Advances in Science, Technology & Innovation

IEREK Interdisciplinary Series for Sustainable Development

Advances in Science, Technology & Innovation (ASTI) is a series of peer-reviewed books based on important emerging research that redefines the current disciplinary boundaries in science, technology and innovation (STI) in order to develop integrated concepts for sustainable development. It not only discusses the progress made towards securing more resources, allocating smarter solutions, and rebalancing the relationship between nature and people, but also provides in-depth insights from comprehensive research that addresses the **17 sustainable development goals (SDGs)** as set out by the UN for 2030.

The series draws on the best research papers from various IEREK and other international conferences to promote the creation and development of viable solutions for a **sustainable future and a positive societal** transformation with the help of integrated and innovative science-based approaches. Including interdisciplinary contributions, it presents innovative approaches and highlights how they can best support both economic and sustainable development, through better use of data, more effective institutions, and global, local and individual action, for the welfare of all societies.

The series particularly features conceptual and empirical contributions from various interrelated fields of science, technology and innovation, with an emphasis on digital transformation, that focus on providing practical solutions to **ensure food, water and energy security to achieve the SDGs.** It also presents new case studies offering concrete examples of how to resolve sustainable urbanization and environmental issues in different regions of the world.

The series is intended for professionals in research and teaching, consultancies and industry, and government and international organizations. Published in collaboration with IEREK, the Springer ASTI series will acquaint readers with essential new studies in STI for sustainable development.

ASTI series has now been accepted for Scopus (September 2020). All content published in this series will start appearing on the Scopus site in early 2021.

Alessandra Battisti • Cristina Piselli •
Eric J Strauss • Etleva Dobjani •
Saimir Kristo
Editors

Greening Our Cities: Sustainable Urbanism for a Greener Future

A Culmination of Selected Research Papers from the International Conferences on Green Urbanism (GU)—6th edition and Urban Regeneration and Sustainability (URS)—2022

Springer

Editors
Alessandra Battisti
Department of Planning Design Technology of Architecture (PDTA)
University of Rome Sapienza
Rome, Italy

Eric J Strauss
Professor Emeritus
Michigan State University
Michigan, MI, USA

Saimir Kristo
Faculty of Applied Science and Creativity
Barelti University
Tirana, Albania

Cristina Piselli
Department of Architecture (DIDA)
University of Florence
Florence, Italy

Etleva Dobjani
Department of Architecture and Design
Barleti University
Tirana, Albania

ISSN 2522-8714 ISSN 2522-8722 (electronic)
Advances in Science, Technology & Innovation
IEREK Interdisciplinary Series for Sustainable Development
ISBN 978-3-031-49494-9 ISBN 978-3-031-49495-6 (eBook)
https://doi.org/10.1007/978-3-031-49495-6

This Springer imprint is published by the registered company Springer Nature Switzerland AG
The registered company address is: Gewerbestrasse 11, 6330 Cham, Switzerland

Paper in this product is recyclable.

Scientific Committee

The Editors warmly thank all the Reviewers who have contributed their authority to the double-blind review process, to ensure the quality of this publication.

Preface

Adaptation of Urban Environments Through Regeneration

It is with great pleasure and anticipation that we present to you the Springer Book of Conference, "Adaptation of Urban Environments through Regeneration." This compilation of articles brings together an array of distinguished researchers and practitioners who have delved into the pressing topic of urban regeneration—an imperative aspect of sustainable urban development.

Cities, with their dynamic and ever-expanding nature, are the epitome of human progress and civilization. However, rapid urbanization has brought about numerous challenges, including environmental degradation, social inequalities and the loss of cultural heritage. In response to these pressing concerns, urban regeneration has emerged as a powerful tool to breathe new life into cities, reinvigorate communities and harmonize the relationship between urban spaces and nature.

This book is not merely an assortment of academic papers; it represents a collective endeavour to understand and address the complexities of urban regeneration. Each article offers unique insights into various aspects of the subject, encompassing green urbanism, nature-based solutions, challenges of regeneration, design methodologies, developmental strategies and preservation of cultural and natural assets.

The articles within these pages transcend the conventional boundaries of urban planning and engineering. They exemplify the interdisciplinary nature of urban regeneration, drawing upon expertise from architecture, environmental science, sociology, economics and beyond. As we navigate through the sections, readers will discover innovative approaches and practical solutions that can foster sustainable urban development across different contexts and regions.

The authors dedication to their respective fields is evident in the depth of research, the rigor of analysis and the clarity of presentation. We express our heartfelt gratitude to each of them for their invaluable contributions, which have enriched this compendium and will undoubtedly stimulate interest in the subject matter.

As editors, it is our hope that this book becomes a source of inspiration for researchers, policymakers and urban planners alike. By assimilating the knowledge shared in these pages, we can collectively pave the way for cities that are ecologically balanced, socially inclusive and culturally vibrant.

"Greening Our Cities: Sustainable Urbanism for a Greener Future" embarks on a journey to reimagine cities as sustainable and resilient hubs of human civilization. Let us embrace this endeavour, fostering a future where urban environments flourish in harmony with nature, heritage and the aspirations of their inhabitants.

Tirana, Albania

Etleva Dobjani, Ph.D. Arch.
Saimir Kristo, Ph.D. Arch.

Acknowledgements The book Editors would like to thank the authors of the chapters of this book for providing high-quality research discussions. They would also like to thank the scientific committee of the reviewers for contributing with their knowledge and constructive feedback to improve the papers and ensure they are of the best quality possible. Finally, the Editors would like to express their appreciation to the IEREK team for supporting the selection of the best research papers submitted to the conference and the successful publication of this book.

Contents

Introduction

Fragments of a World We Would Like to Live In

Alessandra Battisti

Abstract

The recent global transformation is leading us toward an unpredictable future, an unknown territory in front of which there is a need for a global vision that underpins a new humanism and a complex network of objectives in coherence with international and European targets. The goal of sustaining human life on this planet implies intertwined objectives, including improving quality of life through energy, ecological, and digital transitions, limiting resource, land, and energy consumption, ensuring the sustainability of buildings and infrastructure, and adapting and mitigating climate change.

In light of current global events such as climate change, diseases, and conflicts, it is more important than ever to rethink human habitats and carefully reconsider "how we would like to live" in the future. A multidisciplinary approach set at the intersection of exact sciences, human sciences, urban planning, and technology is required to drive research in this direction, in order to study architecture and cities in a global context and to intervene correctly on them by interpreting history and society through the lens of spatial and social justice.

This book seeks possible answers to the question of what happens to the built urban environment and the landscape in times that require urgent and effective responses to global challenges, a time when information and communication technologies not only redesign the transformations occurring in our social system, but also force to rethink the places of politics, to break down borders, and deny the very constraints of space and time. Extremely urgent responses are required, considering that 7 years have already passed since the Paris Agreement on the occasion of COP21 (Moosmann et al., 2017), and that in 7 years time, we extremely urgent responses, given that it has already been 7 years since the Paris Agreement was signed on the occasion of COP21 (Moosmann et al., 2017), and in 7 years time, in 2030, the signatory nations are expected to have reduced emissions of 43% and have change their energy-intensive and polluting lifestyles (IPCC: Summary for Policymakers., 2023).

The essays, divided into six parts, retrace, articulate, and introduce solutions for achieving the ecological, energy, and digital transition in a phase in which downscaling processes, fine-tuning the current vast area strategies, require us to act urgently on the future of the built environment in relation to the capacity for resilience and adaptation to climate change, as well as that of resilience and adaptation to problems related to the progressive limitation and non-renewability of resources. The contributions provide us with a framework in which cities, the highest and most tangible expression of human effort on the planet, are called upon to address these critical challenges immediately in a line with an imperative resilient approach (Lewis & Conaty, 2012).

In this logical framework, the book attempts to provide fragments of visions through which we imagine the future of our living environment, our lifestyles within cities and in rural environments anywhere in the world, and that reflect different cultural, environmental, economic, and social conditions capable of ensuring an architecture of humanity. The essays in this volume deal with ambitious goals, describing practices and experiences in line with the Paris agreements and the European New Deal, ranging from energy and ecological efficiency to digitalization, from culture to sustainable mobility. All local governments are expected to address these topics with the creation of services and policies oriented toward environmental sustainability, circular economy, and social inclusion, through medium and long-term planning.

In relation to the urban regeneration design research experiments illustrated in this volume, several reflections arise. The most important one concerns the governance of complexity, relationships and location, functions and

A. Battisti (✉)
Department Planning Design Technology of Architecture, School of Architecture, Sapienza University of Rome, Rome, Italy
e-mail: alessandra.battisti@uniroma1.it

A. Battisti et al. (eds.), *Greening Our Cities: Sustainable Urbanism for a Greener Future*, Advances in Science, Technology & Innovation, https://doi.org/10.1007/978-3-031-49495-6_1

intended uses, which, in order to become environmentally sustainable and enhance diversity, requires addressing and controlling the systemization of territorial networks at all levels: social, cultural, anthropic, biophysical, and bioclimatic. Taking into account, the territory not only implies paying attention to the context but it also becomes the real driver for enhancing the urban identity in which the settled community may recognize itself. Hence, the importance of building an infrastructural network designed with the inhabitants, involving the reorganization of the mobility network, urban voids, margins, and collective services in order to weave significant relationships, able to influence the inhabitants' forms of life and relationships.

References

IPCC: Summary for Policymakers. (2023). Climate change 2023: Synthesis report. In Core Writing Team, H. Lee, & J. Romero (Eds.), *Contribution of working groups I, II and III to the sixth assessment report of the intergovernmental panel on climate change* (pp. 1–34). IPCC, Geneva, Switzerland. https://doi.org/10.59327/IPCC/AR6-9789291691647.001

Lewis, M., & Conaty, P. (2012). *The resilience imperative: Cooperative transitions to a steady-state economy*. New Society Publishers.

Moosmann, L., Neier, H., Mandl, N., & Radunsky, K. (2017). *Implementing the Paris agreement—New challenges in view of the COP 23 climate change conference*. Brussels: Study for the ENVI Committee, European Parliament, Policy Department for Economic and Scientific Policy.

Green Urbanism

Evaluating Sustainable Urban Mobility Strategies as a Tool to Provide Balance of Centrality Levels in Budapest City

Ola Qasseer and Gábor Szalkai

Abstract

The absence of balance among different urban areas and the centralization/decentralization aspects is one of the challenges that the cities have been facing. The level of centrality affects the distribution of the activities in the city. It also results in the marginality of other areas which face serious problems and may cause an obstacle to the development of a sustainable city. Transportation networks play an essential role in enforcing or weakening the level of centrality. This could be found in the proposed case study of Budapest where the distribution of transportation networks is a major contributor to the different levels of centrality among various urban areas. But, is Budapest city capable of overcoming the different levels of centrality, limiting the centrality problems; congestion, pollution, inequality, etc., and moving toward a more balanced city with strategies of Urban Sustainable Mobility? This will be examined in the proposed research by measuring the centrality levels via transportation networks, specifying the levels of the centrality of the different urban areas, evaluating the situation, and revising the plans of the responsible agencies of transportation for promoting sustainability. Achieving the previously mentioned steps and defining the path of Budapest towards more sustainable transport will help in providing an introduction for developing transportation strategies and improving Sustainable Urban Mobility criteria to provide balance for the different urban areas and reduce the problems of centrality.

O. Qasseer (✉) · G. Szalkai
Department of Regional Science, Institute of Geography and Earth Sciences, ELTE Eötvös Loránd University, Budapest, Hungary
e-mail: ola.qasseer@kti.hu

O. Qasseer
KTI Hungarian Institute for Transport Sciences and Logistics Non Profit Limited Liability Company, Budapest, Hungary

Keywords

Sustainable urban mobility • Centrality • Transportation networks • Transportation strategies • Budapest city

1 Introduction

Cities are complex component of human activities including commercial, touristic, and cultural events (Neal, 2013; Öğdül, 2010). The concentration of activities in urban areas made cities more important and central. This centralization led to the migration of poor people to the city's surrounding areas (Deffner et al., 2012). As a result of the different characteristics of centrality in each urban place (Abdullahi et al., 2017), the centrality level limits the residential use in central areas in some cases (Bunting et al., 2000), and affects the growth of the city (Porta et al., 2009). Replacing the concentrated urban shape with decentralized development after the Second World War affected the distribution of different functions in the city (Garreau, 1991), and encouraged exclusiveness and exclusion (Mace, 2009). In other words, the differences in centralization levels among urban areas caused uneven possibilities and unequal distribution of suitable opportunities (Gutierrez, 2009). One of these opportunities is the access to transportation modes (Hine, 2009). Accessibility and the characteristics of transportation systems affect the centrality level (Abdullahi et al., 2017; Porta et al., 2009) and the attractiveness of a specific place (Banister & Lichfield, 1995). Central places are more likely to be close to street intersections (Porta et al., 2009), due to the important role of road networks in influencing the locations of activities (Miranda et al., 2020). These networks could make a location more central by improving its closeness to other locations (Li et al., 2017). In other words, geographical locations and the development of transportation systems affect the accessibility level by enforcing the central positions for some areas and resulting in the

A. Battisti et al. (eds.), *Greening Our Cities: Sustainable Urbanism for a Greener Future*,
Advances in Science, Technology & Innovation, https://doi.org/10.1007/978-3-031-49495-6_2

peripheral locations for other places (Gutierrez, 2009). Hence, the development and rapid increase of urban transportation has caused many problems for the city. These problems could be summarized as follows; congestion, crowding during peak hours, difficulties in walking, pollution, road accidents, and parking shortage issues (Pacione, 2005). Here, the role of transport planning appears by providing safety, security, equity, environmental protection, and accessibility for all residents (Deffner et al., 2012). Hence, transport planners and urban designers have the responsibility of providing a balanced environment and improving transport for healthier cities (Giles-Corti, 2006), especially with the development of working styles, technological solutions, big data, and the changes in mobility attitudes for the young generations (Stadt Wien, 2014a).

After designing the cities to suit cars, there is a chance now for people to move toward more eco-friendly mobility by using public transportation, cycling, and walking (Stadt Wien, 2014b). Using cars and planning new streets were considered modern during the 1960s, while cycling and walking were considered outdated. It was considered smart for planners to design cities around private cars, remove the cycling and tram lines, and encourage the development of automobility. All that has led to many troubles and consequences in our current cities (Knoflacher et al., 2018). Utilizing land-use planning has a significant role in this regard by reducing traffic, the length of trips, and motorized transportation. It encourages walking, cycling, and developing public transportation facilities which represent the basic concepts for sustainable transportation planning (Deffner et al., 2012). Sustainable Urban Mobility (SUM) could provide equality and improve the quality of life (European Economic and Social Committee, 2022). It improves the accessibility of urban areas and provides more balanced and integrated modes (Interreg Europe, 2018), which could contribute to more balanced and sustainable cities.

2 Budapest City

Similar to what happened in other cities after the spread of the motorized automobile, Budapest also witnessed the removal of tram lines and the planning of its areas according to the movements of cars. This has resulted in a high level of fragmentation, pollution, and lack of cooperation (Budapest Főváros Önkormányzata & BKK Budapesti Közlekedési Központ, 2019). The agglomerations around the city generate a high number of daily car trips. People living in these agglomerations around Budapest are more likely to work, study, and do plenty of activities in Budapest city every day (H2020-Flow, 2015; Budapestvasut, n.d.). The capital city also faces many challenges like congestion and poor level of intermodal interchanges and connections (H2020-Flow, 2015). It has faced real problems regarding pollution, weak cycling services, undeveloped transportation in suburban areas, and the focus on motorized mobility more than the sustainable approaches (Budapest Főváros Önkormányzata & BKK Budapesti Közlekedési Központ, 2019). With more than 1.7 million of population and the daily commuting inside the city and from the surrounding agglomerations, will Budapest be able to control its varied levels of centrality and move to becoming a more balanced and sustainable city?

3 Methodology

The research starts with measuring the centrality of urban areas through public transportation networks. To start the process, all the needed data was requested and collected from the transportation agencies (the public transportation company in Budapest city, the railways company, and the regional buses company); BKK, MÁV-START, and Volánbusz. The data includes the shape files of all routes and stops inside Budapest city and its agglomerations, the number of trips on each route per day, the average number of passengers per day, the average number of boarding passengers per day at each stop, and the average number of alighting passengers per day at each stop. The measurements start with building the network by defining its nodes (the stops and stations of the public transportation) and its edges (the route between every two consecutive stops), then applying the centrality measurements with the help of the GIS environment using Arc Map 10.5 version. To measure the centrality, six measurements are applied. The first measurement includes the visualization of the boarding passengers at each stop (node) and categorization the percentage of the passengers. The second measurement applies the same aforementioned criteria to the number of alighting passengers. The third measurement is concerned with degree centrality to measure the number of connections with each node. This was achieved with the help of Arc Map to produce the count of connections to each node and then to visualize them on the map. The last three measurements are the measurements of Multiple Centrality Assessments (MCA); closeness centrality, betweenness centrality, and straightness centrality. The MCA measurements represent the level of the centrality of each node by its closeness to other nodes in the network, its location between other nodes, and its position on straight routes that connect it with other nodes in the network (Porta et al., 2009). These MCA measurements were measured by calculating the real distances and Euclidean distances between all nodes using network analysis tool in Arc Map, then calculating the level of centrality by applying the needed mathematical equations, and finally visualizing the results of different levels of centrality via maps.

The levels of centrality through the six measurements are visualized and categorized to define the areas with the highest levels among the other nodes of the network. The results are compared with the distribution of land use in the city by observing the connectivity of these nodes, the activities and functions in the surrounding urban areas, and the locations of the nodes in Budapest city. This comparison could represent the related factors causing the differences in centrality level throughout different urban areas and the role of transportation systems in enforcing or weakening this centrality.

The final step in the analysis contains the revision of the suggested published plans by the agencies of transportation and the municipality to propose a roadmap of enhancing the sustainable urban mobility. Defining the main steps and initiating an evaluation of local strategies will help in understanding their role in providing a more balanced urban structure and overcoming the different levels of centrality toward a more sustainable city.

4 Results and Discussion

4.1 Centrality Levels in Budapest

The analyzed network contains 8128 nodes and 57,902,649 edges. The centrality measurements of the built network shows significant differences in centrality levels between different urban areas in the city. The spatial distribution of the nodes with the highest levels of centrality also varied between the six measurements, see Fig. 1. Despite these differences, significant similarities are found in the distribution of the nodes that achieve the highest levels in the number of boarding passengers (Fig. 1a) with the nodes that achieve the highest levels in the number of alighting passengers (Fig. 1b). This could be related to the importance of the locations of these nodes to the daily commuting of passengers to and from these locations. The nodes with the highest levels of degree centrality (Fig. 1c) are also similar in their spatial distribution to the nodes with the highest levels of betweenness centrality (Fig. 1e). As the measurement of degree centrality deals with the number of connections of each node, these connections could support the location of the node between other nodes and increase the level of betweenness centrality. Other similarities are found in the spatial distribution of the nodes with the highest levels of closeness centrality (Fig. 1d) and the nodes with the highest levels of straightness centrality (Fig. 1f). This could be explained by the fact that the nodes that are located on the straightest routes could be closer to other nodes in the network.

It could be noticed that some nodes with a high level of centrality are located in the outer parts of the city. This could be a result of the disconnectivity in some sections of the network and the concentration of nodes in other sections of the network because of the several networks that were considered in the analysis. As a result, some nodes in these parts have a higher level of centrality according to their adjacent urban areas. However, these nodes do not have a great effect when considering the combination of the six measurements to find the nodes that achieve higher levels according to different aspects.

Out of 8128 nodes, the numbers of nodes with the highest levels of centrality according to the applied measurements are as follows; 18 nodes with the highest number of boarding passengers, 17 nodes with the highest number of alighting passengers, 138 nodes with the highest level of degree centrality, 88 nodes with the highest level of closeness centrality, 112 nodes with the highest level of betweenness centrality, and 181 nodes with the highest level of straightness centrality. As a result of the combination of the six measurements, only 7 nodes score the highest level of centrality among several measurements, see Fig. 2.

Two nodes of the final seven nodes, node number 1 and node number 2, represent the main centers in Budapest city. They are located in the central areas of Budapest where the commercial and services sectors are. They also represent important squares that attract transportation and where different modes of several essential routes intersect. Node number 1 is the intersection of three different metro lines out of the four metro lines in the city, one trolley bus line, three tram lines, and six bus lines. The average number of passengers who cross this node exceeds 1,100,000 passengers per day. Node number 2 is an intersection of two metro lines, two trolley bus lines, three tram lines, and four bus lines. This node is also crossed by more than 880,000 passengers per day. On the other hand, the other five nodes are not central in their locations in the city but they scored a high level of centrality according to their position in the network. They are mainly located on the railways that connect the city with other Hungarian cities and with the Budapest agglomerations or close to these routes. Node number 3 is located or close to several public transportation lines that connect the southern parts with the main centers of the city. This area is crossed by two tram lines and 16 bus lines in addition to a regional bus line. Around 165,000 passengers use the lines through this node. Node number 4 is located on one of the main suburban railway lines. This line connects Csepel (the southern island) with the city. It is also close to one of the railways that connect Budapest with the southern cities in Hungary. In addition to one suburban railway line, 12 bus lines, three tram lines, and regional bus lines are crossing this area. Some of these routes also connect the area with the main centers in the city. More than 155,000 passengers cross

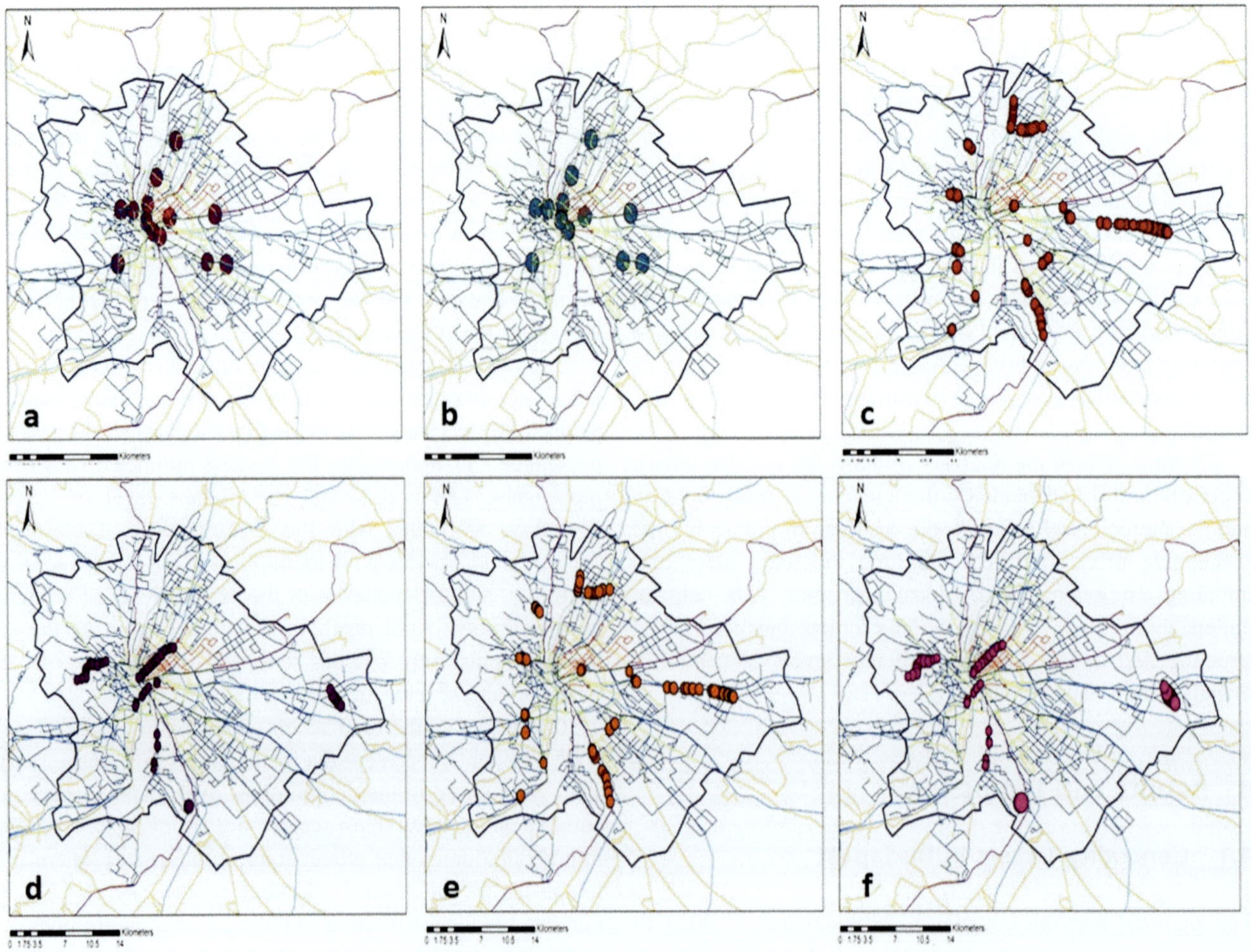

Fig. 1 The spatial distribution of the nodes with the highest levels of centrality according to the six measurements; **a** boarding passengers, **b** alighting passengers, **c** degree centrality, **d** closeness centrality, **e** betweenness centrality, **f** straightness centrality. *Source* Authors

this area per day. Node number 5 is also close to the railway lines and regional buses that connect Budapest with the southeastern cities in Hungary. It is the connection point of five trams and 16 buses. Every day, around 160,000 passengers pass by this area. The last two nodes also represent important nodes along the railway lines to the eastern part of Hungary. They are linked to 10 bus lines and are crossed by more than 150,000 passengers per day.

It was noticed that the highest number of nodes is located on the eastern side of the city (Pest side) where the highest percentage of non-residential activities is located, and the commercial and service functions are concentrated. It was obvious from the analysis that the railways and regional bus networks affected the results of the centrality measurements. The analysis presented the nodes that represent the surrounding connections of the city with its central locations. This shows the important effect of the daily commuting of the suburbs and agglomeration residents to Budapest city and presents the necessity of developing the transportation networks for these areas to reduce the urban problems. Considering only the public transportation network inside the city might highlight more results about the different levels of centrality in the parts of the inner city. This will be presented in the future research to acquire a more comprehensive perspective of the different levels of centrality by the effect of the local public transportation network in Budapest city. Achieving this measurement might help consider all the aspects in the analysis process.

4.2 Sustainable Urban Mobility Strategies in Budapest City

The main strategies for Sustainable Urban Mobility in Budapest city were initiated by the "Balázs Mór Plan" (USER-CHI, n.d.; Budapest Főváros Önkormányzata & BKK Budapesti Közlekedési Központ, 2019). This plan aims to provide integrated network development, green

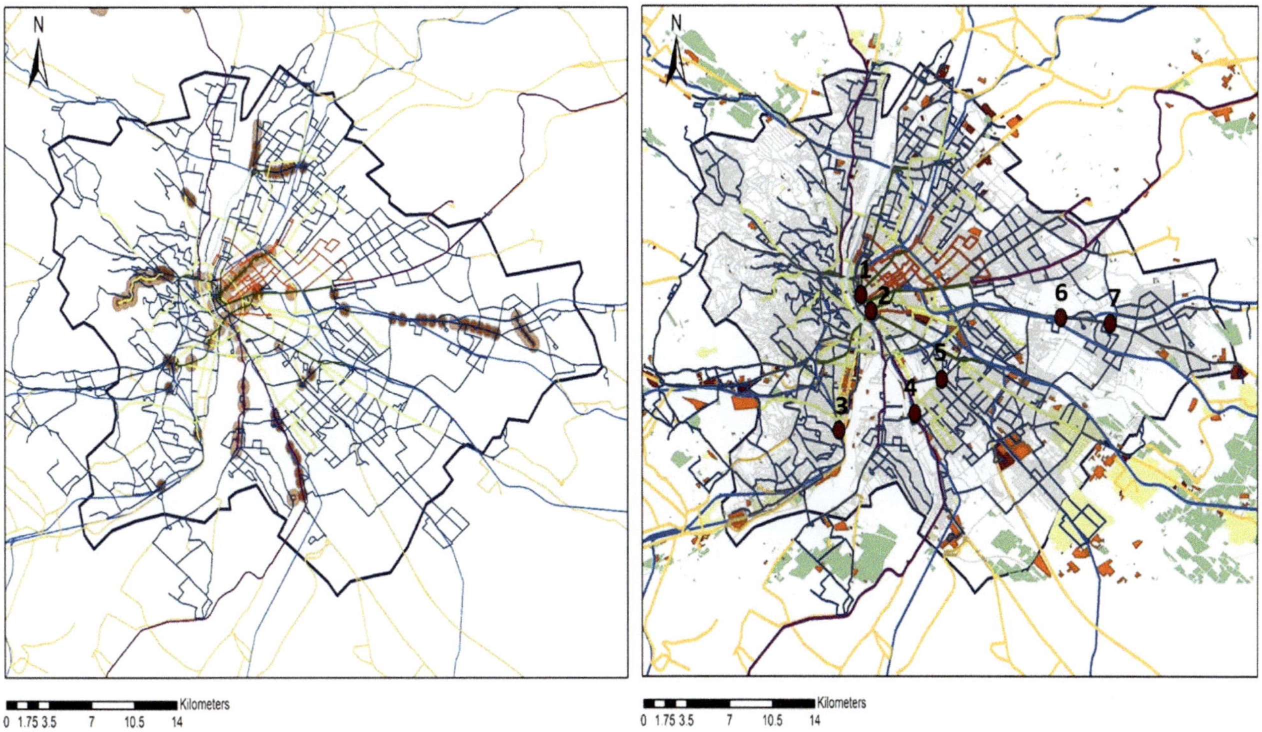

Fig. 2 The nodes with the highest levels of centrality after combining the six measurements and choosing the nodes that achieve the highest level of centrality among several measurements. *Source* Authors

technology, environmental-friendly vehicles, and regional cooperation. It opens a new chapter of transport planning in Budapest to improve urban life and to limit the pollution of the different transport modes. Providing more connectivity, more availability, and a more integrated network are the main directions of this plan. One of the main objectives is to develop the network in suburban areas on an equivalent level to the inner city, to provide more connections via different modes in the outer parts of the city, and to integrate the bike routes with the public transportation network. Another focusing point is to improve accessibility and restore the unity of the city's network. It also focuses on developing the radial connected routes to decrease the movements through the center when it is not needed for the passenger and equipping the city's transport structure with more Danube crossings. Regarding cycling activities, the plan aims to increase the cycling routes, to develop their connections with the transportation network, and to improve their connectivity with regional areas (Budapest Főváros Önkormányzata & BKK Budapesti Közlekedési Központ, 2019). This is what BKK aims to achieve through its service MOL Bubi to provide a fast and eco-friendly alternative by using bike sharing which could be combined with other forms of public transport (MOL Bubi Budapest, n.d.). Another strategy for sustainable urban mobility is the integrated e-mobility concept to reduce the emissions of transport vehicles (USER-CHI, n.d.). Improving cycling networks in Budapest agglomerations connecting them to the city network was also one of the strategies toward more sustainable mobility (Mobilissimus, n.d.). This strategy also aims to connect the cycling network with the railway network, so users can reach close train and bus stops easily (NKK, n.d.). This is accompanied by the development of suburban railways which, started in 2019, and aims to increase the total number of passengers on suburban railway lines by 80% by 2040. It also aims to develop the connections of railways inside the city and increase the capacity so suburban commuters can reach more points than a few limited terminals. Developing the southern rail belt that passes on one of the city's southern bridges is also one of the main goals of this strategy. It aims to increase the capacity of the bridge and to establish more stops inside the city in order to increase the connectivity with other modes of transportation. Another proposal in this strategy was initiating the railway tunnel under the Danube to raise the capacity of railways too and to develop and improve two of the main train stations in Budapest (Budapestvasut, n.d.). Building more connections between Buda and pest sides is also one of the major concerns to increase connectivity and develop accessibility in the city. The project of the new Budapest bridge aims to achieve this goal by improving the connectivity in the southern part of the city (UNStudio, 2018). It is also important to mention the

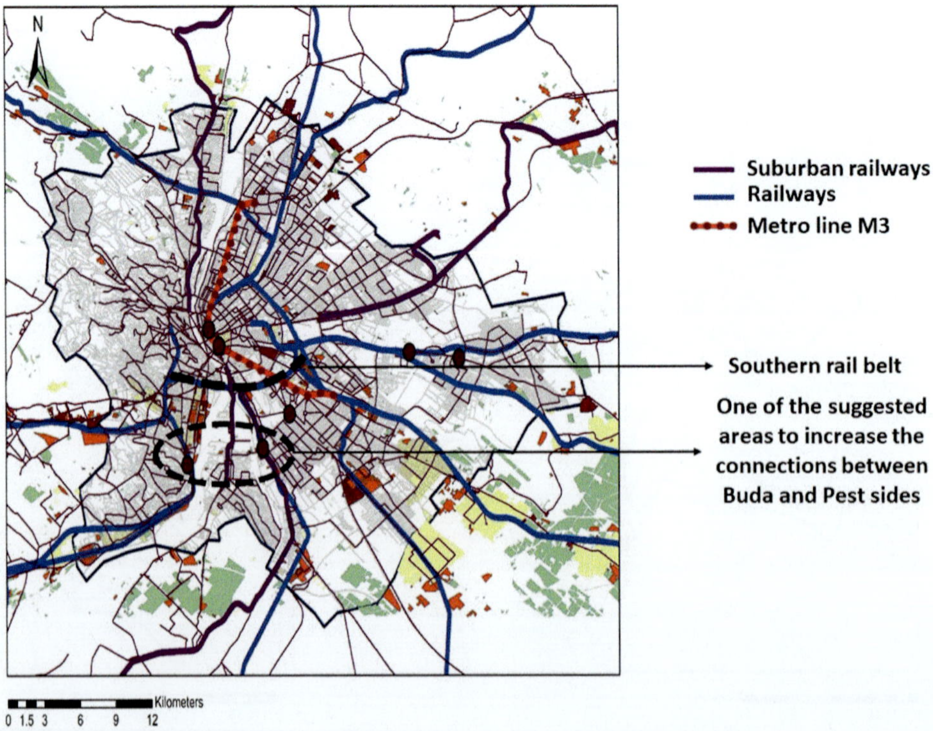

Fig. 3 The relation of the suggested strategies with the network of transportation. *Source* Authors

reconstruction project of metro line M3 that started to develop one of the main lines in the city in order to provide more comfortable service and to improve its connectivity with other routes of public transportation (Official Website of the Reconstruction Project of Metro Line M3, n.d.).

When comparing these strategies with the results of centrality measurements, the importance of developing transportation networks in suburban areas could be noticed. As presented in the measurements, the suburban and outer parts of the city play an essential role in affecting the centrality levels between different nodes because of the high commuting percentage from and to these areas. Developing the transportation networks in these areas could provide more balance to the different parts of the city. In addition, the strategy to develop cycling networks in these areas could also help reduce the high numbers of daily trips by public transportation or traffic. This might reduce the city's congestion and pollution. As presented, most of the nodes in the outer parts of the city are located or close to the railways and suburban trains that connect the commuters with the city. However, these routes are not well connected to the main centers and destinations. Developing and increasing the terminals of these routes, which is suggested in one of the strategies, could also be a helpful factor for more balance and connectivity of all parts of the city. The plans that aim to develop the connections between Buda and pest sides are also essential in providing more balance to the different nodes of the city which is missing between the two sides as the measurements of centrality show. Regarding the central parts of the city, developing metro line M3 which crosses two nodes of the highest node in centrality level (node number 1 and node number 2) is also a good step in the direction of achieving more efficient service across the whole route. Developing the cycling network inside the city is also another aspect for more balance and stability in the central areas of the city, see Fig. 3.

5 Conclusion

It could be concluded that transportation networks have an essential role in affecting the centrality of urban areas. The distribution of these networks could result in more centrality in some areas and less centrality in other parts of the city. The results of the case study of Budapest show that unbalance and different levels of centrality could be a result of the relation between the suburbs or the surrounding areas of the city and the central urban areas. This was obvious when including the railways and the regional bus networks in the analyzed network. The presented strategies and plans show a real concentration on developing the commuting of the residents of agglomerations and suburban areas in addition to applying more sustainable solutions. This could control the problems of centrality and pave the road for Budapest to become a more sustainable and balanced city. However, it is a matter of question if these strategies will be as fast as the continuous changes in commuting attitudes and the new working style, especially with the rapid development of

Information and Communications Technology (ICT) and the recent effects of COVID-19. We are planning to explore these effects in future research and to investigate the changes in commuting patterns. We also suggest that responsible authorities should consider these rapid change into their strategies by conducting regular assessments and developing of these plans to cope with the continuous changes.

Acknowledgements The authors would like to thank the Center for Budapest Transport (BKK) and MÁV-START for the useful data which they provided. The first author would also like to thank Tempus Public Foundation for providing the Stipendium Hungaricum Scholarship.

References

Abdullahi, S., Pradhan, B., & Al-sharif, A. A. A. (2017). Introduction to urban growth and expansion. In B. Pradhan (Ed.), *Spatial modeling and assessment of urban form—Analysis of urban growth: From sprawl to compact using geospatial data* (pp. 3–15). Springer. https://doi.org/10.1007/978-3-319-54217-1

Banister, D., & Lichfield, N. (1995). The key issues in transport and urban development. In D. Banister (Ed.), *Transport and urban development* (pp. 1–16). E & FN Spon.

Budapest Főváros Önkormányzata, & BKK Budapesti Közlekedési Központ. (2019). *Balázs Mór Plan: Budapest Mobility Plan——Volume 1: Objectives and measures*. https://bkk.hu/downloads/6386/DlQmVC-98kefRmUWLj4SQQ==

Budapestvasut. (n.d.). *Budapest suburban railway node strategy*. https://budapestvasut2040.hu/english/

Bunting, T. E., Filion, P., & Priston, H. (2000). Changing patterns of residential centrality: Population and household shift in large Canadian CMAs, 1971–1996. *Cahiers De Geographie Du Quebec, 44*(123), 341–361. https://doi.org/10.7202/022925ar

Deffner, J., Hefter, T., Rudolph, C., & Ziel, T. (2012). *Handbook on cycling inclusive planning and promotion*. Capacity development material for the multiplier training within the mobile2020 project.

European Economic and Social Committee. (2022). *Sustainable and inclusive urban mobility can improve people's quality of life, says EESC*. https://www.eesc.europa.eu/en/news-media/news/sustainable-and-inclusive-urban-mobility-can-improve-peoples-quality-life-says-eesc

Garreau, J. (1991). *Edge city: Life on the new frontier*. Doubleday.

Giles-Corti, B. (2006). People or places: What should be the target? *Journal of Science and Medicine in Sport, 9*(5), 357–366.

Gutierrez, J. (2009). Transport and accessibility. In N. Castree, M. Crang, & M. Domosh (Eds.), *International encyclopedia of human geography* (Vol. 11, pp. 410–417). Elsevier Ltd. https://doi.org/10.1016/B978-008044910-4.01030-0

H2020-Flow. (2015). *Budapest—EN*. http://h2020-flow.eu/flow-cities/budapest/budapest-en/

Hine, J. (2009). Transport and social exclusion. In N. Castree, M. Crang, & M. Domosh (Eds.), *International encyclopedia of human geography* (Vol. 11, pp. 429–434). Elsevier Ltd. https://doi.org/10.1016/B978-008044910-4.01034-8

Interreg Europe. (2018). *Sustainable urban mobility plans: A policy brief from the policy learning platform on low-carbon economy*. European Union, European Regional Development Fund. https://www.interregeurope.eu/sites/default/files/2021-12/Policy%20brief%20on%20sustainable%20urban%20mobility%20plans.pdf

Knoflacher, H., Frey, H., & Leth, U. (2018). Smartness of urban mobility and "quality of life" in Vienna. In *Proceedings of 7th Transport Research Arena TRA 2018*, April 16–19.

Li, J., Qian, J., & Liu, Y. (2017). A novel analysis method of geographical centrality based on space of flows. *ISPRS International Journal of Geo-Information, 6*(5), 153. https://doi.org/10.3390/ijgi6050153

Mace, A. (2009). Suburbanization. In N. Castree, M. Crang, & M. Domosh (Eds.), *International encyclopedia of human geography* (Vol. 11, pp. 77–81). Elsevier Ltd.

Miranda, E., Batista e Silva, J., & Ricardo da Costa, A. (2020). Emergence and structure of urban centralities in a medium-sized historic city. *SAGE Open, 10*(3). https://doi.org/10.1177/2158244020930002

Mobilissimus. (n.d.). *Budapest agglomeration cycling strategy*. https://mobilissimus.hu/en/projects/budapest-agglomeration-cycling-strategy

MOL Bubi Budapest. (n.d.). *Bike sharing in Budapest—Easy bike rental—Everywhere*. https://molbubi.hu/en/

Neal, Z. P. (2013). *The connected city: How networks are shaping the modern metropolis*. Routledge.

NKK. (n.d.). *Kerékpáros Kertváros*. https://nkk.hu/fejlesztesek/kerekparos-kertvaros/

Official website of the reconstruction project of metro line M3. (n.d.). *About the project*. https://m3felujitas.hu/en/information

Öğdül, H. G. (2010). Urban and rural definitions in regional context: A case study on Turkey. *European Planning Studies, 18*(9), 1519–1541. https://doi.org/10.1080/09654313.2010.492589

Pacione, M. (2005). *Urban geography: A global perspective*. Routledge.

Porta, S., Strano, E., Iacoviello, V., Messora, R., Latora, V., Cardillo, A., Wang, F., & Scellato, S. (2009). Street centrality and densities of retail and services in Bologna, Italy. *Environment and Planning B: Planning and Design, 36*(3), 450–465. https://doi.org/10.1068/b34098

Stadt Wien. (2014a). *Step 2025: Urban development plan Vienna—Short report*. https://www.wien.gv.at/stadtentwicklung/studien/pdf/b008379b.pdf

Stadt Wien. (2014b). *Step 2025: Urban mobility plan—Short edition*. https://www.wien.gv.at/stadtentwicklung/studien/pdf/b008444.pdf

UNStudio. (2018). *New Budapest bridge*. https://www.unstudio.com/en/page/11740/new-budapest-bridge

USER-CHI. (n.d.). *Budapest*. https://www.userchi.eu/cities/budapest/

Formal Model for Green Urbanism in Smart Cities

Luca Lezzerini

Abstract

The paper deals with the formulation of the principles of Green Urbanism in a smart city. According to some scholars, the main values provided by a smart city can be summarised as Sustainability, Quality of Life, Equity, Livability and Resilience. Most of these values have the green dimension as a foundation element. Measuring this green dimension can be fundamental to evaluating a smart city's effectiveness in terms of provided values. The research question is how to formally and in a measurable way define the Green dimension in a smart city. Expected results are an improvement of the formal definition of Green Urbanism in the smart city context and, in general, of Green aspects of any built space. Another important result is a first theoretical core to plan and evaluate progress in the development of the Green dimension in a smart city. In this sense, the research results can be considered the starting point to develop a descriptive theory of Green Urbanism. The methodological approach starts by exploring the main business models, in a wider sense, applicable to a smart city, identifying the stakeholders and a framework to define, measure and develop the green dimension of a smart city. KPIs and concepts to be measured are derived from worldwide used standards. A peculiarity of the research is the use of a software engineering standard as a language and an approach to describe such a Green dimension. The standard is an Object Management Group official standard, Essence 1.2, the first formal descriptive theory in software engineering, and it has been applied and customised during the research to enhance the ability to implement, through the software and hardware elements of a smart city continuously keeping into account the Green requirements.

Keywords

Green urbanism • Smart city • Essence • Formal model • Business model • Software engineering

1 Introduction

One of the most famous smart cities is the Korean Songdo. According to its creators, it is one of the greenest smart cities in the world (Townsend, 2015). According to many scholars (for example Mullins, 2017) or Korean activist Edelston (2012), it is not as green as stated above. Simply considering raw figures could be impressive: replacing over 600 hectares of coastal wetlands, populated by shore birds and other life forms typical of such ecosystems, with over 22 thousand apartments and over 460 hectares of commercial space, even if they have been built along a golf course, does not sound green.

This example has been considered to put evidence that it is not easy to objectively measure the "green" dimension of a smart city.

Simply reading the figures above will easily lead to defining Songdo as a complete failure from a green point of view. But considering the opposite solution, leaving the coastal wetlands unchanged, simply cannot be considered a green solution but only doing nothing. Smart cities, whatever definition is used to describe them, are built because they satisfy some needs. Not building a smart city but a normal city will create or leave many problems unresolved.

In a few words, we must consider the global effect of building a smart city, considering the problems it solves. We have to compute a detailed balance, and, in the end, we can state if the smart city is greener or not than other alternatives (and feasible) solutions.

To compute such balance, we must take into account all elements of the smart city development according to a

L. Lezzerini (✉)
Polis University, Tirana, Albania
e-mail: luca_lezzerini@universitetipolis.edu.al

A. Battisti et al. (eds.), *Greening Our Cities: Sustainable Urbanism for a Greener Future*,
Advances in Science, Technology & Innovation, https://doi.org/10.1007/978-3-031-49495-6_3

specific business model, demonstrating that the balance is positive even if some natural space has been destroyed and replaced with built space or, on the contrary, that it is negative because it has not increased the values of the business model. Obviously, the business model must consider the "green factor" as an essential element of its structure.

In this paper, we will search for a possible approach to computing such balance and, also, provide a methodology to state at which point of the development of the smart city we are. This will be done with an interdisciplinary approach derived from the software engineering formal theory named Essence, a formal standard of the Object Management Group, issued in 2014, following the so-called SEMAT Initiative (Jacobson et al., 2012). The Essence 1.2 standard (Object Management Group, 2018) will be used because the smart city must combine urban planning, architecture and other kinds of "soft" requirements with technological systems' "hard" requirements, essentially made by hardware and software requirements. This issue will be further explained in the methodological section but a few words can be spent on it now.

Essence 1.2 is a standard for software engineering defined by the Object Management Group, a worldwide referred authority in the field of software standards. With Essence, it is possible to describe any kind of software engineering process and also state at which point of its completion a project development is. Although a smart city is made of buildings, what makes it smart is the correct and effective application of ICT technologies. So, from this point of view, a smart city is mainly dependent, for its success at implementation time, on software engineering. In this perspective, software engineering is the tool that makes possible what urban planners, architects and urban designers have conceived, often considering requirements that are difficult to be measured.

2 Research Question

The research question is how to formally and in a measurable way define the Green dimension in a smart city, keeping into consideration all the values expected from smart city development, according to the given business model.

Some hypotheses have been done to develop the research:

- The business model is not focused only on economic values but it also considers other values that can be considered "green".
- The approach is followed for one specific business model but it can be applied to any other business model if the previous hypothesis is respected.

3 Expected Results

Expected results are an improvement of the formal definition of Green Urbanism in the smart city context and, in general, of Green aspects of any built space.

Another important expected result is a first theoretical core to plan and evaluate progress in the development of the Green dimension in a smart city. In this sense, the research results can be considered the starting point to develop a descriptive theory of Green Urbanism.

4 Methodological Approach

4.1 Methodological Strategy

The research is a step in wider research that aims to create a formal model for smart cities that can describe, using a formal descriptive theory, urban planning requirements, urban design requirements, architectural requirements and Information and Communication Technology (ICT) requirements, connecting these worlds into a common vision using the same language, each of them for its own competence. In this paper, this main research will not be addressed but it is cited to ease the understanding of the whole picture where the content of this paper is located, underlining the interdisciplinarity of the followed approach.

The methodological approach followed in the research was to define a set of "values", measure them and then calculate the balance. As a reference model for Green Urbanism, Lehmann (2010, 2014) has been chosen but the approach can be also applied to other models (e.g. Sustainable Cities Platform, 1994). Then a smart city framework has been identified. For this framework, having extracted its values, for some of them a specific set of concepts to be measured has been extracted. The selected values have been chosen to aim to consider elements that are difficult to be measured objectively.

Starting from these values, coming from both the Green Urbanism model and smart city model, the "essentialisation" process, defined in Essence 1.2, has been applied to such concepts, transforming them using the language of the standard (named Essence Language) and using the elements of the kernel of the standard. For this paper, only alphas and alpha states elements of Essence will be used, with some exceptions, though also other elements can be used in practice.

After essentialisation, the smart city framework and Green Urbanism model elements (values) have been converted into alphas (sub-alphas) and a set of alpha states, for each of them, has been defined.

Once having defined alphas, alpha states and their relationships, the resulting Essence extension has been evaluated together with the process to obtain it. Using the chosen framework, all possible elements of the smart city implementation have been generated by software but, considering their huge number (over 25 thousand in its basic version), only a few of them have been made explicit in this paper. Nevertheless, the main reasoning remains valid. In further studies, a ranking algorithm will be used to let emerge the most important elements according to some criteria. In research yet in progress, the fitness function of a genetic algorithm has been considered as the possible ranking criterion but results will not be available before the end of 2023.

All these activities have been done concerning Green Urbanism's key concepts.

After the research activity, an additional result has been detected and it will be further investigated in a successive study. This result was about the possibility of using the same Essence 1.2 meta-structure, changing what needed to be, to formally describe the method to develop green urbanism in a specific case, even outside the smart city. This last result can be considered a foundation for a formal descriptive theory of urbanism and other urban planning key concepts.

4.2 Methodological Details

Essence 1.2 is based on 5 "areas of concern" that are the main areas that must be satisfied to have a complete and effective product of a software engineering process. These areas are depicted in Fig. 1.

In the topmost area of concern, there are two elements (they are called alphas): Opportunity and Stakeholders.

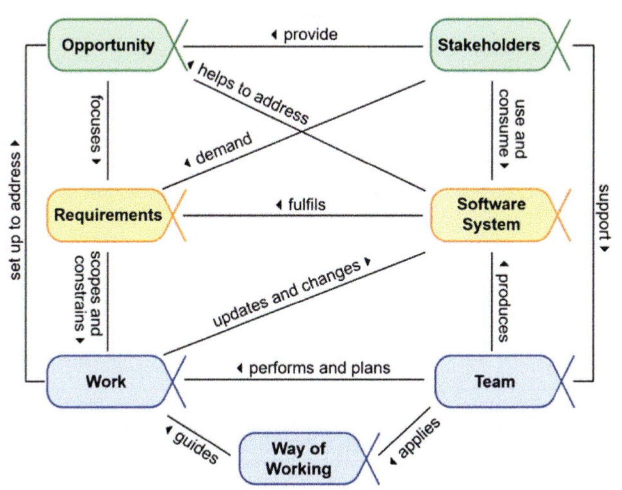

Fig. 1 Essence 1.2 areas of concern. *Source* Object Management Group (2018)

Essence states that to provide an effective software engineering activity, the Stakeholders must be completely identified and involved in the development process. This is represented by the Stakeholders' alpha. In addition, Essence states that also the values that each stakeholder wants to get from the software engineering activity (i.e. the software development process) must be identified and checked throughout the entire process. These values are represented by the Opportunity alpha.

The Requirements alpha represents the requirements to provide the software system (i.e. the smart city).

Stakeholders have been directly taken from the smart city framework, while the Opportunity has been detailed in sub-alphas to support their complete identification. The same sub-alphas detail has been done for the Requirements alpha too. Other alphas have not been considered in this paper because not applicable (Way of Working and Team) or too specific (Work, Software System) so, for the sake of simplicity, they have been omitted. In the wider research cited above, all of these elements have been considered.

5 Research Problem Elements

5.1 The Smart City Framework

Many smart city frameworks have been considered (for example Azizalrahman and Hasyimi, 2019) and other smart city concepts and business models have been evaluated (Townsend, 2015).

The chosen smart city framework was Ramaprasad et al. (2017) because it is flexible and represents an ontology to describe elements of any smart city. In this paper, no changes have been done to this framework. In his vision, Ramaprasad states that the outcomes (i.e. the values provided by the smart city) are:

- Sustainability
- Quality of Life
- Equity
- Livability
- Resilience.

According to this model, the smart city is defined as a function of two elements: "smart" and "city". These two elements will be separately analysed in the following, starting from "city".

The City is defined, in Ramaprasad's framework, by Stakeholders and Values (Outcomes). This is acceptable because The City has the mission to provide some values to the stakeholders. We do not build cities for the sake of building. We build cities to provide what is required by the city's stakeholders.

In Ramaprasad's framework, the Stakeholders are Citizens, Professionals, Communities, Institutions, Businesses and Governments. This set is not important and can be expanded as desired but what is important is that Stakeholders exist and are correctly identified.

In Ramaprasad's vision, the values provided to the Stakeholders are the outcomes described above. In our analysis, we will later add new outcomes (i.e. new values) further detailing those defined by him.

In the framework, the "smart" element of The City, i.e. the Smartness of The City, can be defined as "being able to intelligent sense and respond through semiotics".

Semiotics is defined by Ramaprasad as "an iterative process that generates and applies intelligence". In his idea, Semiotics is Data, Information and Knowledge. Data is the raw, symbolic representation of sensations and measurements. It is a mere recording of something that has been sensed. Data can be considered the raw material from which information is extracted. Information, in fact, in Ramaprasad's vision, is made by the relationships that exist (or are discovered) among the (raw) data. As a general definition from Information Theory, the concept of Information can be considered as the ability to remove uncertainty thanks to the analysis of the relationships among data elements. If before something was not known, after having analysed data relationships, it becomes known. Understanding the meaning of this information is knowledge. Knowledge is, then, the meaning, i.e. the semantic value, of the Information, which is the semantics of the relationships discovered among the data elements. Knowledge is giving a sense of what, being uncertain before, is now known.

According to Ramaprasad, Smartness is defined again as a function, a function of four variables:

$$\text{Smart} = f(\text{Structure, Functions, Focus, Semiotics})$$

where structure is the "artefact" needed to manage the semiotics. To handle data, information and knowledge, some artefacts (physical, like computers or abstract, like processes) are needed.

Architecture (the overall organisation of the space) is the first type of Structure. Architecture is needed because the Smartness of a City needs to shape the space intelligently.

Infrastructure is the structure that relates to physical and virtual connected elements. It provides physical relationships among the components of the City. Infrastructure can be used to support physical communications (e.g. roads, pipelines) or digital communications.

Another infrastructure is called Systems and describes all elements that have their dynamics to process some inputs providing some outputs. Systems can be of three types: computer systems, social systems, and paper-based. They are focused mainly on internal dynamics and represent the intrinsic behaviour that complex aggregations of elements (including stakeholders instances) can generate.

Services are another type of Structure and are used to describe the functions provided by the Systems just described. Services are the general relationship between input and outputs provided by systems and do not consider systems' internal dynamics but only input–output relationships.

To represent regulation activity, an important Structure is one related to Policies. They are, essentially, laws, guidelines and/or rules, and are the result of a governance framework.

Processes and Personnel are two other Structure types. The people are responsible for the processes execution, i.e. they are involved in the processes. Processes are the description at the organisational level of what happens in the city. Often processes involve many systems. When describing a smart city, processes and persons (roles) are essential to its smartness.

In this vision, Structure is what is needed to provide the functions, which are the second variable.

Functions are the functions that are provided by the structure and required to handle the semiotics. They are related to a typical life cycle of semiotics. Semiotics have to be, detected/measured, supervised in their dynamics during the time, processed according to some algorithm, converted into actions by actuators or other forms of control and transferred from one side to another. According to this life cycle, Ramaprasad defines five types.

The first type is the capability to measure something. This first type is called Sense because represents the ability to "sense" what is happening. Sense is the Function that generates the semiotics.

Another function is to observe the semiotics over time to detect patterns in them. Pattern detection is the basis of a reaction schema. This function has been called Monitor.

All the semiotic elements are gathered and monitored to apply some sort of transformation over them. This function has been called Process. It is essential because it allows the generation of added value from the semiotics to get the outcomes.

Many processing need to move out from the semiotics domain (i.e. data, information and knowledge) and become actions to provide results in the physical world or control some variables. This kind of function is called Translate because translates the semiotics (that are intangible) into an interaction with the physical world.

A last function is an ability to exchange semiotics. This function is called Communicate.

Each function applied to semiotics has a specific domain which is the Focus. It targets a specific dimension of the city's dynamics.

One dimension is the Cultural one. It relates to semiotics that is related to culture, including art.

The Economic focus is related to economic semiotic elements and is the most obvious to be managed. They include semiotics about energy, raw materials, basic supplies and similar that influences economics.

In Ramaprasad's vision, also Demographic focus is important. This focus also includes health data (both physical and mental) and probably the name should be changed but, in this research, we have kept it unchanged.

The focus on the environment is another one that must be considered. Environmental focus means considering elements that are related to environmental management and preservation. This focus has gotten a growing interest in the last decades, arriving to be considered one of the important and peculiar of a smart city.

The political focus should be considered in the etymological sense, i.e. "solving problems of the polis", but really can focus even on other elements, like personal opinions.

The focus on Social context is necessary to consider the sociological sphere of the smart city.

The technological focus is on all elements that are strictly related. It is semiotics serving applied technologies. It is to be underlined that new technology trends and general knowledge about technology, including research and innovation, should be included in the culture focus.

Focus is the centre of convergence of the intelligent sense and response, which Ramarasad calls "smartness".

All the framework elements can be used as an ontology to describe the smart city. Combining the various elements of the framework, it is possible to generate a high number of "illustrative elements" that represent, each, single aspect of the smart city.

A single illustrative element can be generated by combining the framework's elements according to the following schema:

$$
\begin{aligned}
&< \text{Structure} > \text{ to } < \text{Functions} > + < \text{Focus} > \\
&\quad + < \text{Semiotics} > \text{by/from/to} < \text{Stakeholders} > \\
&\quad \text{for } < \text{Outcomes} >
\end{aligned}
$$

An example of an illustrative element obtained by applying the above schema could be "Services to Monitor Demographic Data from Citizens for Resilience".

5.2 Smart Cities Standards

International standardisation bodies have proposed various standards for sustainable cities and communities and many of them can be applied (and are applied) to smart cities (Huovila et al., 2019; Midor & Płaza, 2020), even if a common, definitive, definition of smart city has not yet been achieved.

According to BSI (The British Standards Institution, 2014), standards for smart cities can be defined at three different levels in its 2014 view. These levels are strategic, process and technical specifications.

At the strategic level, standards provide high-level guidelines aimed to support governments and other organisations on how to conceive and develop smart city strategies. At this level, priorities are identified, roadmaps developed and progress can be monitored in its evolution along the roadmap.

Standards at the process level provide good practices in project procurement and management for projects that are achieved in smart cities. They contain best practices and guidance also about funding and financing.

The last level, the technical, contains all specifications (i.e. technical requirements) for both products and services needed to implement smart cities projects.

Following such schema, the standards have been grouped and classified in Fig. 2.

The three levels talk different languages and do not have a formal descriptive common language that, indeed, could be very useful to allow easy tracing of requirements implementation from one level to another. Even the risk management group, which spreads over all three levels, is described in different ways through the layers.

The technical committees' framework proposed by BSI in 2014 was organised as in Fig. 3.

Let's consider, as an example, the ISO/TC 268 managed standards. The typical applicable ISO standards coming from ISO/TC 268 and suitable for smart cities are shown in Fig. 4.

ISO 37120 is the central, key, standard and, even if not specifically designed for smart cities but for a wider concept (sustainable cities and sustainable communities). It defines methodologies for a set of indicators that can be used to drive and measure the sustainability of the city/community. Sustainability is defined, in this case, in terms of the city services' performance and quality of life.

ISO 37122 and ISO 37123 are complementary sets of indicators that provide a wide range of capabilities to measure the "smartness" of a city and its "resilience". ISO 37122 goes further than the simple KPIs list and also provides important methods and practices that can have a serious impact on the smart city context (i.e. its social environment and its sustainability from economic and environmental points of view).

ISO 37101 provides requirements for the development of a management system to continuously improve the sustainability dimension of the community. As with any other ISO management system requirements set, it contains the foundation principles that must be satisfied in such a management system. Using this standard with the other three

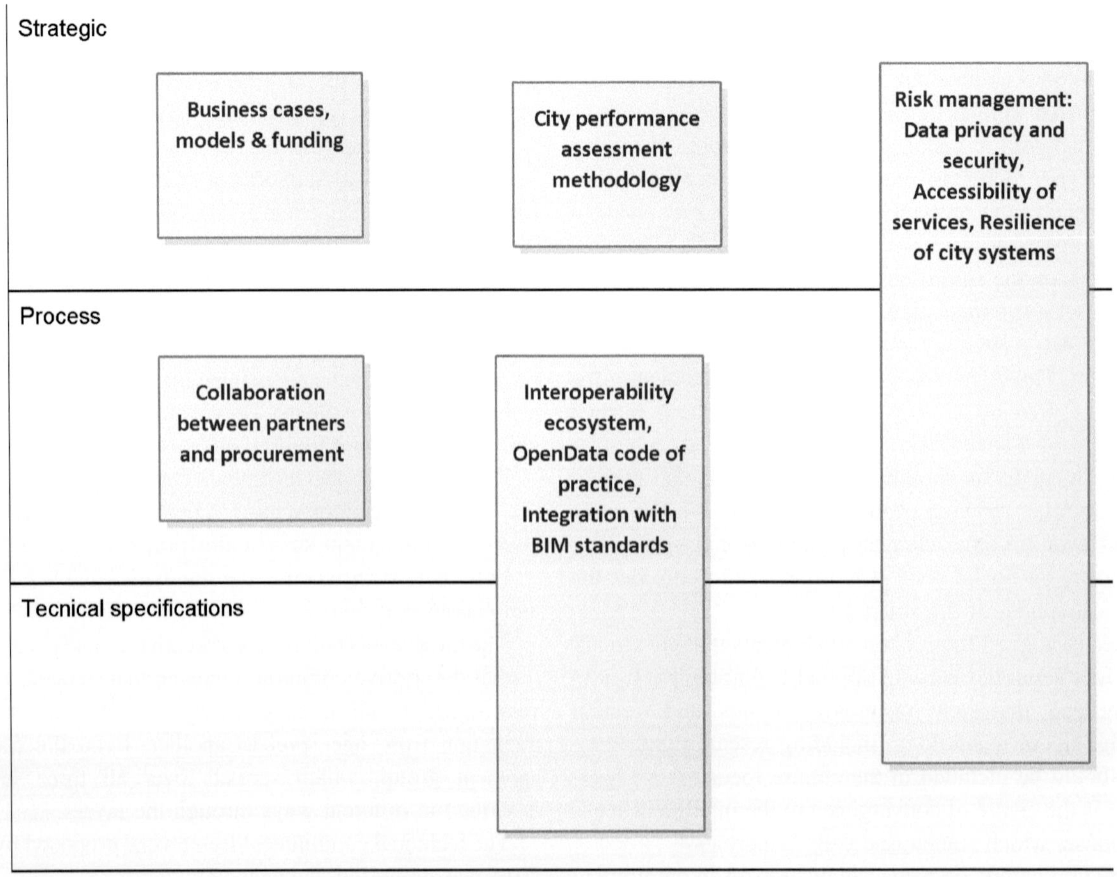

Fig. 2 Level of standards according to BSI 2014 (image of the author)

described above usually gives the ability to implement it effectively and efficiently. ISO 37101 can also be used not only to improve sustainability, smartness and resilience through an interdisciplinary and holistic approach but also to create clearness and consensus about the need (and the opportunity too) to implement sustainable development in a community.

The standards above spread from strategic to process level but they have some common issues.

The first issue is about ISO 37101, i.e. at the strategic level. Its requirements are in a general form and do not provide enough details to implement and measure the progress other standards have been used for doing this task. ISO 37101 is too general and does not take care of the existence of a custom or peculiar requirements like democracy or social engagement.

The other standards define the level of the implementation through indicators that are correlated with the required feature (e.g. sustainability) but that do not measure it. In a few words, KPIs have been defined to measure, in a quantified way, something that, often, is not quantifiable. Some examples of these KPIs are reported below with explicit criticism of their limits.

For example, ISO 37120 indicators are partitioned into two levels: core and support indicators.

Environment core indicators are particulate matter concentrations and greenhouse gas emissions per capita. Environment support indicators are concentrations of NO_2, SO_2, O_3, noise pollution and relative changes in native species (% of number). Although these parameters are important from the environmental point of view, they are not enough to quantify the environmental dimension of the community.

But the real issue comes when a more complex (and effective) model is considered. If the discourse is moved towards Green Urbanism, it becomes clear that something is missing. Not only in the indicators that can be updated or increased in number and detail but in the real approach.

5.3 The Green Urbanism Principles According to Lehmann

The green principles have grown in importance in the last decades, especially in the first years of the XXI century. This importance has been reflected in both urban planning and urban design, becoming essential to any study done in these

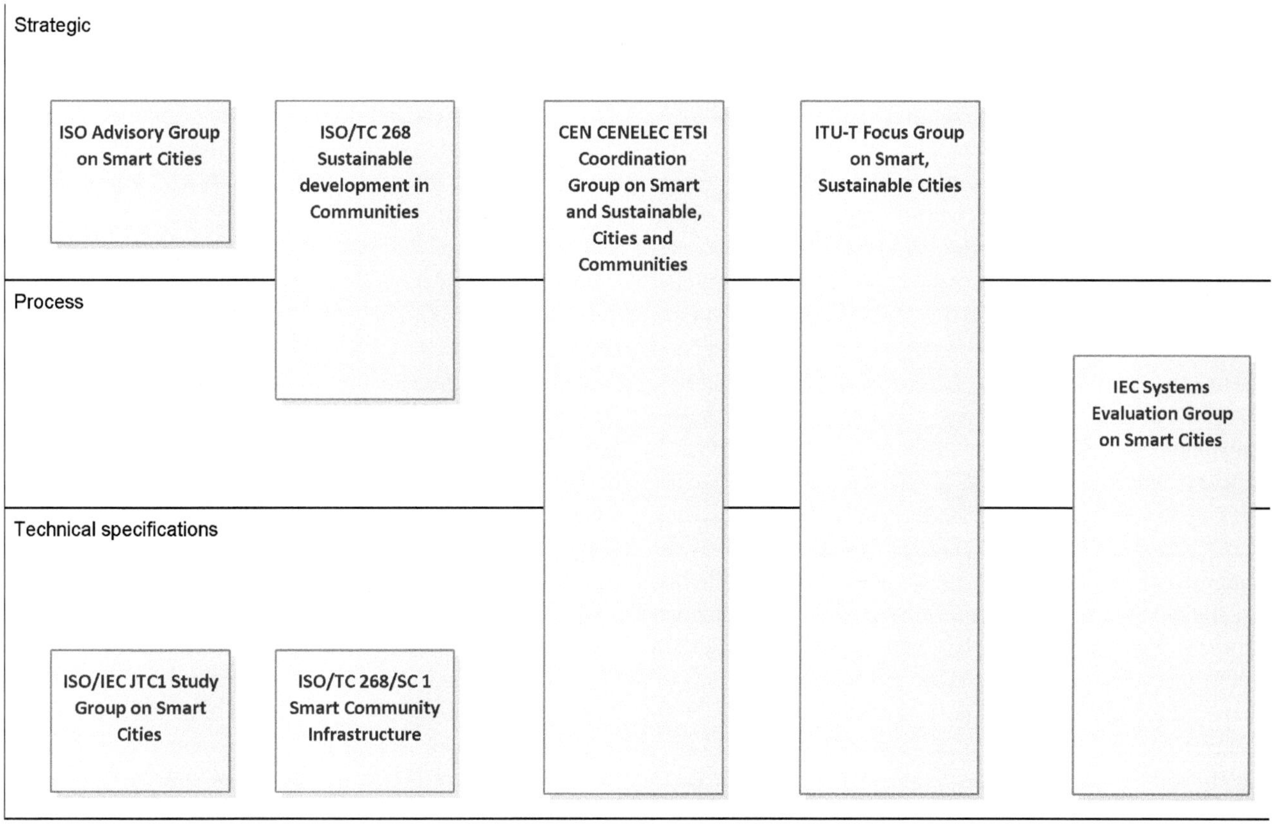

Fig. 3 Technical committees' framework according to BSI 2014 (image of the author)

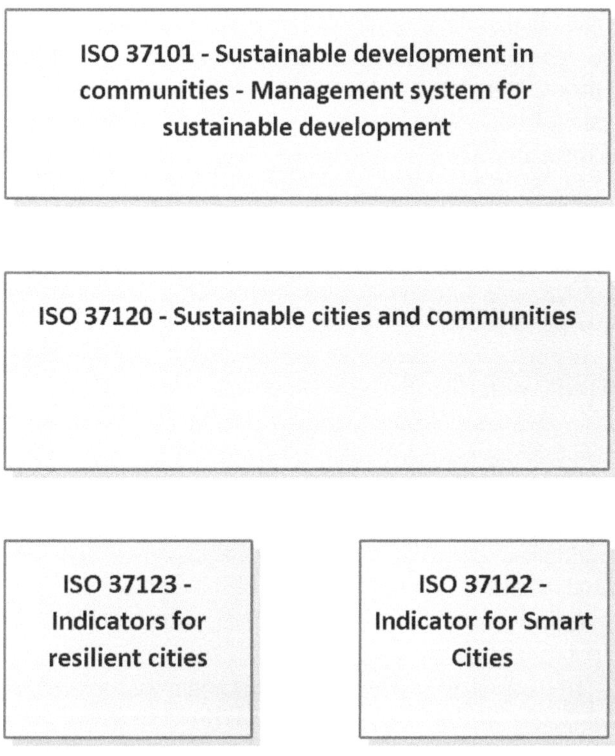

Fig. 4 ISO/TC 268 suitable standards for smart cities (image of the author)

disciplines. Many movements have arisen along this time and have analysed the change in urbanism, moving towards green principles and sustainable growth. These movements can be considered a new, wider movement that has evolved to converge into green urbanism. This "green urbanism" was aimed to develop human communities that had qualities like sustainability and a lower impact on the environment (Beatley, 2000). Since its beginning, this new kind of urbanism, exposed concerns about the environment, starting from land and energy consumption, and moving towards a more generalised perspective. One approach that these new generations of urbanists proposed was a regional approach to development management. Katz (1994) proposed to move people into more liveable communities (more precisely "small towns"). Ewing (1996), considering the case study of Florida, proposed a set of best practices to both assess the quality of existing or guide new development plans. These guidelines were focused on four main axes: land use, transportation, environment and housing but founded on the concept of "green fields". Calthorpe and Fulton (2001) studied the most diffused form of settlements in the USA which was an aggregation of cities and suburbs defining them as "regional cities".

But not only urban planning issues have been developed. Also, the background philosophy and best practices of this new "green" trend in urbanism have been explored by many scholars. In addition to the already cited Calthorpe and Fulton (2001), Duany et al. (2000) considered the need to move from automobile-centred urban planning to a more human-friendly and sustainable vision. In Steuteville and Langdon (2003) there is an important effort to try to change settlement evolution from a "natural" chaotic sprawl into a planned approach, considering many case studies and organising a collection of best practices and philosophical concepts behind them. With Grant (2006) the green urbanism movement got a strong analysis of both theory and practice in green urbanism. To create good communities, Grant criticises in a constructive way current approaches in many parts of the Western countries, Japan included. He also analyses and opens a debate about the concerns that are being applied to new urban planning philosophies, focusing on aesthetics topics too. In other cases, these strategies are more tied to the so-called "smart growth policies" that, essentially, require considering efficiency in using natural resources. Examples of these cases are the studies of Gordon (2003) that even propose preemptive planning focusing on small communities. Their most important concern is to conjugate the New Urbanism movement they were experiencing everywhere with the preemptive planning to protect and preserve the ecosystem. So, their focus is to improve the New Urbanism adding to it a stronger dimension along the environment, letting it become a Green New Urbanism. Another example can be found in White and Ellis (2007).

In these documents, we can see how the New Urbanism, thought to reduce sprawl and increase social-well being evolves into Green Urbanism where the "Green Agenda" becomes prevalent.

While this movement was arising in the USA, the same happened in Europe during the '80s and '90s. An example could be the Brundtland Commission which defined sustainable development as the result of "balancing economic, social, and environmental objectives to preserve options for future generations" (WCED, 1987). This definition contributed to spreading, at the European level, these ideas and increased their relevance in the New Urbanism trends.

In the 1990s, a strong movement started opposing growth evidencing the issues caused by chaotic sprawling. These pressures to avoid, for example, traffic congestion, or air and water pollution forced some local governments to try to even decide to brake or stop development. To solve this conflict, alternative growth strategies were proposed. An example could be the already cited Ewing (1996) with its 12 principles, another is Cervero (1986) that focused his analysis on business complexes and proposed solutions to suburban congestion, focusing on suburban transportation planning. In Downs (1994) the topic of collaboration between suburbs and city centres to find a common "win–win" strategy is analysed defining a very detailed and persuasive proposal of a regional approach to planning.

Also, organisations like the American Planning Association, the Urban Land Institute, the National Governors Association, and the Congress for the New Urbanism quickly became promoters of a new form of growth, that they called "smart growth".

The results of this process have led to many different frameworks for planning. One is Ewing's (1996), another is the one from Gordon and Tamminga (2002). Jabareen (2006) and White and Ellis (2007) are two other examples.

Unfortunately, these frameworks were conflicting. Often, they gave opposite evaluations of the same result. Often some of them proposed strong criticism of others' work, for example affirming that the results of new urbanism trends were poor from the environmental perspective. Most of these critics were focused on missing a real, true, holistic and green vision. For example, Audirac et al. (1990) evaluated some neo-traditional projects in Florida and observed that a compact development policy was an excessive simplification of reality and led to poor results. Reasons underlined by them were the unsatisfactory protection of the environment or the slowness of change adaptation of the local development regulation. Grant et al. (1996), as already cited above, criticise the real application of the most important values for urban stakeholders. In their vision elements like democracy, globalisation, spatial patterns and more are checked against their impact on the New Urbanism movement and they conclude that there is yet a lot to do, especially considering the need to build with high urban densities. Many studies (Durack, 2001; Till, 2001; Zimmerman, 2001) underlined, demonstrating that, that New Urbanism often used environmental issues as a marketing strategy, not based on a real sensitivity to the problem but only targeted to "sell" their projects. Even scholars that favoured the high-density strategies proposed in urban planning, like Gordon and Vipond (2005), evidenced that biodiversity preservation and land preservation were a requirement to be satisfied.

In this paper, the Green Dimension of a smart city is defined as having implemented, in the smart city, the concepts of Green Urbanism. To have a reference framework for it, in this paper we will refer to Lehmann's model for Green Urbanism in the 2010 version. The reason is that this study is well-structured and, for this paper, has a high coherence and a completeness level that allows a comparison with ISO standards. In his studies, Lehmann proposed some principles organised in groups that cover all the elements of Green Urbanism.

The fifteen principles of Green Urbanism according to Lehmann, in its 2010 formulation (Fig. 5), are a good example of the mismatch between them and the ISO standards referred to before.

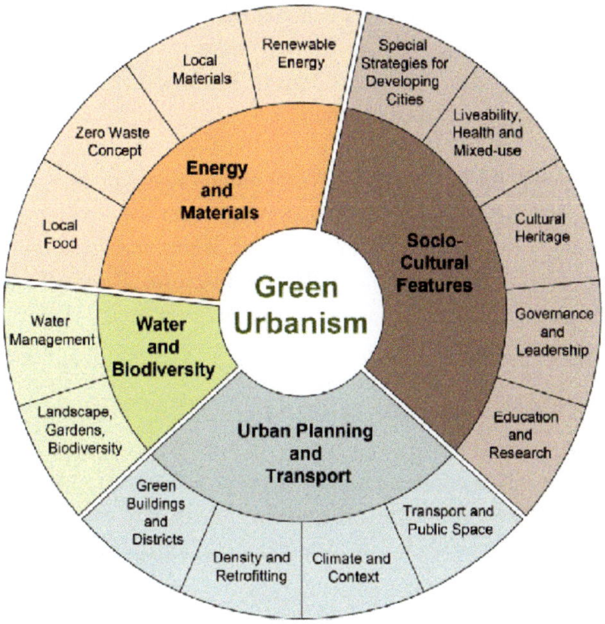

Fig. 5 Principles of Green Urbanism according to Lehmann (image from Lehmann, 2010)

Lehmann divides his Green Urbanism model into four sections: sociocultural features, urban planning and transport, water and biodiversity, energy, and materials. He defines a holistic vision where these four areas (and their fifteen subsections, The Principles) are interconnected and influence each other. In his formulation, Lehmann also considers some actions to be taken for each principle. An example (revised and summarised) of these actions is shown in Table 1.

Matching these actions with the ISO standards is not easy and, often, even not possible. But also defining a new set of KPIs cannot be considered a solution because, again, these KPIs will quite always be correlated to the feature but not a real measure of the feature itself because it is not quantifiable.

What is missing, is a structured approach, feasible in any case, that can allow, at least the:

- Description of the elements to be monitored to state the "health" of the smart city "greenness" (the alphas)
- Definition of milestones to both define a roadmap and assess where the project state is during time
- Definition of relationship among these alphas
- Definition of the relationship between these alphas and real-world results (e.g. a building, an infrastructure)
- Definition of best practices for the implementation of objectives.

5.4 The Application of Essence 1.2

Since the start of software development, a discipline called software engineering arose among all other elements of the software industry and research. But over decades it became more and more evident that a large gap was between the developers' world and the researchers and the academic world in this discipline. In fact, in 2009, Ivar Jacobson, Bertrand Meyer and Richard Soley launched an initiative called SEMAT. It was a call for action and a vision statement. SEMAT aimed to bridge such a gap between the developers' community and the academic community, with the target to create a new community to solve the big issues of immature practices in software development. In its paper, Jacobson (2010) explicated the results of this initiative and provided a framework for its use and evolution. The results of SEMAT during its evolution have been analysed by other scholars (e.g. Dwolatzky, 2012) and they converged into the Essence standard (Jacobson et al., 2013). The Essence

Table 1 Examples of Lehmann's recommendations

Principle	Recommendations
Principle 6: Transport and space Create a good network of public transport reducing its fossil fuels usage	5% GDP investment in public transport Grow tramways and free hybrid buses Multi-modal public transport systems with a high number of connections and choices Implement safe pedestrian and cycle networks with priority for them
Principle 12: Cultural heritage Promote cultural identity	Protect existing elements of cultural heritage Facilitate more adaptive reuse through relaxed building code Increase communities' participation
Principle 14: Education, research and knowledge-sharing Promote sustainability at all educational levels, especially at university	2% GDP in research and innovation Empower university education programmes including sustainability topics Ease sustainable behaviours, especially for long-term behaviour change

standard is now at release 1.2 (Object Management Group, 2018) and it is an officially approved and maintained standard from Object Management Group (OMG) that consist of a formal descriptive theory for software engineering. This standard, although in its recent development, has been demonstrated to be well-formed and practically usable by many scholars (Graziotin & Abrahamsson, 2013; Syynimaa, 2018). Essence provides, in its specification, also a customisation process to create new elements beyond the original ones (The Kernel), as clearly shown by Ivar Jacobson (Jacobson et al., 2019).

To apply a customised form of Essence 1.2 to the smart city, a new organisation of areas of concern has been done. For the sake of simplicity, in this paper, the endeavour area of concern has been omitted, because it goes beyond the scope of the article. In Fig. 6, the new organisation of areas of concern is shown.

The change to be evidenced is in the Software System alpha which has been replaced by the smart city alpha. The other alphas have remained unchanged. This change, according to Essence 1.2 specifications, was not needed because its extension mechanism allows the creation of sub-alphas but, in this case, it seemed to be more opportune to define a new alpha to make the diagram more understandable.

In this customisation, some sub-alphas have been added to implement the chosen Ramaprasad framework. The first sub-alpha set is about stakeholders alpha and is shown in Fig. 7.

The second sub-alpha set is about the Opportunity alpha: the outcomes (values) proposed in Ramaprasad's framework have been mapped as sub-alpha as shown in Fig. 8.

The third sub-alphas set is about Requirements. In this set the Green Urbanism principles have been arranged because they are the "requirements" to be satisfied (after their gathering and discovery) by the smart city alpha, to provide Opportunity (outcomes) needed by the stakeholders, as depicted in Fig. 9.

The last sub-alphas set is about the smart city alpha. This alpha has been put all the elements of the framework "smartness" element, according to Ramaprasad framework, as represented in Fig. 10.

This last sub-alpha set is one that directly controls and measures the smart city development stage. In this paper, this last sub-alpha definition, coming from the Ramaprasad framework, has not been considered useful and the research has moved into using it as an ontology (as Ramaprasad's idea was) to describe all the illustrative elements. Anyway, this representation has been reported because the further investigation will be done in future research about a formal descriptive theory of Green Urbanism.

To effectively apply Essence, for each alpha, a lifecycle composed of some alpha states must be defined (Jacobson et al., 2019). The alpha states are evidenced by checking the evolution of some Work Products, that are used to organise the Alphas and give visibility to their progress. To progress the Alpha States, some activities are performed and their outcomes (changes in Work Products) provide the Alpha states evolution.

In this article, not all the elements will be reported because they are too many for the scope of the paper but the methodology should be clear. Considering only Ramaprasad's elements they are 25,200 possible "illustrative components", but each of them can exist in different forms so the possible number of cases is even higher.

There are some other elements, according to Essence, that must be designed. They are:

- Activities, the task to be accomplished to progress with the smart city development.
- Competencies, to explicitly declare needed skills
- Patterns, to provide qualified solutions to common problems.

Also, these elements will not be detailed to keep the article focused on methodology demonstration.

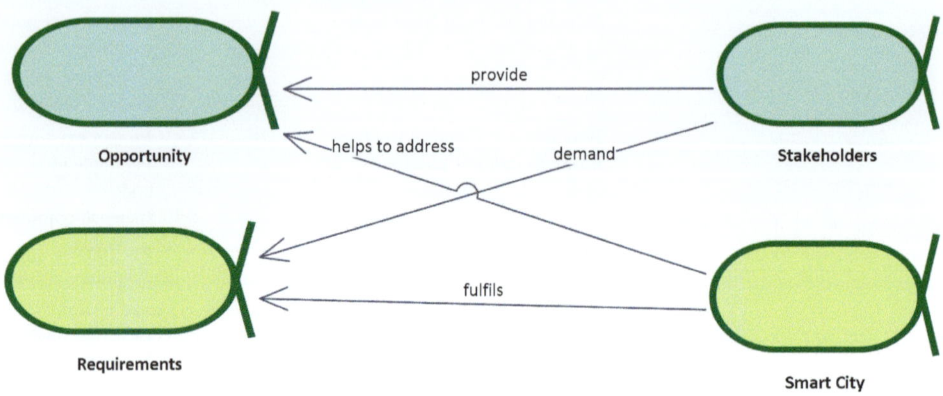

Fig. 6 Essence 1.2 areas of concern adapted to a smart city (image of the author)

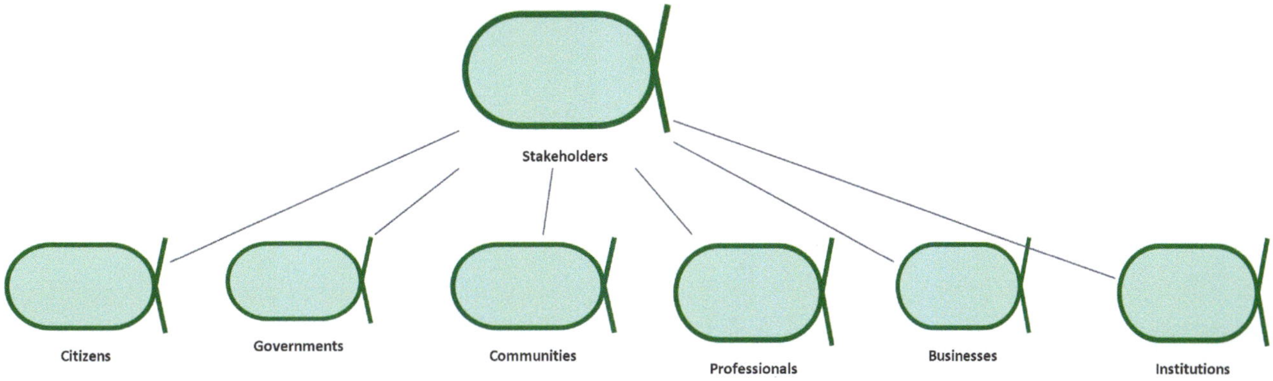

Fig. 7 Sub-alphas for smart city's stakeholders (image of the author)

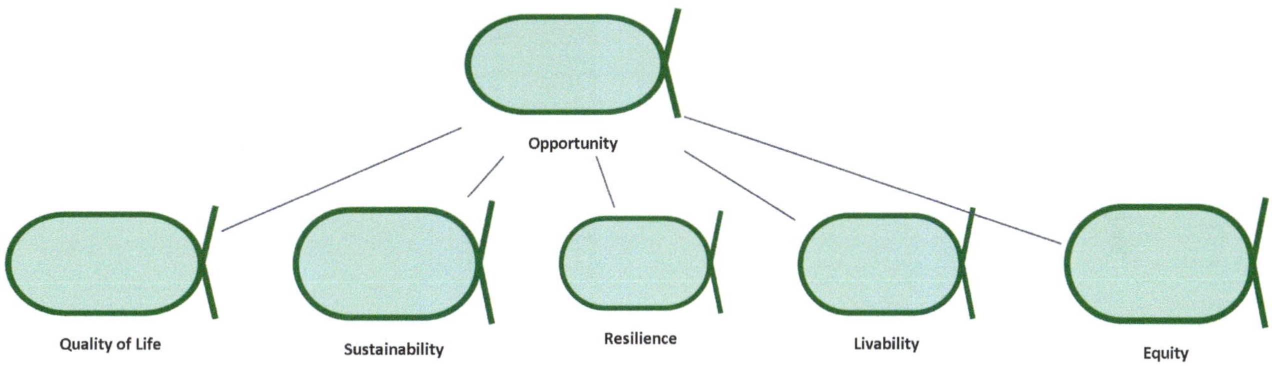

Fig. 8 Sub-alphas for smart city's opportunity (image of the author)

6 Connecting Model, Framework and Essence

6.1 Introduction

Up to now, the Green Urbanism model has been essentialised (i.e. translated into Essence language), and Ramaprasad's framework has been essentialised too (Ng & Huang, 2013). To complete the approach and get an applicable result, the two elements, model and framework, must be connected through Essence.

Connecting the two elements will provide the following results:

- Trace all requirements from their most general form up to their physical implementation
- Represent a traced interdisciplinarity starting from Urban Planning, Urban Design and Architecture set of concepts that are mapped onto Engineering (Civil, Construction, Electronic, Telecommunications, Organisational)
- List all Work Products (i.e. the real-world artefacts that are created or modified during the smart city realisation process)
- Identify all needed activities that must be accomplished to build the smart city.

Further results, not considered in this article, are:

- Identify competencies needed for the smart city development
- Describe all patterns applicable to smart city development
- Map the identified activities onto the Essence Activity Spaces to check that the no activity is missing. This subject will be considered in a further paper due to its importance from the development process point of view.

6.2 Connection Example: Cultural Heritage Connection

As a first application example of connecting the elements, it will be considered the Cultural Heritage Principle in Lehmann's model. Putting this alpha into the framework means combining the applicable elements of the framework to get the expected outcomes for cultural heritage.

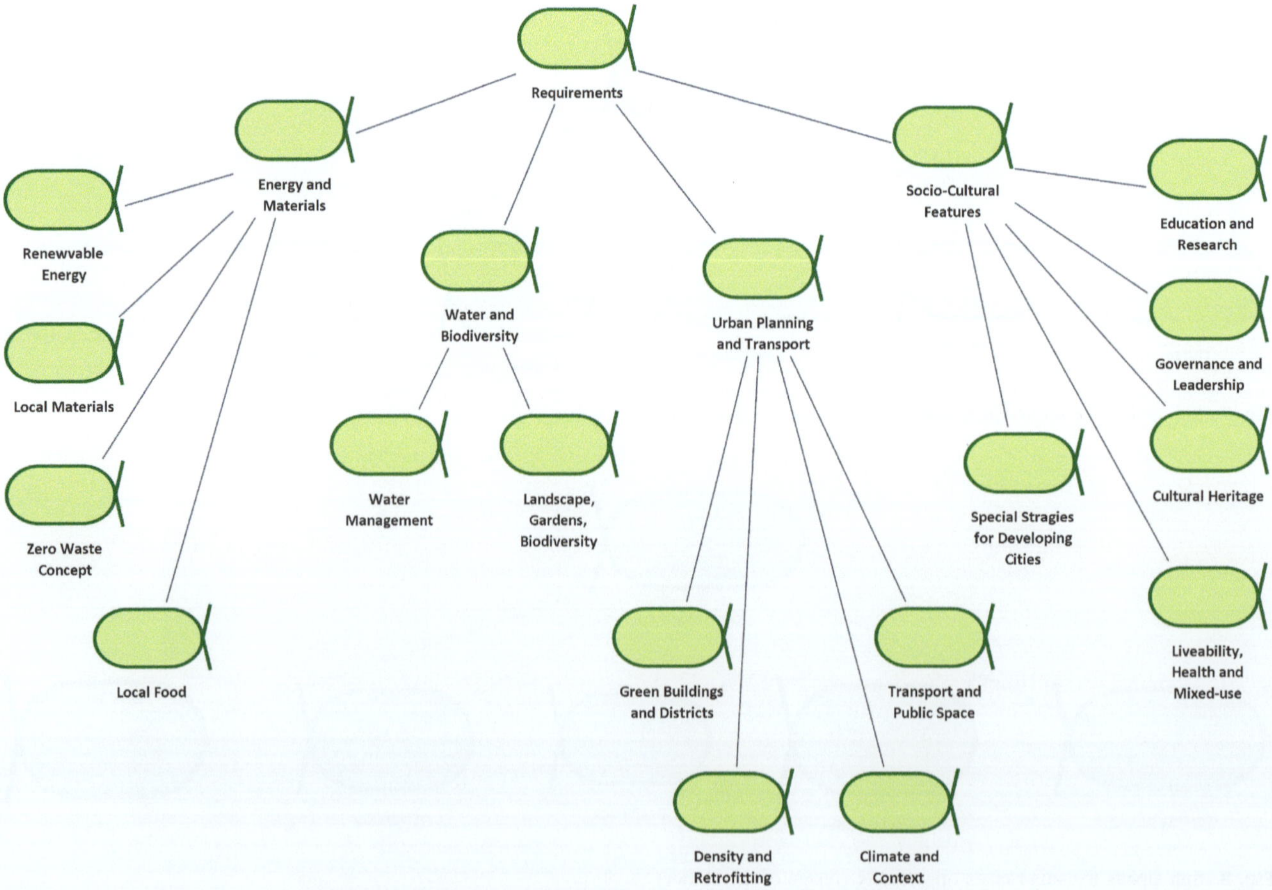

Fig. 9 Sub-alphas for smart city's requirements (image of the author)

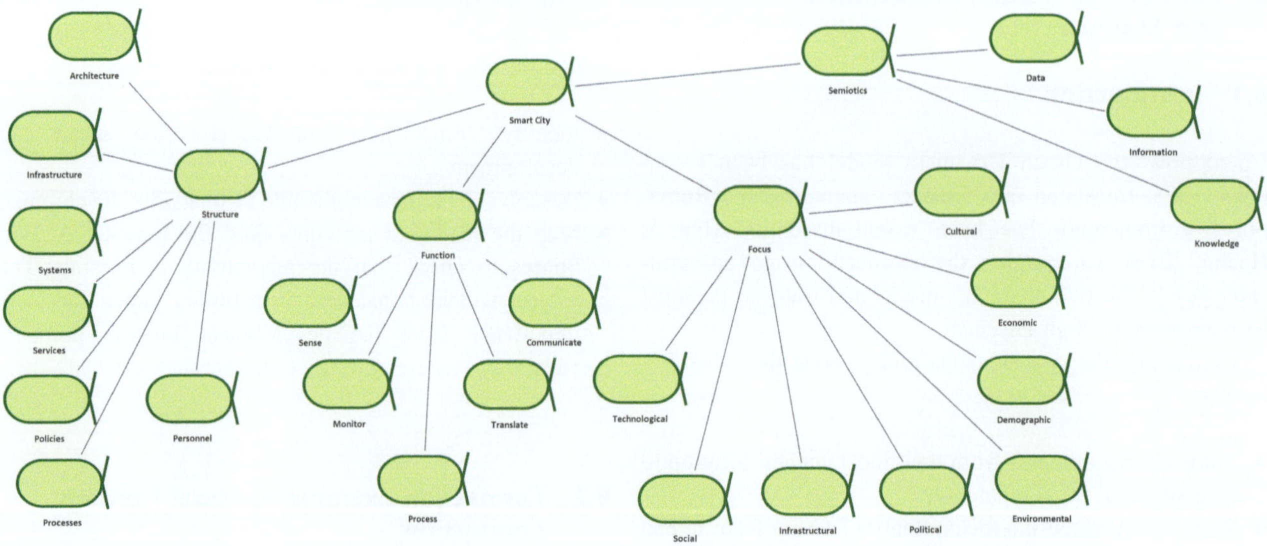

Fig. 10 Sub-alphas for smart city alpha in Ramaprasad's framework (image of the author)

Considering the quality of life definition given by WHO "an individual's perception of their position in life in the context of the culture and value systems in which they live and about their goals, expectations, standards and concerns", Cultural Heritage alpha influences the quality of life alpha, and this connection must be described in terms of illustrative components of the framework. To serve as an example of application it will be considered the following simplified set of illustrative components, each of them identified by a unique identifier:

- IC01: Architecture to communicate cultural data to citizens for quality of life
- IC02: Infrastructure to sense social data from citizens for quality of life
- IC03: Systems to process cultural information to citizens for quality of life
- IC04: Policies to communicate technological knowledge to professionals for the quality of life.

Other illustrative components are omitted for simplicity.

The illustrative components are Work Products (according to Essence language) that must be built to implement the QoL, that is to progress the alpha state of QoL alpha up to the desired configuration. In the diagram depicted in Fig. 11, the QoL alpha is described by the four illustrative elements generated through the framework.

This diagram shows that the QoL alpha is described by the Work Products that are the illustrative components, as Essence requires. To improve the diagram, it is necessary to describe the relationship between the illustrative components and the alpha states of QoL. But in Essence one element can have only one life cycle so it is needed to define four sub-alphas, one for each illustrative component. In Essence, Work Products do not have states but a level of detail so, for each component, it is also needed the relationships with all the activities that contribute to the creation or evolution of the Work Product through its detail levels. The resulting diagram is the following Fig. 12 (only one sub-alpha has been detailed and some relationships have been grouped in a single line):

The same essentialisation process must be done for each work product and sub-alpha and is omitted in this article.

The Work Product detail level is part of the definition of the work product itself. For the IC04 element, the level of detail is given in Table 2.

The activity QoL policies development will influence both the IC04 detail level and QoL Policies alpha state. The last step is then to define the alpha states (the life cycle) for QoL policies, using all the available information (from the Green Urbanism model, from the framework and ISO standards).

The result is described in Table 3 (in this case, they are very similar to those of the Opportunity alpha directly from Essence kernel):

For the Cultural Heritage alpha, an alpha states life cycle has been developed rearranging the Essence typical life cycle for requirements alpha as represented in Fig. 13 and in Table 4.

The same has been done for the quality of life alpha, with a completely new alpha life cycle. In this case, it is possible to identify some work products and some activities that are omitted for the sake of simplicity.

7 Conclusions

Applying Essence 1.2 to a smart city framework and a set of smart city indicators can lead to an explicit, formal representation in form of diagrams that provides many advantages.

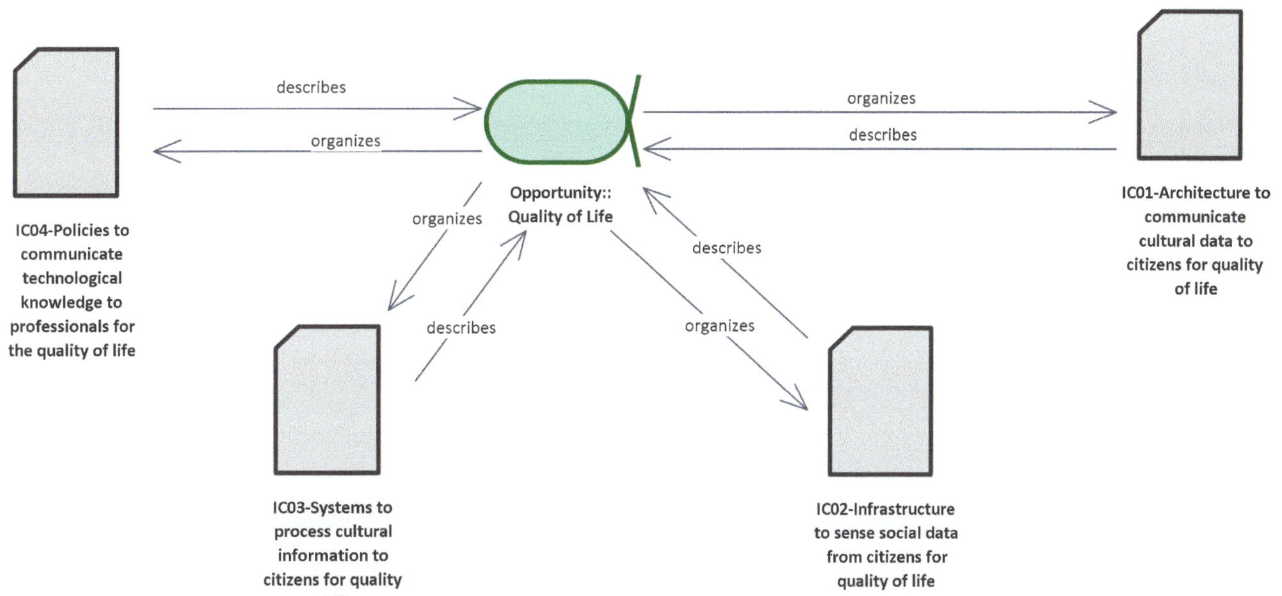

Fig. 11 Essentialised illustrative components for QoL (image of the author)

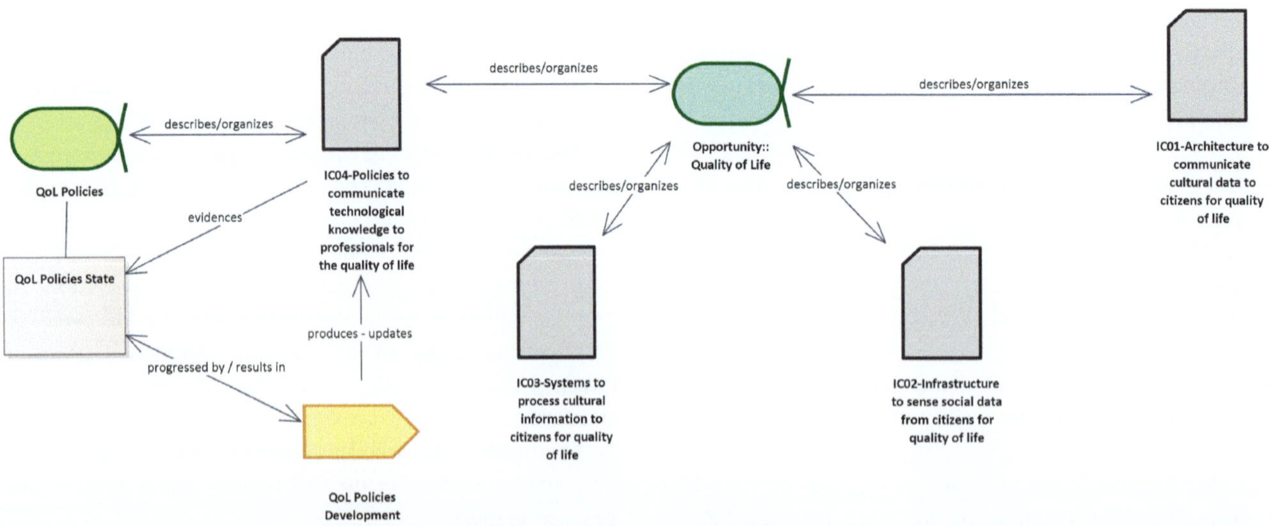

Fig. 12 Complete Essence diagram after essentialisation of QoL (image of the author)

Table 2 Product detail levels for the illustrative element IC04

Detail level	Description
Initial	The main concepts of the policies have been defined
Structured	All sections of the policies have been detailed and rearranged in a discursive flow
Approved	The policies have been approved by the interested stakeholders
Deployed	The policies have been distributed, communication has been done, and training has been provided
Effective	The policies have been adopted in practice and are followed

Table 3 Alpha states for QoL policies

Alpha state	Conditions
Identified	The main ideas of policies have been identified At least one investing stakeholder is interested Other stakeholders identified
Solution needed	Draft schema of policies has been identified Stakeholders' needs identified Issues that are solved by policies have been identified Need for a policy confirmed by all stakeholders
Value established	QoL quantified in terms of KPIs (from standards) and in terms of sub-alphas to be accomplished (if available) QoL impact understood Smart city value in terms of QoL understood Success criteria for QoL defined and understood Outcomes for QoL are clear and quantified
Viable	A QoL solution has been outlined Possible constraints to the solution have been identified Main risks identified, quantified and considered acceptable or manageable The solution is cost-effective The reasons to develop the solution have been clearly understood and explicitly stated The target is viable (in the sense of sustainable)
Addressed	The QoL values have been fully addressed The selected solution is considered worthy to be deployed Stakeholders are satisfied
Benefit accrued	The solution implemented for QoL has been considered in line with the expected benefits Return on Investment (ROI) is considered acceptable

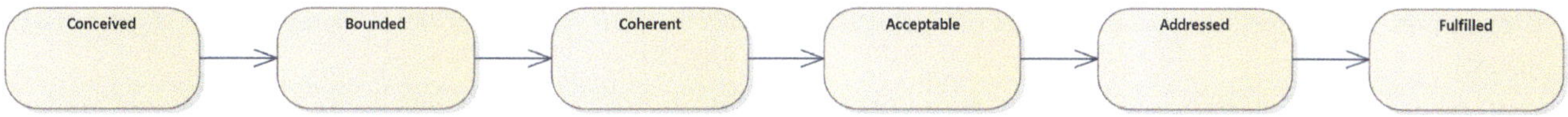

Fig. 13 Alpha states for Cultural Heritage alpha (image of the author)

Table 4 Alpha states for Cultural Heritage alpha

Alpha state	Conditions
Conceived	Stakeholders agree about the need to preserve CH Targets of CH identified Funding stakeholders identified QoL opportunity clear
Bounded	Acquisition stakeholders identified CH borders agreed CH acquisition success criteria agreed Shared roadmap exists Priority scheme clear Constraints and risks identified Hypothesis and assumptions are clear Acquisition requirements defined
Coherent	Conflicts addressed Roadmap shared Key stakeholders identified Priorities are clear and defined CH acquisition team identified Acquisition requirements are completely defined
Acceptable	Roadmap considered acceptable Financial issues have been solved The acquisition management process defined CH value is clear How QoL is influenced is clear Acquisition requirements are considered acceptable
Addressed	The acquisition has gone enough far to be considered accepted Initial acquisition requirements and final results match Realised value is clear Acquisition requirements have been addressed
Fulfilled	Stakeholders accept final results No missing requirements All requirements have been fulfilled

Using the framework as an ontology to represent all the possible elements in a smart city allows urban planners to correctly identify all possible elements to be monitored (i.e. the alphas). This is valid in general and more specifically in the case of Green Urbanism.

All these alphas are monitored using alpha states that are based on checklists that can be fully customisable, putting into them both qualitative and quantitative measures. Mixing both elements into a single state definition allows a more precise evaluation of the progress of the realisation of the smart city project. With this approach, standards based on exact measures and concepts expressed in "soft" form can be considered.

In addition, both direct and indirect measures can be done at any time, giving more reliability to the results.

The essentialised model of the Green Urbanism application to a smart city is, essentially, a description of the smart city Green Urbanism design and, at the same time, a way to define a methodology to implement it measurably, meaning that during the development it is possible to state the health of the project. This progress measurement is possible because Essence allows (Jacobson et al., 2019):

- To define checkpoints (i.e. milestones in the development) using alpha states
- To evaluate, at any time, these states, defining where the project is against checkpoints
- Discover and explicit all the needed activities to move from one state to another, providing elements (tasks) directly usable in a project management methodology

(sometimes with some refinement in terms of the level of detail)

- List all the work products, that are the real, concrete elements to be built, ranging from documents to complete districts, providing effective outcomes at a very fine level
- Check if all the needed activity types (in Essence are called activity spaces) have been addressed to verify if something is missing from the methodological point of view.

One important result of the essentialisation of a Green Urbanism project is the ability to define a way of working in terms of practices (focused ways of doing something) and methods (collections of practices to build something more complex) (Jacobson et al., 2019). This means that it is possible to:

- Define from scratch or reuse best practices to create or improve something (both for practices and methods)
- Compose practices and methods coming from different projects or contexts to define a better way of working (that means continuous improvement of the way of working)
- Clearly state the methodology to apply Green Urbanism to a smart city
- Refine and improve the way of working even during the development of the smart city
- Shape and adapt best practices to specific cases in an effective and clear way.

All these results come directly from the Essence application.

Another aspect that must be considered is the possibility to use the approach shown in this paper to develop a specific formal descriptive theory of Green Urbanism applied to smart cities. This aspect will be investigated in a further paper on the matter, but it is clear from the results of the research described above. Considering this opportunity, it is possible to also imagine a Green Urbanism formal descriptive theory that can be applied to any case, not only the smart city, but this is a further step that will be investigated after the one just cited.

References

Audirac, I., Shermyen, A. H., & Smith, M. T. (1990). Ideal urban form and visions of the good life: Florida's growth management dilemma. *Journal of the American Planning Association, 56*(4).

Azizalrahman, H., & Hasyimi, V. (2019). *Towards a generic framework for smart cities*. INTECHOPEN.

Beatley, T. (2000). *Green urbanism: Learning from European cities*. Island Press.

Calthorpe, P., & Fulton, W. (2001). *The regional city: Planning for the end of sprawl*. Island Press.

Cervero, R. (1986). *Suburban gridlock*. Rutgers University Press.

Downs, A. (1994). *New visions for metropolitan America*. The Brookings Institution and the Lincoln Institute of Land Policy.

Duany, A., Plater-Zyberk, E., & Speck, J. (2000). *Suburban nation: The rise of sprawl and the decline of the American dream*. North Point Press.

Durack, R. (2001). *Village vices: The contradictions of new urbanism and sustainability. Places, 14*(2).

Dwolatzky, B. (2012). *Re-founding software engineering practice—The SEMAT initiative*. IEE Xplore. https://doi.org/10.1109/SE.2012.6242347

Edelston, T. (2012). Still time for Songdo City to protect biodiversity. *Korea Times*, last modified January 8.

Ewing, R. (1996). *Best development practices: Doing the right thing and making money at the same time*. APA Planners Press.

Gordon, D. L. A. (2003). *New urbanism and smart growth: Twins separated at birth?* Places.

Gordon, D. L. A., & Tamminga, K. (2002). Large-scale traditional neighbourhood development and pre-emptive ecosystem planning: The Markham experience, 1989–2001. *Journal of Urban Design*.

Gordon, D. L. A., & Vipond, S. (2005). Gross density and new urbanism: Comparing conventional and new urbanist suburbs in Markham, Ontario. *Journal of the American Planning Association*.

Grant, J. (2006). *Planning the good community: New urbanism in theory and practice*. Routledge.

Grant, J., Manuel, M., & Joudrey, D. (1996). A framework for planning sustainable residential landscapes. *Journal of the American Planning Association*.

Graziotin, D., & Abrahamsson, P. (2013). A web-based modelling tool for the SEMAT essence theory of software engineering. *Journal of Open Research Software*.

Huovila, A., & Bosch, P., & Airaksinen, M. (2019). *Comparative analysis of standardized indicators for smart sustainable cities: What indicators and standards to use and when?* Cities Association of America Geographers conference in Seattle.

Jabareen, Y. R. (2006). Sustainable urban forms: Their typologies, models, and concepts. *Journal of Planning Education and Research*.

Jacobson, I. (2010). A smarter way: The software engineering method and theory initiative (Semat). In Nordio, M., Joseph, M., Meyer, B., & Terekhov, A. (eds.), *Software engineering approaches for offshore and outsourced development. SEAFOOD 2010*. Lecture Notes in Business Information Processing, Vol. 54. Springer, Berlin, Heidelberg. https://doi.org/10.1007/978-3-642-13784-6_1

Jacobson, I., Ng, P., McMahon, P., Spence, I., & Lidman, S. (2012). The essence of software engineering: The SEMAT kernel. *ACM Queue, 10*(10), 40–51. https://doi.org/10.1145/2381996.2389616

Jacobson, I., Ng, P., McMahon, P., Spence, I., & Lidman, S. (2013). *The essence of software engineering: applying the SEMAT kernel*. https://www.ivarjacobson.com/sites/default/files/field_iji_file/article/scott_ambler_interviews_ivar_jacobson.pdf

Jacobson, I., Lawson, H., Ng, P., McMahon, P. E., & Goedicke, M. (2019). *The essentials of modern software engineering: Free the practices from the method prisons!* Morgan & Claypool.

Katz, P. (1994). *The new urbanism: Toward an architecture of community*. McGraw Hill.

Lehmann, S. (2010). *The principles of green urbanism: Transforming the city for sustainability*, Earthscan, ISBN 978-1-8440-7834-9.

Lehmann, S. (2014). *Low carbon districts: Mitigating the urban heat island with green roof infrastructure* (p. 5). Culture and Society.

Midor, K., & Płaza, G. (2020). *Moving to smart cities through the standard indicators ISO 37120, Multidisciplinary aspects of production engineering*—MAPE, Vol. 3, issue 1.

Mullins, P. D. (2017). The ubiquitous-eco-city of Songdo: An urban systems perspective on South Korea's green City approach. *Urban Planning*, *2*(2), 4–12. https://doi.org/10.17645/up.v2i2.933

Ng, P., & Huang, S. (2013). Essence: A framework to help bridge the gap between software engineering education and industry needs. *Software Engineering Education Conference, Proceedings*. https://doi.org/10.1109/CSEET.2013.6595266

Object Management Group. (2018). *Essence—Kernel and language for software engineering methods, Version 1.2*. Retrieved October 18, 2022, from https://www.omg.org/spec/Essence/1.2/PDF

Ramaprasad, A., & Sánchez-Ortiz, A., & Syn, T. (2017). A unified definition of a smart city. In *16th International conference on electronic government (EGOV)*, Sep 2017. St. Petersburg, Russia.

Steuteville, R., & Langdon, P. (2003). *New urbanism: Comprehensive report and best practices guide* (3rd ed.). New Urban Publications Inc.

Sustainable Cities Platform. (1994). *Aalborg Charter—European sustainable cities platform*. Retrieved from https://sustainablecities.eu/fileadmin/repository/Aalborg_Charter/Aalborg_Charter_English.pdf

Syynimaa, N. (2018). Essence: Reference architecture for software engineering. In Proceedings of the 20th International conference on enterprise information systems (ICEIS 2018), pp. 345–350. ISBN: 978-989-758-298-1. https://doi.org/10.5220/0006793603450350

The British Standards Institution. (2014, August). *The role of standards in smart cities*. Retrieved October 18, 2022, from https://www.bsigroup.com/LocalFiles/en-GB/smart-cities/resources/The-Role-of-Standards-in-Smart-Cities-Issue-2-August-2014.pdf

Till, K.E. (2001). *New urbanism and nature: Green marketing and the neotraditional community*, Urban Geography

Townsend, A. M. (2015). *Smart cities*. Norton.

WCED (World Commission on the Environment and Development). (1987). *Our common future. Report of the Brundtland Commission*. Oxford University Press.

White, S. S., & Ellis, C. (2007). Sustainability, the environment and new urbanism: An assessment and agenda for research. *Journal of Architectural and Planning Research, 24*(2).

Zimmerman, J. (2001). *The "nature" of urbanism on the new urbanist frontier: Sustainable development, or defense of the suburban dream?* Urban Geography.

Challenges in the Implementation of Bike Lane Networks in Intermediate Latin American Cities. The Case of Arequipa Peru

Rocío Toni-Llerena, Eliot Loayza-Muñoz, and Carlos Zeballos-Velarde

Abstract

Cycle lanes are transport systems recommended for large and intermediate cities because they desaturate vehicular traffic and at the same time promote non-polluting transportation; however, for this system to work, it must be linked to a public transport system that allows an integrated and multimodal model. As a result of the confinement caused by the covid-19 pandemic, a network of bicycle lanes was implemented in the city of Arequipa, Peru; prior to this circumstance, it was difficult for the municipality to implement them, due to the reluctance on the part of vehicular transport users. However, this proposal was developed with limited technical criteria, without properly studying the capacity of the roads, the volume of traffic, and the behavior of drivers and cyclists, which brought many problems that have resulted in the poor performance of the network. This study proposes a comprehensive model of bike lanes according to the principles of the World Bank, among other institutions, using an integrated multimodal system. For this purpose, in the first place, a general diagnosis of the current network of cycle paths and the complementary measures applied has been carried out, identifying their main problems. Secondly, the perception of the users was measured, using semi-structured questionnaires. Finally, urban design proposals that can help to develop a conceptual model of a cycle system are proposed.

R. Toni-Llerena · E. Loayza-Muñoz · C. Zeballos-Velarde (✉)
Universidad Nacional de San Agustín de Arequipa,
Arequipa, Peru
e-mail: czeballos@unsa.edu.pe

R. Toni-Llerena
e-mail: rtoni@unsa.edu.pe

E. Loayza-Muñoz
e-mail: eloayzam@unsa.edu.pe

Keywords

Cycle paths • Cycle network • Green transportation • Intermediate cities • Arequipa

1 Introduction

The changes generated in society from the situation that was faced worldwide due to the covid-19 pandemic comprise a series of activities, from the adaptation to new habits to the way the population works and mobilizes. The combination of the two main measures adopted, which were "stay at home" and "maintain social distancing", totally reformulated our daily practices, among them avoiding public transport. Therefore, the mandatory socio-spatial practices resulting from the outbreak of the pandemic unleash and raise some critical questions about the ways and conventions that shape our cities (Jabareen & Eizenberg, 2021). Already with the restricted confinement, the fear of getting infected changed the way people travel: walking or cycling began to be preferred instead of going by bus, choosing a private transport car over public transport (Jasinski, 2022).

This situation exposed several vulnerabilities in urban and territorial planning that can respond and be resilient to a social emergency, and at the same time provided the opportunity to implement alternative infrastructure, especially for intermediate cities (Imbert, 2020). To attain more open public spaces necessary for social distancing, it was important to make adjustments in the design and use of the urban fabric by developing new design standards that allow for flexibility and variability of use while also taking social wellbeing into account (Abusaada & Elshater, 2022).

The covid-19 situation has forced national and local governments to develop a series of measures and allocate emergency funds to mitigate the effects of the pandemic. Taking advantage of this circumstance, many cities have carried out projects to reshape traffic, reducing car access

A. Battisti et al. (eds.), *Greening Our Cities: Sustainable Urbanism for a Greener Future*,
Advances in Science, Technology & Innovation, https://doi.org/10.1007/978-3-031-49495-6_4

and expanding bicycle and pedestrian paths. Taking into account that "opportunities arise from crises" and in the search for a more humane and inclusive city model (Díaz-Gamboa, 2020), a pilot plan for bicycle lanes was implemented in the city of Arequipa. The intervention had to happen quickly and without much planning, taking advantage of the lack of cars due to confinement. This study examines the effects of this intervention two years after it was implemented, now that the state of emergency has been greatly reduced and people have returned to normalcy.

2 Literature Review

One of the sectors that suffered the most from the effects of the forced quarantine was public transport, which after the confinement ordered by the authorities, had a drop of up to 50% since people experienced fear of getting infected. Therefore, people chose transport alternatives that somehow limit contact with others; these means of transport are reduced to private cars, linear motorcycles, and non-motorized vehicles such as skateboards, scooters, skates, bicycles, and any vehicle that works with the propulsion force that does not come from an engine, in this case, the human force.

In this regard, Vicuña Mena et al. (2022) affirm that, in terms of the socioeconomic impact and its interaction with public transportation, there was a significant difference in the drop in public transportation use during the pandemic. Also, Kolarova et al. (2021) noted an increase in car use and a decrease in public transportation use, as well as a more negative impression of public transportation as a transportation choice. Traveler behavior changed substantially as the world reacted to covid-19. Governments and transit authorities have planned substantial mitigation measures, but education campaigns will be required to maintain the trend of sustainable usage of public transportation after this pandemic (Labonté-LeMoyne et al., 2020).

Several studies analyze the behavior of users in bike lanes, associating data-driven urban planning schemes in smart city operations management. According to Zhuang et al. (2021) to achieve the objective of improving the accessibility, flexibility, functionality, and connectivity of the cycling traffic system, grouping methods were used to identify communities such as the Louvain method. This is a graphical clustering algorithm, applied in two areas in Shanghai, including a campus and an urban area to automatically find through this analysis the typical cyclical patterns in the spatial and temporal dimensions of cyclists. Likewise, the study carried out by Liu et al. (2021) aims to identify sections of the existing road network suitable for building lanes, aiming to achieving the greatest possible coverage in the cycle path road network and its continuity. The method used was to capture the route options of the cyclists integrated with GPS devices, recording the detailed travel pattern of the cyclists, and understanding the route choices made by the cyclists, in this way, they were able to map their travel patterns.

In this sense, Calderon Peña et al. (2017) mentioned the following components that generate an inclusive policy and planning of bike lane network models (Table 1).

Similarly, Targa et al. (2020) propose eight requirements that a complete, connected, and well-designed bicycle network must meet eight infrastructure requirements. These requirements represent items that will be later analyzed in the development of our study (Table 2).

Table 1 Components of an inclusive cycle network policy

Components	Definition
Model of city	Mobility master plan
Infrastructure	Signposted infrastructure to give priority to cycling
Regulation and institutions	Government institutions dedicated to providing cycle-inclusive policies
Communication, education	Citizen initiatives, bike rides, campaigns, or events

Table 2 Requirements for an effective cycle network

Requirements	Definition
Coherence	It must be recognizable by all the stakeholders on the road, without generating improvisation, informality and uncertainty
Direct routes	Direct and optimal paths
Safety	Sustainable road safety
Comfort	Comfortable experience
Attractiveness	Elements that generate added value to the travel experience
Spatial integration	Integration to the urban network
User experience	Positive experience while using the cycling infrastructure

However, things are very different in the Global South, where cities have not implemented bike lanes or a culture of mass cycling and, due to the current situation of the covid-19 pandemic, there is a need to implement these lanes for cyclists. Far from being a perfect solution, this type of implementation produces social and security conflicts and increased traffic. This study analyzes the case of Arequipa, Peru, and the consequences of the implementation of a poorly planned road network.

3 The Local Background

Arequipa is the second city of Peru, located in the southern part of the country and it has a population of around one million inhabitants. Its historic center is a UNESCO World heritage site and the largest number of roads in the city converge toward it, forming a radio-centric urban model. Before the pandemic, there were only 300 m of bike lanes in the whole city, located in front of a mall, which were never used by the population (Zeballos-Velarde, 2020).

On October 30, 2020, the National Program for Sustainable Urban Transport notified a plan for the completion of 380 km of bike lanes in 24 cities in the country. With Arequipa being among these cities, it was initially established a total of 2,900,000 soles (about USD 800,000) for 33 km of this circulation network. So far, only approximately 13 km have been implemented at a cost of 800,000 soles (about USD 225,000) (Zavalaga, 2022).

The office that should have been in charge of making a bike lane plan in the city is the Municipal Office of Transportation (GMT), but they never did so given the resistance of the public transport operators, who opposed the bike lanes because they take away space on the roads. During the pandemic, the GMT claimed not to have the necessary affordability nor the personnel to carry out this plan due to remote working limitations. In order not to lose the budget assigned to the Provincial Municipality of Arequipa, they concluded in transferring the responsibility to the Municipal Office of the Historic Center (GCH), in charge of the UNESCO zone, which was the one that finally executed the cycle paths in Arequipa (Zavalaga, 2022). This alternative transport project was carried out anomalously and within a context of a pandemic and quarantine in 2020, when there was no public transportation available or it was limited. Therefore, rapid planning was necessary to propose a pilot project that should be evaluated in six months, and in this period all data on frequencies of use, incidents, accidents, levels of satisfaction, etc., would be measured, in order to make the corrections and necessary adjustments. However, according to Azpilcueta from the GCH (personal communication, 2022), the project could never be fully implemented nor subsequently evaluated due to political pressure and pressure from the press, since it lacked political support.

In most of the routes reviewed, various design problems were identified, the most frequent being the width of the track, the lack of continuity in the sections, the lack of connectivity in the network, the lack of signage among others. Some of these problems are presented in Fig. 1 derived from the following factors:

(a) Despite the fact that the bike lane was segregated from motor vehicle traffic for the safety of users through elements such as bollards, due to the intensity of vehicular traffic and having less space for cars, the impact on traffic increased exposing the integrity of the cyclists.
(b) Some of the routes were located in areas where there is a conflict with industrial parking lots and in some cases the entrance to private parking lots was not taken into account, so some vehicles had to invade the bike lane routes and used them as parking lots. See Fig. 2.
(c) The intersections between the routes are conflictive points because their design does not contemplate all the possible movements to be carried out by the cyclists, conflicting with their sense of security.
(d) The layout of bike lanes was carried out mostly along the road, including narrow streets in the center of the city, which causes interference with motorized vehicles, including public transport, causing the bike lanes to be invaded by motorized vehicles, deteriorating their infrastructure (Fig. 3).
(e) Because earlier coordination with transportation inspectors and the Traffic Police was not adopted as a complementary step, the right use of Arequipa's bicycle routes was not monitored, nor were sanctions applied to highlight the intangibility of these places.
(f) The project was implemented without a complementary plan for prior awareness directed at urban transport companies and other types of drivers, causing resistance to the implementation of bike lanes in Arequipa.

In addition, once they were implemented, the bike lanes were left to their own devices (Zavalaga, 2022). Currently, the relocation of damaged bollards and rubber stoppers along the 13.5 km of temporary bike paths located in the center of Arequipa is not a priority of the GMT. In fact, according to the last inspection carried out by management personnel, more than a thousand broken bollards and 120 destroyed rubber stoppers were identified (Morales, 2022).

In addition, short after this study was conducted, the cycles lanes were removed due to political reasons, returning to the situation prior to the pandemic.

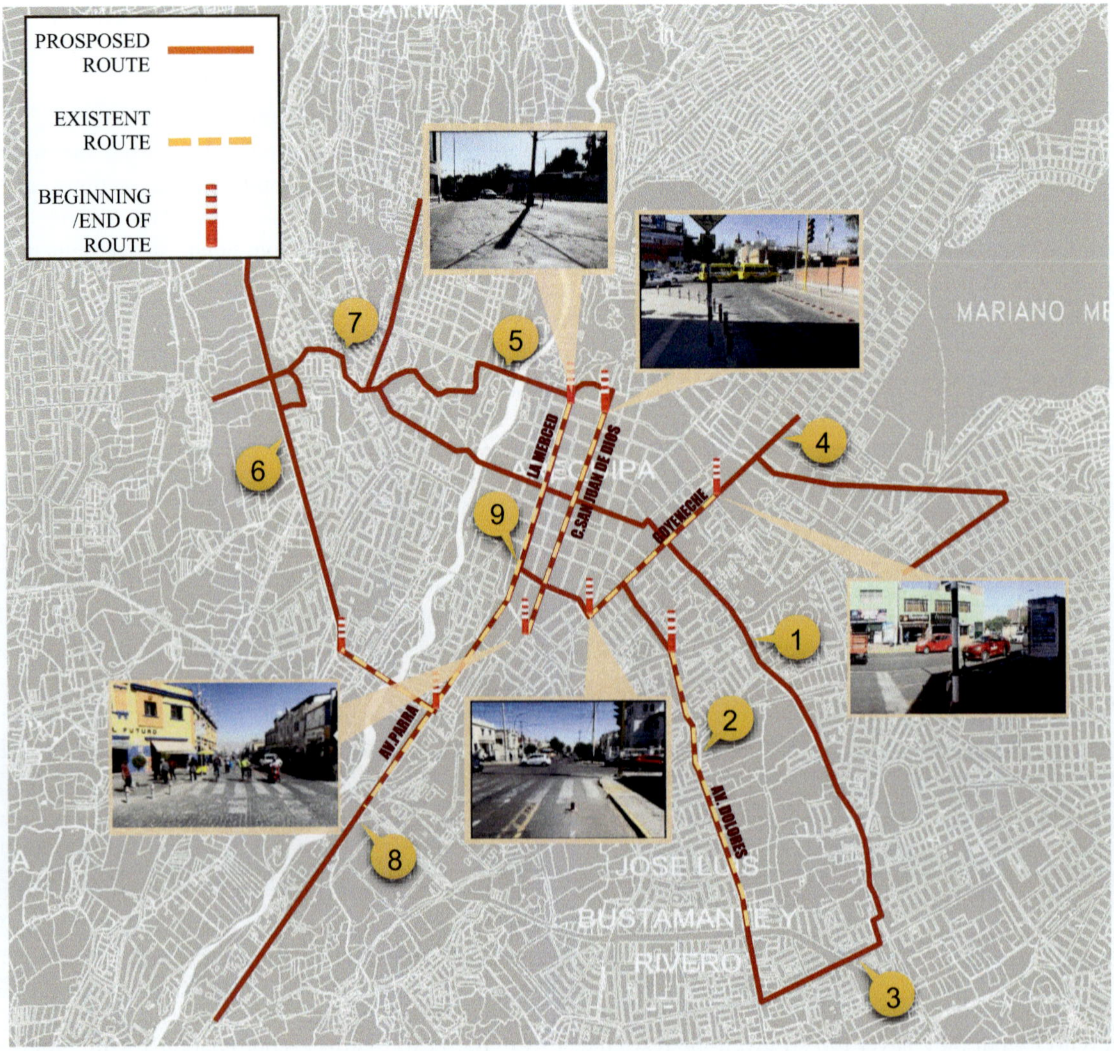

Fig. 1 Existing and planned routes. *Graph* R. Toni, E. Loayza

4 Methodology

4.1 Methods

This study aims to comprehend the peculiarities of a one-of-a-kind urban transportation experiment conducted in Arequipa in a short period of time and under unusual conditions. We also want to understand what caused the inability to create a bike lane network despite having the funds and being recognized as a sustainable mode of transportation suggested by many academics and urban planning professionals.

For this, in the first place, some indicators that a bike lane should have often discussed in the scientific literature, were taken into account. Subsequently, a structured survey was carried out on four types of users: cyclists (25), transport carriers, both public transport and private drivers (13), and urban planning experts (3). The interviews were carried out in September 2022.

Based on the literature review, five main categories were considered for the conception of a bicycle infrastructure network: planning, governance, ease of use, integration, and user experience. A group of questions were asked for each category in order to measure the perception of the three groups of users (cyclists, private motorized users, and public transport operators). These questions were organized in the following semi-structured interview questionnaire (Table 3).

4.2 Limitations

Due to the devastating impact of the covid-19 pandemic in Arequipa (Peru had the highest number of deaths per million in the world during the pandemic), the constraints imposed

Fig. 2 Deterioration of equipment in bike lanes. *Photos* R. Toni, E. Loayza

Fig. 3 Conflicts in the bike lane routes with industrial and private parking lots. *Photos* R. Toni, E. Loayza

Table 3 Semi-structured interview questions

Items	Question
Planning	Do you consider that the implementation of the cycle path motivated the use of the bicycle as a means of transport? Why?
	Do you consider that the implementation of the bike lane improved or harmed urban mobility?
	What significant changes did the pandemic bring to public transportation?
Governance	What are the necessary measures that should be carried out by the authorities concerning bicycle lanes?
	What measures are required to improve the use of bicycles?
Ease of use	Do you consider that there was sufficient surveillance and/or signaling of the correct use of the cycle paths?
	What do you consider to be the reasons why motorized vehicles invade bike lane areas?
	What difficulty do you think a cyclist may have if he goes to work or study by bicycle?
Integration	What aspects could the current bike path change so as not to interfere with the traffic of cyclists and motorized vehicles?
User experience	Do you consider that the cycle path is safe along its route?
	Do you consider that the cycle path is comfortable and easy to use?
	Do you consider that the bike lane offers continuous routes and connections?
	Do you consider that the cycle path is suitable for both novice and experienced cyclists?

by the health authorities made a thorough statistical analysis for this study extremely difficult. As a result, it was decided to take a broad approach to the problem, which enabled us to study user perceptions of the deployment of bicycle lane infrastructure, with the objective of expanding these studies once the sanitary restrictions were released. Unfortunately, because the bike lanes were removed by the end of 2022, statistically deepening these analyses was not possible. However, this study is one of the few academic records on this experiment in urban transportation in Arequipa.

5 Results

The following statistical graphs represent the views of each of the user groups interviewed, differentiated both by group and their overall participation. This allows us to have an idea of the specific interests of each group as well as their differentiated perception of the experimental proposal of bicycle paths in the city.

5.1 Planning

In the planning category, three aspects have been analyzed. Regarding the motivation to use the bike lanes, half of the interviewees were not motivated. However, this opinion varies considerably in the different groups, as 77% of cyclists were motivated, in contrast to private (35%) and public carriers (30%). Regarding the perception about the improvement of the city transport after the implementation of the bike paths, the majority of users (53%) think that transport was rather affected after the installation of bike lanes, reaching 90% of this perception among public transport carriers. Even among cyclists, only 26% think that the installation of this infrastructure constituted a significant improvement for the city transport. On the other hand, regarding the perception of the changes that this implementation produced, there are also different perceptions, because although 60% of cyclists valued the possibility of traveling by bicycle, both private (60%) and public carriers (70%) pointed out their economic losses due to the increase in vehicular traffic. However, it is important to note that 22% of the interviewees valued the reduction of pollution as a positive change during the implementation of this project (Fig. 4a–c).

5.2 Governance

Regarding governance, a question was asked about what measures should be implemented to improve the infrastructure of the bike lanes and the majority of users (72%) thought that it was best to remove them. Here, there is also a differentiated opinion, since the majority of cyclists (83%) thought that it was not necessary to remove the bike lanes, but to improve their planning. There is also a differentiated perception among carriers, because while the majority of private motorized users (83%) wanted the bike paths to be removed but could accept their presence with better

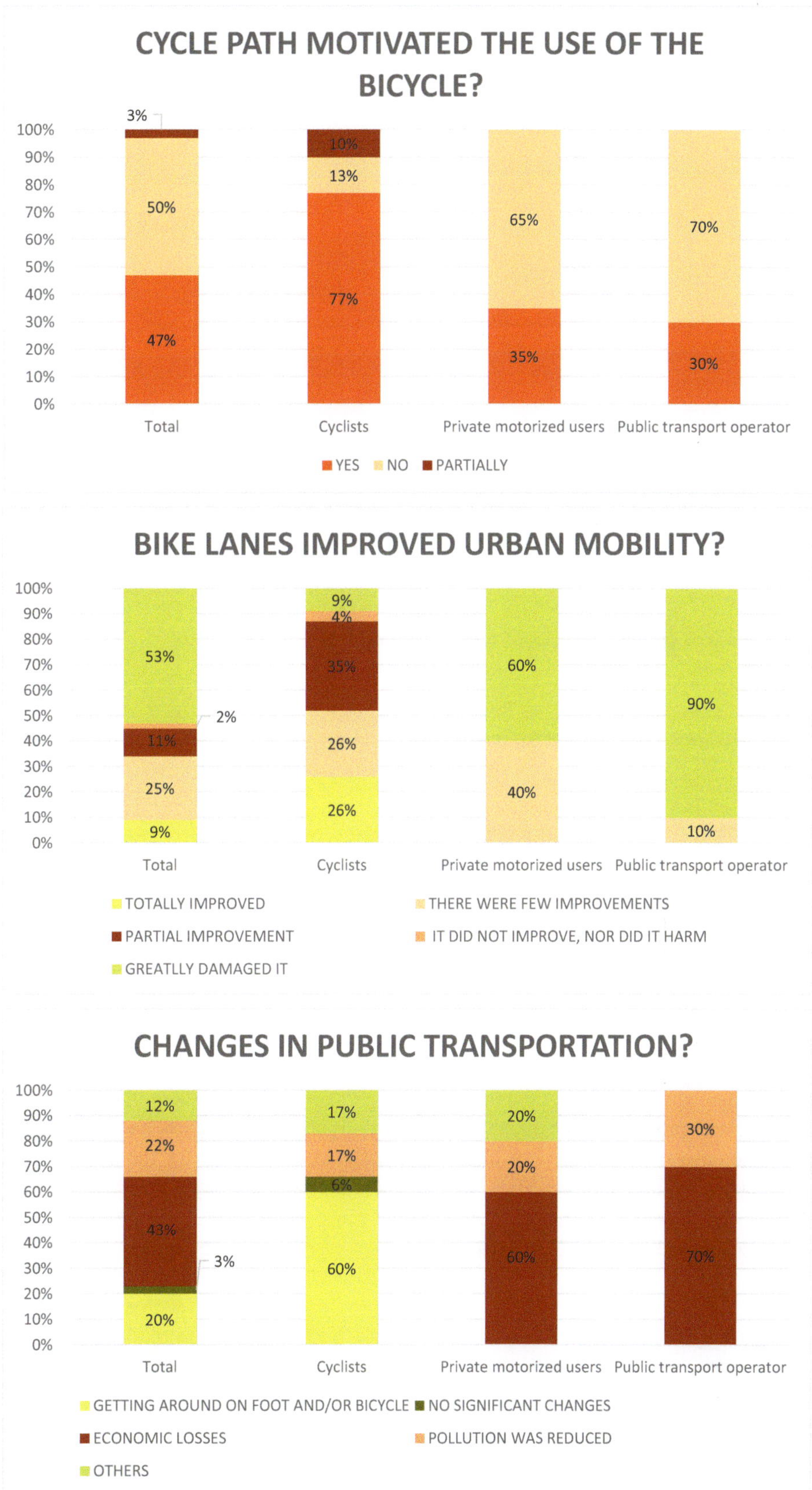

Fig. 4 Answers regarding planning. **a** Motivation. **b** Perception of improvement. **c** Changes in city transport. *Source* Authors

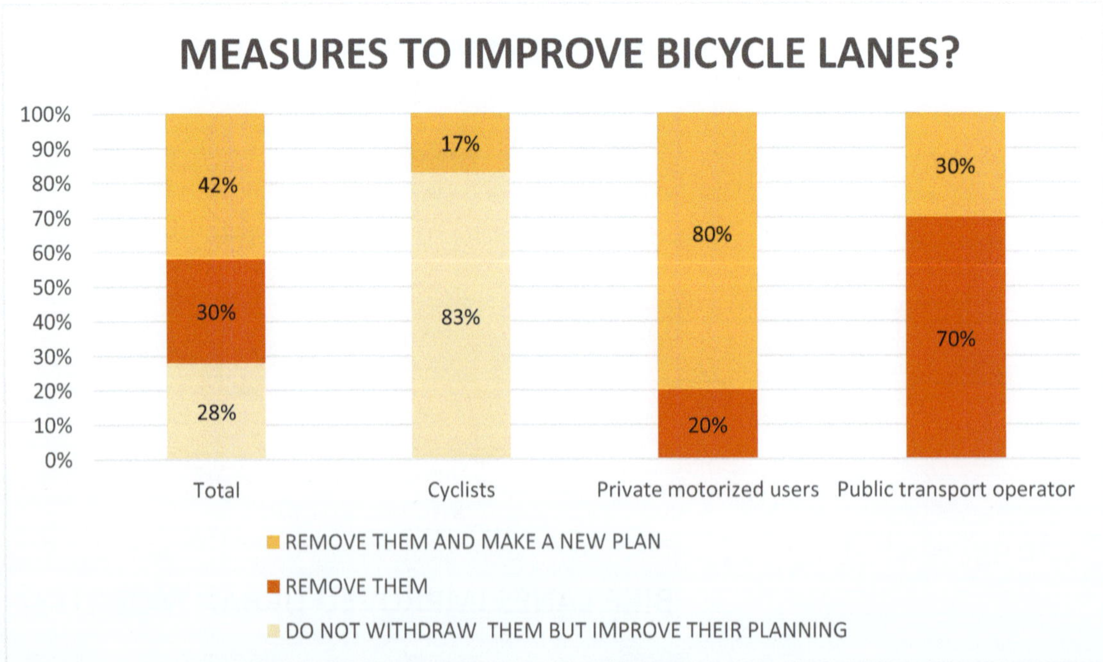

Fig. 5 Answers regarding governance. *Source* Authors

planning, most public carriers (70%) were not interested in the presence of bike lanes at all (Fig. 5).

5.3 Ease of Use

To measure the ease of use of the bike lanes, three questions were asked. As for the signage used, there is a large majority of users (84%) who state that the bike paths were poorly signposted. This is one of the few aspects in which there is total agreement between the three groups of users, since both cyclists (83%), private users (80%) and public carriers (90%) underline the bad or insufficient signage. There is also general agreement (92%) in the perception that the cause of the frequent problem of encroachment on cycle paths by motorized users was due to poor planning. Finally, in terms of the greatest difficulties perceived by cyclists in the use of these roads, it was their frequent fear of being run over by motorized vehicles (46%) (Fig. 6a–c).

5.4 Integration

In the case of integration, an open question was asked about the aspects that should be changed to avoid existing interference between bicycles and motorized vehicles. The majority of users (39%) agreed that the quality of the bike lanes should be improved, followed by a claim for a greater presence of traffic police officers (21%). Other responses included improving the infrastructure (an aspect not considered by private motorized users) and the bicycle inclusiveness (an issue not addressed by the public transport operators) (Fig. 7).

5.5 User Experience

Four questions were asked in order to understand the user experience (Fig. 8a–d). Regarding the safety of the cycle paths, the majority of users (43%) perceived them as very unsafe, followed by 36% who saw them as unsafe. Curiously, it was the private and public carriers who mostly consider them as very unsafe (60% in both cases) while cyclists had different degrees of perception regarding safety. As for the comfort and ease of use of the bike paths, the majority of interviewees considered them very uncomfortable, while 24 percent considered them uncomfortable. Again, this perception was felt more by motorized transport users who do not include any favorable perception. In contrast with cyclists who were 48% neutral, 17% consider them comfortable and 17% very comfortable.

Regarding the connectivity of the bike paths, the majority of users (44%) consider them very poorly connected followed by 42% who view them as poorly connected. Again, most negative perceptions came from motorized carriers, particularly those of public transport, who reach 80% of perception of a very poor connectivity. On the other hand, 25% of the cyclists considered it neutral and 16% thought

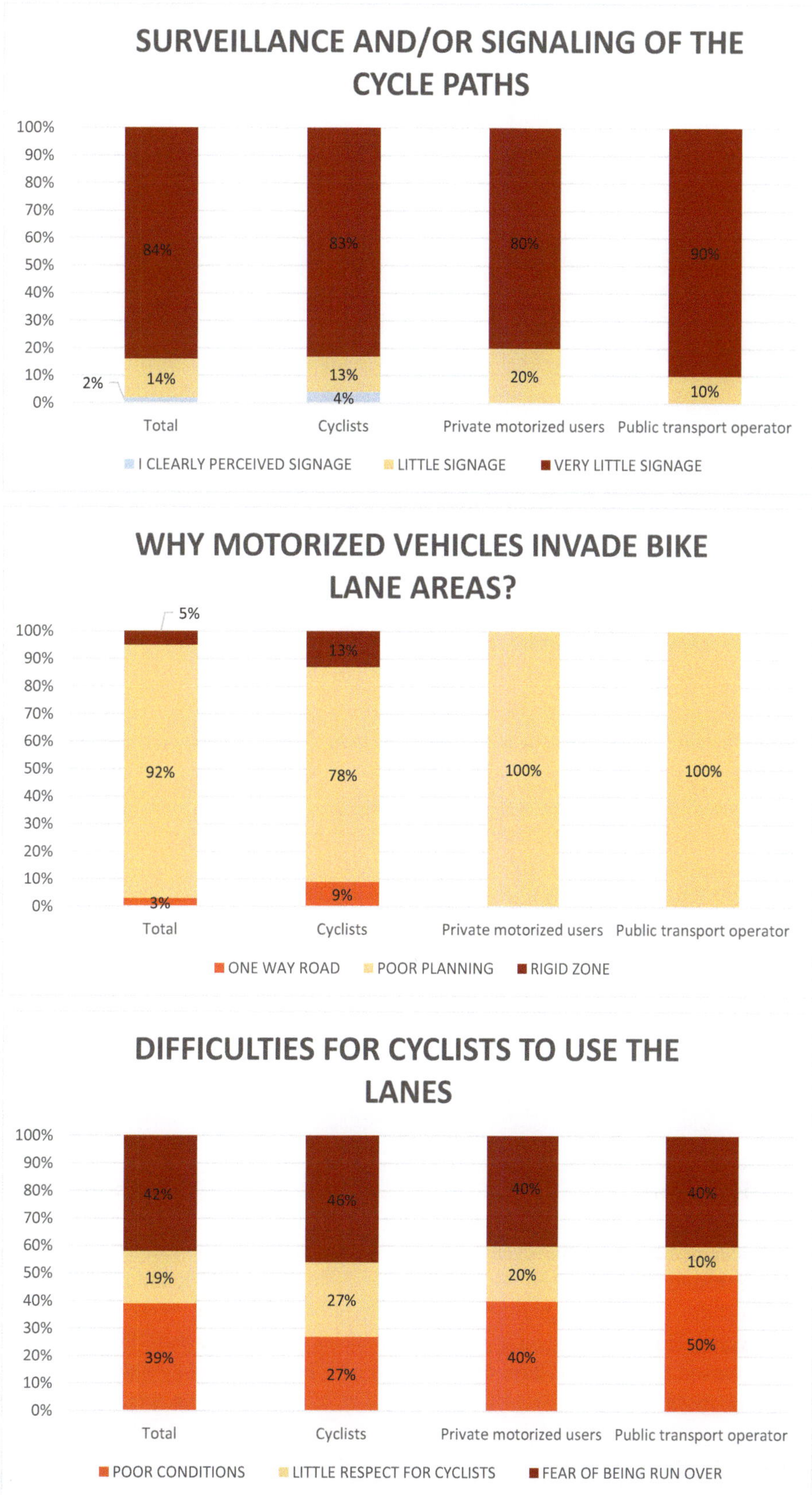

Fig. 6 Answers regarding ease of use. **a** Lanes signaling. **b** Reason for invading bike lanes. **c** Difficulties of cyclists. *Source* Authors

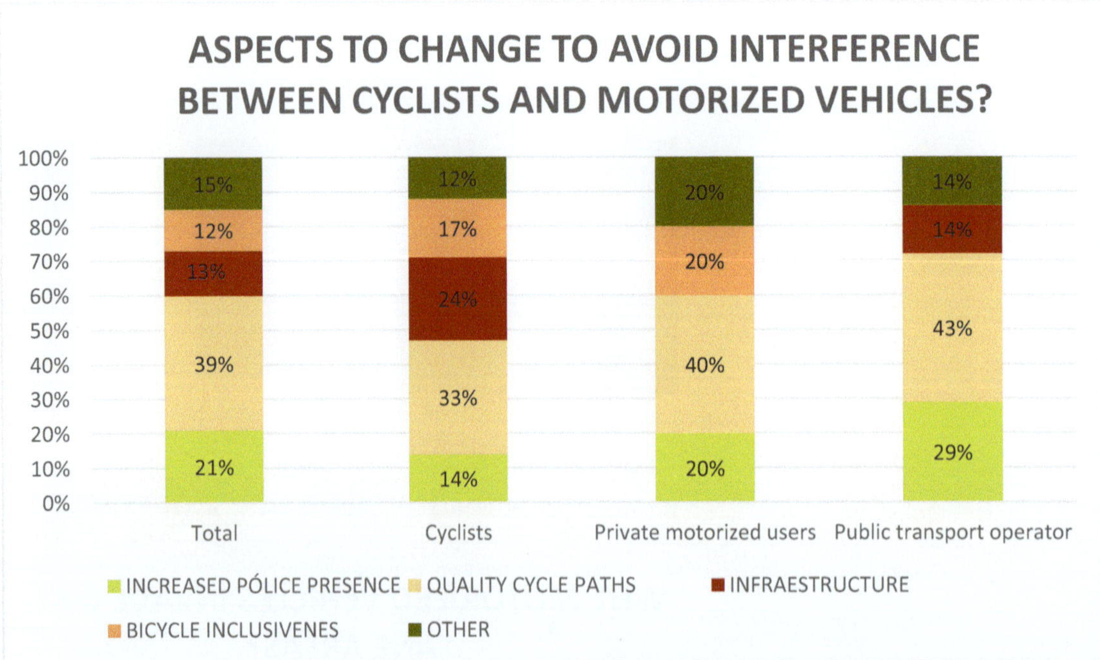

Fig. 7 Answers regarding integration. *Source* Authors

that there was some connectivity; however, nobody thought that the network of cycle paths was well connected.

Finally, regarding the perception that cycle lanes are appropriate for both novice and experienced cyclists, a majority of users (32%) maintained a neutral opinion; however, this point of view varied substantially by group. Most of the cyclists totally agreed (35%) or agreed (30%) that the bike lanes could be used without problem by novice users as well. Regarding private motorized users, 60% maintained neutrality while half of the public carriers were in total disagreement. Once again, none of the carriers, private or public, had positive opinions regarding the ease of use of the bike lanes by novice users.

The results are summarized in Table 4.

6 Discussion

The implementation of the cycle paths in Arequipa had positive results at first, when the confinement restrictions prevented the use of public transport and the use of bicycles was favored. However, as soon as the use of vehicles on the streets moved toward normality, the problems of a lack of planning arised.

This is evident in the interviews conducted, since the vast majority of users criticized various aspects of the implemented bike paths. However, a difference in perception is also evident between the various groups of users interviewed, cyclists, private motorized users, and public transporters a divergence that corresponds to their particular interests. Finally, the interests of the public transporters prevailed, who pressured the newly elected authorities to remove the bike lanes, which occurred in January 2023 when a new mayor took over the leadership of the municipality.

Sanitary restrictions due to the covid-19 pandemic, as well as the subsequent removal of bicycle lanes in Arequipa, made it impossible to carry out a more in-depth statistical study of user perception, however, this study made it possible to measure, even in a general way, the impact that this urban experiment had on the population of Arequipa. Therefore, we propose some strategies that can be implemented in the future plans that include concepts of multimodal and inclusive mobility.

6.1 Planning

The results showed a mostly negative perception regarding cycle life planning, which resulted in little motivation for its use. Well, the idea that it is in fracture ended up being more harmful than beneficial for users. The results showed a mostly negative perception regarding the planning of cycle paths, which translated into little motivation for their use and the idea that this infrastructure ended up being more harmful than beneficial for users. This was due to the short time it took to plan this network, using government funds given to support municipalities to implement sanitary measures to prevent the contagion of covid-19; that money had to be

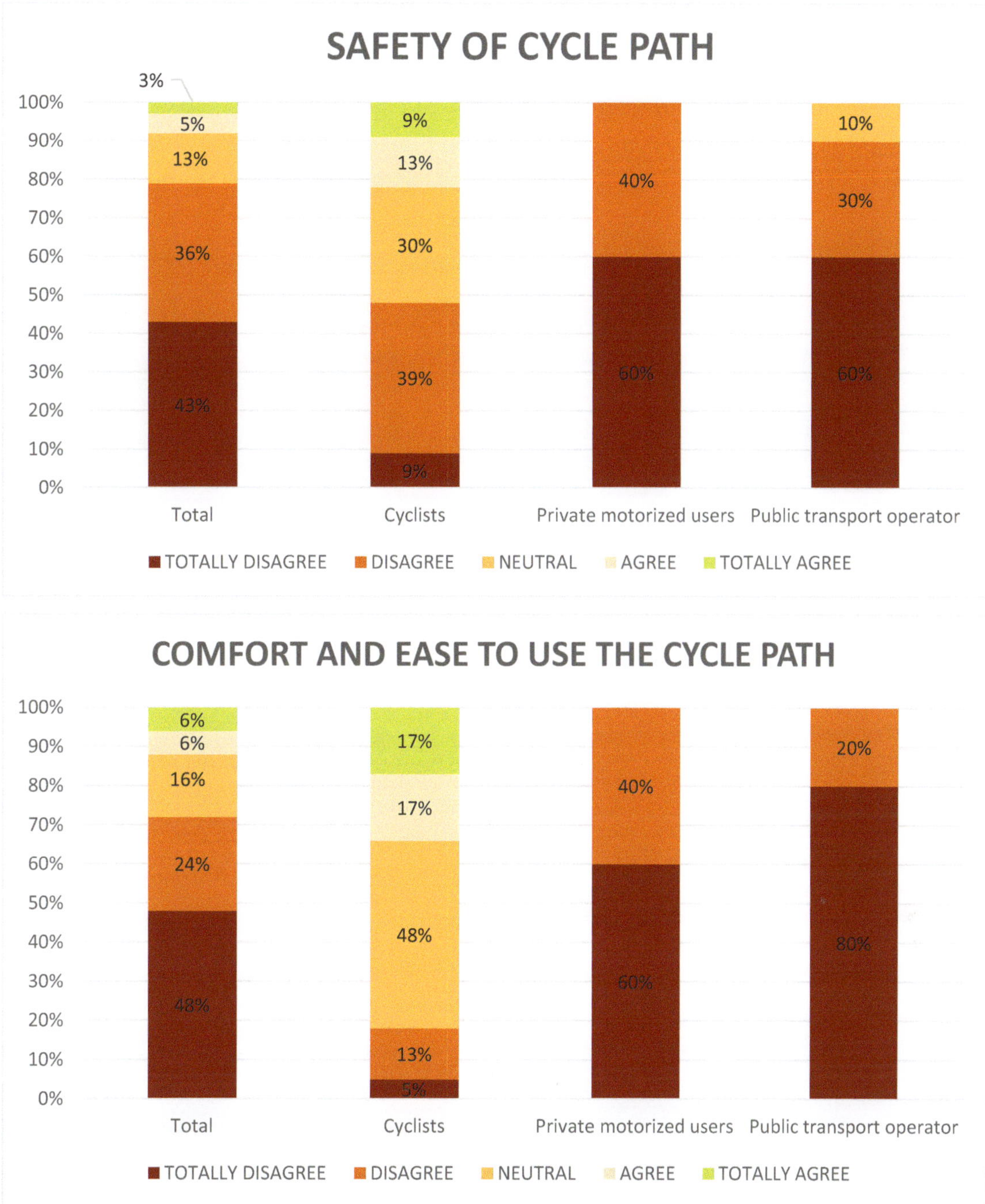

Fig. 8 Answers regarding user experience. **a** Safety. **b** Comfort. **c** Connectivity. **d** Suitability for unexperienced cyclists. *Source* Authors

used promptly or it should be returned to the central state. Another factor for the poor planning was because many transport experts were not willing to carry out field work during the pandemic, therefore were the municipal technicians themselves who assumed this challenge even without having much experience in transport planning, and they did so prioritizing idealistic and non-technical aspects in the conception of the bike paths. Therefore, the bike lane network should have been traced along roads that have a low rate of vehicular traffic, in order not to take space away from public transport vehicles.

Some of the users interviewed, especially cyclists, recognized that the pandemic brought temporary benefits in terms of transportation because interest was taken in a type of sustainable transportation, however, this implementation brought later problems due to its lack of resilience. This was

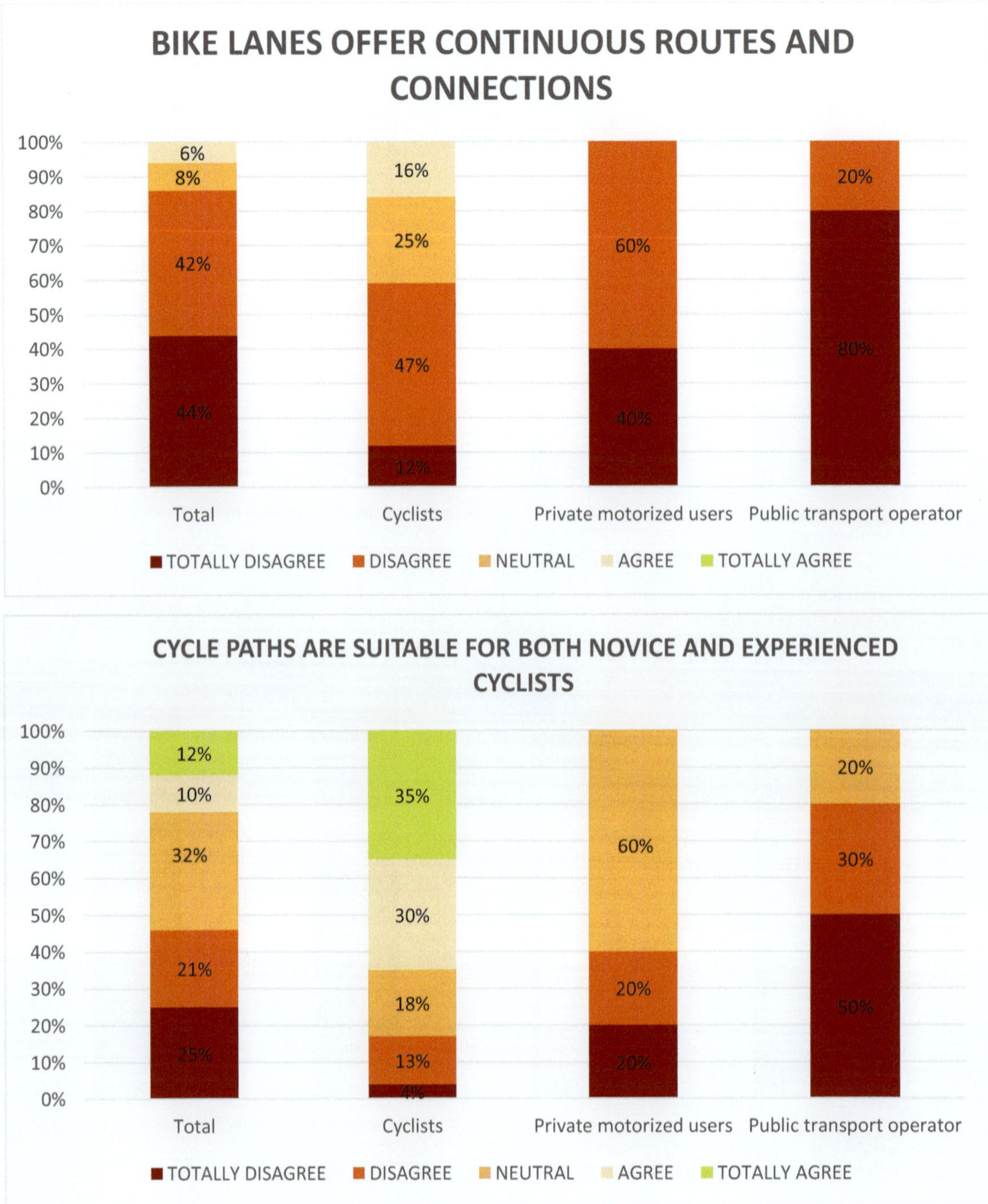

Fig. 8 (continued)

because that priority was given to only one type of user, which is the non-motorized transport operator, who has the option of changing their route to reach his/her destination, something that could not happen with the public motorized transport operator.

It is necessary to provide adequate infrastructure for the circulation of bicycles, so the cyclist must be considered as the center of all decisions in planning and design. For this, the first thing is to know what characteristics of this mode of transport and to know the different types of inclusive roads for bicycles, proposing three categories according to their physical characteristics. These are: (a) shared (b) segregated (c) green (Fig. 9) (Ministerio de Vivienda y Urbanismo [MVU], 2015).

Table 4 Summary of the results

Question	Experts	Cyclists	Private motorized users	Public transport operator
Cycle path motivated the use of the bicycle	Partially, during the emergency	Yes, safe spaces, segregated, accessibility	No, it is not very practical and there is no habit of using it	Yes, the cyclist needs a place to ride
Bike lanes improved urban mobility	Greatly damaged it, lack of a technical proposal, inappropriate routes	Partial improvement. They were necessary despite their poor design	They damaged it, the number of vehicles increased in proportion	There were few improvements, lack of driver education, awareness
Changes in public transportation?	No significant changes	Getting around on foot and/or bicycle	Economic losses, pollution was reduced	Economic losses
Measures to improve bicycle lanes	Do not withdraw them but improve their planning and maintenance	Do not withdraw them but improve their planning and maintenance	Remove them and make a new plan because they did not work	Remove them and make a new plan; only considered the cyclist and not public transport operators
Measures to improve the use of bicycles	Promotion and implementation of road safety	Development of bicycle infrastructure, security control, presence of police	Development of bicycle infrastructure, differentiate types of cyclists	Implementation of a public bicycle system
Surveillance and/or signaling of the cycle paths	Incomplete signage-no surveillance	Little signage-no surveillance	Very little signage	Very little signage
Why motorized vehicles invade bike lane areas	Not all users were not considered	Poor planning, irresponsibility, little road education	Poor planning due to their location on high-traffic avenues	Poor planning, one-way streets
Difficulties for cyclists to work or study	Poor conditions, no parking and safety	Little respect for cyclists, recklessness, intense car traffic	Little respect for the cyclists	Fear of being run over, road safety
Aspects to change to avoid interference between cyclists and motorized vehicles?	Non-overlapping routes	Bicycle inclusiveness, parking, intermodality	More infrastructure for cyclists, more routes and dissemination of their use	Infrastructure and quality cycle paths
Safety of cycle path?	–	Disagree	Totally disagree	Totally disagree
Comfort and ease to use the cycle path	–	Disagree	Totally disagree	Disagree
Bike lanes offer continuous routes and connections	–	Disagree	Disagree	Disagree
Cycle paths are suitable for both novice and experienced cyclists?	–	Agree	Disagree	Disagree

6.2 Comfort

The signposting of the bike paths was not perceived in the main streets or avenues due to the low awareness and poor road culture of pedestrians, cyclists, and motor vehicle operators. However, it is not enough just to place the signs, there must be a strategy so that the signage is perceived by each of the users. Therefore, it is recommended that for each horizontal signage a vertical signage be placed as well. The support of the National Police of Peru is also necessary to regulate and control traffic in critical areas or areas of conflict between the same users, safeguard the safety and, in addition, guarantee compliance with traffic rules.

6.3 Safety

Both motorized and non-motorized users share the fear of suffering an unfortunate traffic accident with fatal consequences. This indicates an empathy with cyclists and their fear of riding on bike paths precisely because they are located on avenues with a very high demand for public transport. Bike paths should not only be able to be traveled by experienced people but also by beginners because novice cyclists would not dare to use the current bike path if they do not feel safe riding on it. Therefore, bike lanes should follow a safer route with very little vehicular presence, without neglecting the personal safety of the cyclist or pedestrian.

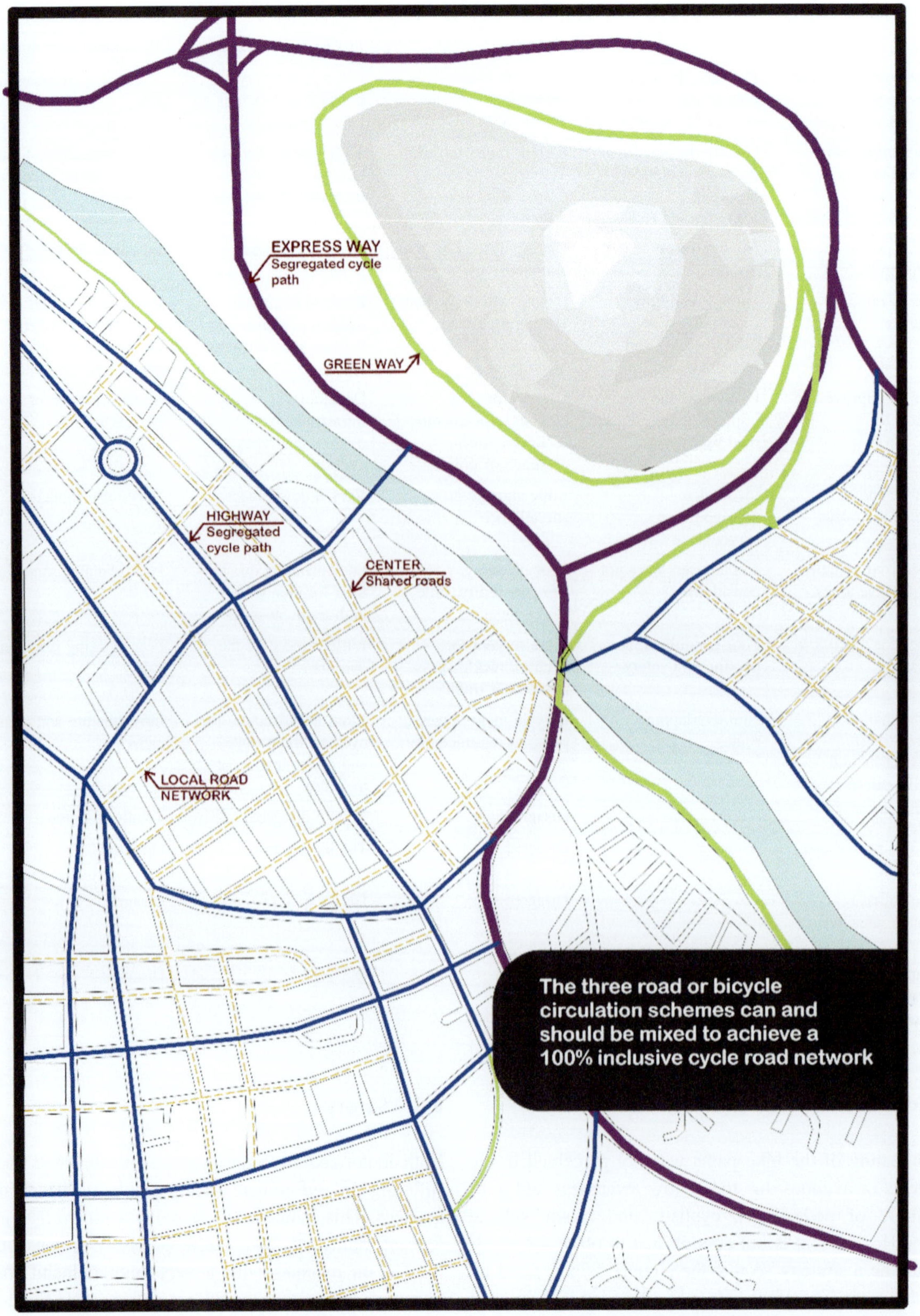

Fig. 9 Example of inclusive cycle road mesh planning. In red, expressways supporting segregated bike lanes. In blue, major roads supporting segregated bike lanes. In yellow, greenways. In gray, local roads supporting shared lanes. *Source* Prepared by the authors based on data supplied by the Ministry of Housing

Fig. 10 Attractiveness in a cycle path network. *Source* Prepared by the authors based on data supplied by the Ministry of Housing

6.4 Governance

The construction of new bike lane network has to be coupled with the improvement and maintenance of current bike lanes. The local government also has an active participation so that the current bike paths work, by promoting this means of transport and inviting more users to join them. The use of bike lanes must be regulated and communicated through promotion and awareness events, creating road safety policies that improve the circulation of cyclists, motorized vehicles and pedestrians.

There must be government institutions that contemplate and consolidate strategies for the education and promotion of bicycle use, in addition to compiling indicators such as travel demand, citizen perception, and economic value, to name a few.

6.5 Spatial Integration

Since the network of bike lanes was proposed only thinking of a single type of user, the cyclist and other motorized means of transport were not taken into account, discontent was generated both in public and private motorized users. On the other hand, if the bike lanes are removed in their entirety, it will generate discontent among cyclists, which is not the most convenient situation. Therefore, intermodality should be considered, linking cycle paths to the new Integrated Transport System (SIT) of Arequipa and not thinking of just one mode of transport implemented solely with signaling.

6.6 User Experience

The user experience was initially successful due to the lockdown due to the covid-19 pandemic, however as soon as it returned to "normal" the problems and improvisation became apparent. It is for this reason that more than 70% of those interviewed do not agree with the insertion of bicycle paths in the city.

Both cyclists, motorized users and urban planners show their disagreement regarding safety, comfort, and continuity throughout the route of the cycle paths. It is necessary to have an infrastructure design that promotes the integration of the bike lanes and a consistent continuity, with adequate pavement conditions, to achieve a comfortable and pleasant travel experience. It must also have travel routes that lead to attractive areas of the city and that offer interesting activities and amenities along the route (Fig. 10).

7 Conclusions

The economic incentive and the financing of a network of cycle paths due to the pandemic, coupled with the need for mobility with social distancing, demanded the planning, and implementation of cycle paths. However, because of the emergency, in the case of Arequipa, this was carried out without a study referring to the demands of the different types of users. This poor planning was evident as the city returned to "normality", generating discomfort in users of motorized transport.

It must be taken into account that for the planning of bicycle paths, the volume of passability of both urban passengers and cyclists must be evaluated to not generate a conflict between the different users. Another important but not considered factor was an origin–destination analysis in the planning of the bike lanes, which provides an understanding of how the cyclists move throughout the city.

The infrastructure is also a determining factor for the planning of the bike path because it must be comfortable and safe for circulation and properly vertically and horizontally signaled. The local governments must work on the promotion and awareness of the use of non-motorized transport. Also, they must integrate cycle paths into the urban transport network, through intermodality so that the user's change of transport without being affected by the topography, vehicular traffic, or road safety.

On the other hand, although there is widespread discontent among the three types of users regarding the quality of the bike lanes and their little planning, most cyclists agree that an improvement should be made to the bike lane infrastructure and a safe route that does not harm the traffic of motorized vehicles.

Also, it is concluded that the experience of the stakeholders themselves regarding the use of cycle paths is contradictory, although cyclists see this new implementation with more positive eyes while some private and public vehicle operators perceive it as an obstacle both for them as for the cyclists themselves.

Finally, the contribution of this study lies in making it clear that, contrary to what is stated in many texts on urbanism, it is not enough to build bicycle lane networks in a city for it to function sustainably, but rather they must be accompanied by comprehensive and inclusive planning strategies that consider the different types of users. The failure of the implementation of cycle paths in Arequipa should not be seen as evidence of the inconvenience of including the bicycle in the city transport, on the contrary, this study analyzes the factors that must be corrected for a healthy and efficient multimodal transport system.

References

Abusaada, H., & Elshater, A. (2022). COVID-19's challenges to urbanism: Social distancing and the phenomenon of boredom in urban spaces. *Journal of Urbanism: International Research on Placemaking and Urban Sustainability, 15*(2), 258–260. https://doi.org/10.1080/17549175.2020.1842484

Calderon Peña, P., Arrua, J. J., & Pardo, C. F. (2017). *Manual de Criterios de Diseño de Infraestructura Ciclo-inclusiva y Guía de Circulación del Ciclista* [Manual of design criteria for cycle-inclusive infrastructure and cyclist circulation guide]. Municipalidad Metropolitana de Lima.

Díaz-Gamboa, L. B. (2020). Enseñanzas del covid-19 [Lessons from covid-19]. *Derecho y Realidad, 18*(35). https://doi.org/10.19053/16923936.v18.n35.2020.11045

Imbert, N. (2020). *COVID 19 - 10 Propositions Pour Des Territoires Plus Résilients* [COVID 19—10 proposals for more resilient territories]. GREEN CROSS France & Territoires.

Jabareen, Y., & Eizenberg, E. (2021). The failure of urban forms under the COVID-19 epidemic. *Town Planning Review, 92*(1), 57–63. https://doi.org/10.3828/tpr.2020.42

Jasinski, A. (2022). COVID-19 pandemic is challenging some dogmas of modern urbanism. *Cities, 11*, 103498. https://doi.org/10.1016/j.cities.2021.103498

Kolarova, V., Eisenmann, C., Nobis, C., Winkler, C., & Lenz, B. (2021). Analysing the impact of the COVID-19 outbreak on everyday travel behaviour in Germany and potential implications for future travel patterns. *European Transport Research Review, 13*, 27. https://doi.org/10.1186/s12544-021-00486-2

Labonté-LeMoyne, É., Chen, S.-L., Coursaris, C., Sénécal, S., & Léger, P.-M. (2020). The unintended consequences of COVID-19 mitigation measures on mass transit and car use. *Sustainability, 12*(23), 9892. https://doi.org/10.3390/su12239892

Liu, S., Shen, Z.-J., & Ji, X. (2021). Urban bike lane planning with bike trajectories: Models, algorithms, and a real-world case study. *Forthcoming in Manufacturing & Service Operations Management, 24*(5), 2500–2515. https://doi.org/10.1287/msom.2021.1023

Ministerio de Vivienda y Urbanismo [MVU]. (2015). *Vialidad Ciclo-Inclusiva: Recomendaciones de Diseño* [Cycle-inclusive road: Design recommendations]. MVU.

Morales, S. (2022, September 6). *Arequipa: Ciclovías siguen abandonadas y destrozadas en la ciudad* [Arequipa: Cyclepaths remain abandoned and destroyed in the city] (p. 1). Diario Correo Arequipa.

Targa, F., Buis, J., Olivares, C., Flores, J., & Peña, J. (2020). *Propuesta de actualización del Plan de Infraestructura Cicloviaria para Lima y Callao* [Proposal to update the Bicycle Infrastructure Plan for Lima and Callao]. Banco Mundial Región de América Latina y El Caribe.

Vicuña Mena, P. A., Pillaca Samora, P. A., & Jauregui Moron, M. C. (2022). *Una mirada al transporte público en tiempos de pandemia: revisión sistemática* [A look at public transport in times of pandemic: Systematic review]. Ciencia Latina Revista Científica Multidisciplinas.

Zavalaga, M. (2022, June 5). *Ciclovías en Arequipa: el otro fracaso en la gestión de Omar Candia* [Cycle paths in Arequipa: The other failure in the management of Omar Candia] (p. 1). El Buho.

Zeballos-Velarde, C. (2020). *Atlas Ambiental de Arequipa [Environmental Atlas of Arequipa]*. Universidad Católica de Santa María.

Zhuang, D., Jin, J., Shen, Y., & Jiang, W. (2021). Understanding the bike sharing travel demand and cycle lane network: The case of Shanghai. *International Journal of Sustainable Transportation*, 111–123.

Sustainable Behaviour Assessment for Building Highways Inside Existing Neighbourhoods: A Case Study of Heliopolis–Cairo–Egypt

Islam Ghonimi and Gehad Ghonemy

Abstract

In the last two years, Egyptian government started a huge urban regeneration and road network rehabilitation projects in existing neighbourhoods; a massive establishment of highways, elevated highways and bridges are erected cutting through existing urban fabric, especially in Nasr city and Heliopolis. These projects aim to avoid traffic jams, increase trafic flow, reduce travel time and enhance urban mobility through the existing urban fabric. Establishing highways inside existing urban fabric is an international old phenomena in developed countries who first advocated and established highways a century ago followed with a scientific conscious of their negative impacts and governmental calls and actions for removing highways from cities especially American and Japan cities. On the contrary, Egyptian government started a huge highway erection, creating a new border that expected to highly change the existing continuity in socio-spatial fabric of current residential areas without taken into consideration that once the urban structure has changed, it could change neighbourhood behaviour including functional efficiency, urban mobility behaviour, social interactions and relations, and safety and security in somehow and could cause great impacts on existing neighbourhoods in terms of functionality, sustainability and satisfaction. Most local and international studies focused on the impacts of highways on macro-scale in terms of experts and resident's satisfaction and perception of the impacts, whilst this research aims to test how far the impacts of such change on changing social and urban behaviour of micro-community residents and accordingly impacts sustainability of existing neighbourhoods, in order to help future research in building design guidelines taking into consideration the future plans for road rehabilitation. The result of the research demonstrates how establishing highways inside existing neighbourhoods of Heliopolis records low services efficiency, unsustainable urban mobility, unsustainable social bonds, and reduced sense of security and safety and accordingly negatively impacts functionality and sustainabilty.

Keywords

Road network regeneration • Highways • Existing neighbourhoods • Street network • Sustainability performance • Behaviour assessment

1 Introduction

The last two years Cairo witnessed a process of urban reformation by introducing a process of highway erection and bridges establishment projects to enable shortcut paths through the urban fabric of east Cairo to link old capital to the new capital to the east. These projects aimed to establish a new paradigm of less traffic congestion and reduced travel time by increasing roads' width and increasing the number of lanes; it came over different methods, some times by the demolition of buildings others by reducing the wide islands with greenery spines, and in other way by removing the islands and converting the old two-way streets to wider highway, other times by removing the tram lanes.

The new paradigm of planning process affects the formulation of road network inside existing towns urban fabric, a wide process of expansion of highway projects cut through high density residential areas inside the existing fabric that makes many transformations without taking proper cautions of their impacts. The government allowed to make changes in urban fabric and structure that can negativly impacts the sustainability and quality of life of existing neighbourhoods.

I. Ghonimi (✉)
Shoubra Faculty of Engineering, Benha University, Banha, Egypt
e-mail: Islam.ghonimi@feng.bu.edu.eg

G. Ghonemy
Institute of Aviation Engineering and Technology, Cairo, Egypt
e-mail: Gehadghonemy@gmail.com

A. Battisti et al. (eds.), *Greening Our Cities: Sustainable Urbanism for a Greener Future*, Advances in Science, Technology & Innovation, https://doi.org/10.1007/978-3-031-49495-6_5

Erecting highways as a traffic solution inside urban fabric of existing cities is an old worldwide phenomena in developed countries. It was initially advocated in developed conutries since the beginning of the twentieth century, then after few decades, criticized by theories. Traffic planners advocated for establishing highways to face city problems as cause of survival of downtown cities, to solve the increase of traffic demand, to pass through the crowded neighbourhoods, to split traffic flow from pedestrian flow and to reduce traffic cognition as the old crowded, narrow, slow streets cannot meet the new urban mobility requirement of fast and high traffic flow through the existing crowded cities (Bauman, 1991; MacKaye & Mumford, 1931; Mohl, 2004).

After few decades, urban planners in developed countries criticized the impacts of highways on sustainability of cities and called for removing highways from cities. Jacobs pointed to the negative consequences on environmental degradation, change in land uses and death of pedestrian life (Jacobs, 1961). Mumford points to the negative impacts on reducing sociability and communication of American families as it reduces mass transportation and relies on private mobility and reduces pedestrian mobility (Mumford, 1964). Saiz added that highways serve private transportation system and neglect public transit systems (Saiz, 2006). Cervero added the negative impacts on urban livability, urban quality of life and placemaking (Cervero, 2009). Ebeling pointed that highway limits the number of passers driving in front of the shops, therefore limiting the potential shopping and reducing economy return (Ebeling, 2013). Levin added the negative environmental impacts it can cause (Levin & Duke, 2012). Schindler discussed how such change could impact residents' exclusion and social segregation (Schindler, 2014).

A recent worldwide scientific conscious found that the old paradigm of erecting highways inside cities to solve urban mobility is failed (Khalaj et al., 2020; Popovich et al., 2021). They suggest that removing highways can face the social and economic ramifications of highways. A leading article by Alana semules "The Atlantic" discussed how the removal of highways can revitalize uban cores of existing urban fabric by replacing motor traffic infrastructure with walkable, green and mixed use streets that can promote sustainable, livable and healthy urban environment. As response for the academic push, most of the world governments start to regulate removing their highways inside cities as part of their agenda; USA proposed new urban legislations and proposed to spend 20 billion dollar to reconnect neighbourhoods divided by highways. New urbanism tracked 33 proposed projects for removing highways in different American cities to make better life for local residents (Popovich et al., 2021).

Regardless, the worldwide scientific conscious for the negative impacts of highways inside cities followed by a massive government actions for highway removal and shift the planning paradigm from car traffic mobility to transit mobility; on the contrary, recently, Egyptian government as one of the developing countries just started a massive spread of highways in old Cairo, aiming to centre planning decision on private motor mobility without consideration to transit and pedestrian mobility.

Although the topic of highways in Egyptian scientific context is new compared to international scientific context, some local scientific effort started to warn of neglecting the experiment of developing countries and warn of the negative impacts of traffic axes rehabilitation on urban reformation, spatial fragmentation and social segregation in existing neighbourhoods. Some scholars assess the impacts of traffic axes development on urban development of places with heritage values with greater concern to the problems of macro-community concern the reduced green areas, the visual negative impacts on visually reach areas in Heliopolis (Saber & Radwan, 2020; Elkatib, 2020; Mounir, 2019). Other scholars concluded that establishing highways can act as barriers and can foster spatial fragmentation in existing neighbourhoods and can cause social segregation, and they used quantitative methods using space syntax to examine how spatially the internal roads can be connected to the highway (Abozied & Vialard, 2020). On the other hand, some scholars stress how urban reformation can impact the wright to the city, and the importance of public involvement and consultation in decision-making before approving the plans (Almoghazy, 2021). Some scholars concern the evaluation of resident's satisfaction with reformation projects (Elsayed, 2021; Fouad et al., 2023), and they concern the overall satisfaction with the changes happen in the macro-community, as shown in Table 1.

In contrast to most international studies that focused on investigating the impacts of highways based on experts' opinions and residents subjective perception of highways, this paper gives more attention to sustainability behaviour assessment of the impacts of highways on changing residents' repeated social and urban behaviour as objective factor impacting sustainability. Furthermore, most local studies focused on the impacts of highways on overall macro-urban form of cities, and the wright to the city, whilst this paper extends more attention to investigate the impacts of highways on changing the urban and social behaviour of local residents inside their micro-community. It aims to trace any significant differences in response to urban sustainability before and after the establishment of highways, as a step to develop suatainable development guidelines for the future process of road rehabilitation.

This research suggests that the new road rehabilitation in Heliopolis rather than support urban connectivity leading to sustainable urban mobility and sustainable social behaviour; on the contrary, it can divide the city urban fabric to isolated separate islands leading to change neighbourhood behaviour to unefficient service provision, unsustainable urban mobility behaviour, unsustainable social behaviour, and less sense of safety and security and accordingly could cause great negative impacts on existing neighbourhoods in terms of functionality, sustainability and satisfaction.

Table 1 Deduced impacts of highways on macro-urban form

	Jacobs (1961)	Mumford (1964)	Saiz (2006)	Cervero (2009)	Schindler (2014)	Ebeling (2013)	Khalaj et al. (2020)	Popovich et al. (2021)	Saber and Radwan (2020)	Abozied and Vialard (2020)	Almoghazy (2021)	Elsayed (2021)	Fouad et al. (2023)
Environmental degradation	√												
Land use change	√												
Death of pedestrian life	√	√		√				√		√			
Reduce sociability		√											
Neglect mass mobility		√	√				√						
Concern private mobility		√	√				√						
Reduced green areas								√	√	√	√	√	√
Visual negative impacts					√					√	√	√	√
Reduce property value				√	√								
Reduce resident's satisfaction						√						√	√
Social exclusion					√								
Economic oppourtunities						√							

2 Methods

In order to test the research hypothesis, an inductive method is used, based on theoretical review and empirical analysis of a case study of Helioplis neighbourhoods. Different data collection methods are used including the literature review, observation, questionnaire and behavioural mapping to deduce the impacts of road rehabilitation on the sustainability of existing neighbourhoods, and the method depends on three stages:

- Theoretical exploration of the literatures concerning the impacts of road rehabilitation on physical change of city is introduced using spatial analysis method to explore the impacts of such changes on the functionality of urban fabric.
- Theoretical exploration of the deduced sustainability factors including efficient service provision, sustainable urban mobility, social sustainability, and safety and security is introduced, providing measuring tools to investigate their change before and after erecting highways.
- A case study of Heliopolis will give explanatory qualitative approach of how highways impact people behaviour using ethnographic observation, behavioural tracing maps and questionnaire to explore the impacts of highways on changing the local residents urban and social behaviour and to deduce the impacts on sustainability of the community.

2.1 The Impacts of Highways on Changing City Form

A new paradigm shift is developed in planning urban mobility from the old paradigm of sustainable walkability with reduced motor traffic flow to the new paradigm of high traffic flow depending on a process of road rehabilitation. The new paradigm developed different strategies for erecting highways inside urban fabric, some times by demolitioning of some buildings, reducing or removing the wide greenery islands, removing the tram lanes, and finally by converting the old two-way traffic flow to wider one-way street. Furthermore, to increase traffic flow and avoid intersection bridges, turnovers and elevated highways are used.

Eventhough the new rehabilitation strategy of road network aimed to increase traffic flow, speed and cause less traffic cognition, it caused important challenge for urban development. These roads become hard, unsafe and consuming more time, effort and cost to cross, and accordingly, these roads rather become connectors to link city parts; they created divisions of loyalties, cut the connection between neighbourhoods to the surrounding residential area of its borders and become cutters of the connection between different parts of inner urban fabric and can impact the sustainability and satisfaction of the existing neighbourhoods.

These projects not only caused losing 90 feddans of green spaces, removing tram way as popular public trnasportation, encouraging private mobility over public one, centring concern to the car instead of walking and cycling, causing high speed and high traffic accidents, but also it can have impacts on the behaviour of people living in the existing neighbourhoods. This paper aims to investigate the impacts that these projects can imprint on the spatial fabric of the existing urban fabric and deduce their expected impacts on the sustainability of these areas.

The new condition imprints some changes on the structure of urban form that changed from traditional concept of being an integrative organic part of the city that react with the city and constitute an important part of it; and turned it to be externally isolated and internally focused, and restricted to its own resources, which may be limited and insufficient to provide true community life, and it became the focus of the community social life, neglecting sense of the macro-city. It became a cellular cell isolated and separated element from the city. Furthermore, the spatial separation was associated with excluding less desirable city life and became self-contained internally focused unit of the city that changed the way of looking to the micro- to macro-relationship.

An important impact of this change is that although they tend to increase the traffic flow through the macro-community and change micro-community to inward oriented and keep out unwanted, rather they kept services area in the boundary face in arterial streets, which promote city public life and ensure modernist superblock ideas, where housing units are arranged in clusters with series of loops surrounding an interior park and open space. Furthermore, neighbourhoods showed a paradigm shift from an integrated section of the city to independent and self-contained unit. This could make neighbourhoods lack the necessary resources to function as a true self-sufficient sustainable community.

To understand the impact of new road network on the behaviour of residential areas and explore its impacts on urban development, it is important to consider the new conditions created by the new features of road rehabilitation and its relevant impacts on micro- and macro-community that make it explicitly and intentionally create division of loyalties between its residents and the town. So establishing highways is not just about the barriers; rather, it is about what is between different territorial systems. It is about the relation between the system of micro-community on one hand and the system of adjacent macro-community, and the system of the city on the other hand.

Hence, erection of highways as borders not only can separate a specific urban area out from its contextual public urban fabric but also can change the relation between micro-community and macro-community, and change the two-way relationship between adjacent territories producing new behaviour for micro-community, their neighbours and the total macro-urban community. In its broader definition, once the structure of urban form has changed, a relative behavioural change is expected to happen in micro- and macro-community and in the in between city life, as shown in Table 2.

The micro-urban form becomes independent, isolated from the context, and restricted to its own resources, which create spatial restriction that could extend to the social manner, as micro-neighbourhood could be planned with homogeneous separated housing type. In addition, the adjacent community neighbour, and due to the new borders, cut the continuity and become excluded from other residents.

2.2 Sustainabile Behaviour Assessment for Road Rehabilitation

Most local and international studies give more attention to satisfaction assessment of the impacts of highways on quality of life to ensure resident's satisfaction about their community after erecting highways. They depend on subjective indicators depending on attitudes, opinions and hopes regarding their personal satisfaction, perception and expectations concerning what people hope to have.

This research concerns sustainable behaviour assessment of the impacts of highways on neighbourhood performance, and it concerns objective measures related to urban performance that sustains well performance behaviour of sustainable communities. It aims to develop report of activities to evaluate the change in what residents used to act of typical and repeated way of behaving between each other and with their community. It measures residents' social and urban behaviour depending on ethnographic observation to trace any significant change of residents' social behaviour and daily life activities, regardless of resident's opinions and attitudes about their community.

In order to assess the impacts of highways on the sustainabiliity of existing neighbourhoods, an extensive literature review in the academia, including paradigms, theories and approaches of sustainability, different dimensions and

Table 2 Schematic representation of the impact of highways on changing urban structure in terms of micro- and macro-scale

	Before road rehabilitation	After road rehabilitation
Micro-community	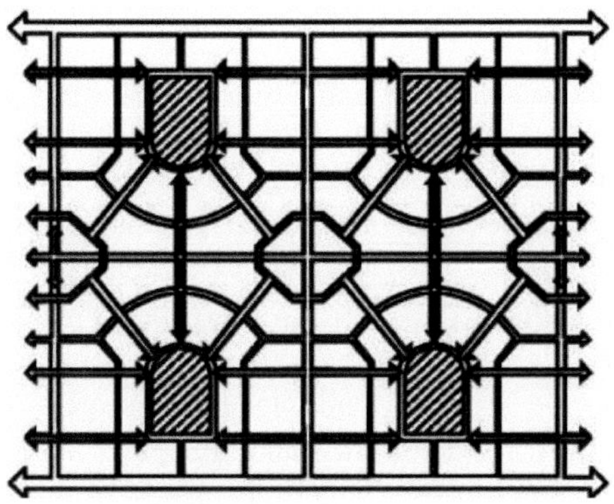	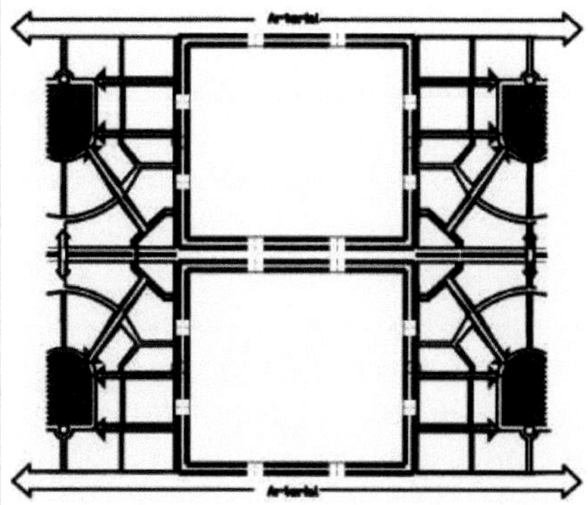
	Existing road network as public fine grain	Transformed from fine grain to super
Macro-community	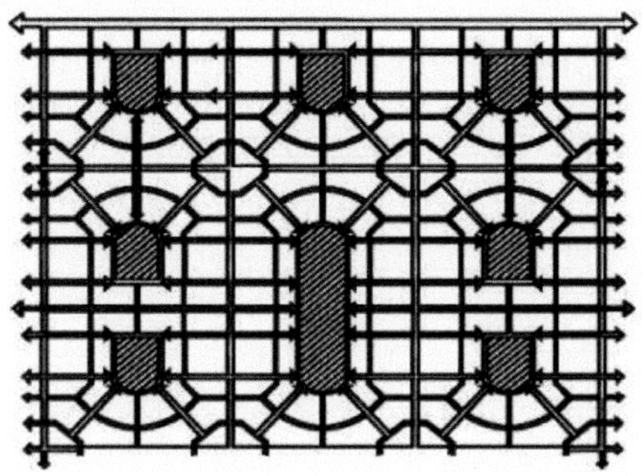	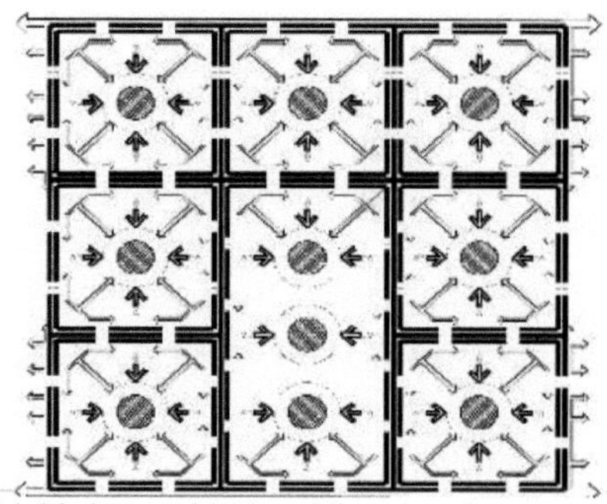
	Connected fine-grained public urban form	Divided super block public urban form

indicators of urban sustainability is developed and summarized. The research concluded four proudly dimensions of functional efficiency, urban mobility, safety and security, and social ties with neighbours. Each dimension consists of domains, and each domain contains its indicators (Al-Qawasmi et al., 2021; Levett, 1998; Wiesli et al., 2021, 2022; Taqi et al., 2021; Vallance et al., 2012; Correia et al., 2022; Turkoglu, 2015; Ghonimi & Awad, 2017a; Ghonimi & Elzamly, 2017b; Ghonimi, 2021; Ghonimi, 2022).

- The Change in Service Efficiency

Erecting highways is expected to change service efficiency to less efficiently function and unsustainable behaviour. Highways cut the conection between neighbourhoods and accordingly impact the service distribution that was previously functioning as an integrative way with the surrounding parts. Its impacts can be measured using three key indicators including service sufficiency, service efficient distribution and parking requirements to probe the change in resident service behaviour (Ghonimi, 2022).

- "Sufficient provision of services" refers to the quantitative sufficient provision of services for daily, weekly and monthly needs within walking distance.
- "Efficient provision of services" refers to the qualitative provision of wide range of choices of services.
- "Efficient distribution of services" refers to the efficient distribution within safe walking distance and the diversity and wide range of services allowed to meet the diversity of resident's needs.
- "Efficient parking provision" refers to the qualitative provision of parking area in range of car ownership, within walking distance.

- The Change in Travel Behaviour

Erecting highways can change local residents' patterns of urban mobility, to less sustainable travel behaviour. Its impacts can be measured using travel mode choices, trip frequency, trip distances, trip duration, trip time and overall trip distances to probe the change in resident willing to sustain urban mobility patterns that encourge public transits and pedestrian mobility rather than rely on private motor mobility (Ghonimi & Elzamly, 2017b; Ghonimi, 2021; Ghonimi, 2022).

- "Mode Choices of urban mobility" refers to potential encourage for alternative transportation options and movement systems (pedestrian, cycling, public transport), including choices for passive motorized mobility and active motorized mobility and single mobility ride, and public mobility ride.
- "Frequency of urban mobility" refers to the times that certain trips could happen per day, and it includes high, medium and low frequent trips.
- "Distance of urban mobility" refer to the types of trips in terms of distance that can happen including short-distance, medium-distance and long-distance trips.
- "Duration of urban mobility" refers to the type of trip in terms of time spent in each trip per day including short-stay, medium-stay and long-stay trips. It measures residents feel safe to spend more time in such mobility mode at certain times of day.
- "Time of urban mobility" refers to the time residents are willing to do their mobility including morning, afternoon and evening trips. It measures residents feeling safe for commuting at certain times of day.

- The Change in Social Bonds

Erecting highways can change social bonds to less socially sustainable behaviour. Its impacts can be measured using four proudly key indicators: social interaction, strength of social relations, social inclusion and social participation that can probe the change in resident willing to sustain social relationship and bonds with their neighbours and encourge them to interact, recognize, know, engage to neighbours, help each other, trust each other, build social ties and define how deep their relations (Forrest & Kearns, 2001; Hamiduddin, 2015; Ghonimi & Awad, 2017a; Ghonimi 2021; Ghonimi 2022).

- "Social interaction" refers to the essential cause for building social relation; it can be justified using three sub-measures of degree of interaction, modes of interaction and actors of interaction:
  - "Degree of interaction" refers to the quantitative measure of interaction; it measures the degree of interaction happens with friends, neighbours and integrators.
  - "Modes of social interaction" refers to the potential encourage for accidental, and intentional modes of interaction; it provides qualitative measure for the type residents used to meet neighbours.
  - "Actors of social interaction" refers to the type of persons involved in social interaction; it involves the integrators, residents, passers and strangers; it also can happen with similar or dissimilar social income people.
- "Strength of social relation" refers to the type of social bonds exists between neighbours; it can be justified through four sub-measures: recognize, know, engage, build deep relations and feel trust:
  - "Recognize neighbours" refers to the degree residents able to know neighbours in the same, floor, building, street and neighbourhood.
  - "Know neighbours" refers to the degree residents know their neighbours in the same floor, in the same building, in the same street and in the same community and how they able to recognize neighbours in streets accidently or intentionally.
  - "Engage with neighbours" refers to how deep are social relations with friends, family and neighbours; and to how far residents know their neighbours. How many times residents like to share their neighbours in trips, invite their neighbours to their homes and visit their neighbours in their homes.
  - "Neighbours social support" refers to how far residents feel trust in neighbours, thought them fair, can support them and thought they are helpful.
  - "Strangers familiarity" refers to how far residents feel trust with strangers, can support and help them, think they are helpful.

- The Change in Safety and Security

Erecting highways can change resident's feel of safety and security to less sense of safety and security. It depends on some investigation tools to probe how far residents percept safety and security; how far they can feel safe for their families, children, females and elders to move freely in the community, how far they feel safe for their properties; how far they can feel secure from traffic accidents and fast moving traffics'; how far they can define neighbours and accordingly distinguish strangers in their community and can define offenders or criminals. This criterion rests on some sub-indicators:

- "Feel safe" refers to the degree residents feel safe from traffic accidents and fast moving traffic whilst moving in streets and urban spaces, feel safe for females, kids and aged groups during day hours, night hours and till late night.
- "Feel secure" refers to the degree residents feel secure from offenders and criminals, feel secured in homes, streets and urban spaces for self, properties and family during day hours, night hours and till late night.

- "Feel of trust measure" refers to the degree residents felt that their neighbours could be trusted, whether they thought people were fair, and whether they thought most people try to be helpful.
- "Surveillance measure" refers to the degree residents use passive inspection to allow clear sight line for the external surroundings when they are inside their house.
- "Target hardening measure" refers to the degree residents use active tools like alarms and technical instruments to control and observe security purposes in home and for their properties.
- "Define and trust strangers" refers to the degree residents can recognize and define neighbours, strangers and offenders that pass infront of their houses; it also includes how to trust strangers, help them and found them as helpful?

2.3 A Case Study of Heliopolis City

Heliopolis city was selected to present one of the leading road rehabilitation projects in Cairo. Table 3 includes a representation of the selected case study showing an extended highway projects cutting through the existing urban fabric of Helipolis, and through the old fabric of the early developed neighbourhoods of Mataria, Zyton, to connect the adminstrative capital in the east to the west of Cairo.

Heliopolis was initially planned and developed by the end of nineteenth century as garden city that is mainly pedestrian oriented. Urban form of Heliopolis was seen as structurally integrated and organically linked to existing city structure, and civic function. It does not miss the basic function and nature of being an entity of a larger urban area; it mentioned its role in the life of the city, by connecting to the large urban environment and surrounding community life. In addition, it was interconnected permeable communities that regularly interact with the suronding macro-community.

By the end of 2019, Cairo started a new paradigm in rehabilitation planning aiming for less traffic cognition, less travel time, more motor traffic flow and increased motor speed, by establishing a wide projects of highway. An important part of heliopolis is Almaza zone that was develped to increase motor traffic flow and car speed. Part of the actions was increasing roads' width, erecting more than 10 bridges, removing the green islands, displacment of traffic direction by changing some of the street directions to become one-way streets, such as Thawra and Omar Ibn Al Khatab streets, and removing squares.

Table 3 The transformation of urban fabric as response to rehabilitation of road network in Heliopolis (Shafik et al., 2021)

Heliopolis before road rehabilitation	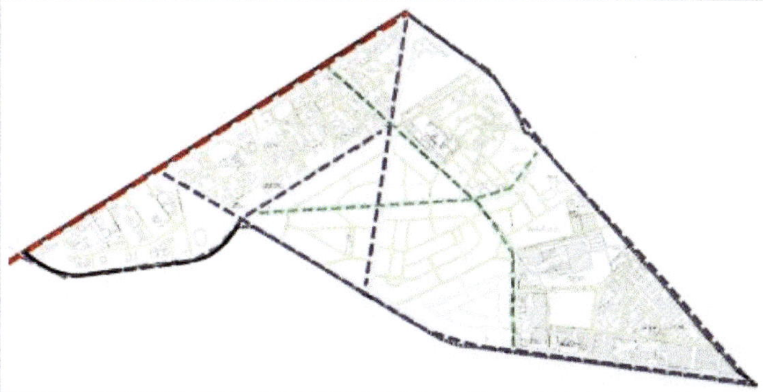
Heliopolis after road rehabilitation	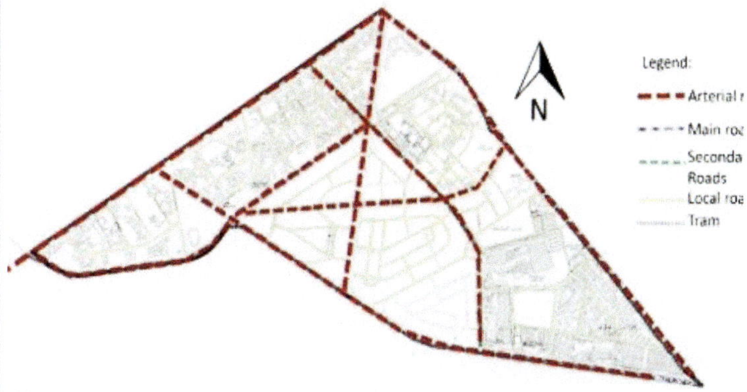

As a reflection to these actions, the connected neighbourhoods were divided by different transformations of roads and transform urban from being interconnected to be a standalone cell. All the preceding actions aimed to increase the connection and increase the traffic flow and speed. For example, Hussein Kamel street, Al Nozha street, El Merghani street, El Thawra street, Omar Ibn El Khatab street, El Horia street; all these streets have transformed from lower levels of local road to become arterial roads.

- Data Collection and Classification

Table 4 represents behaviour data that were collected to explore the influence of establishing highways on residents' sustainability behaviour in terms of the four variables: service provision behaviour, urban mobility behaviour, social relation behaviour and sense of security and safety behaviour as well as to reciprocity, familiarity and trust in neighbours and strangers. These variables represent resident's urban behaviour and social attitudes that concern their typical and repeated behaviour they are more likely to do. Measuring these variables depends on objective investigation behaviour survey tools:

- Spatial analysis is used to deduce social behaviour depending on walk, transit and accessibility score.
- Questionnaire administered to district residents: Sample selection: 40 residents are randomly selected that represent different gender, age, education and income. The questionnaire first explores resident's socio-economic characteristics and then measures the change in social and urban behaviour and the key factors of sustainability indicators. Likert scale was used and has been converted into percentage scale.
- Ethnographic observation of residents' social behaviour is used based on field notes with photographs which are used to extract records. It aims to double check the reported scores by what people say they do in the questionnaire.

3 Results

This part aims to discuss the impacts of road rehabilitation on changing resident's urban behaviour in selected case study in terms of four main factors including service efficiency, urban mobility, depth of social relations, and safety and security.

3.1 Service Efficiency

Figure 1 represents the change in functional efficiency before and after road rehabilitation. It revealed a lack of service provision for daily, weekly and monthly needs in terms of sufficiency 30, 25, 40 and 95 to 15, 10, 5 and 30, respectively; and in terms of efficient diversity from 35, 30, 20 and 85 to 10, 10, 5 and 25, respectively; and efficient distribution and access from 40, 45 and 50 to 15, 10 and 10, respectively. This can be explained that the new highways work as barriers for physical accessibility to the surrounding area and limit access to the services that were in front of residents and can reach with limited effort before highway. Highways make neighbourhood limited to its own resources and lose the mutual relation and exchange between neighbourhood and the surrounding area and make people found it hard to get services, especially the weekly and monthly one; they are forced to move long-distance and high frequent trips that consume more travel time, cost and effort to reach services that were located on the other side of the street that could be met in short time, distance and effort which become very hard to be reached after establishing highways.

3.2 Urban Mobility

Figures 2 and 3 trace the change in patterns of urban mobility before and after road rehabilitation. Figure 2 reveals that highways changed the travel mode choices for short-, medium-, and long-distance trips; walkability is reduced from 60, 40 and 20 to 15, 10 and 5, respectively; public mobility is reduced from 65, 50 and 35 to 20, 15 and 10, respectively; private mobility is increased from 10, 40 and 60 to 50, 70 and 80, respectively. It seems that the new conditions created great concern to private mobility for all types of trips including not only the long-distance trips but also extends to the medium- and the short-distance trips. People are encouraged to use private transportation instead of walkability for the short- and medium-distance trips. This can be explained that depending on public transportation and walkability become unsafe with the massive through traffic.

Figure 3 indicates how the frequency of different types of trips is changed after road rehabilitation. Total frequencies of all types of trips are reduced from 80, 45 and 25 before road rehabilitation to 30, 25 and 60 after road rehabilitation. This indicates how people become unwilling to do most of their daily used activities; they are not willing to do necessary, optional and social activities. In the other

Table 4 Summarized scores for the variables of behaviour performance

Key indicators	Sub-indicators	Indicators	Sustainability performance	
			Before	After
Service efficiency	Quantitative	Sufficient provision for daily needs	30	15
		Sufficient provision for weekly needs	25	10
		Sufficient provision for monthly needs	40	5
	Qualitative	Efficient diversity for daily needs	35	10
		Efficient diversity for weekly needs	30	10
		Efficient diversity for monthly needs	20	5
	Accessibilty	Efficient distribution access in walking distance	40	15
		Efficient distribution access in walking distance	50	10
		Efficient distribution access in walking distance	45	10
Trip distance	Inside the borders	Long-distance trips	60	20
		Medium-distance trips	70	30
		Short-distance trips	80	40
	Outside the borders	Long-distance trips	60	40
		Medium-distance trips	50	10
		Short-distance trips	40	5
Mode choices	Short-distance trips	Private transportation	10	50
		Public transportation	20	25
		Walkability and cyclability	60	15
	Medium-distance trips	Private transportation	40	70
		Public transportation	30	30
		Walkability and cyclability	40	10
	Long-distance trips	Private transportation	60	80
		Public transportation	50	35
		Walkability	20	10
Trip frequency	Short-distance trips	High frequent trips	80	30
		Medium frequent trips	15	10
		Low frequent trips	5	5
	Medium-distance trips	High frequent trips	45	35
		Medium frequent trips	20	15
		Low frequent trips	35	25
	Long-distance trips	High frequent trips	25	60
		Medium frequent trips	25	25
		Low frequent trips	50	15
Social relations	Social inclusion	Necessary activities (work, educate, shop)	4.50	2.50
		Optional activities (entertainment, walking)	4.50	2.50
		Social activities (social, religion, meet neighbours)	4.00	2.50
	Social interaction	Intentional Interact with neighbours and friends	1.00	0.75
		Accidental Interact with neighbours and friends	1.50	2.00
		Collective interaction (meeting in groups)	2.50	2.25
		Diversity interaction with diverse social groups	2.50	2.00
		Diversity interaction with integrator types	0.25	1.00
	Social bonds	Recognize neighbours	1.25	0.25
		Know neighbours	4.8	5
		Engage with neighbours	3.2	4.5
		Social relation	1.00	0.75
	Social relation	No connection	1.50	2.00
		Greetings	2.50	2.25
		Cheap chat	2.50	2.00
		Deep conversation	0.25	1.00

(continued)

Table 4 (continued)

Key indicators	Sub-indicators	Indicators	Sustainability performance	
			Before	After
Safety and security	Security	Through traffic	1.25	0.25
		Accidents	4.8	5
		Intrusions and strangers	3.2	4.5
		Offenders and criminals	1.00	0.75
	Feel safety	Feel safe for movement in streets and parks?	1.50	2.00
		Feel safe in homes	2.50	2.25
		Feel safe for properties	2.50	2.00
		Feel safe for women, kids, elders	0.25	1.00
		Day hours	1.25	0.25
		Night hours	4.8	5
		Late night hours	3.2	4.5
	Target hardener and surveillance	Degree streets are inspected by residents?	1.00	0.75
		Active tools like alarms, camera and guards?	1.50	2.00
		Using guards and security members	2.50	2.25
	Strangers familiarity	Recognize strangers in streets	2.50	2.00
		Trust strangers?	1.50	1.00
		Can help strangers?	1.00	0.50
		Found strangers as helpful?	4.50	2.50

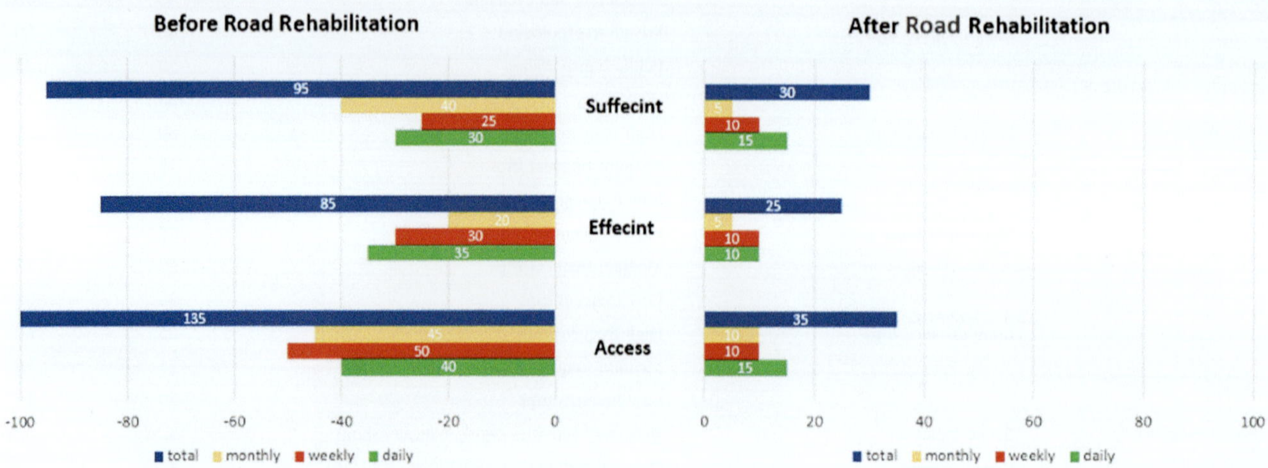

Fig. 1 The change in service provision (by Author)

hand, the frequency of short- and medium-distance trips is highly reduced; in other words, short-distance trips are reduced and most of the trips are changed to medium- and long-distance trips, the short-distance trips become limited, and this cause increases that rely on private motor mobility and reduced non-motorized environmentally friendly mobility.

3.3 Depth of Social Relations:

Figures 4, 5, 6 and 7 trace the change in patterns of social relations before and after road rehabilitation. Figure 4 explores a total reduction in participation to necessary, optional and social activities; this can be explained that most of the social activities can be reached before erecting

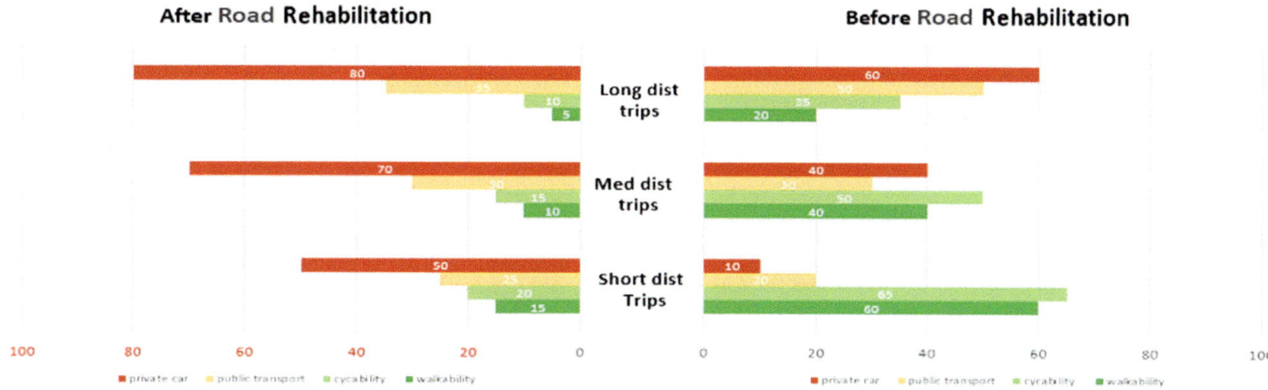

Fig. 2 The change in mode choices in different types of trips (by Author)

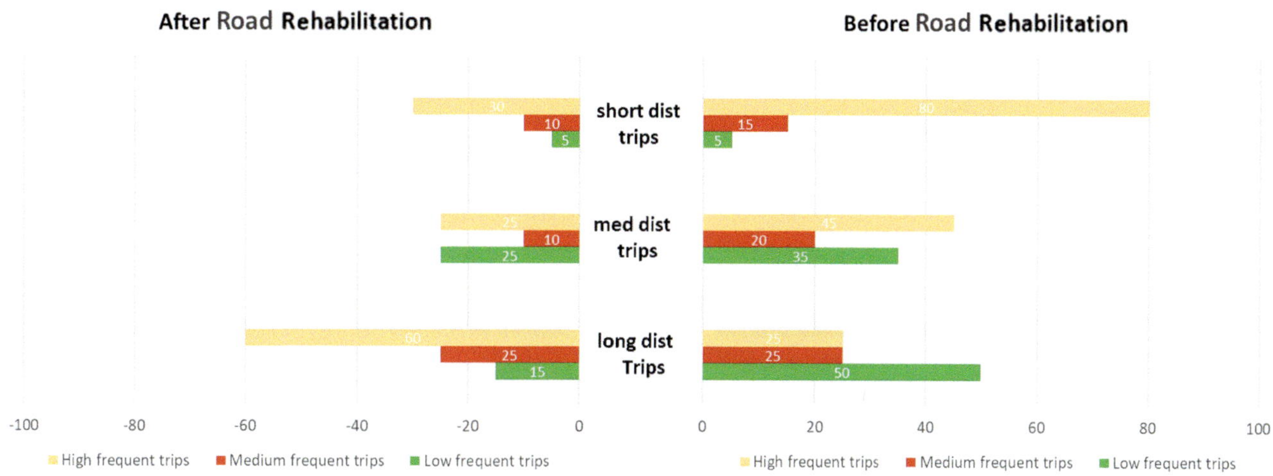

Fig. 3 Change in travel behaviour (frequency of short-, medium- and long-distance trips) (by Author)

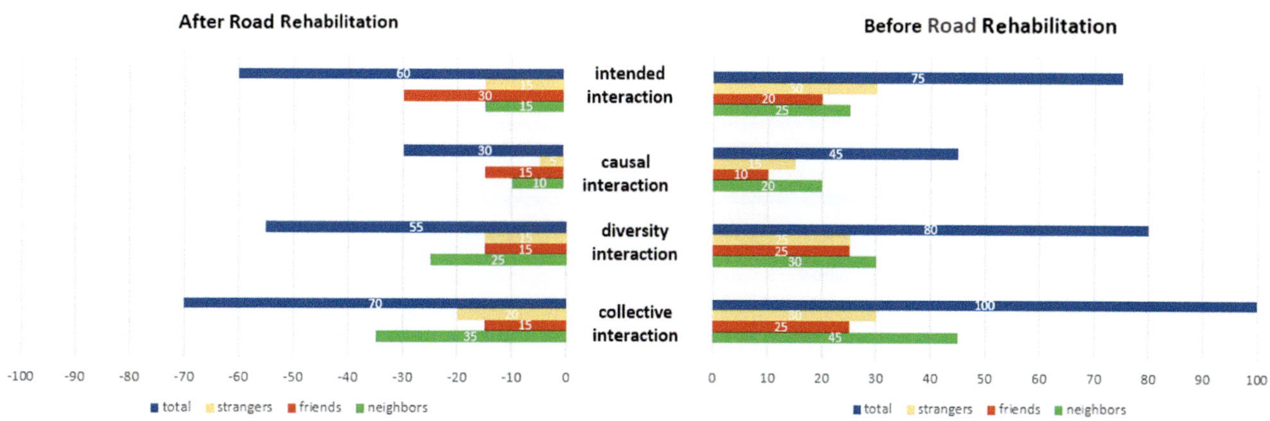

Fig. 4 Tracing the change in patterns of social interaction in term of situation (by Author)

highways, and now it consumes long-distance trips that are not safe and consumes more time, effort and cost with private mobility. It also explores the change in patterns of social interaction before and after the change; it illustrates how the new conditions imprint negative impacts on both intended and causal interaction as people hardly meet their neighbours, and the micro-community are crowded with strangers that make people lose the comfort to be in urban spaces.

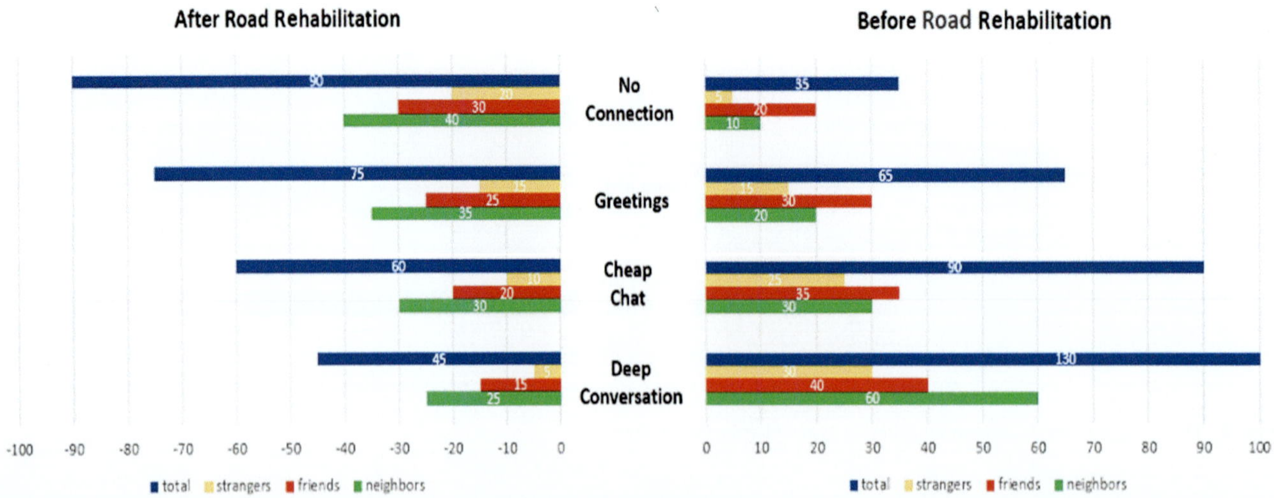

Fig. 5 Tracing the change in patterns of social interaction in term of action (by Author)

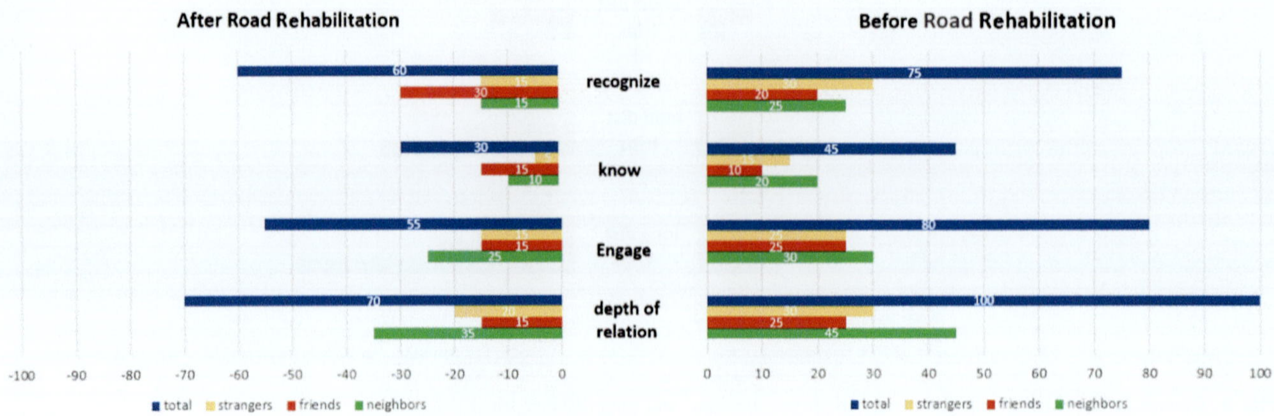

Fig. 6 Tracing the change in depth of social relations (by Author)

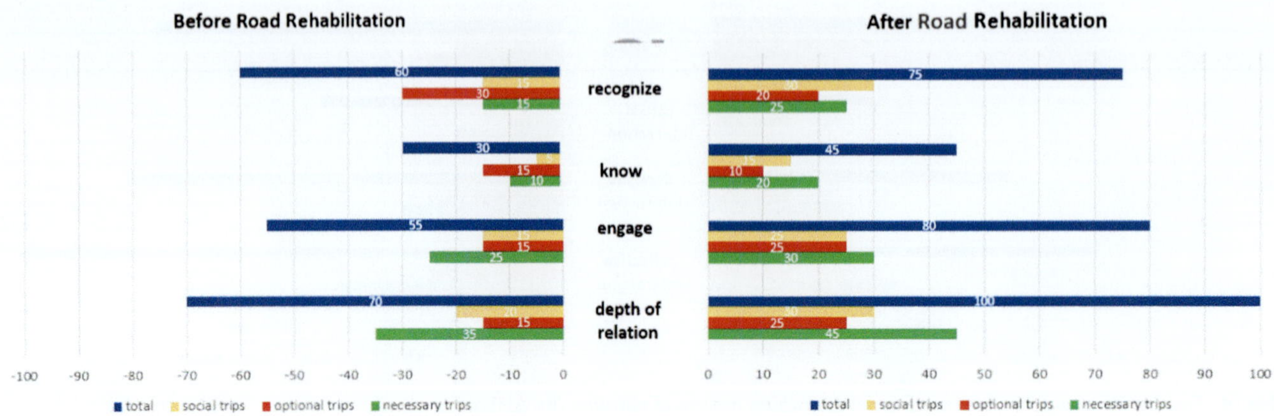

Fig. 7 The change in sense of social relations (by Author)

Figure 5 explores reduced patterns of social interaction before and after the change, it illustrates reduced interaction type from deep conversation to cheap chat to greetings with total gradually increased values of 35, 65, 80, 130 before road rehabilitation to gradually reduced values of 90, 75, 60, 45 after road rehabilitation. This can be explained that the micro-community are crowded with strangers that make people lose the comfort and trust to interact in urban spaces and lack deep conversation and accordingly deep relations.

Figure 6 explores how the social ties with neighbours are reduced, people hardly recognize and know their neighbours, and they are not connected to them as before the road rihabilitaion. The relations are reduced from deep social relation to only greetings; this is because they do not have enough space for gathering, and as it costs time, effort and money to meet friends and neighbours. The physical separation caused by the transformed roads caused a social separation between people in different neighbourhoods; in the other hand, it increased the strangers that reduce safety and quietness and accordingly reduce connection with neighbours in the same street and neighbourhood.

Figure 7 traces the change in sense of community before and after erecting highways; they loss their relation with the people and the place. People lost their sense of micro-community as it was mainly based on the sense of macro-community of Heliopolis. This can be explained that they found the community are crowded and full of strange integrators, high through traffic that makes it unsafe, uncomfortable and lack the privacy, quite and uniqueness that was existing before. Furthermore, they lack the relation with their neighbours.

3.4 Safety and Security

Figures 8, 9 and 10 trace the change in patterns of safety and security before and after road rehabilitation. Figure 8 traces the reduced sense of safety before and after establishing highways, where an increase through traffic is recorded from 30 to 60 that increased the chance for accidents. On the other hand, increased strangers are recorded from 25 to 45 that reduce people sense of security and caused increased concern for more target hardener from 25 to 50.

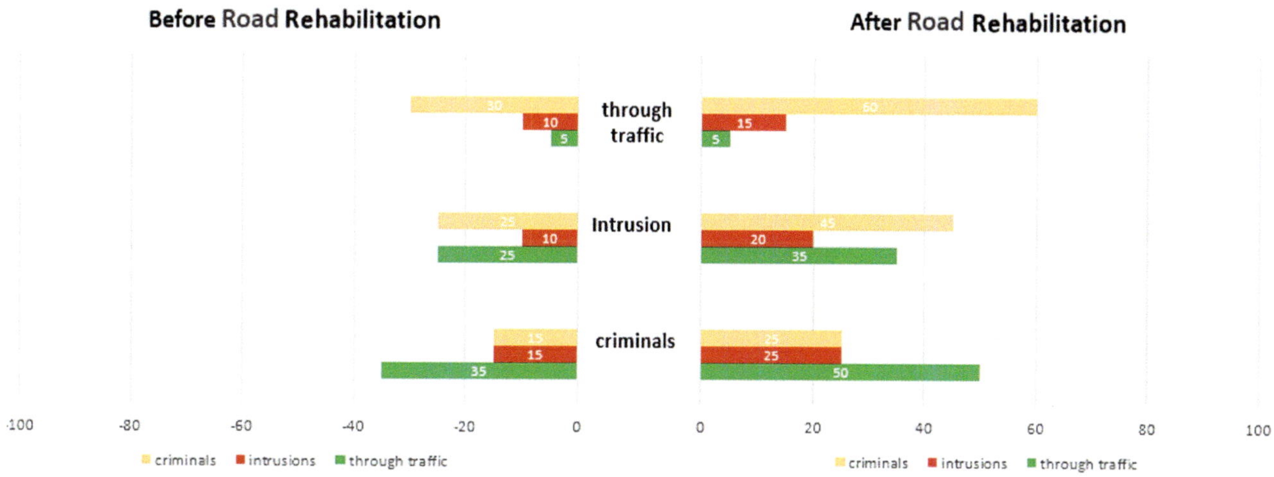

Fig. 8 Tracing the change in sense of safety and security in term of type of offenders nse of safety and secur(by Author)

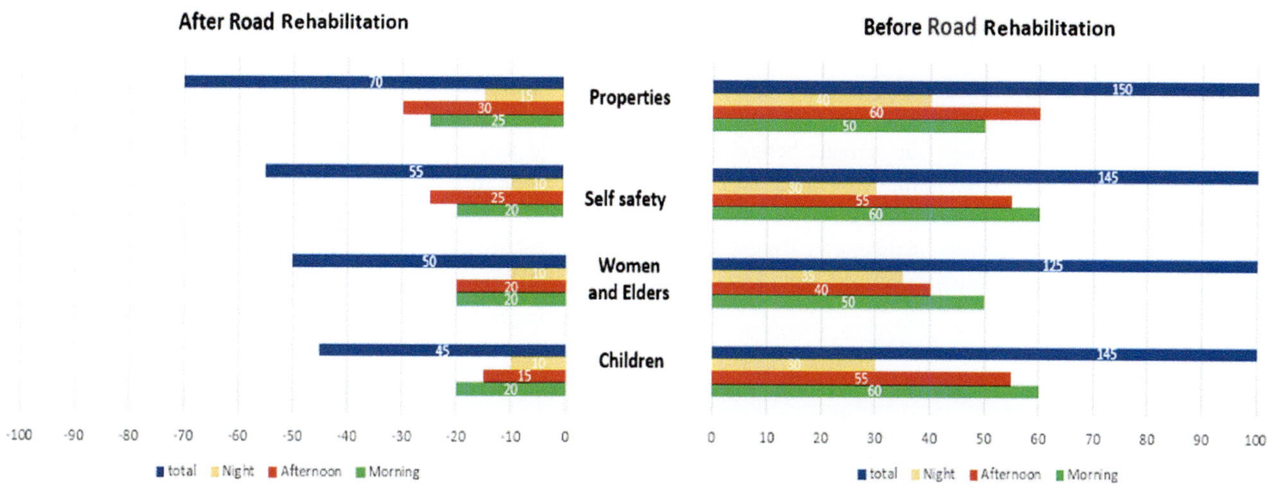

Fig. 9 Tracing the change in sense of safety and security in terms of properties (by Author)

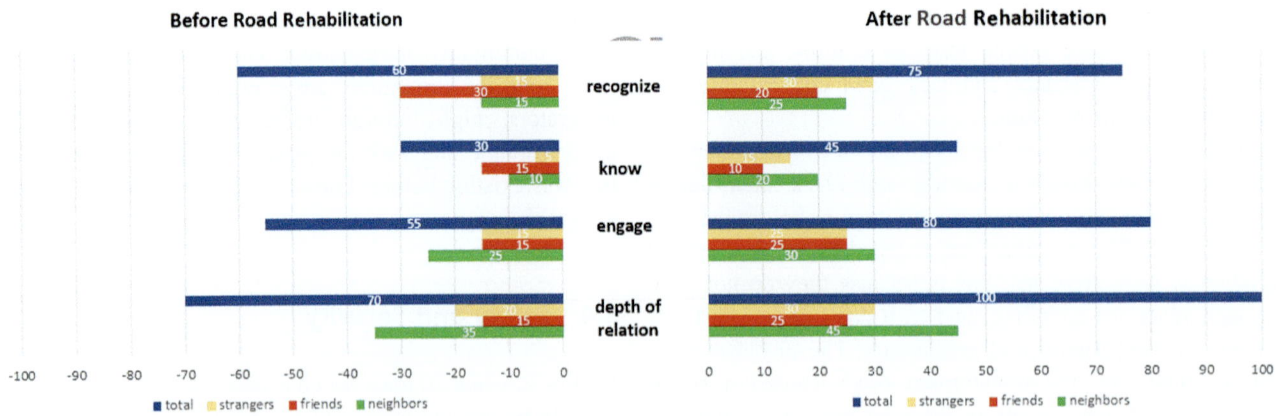

Fig. 10 Tracing the change in sense of safety and security in terms of degree of configuring intrusions (by Author)

Figure 9 traces the impacts of highways on people sense of safety for properties, self-safety, for elders and women, and for kids in morning, afternoon and night hours. People lose their total sense of safety from values of 150, 145, 125 and 145 before the road rehabilitation to the values of 70, 55, 50 and 45 after road rehabilitation.

Figure 10 traces the change in social behaviour, people hardly recognize or now the neighbours, and hardly engage to them, they do not trust on them, and they also found a lot of strangers in the streets so they cannot trust in them or have faith that they are helpful. People are encouraged to increase target hardener tools to meet the new conditions that different strangers and offenders have occupied the urban spaces as they path through the community to get shorter distance to the opposite side of the roads.

4 Discussion

This paper introduces a different perspective for examining the sustainability impacts of erecting highway projects concernning existing neighbourhoods in micro-scale. Urban fabric of micro-community is ruptured in favour of solving the traffic problems of wider community; accordingly, once the urban structure has changed by erecting highways, a relevant change in urban behaviour of micro-community residents, including functional efficiency, urban mobility behaviour, social interactions and relations, and sense of safety and security, happens in somehow and become interesting to compare before and after road rehabilitation.

Results of the field study in Heliopolis supports the international review and contradict with the expected aims of government and the results achieved after application of establishing highways inside existing urban fabric in different dimensions.

- In terms of service efficiency, government sought that encouraging highways would build boundaries and increase roads hierarchy that defines neighbourhoods and shift existing neighbourhood model from existing traditional undefined boundary to modern defined boundary model that can provide residents with the required functional efficiency of having its own services in quantitiative fulfilment and qualitative distrbution. However, empirical evidence of Heliopolis revealed a lack of service provision in terms of quantity and distribution.
- In terms of urban mobility, government thought that highways can encouraged traffic flow and reduce travel distance, time and cost. In contrast, it shifted the traffic jams to the inner community. On the other side, it was expecting to increase local community dependency on walkability. However, empirical evidence of Heliopolis revealed that the modifications make residents depend more on using private car dependency and reduce walkability. A more private car-driven environment increases traffic jams and increases insecurity of accidents and unsafety of intrusions, criminals and offenders.
- In terms of privacy, safety and security, government thought that highways can achieve privacy by diverting the high traffic away from the inner communities and limit the access as positive notion to achieve privacy as it would split the neighbourhood and can force to exclude the through traffic and exclude the strangers and accordingly can provide the neighbourhoods with calm, safety and security. However, empirical evidence of Heliopolis revealed inclusion of through traffic and strangers who are seeking to find shortcut routes to avoid long-distance trips that cause reduced safety and security and increase traffic displacement to inner neighbourhoods.
- In terms of social relations, government thought that highways can move the through traffic away from the neighbourhoods and, accordingly, can provide residents

with safe and defined community that all neighbours know each other without strangers. However, empirical evidence of Heliopolis revealed the opposite, where the community full of strangers and through traffic that make community residents found it unsafe to communicate with each other and fair to build social relations. On the other hand, a limitation of accessibility to surounding area mimnimizes the available social urban spaces that can discourage people involvement to urban spaces and reduce their chances for socal interaction.

The paper introduces some precautions for future projects of road rehabilitation, studying first the impacts in the community and just them choosing to proceed or not with the modifications. Some precautions need to be considered to minimize the negative sustainability impacts on existing neighbourhoods.

- Highway should be considered as a physical barrier for accessibility that cuts the connection of different parts of the existing urban fabric and can cause physical and social barriers for service and activities, instead of being self-sufficient and independent community depending mainly on its own resources which could be limited due to the wrong choice of the highway and insufficient to provide true community life.
- Highways can cause traffic problems to the macro-community that can displace some of the unwanted traffic flow inside micro-community for those who aim to escape the long-distance trips to avoid commute cost and time that not only can cause crowdedness and traffic jams inside micro-community but also can bring strangers who can include offenders and criminals.
- Highways can transform the nature of arterial streets surrounding neighbourhoods, which convert these streets to dead areas with no activity.

Based on the preceding findings and discussion, some development guidelines are provided to assist decision-making process for establishing future process of sustainable road rehabilitation inside existing urban fabric:

- Location of Highways and Area of Divided Communities:

Planning highways should consider sufficient community area with considering community density and size that can provide dense concentration of residents and promote the viable size that could promote a critical mass of activities to animate streets and support feasible range of required daily services within walking distance of divided neighbourhoods. If it is mandatory to erect highways with out considering the area, density and size of divided neighbourhoods, it is important to displace the longitudinal traffic above or below from cutting the horizontal pedestrian traffic between neighbourhoods using elevated or underground highways to facilitate connecting the low self-sufficient areas to connect more areas and services to replace their low sufficiency.

- Displace Civic Life to the Boundary

Planning highways should consider establishment of commercial boundary to reduce negative affected area and paths so it makes streets full of use and full of live through variety of activities that increase natural survilliance to streets and increase the sense of safety, and enhances the visual context.

- Pedestrian Design Between Neighbourhoods

Planning highway should be associated with concern to planning a pedestrian circulation and mass transportation for connecting inner cities. Planning highways should be associated with encouraging pedestrian bridges over the highways since they reduce walking distance between access points and increase external connectivity between micro- and macro-community to encourage pedestrian life between neighbourhoods. Also shifting the highway traffic above or below the connection between neighbourhoods can encourage more space and chance for pedestrian accessibility between neighbourhoods.

- Controlled Motor Access and Elevated Highways

Using elevated highways with controlled motor access and establishing treed street network pattern could minimize the through traffic though the neighbourhoods and encourage privacy, safety and security for inner neighbourhoods.

5 Conclusions and Recommendations

This research suggests that the way we imprint the process of road rehabilitations and urban regeneration in our neighbourhoods affects people social behaviour and accordingly impacts urban sustainability. Different strategies of transformations of roads are used by increasing their width, removing the islands, moving the traffic direction to one way to increase the road width by removing squares and erecting bridges to increase the connection and increase the traffic flow and speed. Even though it caused a very important challenge for urban development, these roads become hard, unsafe and consuming more time, effor, and cost to cross, and accordingly, these roads rather than become connectors to link city parts; they become cutters of the connection between different parts of inner urban fabric and cut the connection to the surrounding residential area. They become cutters for the connection between city parts and can impact the sustainability and satisfaction of the existing neighbourhoods. The research concluded three conclusions:

First: The process of road rehabilitation imprints impacts on the structure of the city; it created new conditions that limit accessibility, connectivity and mutual benefits between neighbourhoods and the surrounding area. This new condition impacts not only the functionality but also the sustainability and satisfaction in the neighbourhoods.

Second: The rehabilitation of road network was expected to divide the loyalty of the macro-community to clear shape neighbourhoods that can provide its residents with the required privacy and local sense of community. Rather it is deduced that people lost their sense of micro-community as it was mainly based on the sense of macro-community of Heliopolis.

Third: The change in urban structure structure reformation cause a change in urban behaviour and social interactions and cause new urban behavioural patterns after road rehabilitation that are deduced to be less sustainable than before urban reformation with specific reference to the case of Heliopolis:

- Limited resources with unsafe accessibility to the services are recorded, accordingly caused low efficiency and sufficiency of urban functionality and cause less sustainable provision of services.
- Unsustainable urban mobility patterns are recorded, with more concern to the private mobility than walkability and cycling, increased medium- and long-distance trips than short-distance trips, and increased frequency of long- and medium-distance trips.
- Unsustainable social impacts are recorded with reduced social ties and reduced sense of micro- and macro-community.
- Reduced sense of safety and security is recorded, where an increase through traffic is noticed that increased strangers in the neighbourhood and increased the chance for accidents; at the same time, a great concern for more target hardener is noticed.

The paper introduced some development regulations that should be regulated to minimize the negative impacts of highways:

- Planners should promote micro-community self-sufficiency and diversity by supporting keeping sufficient community area and sufficient density of population that can keep self-sufficiency inside the isolated neighbourhoods to perform well in providing sufficient and efficient distribution of services and urban spaces that can keep urban form perform well.
- Planners should provide save pedestrian bridges over the highways to facilitate well connection and integration to macro-community surrounding it.
- Planners should provide traffic solutions to avoid forcing the through traffic inside existing neighbourhoods to avoid long trips and avoid increased cost and time.
- Planners should support exclusion of strangers to the existing neighbourhoods to avoid offenders and criminals without been recognized and observed well by residents.
- Planners should encourage integration, through enhancing boundary land use, encouraging activities and keep services area in the boundaries face in arterial streets, which can promote city public life.
- Planners should displace some of the services and facilities to the borders to encourage civic life of macro-community and to work as borders to exclude through traffic.

The paper introduces some recommendations for highways policy makers, planners and future research to better change the impacts of highways on existing urban fabric:

- The traffic planning inside urban areas should be associated with considerations of how the community interacts with and across the highways.
- The traffic planning should be associated with pedestrian and public transit planning.
- More investigation for the impacts of different types of highways is required, including a comparison between the potential impacts of normal highways, elevated highways and flyover highways.
- More investigation for the impacts on different categories of neighbourhoods is required including a comparison between the potential impacts on early developed, early planned and new planned communities.
- More investigation for the impacts on different socioeconomic neighbourhoods is required, including a comparison between the perception of the impacts of low-, middle- and high-income communities.

References

Abozied, E., & Vialard, A. (2020). Reintegrating informal settlements into the Greater Cairo Region of Egypt through the regional highway network. *Regional Studies, Regional Science, 7*(1), 333–345. https://doi.org/10.1080/21681376.2020.1804443

Almoghazy, A. (2021). New roads, bad connections: How new road projects impact the right to the city. Altern. Policy Solut shorturl.at/pzCIZ. Accessed, 25. https://aps.aucegypt.edu/en/articles/608/

new-roads-bad-connections-how-new-road-projects-impact-the-right-to-the-city

Al-Qawasmi, J., Saeed, M., Asfour, O. S., & Aldosary, A. S. (2021). Assessing urban quality of life: Developing the criteria for Saudi cities. *Frontiers in Built Environment, 7*, 682391. https://doi.org/10.3389/fbuil.2021.682391

Asian Development Bank (ADB). (2005). Road Rehabilitation Project (Loan 1455-KAZ) in Kazakhstan. Performance Evaluation Report, Operations Evaluation Department, Asian Development Bank. https://www.adb.org/sites/default/files/evaluation-document/35135/files/28403-kaz-pper.pdf

Bauman, J. F. (1991). The expressway "motorists loved to hate": Philadelphia and the first era of postwar highway planning, 1943–1956. *The Pennsylvania Magazine of History and Biography, 115* (4), 503–533.

Cervero, R. (2009). Transport infrastructure and global competitiveness: Balancing mobility and livability. *The Annals of the American Academy of Political and Social Science, 626*(1), 210–225.

Correia, D., Marques, J. L., & Teixeira, L. (2022). City@ Path: A collaborative smart city planning and assessment tool. *International Journal of Transport Development and Integration, 6*(1), 66–80. https://doi.org/10.2495/TDI-V6-N1-66-80

Ebeling, W. (2013). Urban entomology. *Annual Review of Entomology, 16*, 45–66.

Elkatib, M. (2020). Bridges is coming—How city planners continue to tear apart the urban fabric of Cairo's neighborhoods by means of highways and bridges. https://doi.org/10.13140/RG.2.2.24063.64164

Elsayed, E. (2021). Rehabilitation of road networks inside inherent neighborhoods and their impact on residents' urban quality of life case study: "Heliopolis" neighborhood-Cairo. *MEJ. Mansoura Engineering Journal, 46*(2), 1–14. https://doi.org/10.21608/bfemu.2021.175971

Forrest, R., & Kearns, A. (2001). Social cohesion, social capital and the neighborhood. *Urbanstudies, 38*(12), 2125–2143. https://doi.org/10.1080/00420980120087081

Fouad, T. M. T. F. A., Abdou, A. A., & Morsi, A. A. (2023). Urban quality of life in Nasr city: An assessment of the reformation project of the built environment. *Environment-Behaviour Proceedings Journal, 8*(23), 263–270. https://doi.org/10.21834/ebpj.v8i23.4454

Ghonimi, I. (2021). The role of urban space configuration in maintaining social inclusion: A comparison of four types of neighborhoods in Greater Cairo region-Egypt. *Journal of Urban Research, 39*(1), 75–104. https://doi.org/10.21608/JUR.2021.30232.1002

Ghonimi, I. (2022). Sustainable urban quality-of-life in neighborhoods: Bridging the gap between satisfaction and performance in four types of neighborhoods in GCR-Egypt. *Journal of Engineering Research, 6*(4), 105–121. https://doi.org/10.21608/erjeng.2022.265458

Ghonimi, I., & Awad, A. (2017a). Socially sustainable neighborhood in Egypt: Assessing social capital for different neighborhood models in Greater Cairo Region. *Journal of Engineering Sciences Assiut University, 46*. https://doi.org/10.21608/jesaun.2018.114463

Ghonimi, I., & Elzamly, H. (2017b). Sustainable urban mobility: Assessing different neighborhood models in Greater Cairo Region, Egypt. In *22th International Conference on Urban Planning Regional Development and Information Society*.

Hamiduddin, I. (2015). Social sustainability, residential design and demographic balance: Neighbourhood planning strategies in Freiburg, Germany. In *Town planning review* (pp. 29–52).

Jacobs, J. (1961). *The death and life of great American cities*. Random House.

Khalaj, F., Pojani, D., Sipe, N., & Corcoran, J. (2020). Why are cities removing their freeways? A systematic review of the literature. *Transport Reviews, 40*(5), 557–580. https://doi.org/10.1080/01441647.2020.1743919

Levett, R. (1998). Sustainability indicators—Integrating quality of life and environmental protection. *Journal of the Royal Statistical Society: Series A (Statistics in Society), 161*(3), 291–302. https://doi.org/10.1111/1467-985X.00109

Levin, N., & Duke, Y. (2012). High spatial resolution night-time light images for demographic and socio-economic studies. *Remote Sensing of Environment, 119*, 1–10.

MacKaye, B., & Mumford, L. (1931). Townless highways for the motorist. *Harper's, 163*, 347–357.

Mohl, R. A. (2004). Stop the road: Freeway revolts in American cities. *Journal of Urban History, 30*(5), 674–706.

Mounir, C. (2019). Trees razed for bridge construction in Heliopolis. Available: https://egyptindependent.com/trees-razed-for-bridge-construction-in-heliopolis/

Mumford, L. (1964). *The highway and the city* (p. 175). New American Library.

Popovich, N., Williams, J., & Lu, D. (2021). *Can removing highways fix America's cities*. The New York Times (p. 27). https://www.nytimes.com/interactive/2021/05/27/climate/us-cities-highway-removal.html

Saber, A. M., & Radwan, A. H. (2020). Assessment of the urban and environmental impact of the traffic axes development project of Heliopolis. *MEJ. Mansoura Engineering Journal, 45*(4), 42–57. https://doi.org/10.21608/BFEMU.2020.122375

Saiz, A. (2006). Dictatorships and highways. *Regional Science and Urban Economics, 36*(2), 187–206.

Schindler, S. B. (2014). Architectural exclusion: Discrimination and segregation through physical design of the built environment. *Yale LJ, 124*, 1934.

Shafik, N., Mansour, Y., Kamel, S., & Morcos, R. (2021). The impact of the Cairo Streets Development Project on the independent mobility of children: A field study on the streets of Heliopolis, Egypt. *Infrastructures, 6*(7), 98. https://doi.org/10.3390/infrastructures6070098

Taqi, O. S., Farid, M., & Radwan, A. H. (2021). The importance of quality of urban life indicators in developing new urban communities in Cairo. *MEJ. Mansoura Engineering Journal, 46*(2), 58–66. https://doi.org/10.21608/bfemu.2021.184361

Turkoglu, H. (2015). Sustainable development and quality of urban life. *Procedia-Social and Behavioral Sciences, 202*, 10–14. https://doi.org/10.1016/j.sbspro.2015.08.203

Vallance, S., Perkins, H. C., Bowring, J., & Dixon, J. E. (2012). Almost invisible: Glimpsing the city and its residents in the urban sustainability discourse. *Urban Studies, 49*(8), 1695–1710. https://doi.org/10.1177/0042098011417903

Wiesli, T. X., Hammer, T., & Knaus, F. (2022). Improving quality of life for residents of biosphere reserves and nature parks: Management recommendations from Switzerland. *Sustainability: Science, Practice and Policy, 18*(1), 601–615. https://doi.org/10.1080/15487733.2022.2100128

Wiesli, T. X., Liebe, U., Hammer, T., & Bär, R. (2021). Sustainable quality of life: A conceptualization that integrates the views of inhabitants of Swiss Rural Regions. *Sustainability, 13*(16), 9187. https://doi.org/10.3390/su13169187

Sustainable Praxis on Istanbul's Residential and Green Areas by the Henri Prost's Development Plans

Hülya Coskun

Abstract

The idea of sustainability has not found place in the agenda of the city in Istanbul long years, which has rapidly developed and vastly overpopulated with immigration. Today on sustainability and green urbanism not only new and innovative design and planning ideas have been used, but also in old urban planning methods recently come-back. Istanbul's first Master plans planned by French architect-urbanist Henri Prost with "residential areas," and "green areas" in balanced. Almost after a century, his Master plan and these green areas still significant in the green planning of the city as nuclei. The founder of *l'urbanisme* (urbanism) theories of the twentieth century, French planner Henri Prost's Istanbul city's modernization plans considered vague and controversial many years. Although Prost plans always been popular in every period as the product of the doctrinal structure of the French Urbanism School (*l'ecole*), its effects have lasted from the 1950s to the 70s to 80s until today. The twentieth century's vehicle-based urban planning theories and discourses of CIAM School evolved to the non-vehicle, green and human-oriented plannings with the climate concerns. By the last quarter of the twentieth century, with concern on increasing uncontrolled housing areas, road network, heavy traffic, high carbon emission and less green areas, the cities had to confront being as harmful places for human life. Today, world cities like Paris and London searching the solutions with the specifically arranged *zone* zoning regulations like the twentieth century planners Henri Prost. Hence, this research aimed to present solution of today's cities problematics via re-examination Prost plans. Consequently, findings achieved after the re-examinations on Prost Master plans the purpose of the old "zoning" applications to separate cities into functions such as residential areas and green areas recently with the sustainability entered the city planning agenda the "zoning" planning for "sustainable protection" of the cities to develop balanced relations between "people" and "nature." Also, recently many cities expanded the "zoning" regulations, initiated to maintain climate-change, sustainability, green architecture, pollution, even pandemic, etc.

Keywords

Henri Prost • Development plans • Istanbul • Housing • Sustainability • Green areas

1 Introduction

Since the last quarter of the twentieth century, sustainability has not been on the agenda in İstanbul one of the worlds' giant mega city having the 16 million population today. However, the recent pandemic has made us understand once again the importance of sustainable ideas and necessity to green planning, with "green areas" and "residential areas" arranged as balanced in the cities. Recently, we witnessed the come -back of the some of the old architectural and city planning theories and discourses like E. Howard in the world. Smilarly in this study, Henri Prost İstanbul Master plans and his urban planning theories are re-examined in the possibility of transferring to the present as precious data of this old methods in city planning in future use. İstanbul Prost Master plans even though considered as a modernization plan it was prepared according to specific laws and regulations of *zone* "zoning" (French originated) by Henri Prost previously for Paris and Northern Africa, finally for İstanbul. It is searched the recent status of the Prost İstanbul plans and specifically the "zoning" applications for residential areas, and green areas, within the scope of climate-change and sustainability, the possibility of using

H. Coskun (✉)
Faculty of Architecture, MSGSU, Mimar Sinan Fine Arts University, Istanbul, Turkey
e-mail: her_222@yahoo.com

A. Battisti et al. (eds.), *Greening Our Cities: Sustainable Urbanism for a Greener Future*, Advances in Science, Technology & Innovation, https://doi.org/10.1007/978-3-031-49495-6_6

this valuable data both in Istanbul and in the world cities for future. Although old "zoning" regulations were initially considered modern architectural application from the early twentieth century, recently it was found that these old ideas might be used in the many cities adopted and developed according to today's circumstances to maintain climate-change, sustainability, green architecture, decreasing pollution, even solution for pandemic, etc.

The purpose of the old "zoning" applications to separate cities into "functions" today, the sustainability entered the city planning and zoning literature by *İstanbul Bölge Koruma Kurulu* (Istanbul Regional Conservation Commission) is deemed appropriate to expand the areas of "sustainable protection" and controlled use of the cities to develop more balanced relations between "people" and "nature." According to recent UNEP reports the cities need to be planned to achieve a zero-carbon living, with green planning significant for to make more livable places them (UNEP, 2019). Nowadays, urban planning is based on the values life quality substantiated on such as the clean air, travel times, green planning, and protect natural environment. Today some specifically arranged "zoning plans" laws and regulations put into effect by municipalities and governments in the world to mitigate the climate-change effects like the Henri Prost did in the early twentieth century. Some old *zones* (zoning plans) and (zoning bands) to protect green lands, woods and forests, refrain the uncontrolled urban sprawl and the harmful effects of the environment, pollution, even pandemic recently started the applicate cities like London, Paris, Europe, etc. Thus, like old "zoning" regulations put into as new application its ultra zone to limit low carbon emission in London also Paris plan to remove belt way lanes to cut pollutions (Citylab Bloomberg, 2022). Although existing old "zoning" regulations for the protection of green areas today, Istanbul do not have a more intrusive planning strategy like Paris and London yet. In this study, Prost Master plans, his theories, and regulations like "zoning" are re-examined. Without specifically arranged "zoning maps" uncontrolled urban sprawl, unplanned residential areas became critical threat of the cities' eco-system: green lands, forests, woods, even vital for water supplies. In the cities, a reconciliation aimed between ecology and economy the notion of urban growth of green as inseparable concept of sustainable development (Ferry, 2021, p. 125). Unlike the early twentieth century, the green planning now considered in a multidisciplinary perspective within the scope of sustainability, ecology, economy, urban, and architectural planning. Today, green planning in the cities ensures the stabilization of the temperature and therefore contributes to the improvement of the urban ecosystem where they becoming more harmful places. Likewise the world cities growth continuing a size of Manhattan added to the existing world lands almost every day, doubling number of the vehicles 1.5 million in the world (Calderon & Stern, 2015, p. 79).

Decreasing green planning led to increasing carbon emission and high-level temperature made cities unlivable places as a potential threat climate crisis. The comfort values of cities are now considered equivalent to number of green areas and parks, carbon emissions, so Istanbul was the last places in Green Score in the European Green City Index behind the cities like New York, London, etc. (Forum, 2020). According to recent study, the rate of green spaces in Istanbul is very low; only 2.20% of total surface area comprises of public (Greenpeace, 2022) (Fig. 1).

Although Istanbul today known as with less parks and green, however, the city was well planned with the balanced "green areas," and "residential areas" previously at the beginning of the twentieth century. The urban dynamics that have emerged from the middle of the twentieth century adversely affected the uncontrolled development of the city. At the mid of the twentieth century, after the 1960s to 70s with the mass migration, and unplanned urban sprawl was the critical issue of green lands, and forests were neglected long years. Since the 1960s, Istanbul endured immense urban population and lack of controlled urban planning and infrastructure (The Economist Intelligence, 2009). Furthermore, this research aimed to contribution of the creation of new and innovative urban planning theories for the future plans of the İstanbul as well as the other cities in the world through the reuse of past theories. Thus, research aims to

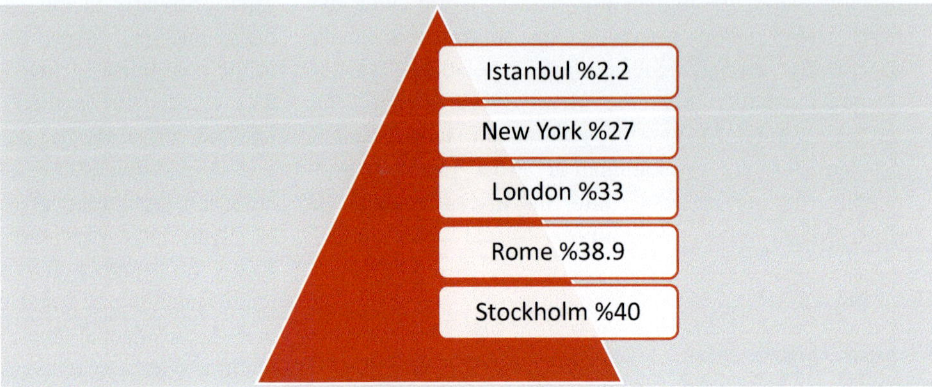

Fig. 1 Istanbul and the other cities in the World Green Areas and Parks. Graphic, designed by H. Coskun according to source of data from Greenpeace (Greenpeace, 2022)

re-examine the Istanbul city's "green areas," and the "residential areas" presented updated research concentrated on French architect-planner Henri Prost's development plans and "zoning" regulations also including post-Prost period and the today.

2 Problem Statements

The main problematic is approached primarily in the context of the climate issues with the aim of the uncontrolled urban sprawl in Istanbul, and urbanization, decreasing the green areas in the city we lived in today.

1. The primary proposition is established examination of old planning theories, laws, and regulations like used by French planner Henri Prost and other urban planners. Although urban planning theories have changed and transformed in a century, these old urban laws and practices could still put into effect to climate change and sustainability for the future. Thus, Henri Prost Master plans are always significant for the İstanbul city as well as the old "zoning" regulations developed and used by him. His Istanbul Master plans based to dividing the green areas, residential areas or industrial areas limited them inspired by the French Cornudet laws originally re-acted by Paris municipality. Indeed, this regulation is obtain similarity to our goals today on the issues climate-change and sustainability. In this context, this research involved the re-examination of old theories, law, and practices in the planning to use to climate-change and sustainability. Also, related to green planning these old theories could be transformed and adapted today's conditions particularly in the sustainability as a newly emerged problematic. The research aimed to contribute for the future development of the city and the other cities through the inferences of old theories and by re-examining this old and notion's planning methods and principles in the past.
2. The secondary proposition is established examination of Henri Prost's Istanbul plans prepared by old French *l'école* (school)planning tool; *l'aménagement* (regulatory) and *l'embellissement* (beautifying) as well as the *l'hygiène* (hygiene) aimed to modernization and traffic transportation of cities.
3. However, this old method based on green planning almost a century ago used as *l'embellissement* (beautifying) in the city of Paris purpose of the "esthetics" rather than the today's environmental concerns and sustainable ideas. At the middle of the twentieth century, a modern planning tool CIAM School become significant, plan the cities with green areas. Today, by the twenty-first century, the green planning has changed into a completely different path with recent environmental and climate concerns.
4. Third proposition is about the recent urban planning method and theories and its effects on Istanbul's green areas, woods, and forest since the beginning of the twentieth century. Today, in Istanbul, planning evolved to a completely a new dimension from previous centuries due to climate crisis, and urban planning has new concerns on to mitigate the harmful effects thinking about the cities future and creating a greener, more livable environment as UN Habitat declarations.

3 The Methodology

The research focuses on the re-examining of the planning Istanbul city green planning in balanced and controlled started from the early twentieth century and later in the post-Prost period and the 1960s to 1970s also comparing today's methods. Hence, this research purposed to find solution of today's cities problematics via re-examination plans and applications specifically Prost zoning plans. Henri Prost's methods and his planning tools via re-examined design principles for the future planning of cities as a unique research topic that researchers have not addressed formerly. The Prost plans were basically divided the city according to its functions such as residential areas, green areas, parks, educational areas, and industrial areas. Henri Prost's planning methods some of the old planning theories recently come-back to create solutions for cities problems. So, this study approached to problematic concentrated on his urban planning methods that he used at the early twentieth century still significant solving the cities problems today like Paris, London, etc. Although the purpose of the old "zoning" applications to separate cities into "functions" of settlements, today, the sustainability has entered the city planning and zoning literature by *İstanbul Bölge Koruma Kurulu* (Istanbul Regional Conservation Commission) was deemed appropriate expanding the areas sustainable protection and controlled use of the cities in order to develop more balanced relations between "people" and "nature." Today İstanbul Municipality and other cities municipalities work on more specific green "zoning" maps such as controlled zones or buffer zones for planning more controlled housing areas to protect the nature. The newly specified "zoning" applications, such as zoning areas for urban pollution, currently have used in cities such as London and Paris. On the basis of research, the primary research methodology depends on re-examination of Henri Prost Master plans, and *zones* "zoning" applications and suggestions for future use:

1. In the first part of the research is included that re-examination and analysis of the *zone* "zoning" practices applied in urban planning by Henri Prost at the early twentieth century.

2. Implementation of Henri Prost's Master plans on Istanbul city and his specifically produced *zones* "zoning" plans their recent situation and possibility to use of these applications after the uncontrolled development of Istanbul.
3. Finally, suggestions are proposed for future cities result from the assumptions from the Prost Master plans and zoning applications and transformation of these data to the practice for the future.

The research aims to re-examine this subject and present an updated study on both the green-areas and parks of the Prost period and post-Prost period and their current status, that have come to the fore in recent years in the city.

4 From Twentieth Century Old Planning Theories; to the Today's Sustainable Planning and Applicability of Old Planning Methods and Zones "Zoning" Plans

By the twenty-first century and toward the end of the twentieth century, the idea of green planning has became significant again due to sustainability with the recent environmental and climate issues. Today, Istanbul urban planning has specifically evolved to different path from previous centuries newly changed urban planning, and green ideas with the concern of the climate crisis to create a greener, more livable environment according to UN Habitat's recent declarations (United Nations, 2016).

Nowadays, some old planning theories in the context of the sustainability or to find solution to climate-based issues recently come -back. Also, the green plannings might be considered a base for a model development for future sustainable city planning. How to make a city climate resilient and what form can this take? Indeed, the first, planners need the most recent data on climatological risks and projections from computer models that predict the impact of climate (Gates, 2021, p. 287).

Even, the some early urban planning theories on green urbanism considered to use in the sustainable ideas like E. Howard's the "garden cities" in the early twentieth century come-back (Lehmann, 2011). Furthermore, just after the recent Covid-19, the housing and city planning theories changed due to people's newly emerged demand and the search with the houses with gardens in the green areas and out of the city or near towns. The Welvyn city designed previously by E. Howard in UK started to re-planning again recently as a sustainable city (Hertslife, 2021; Smith, 2021, p. 3). In İstanbul, after the Covid-19, people changed their requirement and searching houses with gardens or houses on the ground with garden or with large green terraces and balconies out of the city on this way similar to Welvyn garden-city planning. Thus, although these theories first came to the fore as sustainability in recent years, it became more significant, especially after Covid-19. Recently, in the city, and housing planning approach in which arranged with green spaces has become widespread in the world cities.

Even though many European cities have already been planned with considerable "garden-cities" in the out of the city along the green-belts since the early twentieth century, Istanbul city developed with the modernization plan arranged with green areas and parks which was made by Henri Prost. Today when compared with the European cities' norms and standards, Istanbul noted as with less parks and green spaces placed, the last, 24th on the Green Score in the European Green City Index, is behind the cities like New York, London, etc. (Forum, 2020). According to another recent study, the rate of green areas in Istanbul is considered very low only 2.20% of total surface area comprises of public-spaces (Greenpeace, 2022). By the 1990s, the world's agenda completely changed with the first summit in the 1992 (United Nations, 2020) and one of the climate summit, which was held in Istanbul, in 1996 (United Nations, 1996); however, sustainability did not arise the expected interest on and foreseen that it would take many years (United Nations, 2015, 2021). Finally, Istanbul ensured the acceptance of the Local Agenda and the strong international substances of the recent process of aforementioned (Arar, 2021). According to the new declaration in 2019, the UNEP urged the cities needed to balanced planning of the green areas and housing areas in future (UNEP, 2019). This made significant to prepare comprehensive and updated "zoning" laws in cities as well as the Istanbul. According to recent recommendation of UNEP in the context of sustainability, these laws had already arranged by architect-planners such as Henri Prost, with the controlled regulations of housing, green areas, parks, etc. This also made stil significant and the valid in use of Prost plans and his planning methods which is explained the need to re-examination in accordance with today's rules and design methods.

Also, some specific "zoning plans" and "green bands" laws and regulations to protect green spaces or to mitigate the climate-change affects acted by many municipalities and governments in the world likewise the Henri Prost and urban planners did in the early twentieth century. Henri Prost's plans and regulations like zoning plans re-examined to find solution climate-based issues. Thus, today, without "zoning maps" uncontrolled urban sprawl, unplanned residential areas lead to destruction of green lands became critical threat of the northern forests of Istanbul, vital water supplies.

Indeed, the İstanbul city was well planned with the balanced green areas, and parks, previously in the early twentieth century by the French architect-planner Henri Prost. However, the rapidly changing urban dynamics that started from the middle of the twentieth century adversely affected the uncontrolled development of the city (Figs. 2, 3, and 4). At

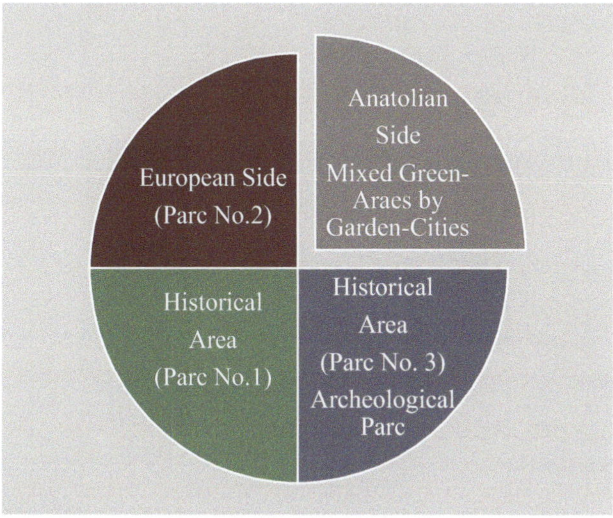

Fig. 2 Istanbul, Henri Prost Master plans, green areas and parks per regions. Graphic design and data by H. Coskun, 2022

the mid of the twentieth century, after the 1960s to 70s with the mass migration, and unplanned urban sprawl the vital issue of green areas, forests were neglected long years. Since the 1960s, Istanbul endured immense urban population and lack of controlled urban planning and infrastructure (The Economist Intelligence, 2009) (Fig. 5).

Hence, Prost Development plans which were founded on a well-balanced planning dynamism interwoven among the green areas, and residential areas considered as a respectable example for plannings in the near future of the cities. However, his old, development plans consist of explicitly green spaces, also implicitly included old, as well as the new residential areas. Henri Prost's urban design technique was varied on the creation of urban areas derived from the specifically arranged *zone plan* (zoning plan). Moreover, it has become vital importance to planning a city with green areas considered to the old *zones* (zoning plans), and urban planning techniques by Henri Prost and colleagues in the early twentieth century which specifically produce to protect cities harmful effects of uncontrolled urban sprawl and aimed to protect green lands, parks, forests, woods started the applicate today's cities such as London, Paris, Berlin, Europe, etc.

4.1 Henri Prost Planning Method and Tools in the Context of the Sustainability

Due to the changing urban theories and his main idea of green planning is regarded as indispensable perquisite of designing concept primarily "healthy cities" by Henri Prost in the beginning of the twentieth century. The idea of the

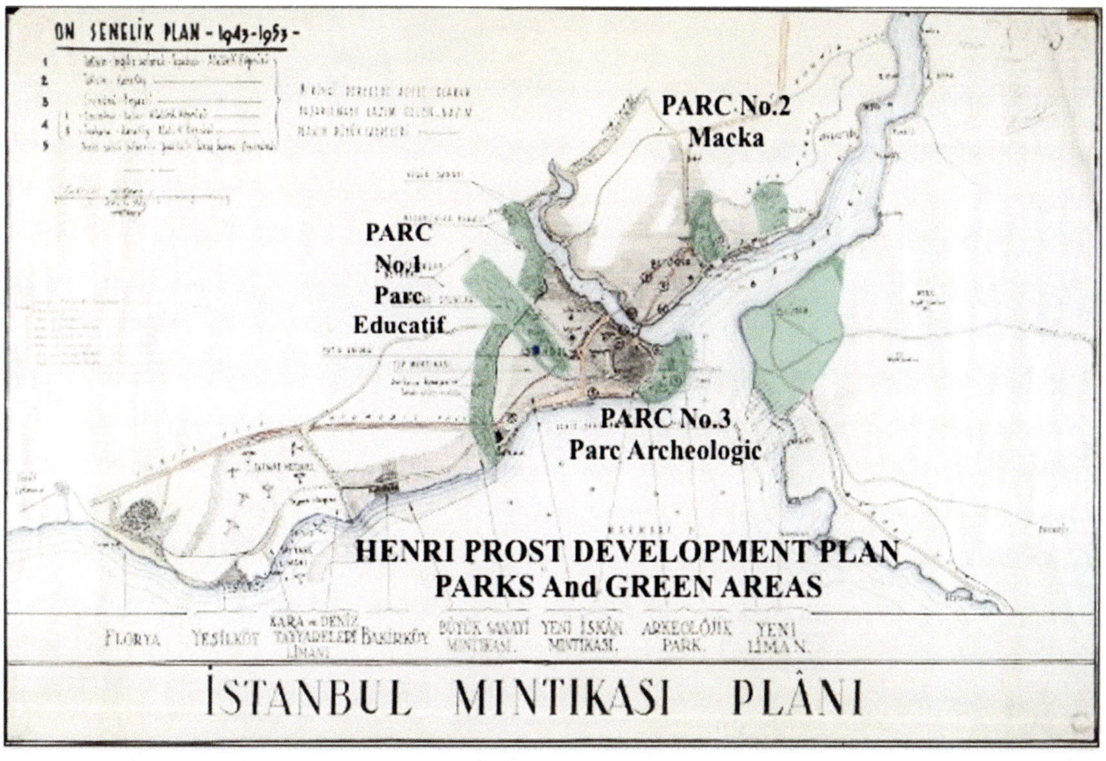

Fig. 3 Henri Prost Istanbul, European side development plans; "zoning plan" for green areas and parks. Plan, drawn by H. Coskun original from Prost plan. Original map, source, IFA, archives *(Académie d'Architecture), Paris*

Fig. 4 Henri Prost Istanbul 10-year plan, "zoning plan" for housing areas distributed the city's regions. Plan, drawn by H. Coskun original from Prost plan. Original map, source, IFA, archives *(Académie d'Architecture)*, Paris

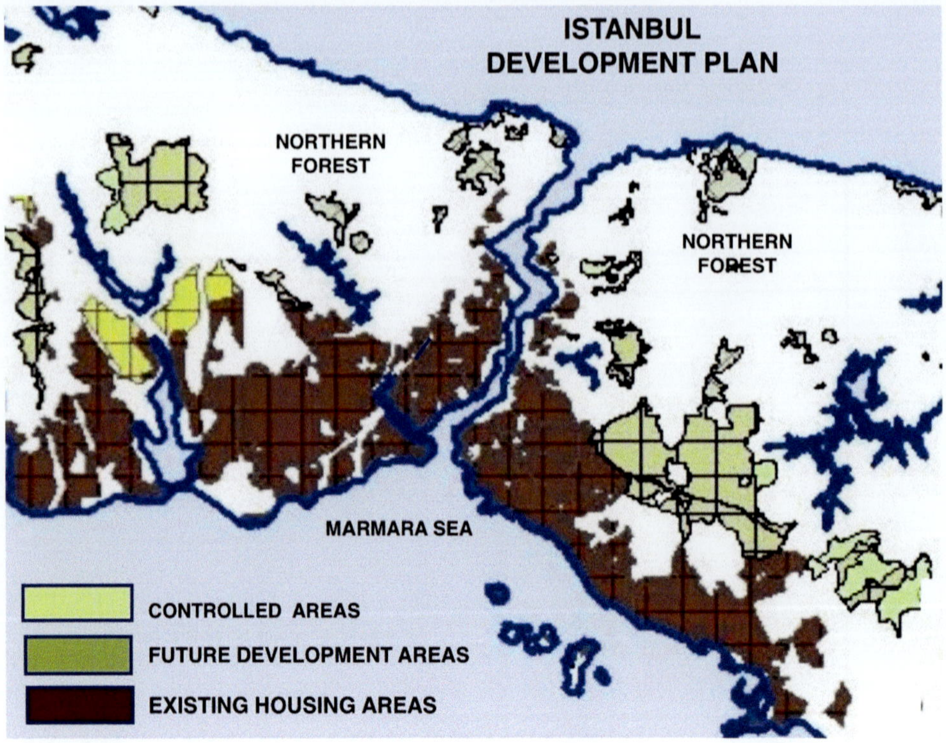

Fig. 5 Istanbul, recent newly arranged "zoning" map, existing residential areas (brown). Future development (green), and controlled areas (yellow). *Map source* Environment Reports of İstanbul Municipality. http://www.ibb.gov.tr/tr-TR/kurumsal/Birimler/SehirPlanlamaMd/Documents/100.000%ÖlçekliÇevreDüzeniPlanı

green planning was used as well as the French urbanism's old planning tool *l'embellissement* (beautifying), *l'hygiène* (hygiene), and *l'aménagement* (regulatory) a century ago (Prost, 1949). These urban planning principles of the early twentieth century were depended on modernization ideas by more green spaces and modern parks which was evoked by

E. Henard in Paris modernization (Bruant, 2011, p. 246) and Cerda's plan Barcelona the early twentieth century (Merlin, 1991, p. 31).

Until the late 1970s, the Istanbul city continued to develop with the newly arranged green areas around the nucleus first started by Henri Prost and enhanced the urban network of traffic roads in accordance with the planning discourse of the CIAM School. Although heavily criticized today, the vehicle-based urban planning was advocated in the mid twentieth century by CIAM, *Congrès International d'Architecture Modern* (International Congress of Modern Architecture) (Le Corbusier, 1994, p. 166). By the middle of the twentieth century, a planning tool became significant, aimed to plan the modern cities with parks and gardens by CIAM (Choay, 1979, p. 31). This urban planning approach was seen as indispensable prerequisite urban planning idea for "healthy cities" that developed in the early twentieth century, the Henri Prost planned. Furthermore, like the green spaces and other terms used *les espaces-publics* (squares-free spaces) Henri Prost frequently used in his planning reports and terminology encompassed all public spaces and open spaces including parks (Bilsel, 2010b, p. 349).

Although, urban planning notion evolved and transformed to sustainable ideas however Henri Prost's modern urban planning methods not far from today's urban planning principles and understanding, known as "realistic" urbanism identified by French F. Choay's latest urbanism theories and discourse (Doğrusöz, 2016) in F. Choay's book *L'Urbanisme Utopies Et Réalities* (Urbanism Utopias and Realities) and P. Merlin's urbanism theories (Choay, 1969; Merlin, 1991). F. Choay's urban theories are considered among the principle sustainable theories referenced as "naturalist" ideas as a pre-ideology in her book *L'Urbanisme Utopies Et Réalities* (Choay, 1979; Carriou & Ratouis, 2014, p. 2). Connected with the ideologies of F. Choay, the sustainable ideas also thought to pioneered by Henri Prost adopting a different notion from CIAM in the context of the F. Choay's progressivist and naturalist model. When the Prost Master plans examined in architectural scale, he planned the İstanbul well-balanced residential areas, and green areas as well as the public spaces, axis, roads, boulevards, etc. (Coskun, 2017a, pp. 34–36).

5 Re-examination of the H. Prost Master Plans and *Zones* (Zoning Plan): Dividing "Green Areas" and "Residential Areas"

Henri Prost's Master plans were known as Istanbul's first modernization plans which organized in well balanced green planning along the newly opened and designed housing settlements. When Prost İstanbul Master plans examined, it is observed that he planned the green spaces or parks in every neighborhood as well as the controlled residential areas. Likewise, the arrangement of the French capital Paris city plan (PARP, Plan *d' Arrangement Region Parisienne*) the project envisioned the planned parks and green areas (Merlin, 1991, p. 60) However, on the disagreeing the previous idea so far, Henri Prost did not interfere historical and green planning as well as the civil, vernacular architecture, he purposed the protect city with the specifically arranged *zones* (zoning) plans (even though, H. Prost's zoning notion different from CIAM) (Doğrusöz, 1981) targeted to increase the city's green regions planned many new parks (Figs. 2, 3, and 4).

Parc No. 1 which was planned in the middle of the Historical Peninsula, and Parc No. 2 in the Pera Region, Maçka-Beşiktaş Valley (known as İstanbul city's *Rive Gauche*) and the other Parc No. 3 as Archeological Park approved (Coskun, 2017b, p. 358; Coskun, 2020). These three large park projects (likewise the previous models proposed by E. Hénard) in Paris city, squares, and green spaces (*espaces publics*) arrangements were considered landmark in new plans for the city in the following years, and in later projects. Even though, the doctrines and dynamics of urban planning have altered in the years; however, as a one of the prominent urban theorist's Henri Prost's, old planning methods and practices were considered still significant, in the early twentieth century.

- **Historical Peninsula: Parc No. 1, *Parc Educatif* (Educational Park) and Housing Areas**

Prost Master plans examined, the residential areas in the Historical Peninsula were arranged around a large-scale *Parc No. 1*, and his 1/5000 Fatih Plan consisted of the majority of the Historical Peninsula (1/3, nearly one-third of the total surface) of the Istanbul city on that era was most notable planned part of the Prost Development plans (Figs. 2, 3, and 4).

Henri Prost explained the Historical Peninsula plannings notion: The consolidation of parcels in the old Istanbul, the re-planning and rationalization of the street pattern, and the creation of a new settlements order by creating green spaces through expropriations where necessary, adopted as a principle (Bilsel, 2011). Henri Prost's strategy of balancing the residential areas designed along the Historical Peninsula with green areas through this large *Parc-Educatif* (educational park) that crossed from one end to the other. This area arranged as cultural and educational park likewise the European examples. As a large-scale park along the Bayrampaşa Valley similar to the *jardin des plantes* (great park with botanical gardens and zoos and a museum, in Paris),

Table 1 The Istanbul housing areas decentralization in the years

	The years	1950s to 1960s (regions including Prost plans)	1970s to 2000s (post-Prost developments)	After 2000s to 2020
City center	Historical Peninsula (center)	(Fatih, Kocamustafapasa)	Out of Historical Peninsula (Bakırköy-Zeytinburnu)	Beylikdüzü, Esenyurt,
North	European side (north)	(Şişli-Levent axis)	Levent, Etiler, Akadlar, Sarıyer	Sarıyer, Kemerburgaz, Zekeriyaköy
East	Asian-Anatolian side (east)	(Kadıköy, Acıbadem, Koşuyolu)	Kadıköy, Maltepe. Kartal, Pendik	Sancaktepe, Çekmeköy
East–west	East–west axis	(Florya, Yeşilköy)	Beylikdüzü-Pendik	Avcılar-Tuzla

Table by H. Coskun

various European cities as cultural activities arranged with the houses with garden for the workers' families resided and cultivated the area (Bilsel, 2010a, p. 122; Tables 1 and 2). So, H. Prost planned strategically divided the Historical Peninsula by opening Parisien style two broad boulevard: Vatan and Millet boulevards and between them he planned a large-scale Parc No. 1 specified as *Parc-Educatif* (educational park) Historical Peninsula consisted of houses with gardens, vacant and burnt city areas on Fatih, Findikzade, Edirnekapi, Kocamustafapasa districts, etc.

The municipality of the time decided the planning of a cultural park as shortly after Henri Prost leaves the Istanbul and also this planned park was not realized. Later, during the DP, Democrat Party reconstruction plans his plans implemented uncontrolled by the authorities responsible of the time (Akpınar, 2010, p. 174). After H. Prost, with the mass migration and population growth demand of the number of the new houses increased. The Historical Peninsula was occupied by the uncontrolled buildings by the private contractors of the time using the Prost plans. So, the park No. 1 area was occupied by apartments and tall buildings in a very short time with *yapsatçı* (built and sell) system for middle-class people (Bozdoğan & Akcan, 2012, p. 163; Figs. 5, 6, 12, 13, 14, 15, 16 and 17; Tables 1 and 2). Henri Prost, 1/5000 Scale Fatih Region Master plan, it was a well-prepared "zoning plan," made according to the urban planning ideologies of the period, depend on the idea divided the green areas, and the residential areas from each other united them around a large park, Parc No. 1.

- **The European Region: Parc No. 2 Maçka-Beşiktaş Valley and Housing Areas**

In the Istanbul Planning Reports, H. Prost adopted the principle of arranging the new "residential areas" to be opened along with "green areas." Indeed, this planning principle was a *zone* "zoning plan" that separated and determined previously planned by Henri Prost in France as well.

In the European Region, Henri Prost planned another large-scale park, Parc No. 2 in the Beşiktaş-Maçka valley between the old and housing areas in Taksim and Harbiye (Figs. 3, 4, 5, 6, 7, 8, 9, 10, 11, 12, 13, 14, 15 and 16) where he planned some building- blocks and apartments. This new park area, Parc No. 2, with the spectacular view of the Bosphorus explained by him like *Parc de Boulogne* (Boulogne Park) in Paris with the new housing settlement (Bilsel, 2010b, p. 361). The park would be established in a large valley between Macka-Beşiktaş on the slope descending from the Bosphorus Heights. This park along the Macka-Beşiktas valley was the also the largest park in the district planned for balancing the green spaces and residential areas in the European Side (Figs. 3, 4, 5, 6, 7, 8, 9, 10, 11, 12, 13, 14, 15 and 16; Tables 1 and 2).

Henri Prost explained his new planning on European Side: First Housing areas planned instead of the Taksim Barracks, and the Second district was planned between the Mecidiyeköy Liquor Factory in Maçka-Beşiktaş, "public parks" and sports venues are established in the valley and houses should be distributed on the high parts of the slopes (Prost, 2007, p. 86). Henri Prost planned the new houses in

Table 2 The housing models in the mid of the twentieth century, 1970–2000 period

Housing regions	Historical Peninsula	Out of the historical Peninsula	European region (Levent, Etiler, Bebek)	Asian-Anatolian region (Kadıköy, Koşuyolu, Maltepe, Pendik)	East–west axis (Yeşilköy, Ataköy, Florya)
Housing models	Property houses	Property houses	Apartment blocks	Garden houses	Apartment blocks
	Building-blocks	Building-blocks	Modern blocks	Apartment blocks	Modern blocks

Table by H. Coskun

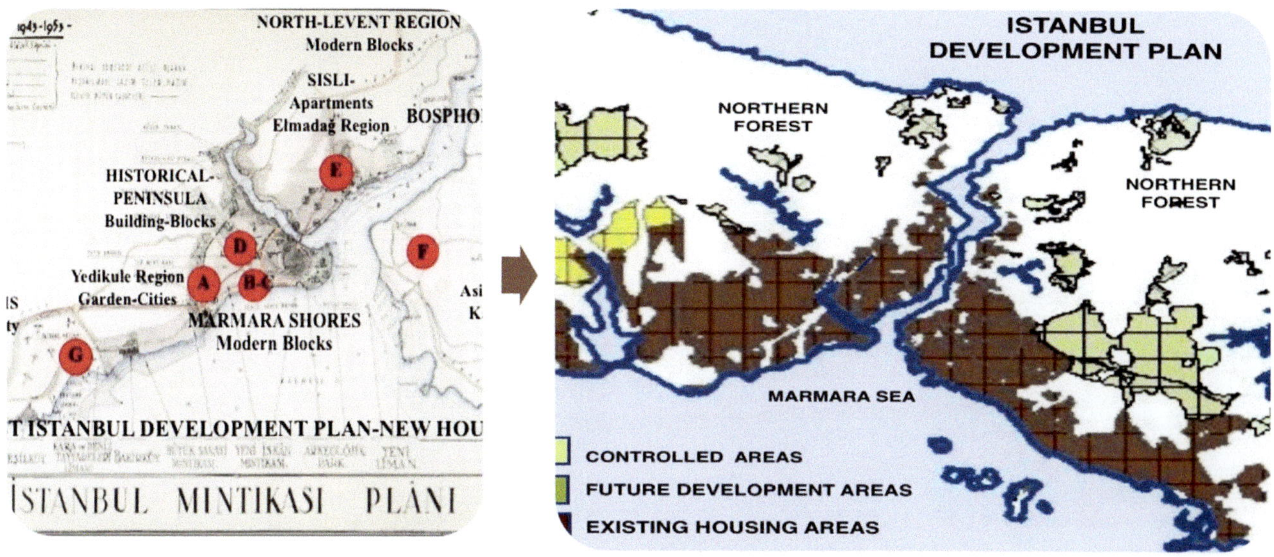

Fig. 6 Istanbul city's development of the housing areas in years. (Left) Map, designed by H. Coskun original from Henri Prost Master plan. *Source* IFA, Archives *(Académie d'Architecture), Paris.* (Right) İstanbul Housing Zoning Plan. *Map source* Environment Report of İstanbul Municipality. http://www.ibb.gov.tr/tr-TR/kurumsal/Birimler/SehirPlanlamaMd/Documents/100.000%ÖlçekliÇevreDüzeniPlanı

Fig. 7 Istanbul, in the early twentieth century, Cihangir, European side, burnt old housing areas. *Photograph source* Henri Prost own photographs, H. Prost, Photography, IFA, Archives *(Académie d'Architecture), Paris*

the Bosphorus Heights and slopes in Taksim and Harbiye, also new park in the Maçka-Beşiktaş valley with public parks and sports areas in the region. Maçka park was implemented in the district, and new building blocks were built; these would be the only blocks that Henri Prost realized in Istanbul. The park and its surroundings still preserve the appearance of being built during the Prost period. Henri Prost and A. Angel also planned some new (Le Corbusien style) modern blocks in newly developed Levent District in the northern axis of the city; however, the project later hand out to Turkish architects and urban planners K. A. Aru and R. Gorbon (Figs. 6, 7 and 8).

- **Historical Peninsula: Sarayburnu Region, *Parc Archéologique*, Parc Archeologic**

Henri Prost has given special importance to the protection of historical sites in Istanbul and declared some specifically protection zoning regulations for these areas. In this context, the ancient Sarayburnu region, which is the oldest inhabited area of the city, he provided the protection of the historical buildings the main Cathedral St. Sophia and Topkapı Palace, by planning an Archeological park. This park was a specifically protected area rather than a park arrangement. (Figs. 6, 7, 8, 9, 10, 11, 12, 13, 14 and 15) In the Historical

Fig. 8 Istanbul, 1950–60s, mid of the twentieth century, European side, Levent District, newly developed northern axis with (Le Corbusien Style Modern Blocks by K. Ahmet Aru and R. Gorbon). *Photograph* www.Levent_Emlak_Bankası web page. Right, today Levent as Business District, with high-rise buildings. *Photograph source* M. Sümer

Fig. 9 Istanbul, today, Historical Peninsula old and new housing areas in Prost plans. The View from Haliç (Golden-Horn) to the Sarayburnu, St. Sophia, Topkapı Palace. *Photograph source* Istanbul.com

Peninsula, including the old, archeological site area from Roman, Byzantine, and Ottoman period extended from Sarayburnu, St. Sophia placed on the top of the Sarayburnu Hill to the other older St. Sophia (known as KüçükAyasofya) in the down shore of the Bosphorus was entirely allocated the archeological park.

- ***Le Plan de Côté d'Asie* (The Development Plan of Anatolian Side)**

Henri Prost planned a specifically designed plan for Asian-Anatolian Side, *Le Plan de Côté d'Asie* stated in his reports as the old part of the city Anatolian Side of Istanbul consisted of houses with gardens, so recommended that this region planned with "garden cities" to preserve old characteristic of this historical region. He also planned and opened new housing areas with gardens accordance to Anatolian Side's old, historical distinguished region and to generate balanced areas arranged with green spaces, and houses with gardens on the Anatolian side; Kadıköy, Acıbadem, Koşuyolu (Coskun, 2021a; Fig. 14). The Anatolian Side region well planned by house of gardens, equally the early European "garden city" model. In his Le *Plan de Côté d'Asie* (Anatolian Side Development plan), housing areas planned

Fig. 10 Istanbul, today, Şişli-Levent development of business areas after the Bosphorus bridge developed between 1970 and 2000s. *Photograph source* Istanbul.com

Fig. 11 Istanbul, today, Şişli-Levent axis, the front Anatolian side garden-city houses along the Bosphorus, planned by Henri Prost, developed after him with the opening Bosphorus Bridge 1970s to 2000s. *Photograph* Istanbul.com

by *cité-jardins* (garden cities) analogous to Ebenezer Howard's British model (Howard, 2008) (Tables 1 and 2).

Known as *cité-jardins* (garden cities) previously in Paris, France, the English housing were transmitted by his colleague Léon Jaussely directly from E. Howard's specific models, later would be implemented by Henri Sellier Paris city's outer quarters (Sellier, 1998, p. 22). Henri Prost envisioned the new houses for specifically planned unique to Istanbul's Bosphorus Heights known as *cité-parcs* (park cities) aimed to controlled planning for picturesque woods of Bosphorus and green spaces arranged as public garden (Bilsel, 2010b, p. 369; Coskun, 2021b) (Figs. 3, 4, 5, 6, 7, 8, 9, 10, 11, 12, 13 and 14). The city continued to develop with the French model new *banlieus* (suburbans) along the Kadıköy, Göztepe, Bostancı Regions next to Marmara Shores in the East-Axis Anatolian Side while developed

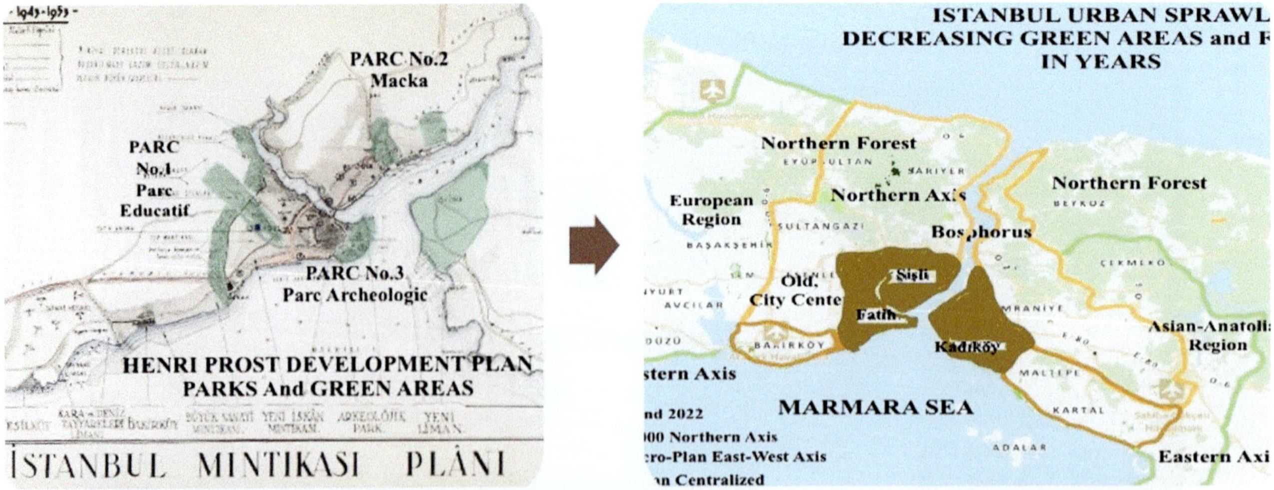

Fig. 12 Left, Istanbul, the early twentieth century, Prost plans green areas, and parks. Map, H. Coskun, original, IFA, Archives (*Académie d'Architecture)*, Paris. Istanbul, the recent green areas. Map, H. Coskun. *Original map source* https://yesil.istanbul/ekosistemharitasi

Fig. 13 Left, early twentieth century, Istanbul, Prost plan, in the mid of historical area Park No. 1 (*Parc Educatif*) not implemented. Map H. Coskun, original from, IFA Archives (*Académie d'Architecture*), Paris. Right, today, historical Peninsula, Parc No. 1 occupied by tall buildings after the 1960s. Photograph, anonym. *Photograph source* https://Istanbul.resimleri.com

with new *cité-satellites* (Satellite-cities) in far West-Axis, Bakırköy, Ataköy districts by the project of L. Piccinato and E. Menteşe (Malussardi, 1993, p. 55) (Figs. 3 and 4).

5.1 The Other Planning Regulations

In the developing urban planning terms and the literature in the context of the sustainability, there are some planning tools that Henri Prost had previously applied in Paris and N. Africa. These applications are now widely used again in the context of sustainability. He applied some *zones* (zoning) regulations for Historical regions in his Istanbul Development plans to protect them.

- The *zones* (zoning), *zone non-ædificandi* (specially protected areas) indeed no-construction zone; Henri Prost applicated this *zones* "zoning" method in the İstanbul to protect the old city wall of İstanbul dated to Byzantinian period.
- The *cordon sanitaire* (sanitary band) he used this zoning application especially in the his N. African plannings (Cohen, 2010, p. 55).

The previous planning methods, and old techniques of French architect-planner Henri Prost affiliated to old, French urbanism school emphasized in this research that these were still significant for future planning of the cities. Related to both "residential areas" and "green areas," these old theories

Fig. 14 Left, in the early twentieth century, Istanbul, Bosphorus Heights, *Parc-Cities* planned by Prost specifically to protect old woods. Photograph by H. Prost's own objective. H. Prost Archive, IFA Archives (*Académie d'Architecture*), Paris. Right, today, Bosphorus Heights, after uncontrolled developments. *Photograph* H. Coskun

Fig. 15 Istanbul, today, Haliç (Golden-Horn) looking Sarayburnu, St. Sophia, Topkapı Palace, Park No. 3 (*Parc Archeologic*), planned by Henri Prost. Archeologic area was well protected today. *Photograph source* https://yesil.istanbul/ekosistemharitasi

transformed and adapted today's conditions particularly in the content of sustainability as a newly emerged problematic. Likewise, his other zoning applications like *zone nonædificandi* (specifically protected areas) or *cordon sanitaire* (sanitary band) developed versions of this band implementations today put into effect. So, similar applications of this planning method re-acted by municipalities and government in the world as aimed to decreasing carbon emission to protect effects of the climate change some of the cities like Paris, London (Citylab Bloomberg, 2022). Hence, like the old regulations applicated by Henri Prost and Ebenezer Howard, today in London, and Paris cities put into a new act and regulations related the re-using old, "zoning" applications pushes its ultra zone to limit low carbon emission limits in London also Paris plan to remove two belt way lanes to cut pollutions.

6 Conclusion

The urban hinterland envisaged at the early twentieth century, even though the Prost İstanbul Master plans considered still significant and valid today; however, this plan has lost its holistic value and urban features, and functionality within

Fig. 16 Istanbul, today, Macka Park No. 2, drawn by Henri Prost in the early twentieth century. Green valley between Maçka-Beşiktaş only green area to breath. *Photograph source* https://yesil.istanbul/ekosistemharitasi

Fig. 17 Istanbul, today, Park No. 1 (*Parc Educatif*-Educational Park), Fatih Region, mid of the historical Peninsula and Atatürk Boulevard and bridge drawn by Henri Prost. *Photograph source* https://Istanbul.resimleri.com

years with uncontrolled implementations. Until 1980s, İstanbul changed and developed with the population of from the early twentieth century 1 million, to the 1980s 2.5 million, and today nearly 15 million. The housing stock of the city increased from in the early twentieth century 5000 to the 250 thousand until 1980s, today its 1.5 million. The inappropriate implementation of Prost Istanbul Master plans from the 1950s to 1960s by the governors of the period, rapid population growth and uncontrolled developments were prevented the healthy development of the city. In the 1960s to 70s, mass migration led to densely urbanization first in Historical Peninsula the residential areas, green spaces, and parks previously proposed in the Prost plans accordance to his "zoning" plans were occupied in a very short period by the apartments, and tall buildings without his permission so, the holistic traces of his plans nearly lost today (Figs. 3, 4, 5, 6, 7, 8, 9, 10, 11, 12, 13, 14, 15, 16 and 17).

In the 1980s, the city's planning worsened by densely urbanization spreading to Bosphorus Heights including European and Anatolian (Asian) Sides. The increase with the popped slum structures (*gece-kondu*) was the catastrophic milestone with the subsequent zoning amnesties led to uncontrolled development (Fig. 14). After the Bosphorus Bridge, in 1970s, the slum areas occupied very short period the last remaining green areas, around the Bosphorus Heights and Anatolian side developed by garden cities (Figs. 11, 12,

13 and 14), the European Side developed as business district toward the old woods and groves to northern forests (Figs. 12, 13 and 14). Although the appearance of the city did not change widely within the framework of the Bosphorus Bridge already envisaged in the Prost plan also the newly opened residential areas as garden cities on the Anatolian side developed (Figs. 11, 12, 13 and 14).

The post-Prost period, today the "housing areas"

Since the post-Prost period, policies of Housing and determining the "housing areas" to be allocated for newly opened settlements in the city have changed radically as well as the city's planning theories and discourse. Today, *İMB, İstanbul Metropolitan Belediyesi*, İMMP (İstanbul Metropolitan Municipality) which carries out the planning of Istanbul to urban-scale, within the scope of developed "zoning" maps that have been determined specifically (Fig. 6). The Prost city maps were basically divided the city certain functions such as residential areas, green spaces, educational, industrial areas, etc. However, today İstanbul Municipality and also other cities' municipalities work on more specific "zoning" maps such as "controlled zones" or "buffer zones" for planning housing areas to protect the natural landscapes. Furthermore, İstanbul Regional Commission's new regulations stated that the decision of the *İstanbul Bölge Koruma Kurulu* (Istanbul Regional Conservation Commission) was deemed appropriate to expand the areas of sustainable protection and controlled use in order to develop more balanced relations between "people" and "nature" (Sözcü, 2022).

İstanbul Metropolitan prepares some specific "zoning" plans for Istanbul, which was expected to develop mostly on the east–west axis and partially toward the north, has now merged with the neighbor cities such as Kocaeli and Tekirdağ in an almost east–west direction, today it has come to a level that threatens the northern forests, which is the only development axis.

The opening of new housing settlements today specified according to housing policies in general by governmental side by *İklim Değişikliği, Çevre ve Şehircilik Bakanlığı* (Ministry of Climate-Change, Environment and Urbanization) and the large-scale and mass-housing projects by *TOKI, Toplu Konut İdaresi* (Mass Housing Administration) middle-scale housing projects by *İMB, İstanbul Metropolitan Belediyesi*, İMMP (İstanbul Metropolitan Municipality) and small-scale and single buildings by local municipalities. In this context, large-scale housing projects and mass-housing projects mostly carried by governmental side by TOKI creating new housing areas out of the İstanbul city toward to Northern forest, and green spaces like Esenyurt, Beylikdüzü, Tuzla, Pendik, etc. The small-scale projects mostly planned in city center in the lands still vacant also between city periphery and TOKI mass-housing areas.

The post-Prost period, today "green areas" and parks

The newly planned green spaces and parks by Henri Prost were examined; except for the Park No. 1, *Parc Educatif* (the Educational Park) (Mid of the Historical Peninsula) (Figs. 12, 13, 14, 15, 16 and 17), Park No. 2 (Maçka Valley) (Figs. 10, 11, 12, 13, 14, 15 and 16), and Park No. 3, *Parc Archeologic* (the Archeologic Park) (Figs. 9, 10, 11, 12, 13, 14 and 15) still exist. Today, the planning and arrangement of new green spaces in the city is planned by both the *İMB, İstanbul Metropolitan Belediyesi*, İMMPO (İstanbul Metropolitan Municipality) and local municipalities according to the requirements of the regions (Fig. 5). These new parks created by using the existing green spaces and vacant areas. Although new "housing areas" and "green areas" continued to open by the government side and İstanbul Municipality, these could not include a holistic urban planning approach like the Henri Prost Master plans or the Italian architect Luigi Piccinato's Metropolitan plan considered as certain extension patches made over the years (Figs. 3 and 4). Henri Prost's plans and planning housing and green areas for each neighborhood became obsolete with the rapid migration after the 1960s to 1980s, and unfortunately this planning method has turned into a forced model that injected into existing neighborhoods (Figs. 5, 6, 7, 8, 9, 10, 11 and 12) mostly next to the large-scale projects or mass-housing areas out of the city.

The post-Prost period today, application of "zoning" regulations in İstanbul

Henri Prost's "zoning" regulations was depended on regulations dividing the cities green areas, residential, industrial, etc. These regulations were valid laws that could still be used by being further developed and adapted to current practices or might be use as new proposals for future cities. The purpose of "zoning" applications to separate cities into certain functions in terms of settlements as a new term created in the early twentieth century as main principles of modern architecture and urban planning. Henri Prost's "zoning" applications divided the cities functions like housing areas, green areas, educational, industrial areas, etc. varied.

Even though known differently the many of the Prost "zoning" regulations was also purposed to protect cities' architectural heritage, civil architecture, vernacular areas. These were also aimed to protect green lands, woods, and forest, etc. Thus, he created specifically arranged "zoning" regulations to protect the city's such as green areas, historical areas, even archeological site areas, etc. Furthermore, the "zoning" regulations prepared specifically for "archeological site areas" proposed for Istanbul city as *zone non-aedificandi* (specifically protected areas) in Istanbul or *cordon sanitaire* (sanitary bands) also specific for Northern Africa. However, these were the laws restricting the new construction of completely new buildings likewise the laws of Cornudet previously made in France, and there is no issue of sustainability and sustainable protection, the idea which will often come to the fore after the 1980s at the beginning of the twentieth century.

In Türkiye today, the term of "sustainability" recently have included in the Governmental Housing Policies and Development Reports *Onbirinci Kalkınma Raporları ve Konut Politikaları* (State Development Reports Including Housing) as a significant matter as well as the earthquake concern (On Birinci Kalkınma Planı, 2018). The vision of UN Habitat's widely emphasized in Turkish Government Reports; "providing everyone's access to adequate, livable, healthy, durable, safe, inclusive, affordable and sustainable housing and basic city services" constitutes the framework of this plan/policy document created for the solution of the housing problem in our country.

Although "zoning" applications evaluated differently from their previous purpose, in the early twentieth century that examined in the old Prost plans today and they are developed in the context of sustainability, considered of nature. Thus, İstanbul Regional Commission's new regulations stated that with the decision of the *İstanbul Bölge Koruma Kurulu* (Istanbul Regional Conservation Commission) was deemed appropriate to expand the areas of "sustainable protection" and controlled use to develop more balanced relations between "people" and "nature" (Sözcü, 2022). With the decision of the Commission of District Conservation dated July 4, 2018, it was determined that "sustainable protection," controlled use, and partially qualified natural protection areas were created for the parcels where the site rating was made.

In Istanbul, today, both the State, the *İstanbul Büyükşehir Belediyesi* İMM (İstanbul Metropolitan Municipality) and local governors have realized that to protect the green areas, woods, forests, and even water basins should be via arranging new strict "zoning" laws and regulations, also limited the "residential areas." As recent application of governmental side in the context of the as an old method the "zoning" (like first applied by Henri Prost in Paris), continued to implement broader their scope. Thus, these old applications developed and varied adding some newly created regulations according to today's circumstances which were not existed previously "specifically protected areas" or "controlled use areas" by government.

As stated by İMMPO, *İstanbul Metropolitan Planlama Ofisi* (Istanbul Metropolitan Planning Office's) in their recent planning proposal some concerns on global warming and climate crisis still exist in as an extensively prepared plan on sustainability.

In İstanbul, today, some new regulations and "zoning" practices have already entered the city planning agenda. According to a new urban declaration, the characteristics and principles of "sustainable conservation" and "controlled use areas," considered among the natural sites, are explained as follows within the scope of the "Regulation on the determination, registration and approval of protected areas" (Emlakkulisi, 2016).

Examining the recent "zoning" applications (which were first realized by Henri Prost in İstanbul, revised and re-arranged today by the Municipalities), the scope of today's is generally as follows.

(a) It contains human settlements compatible with its landscape.
(b) It includes or contributes to the protection of natural resource management systems and related cultural values, ecosystems, and habitats.
(c) It provide to the development and maintenance of balanced relations between "people" and "nature."
(d) Contributes to the social and economic gains of the local people where applicable.
(e) It supports sustainable use of natural resources and development at national, regional, and local levels.

Although some zoning practices for the protection of green areas in the city on the agenda, they do not have a more intrusive planning strategy like Paris and London of other world cities. *IMMPO* (Istanbul Metropolitan Planning Office) declared in its newly regulated Development Plan appealed the requisite of avoiding the planning of the city toward the northern forests, as well as the existing the water basins that encourage the sprawl of the city toward the west and east directions (IMMPO, 2022). Recently İMMPO, Istanbul Metropolitan Office, puts into effects on newly produced regulations considering to new problematic sustainability as well as the to protect green areas especially northern forests (Figs. 5, 6, 7, 8, 9, 10, 11, 12, 13 and 14).

Nowadays, "zoning" applications are becoming widespread, and its scope is being developed and varied. Also, they have become vital importance to planning cities with green areas, and residential areas according to the old *zones* (zoning plans) and techniques like by Henri Prost and colleagues in the early twentieth century. These were specifically produced laws limited the housing settlements to protect cities harmful effects of uncontrolled and unplanned urban sprawl and aimed to more protected green areas, parks, forests, woods started to applicate the cities like London, Paris, İstanbul, Europe, etc.

In the twenty-first century, the cities are under the different circumstances and problems than the urban dynamics of the twentieth century. The scope of the old "zoning" applications has been further re-regulated, developed, and adapted to today's conditions in the context of climate change issues, and sustainability. In the world, the comfort values of cities equivalent to number of green areas and parks and how little carbon emissions are related to city's heat regulations (Greenpeace, 2002).

In the early twentieth century, with the uncontrolled urban sprawl, urbanization, road network, heavy traffic, and the increasing number of vehicles, and high carbon emission less green areas the cities had to confront being as harmful places for human life. Today, England put into a new act and regulations related the re-using old, "zoning" application pushes its ultra zone to limit low carbon emission limits in London.

References

Akpınar, İ. Y. (2010). İstanbul'da Modern Bir Pay-i Taht, Prost Planı Çerçevesinde Menderes'in İcraatları, İmparatorluk Başkentinden Cumhuriyet'in ModernKentine, H. Prost istanbul Planlaması (1936–1951), İstanbul Anadolu Araştırmaları Enstitüsü.

Arar, A. (2021). TC. Dış İşleri Başkanlığı, Daire Başkanı, Uluslararası Ekonomik Sorunlar Dergisi, Sayı VI, Yerel Gündem 21.

Bilsel, C. (2010a). Henri Prost'un İstanbul Planlaması (1936–1951), Nazım Planlar ve Başkentinden Kentsel Operasyonlarla Kentin Yapısal Dönüşümü, İmparatorluk Cumhuriyet'in Modern Kentine, Henri Prost İstanbul Planlaması (1936–1951), İAE.

Bilsel, C. (2010b). Serbest Sahalar: Parklar, Geziler, Meydanlar. İmparatorluk Başkentinden, Cumhuriyet'in Modern Kentine H. Prost İstanbul Planlaması (1936–1951). İAE.

Bilsel, C. (2011). Les Transformations d' İstanbul: Transformations of İstanbul by Henri Prost. *AIZ, Journal of Faculty of Architecture, 8* (1, Spring), 100–116.

Bozdoğan, S., & Akcan, E. (2012). *Turkey, modern architectures in history*. Redaktion Books (First Published, 2012). ISBN-13: 978-1861898784.

Bruant, C. (2011). Eugene Henard, Paquot, Thierry, Les Faiseur De Villes, İnfolie Editions.

Calderon, F., & Stern, N. (2015). La Nouvelle Economie Climatic, Une Meilleure Croissance Un Meilleure Climatique, Pour le traduction Française. Published by Les Petit Matins. ISBN: 976-2-36383-178-1.

Carriou, C., & Ratouis, O. (2014, November 27). *Quels Modeles Pour l'Urbanisme Durable* [O. Waine, trans.]. Metropolitque.

Choay, F. (1979). L'Urbanisme, Utopies et Réalities, Une Anthologie. Edition du Seuil.

Choay, F. (1969). *Pre-urbanisme and Urbanisme, The progresist Model* (The modern city: Planning in the 19th century) [M. Hugo & G. R. Collins, Trans.]. George Braziller.

Citylab, Bloomberg. (2022). Paris plans to remove two beltway to cut the pollution. Page published in May 2022, Instagram. https://www.instagram.com/p/Cd07937pLEI/?ref=chowhound&hl=nl. Accessed September 25, 2022

Cohen, J. L. (2010). Büyük Peyzajlardan, Metropollere : H. Prost., İmparatorluk Başkentinden, Cumhuriyet'in Modern Kentine, H. Prost İstanbul Planlaması (1936–1951). İAE.

Coskun, H. (2017a). *In the beginning of the 20th century, analyzing methods of the housing problem and an example: Henri Prost's Istanbul plannings* (PhD Thesis). MSGSU, Mimar Sinan Fine Arts University, The Institute of Science, Faculty of Architecture, Building Design.

Coskun, H. (2017b). ISTANBUL: 19th and 20th century transformations in the historical Peninsula and after the 1950s reflections of the globalization. In *ISUF, 2017 Conference*, Universitad Valencia, Spain, September 2017. E-Press.

Coskun, H. (2020). Henri Prost's Paris and Istanbul plannings, zoning regulations urban planning tools: Housing, green areas, parks, axis. In *GU—Green Urbanism*, Rome Tre University, Rome, Italy.

Coskun, H. (2021a). The housing models for green architecture; The come-back of garden-cities first known as bank-houses in Istanbul. In *International Conference, GU, Green Architecture*, 5th ed., Rome La Sapienza University, Rome, Italy, December 06–07, 2021.

Coskun, H. (2021b). Sustainable and resilient plannings, developed housing models for Istanbul. In *International Conference, RRAU, Resilient& Responsible Architecture and Urbanism*, 4th ed., Ximanen University Malaysia, Southwest Jiatotong University China, Malaysia, December 03–05, 2021.

Doğrusöz, U. (1981). *Henri Prost* (Unpublished Master Thesis).

Doğrusöz, U. (2016). Interview, Istanbul, 2016.

Emlakkulisi. (2016). *Sürdürülebilir koruma ve Kontrolllü Kullanım Alanı Nedir?* (What is sustainability and controlled using of areas). https://emlakkulisi.com/surdurulebilir-koruma-ve-kontrollu-kullanim-alani-nedir/497413. Accessed October 13, 2022

Ferry, L. (2021). Les Sept Ecologies. l'Edition de Observatoire/Humensis.

Forum, W. C. (2020). *World Cities Culture Forum*. Retrieved September 20, 2022, from http://www.worldcitiescultureforum.com/data/of-public-green-space-parks-and-gardens

Gates, B. (2021). *Climat Comment Eviter Un Desastre*. Flammarion.

Greenpeace. (2022), Büyük kentlerde Park ve Bahçelerin Kent İçindeki Yeri. Instagram. https://www.instagram.com/p/Clihuq3Orqt/. Accessed November 22, 2022

Hertslife. (2021). Transformation of Welwyn Garden City Town Center Spot to Begin, M. Smith. https://www.hertfordshiremercury.co.uk/news/hertfordshire-news/transformation-welwyn-garden-city-town-5603085. Accessed September 05, 2021

Howard, E. (2008). (1898; reprint 1902/1989). *Tomorrow, a peaceful path to real reform. Garden cities of tomorrow*. ATC Books/Faber and Faber.

IMMPO. (2022). Istanbul Metropolitan Municipality Planning Office, Open Data Portal. https://data.ibb.gov.tr/dataset/ilce-yıl. http://www.ibb.gov.tr/tr-TR/kurumsal/Birimler/SehirPlanlamaMd/Documents/100.000%20Ölçekli%20Çevre%20Düzeni%20Planı%20Sunum%20TR.pdf. Accessed, October 10, 2022

Le Corbusier. (1994). L'Urbanisme, Edition, Champart.

Lehmann, S. (2011). *What is green urbanism? Holistic principles to transform cities for sustainable* (Chap. 14). Researchgate. https://www.researchgate.net/publication/221915598_What_is_Green_Urbanism_Holistic_Principles_to_Transform_Cities_for_Sustainability. Accessed May 2021

Malussardi, F. (1993). L'Azione per Una Cultura Urbanistica Senza Frontiere, Luigi Piccinato e L'Urbanistica Moderna. Edizioni Officina.

Merlin, P. (1991). *L'Urbanisme*. Presses Universitaires de France.

On Birinci Kalkınma Planı. (2018). Konut Politikaları Özel Komisyonu Raporu. https://www.sbb.gov.tr/wp-content/uploads/2020/04/Konut PolitikalariOzelIhtisasKomisyonuRaporu.pdf. Accessed September 09, 2022

Prost, H. (1949). İstanbul Belediyesi Şehircilik Mütehassısı, İmar Planlarından Doğan, Gayrimenkul Mükellifiyetlerinin (Servitudes) Tatbiki Hakkındaki Fikirler. Çeviren; Z. Feran, 1949, Arkitekt, C.18, S.39.

Prost, H. (2007). İstanbul Hakkında Notlar, Cumhuriyet Dönemi, İstanbul Planlama Raporları, 1934–1995, Derleyen Ş. Özler, TMMOB, Mimarlar Odası, İstanbul Şb.

Sellier, H. (1998). Une Cité Pour Tous. Edition de Linteau.
Smith, M. (2021). Transformation of the Welwyn Garden City Town, Center Spot the Begin. Hertslife.com
The Economist Intelligence. (2009). *European Green City Index*. Retrieved October 18, 2020, from http://sg.siemens.com/city_of_the_future/_docs/greencityindex_report_en.pdf
Sözcü. (2022). *Kuzey Ormanları Kuşatması* (Northern Forest's Planned 657 More Housing). https://www.sozcu.com.tr/2022/emlak/kuzey-ormanlari-kusatmasi-153-blok-657-konut-daha-7245051/. Accessed October 20, 2022
United Nations. (1996). *Report of the United Nations Conference on Human Settlements (Habitat II)*, Istanbul, June 3–14, 1996.
United Nations. (2015). *Transforming our world: The 2030 agenda for sustainable development*. United Nations.
United Nations. (2016). *Housing is at the Center of the Sustainable Agenda*. https://blogs.worldbank.org/psd/housing-center-sustainable-development-agenda. Accessed November 20, 2021
United Nations. (2020). The future we want: Final document of the Rio +20 Conference. In *Proceedings of the Rio20*.
United Nations. (2021). *Conference Sustainable Development*, Rio de Janeiro, Brazil, June 20–22, 2012.
UNEP. (2019). *UNEP Reports, Global Environment Outlook-GEO6Healty Planet Healthy People*.

Nature Based Solutions in Mediterranean City

An Ecological Study of Garden Cities (Case Study: Iranian Garden Cities—Isfahan)

Ali Moazzeni Khorasgani, Maria H. Villalobos, and Ghazal Asadi Eskandar

Abstract

Cities' growth and rising demand for energy resources have created problems and hazards for human life due to resource depletion. Therefore, it is crucial to establish a framework that promotes resource efficiency to achieve sustainable development and energy preservation, which is now a vital daily requirement. Developing ecosystem services in today's cities can guarantee resources for future generations, urban health, and the realization of a healthy city and life. Research on Iranian–Islamic garden cities has revealed their structure shares similarities and adaptations with the overall ecosystem services framework. These garden cities were designed to create healthy urban environments in their time and can serve as a model for developing sustainable towns today. Meanwhile, the Isfahan Garden City in the Safavid period is a perfect example of various sustainability and ecological functions. The analysis of the Isfahan city garden from the perspective of different sub-structures, including water, green, social, and economic, makes it possible to achieve this goal. This research aimed to compare Iranian–Islamic garden cities with typical urban areas, focusing on the various aspects of ecosystem services, their classification, and the strategies for developing sustainable and green cities. The study also analyzed the infrastructure services of Isfahan, a prominent Iranian–Islamic garden city, and explored how these services align with the ecosystem services of green cities. The research aimed to identify which ecosystem services present in Iranian–Islamic garden cities are compatible with the strategies for creating green cities. Or the answer to this year, which ecosystem service items have been considered in the design of Iranian–Islamic garden cities, especially Isfahan? The Iranian garden cities have established essential regulations and ecosystem support infrastructure, instrumental in their sustained existence as green and sustainable cities. These cities have promoted the creation of healthy urban environments and contributed to the ecosystem's overall health. As a result, these garden cities have become crucial for the continuity of life and the sustainable development of both themselves and their residents.

Keywords

Ecological study • Garden cities • Isfahan • Sustainable • Ecosystem services • Infrastructures

1 Introduction

The rapid development of technology and industry caused cities' rapid growth, the suburbs' story, and the destruction of green spaces inside cities. In the past, the buildings had a place identity, and connection with the alley, street, and market completed this identity. The urban communication networks had different and original views, and their spaces were used differently. These communication spaces brought nature along with them in different ways; with the advent of the new era, the buildings and structures took on a rude appearance, creating a soulless environment that tired and annoyed the human spirit (Eslamian & Maleki, 2021). In the cities and cultures of the past, different types of the formation of nature can be observed, one of the valuable examples of which can be seen in the Iranian–Islamic cities of the past. The Iranian garden is a brilliant example of the connection between nature and architecture. Moreover, the garden city

A. M. Khorasgani (✉)
Landscape Architecture + Urbanism Program, Illinois Institute of Technology, Chicago, IL, USA
e-mail: Amoazzenikhorasgani@iit.edu

M. H. Villalobos
Illinois Institute of Technology, Chicago, IL, USA
e-mail: mvillaloboshernandez@iit.edu

G. A. Eskandar
South Dakota State University, Brookings, SD, USA
e-mail: Ghazal.AsadiEskandar@sdstate.edu

A. Battisti et al. (eds.), *Greening Our Cities: Sustainable Urbanism for a Greener Future*, Advances in Science, Technology & Innovation, https://doi.org/10.1007/978-3-031-49495-6_7

is another manifestation of the Iranian–Islamic town that appeared in different forms in different periods of Iran's history. During the Safavid era, the development of the capital and important cities such as Isfahan was built with the model of a city garden, which is very helpful for the development of today's cities. Now, the question is whether, by studying and examining these garden cities, it is possible to achieve sustainable principles and principles regarding healthy cities concerning ecosystem services in them. Iranian–Islamic garden cities in their time and even in the current time, despite the changes that have taken place in them, what principles and foundations are they based on in terms of ecosystem and sustainability?

Many articles and studies have concerned healthy, sustainable, and green cities and their ecosystem services and considerations. In the meantime, many researchers in this field have considered the investigation of Iranian–Islamic garden cities, and results have been obtained in this regard. This research attempt has been made by examining and reviewing studies on green and sustainable cities, the principles and basics of design and ecosystem services, and their adaptation and comparison with Iranian–Islamic garden cities, especially the historical city of Isfahan. The primary, critical, and practical components in the design of sustainable cities.

2 Research Method

The studies in this research are related to the previous research and articles and the results obtained about garden cities and ecological and sustainable cities in a library form and considering the lack of resources regarding the ecosystem studies of Iranian garden cities, especially the case example of this article—historical city. Therefore, this research will be presented in an ecosystem services survey and an Iranian city garden–Isfahan city garden.

The research method in the present article is based on a document study, based on the review of documents related to prominent and prominent research in the field of ecosystem services and gardens of Iranian–Islamic cities and their different levels. It has been carried out by the descriptive-analytical research method. The method of collecting information is in two forms, documentary, and field.

3 Ecosystem Services

The term ecosystem service has many meanings. The United Nations Millennium Assessment of Ecosystem Services defines ecosystem services as "benefits humans derive from ecosystems." The "Economics of Ecosystems and Biodiversity" initiative defines ecosystem services as ecosystems' direct and indirect contributions to human well-being. In contrast, in an article that critiques ecosystem services, Fisher and Turner describe them as dimensions of the ecosystem that have been utilized actively or passively in creating human well-being. All these definitions are related to how nature is valuable to humans. This is an anthropological concept. Ecosystems are often very complex and depend on all the components that interact and influence each other. A theoretical framework is a way to conceptualize such complex processes. Valuing ecosystem services is a relatively new approach. Since it helps urban policymakers make better decisions in line with sustainable urban planning, it is mentioned as a suitable method in reliable sources (Dekay & Brien, 2001). In general, ecosystem services are categorized into four main groups.

3.1 Supplier Services

Ecosystem services include all materials extracted from ecosystems, such as food, fibers, fresh water, or healing resources.

3.2 Regulatory Services

Regulating ecosystem services includes all the methods to adjust the environment, including handling weather conditions, extreme events, erosion control, or biological control.

3.3 Cultural Services

The intangible outputs of ecosystems that affect humans' physical or mental conditions are included in the category of cultural ecosystem services. For example, we can list spiritual experiences, entertainment, aesthetic appreciation, or a sense of place.

3.4 Support Services

Finally, supporting or habitat-dependent ecosystem services are defined as ecological processes and functions necessary for producing "final or final ecosystem services," which include the habitat of species and the care of genetic diversity.

On the other hand, in the last decade, ecosystem services have been well equipped as an "educational tool" or a "communicative metaphor" that supports biodiversity and ecosystem conservation. This approach has also been widely used to understand and analyze the relationship between society and nature. A sustainable city can be built on three interrelated mental models, each dependent on a different count as success.

Dekay and Brien (2001) have proposed three mental models for a sustainable city in their article titled "Grey City, Green City";

- The city as a living system: This type of thinking asks us what form the city takes if we consider it a manifestation of a natural process.
- The city as an experience of nature: This type of thinking asks us what form the city takes if we consider spiriting an essential principle for human development.
- The city as a specific place: This type of thinking asks us what form the city takes if we consider it part of a larger whole.

According to Dekay, ecosystem services only meet our basic needs. The biosphere nurtures our mind, spirit, stomach, and lungs. Restoring a healthy ecosystem in the city helps restore human health and supports the development of our human and cultural potential. Nature is used positively in the green town (Dekay & Brien, 2001). Interaction with nature can restore physiological and psychological health (Kaplan & Kaplan, 1998). Patients recover faster when they can see natural scenery; stress is reduced, and psychological relaxation occurs (Ulrich, 1993).

However, in the gray city, humans are imprisoned in a concrete jungle, like a lion in a zoo, pacing furiously, grappling with its dark emotions. In the green city, man frequently interacts closely with nature, which provides freedom of movement. In Table 1, Dekay and Brien (2001) presented the green city and gray strategies for using ecosystem services.

4 Iranian–Islamic City Garden

In recent times, cities have become increasingly crowded due to a rise in population and migration from rural areas, leading to higher population density and pollution (Khorasgani & Haghighatbin, 2023). Amidst urban residents' busy and stressful lifestyles, green spaces offer a calming and rejuvenating effect, crucial for maintaining physical and mental well-being. Research shows that 43% of people engage in recreational activities, travel, and intellectual rest, indicating the importance of green spaces (Azani, 2002). Green spaces not only provide a connection with nature, but they also help to reduce environmental pollution. The Safavid era witnessed the development of cities like Isfahan, which were designed based on the Baghshahr model. The Safavid garden design established a connection between the interior and exterior spaces, with examples such as ponds and water features with fountains inside buildings. The Timurid style influenced the streets, squares, and public passages in the Safavid era, and regular-shaped roads were designed for promenades and urban communication routes. The Chaharbagh of Isfahan is a prominent example of this style, with ornamental shrubs, red flowers, jasmine, and shaded rows of trees lining the streets. The water features in these gardens were adorned with blue-painted tiles. This research seeks to explore the influence of Iranian and Chinese urban planning on each other and the role of natural elements in Iranian–Islamic urban planning. It recommends implementing the Safavid garden city model for urban development in the western country's western region has similar climatic and geographical potentials.

4.1 Isfahan Garden City in the Safavid Period

Isfahan's gardens and mosques present a beautiful and memorable scene to travelers approaching the city from the south, with the gardens acting as a connective element between the government house and the city's three main axes. The plan for Isfahan's development during Shah Abbas's time was based on a unified structure consisting of the existing Zayandeh Rood River and the artificial Chaharbagh axis, both perpendicular to each other and in line with the Iranian garden model. Isfahan's gardens were developed on two levels, man-artificial natural, and two scales, the whole city and the micro-scales of the gardens themselves. The gardens were situated on the west side of Naqsh Jahan Square, around the axis of Chahar Bagh, and along the Zayandeh Rood River, forming the main structure of the city and serving as the foundation for its garden city expansion (Muqtada, 2003). The gardens' importance lies not in their vastness but in their role in the city's urban planning system. Sharden described Chaharbagh as the most beautiful street he had ever seen. At the same time, Donald Wilbur wrote about the gardens' rows of sycamore trees, gold and silver coins planted among the roots, and the fragrance of flowers that filled the space in the evenings. The promenade was dotted with gardens named after trees, birds, and dervishes, and each had two Kolah Ferangi mansions, one above the gate and one inside the garden. The remaining Hashtbehesht building is a testament to the grandeur of the gardens, located in the middle of Bulbul Garden and featuring an octagonal design with a central courtyard, pool, and fountain (Haghighatbin et al., 2007).

The gardens of Isfahan were indeed a marvel of their time and continue to be a popular tourist attraction today. The Persian tradition of building gardens designed to be an earthly representation of paradise influenced the city's garden design. Isfahan's gardens were built to provide a serene and peaceful retreat for the city's residents, as well as a place for social gatherings and celebrations.

Table 1 Urban strategies for ecosystem services (Dekay & Brien, 2001)

Green city strategy	Gray city strategy	Ecosystem services
Multiple regional renewable energy: solar, wind, hydro, biofuels, energy production from a type of fuel, conservation	Concentrated production: fossil fuels, nuclear energy	Energy production
Diversifying urban, suburban, and regional agriculture: a collection of gardens, producer parks	Remote integration: industrial farms	Food production
Wind planning: green corridors, scattered buildings, windbreaks	Mechanical systems in the building; lack of urban strategy	Air source
The catch basin in the place and building; natural pits, streams, and streams; restoration of the creek	Wastewater pipes and urban sewage, irrigation, sewerage, and burial	Water source
Local production; consecutive use: restaurant to garden complex, sewage to fertilizer, neighborhood compost	Centralized production; food, gardening, and construction industry; distribution of national exchange	Nutrition/supply of materials
Composting, recycling, industrial parks, composting toilets, improving neighborhood sewage	Trucks and landfills, incineration, sink drains, sanitary sewers	Removal of waste
Landscape ecology model: Corridor, patch, matrix	Separation of critical habitat from land development	Biodiversity
Water filter/reuse; living filters, soil infiltration, and loading; constructed wetlands	Chemicals, mechanical filters	Water refinery
Constructed wetlands, compost	Settling ponds, chlorine, chemicals, landfills	Waste processing
Determining the location outside the flood zone; green roofs, permeable cover; renewed areas: wetlands, site protection	Discharge downstream: floodwall, embankment, channelization; concentrated storage: dams and reservoirs	Flood mitigation
Protection; reservoirs, detention centers, ponds, and local lakes; xeriscape, native plants	Central reservoirs; interbrain pumping extraction from the underground water table; irrigation dam	Reduce drought
Acceptance of natural changes; planting, ground cover, avoiding the development of erosion in the area	Physical barriers, sea walls, erosion fence	Erosion control
Demand/supply cycle; the structure of buildings; ice, phase change of substances; reservoirs, forests, biofuels	Gas/oil reserves	Energy saving

Chelsetun Garden, also known as Chehel Sotoun, is one of the most famous gardens in Isfahan. It was built in the seventeenth century during the Safavid era and was used as a royal pavilion for receptions and festivals. A large reflecting pool surrounds the garden and features a grand hall decorated with beautiful frescoes and murals.

On the other hand, Saadat Abad was a residential neighborhood in Isfahan, known for its beautiful gardens, as depicted in Fig. 1. Some of the famous gardens in this area include the Garden of Hajarjarib and Farah Abad. The Garden of Hajarjarib was famous for its KolahFerangi buildings, which were European-style pavilions popular in Iran during the Qajar period. Farah Abad, on the other hand, was a more extensive complex featuring several gardens, including the Garden of Paradise, its beautiful flowers, and fountains.

Overall, the gardens of Isfahan were a testament to the creativity and skill of Iranian architects and landscapers of the time, and their legacy continues to be appreciated by visitors from around the world.

4.1.1 Urban Infrastructures in Bagh Shahr of Isfahan

The infrastructure of Isfahan's urban area can be categorized into four different types, namely: green infrastructure, water infrastructure, sociocultural infrastructure, and economic infrastructure:

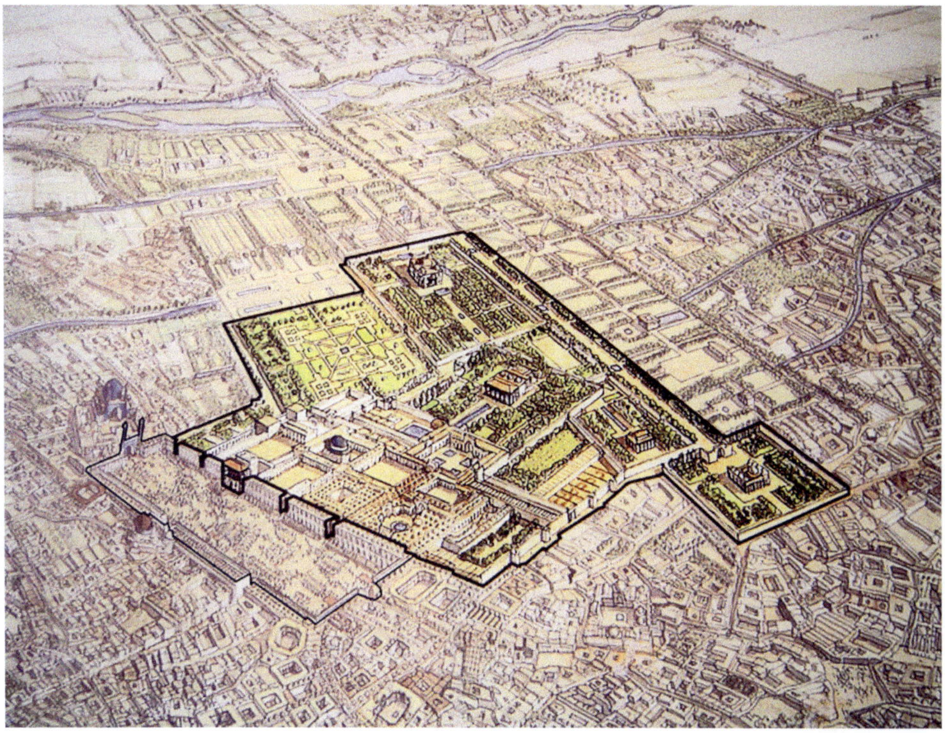

Fig. 1 Bird's eye view of Isfahan in the seventeenth century J.C Goluin-Reference: director of research at the C.N.R.S

4.1.2 Green Infrastructure

The city of Isfahan is considered one of the oldest cities in Iran, with a history that spans several thousand years, as indicated by many writings. The development of Seljuk gardens and promenades within the city walls (such as Naghsh Jahan Garden) and beyond the city boundaries toward the river (including Baghkaran, Ahmedsiyah Garden, Bakr Garden, Falasan Garden, etc.) played a crucial role in connecting the central city (Yehudieh) to the surrounding scattered points, facilitated by the presence of the city and the river. This also reinforced the axis of Zayandeh Rood as a recreational area outside the town. In the early tenth century, the establishment of the Safavid dynasty by Shah Ismail marked a significant and novel event in Iranian civilization. In the 11th year of his reign, Shah Abbas I, the fifth king of this dynasty, chose the old city of Isfahan as the new capital. Shah Abbas aimed to create an earthly paradise in Isfahan and devoted great care and attention to the renovation and development of the city, as shown in Fig. 2. The objective of the new city was to establish a green network in the southwest of the old city, and its layout is cleverly interconnected with the ancient city. Thanks to the attention and encouragement of Shah Abbas, green spaces, tree plantations, and the number of gardens flourished to such an extent that most Western tourists visiting the city in the seventeenth and eighteenth centuries went to Isfahan and described the city's landscape as a garden, where among the abundance of trees, only the gleaming walls of mosques and minarets could be seen. Even the palaces of the king and politicians paled compared to the vastness and grandeur of the city's extensive gardens. The new city's central axis, considered one of the first urban planning schemes of the capital, was Bagh Moshjiri, where the mainstream and side streams cascaded through the waterfalls and numerous ponds.

Based on the writings of Della Valle et al. (1892), who were travelers during the time of Shah Abbas I, it can be concluded that the Safavid city was divided into four main parts and formed thirty gardens with secondary walls. The king gave some of these gardens to the elders and nobles, while others were allocated for personal or public use. These gardens not only added to the city's prosperity and appearance, but they were also like paradise themselves with their charming small buildings and palaces, clean corridors, blooming gardens, rare hand-picked bushes and flowers, and various water tanks and fountains. The city's recreational, public, private, and royal functions emerged from the two main axes: the artificial axis of Cheharbagh and the natural axis of Zayandeh-Rood, which were connected. The Safavid government, in developing the city and creating a new town, took advantage of the potential local natural factors, especially the Zayandeh River, without destroying or damaging the old Seljuk city. The development of gardens and gardening, including creating the central artificial perpendicular Chaharbagh, which was entirely in harmony with the natural environment and the river, laid the groundwork for transforming Isfahan into a garden. This process was so successful that Tavernier, in his travel book, described Isfahan as a city with a small population, where each family had a

Fig. 2 Garden City of Isfahan—Isfahan engraving from the north side. 1995 *Source* Drower Flan ()

separate house and a unique garden, giving the city a spacious and spacious and green appearance (Inanlou, 1997). The order of plants was essential in defining and strengthening the physical and visual boundaries of the space, components, and building elements, as well as being the visual identity of the garden city. In Isfahan Garden City during the Safavid era, the order of plants and the presence of green elements and gardens in the physical structure of urban spaces were among the most important factors that gave identity to the newly built parts of the city by foreign visitors. The order of plants can be seen on both micro- and macro-scales in the gardens and the space inside all urban buildings in the form of a garden or body-like tiles where a garden is impossible.

According to the text, the plant order in Isfahan Gardens during the Safavid era significantly impacted the city's urban landscape. It has become its visual identity, as shown in Fig. 3. This is particularly evident on a macro-scale, where the arrangement of plants in rows of trees and gardens has strengthened the physical and visual boundaries of the gardens' spaces, components, and elements. The plant order used in Isfahan Gardens has caused a massive change in the urban landscape and has been instrumental in defining the new texture of the city.

According to Sharden, when observing the city of Isfahan, one can only see it through the trees of the blue domes and tall minarets that resemble a forest or a grove from all directions.

The design of Isfahan's garden city incorporates principles of Iranian gardening, using a combination of natural and artificial landscape factors and right-corner geometry to create a virtual expanse along the paths and gardens and to direct the main view (Ulrich, 1993). The paths in new neighborhoods and the Chaharbagh axis feature a compound use of right-angled geometry and row planting of trees on the edges to enhance their visual impact, increase their length, and focus the viewer's attention on a specific point. These principles are consistent with those used in Iranian gardens, including the design of the garden's central axis (Bahjat, 2010; Haghighatbin et al., 2012).

According to Eskandar et al., (2022), the research results indicate that mixed organization areas have a higher thermal comfort than those with tributary organizations. Furthermore, the thermal performance is better in streets with radial-tributary, grid-tributary, and tributary-tributary organizations, particularly with Northeast-Southwest and Northwest-Southeast orientations, compared to those with North–South and East–West orientations. The study also reveals that the thermal conditions of roads adjacent to old plots, mixed plots, and plots with new arrangements are better, respectively. Additionally, the thermal stress is slightly different in streets with a lower aspect ratio but almost the same as those with a higher aspect ratio. These findings highlight the significance of such considerations in traditional Iranian urban planning, which is a distinct characteristic of the planning process.

4.2 Water Infrastructure

In the Safavid era, Isfahan's gardens were constructed in different ways. The main design and organization elements of Isfahan's garden were the water flowing and its infrastructure, shown in Fig. 4. The river and its branches provided the organic water order on a macro-scale for the entire garden city. In contrast, its geometric order on a micro-scale

Fig. 3 The green infrastructure of Isfahan. *Source* Authors

in the constituent elements of the garden city, such as gardens and access routes, was organized using two forms of flowing and stagnant water (Haghighatbin et al., 2007).

The word "Madi" in the Isfahani dialect refers to the broad rivers that have branched off from the Zayandeh River to supply water around this city, depicted in Fig. 5. It is not used anywhere else except in this city, and it is a sign of the historical antiquity of this word. On the other hand, Roshni is one of the interests and efforts of people in developing and settling their land. The complex and fantastic method of dividing water based on the two principles of time and space by these people was the result of a deep thought that brought water to all the villages in proportion to the need, created such a lushness and abundance that throughout history has forced all the people of "Qalam to confess" (Jafari, 1993). The materials are separated in the form of wide channels upstream from the river, with a slope of one in a thousand; after a few tens of kilometers, they have reached the watershed, and in this part, they have flowed on the surface of the earth with small streams. Madis eventually fed and irrigated many villages along their route, and downstream their wastewater flows into springs or other Madis. The most important effect of materials has been to create sustainability in the green spaces (gardens) of the green border of Chaharbagh Street and numerous parks in the city (Hekmati, 1987). Supplying water was an essential factor in developing gardens. It provided the possibility of dispersing water-related jobs, such as mills, etc., from the riverside in different parts of the city and the transformation of the city of Isfahan in the Seljuk period with the characteristics of the towns into hot and dry places. Moreover, the narrow alleys leading to the garden city, with numerous paths and gardens, played an essential role (Jafari, 1993).

The Isfahan School of Garden shaped the Isfahan Garden City's physical-spatial structure and was influenced by intellectual and Shiite wisdom and teachings. The water order in Baghshahr of Isfahan on a macro scale is not only related to the order of the earth in terms of shape and slope but also embodies Shia intellectual wisdom, particularly the essential movement of the Mullasadra.

The Safavid urban designers emphasized the display of water order in various urban spaces based on the principles of the Isfahan school of gardening and urban planning. This was achieved by utilizing the physical features of the earth,

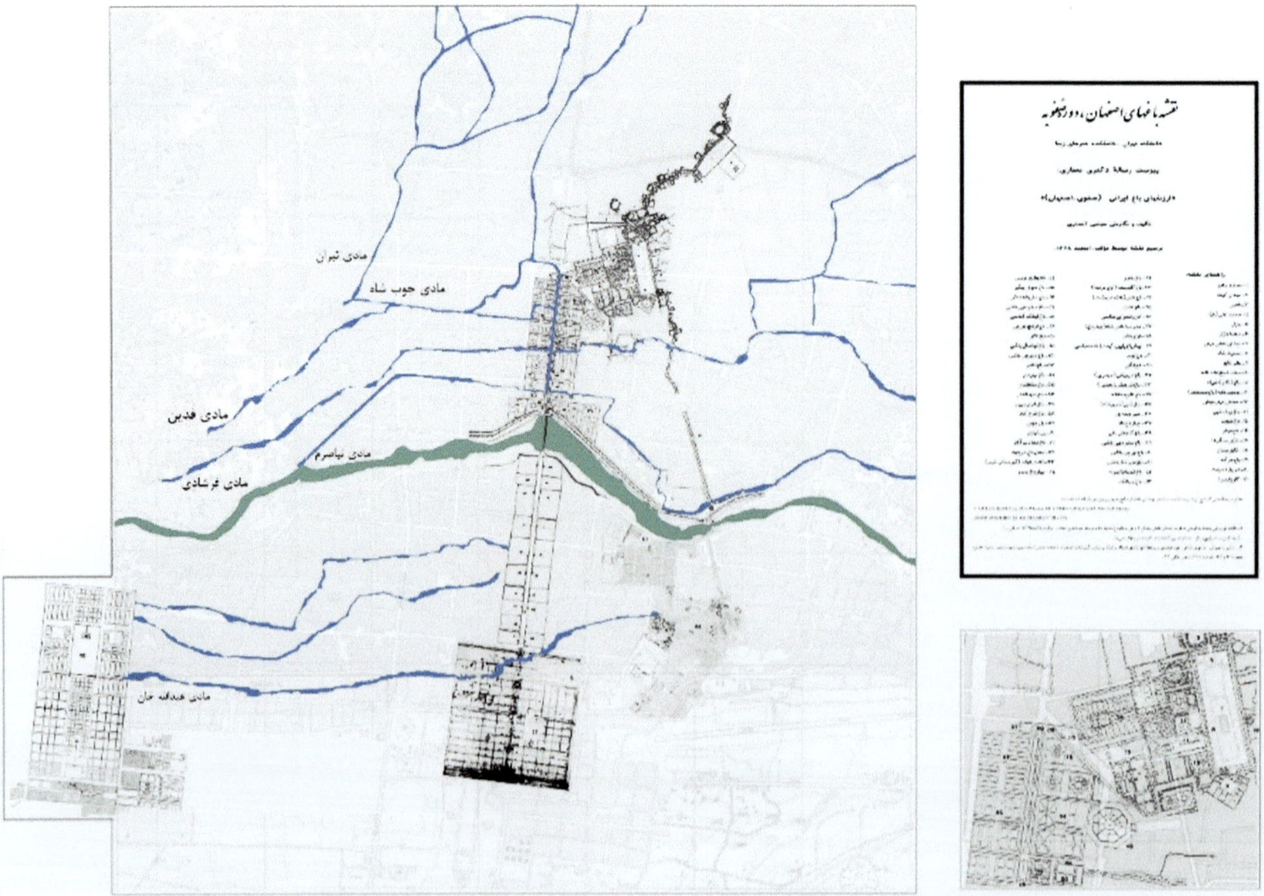

Fig. 4 The water infrastructure of Isfahan. *Source* Authors

Fig. 5 Isfahan Madis—*source* Authors

such as the Zayandehroud river and its branches, which were organized into materials and degrees of units of materials in the form of atmosphere and joule flow throughout the city (Haghighatbin et al, 2007; Hosseini, 2002).

This approach to historical landscape provides a platform for exploring the complex relationships between humans and their environment and the diverse ways in which people have interacted with and transformed the landscape over time. By incorporating narratives and multiple perspectives, this approach can offer a more nuanced understanding of historical contexts and enable a deeper appreciation of landscapes' cultural and ecological values. Additionally, this perspective can inspire activism and encourage a more active engagement with historical landscapes, promoting their preservation and sustainable management for future generations. This represents a possible step towards promoting the development and preservation of ancient cities. The idea behind it is that it could provide a fresh interpretation of historical backgrounds, which may reinforce the feeling of attachment and human identity in urban areas through people's experiences of their surroundings (Khorasgani et al., 2023).

4.3 Social Infrastructures

The emergence of new concepts in native architecture and urban planning in a society without a cultural context and the people's input as builders and users is unlikely. In the

Safavid era, cultural development was based on religious and spiritual beliefs, resulting in achievements in literature, ethics, education, training, philosophy, mysticism, and art. Alongside this, the development of material culture, including science, industry, production, and commercial exchanges, was also significant. Different peoples and clans formed the Safawi community. They were the outcome of successful socioeconomic planning by the authorities, including a powerful central government and unity among the people under the banner of the Shia religion.

Regarding Fig. 6, during the Safavid era, the collective spaces of Isfahan played an essential role in fostering mechanical solidarity based on ideological similarities rather than class or gender differences. The government needed to maintain unity and solidarity among the people, so they organized magnificent national ceremonies and religious celebrations to create a collective identity. The social structure of Safavid society was a pyramid with the king at the top and ordinary people at the base, including peasants, artisans, shopkeepers, and small traders. Military and country nobles and group officials were between these two classes. Different classes had guilds and social and religious associations with representatives in every important city. The Safavid society attempted to create a link between the Iranian and Turkmen Qazalbash organizations and the union between scholars and marketers. The clerical authority managed many properties under their control through endowments or donations from individuals for religious purposes.

4.4 Economic Infrastructures

Undoubtedly, the economic life during the years of Safavid rule is one of the new research topics that has recently attracted the attention of researchers. One of the reasons for this was the nature of traditional historiography and the tendency of Iran's science toward sovereignty and engagement with political figures (Romlu, 1997). On the other hand, the long-standing tradition of historiography in Iran has long served as a tool for governments and has expressed the events in a favorable intellectual format of the ruling class. In this way of thinking, artisans, and merchants were considered level. Their social expression and protests defending their rights were sometimes deemed worthless and

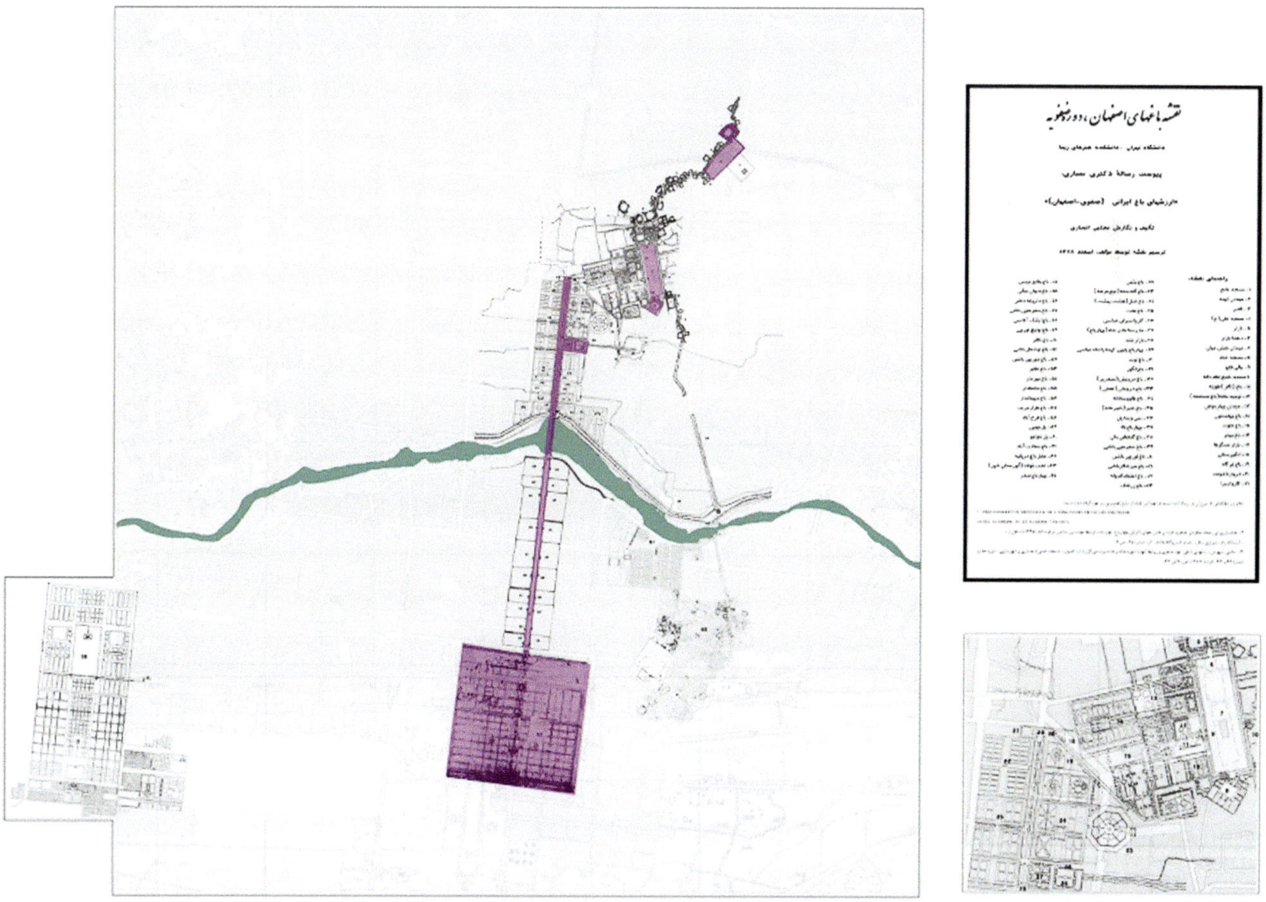

Fig. 6 The social infrastructure of Isfahan. *Source* Authors

humiliating. They were called insulting nicknames such as "Ajlaf" (Romlu, 1997). This issue can be examined from several perspectives. Firstly, economic components could have political and cultural power due to their importance and influence in the political developments of this period, as they were able to leave their artistic effect on other groups and organizations. In this context, the impact of these components on the rebellions with the institutions of this period is evaluated based on what is presented in Fig. 7. Another importance of this issue is in examining the influence of Safavid political-religious rule on the economy of this period. Iran's Sunni society before the Safavids gradually transformed into a Shia-majority society organization in the Safavid era. In this development, the politically instituted religious institutions have a familiar role. The religious institution, which was integrated and mixed with the political institution at the beginning, was gradually separated from it and found a clear identity and a stable position during Shah Tahmasb and after undergoing a temporary crisis during the reign of Ismail II and Muhammad Khodabandeh. Abbas first found his stability and strength. This process continued until the end of the Safavid rule. About this issue, it can be said that the relationship of the Safavid government with the strata of society (including the people, traders, markets, guilds, etc.), as a part of the body of the political institution, was a direct relationship. The ruling political supervision was the same as in the previous era. It was applied during the Safavid period, but its religious leadership took on more water and color with Shi'ite color and margins (Romlu, 1997).

Based on the statement, it can be inferred that during the early period of the Safavid government, militarism was a key driving force for the government's institutions and goals. The government's judicial structure mainly focused on supporting the political establishment, further reinforcing the emphasis on military and war objectives. The economic affairs of Iranian society were also primarily driven by ethnic and tribal considerations rather than religious, economic theories. The government did not prioritize the various aspects of social and economic life as part of its social responsibilities, although it did make efforts to improve energy. The main task of the government was to establish political power. Therefore, the main components that dominated the interaction between political governance and

Fig. 7 The economic infrastructure of Isfahan. *Source* Authors

economic life in Safavid society during the early period were militarism, political establishment, and tribal considerations. Overall, the Safavid government's focus on militarism and political power during the early period may have contributed to neglecting social and economic development, which may have had long-term implications for Iranian society. (from the era of Shah Ismail to the beginning of the reign of Shah Abbas I).

Each society's market and economic structure operates on its principles, prioritizing work and financial stability over the disciple and mastery relationship. However, if the relationships between Merid and Moradi overshadow this logic, there may be a lack of motivation to develop the market and promote economic growth. In such relationships, the mentor holds all the power, and the disciples are expected to follow without questioning. This mentality places greater emphasis on the mentor rather than on economic prosperity for society. As a result, individuals striving for economic success may feel deprived of motivation, as their interests are not considered.

The Sufi worldview emphasizes a disregard for the world, at least as defined by powerful Sufis. In this worldview, there is little consideration given to society's development from political, cultural, and economic perspectives. Instead, everything is seen as being in service of the guru, with little responsibility towards the wider community's interests and economic well-being. The Sufi world is a small, simple, and benign place where concepts like capital, economic growth, and wealth accumulation based on private interests hold no meaning (Nelson et al., 2014). As such, it cannot fully understand the complexities of a vast world and cannot endure for long periods, as was the case with the Safavid society during the reign of Shah Abbas I. Shah Abbas I sought to challenge and erase this thinking by introducing other elements in the construction of power. The fragility of power and its changing hands in previous periods showed that Sufi thinking was incompatible with the world that Shah Abbas was trying to define. While this thinking may have worked during the founding period, it proved ineffective during the establishment period (Haghighi-Borujeni, 2017) (from Abbas onwards).

The core principle of monastic thinking is to offer water and bread to anyone who enters the house, regardless of their faith. While this may seem like a commendable moral stance, it ultimately promotes a sort of parallelism. In this religious mindset, the individual is not regarded as a productive force and is not encouraged to contribute to economic growth. This outlook is at odds with the incentives of the market, where individuals who work harder and play a more significant role in capital circulation tend to earn more and significantly impact society (Aghili, 2015).

5 Adaptation of Ecosystem Services and Infrastructures of Bagh Shahr, Isfahan

The adaptation of the green, water, social, cultural, and economic infrastructures of Isfahan Garden City shows the overall coherence of the city and the coordination of the infrastructures above as parts of the whole. In the design of Isfahan Garden City, compliance with the principles of sustainability and achieving a healthy town, considering that the city is an allegory of heaven, is the idea and goal of the design, and it seems that in this regard, Isfahan has met the expectations of its users well.

As mentioned in Table 1, Mark Dekay has compared green and gray cities from the perspective of ecosystem services, resulting in these services in green cities being ultimately sustainable and leading to a healthy city. According to its infrastructure, Isfahan Garden City has been a kind of answer and supplier of these ecosystem services, as shown in Fig. 8.

A significant portion of Isfahan's food production is sourced from the city's gardens and the agricultural lands in the suburbs. The city's material paths, designed as water arteries, not only consider the natural order of slope and form but also consider the placement of water and green spaces to align with the prevailing and favorable wind directions. This design functions as a natural filter (Khorasgani et al., 2023).

In Table 2, an attempt has been made to compare ecosystem services and their components and explain and examine them with Isfahan's Safavi garden and green urban strategies.

6 Summary and Conclusion

The ideal city of Iranian garden cities is rooted in the religious beliefs of the builders. Research on remaining historical sources shows the significance of gardening and gardens in the Safavid era, which established a close relationship with the Shiite religion and intellectual life. This close semantic link was based on verses of the Qur'an, words of the Prophet, and pure imams about heaven. The thinkers and artists of the Safavid era viewed the world as an example of the hereafter. They believed in a balance between the tangible and the sensible, or the external and the internal, referred to as a balance between the world and the afterlife. Iranian garden cities were designed based on physical and semantic principles of Iranian gardening, to visualize the city as a shadow of paradise. Success in Safavid era city planning was due to factors such as relying on teachings of the Shia religion

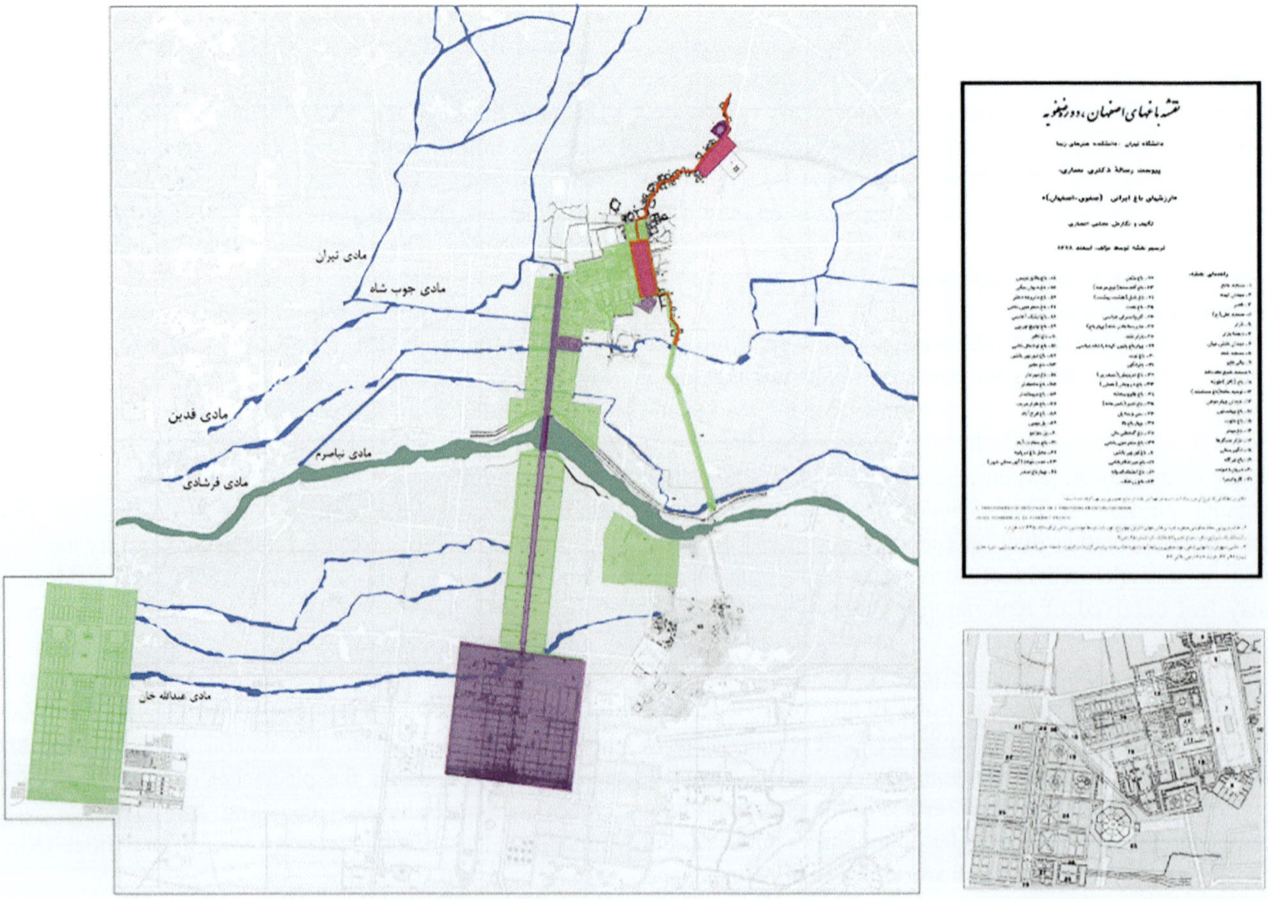

Fig. 8 The infrastructure of Isfahan. *Source* Authors

and Isfahan school principles in city planning, combined with the physical and semantic model of the Iranian garden. Isfahan City was developed as a combination of gardens with the construction of royal and government residences on the outskirts of Seljuk City. This embodiment of symbolic concepts and the massive transformation of the city's urban landscape and visual quality caused by the builders is confirmed by descriptions from foreign travelers such as those visiting Isfahan then.

According to Table 2 examining the factors of ecosystem services in green cities and comparing them with similar features in Iranian garden cities shows that most factors are consistent in supplier services. Still, the emphasis is on agriculture for food production in Iranian garden cities and using air resources (natural wind direction) has played an essential role in the overall design and orientation of towns. In Iranian garden cities, the design and implementation of water diversion routes such as streams and dams have also influenced the distribution and optimal use of water resources (Mukheibir, 2007).

In regulatory services, despite the commonalities of the garden and green cities, it seems that Iranian garden cities need to be reviewed and updated in this area. Comparative studies have been conducted according to the daily needs of cities.

In social services, according to the oriental culture that dominates Iranian garden cities, more attention has been paid to spiritual experiences and metaphysical fields. Therefore this is the adaptation of the features of the green city to the culture and civilization of the users of Bagshahr. Regarding habitat services, there are similar actions and results in green and garden cities (Smit & Wandel, 2006).

The study of Iranian garden cities, especially the garden city of Isfahan, and ecosystem services shows that ecosystem services have been considered experimentally in their formation. Using the principles of traditional Iranian architecture and urban planning, they can be classified as green cities. In addition to the mentioned functions, Isfahan Garden City, a symbol of an Iranian–Islamic garden city, has provided the necessary infrastructure to regulate and support the ecosystem. Therefore, this garden city has been influential in the continuity of life and sustainable continuity of itself and its users as a green and sustainable city, as well as creating a healthy city and ecosystem health.

Table 2 Urban strategies for ecosystem services and Isfahan Garden City (Authors)

Green city strategy	Isfahan garden strategy	Ecosystem services	General categories of ecosystem services
Multiple regional renewable energy: Solar, wind, hydro, biofuels, energy production from a type of fuel, conservation	Multiple regional renewable energy: Solar, wind, hydro, biofuels, energy production from a type of fuel, conservation	Energy production	Supplier services
Demand/supply cycle; the design of buildings; ice, the phase change of substances; reservoirs, forests, biofuels	Demand/supply cycle; construction structure, material phase change; reservoirs, forests, biofuels	Energy saving	
Diversifying urban, suburban, and regional agriculture: a collection of gardens, producer parks	Diversifying urban, suburban, and regional agriculture: collection of gardens, agriculture	Food production	
Wind Planning: Green corridors, scattered and scattered buildings Windbreakers	Wind planning: green corridors, use of favorable and prevailing winds of the city, and urban design according to the direction of wind flow	Air source	
The catch basin in the place and building; natural pits, streams, and streams; restoration of the creek	Zayandehroud natural river; materials; streams and tables; canals and wells	Water source	
Local production; consecutive use: Restaurant to garden complex, sewage to fertilizer, neighborhood compost	Local production; consecutive use: Sewage to fertilizer	Nutrition/supply of materials	
Composting, recycling, industrial parks, composting toilets, improvement of neighborhood sewage	Recycle	Removal of waste	Regulatory services
Landscape ecology model: Corridor, patch, matrix	Landscape ecology model	Biodiversity	
Water filter/reuse; living filters, soil infiltration, and loading; constructed wetlands	Reuse; living filters, soil infiltration, and loading; Made straps	Water refinery	
Constructed wetlands, compost	Garbage landfill and conversion to gas methane	Waste processing	
Choosing the place outside the flood zone; green roofs, permeable cover; renewed areas: wetlands, site protection	Determining the location outside the flood zone; clauses, site protection	Flood mitigation	
Protection; reservoirs, detention centers, ponds, and local lakes; xeriscape, native plants	Protection; local ponds and lakes, embankments and embankments; native plants	Reduce drought	
approval of natural changes; planting, and ground cover, preventing the result of the decline in the area	acceptance of natural changes; planting, ground cover, avoiding the development of erosion in the area	Erosion control	
Urban green areas provide facilities for recreation, leisure, and relaxation; forest recreation areas, fishing activities, leisure activities	Urban green areas, recreational facilities, leisure time	Recreation and ecotourism	Social service
Beautiful views that city parks create for buildings; landscape aesthetics, landscape quality	Beautiful views created by urban gardens; landscape aesthetics, landscape quality	Aesthetic values	
Actual gardening to protect ecological-social knowledge	Dedicated and public gardening to protect ecological-social knowledge	Cognitive development	
–	An allegorical garden of paradise	Spiritual experiences	
Urban green spaces are a habitat for birds and other animals that people enjoy watching	Zoos and game parks	Biodiversity habitat	Habitat-related services

The characteristics of Iranian garden cities their knowledge, principles, and traditions and their compatibility with the studies conducted on sustainable and green cities and their design management, as well as ecosystem services, can provide solutions in line with the design of sustainable cities.

References

Aghili, S. A. (2015). The relationship between political sovereignty and economic life in the first period of Safavid Government. *Britannica, 17*(4), 47–86

Azani, M. (2002). Isfahan landscape and the sustainable development trend. *Quarterly Journal of Geographical Space, 1*(8), 65–82.

Bahjat, F. (2010). Iranian gardens from the past to the present (emphasis on tourism in Isfahan). *Journal of Geographical Research, 8*.

Dekay, M., & Brien, M. O. (2001). Gray City, Green City. *Forum for Applied Research and Public Policy Journal, 16*(2), 19–27.

Della Valle, P., Havers, G., & Grey, E. (1892). *The travels of Pietro Della Valle in India: From the Old English translation of 1664*. Hakluyt Society.

Drower, M. S. (1995). *Flinders Petrie: A life in archaeology*. University of Wisconsin Press.

Eskandar, G. A., Zamani, B., Kariminia, S., & Sichani, M. G. (2022). *Thermal comfort in public open spaces: The historical core of Isfahan*.

Eslamian, S., & Maleki, M. (2021). Disaster resilience and computational methods for urban infrastructures. In S. Eslamian & F. Eslamian (Eds.), *Handbook of disaster risk reduction for resilience* (1st ed., pp. 279–298). Springer International Publishing.

Haghighatbin, M., Ansari, M., & Pourjafar, M. (2007) . Sustainable design in isfahan urban green space during the safavid period and its destruction during the contemporary period. *Modares Honor Quarterly, 2*(1).

Haghighatbin, M., Ansari, M., & Zabihian. (2012) . We are examining the principles of Howard's garden and comparing it with Safavi's garden. *Naghsh-e-Jahan Magazin, 2*(2).

Haghighi-Borujeni, S. (2017). *Interventions of the power class and its effect on the city of Isfahan's structure during Shah Abbas Safavid's reign*

Hekmati, J. (1987). *Garden and park design* (1st ed.). Jamshid Hekmati Publishing.

Hosseini, S. H. (2002). The traditional water management of Zayandehroud River, a debate in Iranian indigenous knowledge. *Quarterly Journal of Faculty of Literature and Human Sciences (isfahan University), 15*(4), 101–120.

Inanlou, S. (1997). The impact of tehran population growth rate on landscape per capita. *Quarterly Journal of Sociology and Social Sciences, 19 and 20*.

Jafari, N. (1993). *We are improving the role of an urban element of material passage as an urban pedestrian axis, material Fedan Isfahan* [Master's Thesis in Urban Planning]. University of Tehran.

Kaplan, R., & Kaplan, S. (1998). *With people in mind: Design and management of everyday nature Washington*. Island Press.

Khorasgani, A. M., & Haghighatbin, M. (2023). Regeneration of historic cities: Reflections of its evolution towards a landscape approach. *ISVS e-Journal, 10*(5), 60–78.

Khorasgani, A. M., Villalobos, M. H., & Eskandar, G. A. (2023). Sustaining historic cities: An approach using the ideas of landscape and place. *ISVS e-Journal, 10*(1), 320–332.

Mukheibir, P. (2007). Qualitative assessment of municipal water resource management strategies under climate impacts: The Northern Cape South Africa case. *Water Sa, 33*(4), 575–581.

Muqtada, R. (2003). *Isfahan: A picture of Chahar Bagh*. Irannameh Quarterly (Year 21)

Nelson, G. C., Valin, H., & Sands, R. D. (2014). Climate change effects on agriculture: Economic responses to biophysical shocks. *Proceedings of the National Academy of Sciences, 111*(9), 3274–3279.

Romlu, H. (1997). The economic life during Safavid rule: A new research topic [In Persian]. *Iranian Journal of Historical Studies*, 29–30.

Smit, B., & Wandel, J. (2006). Adaptation, adaptive capacity, and vulnerability. *Global Environmental Change, 16*(3), 282–292.

Ulrich, R. S. (1993). Biophilia, biophobia, and natural landscapes. In: S. Kellert & E. O. Wilson (Eds.), *The Biophilia Hypothesis*. Island Press.

Minimizing Energy Consumption of Educational Buildings by Testing Alternatives of Green Envelopes in Alexandria

Ahmed Soliman

Abstract

A green building is a building that is ecologically responsible for using eco-friendly building materials and construction practices. By enhancing the process of design, construction, building running, maintaining, and removal, architects can preserve natural resources, contribute significantly to the decarbonization of the built environment and consequently improve quality of life. In that respect, building envelopes are not just physical separators between indoor and outdoor environments, but they have a significant impact on reducing the building's overall energy consumption and in improving the indoor environmental quality. To make better decisions, technology can significantly assist architects in the pre-design stage or retrofit interventions to select the appropriate parameters of green buildings. In that matter, modeling and simulation tools for integrating the parameters and materials of green architecture in building envelopes are useful interfaces. Based on relevant bibliography, the research developed the parameters of green building envelopes, which were experimentally and numerically investigated one by one through a case study conducted on an educational building in Alexandria. Autodesk Revit Architecture software is used to create the building model, and energy simulation is performed using DesignBuilder to assess the indoor environmental quality and energy consumption. Hygrothermal comfort (interior air temperature and relative humidity) and greenhouse gas emission (indoor environmental quality), and cooling electricity (energy consumption) are assessed along the year to measure the difference between the performance of the building's envelope before and after modifying the parameters of green buildings. The results show that by setting the interior air temperature at 25 °C in the simulation process, integrating parameters of green envelopes will decrease the energy consumption and CO_2 emissions by 20% at June and by 19% at September.

Keywords

Green building • Envelope-energy • Consumption • Computer simulation • Egypt

1 Introduction

The buildings and construction sector are responsible for 36% of global energy consumption and 37% of energy related CO_2 emissions. By 2020, due to the COVID-19 pandemic, greenhouse gas emission were reduced to 10%. However, long-term progress remains largely inefficient (United Nations Environment Programme, 2021). Since the past few decades, terms such as "sustainable", "ecological", and "green" are often being used interchangeably in the body of literature related to the construction industry to address climate change and global warming and to reduce the impact of the building sector on the environment. The term "green architecture" refers to a combination of values that are environmental, social, political, and technological. Using the term "green" implies the inclusion of terms such as sustainability, ecology, and performance. As a result, a building can be sustainable but not ecological or green, whereas a green building must be a combination of sustainable, ecological, and performative characteristics (Attmann, 2010). To measure the "greenness" of architecture, it is essential to define and isolate the different activities or operations. As a result, green architecture is the profession of designing buildings and employing ecologically responsible methods throughout a building's life cycle, such as "site selection," "design," "construction," "operation and maintenance," "renovation," and "demolition" (Attmann, 2010).

A. Soliman (✉)
Department of Architectural Engineering and Environmental Design, Arab Academy for Science, Technology, and Maritime, Alexandria, Egypt
e-mail: Ahmedsleman87@gmail.com

A. Battisti et al. (eds.), *Greening Our Cities: Sustainable Urbanism for a Greener Future*, Advances in Science, Technology & Innovation, https://doi.org/10.1007/978-3-031-49495-6_8

This practice complements the conventional architecture concerning environmental, economic, and social aspects including utility, durability, and comfort. According to Dusek (2008), sustainability has increasingly become the dominant metaphor and language for expressing ecological concerns about the economy and technology. Hence, green buildings minimize or eliminate negative impacts on the environment and shared ecosystem while retaining the potential for positive effects. Green buildings consequently encompass passive strategies that maintain a comfortable temperature within the building using the climate and natural elements to get the optimum benefit without using the artificial energy. Passive solar design involves nontechnical design methods that affect the building energy consumption: site conditions, local climate, sun angle, massing and orientation, and daylight (Attmann, 2010).

In design, architects have to steward the biological environment and the holistic sum of the parts, including building materials, mechanical and electrical systems, controls, envelope (walls, roof, and slabs), and fenestration (glazing material and shading elements) to use energy efficiently. Using building energy modeling to simulate a building's operational use enables designers to make better decisions (Hemsath & Bandhosseini, 2018). In Egypt, it is nearly impossible to fully climate-control the whole building to attain human comfort only by passive strategies especially during the summer. Nevertheless, it is totally feasible to decrease the utilization of mechanical cooling systems to the bare minimum if the appropriate parameters of green buildings are applied. During the building life cycle, air conditioning is the biggest factor in electricity demand. Furthermore, during the operation stage, there are significant energy losses through the building envelope elements: external wall, roof, and openings, which raises the demand for energy to return the indoor temperature to the thermal comfort zone (Bachrun et al., 2019). Comparisons between different design decisions using metrics and measurements, within energy modeling, show how different elements have the potential to reduce the energy footprint of the building and to minimize the initial costs of the heating, ventilation, and air conditioning (HVAC) system in addition to the long-term costs associated with operating it.

So based on the relevant literature Table 1, the researcher has identified the following parameters that impact the thermal performance of green envelopes (Bachrun et al., 2019; The Government of the Province of Jakarta Capital Special Territory, n.d.; Motawa et al., 2021):

- Building orientation
- Building shape and massing
- Windows orientation and shading devices
- Window-to-wall ratio (WWR)
- Glazing
- Vegetation (green walls and green roofs)
- Insulation (including PCM materials) in walls.

2 Methodology

The research methodology is based on data collection and review of literature related to green buildings to define the parameters of the envelope. Then, an experimental and numerical study is conducted to evaluate the impact of each of the different parameters of the green envelope on the indoor environmental quality and the overall energy consumption (cooling electricity). Building E of the Faculty of Engineering and Technology in the Arab Academy for Science, Technology and Maritime Transport (AAST) at Abu Qir campus in Alexandria was selected. This as-built base case AAST building is compared to the simulation model with different alternatives of green parameters. The research focused on four parameters: shading devices, glazing, vegetation in walls and the roof, and thermal insulation for walls (Fig. 1). The researchers did not change the building massing orientation or its shape and massing, as these parameters of green envelopes cannot be changed in retrofitting interventions.

Likewise, the WWR in the AAST base case could not be increased as it already reached approximately 54%, in east, west, south facades and 35% in the south and north facades. So, this parameter was not put into consideration in the simulation. The research applied a spilt system as air conditioning for the whole building simulation as it is the current situation of the AAST building, which is depending on spilt unit as air condition for offices, laboratories, and meeting rooms and fan system for classrooms and studios. The researchers compared between the numerical simulations of indoor air comfort (temperature and humidity levels), greenhouse gas emission and energy consumption in the base case, and the same numerical simulations in the case of modifications of the green parameters of the envelope to verify and measure the effectiveness of the design alternatives. Each strategy' effect is studied separately, so when studying the effect of thermal insulation on reducing energy consumption, in the base case, the rest of the envelope components were left unchanged. However, in the case of the combined alternative, the parameters were selected based on the best results from among the building envelope options presented in each strategy. The parameters for the combined alternative, on the other hand, were selected based on the best results of each strategy's building envelope options (Al-Tamimi, 2022).

Table 1 Shows the different parameters of green architecture (the author)

Author	Findings
Ariff et al. (2019)	The green envelope variances are characterized based on the following; • Strategies of passive design • Thermal insulation for walls • The materials of windows glazing • Vegetation over the roof • Shading devices
Kibert (2013)	The green building is the design that implants the concept of net zero energy with emphasizing on the following strategies of passive design; • Building orientation and geometry • Glazing materials • Windows shading devices
Visser and Yeretzian (2013)	Energy-efficient building design depends on the reduction of the cooling demand and the related energy consumption, thus reducing CO_2 emissions, and this can be accomplished by proper design concerns for the following guidelines of passive design; • Building orientation • Building form and typology • Size and locations of windows • Shading devices • Material selection • Landscaping over the building envelope Also, the paper presents two different examples for green buildings, which using the following; • Double glazing windows • Window, roof, and wall insulation • Thermal bridge protection • Shading devices
Bachrun et al. (2019)	Sustainable architecture and design (green construction and building) is often found in traditional architecture which is responsible in reducing greenhouse gas emission (greenhouses effect) by adapting many passive architectural strategies in building envelope which have also a substantial impact on total energy consumption by emphasizing on the following; • Window-to-wall ratio (WWR) • Building shape and orientation • Windows glass materials • Shading devices • Thermal characteristics of envelope materials (external walls and roof)
BCA (2010)	Green or high-performance building relies on the fundamental to optimizing passive design by emphasizing on the following; • Massing and site planning • Building orientation • Windows glass materials • Wall insulation • Window-to-wall ratio (WWR) • Shading devices • Double skin facades • Material selection • Green wall Also, the paper present different examples for green buildings, which using the following; • Optimized orientation • Sun shading devices • PV panels over the roof • Vegetation over the roof (green roof)

(continued)

Table 1 (continued)

Author	Findings
The Government of the Province of Jakarta Capital Special Territory (n.d.)	The design of green building is emphasizing on the following principles; • Building forms and orientation • The proportion of window area (WWR) • Glazing materials • External shading • Light shelf • Internal shading
Perini et al. (2017)	Green facades, like other forms of green infrastructure, are increasingly being considered as a design feature to reduce building energy consumption by cooling internal building temperatures • Vertical greening systems (VGS) have a positive impact on the thermal performance of the building envelope, particularly during cooling periods
Addo-Bankas et al. (2021)	Green walls are known as the soul of green buildings because of their role in promoting energy efficiency, reducing sound transmission into buildings, improving indoor air quality, increasing building longevity, grey water treatment, and recycling
Shams (2017)	Green or sustainable design strategies are characterized into three groups; sustainable design in architecture, sustainable structural design, and passive design which are as follows; • Incorporation of nature • Building orientation • The ratio of building's length to its width • Roof materials with the consideration of vegetation elements • The colors of envelope materials • Glass materials • Shading overhangs

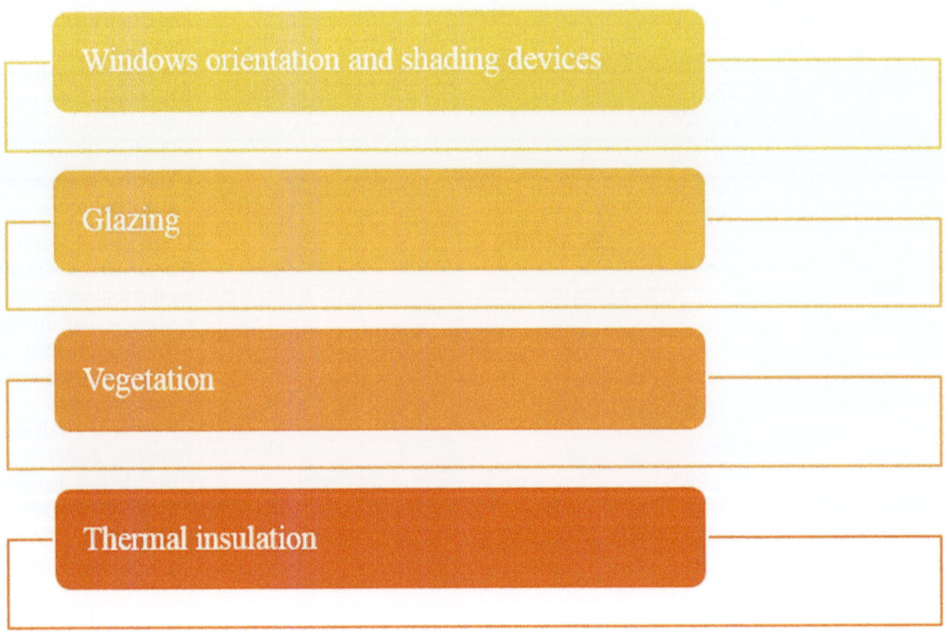

Fig. 1 Used parameters in the case study (the author)

3 Simulation Programs

The base case model was rendered using Revit Software and DesignBuilder software, as well as Energy Plus to carry out energy simulations. DesignBuilder provides access to all of the most commonly required simulation capabilities covering building fabric, thermal mass, glazing, shading, renewable energy, heating, ventilation, and air conditioning (HVAC) system (Designbuilder Software Ltd—Simulation, 2022), as well as financial analysis, making it the first comprehensive user interface for Energy Plus dynamic thermal simulation engine. The researchers used the simulation tool Energy Plus (E+) version 9.4, with an interface to the software modeling tool DesignBuilder, version 7.0.0.116. The simulation tool E + has been selected since it is widely used in the scientific communities as a tool for realistic thermal and energy performance simulations. Moreover, the process has been validated by Weather files of Alexandria (Weatherbase, n.d.).

DesignBuilder evaluates the indoor environmental quality: interior air temperature, relative humidity greenhouse gas emission, and the energy consumption from cooling electricity for the whole building. To evaluate energy consumption, the simulation must be set with the use of the heating, ventilation, and air conditioning (HVAC) system for cooling. It is not possible to measure solar heat gains for each façade individually, but the energy consumption and CO_2 emissions are measured for the whole building.

Simulation process

Simulating involves a number of steps that must be followed in order to ensure that the process is well thought out and reliable results can be obtained Table 2. The simulation started with four parameters of green architecture one by one, while the others parameters remain constant.

Step 1: The model of the building is drawn using DesignBuilder, including floor plans, elevations, and sections. The total floor area is 1075 m^2 for each floor, the building is located in Alexandria, Egypt, and the months for the simulation are June and September (as they are considered the hottest months of the educational year) to evaluate energy consumption and greenhouse gas emission, while the interior air temperature remains controlled at 25 °C with the use of the heating, ventilation, and air conditioning (HVAC) system (educational building).

Step 2: The simulation is run for the whole building without any changes and the results are recorded.

Step 3: The parameters of green architecture are modified one by one, while the other parameters remain constant and the results are recorded.

Step 4: The parameters are modified and are applied onto the building, and the results are recorded considering energy consumption and CO_2 emissions.

Step 5: The results are tabulated and translated into a bar chart, then the results are analyzed and discussed in both cases: before and after the modifications of the parameters of the green envelope.

The model is simulated under the weather file of Alexandria city (EGY_AL ISKANDARIYAH_ETMY) (Weatherbase, n.d.).

4 Case Study: Building of Faculty of Engineering and Technology, Arab Academy for Science, Technology and Maritime

4.1 Building Information

The case study selected for the simulation is the Faculty of Engineering building of the Arab Academy for Science,

Table 2 Shows the steps of simulation process for the case study (the author)

Simulation steps	Process
Step.1	Drawing a model for the case study by using DesignBuilder program
Step.2	Apply the simulation process for the case study and record the results for greenhouse gas emission and the energy consumption (cooling electricity)
Step.3	Apply the following parameters one by one and record the results for greenhouse gas emission and the energy consumption (cooling electricity)
	Shading devices (0.5, 1.0, and 1.5 m)
	Glazing (triple glass material)
	Vegetation (green wall and green roof)
	Insulation material (PCM material)
Step.4	Apply all the above parameters onto the case study to determine the best alternative in reducing energy consumption (cooling electricity), and greenhouse gas emission
Step.5	Comparison between the results form Step.2 and Step.4 to determine reduction percentage in energy consumption (cooling electricity) and greenhouse gas emission

Fig. 2 Base case: AAST building e showing north-east facade (the author)

Technology, and Maritime Transport at Abu Qir campus located in Alexandria, Egypt, Building E, (Figs. 2, 3, 4 and 5).

According to the Köppen-Geiger Climate Classification subtype for the climate in Alexandria is "Bwh" (tropical and subtropical desert climate) (HBRC, 2008). The average temperature for the year in Alexandria is 20.6 °C. The warmest month, on average, is August with an average temperature of 26.7 °C. The coolest month on average is January, with an average temperature of 13.9 °C. In May, Alexandria recorded its highest temperature of 42.2 °C. While the lowest recorded temperature was 2.8 °C in January (Weatherbase, n.d.).

Housing and building research center (HBRC) in Egypt developed another climate classification that divides Egypt into eight climatic zones, and Alexandria is classified in the Northern Coast zone: The city receives only about 200 mm of precipitation per year (Ministry of the Environment, 2005).

The general characteristics of the base case are described in Table 3, and the construction specifications are mentioned in Table 4.

The air gap between the walls works is an efficient low-cost insulating material and can reduce significantly the heat transferred to internal living space (Avila & Ismail, 2017).

4.2 Results and Discussion

The following part will discuss the effect of using four parameters of green architecture by modifying: shading louvers, glazing materials, vegetation elements on roofs and façades, and insulation materials (phase change material) on exterior walls. Other parameters: orientation, massing, and shape are discarded, as they are impossible to modify in retrofit interventions. WWR was also disregarded as the windows occupy nearly the maximum surface of wall surfaces in the base case and any change would have been minimal and insignificant in the results.

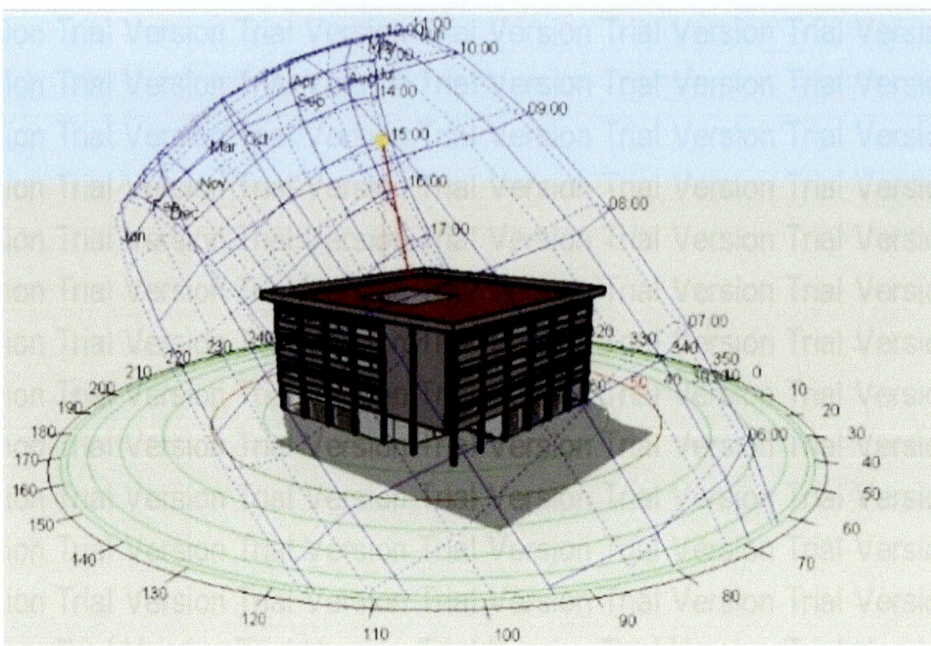

Fig. 3 Rendering of the AAST building (base case) using designbuilder showing the facades' orientation (the author)

Fig. 4 Plans of the AAST building using Revit (the author)

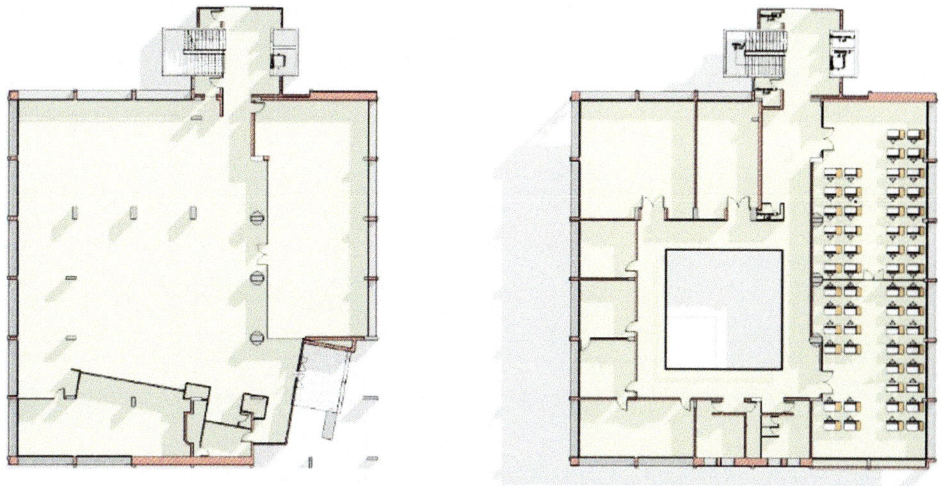

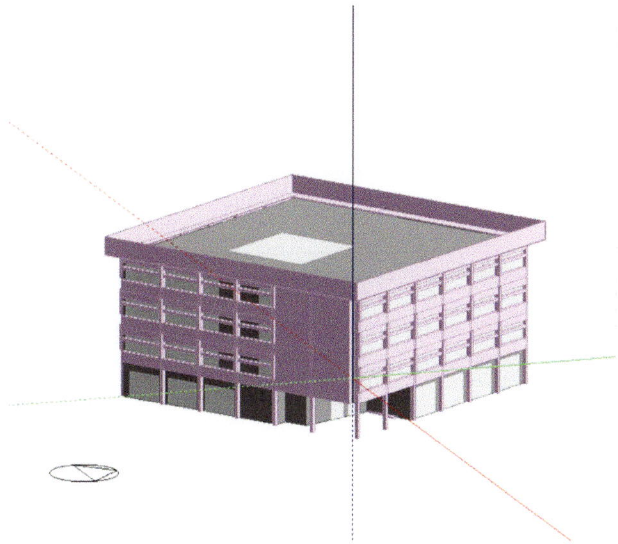

Fig. 5 Model generated by DesignBuilder (the author)

Table 3 General characteristics of case study (by researcher)

Climate	Mediterranean climate
City	Alexandria
Latitude	31.33° and 22° North
Longitude	26° and 35° East
Type of building	Educational building
Orientation	35° from the north
Building shape and massing	The building is a compact cube: It has the lowest surface of thermal exchange per conditioned volume
Area of one floor	1075 m^2
Number of floors	5 floors
HVAC system	Split unit

The indoor environment quality (interior air temperature)

The aim of the research is to measure energy consumption and CO_2 emissions while maintaining a stable hygrothermal (temperature and relative humidity) comfort inside the building (Fig. 6). According to relevant literature and ASHRAE standards (Rodriguez-Ubinas et al., 2014; ASHRAE Standard 55, 2010), the best interior temperature for human comfort should be between 23 and 25 °C, and relative humidity level should be between 40 and 55% (Rodriguez-Ubinas et al., 2014) Figure 6. So, the simulation is set to 25 °C.

4.2.1 Shading Devices

Window orientation and shading device design are two main components of passive design that regulate daylighting integration in buildings (Altan et al., 2016). Building exterior shading devices improved both visual and energy-saving measures (Sari & Chiou, 2018). The primary function of using shading devices is to reduce solar heat gains, the secondary functions are to control views into and out of a building, reduce solar glare, provide rain protection for opening windows, and serve as part of a maintenance strategy (BCA, 2010).

High windows, in the base case, allow hot air to escape, as well as a window recess of 70 cm that is a practical self-shading technique easily integrated without reducing natural light and obscuring views. The following Table 5 will show the comparison between the current situation of the as-built case with the 70 cm recess (Fig. 7) and in the case of adding to the windows shading louvers in the modified model (0.5, 1.0, and 1.5 m).

From the Table 5, it is noted that using 1.5 m shading louvers has a significant impact on decreasing greenhouse gas emission and the energy consumption (cooling electricity) by 5% at June and by 4% at September.

Table 4 External walls and roof of the as-built AAST base case (the author)

Element	Material layers	Thickness (mm)	U-value (W/m^2 K)
External walls	External plaster	20	0.829
	Bricks	200	
	Cavity layer	300	
	Bricks	200	
	Internal plaster	20	
Fenestrations	Generic PYR clear (double glazing)	60	2.716
Roof	Mosaico tiles	30	1.298
	Mortar	20	
	Sandstone	60	
	Cast concrete	70	
	Bitumen, felt/sheet	3	
	Cast concrete	50	
	Glass fiber slab	50	
	XPS extruded polystyrene-CO_2 blowing	50	
	Concrete, reinforced (with 1% steel)	250	
	Plaster	0.02	

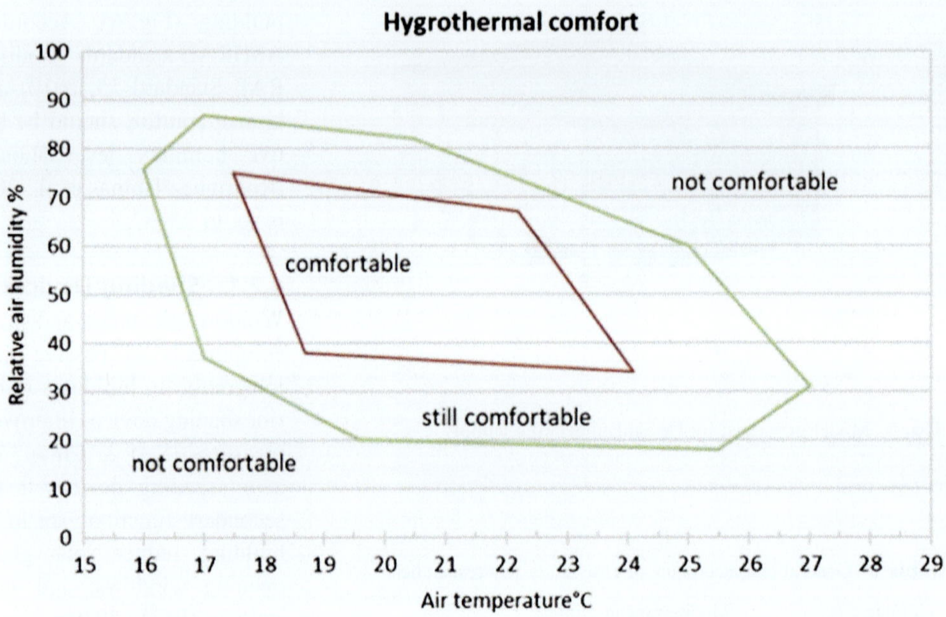

Fig. 6 Hygrothermal comfort in buildings (Frank et al., 1975)

Table 5 Effect of using shading louvers on greenhouse gas emission and energy consumption (the author)

The indoor environment quality	
Greenhouse gas emission	Using the shading louvers 0.5, 1.0, or 1.5 m will decrease CO_2 emissions from AC by the following; • 0.5 m louvers: by 611 kg and by 706 kg for June and September, respectively • 1.0 m louvers: by 1172 kg and by 1259 kg for June and September, respectively • 1.5 m louvers: by 1314 kg and by 1344 kg for June and September, respectively
Energy consumption	
Using the shading louvers 0.5, 1.0, or 1.5 m will decrease energy consumption by the following; • 0.5 m louvers: by 1009 kWh and by 1164 kWh for June and September, respectively • 1.0 m louvers: by 1934 kWh and by 2077 kWh for June and September, respectively • 1.5 m louvers: by 2168 kWh kg and by 2218 kWh for June and September, respectively	

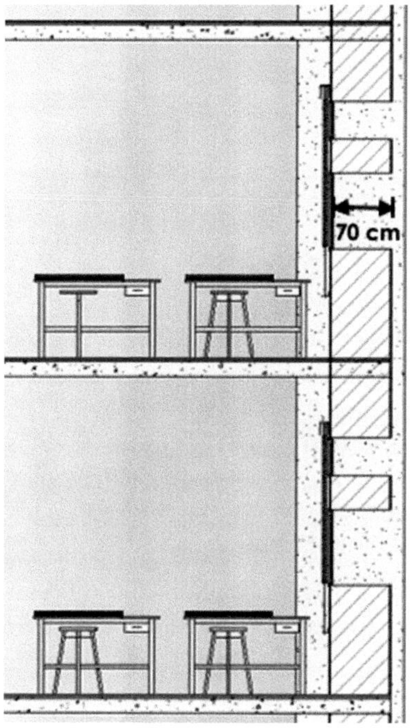

Fig. 7 Existing 70 cm recess of the window in the base case (the author)

4.2.2 Glazing

According to body of scientific research, glass properties have a significant impact on reducing cooling load (Bachrun et al., 2019; Motawa et al., 2021; Rodriguez-Ubinas et al., 2014). Thermal properties of glass materials vary depending on solar transmittance, solar absorption, solar reflectance, and visible transmittance VT. The thermal transmission characteristics of glass are determined by U-values for conduction and by solar heat gain coefficients (SHGC) or shading coefficients (SC) for radiation (Bachrun et al., 2019). Therefore, there should be a balance between U-value and SC value with visual transmission to allow daylighting, while keeping heat out. Table 6 shows U-values, visible light transmission, and SHGC for a variety of glass materials. Considering the building envelope as a barrier between the exterior environment and the interior, a low thermal transmittance (U-value) indicates high levels of insulation (Bachrun et al., 2019).

The selection of the type glass for windows depends on the façade orientation and the function of space, considering daylighting and energy saving, with a less amount of heat gains, at the same time. From the Table 6, it is noticed that the best solution for the case study is to use clear glass materials, which has the lowest U-value. Table 7 will show the comparison between the current situation of the case study (using double clear glazing) and with using triple-glazing from total solar transmission (SHGC) and U-value.

From the Table 8, increasing the number of glazing layers with argon layer between them (triple-glazing units) will affect on decreasing greenhouse gas emission and the energy consumption (cooling electricity) by 4% at June and by 3% at September.

4.2.3 Vegetation

The vegetation from green roofs and green facades are key solutions to mitigate building-related energy consumption and greenhouse gas emission (Ahmet & Erdem, 2018). Vegetation is green wall techniques which are called vertical greening/greenery systems, vertical garden, bio-walls, and so on (Manso & Castro-Gomes, 2015). They can be considered as a sustainable strategy that improves thermal comfort by reducing energy demand for heating and cooling (Perini et al., 2017), also they are increasing the value of the real estate and allowing in sound insulation (Hadba et al., 2017).

The paper will focus on the comparison between the current situation of the case study (without vegetation) and

Table 6 U-value, visible light transmission VLT, and SHGC values of typical glass material (Bachrun et al., 2019)

Single glazing, 8 mm	Clear	Color	Reflective	Low
U-value	4.94	5.18	5.18	5.54
VLT (%)	89	55	42–48	35–67
SC	0.95	0.51–0.57	0.42–0.53	0.40–0.69
SHGC	0.82	0.44–0.49	0.36–0.46	0.34–0.59

Table 7 Comparison between glazing materials (the author)

Characteristics	Double glazing	Triple LOE
Layers	Clear 6 mm/13 mm air	Clear 3 mm*3/13 mm argon*2
Total solar transmission (SHGC)	0.568	0.474
Direct solar transmission	0.474	0.358
Light transmission	0.745	0.661
U-value (ISO 10292/EN 673)	1.357 W/m^2 k	0.776 W/m^2 k
U-value	1.493 W/m^2 k	0.780 W/m^2 k

Table 8 Effect of changing glazing materials on the IEQ and the energy consumption (the author)

The indoor environment quality	
Greenhouse gas emission	Using triple-glazing units will decrease CO_2 emissions from AC by 948 kg at June and by 1015 kg at September
Energy consumption	
Using triple-glazing units will decrease energy consumption by 1565 kWh at June and by 1675 kWh at September	

in the case of using green walls in the first part, then in the case of using a green roof in the second part. Vegetation integrated in façades and green roofs are greenery systems that are passive solutions to mitigate building-related energy consumption and greenhouse gas emissions (Ahmet & Erdem, 2018).

Green walls

Green walls systems support selected plant species on a wall or non-horizontal surface of a building. They provide environmental, economic, social, and health benefits. Their environmental benefits include improved air quality and energy and natural resource conservation. Green walls are also capable of creating thermally comfortable indoor and outdoor environments (Addo-Bankas et al., 2021). Economically, they can minimize the cost of building operations by increasing energy efficiency and conserving water and recycling (Palermo & Turco, 2020). Based on their construction methods, there are two kinds of green walls: green facades and "living" walls. In green façades, plants are rooted in soil on the ground and climb on façade, while in "living" walls, pre-vegetated sheets are attached to a structure or frame on the wall.

A. ***Green facades***

A green facade is characterized by a low level of systemic technology, a limited number of constituent elements, and a limited degree of integration between plants and walls. They are lightweight, simple to install, and generally aim to encourage the natural growth of climbing plants that can reach a height of 25 m and have evergreen or deciduous foliage, but it takes years to completely cover the wall (Pérez et al., 2011). There are two types of green façades: direct and indirect.

A direct green façade consists of self-clinging plants that are rooted directly in the ground and attached directly to the wall. Indirect green façades provide a vertical support structure for climbing plants. Rooted directly in the ground or in planters, these plants grow with support structures such as continuous guides or modular trellises (Pérez et al., 2011; Manso & Castro-Gomes, 2015).

B. ***Living walls***

In this context, it is defined as vegetation that grows on a separate structural system, whether it be freestanding or attached to a wall (Palermo & Turco, 2020). Living walls allow the integration of a wide variety of plant species. Large surfaces can be quickly covered with uniform vertical growth, reaching higher areas and adapting to all types of buildings. Based on the above results from the literature review, the researchers applied green facades and living walls on the external walls. Table 9 illustrates the material layers of the four different kinds of green walls that have been used in the simulation process and their thermal properties.

Table 10 shows that using type.1 of felt system of living wall instead of a bare wall will have significant impact on decreasing greenhouse gas emission, and energy consumption by 6% at June and by the same rate at September. The results showed that the plant cover layer minimized heat gain on facades so that it decreased the cooling loads.

Green roofs

Energy-saving, thermal insulation, shading, and evapotranspiration features highlight the key role of green roofs in the overall thermal performance of buildings and in improving the cooling process and microclimatic conditions of indoor environments (Addo-Bankas et al., 2021). The roof has an exterior insulation layer, an effective way to minimize thermal bridges. Additionally, the thermal properties and thickness of the layers have an effect on a building's heat transfer. Table 11 illustrates the different layers and their thermal properties, which are deducted from the literature review of relevant references of the roof designed for the modified case study, from top to bottom.

From the Table 12, using type.3 of green roof (extensive green roof) instead of a bare roof has an impact of 3% at June and the same percentage at September in decreasing greenhouse gas emission and energy consumption.

4.2.4 Insulation Materials in Walls (Including PCM Materials)

Thermal building insulation is critical for energy retention, which is one of the most pressing and oldest issues in the

Table 9 Layers of green walls (the author)

Type.1: Felt system living wall		Type.2: Panel system living wall		Type.3: Direct Green Facade		Type.4: Indirect Green Facade	
Material layers	Thickness (mm)	Material layers	Thickness (mm)	Material layers	Thickness (mm)	Material layers	Thickness (mm)
Grass/straw materials-straw thatch	150	Grass/straw materials-straw thatch	150	Grass/straw materials-straw thatch	150	Grass/straw materials-straw thatch	150
Cultivated peat soil 366%	200	Cultivated peat soil 366%	200	External plaster	20	Aluminum	50
Felt 3/8 in	50	Aluminum	50	Brick	200	External plaster	20
PVC	10	Air gap	200	Cavity layer	300	Brick	200
Air gap	200	External plaster	20	Brick	200	Cavity layer	300
External plaster	20	Brick	200	Internal plaster	20	Brick	200
Brick	200	Cavity layer	300			Internal plaster	20
Cavity layer	300	Brick	200				
Brick	200	Internal plaster	20				
Internal plaster	20						
U-value: 0.247 W/m^2 k		0.251 W/m^2 k		0.299 W/m^2 k		0.279 W/m^2 k	

Table 10 Effects of using green roof on the IEQ and the energy consumption (the author)

The indoor environment quality	
Greenhouse gas emission	Using vegetation on external walls will decrease CO_2 emissions from AC by the following; • Type.1: by 1369 kg at June and by 1879 kg at September • Type.2: by 1354 kg at June and by 1840 kg at September • Type.3: by 169 kg at June and by 451 kg at September • Type.4: by 780 kg at June and by 1251 kg at September
Energy consumption	
Using vegetation on external walls will decrease energy consumption (cooling electricity) by the following; • Type.1: by 2259 kg at June and by 3100 kg at September • Type.2: by 2234 at June and by 3036 kg at September • Type.3: by 279 kg at June and by 744 kg at September • Type.4: by 1288 kg at June and by 2065 kg at September	

architectural field (Crosbie, 1998). Because of the reduced heat transfer from outside to inside, well-insulated buildings have a lower cooling load in hot climates. This improves indoor environmental quality, increasing people's comfort and productivity by reducing the need for additional cooling (Visser & Yeretzian, 2013).

According to the literature review, increasing the thickness of the wall increases its thermal resistance thereby reducing the heat transfer rate and increasing the time it takes for the temperature to reach the surface of the internal wall (time lag) (Avila & Ismail, 2017). Moreover, considerable energy can be saved when a building envelope is properly insulated. In the AAST building, cavity walls are adopted as an insulation system in the existing building.

Simulations and literature showed that advanced wall systems that consisted of plaster, brickworks, and extruded polystyrene (XPS) foam as insulation material with, air gap, and brickworks can enhance indoor thermal comfort (Avila & Ismail, 2017). In fact, bricks in Egypt are by far the most common typical material used in the building construction due to its availability in the local market and its low cost. Air trapped between the walls acts as a low-cost insulating material and can reduce significantly the transfer of heat.

This study found that phase change materials (PCM) enhance the energy efficiency of buildings when implemented as thermal insulation because they provide an opportunity for improved indoor thermal comfort and reduce energy consumption as a result of load reductions or shifting (Avila & Ismail, 2017). PCM can be implemented in walls, ceiling or floor, and in a separate heat storage system. The researchers focused on the implementations of PCM technique in walls in addition to XPS foam. If PCM are added as an insulation layer, the results of the comparisons between the existing wall and the modified PCM-enhanced insulation walls show positive results in favor of the use of PCM in terms of thermal performance. After comparing between

Table 11 Layers of green roof (the author)

Type.1: intensive		Type.2: intensive		Type.3: extensive	
Material layers	Thickness (mm)	Material layers	Thickness (mm)	Material layers	Thickness (mm)
Green roof material (vegetation)	100	Green roof material (vegetation)	304.80	Cultivated clay soil 12.5% D. W. moisture	50.50
Polypropylene	2.10	Glass fiber quilt	251.20	Miscellaneous materials-polyurethane	5.10
Polyethylene/Polythene, high density	5	Miscellaneous materials-aggregate (sand/gravel/stone)	200	2.0 in depth wood framing, R-11.2 insulation	50
XPS extruded polystyrene-CO_2 blowing	15	Hard rubber	13	Hard rubber	13
Polystyrene	50	Miscellaneous materials-polyurethane, expanded	10	8 in. concrete at R-0.0625/in	200
Floor/roof screed	60	6 in. depth cavity R-19 insulation mineral fiber	152.40		
Concrete roofing slab, aerated	300	0.75 in. plywood/wood panels	19.10		
Air gap	200	Aerated concrete slab	200		
Gypsum plastering	12				
0.283 W/m^2 K		0.073 W/m^2 K		0.478 W/m^2 K	

Table 12 The effects of using green roof on the IEQ and the energy consumption (the author)

The indoor environment quality	
Greenhouse gas emission	Using green roof instead of bare roof will decrease CO2 emissions from AC by the following; • Type.1: by 731 kg at June and by 825 kg at September • Type.2: by 508 kg at June and by 639 kg at September • Type.3: by 820 kg at June and by 872 kg at September
Energy consumption	
Using green roof instead of bare roof will decrease the energy consumption: • Type.1: by 1,207 kWh at June and by 1,362 kWh at September • Type.2: by 838 kWh at June and by 1,055 kWh at September • Type.3: by 1,353 kWh at June and by 1,438 kWh at September	

Table 13 Constituents of the wall simulated (the author)

Material layers	Thickness (mm)	U-value (W/m^2 K)
BioPCM M182/Q25	74.20	0.362
XPS extruded polystyrene-CO_2 blowing	40	
Plaster	20	
Brick	200	
Air layer	300	
Brick	200	
Plaster	20	

different PCM proposed in DesignBuilder, (PCM material-BioPCM M182/Q25) was chosen, as shown in Table 13.

From the Table 14, we note when we used BioPCM M182/Q25 as a type of PCM material, energy consumption, and greenhouse gas emission were decreased by 1% at June,

Table 14 Effects of using PCM materials on exterior walls on the IEQ and the energy consumption (the author)

The indoor environment quality	
Greenhouse gas emission	Using PCM material for exterior walls will decrease CO_2 emissions from AC by 176 kg at June and by 265 kg at September
Energy consumption	
Using PCM material for exterior walls will decrease the energy consumption by 291 kWh at June and by 437 kWh at September	

Table 15 Parameters used in the simulation process (the author)

Walls	Felt living wall (thickness 130 mm)
Roof	Green roof (type.3)
Glazing	Triple LOE clr6mm.13 mm argon
Shading devices	Louvers 1.5 m over the 70 cm recess of windows

and by the same percentage at September. So better results are realized with green walls, better results.

4.2.5 Applying the Parameters of the Green Envelope

After obtaining the above results, the researchers applied all the parameters combined. The parameters selected based on the best results in terms of reduction of CO_2 emissions and energy consumption as shown in Table 15.

As showed in Table 16, after introducing the four parameters, CO_2 emissions and energy consumption are reduced by 20% at June and by 19% at September, to maintain the comfort level of indoor air temperature of 25 °C by the heating, ventilation, and air conditioning (HVAC) system and fans (Figs. 8 and 9).

5 Discussion, Recommendations and Conclusion

As stated previously, during the building life cycle, air conditioning is the biggest factor in electricity demand, so the researchers focused on air conditioning (split units) and fans for the simulations of alternatives and did not put into account artificial lighting and other devices in the calculations. Nevertheless, future research should include all the energy demands of the buildings for more accurate results. The aim of the research was to test the parameters of green envelopes as investigated in relevant literature. According to the results of this research, the following conclusions can be drawn regarding energy-efficient retrofitting of existing educational buildings and for the design of new educational facilities.

It is recommended to introduce external shading devices that provide shade to the external windows and reduce the transition of direct solar heat gains so that cooling loads can be reduced. Shading louvers of 1.5 m over external windows with an existing 70 cm recess realize a 5% reduction of greenhouse gas emission, and energy consumption at June, and by 4% at September.

Based on the study's findings, it is recommended that clear triple-pane low-E windows with a U-value of 0.78 $W/m^2/k$ and 1.22 W/m/k be used: These windows offer good light transmission and a low U-value of 0.78 W/m^2 k and can reduce energy consumption by 4% in June and 3% in September.

The results of the simulation indicated that using felt system in living walls (grass/straw thatch of 150 mm + cultivated peat soil 366% of 200 mm + felt 3/8in of 50 mm) instead of a bare wall will have an impact of 8% at June and by 9% at September, in decreasing greenhouse gas emission and energy consumption.

It is recommended to use an extensive green roof composed of cultivated clay soil 12.5% D.W. moisture + miscellaneous materials-polyurethane + 2 in. depth wood framing, R-11.2 insulation + hard rubber + 8 in. concrete instead of a bare roof as it has an impact of 3% at June, and by the same percentage at September in decreasing greenhouse gas emission energy consumption. Regular maintenance inspections should be a part of the standard green roof installation to avoid leaks.

Table 16 Effects of parameters of green architecture on the IEQ and the energy consumption (the author)

The indoor environment quality	
Greenhouse gas emission	The greenhouse gas emission from AC decreased by 4820 kg at June and by 5686 kg at September
Energy consumption	
The energy consumption decreased by 7954 kWh at June and by 9383 kWh at September	

Fig. 8 Impact of using green building parameters on reducing CO_2 emissions (the author)

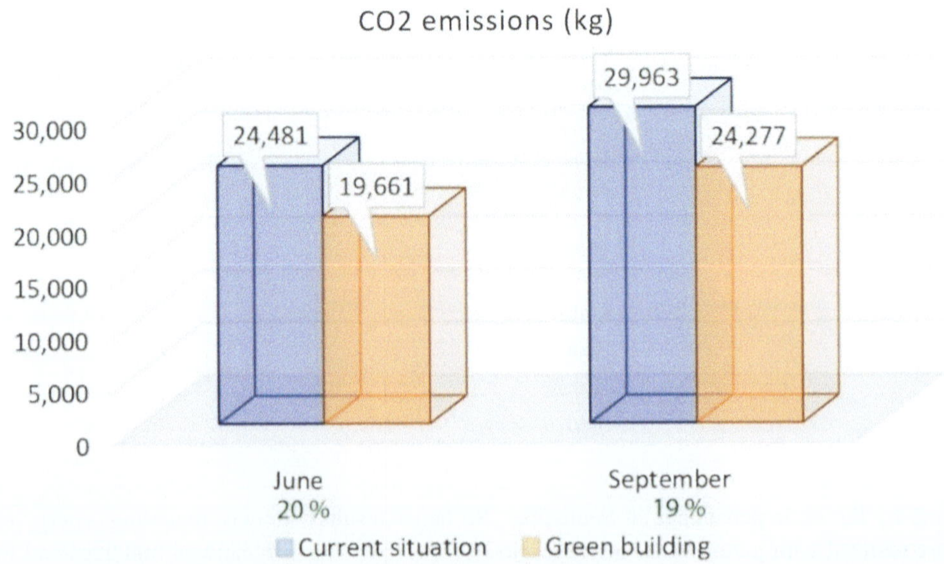

Fig. 9 Impact of using green building parameters on reducing energy consumption (the author)

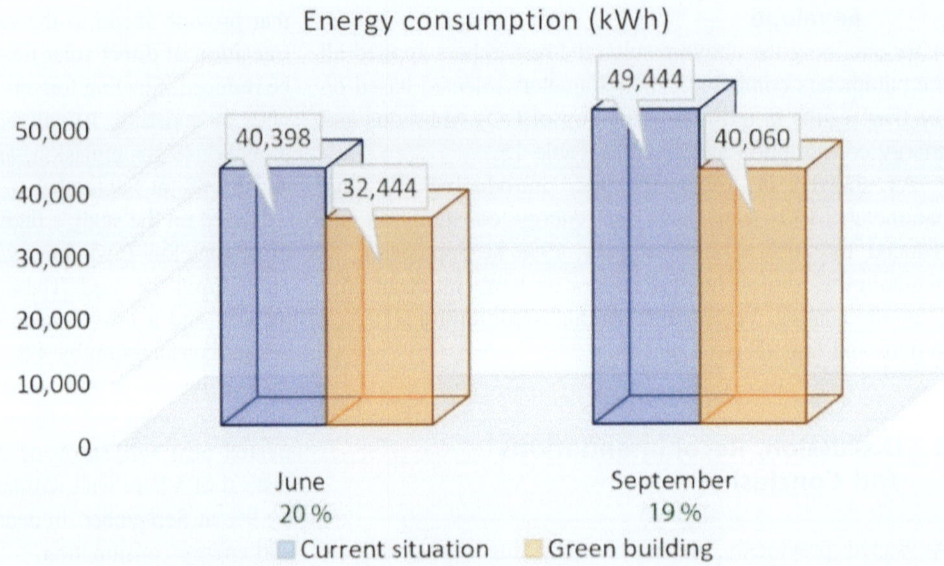

Concerning insulation materials, the researchers tested the as-built base case of AAST which external wall thickness is composed of an air gap of 300 mm between 2 brick walls of 200 mm in comparison with the same layers but with an added layer of thermal insulation. By using BioPCM M182/Q25 material as a type of PCM material, energy consumption and greenhouse gas emission were decreased by 1% at June and by the same percentage at September. So, green walls are more efficient in reduction of energy consumption and CO_2 emissions.

So, by maintaining the comfort level of indoor air temperature by using the heating, ventilation, and air conditioning (HVAC) system and setting the temperature at 25 °C, the results showed a decrease of greenhouse gas emission and energy consumption of 20% at June and of 19% at September by combining the best results of these alternatives. Then, solar heat gains either from walls, windows, or from the building roof were reduced after applying the green parameters.

A combination of different passive design strategies and green building envelope parameters can help reduce energy consumption, nevertheless it is recommended to perform the economic evaluation and the calculation of payback periods to reach best decisions and the amount of energy saved.

The researchers recommend coupling both prescriptive-based and performance-based guidelines and assessment methods using simulations tools into an updated local Egyptian code to enhance energy performance and thermal performance in educational buildings. This updated code should be incorporated to the existing code of 2006 "Building Energy Code for the Residential Sector" (HBRC, 2008).

The limitations of this research are that DesignBuilder can be used to evaluate the indoor environmental quality consisted of interior air temperature for spaces, but for greenhouse gas emission and energy consumption from cooling electricity, they are measured for the whole building and not for a specified façade, because the program shows only the interior air temperature and solar heat gains for different façade elements at the selection of a particular space. So, for more accurate results on each façade of the building, there is a need to use another simulation program to evaluate the parameters of green buildings based on its orientation and the function of its spaces.

References

Addo-Bankas, O., Zhao, Y., Vymazal, J., Yuan, Y., Fu, J., & Wei, T. (2021). *Green walls: A form of constructed wetland in Green Buildings. Journal of Ecological Engineering, 169*. https://doi.org/10.1016/j.ecoleng.2021.106321

Ahmet, B., & Erdem, C. (2018). Green roofs and facades: A comprehensive review. *Renewable and Sustainable Energy Reviews, 82*, 915–939. Available: https://doi.org/10.1016/j.rser.2017.09.106

Al-Tamimi, N. (2022). Passive design strategies for energy efficient buildings in the Arabian Desert. *Journal of Frontiers in Built Environment, 7*. https://doi.org/10.3389/fbuil.2021.805603

Altan, H., Hajibandeh, M., Aoul, K. A., & Deep, A. (2016). Passive design. In *ZEMCH: Toward the delivery of zero energy mass custom homes* (pp. 209–236). Springer tracts in civil engineering. Springer International Publishing.

Ariff, A., Ahmad, S. S., & Hussin, M. A. (2019). Green envelope as an architectural strategy for energy efficiency in a library building. In *Paper presented at MATEC Web Conferences* (Vol. 266(1)). EDP Sciences, January 2019. https://doi.org/10.1051/matecconf/201926601004

ASHRAE Standard 55. (2010). *Thermal environmental conditions for human occupancy*. American Society of Heating, Refrigerating and Air Conditioning Engineers, Inc.

Attmann, O. (2010). *Green architecture: Advanced technologies and materials*. McGraw-Hill.

Avila, J., & Ismail, K. A. R. (2017). Composite thermal walls for low energy Brazilian homes. In *Paper presented at 13th International Conference on Heat Transfer, Fluid Mechanics and Thermodynamics*. Passive thermal comfort in buildings, Slovenia, July 2017.

Bachrun, A. S., Ming, T. Z., & Cinthya, A. (2019). Building envelope component to control thermal indoor environment in Sustainable Building: A review. *SINERGI, 23*(2), 79. https://doi.org/10.22441/sinergi.2019.2.001

BCA. (2010). Building planning and massing.

Crosbie, M. J. (1998). *The passive solar design and construction handbook*. Steven Winter Associates Inc.

Dusek, V. (2008). *Philosophy of technology: An introduction*. Wiley.

Frank, W. (1975). *Raumklima und Thermische Behaglichkeit. Ber. Bauforsch*. Hygrothermal comfort in buildings 104.36.

Hadba, L., Silva, L., & Mendonça, P. (2017). Green walls: An efficient solution for hygrothermal, noise and air pollution control in the buildings. In *Paper presented at Living and Sustainability: An Environmental Critique of Design and Building Practices, Locally and Globally*. London, England, February 2019. https://www.researchgate.net/publication/313852100

HBRC. (2008). Egyptian code to improve energy efficiency in residential buildings. Code No. 306-2005.

Hemsath, T., & Bandhosseini, K. A. (2018). *Energy modeling in architectural design*. Routledge.

Kibert, C. J. (2013). *Sustainable construction: Green building design and delivery*. Wiley.

Designbuilder Software Ltd—Simulation. (2022). https://designbuilder.co.uk/simulation. Accessed September 25, 2022.

Manso, M., & Castro-Gomes, J. (2015). Green wall systems: A review of their characteristics. *Journal of Renewable and Sustainable Energy Reviews, 41*, 863–871. https://doi.org/10.1016/j.rser.2014.07.203

Ministry of the Environment. (2005) (rep.). A study to guide strategies for Japanese cooperation for strengthening environmental management in the early 21st century.

Motawa, I., Elsheikh, A., & Diab, E. (2021). Energy performance analysis of building envelopes. *Journal of Engineering, Project, and Production Management, 11*(3), 196–206.

Palermo, S.A., & Turco, M. (2020). Green wall systems: Where do we stand. *IOP Conference Series: Earth and Environmental Science, 410*(1), 012013. https://doi.org/10.1088/1755-1315/410/1/012013

Pérez, G., Rincón, L., Vila, A., González, J. M., & Cabeza, L. F. (2011). Green vertical systems for buildings as passive systems for Energy Savings. *Journal of Applied Energy, 88*(12), 4854–4859. https://doi.org/10.1016/j.apenergy.2011.06.032

Perini, K., Bazzocchi, F., Croci, L., Magliocco, A., & Cattaneo, E. (2017). The use of vertical greening systems to reduce the energy demand for air conditioning. Field monitoring in Mediterranean climate. *Journal of Energy and Buildings, 143*, 35–42. https://doi.org/10.1016/j.enbuild.2017.03.036

Rodriguez-Ubinas, E., Montero, C., Porteros, M., Vega, S., Navarro, I., Castillo-Cagigal, M., Matallanas, E., & Gutiérrez, A. (2014). Passive design strategies and performance of NET Energy Plus Houses. *Journal of Energy and Buildings, 83*, 10–22. https://doi.org/10.1016/j.enbuild.2014.03.074

Sari, D. P., & Chiou, Y.-A. (2018). Energy saving potential from shading design for residential house in rural area. In *Paper presented at MATEC Web of Conferences* (p. 164). Bali, Indonesia. https://doi.org/10.1051/matecconf/201816401007

Shams, S. (2017). Green building. In *Sustainable utilization of natural resources* (1st ed., pp. 1–600). CRC Press.

The Government of the Province of Jakarta Capital Special Territory. (n.d.). (rep.). *Jakarta green building user guide building envelope* (Vol. 1).

United Nations Environment Programme. (2021). 2021 Global status report for buildings and construction: Towards a zero-emissions, efficient and resilient buildings and construction sector. Nairobi.

Visser, F., & Yeretzian, A. (2013). Energy efficient building guideline for MENA region. Energy efficiency in the construction sector in the Mediterranean (MED-ENEC).

Weatherbase. (n.d.). Travel weather averages (Weatherbase). Weatherbase. Retrieved September 21, 2022, from https://www.weatherbase.com/

Nature-Based Solutions Co-creation and Simulation of Land Use and Socio-economic Impacts in Genova, Italy

Rita Mendonça, Peter Roebeling, Teresa Fidélis, and Miguel Saraiva

Abstract

Urbanization alters urban patterns, which might decrease quality of life as well as its resilience to climate change impacts, namely in European contexts. Nature-based solutions (NBS) are acknowledged as effective tools to support urban sustainability, with environmental, economic and social benefits and co-benefits, having the ability to steer urbanization processes and increase urban resilience. Recent studies have mostly looked at the effects of NBS on flooding, urban heat island as well as air quality, while only few studies assess its socio-economic impacts, such as urban sprawl, gentrification and real-estate value, namely in combination with stakeholder engagement approaches. Hence, the objective of this study is to assess the land use and socio-economic impacts of NBS in an urban context through NBS co-creation and simulation. To this end, the Sustainable Urbanizing Landscape Development (SULD) model is used to assess the impacts of co-created NBS scenarios (i.e. urban park, green roofs and green and blue space) for the case study of Genova in Italy. Results show that NBS have the ability to promote urban compactness, population density and real-estate values while increasing the share of green urban areas and access to attractive amenities, depending mostly on their degree of attractiveness, size, location and the original amount of greenness. Nevertheless, the green and blue space NBS led to some green gentrification. Hence, this NBS, considered the most attractive NBS, had the greatest land use (urban contraction, ratio of open space to build form and per cent built up area) and socio-economic impacts (gentrification, real-estate value appreciation, household density and living space); followed by the urban park and finally green roofs, considered the least attractive NBS, that posed small land use and socio-economic impacts. Overall, NBS do have an impact on the urban form and its dynamics, however, they might cause some green gentrification.

Keywords

Co-design workshop • Nature-based solutions • Simulation modelling • Socio-economic impacts • SULD model • Urban planning

R. Mendonça (✉) · P. Roebeling
CESAM and Department of Environment and Planning, Aveiro University (DAO-UA), Aveiro, Portugal
e-mail: ritam11@ua.com

P. Roebeling
e-mail: peter.roebeling@ua.pt

P. Roebeling
Wageningen Economic Research, Wageningen University and Research (WUR), Wageningen, Netherlands

T. Fidélis
GOVCOPP and Department of Environment and Planning, Aveiro University (DAO-UA), Aveiro, Portugal
e-mail: teresafidelis@ua.pt

M. Saraiva
CEGOT—Centre of Studies in Geography and Spatial Planning, of the Faculty of Arts and Humanities of the University of Porto, Porto, Portugal

1 Introduction

Following the current unsustainable urbanization patterns and climate change impacts, namely in European urban contexts, quality of life and urban resilience is decreasing (Dell'Anna et al., 2022). Besides the biophysical impacts, these urbanization trends lead to soil destruction and urban sprawl—i.e. scattered and unplanned low-density urban development—as well as gentrification, social inequality and low access to quality green spaces and environmental amenities in general (Mendonça et al., 2020). This feeds the need and creates an opportunity for alternative solutions that promote more sustainable urban planning and decision-making processes to be developed.

A. Battisti et al. (eds.), *Greening Our Cities: Sustainable Urbanism for a Greener Future*, Advances in Science, Technology & Innovation, https://doi.org/10.1007/978-3-031-49495-6_9

Nature-based solutions (NBS), such as green roofs, green walls, green parks, street trees and river daylighting, are "solutions that are inspired and supported by nature, which are cost-effective, simultaneously provide environmental, social and economic benefits and help build resilience" (EC, 2021, p. 17), building on other existent green-related concepts by highlighting the social sphere. Due to their multiple and transversal benefits and co-benefits, NBS are considered a better alternative to more traditional solutions to attain urban sustainability. NBS biophysical benefits include air and water pollution control, decreasing air temperature, flood reduction and biodiversity enhancement, among others (Cortinovis et al., 2022; Faivre et al., 2017).

Regarding socio-economic benefits, NBS are acknowledged to promote green connectivity and accessibility as well as social cohesion and urban regeneration. Moreover, by being considered attractive amenities, they have the potential to control land expansion and urban sprawl as households are willing to pay to live near them. Additionally, by creating awareness, they contribute to a better citizen engagement and informed decision-making, while promoting mental and physical health (Cortinovis et al., 2022; Faivre et al., 2017; Mendonça et al., 2021a). By upholding aesthetic values, NBS also contribute to increase land and property values and to create green jobs (EC, 2021). Nevertheless, they can also entail some disservices, such as littering, high costs and green gentrification (Black & Richards, 2020; Bockarjova et al., 2020; Voskamp et al., 2021). Green gentrification is being documented in Europe, accruing from an uneven distribution of access to environmental goods, which might be heightened by increasing property prices near new attractive interventions (Bockarjova et al., 2020; Haase et al., 2017). To better understand and minimize the gentrification process, it is important to create personalized and adapted solutions, thus increasing NBS implementation success. To achieve that, policy instruments, namely information, such as stakeholder engagement initiatives, workshops, and modelling approaches are considered crucial (De Los Rios-White et al., 2020; Hérivaux & Coent, 2021; Mendonça et al., 2021a).

Nevertheless, despite the NBS's potential to steer urban form and influence housing location, gentrification and housing prices by acting as amenities, most studies are theoretical or focus only on its biophysical impacts (Black & Richards, 2020; Cortinovis et al., 2022; Faivre et al., 2017). Accordingly, there are few studies that use simulation methods, creating personalized and case-specific solutions, together with co-creation methods to assess NBS socio-economic impacts, even though both contribute to build a strong evidence base while promoting acceptance and engagement (De Los Rios-White et al., 2020; Mendonça et al., 2021b) and even fewer that assess eventual negative impacts such as green gentrification processes (Bockarjova et al., 2020).

In line with Mendonça et al., (2021b) findings, there is a lack of spatially explicit, economic and simulation models and the model SULD (Sustainable Urbanizing Landscape Development) has already been used to assess the socio-economic, land use and urban form impacts of different green and blue spaces, being able to simulate NBS scenarios (Mendonça et al., 2020; Roebeling et al., 2007; Saraiva et al., 2017). Moreover, the combination of co-design and modelling approaches, from an information policy instrument point of view, aims to provide ex-ante information while incorporating the needs of the case study to foster a more informed decision-making process regarding NBS implementation.

Hence, the objective of this study is to address this literature gap by simulating, from a spatially explicit perspective, the socio-economic and land use impacts of different NBS scenarios using the SULD model. These scenarios were the output of a co-designed workshop setting with the Municipality of Genova, Italy, facilitated through the UNaLab project. The next section (Sect. 2) includes the methodology, that encompasses the case study and NBS co-design workshop (Sect. 2.1) as well as the SULD simulation model information (Sect. 2.2). These are followed by the results (Sect. 3), with the co-design workshop results (Sect. 3.1), the baseline results (Sect. 3.2) and the simulation results (Sect. 3.3) and the last section represents the discussion and conclusions (Sect. 4).

2 Methodology

This section describes the case study and NBS co-design workshop held with the Municipality of Genova, Italy (Sect. 2.1) as well as the description of the SULD simulation model (Sect. 2.2).

2.1 Case Study and NBS Co-design Workshop

Genova is a port city, known for its commerce trades, located in the Liguria Region in northern Italy. It is also one of the biggest and most populated cities in the country. In addition, it is a region known for its high share of impervious surfaces, hence with high NBS implementation potential for different benefits (Di Pirro et al., 2022). The city is one of the UNaLab Project front-runner cities, aiming to serve as example to other cities regarding the implementation and upscale of NBS to improve urban resilience and increase urban greenness, facilitating the co-design workshop (EC, 2021).

To create NBS scenarios from a co-creation and stakeholders' engagement perspective, a co-design workshop was held (25 March 2020) with the Municipality of Genova, via Microsoft Teams due to the pandemic constraints. To adapt the workshop to an online version, the tool UMap was used, an online software that allows georeferenced mapping in real time as well as the download of the designed maps. The workshop followed a sequential discussion, including all the participants with two design phases followed by group discussions. The participants (9) from the Municipality of Genova were first asked to identify and map the urban problems, using the area with the majority of the problems as the case study area to be considered. Following this, with the aid of an NBS benefits list, the most adequate NBS with the ability to tackle each urban problem were identified and mapped. After a group discussion, the final NBS scenarios were created.

2.2 SULD Simulation Model

After the selection of the NBS scenarios through a co-design workshop, the scenarios were simulated using the model SULD (Sustainable Urbanizing Landscape Development), (see Roebeling et al., 2007) to understand its socio-economic and land use impacts. This is a general equilibrium, spatially explicit, economic and hedonic pricing model that simulates urban pattern dynamics, including land use changes, densities and housing prices for different household types, as they aim to enhance living conditions through location according to proximity to environmental amenities (as NBS) and urban centres, subject to income and space constraints. Conversely, developers aim to maximize their profit, according to households' willingness to pay for housing and developing costs. In equilibrium, development takes place. In fact, this model has already been used to simulate green spaces and social dynamics (Saraiva et al., 2017) and policy instruments (Mendonça et al., 2020). Moreover, Mendonça et al., (2021b) findings report that there is a lack of spatially explicit models that have the ability to simulate socio-economic impacts associated with NBS implementation as well as considering policy instruments in order to support the decision-making process.

The model has been developed and adapted to the city of Genova, with socio-economic data collected from the municipality and local statistics (as population, household composition, income and transport costs (Istat Statistics, 2018) while land use data was retrieved from Urban Atlas (Urban Atlas, 2018). The model has been validated and calibrated following Saraiva et al., (2017), including similarity of socio-economic data as well as location and quantity disagreement indicators. The amount of residential cells is overestimated by 1% (quantity disagreement) while the location of the total cells accuracy is 96% and of residential cells is 95% (location agreement). Three different household types with different utility levels were considered: type 1 or low-income (Res1, 34,903 households, average 9599 €/year), type 2 or middle-income (Res2, 76,603 households, average 24,432 €/year) and type 3 or high-income (Res3, 7 757 households, average 85,140 €/year). Transport costs were considered 250 €/Km, construction costs 1500 €/Km^2 and opportunity cost of land of 1000 €/ha/year (data provided by the Municipality of Genova). The model considers a decay function based on the housing distance to the nearest environmental amenity and urban centres. The environmental amenities are divided into type 1 for highly attractive, city scale amenities—Scenario 3/green and blue space; type 2 for medium attractive, neighbourhood scale amenities—Scenario 1/urban park and type 3 for low attractive local scale amenities—Scenario 2/green roofs. Concerning the case study area, the city centre and peripheral area were considered, adopting a 185 × 185 grid of 34 m by 34 m (total area of 1156 m^2) using GIS software (see Fig. 2).

3 Results

This section is divided into co-design workshop results (Sect. 3.1), baseline results (Sect. 3.2) and scenario simulation results (Sect. 3.3) for the different NBS scenarios implementation: urban park (Scenario 1), green roofs (Scenario 2) and green and blue space (Scenario 3).

3.1 Co-Design Workshop Results

To create the NBS scenarios to be simulated (using the SULD model), the workshop included 9 participants from different departments, such as urban space management, green spaces and mobility. In the first instance, the workshop started with the definition of the case study area—i.e. where the most problematic and worrisome problems are located (see Fig. 1). These problems included socio-economic dynamics as urban expansion and housing prices, as well as biophysical, such as flooding, air quality and biodiversity loss that were identified and spatially mapped (see Fig. 1). In the second phase of the workshop, several NBS scenarios were proposed and mapped considering their benefits and ability to target the previously identified urban problems (with the aid of a NBS list and benefits provided by the organizers) (see Fig. 1). After discussing all the proposed NBS scenarios, the final NBS scenarios were selected, including an urban park (Scenario 1), green roofs (Scenario 2) and a green and blue space (Scenario 3)—See Fig. 2 for their location.

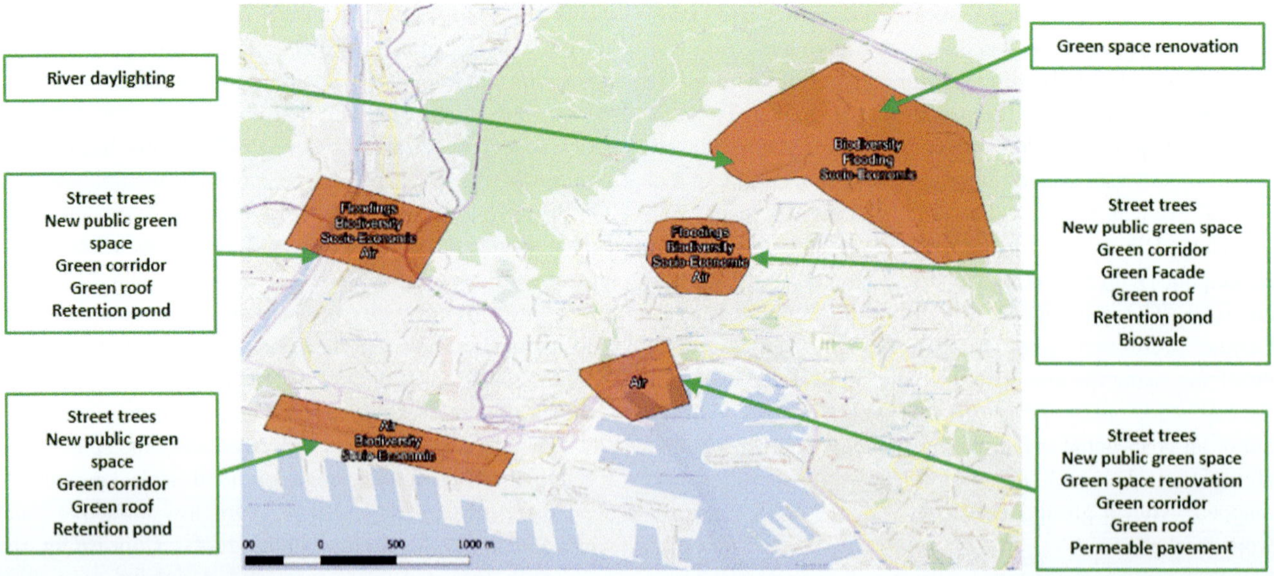

Fig. 1 Co-design workshop map case study area including the problems identified (in orange) and the NBS possibilities (green boxes). *Source* author

3.2 Baseline Results

Baseline results (Fig. 2) show that the majority of land use within the case study area is urban (936 ha), followed by water (846 ha) and industry/commerce (751 ha), with only 161 ha of green urban areas. Considering the urban residential land use, middle-income households occupy 633 ha, low-income households 181 ha and high-income households 122 ha. Regarding the housing quantity (square metres of built area), Res2 are also responsible for the majority of the occupied area, 305 * 103 m^2, followed by Res3, that have 159 * 103 m^2 and Res1, with a total of 42 * 103 m^2. As for the living space, Res3 largely surpass Res2 and Res1 (with 150, 55 and 26 m^2/hh, respectively), with an average value of 53 m^2/hh for the city. Real estate values present higher values near the most attractive environmental amenities (up to 171 €/m^2/yr), where higher income households live in bigger places and household density is lower. Conversely, low-income households live near roads and urban centres, in smaller more dense settlements with less access to environmental amenities. The average housing price is 124 €/m^2/yr, ranging from 157 €/m^2/yr for Res3, 124 €/m^2/yr for Res2 to €/m^2/yr 105 for Res1.

Considering the objective of the present study, with the NBS scenario implementation simulation, it is expected that the urban area experiences a contraction, with an increase in non-built (agriculture and urban green spaces) land use types. Consequently, household density and housing prices are expected to increase while housing quantity and living spaces are expected to decrease as households are willing to pay more and decrease their living conditions to live near these attractive places which, in turn, can create some household type imbalances.

3.3 Scenario Simulation Results

The implementation of an urban park (Scenario 1; + 1.3% in green urban areas) resulted in a contraction of the urban area (− 1%) and increase in agricultural area (+ 4.2%) (see Table 1 and Fig. 2). Res3 (− 3.9%) and Res2 (− 0.7%) were responsible for this contraction in residential area as they already lived near the implementation location. This new medium attractive amenity (type 2) led to these households being willing to decrease housing quantity, hence paying more to live in smaller and more dense spaces (Fig. 2). Hence, housing quantity decreased on average − 2.3% (− 5.4% for Res3 and − 1.1% for Res2) and the living space followed the same trend (average − 0.5% and − 1.6%, − 0.3% for Res3 and Res2, respectively). Regarding the housing prices, with the new amenity, an average city level increase of + 0.3% is visible (around + 7 €/m^2/yr near the implementation area) (Table 1). Low-income households did not suffer any visible alterations in this scenario, with no added green gentrification happening as they do not have the budget to compete with higher income households in the high-priced areas. The urban park implementation (amenity type 2), led to an overall decrease of -6.3% of the distance to amenity type 2, without any changes in the other amenity types distance, as expected. With the urban area contraction, the ratio of open space to build form increases + 0.8% and the per cent built up area decreases − 0.4%.

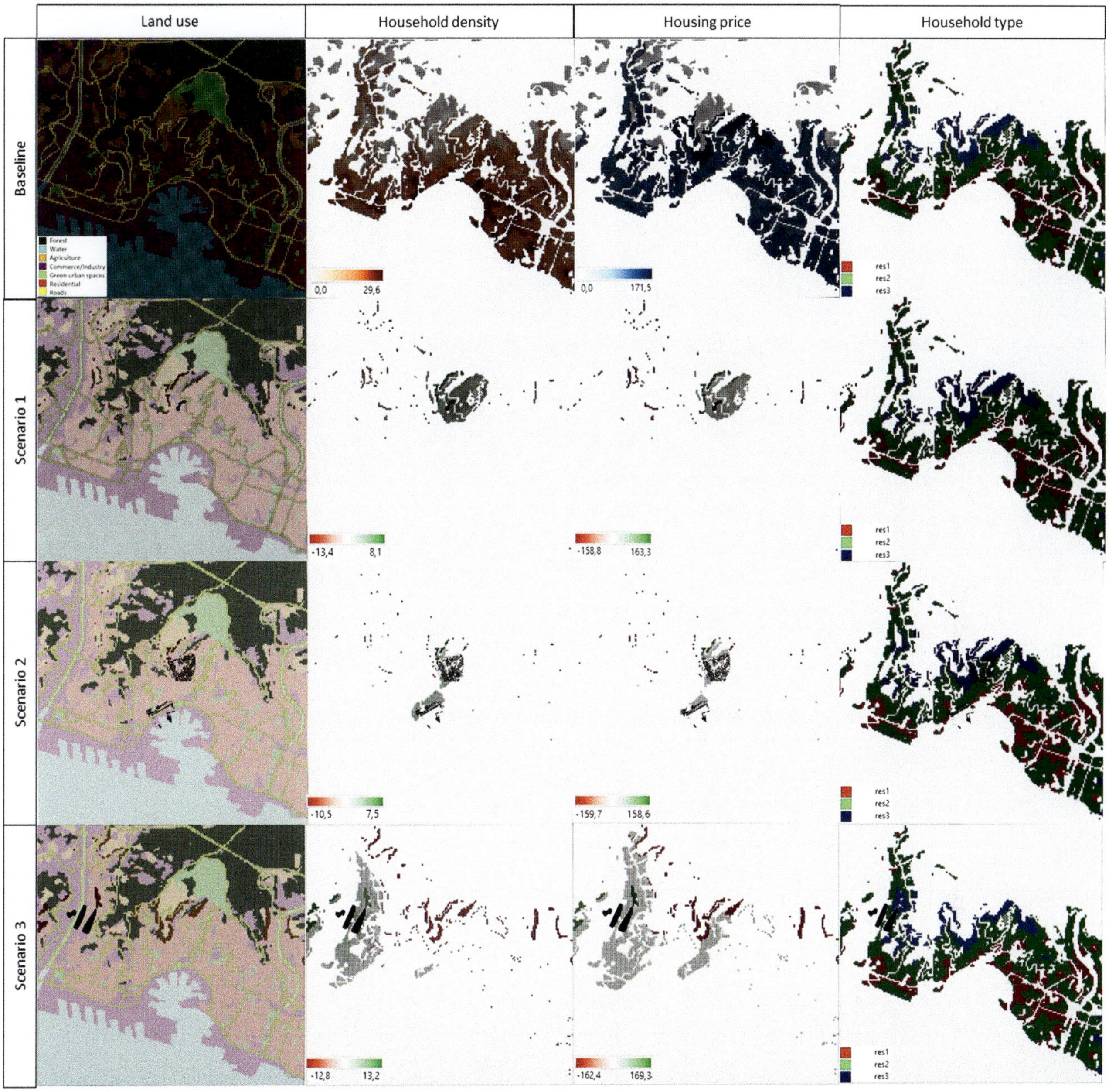

Fig. 2 Baseline and NBS scenario (in black) simulation results (Scenario 1: urban green park; Scenario 2: green roofs; Scenario 3: green and blue space) with land use, household density and housing price difference maps and household type maps. *Source* author

The implementation of green roofs (Scenario 2; resulted in no change in green urban areas as they are not accessible to the citizens apart from house-owners) led to a small contraction of the urban area (− 0.4%) and increase in agricultural area (+ 2%) (see Table 1 and Fig. 2). All households suffer a small decrease in residential area (− 1.1%, − 0.3% and − 0.2% for Res3, Res2 and Res1, respectively), as the large implementation area impacted locations where all households type lived. The new amenity, even if low attractive (type 3), was enough so that some households were willing to pay slightly more to live in buildings with green roofs, decreasing its housing quantity and living spaces while increasing housing prices and household density near the implementation area (Fig. 2). Following this, housing quantity decreased on average − 0.8% (− 1.5%, − 0.4% and − 0.2% for Res3, Res2 and Res1). Living space also decreased, on average − 0.2% (− 0.4%, − 0.1% and − 0.1% for Res3, Res2 and Res1). Conversely, housing prices increased at city level on average + 0.1%, (around + 3 €/m^2/yr near the implementation area) with increases of + 0.1% for Res1 and 2 and 0.4% for Res3 (Table 1). Despite the interest of all household types,

Table 1 Baseline and NBS scenario simulation results at city scale (Scenario 1: urban green park; Scenario 2: green roofs; Scenario 3: green and blue space)

Land use	Baseline	Scenario 1	%	Scenario 2	%	Scenario 3	%
Forest (ha)	693	693	0.0	693	0.0	693	0.0
Water (ha)	846	846	0.0	846	0.0	846	0.0
Agriculture (ha)	184	192	4.2	188	2.0	220	19.4
Industry/Commerce (ha)	751	751	− 0.1	751	0.0	738	− 1.7
Green urban spaces (ha)	161	163	1.3	161	0.0	177	9.6
Roads (ha)	385	385	0.0	385	0.0	385	0.0
Urban (ha)							
Res1 (ha)	181	181	0.0	181	− 0.2	178	− 1.9
Res2 (ha)	633	629	− 0.7	631	− 0.3	608	− 4.0
Res3 (ha)	121	117	− 3.9	120	− 1.1	112	− 7.4
Total (ha)	936	927	− 1.0	932	− 0.4	898	− 4.1
Total (ha)	3956	3956	0.0	3956	0.0	3956	0.0
Housing quantity							
Res1 ($1000m^2$)	42.5	42.5	0.0	42.4	− 0.2	41.3	− 2.8
Res2 ($1000m^2$)	304.8	301.6	− 1.1	303.5	− 0.4	287.2	− 5.8
Res3 ($1000m^2$)	158.6	150.0	− 5.4	156.1	− 1.5	142.4	− 10.2
Total ($1000m^2$)	505.8	494.1	− 2.3	502.0	− 0.8	470.8	− 6.9
Living space							
Res1 (m^2/hh)	26.9	26.9	0.0	26.9	− 0.1	26.7	− 0.7
Res2 (m^2/hh)	55.0	54.9	− 0.3	55.0	− 0.1	54.2	− 1.5
Res3 (m^2/hh)	150.5	148.2	− 1.6	149.9	− 0.4	146.1	− 3.0
Average (m^2/hh)	53.0	52.8	− 0.5	52.9	− 0.2	52.1	− 1.6
Real estate value							
Res1 (€/m^2/yr)	105.3	105.3	0.0	105.4	0.1	106.2	0.8
Res2 (€/m^2/yr)	123.9	124.3	0.3	124.0	0.1	126.1	1.8
Res3 (€/m^2/yr)	157.4	159.9	1.6	158.1	0.4	162.3	3.1
Average (€/m^2/yr)	124.6	125.1	0.3	124.8	0.1	126.7	1.6
Total (€/m^2/yr)	805.0	805.1	0.0	805.0	0.0	806.5	0.2
Distance to amenity 1 (m)	1009.7	1009.7	0.0	1009.7	0.0	804.4	− 20.3
Distance to amenity 2 (m)	1165.9	1092.5	− 6.3	1165.9	0.0	1165.9	0.0
Distance to amenity 3 (m)	633.7	633.7	0.0	618.6	− 2.4	633.7	0.0
Ratio of open space to build form (#)	0.91	0.92	0.8	0.91	0.4	0.96	5.3
Per cent built up area (%)	52.4	52.2	− 0.4	52.3	− 0.2	51.1	− 2.5

no added green gentrification was visible as households maintained their living location, with the NBS not being attractive enough to attract extra higher income households. The green roofs implementation (amenity type 3) led to an overall respective decrease of − 2.4% of the distance to amenity 3, without any changes to other amenity types distance. The small urban contraction resulted in a + 0.8% increase in the ratio of open space to build form and a − 0.2% reduction in the overall per cent built up area.

The implementation of the green and blue space (Scenario 3; + 9.6% in green urban areas) resulted in a significant contraction of the urban area (− 4.1%) and increase in agricultural areas (+ 19.4%) (see Table 1 and Fig. 2). All households suffered a decrease in the residential area (− 7.4% for Res3, − 4% for Res2 and − 1.9% for Res1), impacting mostly locations where Res2 and Res3 lived. The new attractive amenity (type 1) made all households willing to pay more and live in smaller denser spaces to be near the implementation area (Fig. 2). Hence, housing quantity decreased on average − 6.9% (− 10.2% for Res3, − 5.8% for Res2 and − 2.8% for Res1) and living space decreased on average − 1.6% (− 3% for Res3, − 1.5% for Res2 and − 0.7% for Res1). As expected, housing prices suffered an increase, on average of 1.6% (around + 20 €/m^2/yr near the implementation area), with values ranging from + 3.1% for Res3, + 1.8% for Res2 to + 0.8% for Res1 (Table 1). Due to the high-income household's interest and available budget, they can buy out the middle-income households that lived in the area, that are forced to move to the locations left by other households, namely by high-income ones in the northern periphery, leading to some green gentrification. With the green and blue space implementation (amenity type 1), the correspondent distance to amenity 1 significantly decreased (− 20.3%), with no alteration for the other amenity type differences. The urban contraction led to an increase of + 5.3% in the city ratio of open space to build form and a reduction of − 2.5% in the built up area.

4 Discussion and Conclusion

This study simulates the socio-economic and land use impacts of different NBS, namely an urban park (Scenario 1), green roofs (Scenario 2) and a green and blue space (Scenario 3) implementation in the city of Genova, Italy. These scenarios are the result of a co-design workshop that identified the most worrisome urban problems and its locations in order to pinpoint the most suitable NBS to address them. The workshop together with the simulation modelling approach, from an information policy instrument perspective, aims to promote awareness and create a personalized evidence base, being acknowledged to foment the implementation success and a more informed decision-making process (Voskamp et al., 2021).

Comprehensively, all simulated NBS scenarios had an impact on the urban form and socio-economic dynamics, such as household density, housing quantity, living space and housing prices, while also increasing the non-built urban areas and, thus, increasing overall city accessibility to amenities (in line with Adegun et al., (2021) and Mendonça et al., (2021b) findings). In fact, all scenarios led to a contraction of the urban residential area, as households are willing to pay more to live in smaller and more dense spaces to be near the attractive NBS. Hence increasing property prices around implementation areas as widely mentioned in literature (Black & Richards, 2020; Chen et al., 2022; Dell'Anna et al., 2022; Mendonça et al., 2020). These overall impacts in all scenarios Chen, even in less attractive ones, can be attributed to the city originally low share of green spaces and especially of attractive ones (Tzortzi et al., 2022).

As expected, the impacts were greater for the most attractive NBS, also the bigger—green and blue space (Scenario 3), followed by the medium attractive NBS—urban park (Scenario 2), and finally the least attractive one—green roofs (Scenario 1). This highlights the importance of the attractiveness level of the NBS implemented in the socio-economic impacts, thereby noting that large green and blue spaces are recognized to be highly attractive (Cortinovis et al., 2022; Li et al., 2022) decreasing urban residential areas and resulting in the greatest increase in property prices (Tuofu et al., 2021). Alternatively, green roofs are less accessible and less attractive, with smaller and more local impacts, thus with less overall widespread socio-economic benefits recognized (Di Pirro, et al., 2022; Tzortzi et al., 2022). Despite the fact that urban parks are considered attractive amenities with cultural values (Mexia et al., 2018), its smaller size, combined with the medium level attractiveness, might explain the smaller impact in comparison with the green and blue space scenario (Badura et al., 2021).

Nevertheless, the blue and green space scenario led to some green gentrification phenomenon. The high NBS attractiveness and size made higher income households wanting to move towards the new implementation area, buying out middle-income households that already lived in the area. In turn, these were forced to move to locations they can afford left by other households due to the high property prices in the new implementation area that they cannot compete with. With this, the location of the NBS might dictate who suffers or not with green segregation (Black & Richards, 2020; Bockarjova et al., 2020). Policy-wise, this feeds the need to identify households who were mostly impacted and forced to relocate and more studies are needed to better understand this phenomenon, its causes and how to steer it. Moreover, it emphasizes the importance of information policy instruments such as co-creation and simulation tools to understand the case-specific impact of NBS in order to create strategies to minimize potential negative impacts while promoting a more informed decision-making process (Hérivaux & Coent, 2021; Walker, 2021).

Overall simulation results are consistent with previous works (Mendonça et al., 2020; Roebeling et al., 2007; Saraiva et al., 2017), in which an urban contraction leads to a reduction of housing quantity and living space, with a respective increase in housing values and household density. The increased share of green urban areas also increases the ratio of open space to build form and access to amenities while decreasing the per cent built up area. Notwithstanding, it is noteworthy that the results represent trends, as relocations in real life are dependent on other external factors not considered as policies and other developments. New studies could entail policies as an attempt to control the green gentrification phenomenon and simulation of different NBS sizes and locations to fundament the results in different urban settings.

In conclusion, NBS have the potential to impact the socio-economic and land use dynamics of a city, as all simulated scenarios (green roofs, urban park and green and blue space) led to an overall reduction of the residential area, with increased household densities and property prices and decreased living spaces near the implementation areas. In addition, all scenarios improved overall access to environmental amenities and the amount of natural area, increasing the ratio of open space to build form and decreasing the per cent built up area, mostly due to the initial low share of green spaces. Besides that, the impact is mainly dependent on the attractiveness level, size and location, as the green and blue space posed the greatest difference (most attractive and biggest), followed by the urban park (medium attractiveness and smaller) and lastly, the green roofs (least attractiveness, scattered and with limited access). Despite having the most impact on urban contraction, densities and property prices, the green and blue space also created green gentrification around the implementation area, with higher income households moving near the attractive NBS, buying out others that were forced to relocate. This highlights the importance of information policy instruments and case-specific tools, in order to promote a more informed and sustainable urban planning and decision-making processes.

Acknowledgements This work was supported by the UNaLab project (Grant Agreement No. 730052, Topic: SCC-2-2016-2017: Smart Cities and Communities Nature-based solutions). Thanks also for the financial support to the PhD grants of R. Mendonça (SFRH/BD/147005/2019). Finally, thanks are due to FCT/MCTES for the financial support to CESAM (UIDP/50017/2020 + UIDB/50017/2020), through national funds.

References

Adegun, O. B., Ikudayisi, A. E., Morakinyo, T. E., & Olusoga, O. O. (2021). Urban green infrastructure in Nigeria: A review. *Scientific African, 14*, e01044. https://doi.org/10.1016/j.sciaf.2021.e01044

Badura, T., Krkoška Lorencová, E., Ferrini, S., & Vačkářová, D. (2021). Public support for urban climate adaptation policy through nature-based solutions in Prague. *Landscape and Urban Planning, 215*. https://doi.org/10.1016/j.landurbplan.2021.104215

Black, K., & Richards, M. (2020). Landscape and urban planning eco-gentrification and who benefits from urban green amenities: NYC's high Line. *Landscape and Urban Planning, 204*(July), 103900. https://doi.org/10.1016/j.landurbplan.2020.103900

Bockarjova, M., Botzen, W. J. W., van Schie, M. H., & Koetse, M. J. (2020). Property price effects of green interventions in cities: A meta-analysis and implications for gentrification. *Environmental Science and Policy, 112*(June), 293–304. https://doi.org/10.1016/j.envsci.2020.06.024

Chen, S., Zhang, L., Huang, Y., Wilson, B., Mosey, G., & Deal, B. (2022). Spatial impacts of multimodal accessibility to green spaces on housing price in Cook County, Illinois. *Urban Forestry and Urban Greening, 67*(September 2021). https://doi.org/10.1016/j.ufug.2021.127370

Cortinovis, C., Olsson, P., Boke-Olén, N., & Hedlund, K. (2022). Scaling up nature-based solutions for climate-change adaptation: Potential and benefits in three European cities. *Urban Forestry and Urban Greening, 67*. https://doi.org/10.1016/j.ufug.2021.127450

De Los Rios-White, M., Roebeling, P., & Valente, S. (2020). Mapping the life cycle co-creation process of change adaptation. *Resources, 9* (4), 39. https://doi.org/10.3390/resources9040039

Dell'Anna, F., Bravi, M., & Bottero, M. (2022). Urban Green infrastructures: How much did they affect property prices in Singapore? *Urban Forestry and Urban Greening, 68*(January). https://doi.org/10.1016/j.ufug.2022.127475

Di Pirro, E., Sallustio, L., Castellar, J. A. C., Sgrigna, G., Marchetti, M., & Lasserre, B. (2022). Facing multiple environmental challenges through maximizing the co-benefits of nature-based solutions at a national scale in Italy. *Forests, 13*(548), 1–17. https://doi.org/10.3390/f13040548

EC. (2021). Evaluating the impact of nature-based solutions: A handbook for practitioners. European Commission. Retrieved from https://doi.org/10.2777/244577

Faivre, N., Fritz, M., Freitas, T., de Boissezon, B., & Vandewoestijne, S. (2017). Nature-based solutions in the EU: Innovating with nature to address social, economic and environmental challenges. *Environmental Research, 159*(December 2016), 509–518. https://doi.org/10.1016/j.envres.2017.08.032

Haase, D., Kabisch, S., Haase, A., Andersson, E., Banzhaf, E., Brenck, M., Fischer, L. K., Frantzeskaki, N., Bar, F., Kabisch, N., Krellenberg, K., Kremer, P., Kronenberg, J., Larondelle, N., Mathey, J., Pauleit, S., Ring, I., Rink, D., Schwarz, N., & Wolff, M. (2017). Greening cities To be socially inclusive ? About the alleged paradox of society and ecology in cities, *64*. https://doi.org/10.1016/j.habitatint.2017.04.005

Hérivaux, C., & Coent, P. Le. (2021). Introducing nature into cities or preserving existing peri-urban ecosystems ? Analysis of preferences in a rapidly urbanizing catchment, *13*(587). https://doi.org/10.3390/su13020587

Istat Statistics (2018). Available at: http://dati.istat.it/?lang=en

Li, H., Peng, J., Jiao, Y., & Ai, S. (2022). Experiencing urban green and blue spaces in urban wetlands as a nature-based solution to promote positive emotions, *13*(473). https://doi.org/10.3390/f13030473

Mendonça, R., Roebeling, P., Fidélis, T., & Saraiva, M. (2021a). Policy instruments to encourage the adoption of nature-based solutions in urban landscapes. *Resources, 10*(8), 1–14. https://doi.org/10.3390/resources10080081

Mendonça, R., Roebeling, P., Fidélis, T., & Saraiva, M. (2021b). Socio-economic models to assess and policy instruments to steer the impact of nature-based solutions: A review. *WIT Transactions on Ecology and the Environment, 253*, 551–564. https://doi.org/10.2495/SC210451

Mendonça, R., Roebeling, P., Martins, F., Fidélis, T., Teotónio, C., Alves, H., & Rocha, J. (2020). Assessing economic instruments to steer urban residential sprawl, using a hedonic pricing simulation modelling approach. *Land Use Policy, 92*(December 2019), 104458. https://doi.org/10.1016/j.landusepol.2019.104458

Mexia, T., Vieira, J., Príncipe, A., Anjos, A., Silva, P., Lopes, N., Freitas, C., Santos-reis, M., Correia, O., Branquinho, C., & Pinho, P. (2018). Ecosystem services : Urban parks under a magnifying glass. *Environmental Research*, *160*(December 2016), 469–478. https://doi.org/10.1016/j.envres.2017.10.023

Roebeling, P. C., Fletcher, C. S., Hilbert, D. W., & Udo, J. (2007). Welfare gains from urbanizing landscapes in Great Barrier Reef catchments? A spatial environmental-economic modelling approach. *WIT Transactions on Ecology and the Environment, 102*, 737–749. https://doi.org/10.2495/SDP070712

Saraiva, M., Roebeling, P., Sousa, S., Teotónio, C., Palla, A., & Gnecco, I. (2017). Dimensions of shrinkage: Evaluating the socio-economic consequences of population decline in two medium-sized cities in Europe, using the SULD decision support tool. *Environment and Planning b: Urban Analytics and City Science, 44*(6), 1122–1144. https://doi.org/10.1177/0265813516659071

Tuofu, H., Qingyun, H., Dongxiao, Y., & Xiao, O. (2021). Evaluating the impact of urban blue space accessibility on housing price: A spatial quantile regression approach applied in Changsha, China. *Frontiers in Environmental Science*, *9*(May), 1–15. https://doi.org/10.3389/fenvs.2021.696626

Tzortzi, J. N., Guaita, L., & Kouzoupi, A. (2022). Sustainable strategies for urban and landscape regeneration related to agri-cultural heritage in the urban-periphery of South Milan. *Sustainability, 14* (6581). https://doi.org/10.3390/su14116581

Urban Atlas. (2018). Available at: https://land.copernicus.eu/local/urban-atlas/urban-atlas-2018

Voskamp, I. M., de Luca, C., Polo-Ballinas, M. B., Hulsman, H., & Brolsma, R. (2021). Nature-based solutions tools for planning urban climate adaptation: State of the art. *Sustainability (switzerland), 13* (11), 1–17. https://doi.org/10.3390/su13116381

Walker, R. H. (2021). Engineering gentrification: Urban redevelopment, sustainability policy, and green stormwater infrastructure in Minneapolis. *Journal of Environmental Policy and Planning, 23*(5), 646–664. https://doi.org/10.1080/1523908X.2021.1945917

Energy Savings of Green Roofs in Mediterranean Cities

Eman M. Elmazek and Aziza A. Safour

Abstract

Green roofs have shown great benefits to the built environment through both research, and application. One of the most important benefits is to increase the energy performance of the buildings. Green roofs can reduce energy demand due to the heat flow through the roof in the summer by more than 75% in some cases. In addition, they affect the reflected heat in the surrounding, humidity, and air quality in cities and urban areas. Nowadays the market offers many types of green roofs depending on their structure systems and type of vegetation. Moreover, different climates have different environmental and energy needs. One of the most populated areas are the Mediterranean cities. There are 285 Mediterranean cities with more than 126 million inhabitants. One of these cities is Tripoli, the capital of Libya. It has a unique built environment that has been responding to its climate through the ages. However, the city as other cities in the same region is failing to respond to the increase in energy demand due to the industrial revolution. This paper will investigate the contribution of the different types of green roofs in energy performance. To discover what gives Tripoli buildings more energy savings, it is important to recount these types' vegetation and structure. Any differences might affect the roof energy performance. The research will use EnergyPlus simulation program to get results that support the green roof type choice for Tripoli climate. The aim of the research is to assume the best scenario in energy performance between the six types of roofs, namely extensive, semi-intensive, intensive, brown, white, and black.

E. M. Elmazek (✉) · A. A. Safour
Architecture and Urban Planning Department, University of Benghazi, Benghazi, Libya
e-mail: eman.elmazek@uob.edu.ly

A. A. Safour
e-mail: aziza.safour@uob.edu.ly

Keywords

Green roofs • Energy performance • Urban sustainability • Mediterranean climate • Sustainable building • Tripoli

1 Introduction

Energy used by the building sector has been the focus of many sustainable development stakeholders. The specific energy load that the building sector consumes is hard to determine. However, a significant number of researches is estimating it to be around 35%. This is a relatively high contribution to greenhouse gas (GHG) emissions. There are many attempts to lessen building energy consumption and controlling the GHG effect on urban areas. These areas are excluded from the natural environment which sustains it and suffers a lack in urban green spaces.

Numerous sustainable approaches and environmentally responsive energy-efficient technologies were introduced for the sake of low energy design. Green roofs have been proposed for energy saving plans in many areas with different climatic conditions. The cooling and heating energy saving capacity highly depends on the climate, roofing, and building characteristics. Nowadays, the market offers a variety of green roofing systems each of which has its own advantages and disadvantages. It is essential to understand every green feature ability and respond to energy saving in the Mediterranean climate. Tripoli is the case study in this paper: It represents the hot arid Mediterranean climate as it is located in the south of the Mediterranean Sea and in the middle of North Africa shore.

Four green scenarios are presented and simulated to investigate the energy performance of each in this paper. However, there are many other factors to determine the best fit roof system for the target building. These considerations include aesthetic, environmental concepts, economic, social,

A. Battisti et al. (eds.), *Greening Our Cities: Sustainable Urbanism for a Greener Future*, Advances in Science, Technology & Innovation, https://doi.org/10.1007/978-3-031-49495-6_10

and technical benefits. For instance, brown or biodiversity roofs are claimed to reflect local ecology and have a stronger relation with nature. Roof gardens (intensive green roofs) bring aesthetics to the urban environment and support a balanced urban landscape. On the other hand, extensive roofs do not show much success due to financial and weight constraints. Moreover, extensive roofs have hard growing conditions (Castleton et al., 2010).

What are green roofs?

Green roofs or living roofs are any open space that is located on the top of any building and has a growing medium. It is a nature-based solution to environmental problems in urban areas. It combines the abiotic and biotic components to create a green space in urban areas (Magill et al., 2011). Green roof is a roofing system that consists of many layers that differ in type and thickness. These layers are mainly the growing medium, drainage layer, isolation, root barrier, waterproof layer, and the structural support as shown in Fig. 1.

Background and history

The history of green roofs starts with the Hanging Gardens of Babylon built by King Nebuchadnezzar II for his wife, Amyitis who had a passion for green mountains from her homeland (Magill et al., 2011). This technology started in the fifth century BC by the ancient civilizations of the Tigris and Euphrates River valleys. With the development of concrete as a structural material in the 1800s, the flat roofs of building became more common in Europe and America. The World Exhibition in Paris was built in 1868. It has a nature roof as the first several experimental projects in Europe. The over ninety-year-old Moss Water Filtration Plant near Zurich, Switzerland, is considered to be the third oldest green roof. It was built in 1914 to keep the interior cool and prevent bacterial growth in the filtration beds. A drainage layer of gravel and a layer of soil were put over the roofs (Jim, 2017). Others include a restaurant with a roof garden in Chicago in 1914 designed by Frank Lloyd Wright, and a similar project by Gropius in Cologne in the same year. In Paris in 1903, a block of apartments was established with planted terraces and a roof garden. Le Corbusier has used roof gardens since 1920 but for expensive apartments. Frank Lloyd Wright, Gropius, and Le Corbusier adopted the green roof to merge nature with the construction in their buildings (Lepp, 2008). An example of these buildings is the London department store Derry and Toms built in the 1930s. It had a roof garden that contained three themed gardens covering 1.5 acres.

The new improved building techniques in the twentieth century were the first promotion of flat roofs that can carry large loads. This development led to the expansion of green roofs. Especially in the second half of the twentieth century, the construction of plazas over underground garages, roads, and subways became a prominent usage of roof gardens. In the1960s and 1970s, several projects were done with attempts to integrate buildings with plants in German-speaking countries to control winds and rainstorms. These countries are considered the origin of contemporary green roofs. About 10% of roofs in German are green (Stromberg et al., 2007). In the USA, in San Bruno, California, William McDonough designed the headquarters of GAP with a 69,000 square-foot green roof in 1997.[1] In 1998, in Chicago, the mayor Richard M. Daley influenced from what he had seen in Germany, directed municipal funds for green roof development. In the same year, in Washington, D.C., the LEED rating system was put by the USGBC in which using green roofs buildings can get 6 out of 69 points. In 1999, Green Roofs for Healthy Cities, an organization of public and industry groups, was formed in Toronto, Canada (Elmazek, 2020).

Types of green roofs

Nowadays, green roofs come in four types: extensive, intensive, semi-intensive, and brown roofs. They are differentiated mainly by cost, depth of growing medium, and the choice of plants (Luckett, 2009). The extensive green roof is characterized by low weight, low cost, low plant diversity, and minimal maintenance requirements. It is usually not accessible. The nutrient media is shallow; its depths are usually between 5 and 15 cm and weigh between 72.6 and 169.4 kg/m^2 when fully saturated as shown in Fig. 2a (Sopranature, 2023). It usually has sturdy plants such as dry land or indigenous. These types of plants need irrigation and fertilization just at the beginning of installation and some maintenance twice annually to avoid weeding and membrane problems. The intensive green roof is characterized by a deeper soil and a greater weight, higher capital costs, more diverse plantings, and therefore more maintenance requirements. The growing medium is often soil-based, ranging in depth from 20 to 60 cm, with a saturated weight increase of between 290 and 967.7 kg/m^2 as shown in Fig. 2c (Sopranature, 2023). They are always accessible to the general public or building occupants. It contains different types of plantings including some trees and shrubs. This makes it form a more complex ecosystem that requires more maintenance and watering. It demands a developing and challenging watering structure. Moreover, a heavier structural system is important due to the increase of green roof weight. The installation team should hire agricultural and construction experts (Peck & Erickson, 2008). The semi-intensive roof has characteristics between the intensive and extensive roofs. It could be in the form of a hybrid system with different growing

[1] GAP Headquarters, 901 Cherry, 2021.

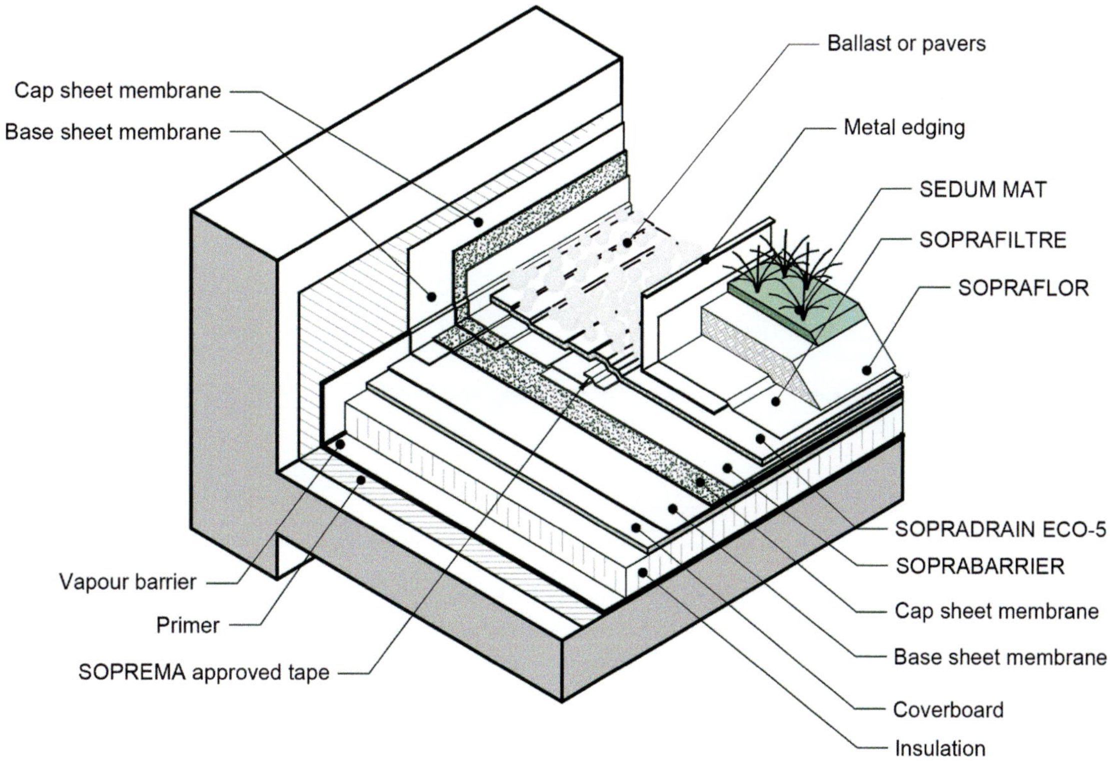

Fig. 1 Green roof parapet detail—sedum mat (Sopranature, 2023)

media depths on the same roof. Each depth forms at least 25 per cent of the total area of the green roof as shown in Fig. 2b (Silva et al., 2014). Semi-intensive roofs are often accessible. Brown roofs or biodiverse roofs are a new concept that follows the same principles of green roof systems of a growing medium, a drainage layer, an isolation, a root barrier, and a waterproof layer. Here, the growing medium is left without vegetation to give the space for a self-vegetation environment that reproduces the local ecological requirements. The proprietary biodiverse substrate encourages the local species and habitats to localize. These roofs should not be trafficked and has the same construction characteristics of the extensive green roofs without any planting.

Benefits of green roofs

Green roofs are used for various reasons depending on climatic site factors and the space functional needs due to the different advantages of applying green roofs. For instance, in wet urban areas they are used for storm water management. It reduces the amount of storm water runoff. Depending on the system design and the depth of the growing medium, a green roof can reduce storm water runoff by 50–90% (Un, 2017). The substrate stores the water and then takes it to the plant. The water is returned to the natural water cycle by evaporation and transpiration of the green roof. Green roofs moderate the temperature of the water and filter it before draining.

Another benefit of green roof systems is the social advantage. Converting unused roof space area to a garden that can be used for public gatherings and supports a healthier lifestyle. A household survey was carried out in February 2016 targeting individuals from different ages and backgrounds in North Africa. 80% of the participants would like to have a roof garden. If they have a roof garden, 54% reported that they would use it for urban agriculture and amenity space at the same time, 31% would keep just as an amenity space, 7.5% would leave for nature, and the rest (7.5%) would use it just for urban agriculture. As for public buildings, 92% of the responses agree that these buildings should have green roofs (Elmazek, 2020). The property's value increases by having an amenity space that can be used for recreational purposes. This garden might be used for urban agriculture to produce convenient food. Moreover, they can improve sound insulation by up to 8 db (Elmazek, 2020). They reduce sound reflection by up to 3 db. Reducing noise level is a significant benefit in impacted areas such as sites near airports, trains, highways, or industries (Peck & Erickson, 2008).

Green roofs help in increasing the life span of the roof outside layer by protecting it from biological, chemical, and physical burdens. They work as buffer from wind, hail, fireworks, or vandalism. They delay the roof aging due to the unstable day temperature and UV radiation. In addition, they act as additional thermal insulation. Experiments by Karen

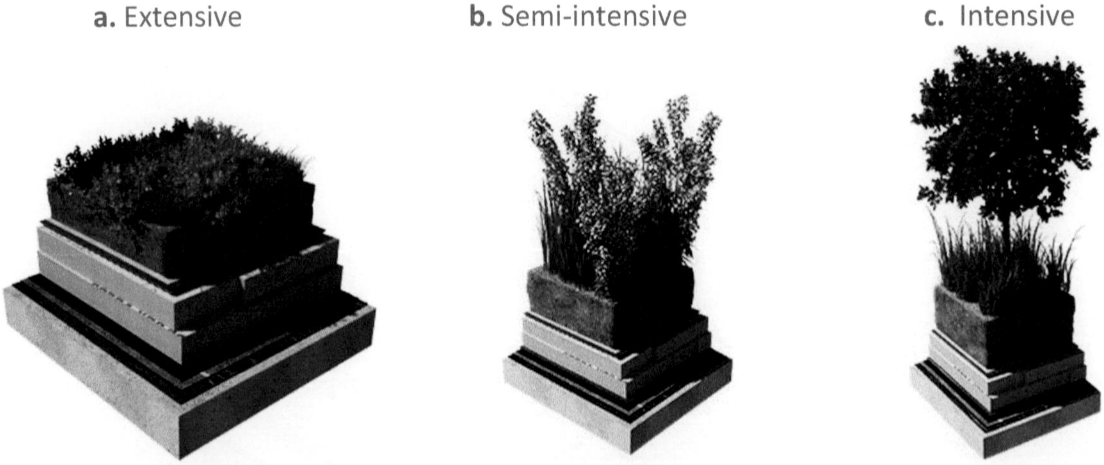

Fig. 2 Three common types of green roofs (Sopranature, 2023)

Liu in Ottawa Canada found that 15-cm extensive green roof reduced heat losses by 26% and heat gains by 95% compared to a conventional roof (Liu & Baskaran, 2003). Green roofs reduce indoor temperatures through transpiration in hotter climates. They also benefit urban areas especially big cities. They lower the city temperature by causing a cooling effect due to the transpiration and shading process, reducing greenhouse gas emissions and urban heat island effect. Furthermore, green roofs reduce dust and smog levels. One m^2 of green roof can filter approximately 180 g of dust and smog particles per year. The roof vegetation converts carbon dioxide, water, and sunlight into oxygen and glucose through photosynthesis offsetting the nature environment excluded by the expansion of urbanism.

2 Methodology

A hypothetical building is used and fully analyzed based on the roof system. The case study was carried out in Tripoli. To determine the best type of roof that should be used in the studied location, Tripoli, which has a Mediterranean climate in terms of energy savings; the study proposed a space with specified characters such as the location, the orientation, the dimensions, and the structural roof and walls materials features. It is also important to know in specific the energy consumption for each roof type in the cooling and the heating seasons to put in an accurate comparison between different types energy performance and balance the results with other economic, social, and ecological factors to decide which roof fits in different cases and buildings functions. The study ran an energy simulation for six types of roofing determined according the global market availability as shown in Figs. 3 and 4. These six types are as follows:

1. Extensive green roofs.
2. Semi-intensive green roofs.
3. Intensive green roofs.
4. Brown roofs.
5. White or cool roofs.
6. Black roofs.

The difference between these types is the structural layers and their depth, the vegetation features, and irrigation required as discussed above in the types of green roofs section and later in Table 1. The boundary conditions of the four vegetated roofs used in the simulation are shown in the simulation table in Fig. 3. Figure 4 shows the boundary conditions for the white roof, the black roof, the floor slab, and the walls.

The study uses a hypothetical space in the Mediterranean city Tripoli as a model to run EnergyPlus simulation program with six different roof scenarios. The program uses Sailor's green roof model and the weather file of Tripoli. A comparison is done between energy consumption for the heating and cooling seasons for the six scenarios to conclude which type of roof gives the best energy performance compared to the other factors such as economic, social, and ecological. The case study building boundary conditions are as shown in Fig. 5, and the thermos set points conditions are shown in Fig. 6.

EnergyPlus

EnergyPlus is a whole-building energy simulation program. It is also known by Eplus or + Energy. It is a tool to simulate energy consumption in a building by inputting the building and weather data. It gives detailed results of how energy is consumed in the building with the given data. It is used to model different building features options to assume which building shape and features lead to more energy savings.

Field	Units	Obj1	Obj2	Obj3	Obj4
Name		EXTESIVE GREEN ROOM	SEMI INTENSIVE GREEN ROOM	INTENSIVE GREEN ROOF	BROWN ROOF
Height of Plants	m	0.1	0.5	1	0.01
Leaf Area Index	dimensionless	1	2.5	5	1
Leaf Reflectivity	dimensionless	0.22	0.22	0.22	0.22
Leaf Emissivity		0.95	0.95	0.95	0.95
Minimum Stomatal Resistance	s/m	180	180	180	180
Soil Layer Name		Green Roof Soil	Green Roof Soil	Green Roof Soil	Green Roof Soil
Roughness		MediumRou gh	MediumRou gh	MediumRou gh	MediumRou gh
Thickness	m	0.15	0.35	0.7	0.15
Conductivity of Dry Soil	W/m-K	0.35	0.35	0.35	0.35
Density of Dry Soil	kg/m3	1100	1100	1100	1100
Specific Heat of Dry Soil	J/kg-K	1200	1200	1200	1200
Thermal Absorptance		0.9	0.9	0.9	0.9
Solar Absorptance		0.7	0.7	0.7	0.7
Visible Absorptance		0.75	0.75	0.75	0.75
Saturation Volumetric Moisture Content of the Soil Layer		0.3	0.3	0.3	0.3
Residual Volumetric Moisture		0.01	0.01	0.01	0.01
Initial Volumetric Moisture		0.1	0.1	0.1	0.1
Moisture Diffusion Calculation		Advanced	Advanced	Advanced	Advanced

Fig. 3 Boundary conditions of EnergyPlus for the four types of green roofs (*source* Authors)

Field	Units	Obj1	Obj2	Obj3	Obj4
Name		WHITE CENCRETE SLAB	CMU	FLOOR	BLACK CONCRETE SLAB
Roughness		Rough	MediumRough	Rough	Rough
Thickness	m	0.25	0.2	0.15	0.25
Conductivity	W/m-K	1.9	1.5	1.9	1.9
Density	kg/m3	2300	1680	2300	2300
Specific Heat	J/kg-K	840	840	840	840
Thermal Absorptance		0.9	0.9	0.9	0.9
Solar Absorptance		0.2	0.7	0.2	0.8
Visible Absorptance		0.6	0.7	0.6	0.6

Fig. 4 Boundary conditions of EnergyPlus for the walls, the roof, and floor roofs (*source* Authors)

Green roof model

The study used the green roof model done by Sailor (2008). The Sailor developed model "ecoroof" is an effective tool to help designers to use a numerical asessment to measure the energy balance of green roofs. Sailor's EnergyPlus green roof model allows the user to specify a vegetated roof as the outer layer of a rooftop in an information data file (IDF). The user can input different features of the structure of his model green roof such as the soil deepness, thermal factors, leaf area index, plant height, moisture transpiration, and irrigation conditions. The model was based on the fast all season soil strength (FASST) model developed by Frankenstein and Koenig for the US Army Corps of Engineers in 2004. FASST tracks the energy and moisture balance (including ice and snow) within a vegetated soil as shown in Fig. 7 (Sailor, 2008). Sailor's model takes in consideration factors that affect the soil and the plants such as radiation exchange between the soil and the vegetation, and the heat flow, storage, and transfer. Figure 7 displays the energy balance of a green roof and the variables discussed.

The Sailor model contains energy balance which is the sum of the energy balance of the vegetation layer (F_f) and the energy balance of the soil (F_g). Equation (1) calculates

Table 1 Characteristics of the six roofs studied

Conventional roofs	White roof	Membrane (5 mm)	Albedo = 0.8
	Black roof	Membrane (5 mm)	Albedo = 0.2
Green roofs	Extensive	LAI	1
		Height of plants (m)	0.1
		Soil depth (m)	0.15
	Semi-intensive	LAI	2.5
		Height of plants (m)	0.5
		Soil depth (m)	0.35
	Intensive	LAI	5
		Height of plants (m)	1
		Soil depth (m)	0.7
Brown roofs	Brown roofs	LAI	1
		Height of plants (m)	0.01
		Soil depth (m)	0.15

Field	Unit	Obj1	Obj2	Obj3	Obj4	Obj5	Obj6
Vertex 1 Z-coordinate	m	2.7	2.7	2.7	2.7	0	2.7
Vertex 2 X-coordinate	m	8	8	0	0	0	0
Vertex 2 Y-coordinate	m	6	0	0	6	6	0
Vertex 2 Z-coordinate	m	0	0	0	0	0	2.7
Vertex 3 X-coordinate	m	0	8	8	0	8	8
Vertex 3 Y-coordinate	m	6	6	0	0	6	0
Vertex 3 Z-coordinate	m	0	0	0	0	0	2.7
Vertex 4 X-coordinate	m	0	8	8	0	8	8
Vertex 4 Y-coordinate	m	6	6	0	0	0	6
Vertex 4 Z-coordinate	m	2.7	2.7	2.7	2.7	0	2.7

Fig. 5 Boundary conditions of EnergyPlus for case study building roofs (*source* Authors)

Field	Units	Obj1
Name		Office Thermostat Dual SP Control
Heating Setpoint Temperature Schedule Name		ALWAYS 20
Cooling Setpoint Temperature Schedule Name		ALWAYS 24

Fig. 6 Boundary conditions of EnergyPlus for heating and cooling set points (*source* Authors)

the leaf energy balance (foliage energy balance F_f) where the total heat exchange to the vegetation layer (F_f) is calculated in (W/m^2).

$$F_f = \sigma_f\left[I_S^{\downarrow}(1-\alpha_f) + \varepsilon_f I_{ir}^{\downarrow} - \varepsilon_f \sigma T_f^4\right] + \frac{\sigma_f \varepsilon_g \varepsilon_f \sigma}{\varepsilon_1}\left(T_g^4 - T_f^4\right) + H_f + L_f \quad (1)$$

where σ_f is calculated as a function F_f of leaf area index (LAI):

$$\sigma_f = 0.9 - 0.7\exp -0.75\text{LAI} \quad (2)$$

LAI in Eq. 2 is leaf area index which is the amount of leaf material in a soil area (m^2/m^2) (Silva et al., 2014). The Sailor Ecoroof model is a useful tool for architects to measure the advantages of using green roofs in energy performance. As shown in Eqs. 1 and 2 and as discussed above, the model takes in consideration many factors such as moisture effect, soil heat conductance, convective heat transfer, and radiance heat (Castleton et al., 2010).

Study case

This study used the software EnergyPlus for building energy simulation. A model of a small built space that has a rectangular shape with an aspect ratio of 3:4. The base case model with a cool roof was used for several simulations to compare the four types of green roofs over energy

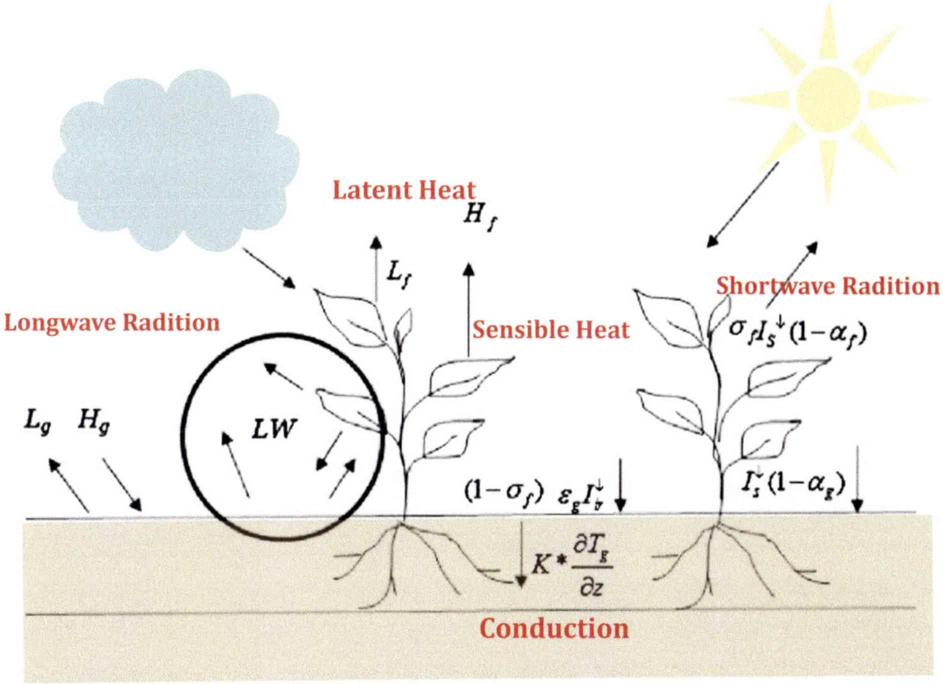

Fig. 7 Energy balance model for a green roof, showing the different methods of heat transfer (Sailor, 2008)

consumption. Another two simulations were run on the same base model to compare between white and black roofs.

The study case used for green roof energy simulation is a space with 6 m wide, 8 m long, and 2.7 m height. The total roof area is 48 m^2, and the whole area is used in the simulation of five different types of roofing system. The north/south facades are 21.6 m^2 each (8 m * 2.7 m). The east/ west facades are 16.2 m^2 each (6 m * 2.7 m) as shown in Fig. 9. It is located in Tripoli, the capital of Libya. Tripoli is in the south shore of the Mediterranean Sea as shown in Fig. 8.

The space is used as a case study to compare four types of green roofs and the conventional white and black concrete roofing systems. The roof support structure was presumed to be a concrete slab with a thickness of 0.25 m in all compared cases with a conductivity of 1.9 W/m-K, a density of 2300 kg/m^3, a specific heat capacity of 840 J/kg-K, a thermal absorptance of 0.9, a visible absorptance of 0.6, and a solar absorptance of 0.2 for white roofs and 0.8 for black roofs. The four walls are presumed to be made of concrete masonry units (CMU) of a thickness of 0.2 m, with a conductivity of 1.5 W/m–K, a density of 1680 kg/m^3, a specific heat capacity of 840 J/kg-K, a thermal absorptance of 0.9, a visible absorptance of 0.7, and a solar absorptance of 0.7. The floor is a concrete slab with a thickness 0.15 m with a conductivity of 1.9 W/m-K, a density of 2300 kg/m^3, a specific heat capacity of 840 J/kg-K, a thermal absorptance of 0.9, a visible absorptance of 0.6, and a solar absorptance of 0.2.

The four green roofs compared extensive, semi-intensive, intensive, and brown roofs. This represents the four main typologies mentioned previously. The extensive green roof has a leaf area index (LAI) of 1, with a soil thickness of 0.15 m, and an average height 0.10 m. The semi-intensive green roof has a leaf area index (LAI) of 2.5, with a soil thickness of 0.35 m, and an average height 0.5 m. The intensive green roof has a leaf area index (LAI) of 5, with a soil thickness of 0.70 m, and an average height 1 m. The brown roof has the same construction of the extensive green roof except the brown roof has an average height of plants 0.01 m as it is not subject to any vegetation installation. Table 1 shows a summary of the characteristics of the six roofs studied (Silva et al., 2014).

3 Results

Model evaluation

The model evaluation data was based on the results from the use of these green roof assemblies and comparing them with each other and with white and black conventional roofs. The simulation results display the energy consumed by heating and cooling for each type of green roof using four models each with different roof features for a 48 m^2 room in Tripoli.

Six different simulations have been carried out on the hypothetical building. All different types were used two of which are for the building without green roof and the four others are for the building with vegetated roofs as described in Table 1. Simulation results are reported in Fig. 10. The simulation results refer only to the room under the conditions mentioned in the case study section. Table 2 and the graphs in Figs. 11 and 12 outline a greater energy consumption

Fig. 8 Tripoli location (*Saudi Aramco World: Tripoli: Crossroads of Rome and Islam, n.d.*)

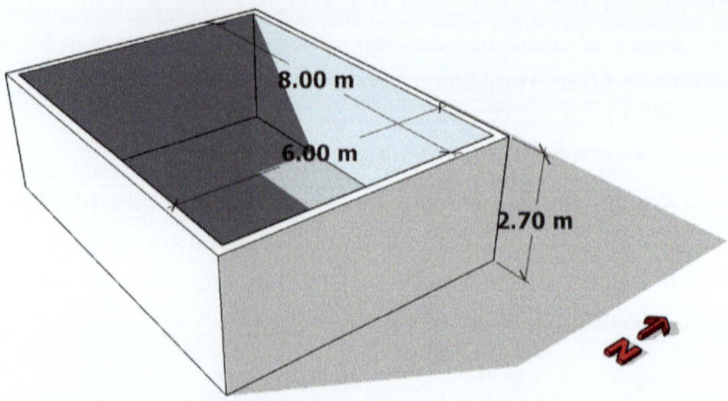

Fig. 9 Case study space dimension (*source* Authors)

during the heating season than in the cooling season. They also show that the intensive green roofs have more energy savings in both cooling and heating seasons. However, the numbers for all scenarios should be compared with other factors that affect picking the suitable roof system.

Table 2 shows the summary of the EnergyPlus simulation results for the six roof models giving different parameters for the roof construction as stated in Table 1.

The simulation results also show that for the cooling season the white roof shows more energy saving than black roofs by 47%. However, it shows minuscule difference in results when applying extensive or brown roofing systems with a slight advantage to white roof application. Semi-intensive green roofs displayed 15% in cooling energy savings compared with the white roof, while the intensive demonstrated 26% cooling energy savings. Intensive green and black conventional roofs simulation results show 60% cooling energy savings when applying intensive green roofs.

For the extensive green roof and brown roof scenarios, the results are similar in both cooling and heating season energy consumptions. As for the heating season there is no clear difference between the two conventional roofs since the white roof functions only to decrease the heat flow to the interiors in summer. Intensive green roofs displayed a reduction in heating energy consumption 14% compared to extensive green roofs as the increase in the growing media thickness gives more insulation and thermal mass. While they showed 40% energy reduction compared to both conventional roofs. On the other hand, extensive green roofs and brown roofs demonstrated 31% energy savings, and semi-intensive green roofs demonstrated 36% energy savings compared both with conventional roofs. Figure 11 shows energy consumptions of the three categories of roofs with no difference between extensive green roof and brown roofs. The thicker the growing media the more energy savings it gives in all cases (cooling and heating).

The results can be summarized in these points:

- Energy consumption is more than twice as high in the heating season as in the cooling season.
- Extensive green roofs and brown roofs show near-identical results in energy consumption.

Fig. 10 Parts of the simulation result spreadsheets for the heating and cooling seasons for the six roof types (*source* Authors)

Date/Time	Environment:Site Outdoor Air Drybulb Temperature [C](Hourly)	ZONE ONE:Zone Air System Sensible Heating Energy [J](Hourly)	ZONE ONE:Zone Air System Sensible Cooling Energy [J](Hourly)	ZONE ONE:Zone Air Temperature [C](Hourly)	DistrictHeating:Facility [J](Hourly)	DistrictCooling:Facility [J](Hourly)
01/21 01:00:00	4.2	14098430.06	0	20	14285970.03	0
08/21 24:00:00	22.043125	0	1550162.862	24	1.02E-10	1550162.862
					1285306531	697649847.6
					8925739.798	2430835.706
					8925.739798	2430.835706
					185.9529125	50.64241054

a. White Roof

Date/Time	Environment:Site Outdoor Air Drybulb Temperature [C](Hourly)	ZONE ONE:Zone Air System Sensible Heating Energy [J](Hourly)	ZONE ONE:Zone Air System Sensible Cooling Energy [J](Hourly)	ZONE ONE:Zone Air Temperature [C](Hourly)	DistrictHeating:Facility [J](Hourly)	DistrictCooling:Facility [J](Hourly)
01/21 01:00:00	4.2	14098430.06	0	20	14285970.03	0
08/21 24:00:00	22.043125	0	4435972.854	24	0	4435972.854
					1281363943	1304689854
					8898360.715	4545957.679
					8898.360715	4545.957679
					185.3825149	94.70745164

b. Black Roof

Date/Time	Environment:Site Outdoor Air Drybulb Temperature [C](Hourly)	ZONE ONE:Zone Air System Sensible Heating Energy [J](Hourly)	ZONE ONE:Zone Air System Sensible Cooling Energy [J](Hourly)	ZONE ONE:Zone Air Temperature [C](Hourly)	DistrictHeating:Facility [J](Hourly)	DistrictCooling:Facility [J](Hourly)
01/21 01:00:00	4.2	10227825.75	0	20	10363878.22	0
08/21 24:00:00	22.043125	0	1798390.007	24	0	1798390.007
					886598057	**718493099.1**
					6156930.952	2503460.276
					6156.930952	2503.460276
					128.2693948	52.15542241

c. Extensive Green Roof

Date/Time	Environment:Site Outdoor Air Drybulb Temperature [C](Hourly)	ZONE ONE:Zone Air System Sensible Heating Energy [J](Hourly)	ZONE ONE:Zone Air System Sensible Cooling Energy [J](Hourly)	ZONE ONE:Zone Air Temperature [C](Hourly)	DistrictHeating:Facility [J](Hourly)	DistrictCooling:Facility [J](Hourly)
01/21 01:00:00	4.2	9333707.345	0	20	9457866.254	0
08/21 24:00:00	22.043125	0	1200781.227	24	0	1200781.227
					815913131.2	592799823.4
					5666063.411	2065504.611
					5666.063411	2065.504611
					118.0429877	43.03134606

d. Semi-intensive Green Roof

Date/Time	Environment:Site Outdoor Air Drybulb Temperature [C](Hourly)	ZONE ONE:Zone Air System Sensible Heating Energy [J](Hourly)	ZONE ONE:Zone Air System Sensible Cooling Energy [J](Hourly)	ZONE ONE:Zone Air Temperature [C](Hourly)	DistrictHeating:Facility [J](Hourly)	DistrictCooling:Facility [J](Hourly)
01/21 01:00:00	4.2	8656674.392	0	20	8771827.105	0
08/21 24:00:00	22.043125	0	890415.2802	24	0	890415.2802
					765744472	515780984.4
					5317669.945	1797146.287
					5317.669945	1797.146287
					110.7847905	37.44054765

e. Intensive Green Roof

Date/Time	Environment:Site Outdoor Air Drybulb Temperature [C](Hourly)	ZONE ONE:Zone Air System Sensible Heating Energy [J](Hourly)	ZONE ONE:Zone Air System Sensible Cooling Energy [J](Hourly)	ZONE ONE:Zone Air Temperature [C](Hourly)	DistrictHeating:Facility [J](Hourly)	DistrictCooling:Facility [J](Hourly)
01/21 01:00:00	4.2	10229076.98	0	20	10365146.09	0
08/21 24:00:00	22.043125	0	1799178.945	24	0	1799178.945
					886810059.1	718659692.1
					6158403.188	2504040.739
					6158.403188	2504.040739
					128.3000664	52.1675154

f. Brown Roof

Table 2 Energy consumptions for the six roof solutions (KJ/h/m^2)

Roof type		Cooling (KJ/h/m^2)	Heating (KJ/h/m^2)
Conventional roofs	White roof	50.64	185.95
	Black roof	94.71	185.38
Green roofs	Extensive	52.16	128.27
	Semi-intensive	43.03	118.04
	Intensive	37.44	110.78
Brown roofs	Brown roofs	52.17	128.30

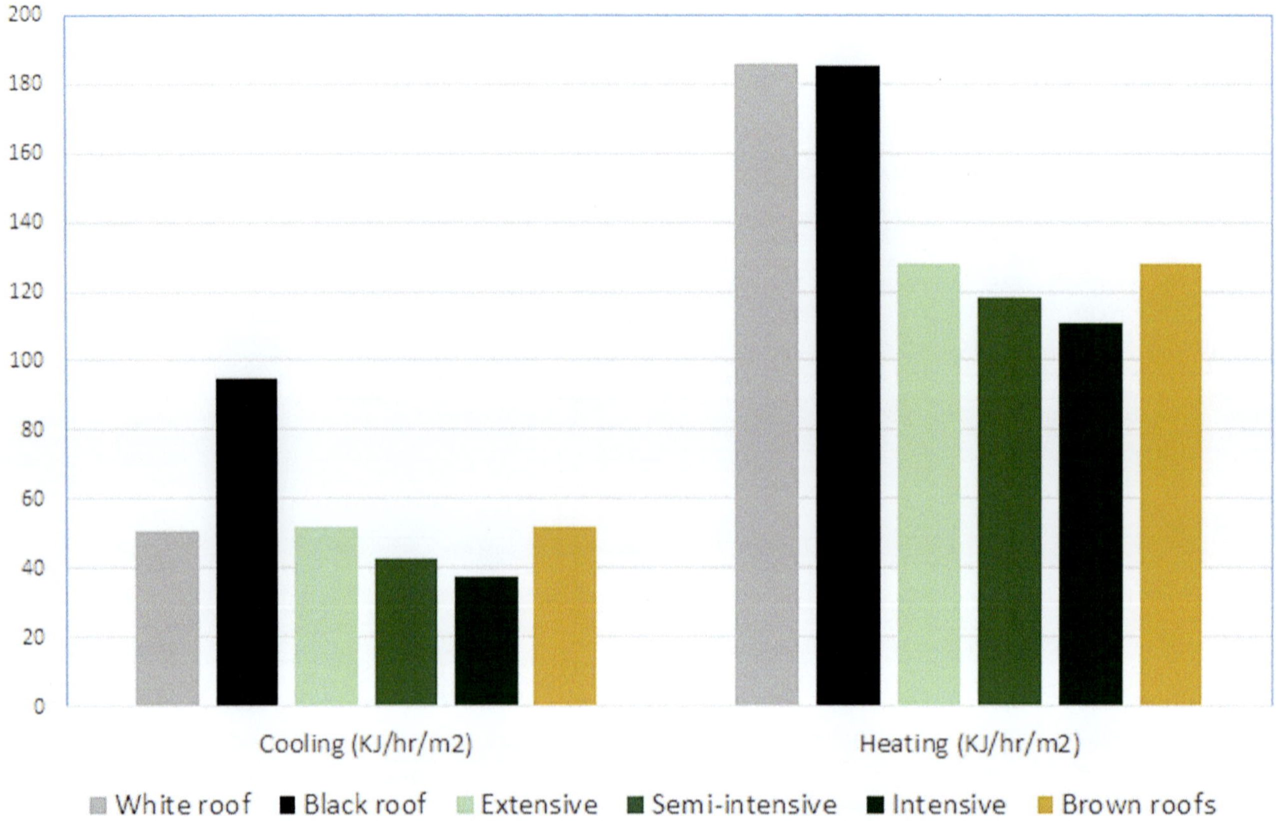

Fig. 11 Energy consumptions for the six roof cases for the cooling and heating seasons in Tripoli, Libya (KJ/h/m^2) (*source* Authors)

- The highest isolation rate is that for the intensive green roofs.
- Green roofs of all kinds provide energy consumption at high rates, especially in the heating season.
- Using white roofs can substitute for the use of green roofs in the cooling season, as the energy savings are low in most cases.

4 Conclusions

Green roofing is always acknowledged as a system with high energy performance. However, different climate conditions can result in different green roof responds to energy savings. Moreover, there are different form of green roofs with different soil depth, leaf area index, and planting heights. All these elements affect the energy performance of the green roofs. It is important to determine how each of these elements supply cooling and heating energy savings in the Mediterranean climate. EnergyPlus is used in the simulation to define every case attitude to energy in the Mediterranean city Tripoli. The results of the simulations show that for cooling seasons green roofs do not grant much energy savings compared to white roofs in the case of extensive green or brown roofs. As for the heating season in Tripoli, white roofs do not grant any energy consumption difference with black roofs. On the other hand, green roofs provide energy savings up to 40%.

The extensive green roofs and the brown roofs show the same results on both heating and cooling seasons.

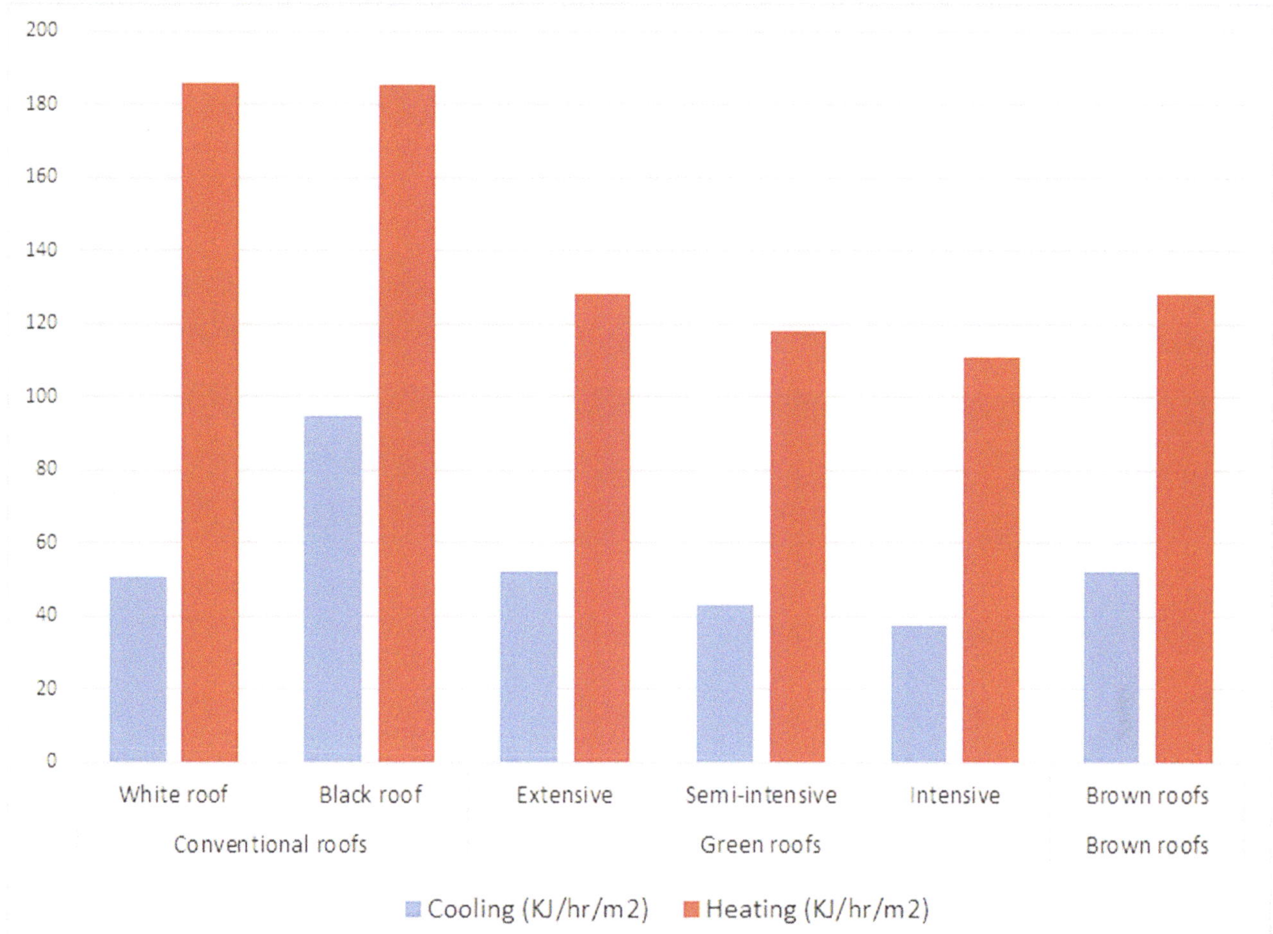

Fig. 12 Energy consumption for the cooling and heating seasons for conventional, green, and brown roofs in Tripoli, Libya (*source* Authors)

This might give favor to the brown roof than the extensive roof, as both are not accessible and cannot be used for amenity purposes while brown roofs have lower installation costs, less maintenance requirements, and more biodiversity benefits. It also has the potential to use substrates with reduced thickness and weight. Although there are different factors in addition to energy performance that play a role to determine which roof to implement such as social, functional, aesthetic, and others. It is recommended to consider these factors since that one of the main intentions to construct green roofs is the energy performance especially in terms of maintaining a reasonable economic payback.

References

Castleton, H., Stovin, V., Beck, S., & Davison, J. (2010). Green roofs; building energy savings and the potential for retrofit. *Energy and Buildings, 42*(10), 1582–1591. https://doi.org/10.1016/j.enbuild.2010.05.004

Elmazek, E. (2020). Agriroofs and agriwalls: Applications of food production in green roofs and green walls. *International Journal of Aerospace and Mechanical Engineering, 14*(12).

Jim, C. Y. (2017). Green roof evolution through exemplars: Germinal prototypes to modern variants. *Sustainable Cities and Society, 35*, 69–82. https://doi.org/10.1016/j.scs.2017.08.001

Lepp, N. W. (2008). Planting green roofs and living walls. *Journal of Environmental Quality, 37*(6), 2408. https://doi.org/10.2134/jeq2008.0016br

Liu, K., & Baskaran, B. (2003). Thermal performance of green roofs through field evaluation. In *First North American green roof infrastructure conference, awards and trade show, Chicago, IL* (pp. 1–10).

Luckett, K. (2009). *Green roof construction and maintenance (GreenSource Books) (McGraw-Hill's Greensource)* (1st ed.). McGraw Hill.

Magill, J. C., Midden, K., Groninger, J. W., & Therrell, M. D. (2011). A History and Definition of Green Roof Technology with recommendations for future research. *Research Papers, 91*. https://opensiuc.lib.siu.edu/gs_rp/91/

Peck, S. W., & Erickson, A. C. (2008). *Award-winning green roof designs: Green roofs for healthy cities (Schiffer Book) (illustrated)*. Schiffer Pub Ltd.

Sailor, D. (2008). A green roof model for building energy simulation programs. *Energy and Buildings, 40*(8), 1466–1478. https://doi.org/10.1016/j.enbuild.2008.02.001

Saudi Aramco World : Tripoli: Crossroads of Rome and Islam. (n.d.). https://archive.aramcoworld.com/issue/201003/tripoli.crossroads.of.rome.and.islam.htm

Silva, M., Silva, C. M., & Gomes, M. G. (2014, May). Green roof energy simulation (in portuguese). *DECivil, Instituto Superior Técnico, Lisbon University, Lisbon, Portugal*. https://www.greenroofs.pt/en/member/tecnico-de-lisboa

SOPRANATURE | SOPREMA. (2023). SOPREMA. https://soprema.us/sopranature/

Stromberg, M. R., D'Antonio, C. M., Young, T. P., Wirka, J., & Kephart, P. (2007). California Grassland Restoration. In *University of California Press eBooks* (pp. 254–280). https://doi.org/10.1525/california/9780520252202.003.0021

Un, K. (2017). *Fact sheet: Green roofs—MAPC*. MAPC. https://www.mapc.org/resource-library/fact-sheet-green-roofs/

Nature-Based Solutions Framework for Wildfire Risk Reduction: Evaluating Governance Recommendations in Girona Province, Spain

Anna Giulia Castaldo and Israa Mahmoud

Abstract

Nature-based solutions (NBS) can have a significant role in wildfire management, since they effectively address key climate-induced impacts and simultaneously provide social benefits improving sustainable land management. However, the NBS framework has so far been underused in wildfire management for different reasons, such as more focus on fire suppression rather than prevention, and insufficient build-up of policy synergies among fire-related sectors. This research aims to contribute to the theme of NBS implementation in wildfire management, by drawing upon a case study in the Province of Girona, on the use of prescribed/targeted grazing practice as wildfire prevention tool. This case study advances wildfire governance research by providing an evaluation model to show up positive spillovers for the future empowering role of stakeholders affected by wildfires in decision-making. Data were collected through semi-structured interviews with local stakeholders, structuring questionnaires according to the phases of the decision-making framework of NBS. Interview data were analyzed through NBS phases to provide evidence of (i) the prescribed grazing project's contribution to the local governance; (ii) components that can improve the effectiveness of such a project for wildfire governance implementation over time. Results point out significant governance supported: shaping of polycentric governance from forestry to peri-urban scale; engaging stakeholders on shared values to be protected; integrating resources and local knowledge in wildfire management; establishing formal and informal agreements to reduce landscape vulnerability; disseminating information on financial resources to employ preventive measures. Such aspects have serious implications for risk prevention and sustainable planning to preserve natural resources.

A. G. Castaldo
Department of Architecture and Urban Studies (DAStU),
Politecnico di Milano, Milan, Italy
e-mail: annagiulia.castaldo@polimi.it

I. Mahmoud (✉)
"Laboratorio di Simulazione Urbana Fausto Curti",
Department of Architecture and Urban Studies (DAStU),
Politecnico di Milano, Milan, Italy
e-mail: israa.mahmoud@polimi.it

Keywords

Nature-based solutions • Wildfire Governance • Socio-ecological resilience • Evaluation models • Governance

1 Introduction

The last fire seasons saw the nation of Spain among the major exponents of the devastion of European wildfires, during documented record-breaking heat waves (Hassan, 2022). Unfortunately, the impacts of extreme wildfire events are not expected to diminish in southern European territories, due to numerous drivers. Wildfire activity is in fact a complex interweaving of human activities, vegetation and landscape patterns, and climate conditions (Duane et al., 2021). Human-induced ignitions, which change the fire regime and, in some cases, take them outside the range of variability (McLauchlan et al., 2020), currently overlap with the effects of climate change—high temperatures, low humidity, and strong wind (Bradstock, 2010). Nature-based solutions (NBS) in wildfire management are measures which intervene on biomass reduction to control the fire regime in order to foster more fire-resistant/resilient environments. They also contribute to biodiversity conservation and the sustainable supply of ecosystem services in socio-ecological systems. Surprisingly, NBS benefits between forest fire mitigation and ecosystem services, and further co-benefits (related to health, wellbeing, and socio-cultural aspects), remain largely unexplored (FirESmart project, 2020). To date, there are many obstacles to NBS implementation in

A. Battisti et al. (eds.), *Greening Our Cities: Sustainable Urbanism for a Greener Future*,
Advances in Science, Technology & Innovation, https://doi.org/10.1007/978-3-031-49495-6_11

different policy contexts (Mahmoud et al., 2022); firstly, fire-related policies focus mostly on fire suppression rather than prevention; secondly, the effectiveness of prevention measures on biomass reduction, such as fuel reduction programs, meets with some skepticism, due to the suboptimal placement of treatments (Barnett et al., 2016; Schoennagel et al., 2009), limited technological advancements, resources and personnel (Mahmoud et al., 2024), as well as an imperfect understanding of fire behavior (Yemshanov et al., 2021). This turns out to be even more problematic if considered the high loaded temperature forests, as in Central Europe, predicted to increase in fire weather conditions in the future (Dowdy & Pepler, 2018; Millar & Stephenson, 2015). Additionally, planning with NBS needs to adapt general solutions to place-based characteristics to ensure resource efficiency and resilience (Albert et al., 2021). Some contexts in Europe show the arise of negative social and management processes leading to increased wildfire risk. Land abandonment, which involves an increase in quantity and continuity of fuel with consequences for fire regime (Pausas & Fernández-Muñoz, 2012), places Mediterranean regions in condition of risk (Di Virgilio et al., 2019; Dowdy & Pepler, 2018; Millar & Stephenson, 2015; Sommerfeld et al., 2018; Terres et al., 2015). The problem could be tackled with the NBS framework, since previous research showed NBS validate small-scale rural producers and indigenous peoples as guardians of the natural environment, and limit the impact of natural disasters on communities and their environments (Ankarfjard et al., 2021).

Some countries have been considering NBS in different environments for wildfire management, such as the promotion of economic value chains of forest products to incentivize sustainable forest management (Verkerk et al., 2018). Other regions advocate for an agricultural landscape mosaic, and silvo-pasture practices to reduce the risk of large and intense fires (Moreira et al., 2020).

Catastrophic events exhibit non-linear behaviors that are often generated by cross-scale interactions (Peters et al., 2004), and these phenomena translate into needs to be managed at the dimension of wildfire governance on the different scales within territories. Agricultural policy, for instance, can contribute to sustainable environmental management by focusing on farmers practices and their crops prosperity as well as addressing natural hazards from climate change including wildfires. The concept of NBS plays a crucial role as an holistic approach used as a means to mitigate risks and for the demanding process that requires a range of methods and tools aligned with all possible procedures (Anderson et al., 2021).

In this regard, we analyzed a case study in Girona province (Spain), and investigated whether this phenomenon enabled the activation of wildfire governance (see the location in Fig. 1). We examined the case of Girona through the lens of the NBS approach. Wildfire risk management requires presently to develop governance approaches that boost social-ecological resilience to wildfires, which often fail shortly for different reasons connected to formal and informal institutions: limitation in scope and scale, insufficient funding, or lack of broad public support (McWethy et al., 2019). The article is organized as follows and is grounded in the following research questions:

- What governance models provide the Prescribed Grazing practice analyzed through the lens of NBS framework?

In this case, the research highlighted the different governance models which were identifiable by interviewing the project's principal stakeholders. The modalities of the interview and the list of participants are presented in the next section.

- Which of these governance models should continue to be increased and which ones should be added for the future of the initiative?

2 Methodology

This research uses qualitative methodology with semi-structured face-to-face interviews. The respondents are the main engaged stakeholders of the fuel reduction program, implemented at the landscape scale in the Province of Girona. The context hosts a fuel reduction program using prescribed grazing, set up by the regional government, and the association of a pilot project to exploit grazing products (see Fig. 2). The participants and the dates of data collection are shown in Table 1. Participants have been classified according to the following four categories, according to: (1) regulatory bodies, (2) intermediaries, (3) suppliers and sellers, and (4) beneficiaries and buyers. The first group includes the members of the firefighters of the Girona province and the technical manager for Wildfire Prevention of the Environmental Department of the province of Girona. In this case, it should also be mentioned that these two institutions fall either in the category of beneficiaries and buyers, since they are directly affected by the wildfire issue and by the loss of any protective function of the NBS. The intermediaries are represented by the members of the Foundation "Pau Costa," who deal with part of the NBS management (the pilot project to exploit grazing products, called "Fireflocks"), while the Butchers' Association of the province of Girona, represented in the interview by the president in charge, belongs to the category of suppliers and seller.

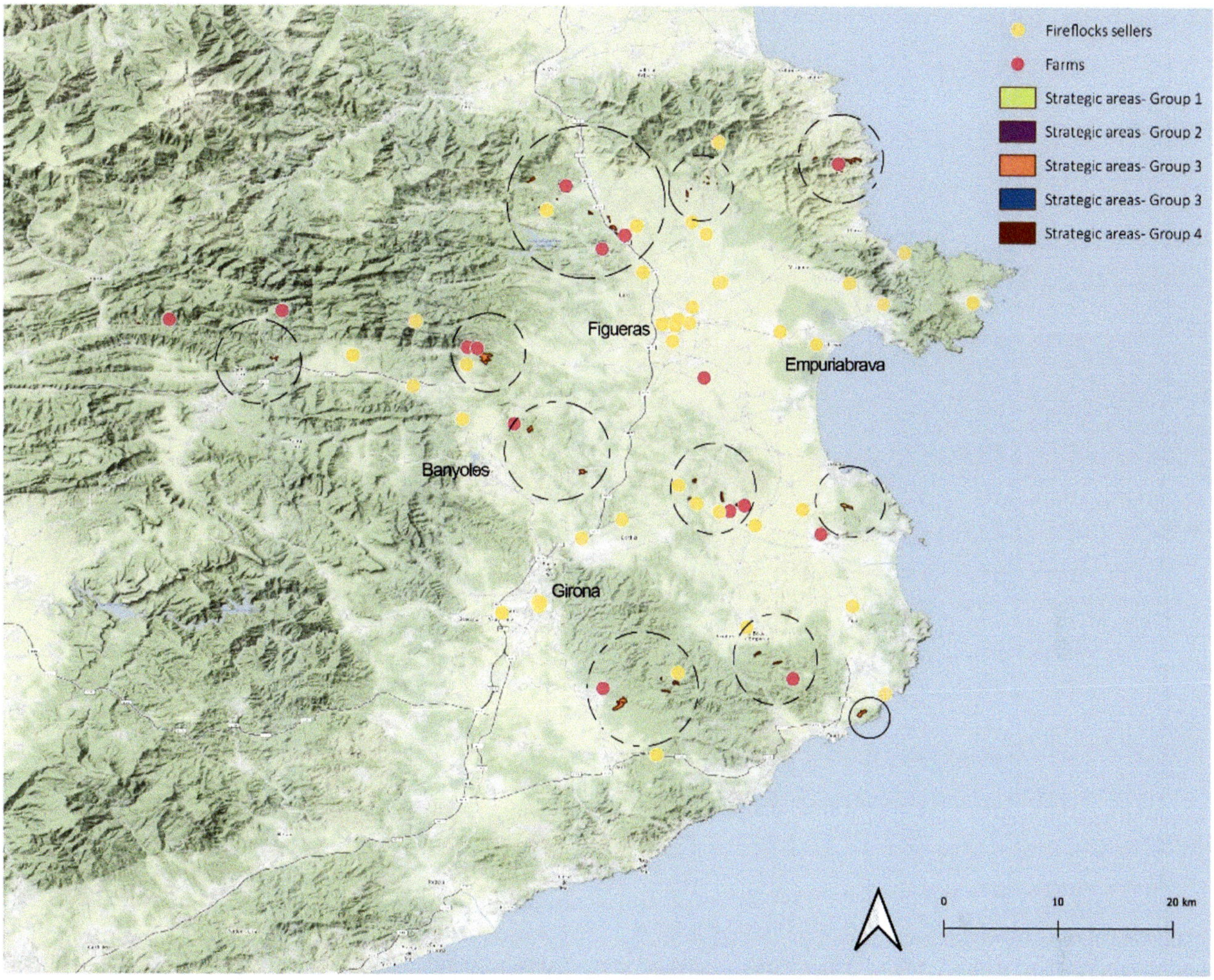

Fig. 1 Fuel reduction program at the landscape scale. Strategic areas are devoted to the reduction of biomass by the animals (prescribed grazing practice). The location points represent the areas of production or sale of livestock products supported by the associate pilot project "Fireflocks" (*source* Authors)

The interview questions were defined based on two aspects, (i) the decision-making frameworks of NBS and (ii) NBS impact indicators. NBS impact indicators were classified as structural, process, and outcome indicators, as reported by the Independent Expert Report published by the European Commission (Cardinali et al., 2021). Their meaning refers to:

- Structural indicators (S): They were considered in the interviews of all participants. The content of these indicators was used to understand the nature of the wildfire issue in the local context and the associated rationale chosen to set up the objectives of the wildfire management.
- Process Indicators (P): Indicators content refers to detect the consistency of the specific procedures related to prescribed grazing practice used to achieve the objectives of wildfire management.
- Outcome indicators (O): These indicators were used to understand the outcomes of prescribed grazing practice and impacts (measured and unmeasured) of wildfire governance.

NBS impact indicators employed refer to the following categories: Natural and Climate Hazards Indicators, Participatory Planning and Governance, Green Space Management, Biodiversity Enhancement, Economic Opportunities and Green Jobs, Knowledge and Social Capacity Building for Sustainable Urban Transformation. This experimental design was employed considering that some respondents witnessed the phenomenon by knowing specific aspects. For example, firefighter interviewees were asked questions about the strategic areas for wildfire management (adopting indicators of Natural and Climate Hazard, Green Space Management, and Participatory Planning and Governance).

Fig. 2 Herds grazing strategic areas in Girona Province. *Source* Castaldo (2022)

Interesting results from this section will be discussed in the final sections. The defined framework of the interview's deployment is shown in Fig. 3.

The analysis of the interview content was then carried out with the NVIVO 12 software. The content has been coded through the software, relying on frameworks and categories defined by the literature related to Wildfire Management and Governance. Following this approach, four governance models were defined according to the qualitative data find out. Also, in this correlation, the NVivo software was particularly useful for the data analysis, as it allowed us to link the content of the interviews to the identified categories. Among the Wildfire Governance models used for reference is Tedim et al. (2020) "Shared Wildfire Governance" (SWG).

3 Results

3.1 Girona Province Case Study and NBS Phases

The present study confirms the findings by Colantoni et al., (2020) for which effective NBS for wildfires, such as extensive grazing, require optimization of available resources and knowledge, as well as pursuing the aim of achieving calculable results. The program phases were defined and aligned with NBS implementation phases, comparing the qualitative results with the NBS principles and phases (Cohen-Shacham et al., 2019; Raymond et al., 2017).

To begin, for the "identify problem or opportunity" phase, it can be stated that the program was initially launched at the regional level in Catalonia as set up by the regional institution, incorporating prescribed grazing in the wildfire prevention system. This supports the notion that the Catalan regional context constitutes a virtuous system of implementation in the European area, considering that forest grazing is still prohibited in other countries of the world (Varga et al., 2020). In these specific cases, several conflicts have arisen between different stakeholder groups (hunters, forestry, etc.) interested in the use of forest products (Öllerer et al., 2019; Varga et al., 2020). Conversely, in other contexts, although there is a willingness to implement this instrument, national and regional institutions lack sectoral policies and references for including grazing in territorial plans (internal communications). The Catalan Regional Policy introduced the grazing practice program through legal instruments from the Rural Development Programme of

Table 1 Interviewees and dates of the interviews, *source* the Authors

Participants	Date of data collection	Number of interviewees
Members of the GRAF Team (Group of Support to Forest Actions) of the Firefighters of the Catalan Government. Section of the Province of Girona	November 12, 2021	2
President of the Butchers' Corporation of Girona County (*GREMI* acronym in Catalan);	November 15, 2021	1
Shepherds included in the *Fireflocks* Project	16, 18 November 2021	3
Members of the Pau Costa Foundation	November 17, 2021	2
Technician of the Environmental Department of the Province of Girona	November 19, 2021	1
Merchants of the Central Market of Girona (sellers of *Fireflocks* project products)	November 23, 2021	3

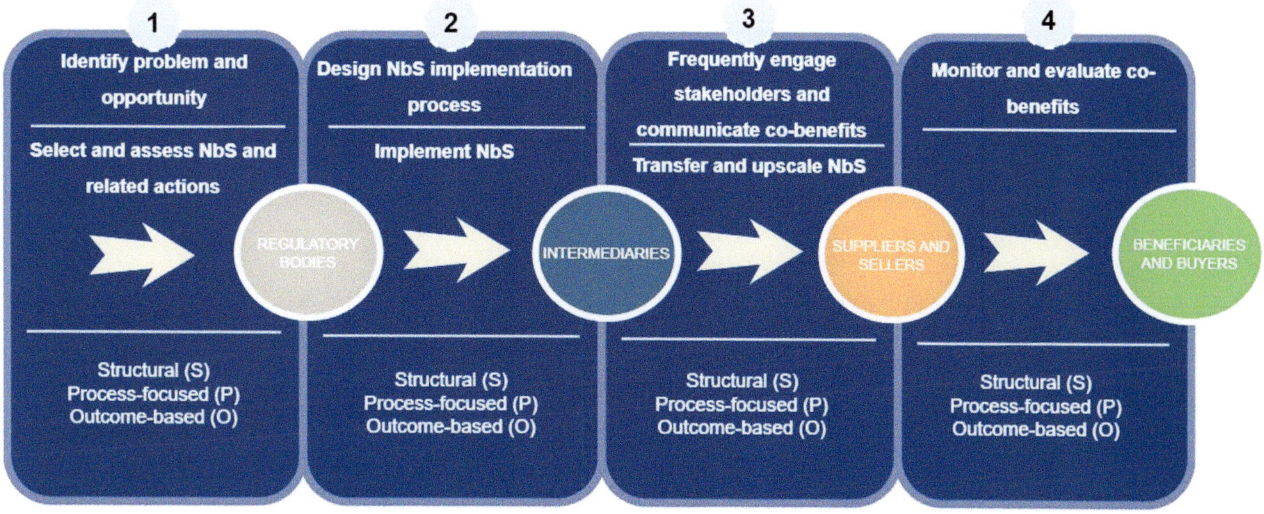

Fig. 3 Methodological model of the interviews based on the NBS decision-making framework and the NBS impact indicators (*source* Authors)

Catalonia 2014–2020 (Ordinance ARP/93/2018). These deliberations take place under EU Regulation No. 1305/2013 on support for rural development by the European Agricultural Fund for Rural Development (EAFRD). Specifically, as reported by the interviews, the process of implementation in Girona Province started by virtue of a collaborative process between the Fire Department (GRAF Section) and the Forest Science and Technology Centre of Catalonia (CTFC). In this case, both identified the practice of grazing within additional dimensions of the wildfire problem by looking for potential blind spots, or, missed opportunities. This collaboration also reviews the possible role of pastoralism in Catalonia compared to what has been affirmed by previous external research on preliminary forms of grazing service payment (Española et al., 2017). The collaborative process in fact provides the experience of the Province of Girona with budgetary stability through the regional legislative framework (Varela Redondo et al., 2017), and concession agreements with shepherds were executed to implement the prescribed grazing not only for public forests and lands, but also for private ones.

In this specific case, we refer to private and public land ownership soils falling within the so-called forested "strategic areas," identified as priority areas for controlling biomass and protecting the surrounding areas from wildfires. It should also be noted that, by observing the NBS aspect of the "ecosystem process utilization" (Albert et al., 2019, 2021), this collaboration allows to consider the acceptable level of human intervention linked to ecosystems due to the presence of CTCF. A number of authors have recognized that can influence plant and soil conditions, but the extent of impacts, namely negative spillovers, depend strongly upon grazing intensity and frequency, local climate, and the type of plant community (Eldridge et al., 2016). This supports the largely approved concept which NBS must protect essential ecosystem processes and resources (Scott et al., 2016).

Secondly, for the "Select and assess NBS and related actions" phase, another important aspect that emerged from the interview with the firefighters is the ability of the local NBS to be implemented alone or in an integrated way, which suggests creating a diversity of management approaches (Cohen-Shacham et al., 2019). In this case, the practice connected with grazing is prescribed burning, a tool used together with grazing to reduce the fire hazard (Diamond, 2009). In this case, the NBS could be shown to be implemented on the ground in an integrated mode, due to the intention expressed by local firefighters and the existence of specific operational capabilities. This integration makes it possible to link project objectives to different actions and to define potentially more indicators to measure the effectiveness of the actions.

Thirdly, for the "Design and implement NBS" phase, the program experienced the definition of two operations. The engagement phase for the marketing and valorization of the livestock products from the program opened up the process to other local stakeholders. It started with an informal meeting at the local level in one of the province's municipalities (Armentera); the firefighters encouraged this proposal adding to the initially involved institutional stakeholders the provincial association of local butchers and the Pau Costa Association members. This process, as evidenced directly and indirectly by the interviews, represented a key element for the project maintenance and recognition, given the participation of diverse groups of stakeholders. It took the name of "Fireflocks" project, supporting the regional-level framework. All this has made it achievable to support wildfire-fighting operations on a larger spatial scale as well as introduced actions belonging to functional groups in urban areas connected with the Fireflocks project. As a result, in the case of the butchers' association in the urban environment, it involved selling in their commercial activities the shepherds products within the program and engaged by the Fireflocks project. From an urban planning perspective, attention needs to be given to these processes since they bridge different knowledge on urban systems (Frantzeskaki & Kabisch, 2016). The second operation consisted of economic assistance for local municipalities in activities of wildfire management for the maintenance of WUI (Wildland-Urban Interface), creating an institutional regulatory economic mechanism that remunerated shepherds for the WUI maintenance. This process observed is usually less subsidized in peri-urban areas, with little or no incentive at the community level in other European contexts (Colantoni et al., 2020).

For the phases "Frequently engage stakeholders and communicate co-benefits" and "Transfer and upscale NBS," a different perspective is needed. Regional implementation and the urban involvement project phases are in their early stages and have limitations. To engage stakeholders, one of the necessary objectives of communication is to ask them about their level of satisfaction with participation and the status of their activities. According to Raymond et al. (2017), communication is effective when it is achieved through processes of top-down and bottom-up levels that can be parallel and overlapping. At the current state, the members of Pau Costa, for instance, conduct extensive communication and dialogue with the shepherds, allowing to bring out positive and negative feedback, such as a lack of alignment of management vision with the impositions of the central regional government.

Concerning the need to upscale the NBS framework, it requires multi-actor partnerships (Frantzeskaki et al., 2022), and in the instance of Girona, Pau Costa members stated that there are additional shepherds on the territory available to manage new strategic areas. However, it is necessary to determine the localization of these new areas from institutional actors. In addition, most forest areas are privately owned, which usually means struggling with tracing back the owners to establish maintenance agreements.

Figure 4 illustrates the impact indicators linked to the NBS phases for developing the interview protocol and the different types of interview questions. First, the table casts a new light on the need for indicators to be directly linked to extensive grazing and/or wildfire management projects in the official NBS catalogue. In addition, interesting results have emerged during the interviews. Some stakeholder groups did not have the ability to respond to questions related to specific impact categories, although their ability for response was assumed to be positive.

For the butchers' association, it was assumed that there was an ability to address the social capacity-building impact of their customers through the "Fireflocks" project. Only entrepreneurs have been claimed to have recognized learning dynamics. Instead, other stakeholders, such as shepherds, have demonstrated their awareness of phenomena influencing biodiversity or recognizing post-fire recovery times for vegetation.

3.2 Wildfire Governance Recommendations Identified

This section of results demonstrated governance features established by the Regional Programme and the Project at the local scale and discusses important recommendations on wildfire governance improvement.

Empowering stakeholders on shared values to be protected

The program resulted in the recognition of a major central theme of interest throughout stakeholders' cooperation, the local forest resources, which where partly already considered by some and underestimated by others before the intervention. According to the Firefighters, whose institutional mission is resource protection, the implementation of grazing has contributed to the improvement of the forestry resource protection by strengthening the interest of shepherds in obtaining spaces such as land and forests, to conduct wildfire management and their subsistence activities at the same time.

Furthermore, the majority of shepherds who responded felt the merit of their work to allow greater access to forestry resources by Firefighters reducing forestry and shrub areas biomass, making their passage easier in case of emergency event.

> [shepherds] are at the end of their long travel. We pay for them. They are doing a service. The objective, behind other objectives, is that we are saving money mixing two sectors in one action: from one side, shepherds need land, forest, for the animals, and for our side we have need of forest managed (GRAF department official).

Firefighters reported that this collaboration, therefore, offers a path of possible integration between wildfire policies and other policy sectors and agreements relating to the sustainable use of forests. In this sense, the interviews revealed that the shared value of the forestry resource protection (which we will define as *primary value for convention*) has been associated in interview responses with other values, which may classified as *second* nature: values related to the production of the Catalan territory, sale of products with a short supply chain, control of disappearance of rural activities. Therefore, grazing could be interpreted as a

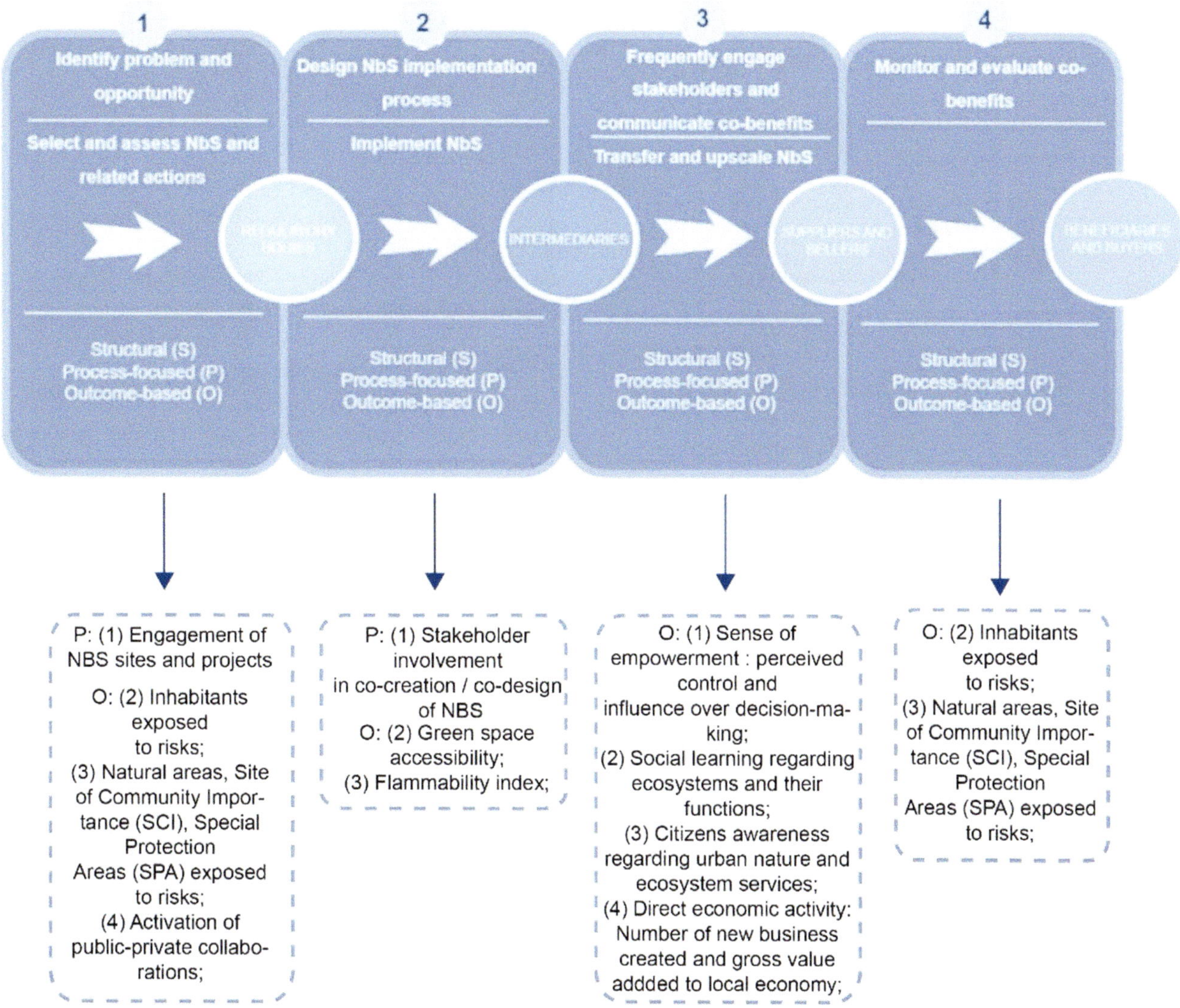

Fig. 4 Indicators used for developing the Interviews Protocol. NBS indicators have been used to set different types of interview questions. *Source* the authors

facilitator of the sharing of these territorial values, and in response to that, the regulatory bodies interviewed indicated that grazing policy, along with the sale of herds products, could be agreeably further structured in the future. In fact, the use of forest resources in the territory, compared to other European contexts, is not linked to production but to recreational and tourist aspects, as affirmed by firefighters, despite the fact that a small minority of shepherds reported the opposite—because some areas they owned produce goods such high-quality cork. It is also important to point out that the theme of cooperation and values was supported by the association Pau Costa Foundation, which, from the interviews conducted, appears to have been a crucial intermediary with a role of engaging most of the new stakeholders, particularly shepherds, during the initiative.

> "Pau Costa Foundation was the intermediary of the [Fireflocks] project and began to develop the idea of Fireflocks—market the lamb meat from the herds—and the meeting with the different actors." (President of GREMI)

From this standpoint, Pau Costa can be regarded as an appropriate intermediary for this type of projects due to the fact that has strategic objectives as well as the possibility of achieving broader goals related to wildfire management, such as the dissemination of knowledge, and interest of contribution in various sectors.

Increase knowledge and use of local resources

Over half of those questioned affirmed that the project has increased the general public's cognition of land management for risk reduction. In this case, the urban communities purchasing the so-called fire products through the Fireflocks project have understood the merit of these goods in representing an agreement among "firefighters, shepherds and butchers" (as reported by one the sellers of the herds' products from the Central Market in Girona city).

> "Fireflocks has given the benefit of giving birth and concreteness to an existing existence but giving it a better history (better management, organization of a piece of knowledge, awareness on the part of the general public)." (President of GREMI)

In addition to promoting an *external learning process for* urban communities, the project also brought new understanding within the stakeholder groups. First, the President of GREMI stated that the project has necessitated several internal communication meetings with association members. Second, association members have established a relationship of trust with the shepherds of the Project. Most were already sellers of shepherds' products, but they never met in person.

The project was also a way to identify the internal needs of individual stakeholder groups. The GREMI association presented through the project that the small slaughterhouses in operation on the territory, which are required for the production and distribution of the short supply chain, are essential tools for achieving the Fireflocks project objectives. This may raise concerns about NBS's long-term maintenance. The respondent stated that the Catalan Region should pay greater attention to the maintenance of such structures. Many slaughterhouses have been closed, forcing farmers and butchers to devote more economical and logistical resources to reach the few that remain. Under certain assumptions, this can be seen for butchers as the need to improve their associative capacities.

Establishing formal and informal agreements for the peri-urban context

A further novel finding is that the project drove the prescribed grazing in the context of peri-urban areas. From the interview with the Environmental Department of the Province of Girona, this experimental procedure relies on the existing need of public officials of the municipalities of the Province to be additionally helped in managing the Wildland-Urban Interface (WUI) areas. Usually, to make an appropriate contribution to peri-urban and urban wildfire management, the Environmental Department is responsible for assisting peri-urban protection zones from wildfires (the so-called WUI) when the Province's municipalities cannot devote enough financial subsidy for the prevention management of WUI.

Thanks to the Fireflocks project, the method of maintenance with the prescribed grazing with shepherds from the Project was introduced for WUI areas. The Province thus started providing grants not only for both machines maintenance and animals. The Department also took the initiative into significant consideration due to settlement and social characteristics related to the territory. The territory has a housing heritage of many villages with few services, often with an elderly population, particularly along the coast. In addition, numerous remote buildings in wooded areas exist, where methods to get them in case of fire are limited. Additional dynamics regarding fire management agreements for these peri-urban contexts and isolated settlements are also reported from interviews with other actors. According to the firefighters, there are informal management agreements in place through grazing maintenance. In some of these isolated settings, where houses are summer residences, agreements have been made between landowners and some local shepherds to keep the space near houses at low risk in case of fire.

Communication and use of information on financial resources

In particular, the Fuel Reduction Program allowed for the proliferation of public calls to create such legally-binding agreements, which is required to support the implementation of the Common Agricultural Policy (CAP) goals. The reiteration of such a practice supports and guarantees the continuity of payment schemes, which the CAP claims demonstrate reduced fire risk, improved biodiversity, sequestration and storage of carbon and attractive landscapes.

Interestingly, there are discrepancies among shepherds regarding their ability to apply to calls for grazing agreements. Some report that they have no issues with the bureaucratic aspects of participating in funding, while others (such as the older shepherds) experience problems. In this case, however, the interviews highlighted the Pau Costa Foundation as fundamental in this role of advisor and supporter.

Interviews even turned up a second funding source for the bottom-up level for the Fireflocks project implementation. As a result, the Girona grazing measures include a top-down economic directive essential for implementing such measures into action. However, it was later reported that the financial ability of the Fireflocks project to participate in the regional-level initiative was made possible to specific funding for innovative pilot projects provided by Catalonia to professionals in the agro-food sector. In order to adopt this financing, it is necessary to set up an association (Group Operatiu in Catalan) with partners with shared common interests.

4 Conclusions and Future Work

The qualitative findings brought out in this study report important considerations for the theme of wildfire governance by establishing a methodology with NBS framework in wildfire management. Firstly, taken together, these results show the ability of NBS implementation framework to bring out distinctive phases of implementation of a Fuel Reduction Program with prescribed grazing at different spatial scales. The experimental design highlights how this case study incorporates elements of a socio-ecological system. In this case, adding local scale dynamics to a regional model makes it more virtu-filled. Through the narrative that interviews provided, the study revealed parts of the dynamics of interaction between actors at the regional and provincial levels that would otherwise go unnoticed in the absence of such a research approach. Processes at the local scale allowed the prescribed grazing practice to advance and make it more sustainable for the socio-ecological system.

Governance processes and recommendations identified reveal benefits of various kinds for wildfire management. The case study has produced, through the regional and community levels, recognized benefits in informing people about the significance of grazing in land management for risk reduction. The work of the shepherds shows visible improvements in gaining importance, although it needs more assistance to not disappear in the future. The results confirm the need for collaboration and multi-level governance for NBS at the forest and mountain scale, as evidenced by previous research which has also shown that a higher number of actors, skills, and management capabilities results in higher resilience to climate change in mountain systems (Palomo et al., 2021). The results indicate that to increase resilience of forest and mountain systems and to foster equitable and long-lasting stakeholder engagement, it is functional to associate the preservation of forest resources with other territorial values. It is the case of the protection of the value of the forest, which was associated to internal necessities of the stakeholders in order to increase their working activities. More than the value of the remuneration of the shepherd by the regional authority of wildfire management has weighted the ability of the project to trace the internal needs of the shepherd's community. For this reason, in the future, it becomes appropriate combining policies belonging to the agricultural sector, or policies safeguarding shepherds' communities with wildfire management systems is a win–win solution. Consequently, understanding the geographical realities is essential, and further work is certainly required to disentangle these complexities.

A further novel finding is that the project drove the prescribed grazing in the context of peri-urban areas. This process, as evidenced directly and indirectly by the interviews, represented an essential phase for its maintenance and recognition, due to the participation of diverse groups of stakeholders. It took the name of "Fireflocks" project, supporting the regional-level framework. All this has made it possible to support wildfire-fighting operations on a broader spatial scale and introduced actions belonging to functional groups in urban areas, coupled with the Fireflocks project. Therefore, in the case of the butchers' association in the urban environment, it implied selling in their commercial activities the livestock products of shepherds within the program and engaged by the Fireflocks project. From an urban planning perspective, attention needs to be given to these processes since they bridge different knowledge on urban systems (Frantzeskaki & Kabisch, 2016).

The research work has presented future dimensions of the initiative, namely the fact that the practice will be associated with the prescribed burning technique, a tool used together with grazing to reduce the fire hazard (Diamond, 2009). Interviews even revealed another financial resource for the bottom-up level for the Fireflocks project implementation. Thus, the grazing measures in the context of Girona contain a top-down economic directive essential for implementing such measures. On the other hand, it was then reported that the economic capacity of the Fireflocks project to take part in the regional-level initiative was also due to specific funding for innovative pilot projects provided by Catalonia to professionals in the agro-food sector.

This may raise concerns about NBS's long-term maintenance. The most important observation relatively is that working in urban communities does not yet seem to have undergone the leap from tactical impact to transformative impact (Malekpour et al., 2021), which is recognized as an important aspect for the NBS collaborative governance. In this case, to provide an impact beyond the business-as-usual, as reported by the case studies on the subject of urban greening and climate change resilience, it is necessary to set various levels of impact. In the case of program established from the Environmental Department for peri-urban areas, they seem to provide what is termed a tactical impact of NBS, which requires the involvement of a small group of actors; however, ongoing engagement with a larger and more diverse group of actors is necessary to provide what is known as a transformative impact. In this case, the findings show that this transformation would require action on the following points: (1) keeping animals within 25 m of the interface areas, e.g., through the use of structures such as fences; (2) setting up additional support structures such as water basins for animals; (3) stepping up dialogue with municipalities where intervention can be carried out for reasons of risk reduction and grassland quality (in some cases, sites host biomass of low quality for animal feed), (4) developing a methodology that simultaneously employs machines and shepherds with different management schedules.

This last case is particularly challenging, since "*the Wildland Urban Interface is a collective band*", according to the Environmental Department, which is in line with similar experiences in other Mediterranean contexts with wildfire issues. According to the provincial technician, people do not learn to work in the forest to reduce risk. With the foundation, meetings have taken place with some municipalities also in *Alta Empordà* province, but agreements have yet been reached large consensus on implementing the practice in the interface areas at the time of the study. Grazing in peri-urban areas requires institutional resources and differentiation of application based on isolated, scattered, and dense housing. Future research should take into account the possibilities of extending the study on this research point, as there is no section of the interviews devoted to a sample of the community interested in the project.

Acknowledgements The authors gratefully acknowledge the Pau Costa Foundation for its helpfulness, the exchange, and the information provided, allowing numerous improvements to the research. The authors gratefully acknowledge Professor Eugenio Morello for his collaboration. Conceptualization: AGC, IM. Methodology: AGC, IM. Formal analysis: AGC. Writing: AGC-original draft; IM writing-review and editing. Supervision: IM. Funding acquisition: IM. Both authors have contributed equally to the results section and agreed to the final submitted version of this manuscript. This research has received funding from the internationalization funds of Department of Architecture and Urban Studies (DAStU) of Politecnico di Milano, Italy under the code Z6D3840342 entitled to Israa Mahmoud.

References

Albert, C., Brillinger, M., Guerrero, P., Gottwald, S., Henze, J., Schmidt, S., Ott, E., & Schröter, B. (2021). Planning nature-based solutions: Principles, steps, and insights. *Ambio, 50*(8), 1446–1461.

Albert, C., Schröter, B., Haase, D., Brillinger, M., Henze, J., Herrmann, S., Gottwald, S., Guerrero, P., Nicolas, C., & Matzdorf, B. (2019). Addressing societal challenges through nature-based solutions: How can landscape planning and governance research contribute? *Landscape and Urban Planning, 182*, 12–21.

Anderson, C. C., Renaud, F. G., Hanscomb, S., Munro, K. E., Gonzalez-Ollauri, A., Thomson, C. S., Pouta, E., Soini, K., Loupis, M., Panga, D., & Stefanopoulou, M. (2021). Public acceptance of nature-based solutions for natural hazard risk reduction: Survey findings from three study sites in Europe. *Frontiers in Environmental Science, 9*(July), 1–22. https://doi.org/10.3389/fenvs.2021.678938

Ankarfjard, R., Subsol, S., & Harvey Williams, F. (2021). *Nature-based solutions: Key results and lessons learned from IFAD's adaptation for smallholder agriculture programme*.

Barnett, K., Parks, S. A., Miller, C., & Naughton, H. T. (2016). Beyond fuel treatment effectiveness: Characterizing interactions between fire and treatments in the US. *Forests, 7*(10), 237.

Bradstock, R. A. (2010). A biogeographic model of fire regimes in Australia: Current and future implications. *Global Ecology and Biogeography, 19*(2), 145–158.

Cardinali, M., Dumitru, A., Vandewostijne, S., & Wendling, L. (2021). *Evaluating the impact of Nature-based Solutions: A summary for policy makers*.

Cohen-Shacham, E., Andrade, A., Dalton, J., Dudley, N., Jones, M., Kumar, C., Maginnis, S., Maynard, S., Nelson, C. R., Renaud, F. G., Welling, R., & Walters, G. (2019). Core principles for successfully implementing and upscaling nature-based solutions. *Environmental Science and Policy, 98*(February), 20–29. https://doi.org/10.1016/j.envsci.2019.04.014

Colantoni, A., Egidi, G., Quaranta, G., D'Alessandro, R., Vinci, S., Turco, R., & Salvati, L. (2020). Sustainable land management, wildfire risk and the role of grazing in mediterranean Urban-rural interfaces: A regional approach from greece. *Land, 9*(1). https://doi.org/10.3390/land9010021

Di Virgilio, G., Evans, J. P., Blake, S. A. P., Armstrong, M., Dowdy, A. J., Sharples, J., & McRae, R. (2019). Climate change increases the potential for extreme wildfires. *Geophysical Research Letters, 46*(14), 8517–8526.

Diamond, J. M. (2009). *Effects of targeted grazing and prescribed burning on fire behavior and community dynamics of a cheatgrass (Bromus tectorum)-dominated landscape*. Utah State University.

Dowdy, A. J., & Pepler, A. (2018). Pyroconvection risk in Australia: Climatological changes in atmospheric stability and surface fire weather conditions. *Geophysical Research Letters, 45*(4), 2005–2013.

Duane, A., Castellnou, M., & Brotons, L. (2021). Towards a comprehensive look at global drivers of novel extreme wildfire events. *Climatic Change, 165*(3–4), 1–21. https://doi.org/10.1007/s10584-021-03066-4

Eldridge, D. J., Poore, A. G. B., Ruiz-Colmenero, M., Letnic, M., & Soliveres, S. (2016). Ecosystem structure, function, and composition in rangelands are negatively affected by livestock grazing. *Ecological Applications, 26*(4), 1273–1283.

Española, S., Sep, D. E. P., Redondo, E. V., López-i-gelats, F., Fabré, F. P., Górriz-mifsud, E., & Lalaguna, B. C. (2017). *Governance and Resilience in Wildfire Prevention Management*, *47*(1), 6–23.

Firesmart Project. (2020). https://firesmartproject.wordpress.com/. Accessed July 2022.

Frantzeskaki, N., & Kabisch, N. (2016). Designing a knowledge co-production operating space for urban environmental governance —Lessons from Rotterdam, Netherlands and Berlin, Germany. *Environmental Science & Policy, 62*, 90–98.

Frantzeskaki, N., Mahmoud, I. H., & Morello, E. (2022). Nature-based solutions for resilient and thriving cities: Opportunities and challenges for planning future cities. *Nature-Based Solutions for Sustainable Urban Planning: Greening Cities, Shaping Cities*, 3–17.

Hassan, J. (2022). *Spain devastated by wildfires amid record-breaking heat wave*. https://www.washingtonpost.com/world/2022/06/20/spain-wildfires-2022-europe-heatwave/

Mahmoud, I. H., Morello, E., Rizzi, D., & Wilk, B. (2022). Localizing sustainable development goals (SDGs) through co-creation of nature-based solutions (NBS). In R. Bears (Ed.), *The Palgrave encyclopedia of urban and regional futures* (pp. 1–17). Springer International Publishing. https://doi.org/10.1007/978-3-030-51812-7_354-1

Mahmoud, I., Morello, E., Bisello, A., & Kolokotsa, D. (2024). Augmented Nature-Based Solutions: A Possible Taxonomy of Technologies "in" and "for" Urban Greening Strategies. In: Bisello, A., Vettorato, D., Bottero, M., Kolokotsa, D. (eds) *Smart and sustainable planning for cities and regions*. Green Energy and Technology. Springer, Cham. https://doi.org/10.1007/978-3-031-39206-1_10

Malekpour, S., Tawfik, S., & Chesterfield, C. (2021). Designing collaborative governance for nature-based solutions. *Urban Forestry & Urban Greening, 62*, 127177.

McLauchlan, K. K., Higuera, P. E., Miesel, J., Rogers, B. M., Schweitzer, J., Shuman, J. K., Tepley, A. J., Varner, J. M., Veblen, T. T., & Adalsteinsson, S. A. (2020). Fire as a fundamental ecological process: Research advances and frontiers. *Journal of Ecology, 108*(5), 2047–2069.

McWethy, D. B., Schoennagel, T., Higuera, P. E., Krawchuk, M., Harvey, B. J., Metcalf, E. C., Schultz, C., Miller, C., Metcalf, A. L., Buma, B., Virapongse, A., Kulig, J. C., Stedman, R. C., Ratajczak, Z., Nelson, C. R., & Kolden, C. (2019). Rethinking resilience to wildfire. *Nature Sustainability, 2*(9), 797–804. https://doi.org/10.1038/s41893-019-0353-8

Millar, C. I., & Stephenson, N. L. (2015). Temperate forest health in an era of emerging megadisturbance. *Science, 349*(6250), 823–826.

Moreira, F., Ascoli, D., Safford, H., Adams, M. A., Moreno, J. M., Pereira, J. M. C., Catry, F. X., Armesto, J., Bond, W., González, M. E., Curt, T., Koutsias, N., McCaw, L., Price, O., Pausas, J. G., Rigolot, E., Stephens, S., Tavsanoglu, C., Vallejo, V. R., … Fernandes, P. M. (2020). Wildfire management in Mediterranean-type regions: Paradigm change needed. *Environmental Research Letters, 15*(1). https://doi.org/10.1088/1748-9326/ab541e

Öllerer, K., Varga, A., Kirby, K., Demeter, L., Biró, M., Bölöni, J., & Molnár, Z. (2019). Beyond the obvious impact of domestic livestock grazing on temperate forest vegetation–A global review. *Biological Conservation, 237*, 209–219.

Palomo, I., Locatelli, B., Otero, I., Colloff, M., Crouzat, E., Cuni-Sanchez, A., Gómez-Baggethun, E., González-García, A., Grêt-Regamey, A., & Jiménez-Aceituno, A. (2021). Assessing nature-based solutions for transformative change. *One Earth, 4*(5), 730–741.

Pausas, J. G., & Fernández-Muñoz, S. (2012). Fire regime changes in the Western Mediterranean Basin: From fuel-limited to drought-driven fire regime. *Climatic Change, 110*(1), 215–226.

Peters, D. P. C., Pielke, R. A., Sr., Bestelmeyer, B. T., Allen, C. D., Munson-McGee, S., & Havstad, K. M. (2004). Cross-scale interactions, nonlinearities, and forecasting catastrophic events. *Proceedings of the National Academy of Sciences, 101*(42), 15130–15135.

Raymond, C. M., Frantzeskaki, N., Kabisch, N., Berry, P., Breil, M., Nita, M. R., Geneletti, D., & Calfapietra, C. (2017). A framework for assessing and implementing the co-benefits of nature-based solutions in urban areas. *Environmental Science and Policy*. https://doi.org/10.1016/j.envsci.2017.07.008

Schoennagel, T., Nelson, C. R., Theobald, D. M., Carnwath, G. C., & Chapman, T. B. (2009). Implementation of National Fire Plan treatments near the wildland–urban interface in the western United States. *Proceedings of the National Academy of Sciences, 106*(26), 10706–10711.

Scott, M., Lennon, M., Haase, D., Kazmierczak, A., Clabby, G., & Beatley, T. (2016). Nature-based solutions for the contemporary city/re-naturing the city/reflections on urban landscapes, ecosystems services and nature-based solutions in cities/multifunctional green infrastructure and climate change adaptation: Brownfield greening as an adaptation strategy for vulnerable communities? Delivering green infrastructure through planning: Insights from practice in Fingal, Ireland/Planning for biophilic cities: From theory to practice. *Planning Theory & Practice, 17*(2), 267–300.

Sommerfeld, A., Senf, C., Buma, B., D'Amato, A. W., Després, T., Díaz-Hormazábal, I., Fraver, S., Frelich, L. E., Gutiérrez, Á. G., & Hart, S. J. (2018). Patterns and drivers of recent disturbances across the temperate forest biome. *Nature Communications, 9*(1), 1–9.

Tedim, F., McCaffrey, S., Leone, V., Delogu, G. M., Castelnou, M., McGee, T. K., & Aranha, J. (2020). What can we do differently about the extreme wildfire problem: An overview. *Extreme Wildfire Events and Disasters*, 233–263.

Terres, J.-M., Scacchiafichi, L. N., Wania, A., Ambar, M., Anguiano, E., Buckwell, A., Coppola, A., Gocht, A., Källström, H. N., & Pointereau, P. (2015). Farmland abandonment in Europe: Identification of drivers and indicators, and development of a composite indicator of risk. *Land Use Policy, 49*, 20–34.

Varela Redondo, E., López-i-Gelats, F., PaunéFabré, F., Górriz-Mifsud, E., & Ciprés Lalaguna, B. (2017). Governance and resilience in wildfire prevention management: the role of pastoralism in Andalusia and Catalonia. *Pastos, 47*(1), 6–23. http://polired.upm.es/index.php/pastos/article/view/3862/3955

Varga, A., Demeter, L., Ulicsni, V., Öllerer, K., Biró, M., Babai, D., & Molnár, Z. (2020). Prohibited, but still present: Local and traditional knowledge about the practice and impact of forest grazing by domestic livestock in Hungary. *Journal of Ethnobiology and Ethnomedicine, 16*(1), 1–12.

Verkerk, P. J., Martinez de Arano, I., & Palahí, M. (2018). The bio-economy as an opportunity to tackle wildfires in Mediterranean forest ecosystems. *Forest Policy and Economics, 86*(September 2017), 1–3. https://doi.org/10.1016/j.forpol.2017.10.016

Yemshanov, D., Haight, R. G., Liu, N., Rempel, R. S., Koch, F. H., & Rodgers, A. (2021). Balancing large-scale wildlife protection and forest management goals with a game-theoretic approach. *Forests, 12*(6), 809.

Challenges of Urban Regeneration

An Integral Sustainable Design Framework to Evaluate Transferability of the 15-Minute City to Developing Cities: The Case of Blantyre, Malawi

Chitsanzo Isaac

Abstract

Worldwide, the COVID-19 Pandemic Tested Policy Frameworks and Urban Systems, Highlighting Systemic Inequities and Vulnerabilities Among Many Communities. City Authorities and Urbanists Have Devised Strategies to Better Facilitate Resilient City Living to Enable Communities to Not Only Survive But Also Thrive in Times of Crisis, Such As the COVID-19 Pandemic. Urban Planning Concepts Such As Chrono-Urbanism Have Become Popular, and the 15-Minute City Has Particularly Shown Promise As a Planning Strategy for Resiliency in the Global North. However, Its Transferability to the Global South Is Unknown. Thus, This Research Posed the Question: to What Extent Is the 15-Minute City Model Integral and Sustainable Design, and Is It Transferable to Developing Cities? Using Secondary Quantitative Data Analysis and Convergence Analysis in the Urban Region of Blantyre, Malawi, this research aimed to test the conceptual approach of the 15-Minute City Against the Framework of Integral Design Theory and Explore Its Transferability to Developing Cities. This Study Used the Perspectives of Experience, Performance, Culture, and Systems to analyze and map the impacts of the 15-Minute City On Urban Dwellers. The Research Findings Showed That the 15-Minute City Model Could Not Be Transferred to Developing Cities Unless Modified to a Framework That Considered Developing Urban Phenomena Such As Informality and Low Digitization. Therefore, a More Suitable Approach Would Be the Integral Developing City Concept. This Model Can Be Deployed in Harmony With Informal Systems Prevalent in Most Developing Cities in the Global South and to steer sustainable development.

Keywords

15-Minute City • Integral Developing City • Chrono-urbanism • Urbanism • Malawi • Blantyre

1 Introduction

The COVID-19 pandemic has tested policy frameworks and urban systems worldwide, highlighting systemic inequities and vulnerabilities among many communities. Confinement and social distancing to contain the COVID-19 virus shocked cities' physical and social fabrics, causing urban dwellers to reassess how they occupy their urban space and maintain social ties. The COVID-19 virus and the subsequent global pandemic also highlighted the overdependence on automobiles, which was already extensively documented to have severe negative impacts on the socio-economic structure of cities (Moreno et al., 2021).

Although the stipulations for lockdown were similar globally, the negative impacts varied from city to city. The quality of outdoor space and social cohesion became a defining factor in how people coped on psychological and physical levels (Silveira et al., 2022). The COVID-19 pandemic taught us that cities need social, economic, and environmental resilience. In addition, there is a pressing need for city leaders and planners to address mental health and welfare in times of crisis and overwhelming stress. Furthermore, resilient measures for climate change must be implemented while addressing urban vulnerabilities.

According to Moreno et al., COVID-19 prompted a need for innovative mechanisms for cities to pursue their economic activities while enforcing strict health protocols. Of the many forms of urbanism tested for resiliency during this period, the novel concept of "Chrono-Urbanism" demonstrated prioritizing building more humane urban fabrics (Moreno et al., 2021). Chrono-Urbanism argues that the quality of urban life is inversely proportional to the amount

C. Isaac (✉)
Thomas Jefferson University, Philadelphia, PA, USA
e-mail: chitsanzo.isaac@gmail.com

A. Battisti et al. (eds.), *Greening Our Cities: Sustainable Urbanism for a Greener Future*,
Advances in Science, Technology & Innovation, https://doi.org/10.1007/978-3-031-49495-6_12

of time invested in transportation, particularly automobiles (Moreno et al., 2021); thus, it provides urban dwellers with a framework to meet their physical, mental, economic, and social needs. Cities like Paris and Barcelona championed this urban concept by implementing the "15-Minute City" concept.

Given this background, this study aimed to test the conceptual approach of the 15-Minute City against the framework of Integral Design Theory (Fleming & Roberts, 2019) and explore its transferability to developing cities. Using the Integral Sustainable Design framework's quadrants of experience, performance, culture, and systems as variables, this research mapped and analyzed the components of the 15-Minute City. It explored whether the 15-Minute City could be applied to cities in the Global South using a case study of Blantyre, Malawi. This approach allowed the author to investigate the transferability of the 15-Minute City concept to examine its merits, applicability, influences on quality of life, and areas for future research.

1.1 Research Methods

To explore the intersection of Chrono-Urbanism, developing cities, and the thematic focus areas mentioned above, this study conducted a thorough literature search that included peer-reviewed academic literature, government documents, and reports from multiple disciplines. This search aimed to identify current trends, impacts, issues, and responses relevant to the subject matter. Then, the experiment defined and analyzed the structure and framework of the 15-Minute City. Its components were divided into density, proximity, diversity, and digitalization. The next step was to determine the viability of these parameters for developing cities using a case study of Blantyre, Malawi. Then, angular segment analysis and metric angular step analysis were used to analyze the proximity dimension and an integrated analytical evaluation was performed using the Integral Sustainable Design Framework.

1.2 Literature Review

1.2.1 15-Minute City

The literature review discusses the relationship between Chrono-Urbanism, developing cities, and the 15-Minute City concept.

The Global North has seen a shift in the thinking of how the resilient post-pandemic city will look. Numerous concepts have been identified as people-centric, sustainable, and resilient for the future of cities, such as Tactical Urbanism, Chrono-Urbanism, and Compact Cities. Tactical Urbanism is an approach to neighborhood building and activation using short-term, low-cost, and scalable urban interventions (Lydon & Garcia, 2015). On the other hand, Chrono-Urbanism has introduced concepts such as the 20-, 15-, and even 1-min Cities.

The 15-Minute City, as coined by Moreno et al., is the manifestation of Chrono-Urbanism, in which the proximity of urban dwellers to amenities is the determining factor of urban organization (Moreno et al., 2021). The social needs of different demographic groups are met through the 15-Minute City. For example, basic amenities for younger people, like children (playgrounds and recreational facilities), may differ from those for the elderly. Furthermore, all these different groups of city users have other walking behavior and speeds.

Moreno et al. also noted that different demographic groups are not provided with sufficient social needs without a walkable distance. At any time, one or more groups were deprived of basic amenities (Moreno et al., 2021). This means that in many cities, urban dwellers lack the basic needs to survive and thrive, especially in times of crisis.

In addition, urban dwellers perceive maintaining close ties with the community as crucial for increased quality of livability (Moreno et al., 2021). The 15-Minute City similarly focuses on other dimensions, such as ecological sustainability, promoting social interactions, citizen participation, and diversity. These dimensions can lead to more closely knitted communities and address the dependence on automobiles by emphasizing the proximity of all essential services.

However, there have been some counterarguments toward the 15-Minute City concept. Harvard professor Ed Glaeser has raised the concern of the concept of reinforcing spatial segregation in communities if not properly planned. It also runs the risk of adding the burden of inequity to some already fragmented neighborhoods (Carlo & Muggah, 2023). He also argues that low-income people must have the ability and flexibility to travel beyond their own communities for employment opportunities in different city locations. Others, such as urban designer and thinker Jay Pitter, have also warned that transplanting the concept to other geographic contexts may seem "presumptive and colonial" (O'Sullivan, 2021). This can add more pressure to the existing post-colonial social tensions of cities struggling with oppressive colonial legacies. Furthermore, to some more seasoned urban planners, the 15-Minute City concept merely takes urban planning strategies that have existed for many years and simply repackages them (Chamberlain, 2022). However, Allam et al. argue that developing a 15-Minute City does not propose overhauling and restructuring cities to erase urban areas' identity and strength but offers new economic and social growth opportunities (Allam et al., 2023).

It should be noted that another comparable example is the Compact City concept. In examining how this urban model is practiced, Bibri et al. showed that similar to the 15-Minute City, the predominant design strategies of Compact City planning encompass compactness, mixed land use, diversity, density, sustainable transportation, and the inclusion of green spaces (Bibri et al., 2020). According to Racki et al., cities with higher density and compactness, along with mixed land uses, have the potential to decrease the energy requirements for community services and transportation (Racki et al., 2014).

However, they also found that to encourage sustainable urban growth, economic objectives often take precedence over environmental and social objectives (Bibri et al., 2020).

1.2.2 Chrono-Urbanism

The 15-Minute City is derived from the concept of Chrono-Urbanism. According to Moreno, six social functions are essential to urban life, and residents will enjoy a higher quality of life when they can effectively fulfill these functions. These include living, working, commerce, healthcare, education, and entertainment (Moreno et al., 2021). He posits that these functions can be attained by restructuring the urban landscape to ensure it complies with proximity, diversity, density, and ubiquity (Moreno et al., 2021). In this way, the city can offer an urban life of high value. Furthermore, Moreno suggests this can be major for those pursuing cities offering a thriving urban life (Moreno et al., 2021).

Despite the increasing acceptance of Chrono-Urbanism, Lu and Diab identify vulnerabilities in what they call "x-minute cities." Even with the popularity of the 15-Minute concept, there has been little effort to understand its determinant. They argue that even though many cities have tried to adopt the model, there have been challenges in defining the parameters of practical implementation. Furthermore, some have deviated from the original concept by adding cars as a mode of transportation (Lu & Diab, 2023).

1.2.3 Developing Cities and the Choice of Blantyre, Malawi as a Case Study

Developing countries are defined by a low gross domestic product (GDP) per person (Kuepper, 2022) (The World Bank Group, 2022). Although the primary definition is economy-based, the socio-economic characteristics of such nations include less industrialized countries, which tend to have a lower per capita income and often rely on agriculture as the primary form of industry (Kuepper, 2022). These countries are found on the United Nations (UN) and World Bank (WB) lists of least developed countries (LDC) (Nations, 2022). A good example of an LDC is Malawi, a country whose economy has not yet matured (Silver, 2010). As the second largest city in Malawi, Blantyre shares many characteristics of developing cities and offers a manageable size for a case study. Developing cities possess the same traits as developing countries and often present with socio-economic phenomena such as: economically marginalized communities, informal settlements, and economic activities, low capacity of electricity and water, and urban agricultural practices (Nations et al., 2018). Due to high urbanization growth rates, developing cities are characterized by high-density urban spaces with overcrowded sidewalks due to many outdoor daily functions (United Nations Human Settlements Programme (UN-Habitat), 2020). Key locations form social nodes such as marketplaces and religious centers. Traffic congestion is prevalent in these cities due to overwhelmed or inadequate infrastructure, which usually does not accommodate bicycles and pedestrians—the highest forms of mobility.

According to Mwathunga and Donaldson (2018), these urban characteristics were formed despite colonialism through planning that has emphasized the production of abstract spaces as its end product (Mwathunga & Donaldson, 2018). These abstract spaces become zones of space-use intensity and reflect how informal settlements are more than just residential areas; they are foremost economic sites of production. Carrizosa highlights that informality helps city function. This is demonstrated by how informal service providers guarantee the flow of goods and services wherever formal conduits do not reach. (Carrizosa, 2022).

1.2.4 Impact of the COVID-19 Pandemic on Developing Cities

In many developing countries, numerous essential daily functions occur outdoors in city space. Cultural, climate, and economic conditions ensure that life in the city significantly impacts living conditions and quality of life (Gehl, 2010). Factors such as overpopulation of existing housing areas can put pressure on cities and overwhelm utility and traffic systems and shared public spaces. According to Gehl (2010), rapid urban population rise in developing countries is the catalyst for numerous urban problems and challenges. In various regions, overwhelming numbers of new urban dwellers have resulted in the spread of large, densely populated, informal housing areas with primitively constructed dwellings lacking most forms of essential services (Gehl, 2010).

Traditional outdoor life featuring handicrafts, street exhibitions, street trade on sidewalks, street kitchens, and small eateries along buildings and street edges are also under pressure. All types of outdoor activity in the city are negatively impacted by noise, pollution, and insecurity (Gehl, 2010).

The COVID-19 virus has disproportionately impacted cities of the Global South compared to those of the North (Sareen et al., 2021). Sareen et al. argue that vulnerability, social transformation, and resilience in the face of COVID-19 depend on various factors. These factors include demographics (such as age, gender, and socio-economic status), the quality, location, and size of homes and workplaces, access to public services, reliable energy, clean water, digital services, and urban governance. Moreover, preexisting disparities have significantly impacted how cities have adjusted to the pandemic's impact on urban life. The degree of resilience shown by city residents in adapting to the "new normal" has been unevenly distributed in the Global North and South (Sareen et al., 2021).

In April 2020, the United Nations Economic Commission for Africa expressed the fear that even with social distancing, the continent of Africa might have nearly 123 million cases of COVID-19 that year and that 300,000 people might die of the disease (Ogunbiyi, 2022). Furthermore, it was projected in 2020 that developing countries already in poverty would be pushed to extreme poverty due to COVID-19, as shown in Fig. 1. Although the virus had hit Sub-Saharan Africa relatively less from a health perspective, early projections suggested that it would be the region hit hardest in terms of increased extreme poverty. One projection suggested that 23 million people living in Sub-Saharan Africa would be pushed into poverty and 16 million in South Asia (Mahler et al., 2020).

However, although public spaces in the Global North became dynamic places of gathering, well-being, and civic expression (Slater et al., 2020), the same was not the case in the Global South. Indicators that the lockdown experience in the Global South proved more challenging included: mental health, physical health, access to amenities and food, access to outdoor space, access to employment opportunities, access to ICT technological resources for employment, access to childcare, access to education and social inclusion (Sareen et al., 2021).

During this time, disparities in housing conditions and energy consumption, as well as vulnerabilities of slums and informal settlements, posed significant challenges in responding to the pandemic. The preexisting digital divides, which are evident even in well-connected cities, resulted in unequal readiness to adapt to lockdown measures. This, as highlighted by Sareen et al. (2021), contributed to the uneven impact of the pandemic on different social groups. (Sareen et al., 2021). Meanwhile, Chirisa et al. (2020) noted that implementing lockdown measures in densely populated settlements has been particularly challenging, as social distancing was virtually impossible due to overcrowding. Consequently, the survival prospects of slum dwellers were severely undermined. (Chirisa et al., 2020). Therefore, planners must pay special attention to how they can achieve resilience in developing cities, particularly in the Global South.

1.3 15-Minute City Framework

According to the Congress for New Urbanism (CNU), the 15-Minute City has been defined as an ideal geography where most human needs and desires are located within a travel distance of 15 min. It emphasizes providing access to all these fundamental needs and amenities by walking or bicycling for that amount of time (Duany & Steuteville, 2021). However, even with the basic framework of this concept being mostly well received, Moreno et al. modified the original model and proposed a "modified 15-Minute City" framework. Figure 2 depicts four identified dimensions to make the model more efficient.

These are (a) density, (b) proximity, (c) diversity, and (d) digitalization (Moreno et al., 2021). The four dimensions identified are based on observations of different cities across the globe, the challenges they endured during the height of widespread cases of the COVID-19 virus, and the subsequent lockdown measures and protocols followed to mitigate the spread of the virus.

These four dimensions express the most critical factors leading to city dwellers' urban living experiences ranging

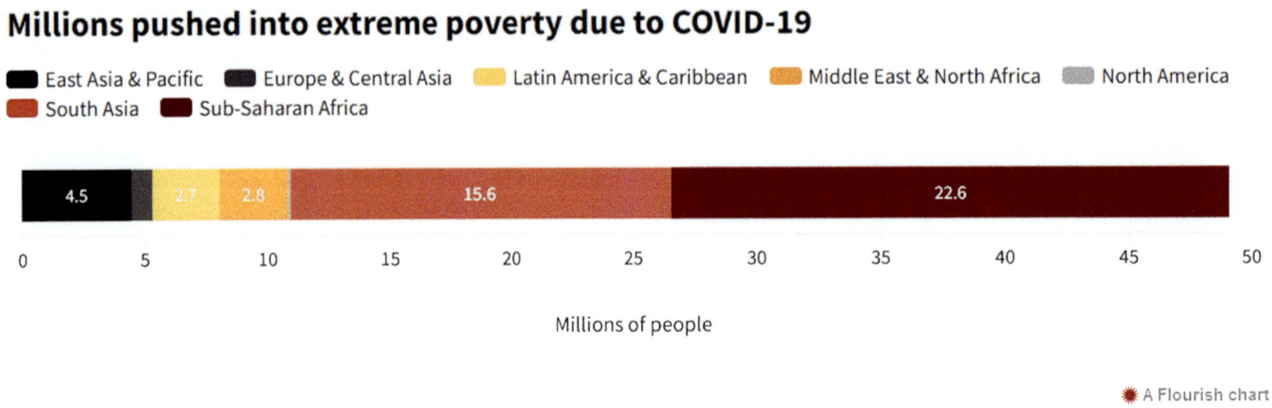

Fig. 1 Millions pushed into extreme poverty due to COVID-19 (Mahler et al., 2020)

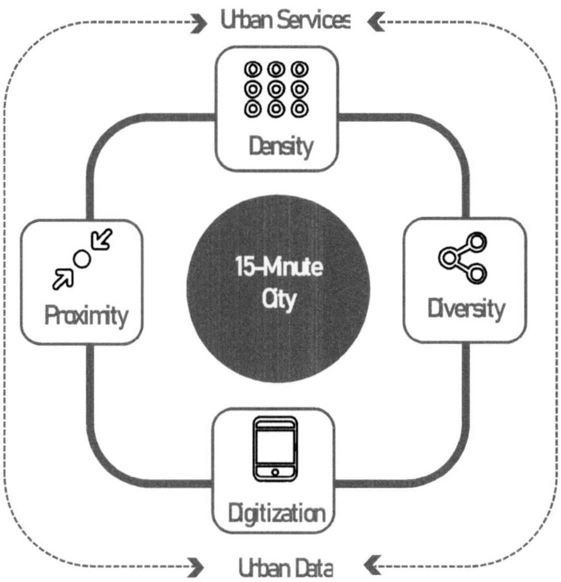

Fig. 2 The 15-Minute city framework. Illustration adapted from Moreno et al. (2021)

from bearable to intolerable. In addition, access to urban services and data also determined people's quality of life while they were confined to their homes.

Density

According to Moreno et al., density is a critical dimension of the city and the built environment. Numerous researchers have noted the direct link between travel and density. In mainstream planning practices, city planners have equated density with high-density buildings; however, in the 15-Minute City model, density is defined in terms of people per square kilometer (Moreno et al., 2021). Therefore, it is crucial that the optimal number of people in a given area be comfortably sustained through urban delivery and balanced resource consumption. (Moreno et al., 2021).

Proximity

Moreno views proximity to be both temporal and spatial. This means that residents can readily and quickly access essential services within 15-Minute radial nodes. According to the CNU, an average individual can comfortably walk approximately ¾ of a mile within a 15-Minute timeframe. This suggests that their accessible area, which includes essential destinations such as grocery stores and amenities, extends to about 1.75 square miles around their residence. Alternatively, using a bicycle allows for covering a distance of two miles, as illustrated in Fig. 3 (Steuteville, 2021).

Diversity

For the dimension of diversity, Moreno highlights two main factors that are crucial to achieving a healthy 15-Minute City: (i) mixed-use neighborhoods that encourage a healthy mix of commercial, residential, and recreational components for entertainment, and (ii) focus on culture and diversity of the city dwellers (Moreno et al., 2021).

Mixed-use spaces sustain economically vibrant urban fabrics and optimize everyday activities such as living, working, commerce, healthcare, education, and entertainment. They also ensure adequate housing and encourage sustainable practices. In addition, embracing diversity also promotes social cohesion, capital, and inclusivity. (Moreno et al., 2021).

Digitization

Digitization is a key tenet of the dimension of proximity. This demonstrates that physical proximity is not necessary to enhance everyday activities. A clear example of this has been found in the Smart City concept, where virtual services such as cashless transactions and online shopping are favored. Further, human interactions, communications, and work are also performed virtually (Moreno et al., 2021). However, they highlight the socio-economic divide of inequitable access to technologies and virtual transactions.

1.4 Integral Sustainable Design Framework

The Brundtland Commission defined modern-day sustainable development in 1987 (Commission on Environment, n. d.): "Sustainable development is the development that meets the needs of the present without compromising the ability of future generations to meet their own needs." This definition birthed the three-ring-sector-view Venn diagram, which intersects the Environment, Society, and Economy to give what is known as the Triple Bottom Line (Fig. 4). Giddings evolved this framework to become nested instead of intersected. Nested sustainable development implies that the economy depends on society, and both are dependent on the environment (Giddings et al., 2002).

Even though the Giddings framework has been recognized as a robust sustainability model, Integral theory has taken further steps to include the underlying complexities of sustainable urban development. Integral theory, as conceptualized by the contemporary American philosopher Ken Wilber, is a philosophical map that integrates over 100 ancient and modern theories from fields such as psychology, philosophy, contemplative traditions, and sociology. Instead

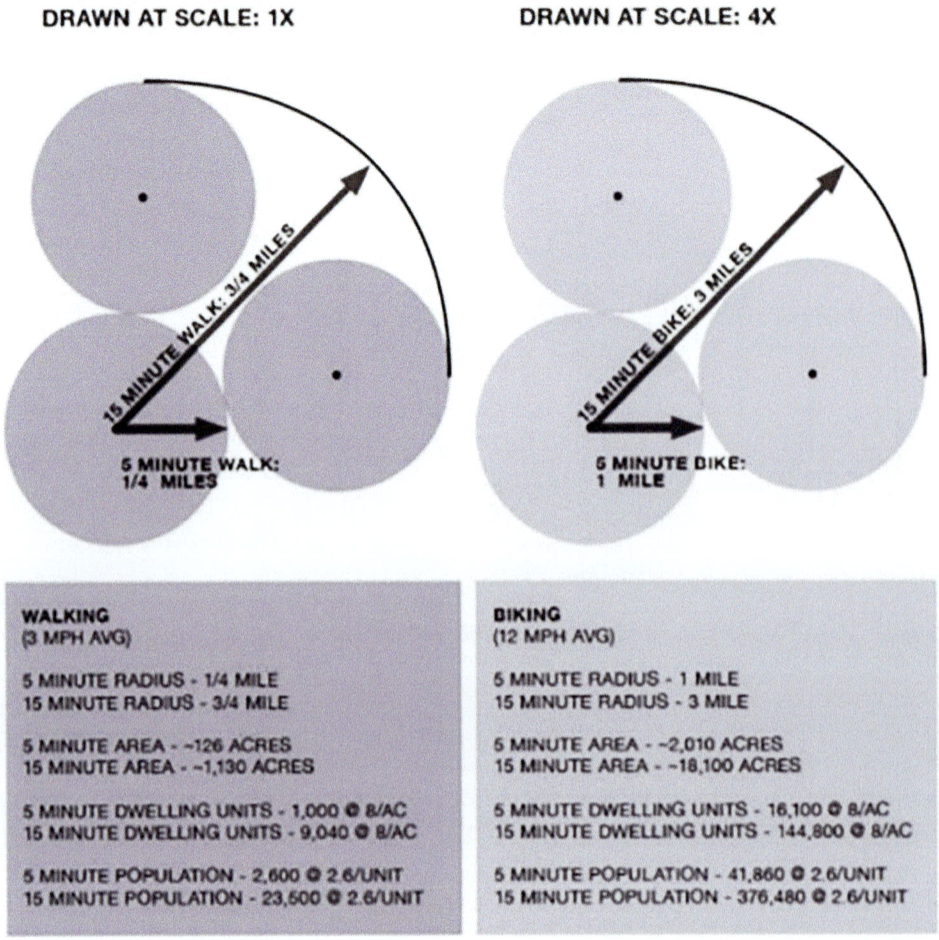

Fig. 3 15-Minute radii compared from walking to bicycling. (Duany & Steuteville, 2021)

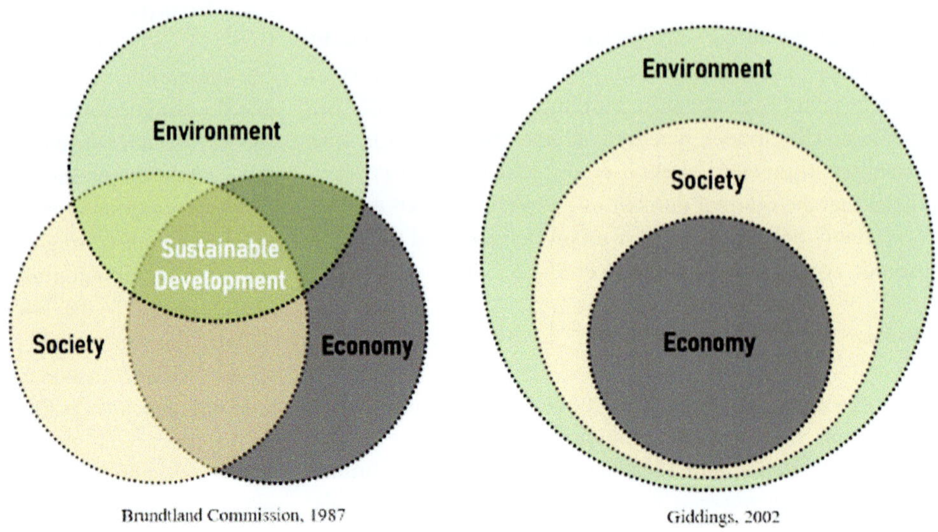

Fig. 4 Triple bottom line by Bruntland Commission and Giddings (Giddings et al., 2002)

of prescribing a single defining perspective, Integral theory aims to provide a framework for comprehending and appreciating the unique insights and perspectives offered by each theory and philosophical tradition and their interrelationships (Duffy, n.d.).

Integral theory is known for the use of the four quadrant model. It divides all sciences and approaches to knowledge into four quadrants by a 2 × 2 matrix. It has evolved through many iterations, including that seen in Fig. 5 (Integral European Conference, 2022).

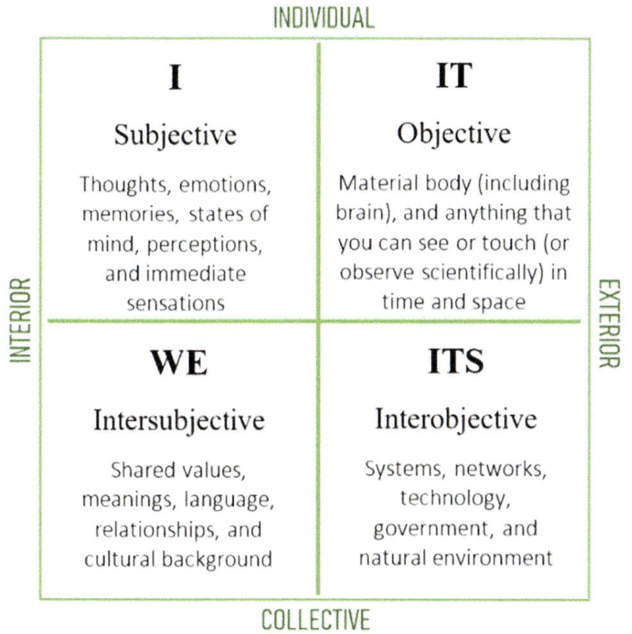

Fig. 5 Integral Theory Maxtrix (Integral European Conference, 2022)

Following this foundation, we see the introduction of the Integral Sustainable Design framework for the built environment by Fleming and Roberts, in which Place substitutes the I Subjective quadrant, Culture with the WE Intersubjective quadrant, Profit with the IT Objective quadrant, and Planet with the ITS Interjective quadrant (Fleming & Roberts, 2019). Figure 6 shows how the framework further evolves to translate to elements required for the holistic design approach of the built environment on many scales. The Experience quadrant refers to beauty, Culture relates to equity, Performance refers to metrics and quantifiable goals, and Systems refers to ecological and human-made systems (Fleming & Roberts, 2019).

Inspired by Marilyn Hamilton's Integral City Inquiry and Action, this framework allows for a holistic view of the sustainable city. Cities are complex systems. They have many interconnected systems like sewers, roads, public transportation, electric grids, people, birds, plants, and animals, all functioning and coexisting in a very limited space (Fleming & Roberts, 2019). Jane Jacobs states, "mixture of uses, if it is to be sufficiently complex to sustain city safety, public contact, and cross-use, needs an enormous diversity of ingredients" (Jacobs, 1993). Therefore, according to Fleming & Roberts and Jacobs, anyone involved in an urban design project, from a city planner to a neighborhood resident or a stakeholder, drives the project's long-term success by offering valuable perspectives toward a holistic solution.

Looking closer at the city through each quadrant, one can break down the sustainable design goals required for holistic design, as shown in Table 1.

1.4.1 Experience

It is vital to work toward inner health and psychological well-being. This can be done by beautifying the city and engaging all the senses. Bio-inspired design can be applied to parks, roadways, and building exteriors. This matters because the sense of place, a critical sustainability value of particular interest to designers, is the last key to a sustainable city (Fleming & Roberts, 2019). Furthermore, the fundamental need for beauty and a sense of connection and belonging to a geographical location urges the designer to integrate existing ecological, historical, and cultural elements into their design spaces. According to Fleming and Roberts, this applies at every scale and creates spaces and

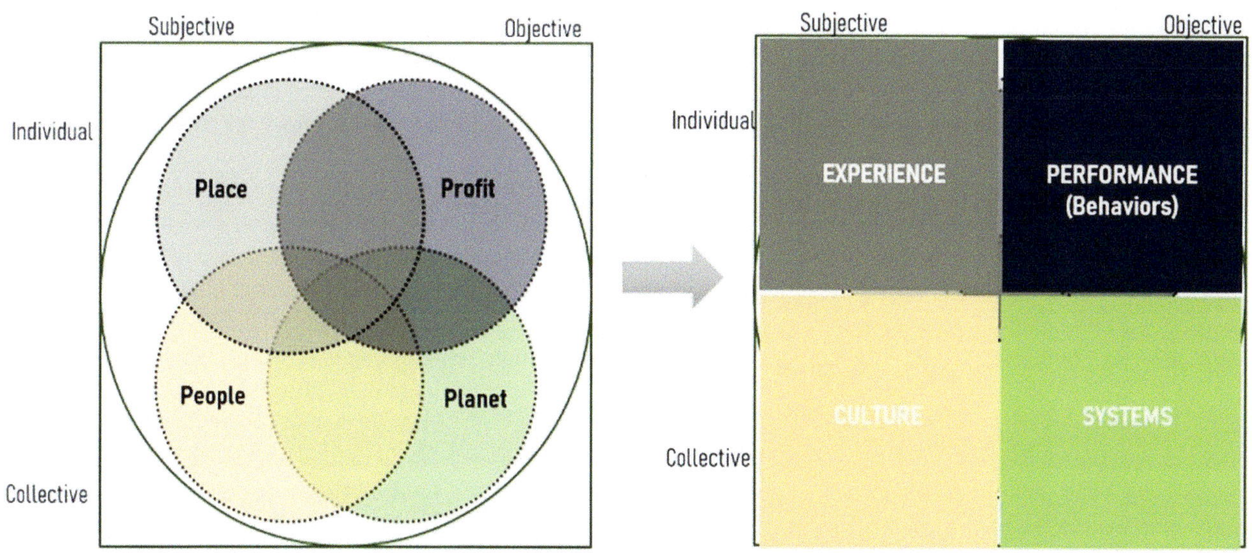

Fig. 6 Integral Sustainable Design Framework (Fleming & Roberts, 2019)

Table 1 Sustainable urban design goals within the Integral Framework (Fleming & Roberts, 2019)

	Subjective	Objective
Individual	*Experience* • **Work toward inner health and psychological well-being** • Make the city beautiful and engage all the senses • Apply bio-inspired design to parks, roadways, and building exteriors	*Performance* • **Work toward "outer health," physical health, and well-being** • Increase availability of healthy food • Increase walkability • Increase bicycle-friendly urban design • Maximize energy efficiency • Maximize water efficiency
Collective	*Culture* • Create **culturally vibrant urban spaces** and places and celebrate the heritage of historic structures • Embed **social equity** into every decision with transparency and accountability • Build a socially resilient urban culture	*Systems* • **Life-cycle Systems**: grow local food, bring back local manufacturing, and process waste locally • **Passive Systems**: organize buildings to maximize potential for solar • **Active Systems**: Use Smart City technology to optimize city systems for resilience • **Living Systems**: Create a regenerative and resilient urban ecology • **Human Systems**: Build a strong economy

places people want to spend time in and care for, creating a rich, vibrant, and diverse urban landscape. Creating viable open-access public spaces interwoven between buildings and informed by ecology is a challenging and rewarding aspect of sustainable design (Fleming & Roberts, 2019).

1.4.2 Culture

Creating culturally vibrant urban spaces and places that celebrate historic-structure heritage is also crucial. Similarly, designers must embed social equity in every decision with transparency and accountability. This allows for the building of a socially resilient urban culture. Fleming and Roberts state that the city is also a place where social equity is constantly on display simply due to the number and diversity of people living in such close proximity. Inevitably, conflict emerges. Truly sustainable cities value access, educational opportunities, and infrastructure and provide these resources for the entire democratic city (Fleming & Roberts, 2019).

1.4.3 Performance

Working toward "outer health," physical health, and well-being is essential. This is achieved by: increasing the availability of healthy food, improving walkability, increasing bicycle-friendly urban design, and maximizing energy and water efficiency.

1.4.4 Systems

Life-cycle systems that grow local food, bring back local manufacturing, and process waste locally are in demand. Passive systems are required to maximize the potential for solar energy and water harvesting in the built environment. Active Systems allow the use of Smart City technology to optimize city systems for resilience, while Living Systems create a regenerative and resilient urban ecology. Finally, human systems can build a strong economy. Table 1. Sustainable urban design goals within the Integral Framework (Fleming & Roberts, 2019).

2 The Developing City Context: A Case Study of Blantyre (Malawi)

Located in the Southeast of Sub-Saharan Africa, Malawi is a developing country that is landlocked and was formerly known as Nyasaland. Zambia forms the western border, Tanzania lies to the north and northeast, and Mozambique shares borders with Malawi to the east, south, and southwest (Fig. 7). With a total area of 118,484 km^2 (45,747 sq mi), Malawi is home to an estimated population of 19,431,566 as of January 2021 (Government of Malawi & Malawi Department of Tourism, 2022). The country comprises four major urban areas, depicted in Fig. 7, namely: the capital city of Lilongwe, located in the central region; Mzuzu in the northern region; the former capital city of Zomba; and the commercial city of Blantyre situated in the southern region.

According to UN-Habitat, Malawi demonstrates rapid urbanization, ranking among the world's fastest-growing countries in this regard, with an annual urban growth rate exceeding 5%. Urban growth in Malawi, which is estimated at 214,000 individuals annually between 2020 and 2025, will outpace rural growth (193,000) before 2025 (UN-Habitat, 2022).

The findings of the 2008 Population and Housing Census revealed that 15.3% of the total population resides in urban

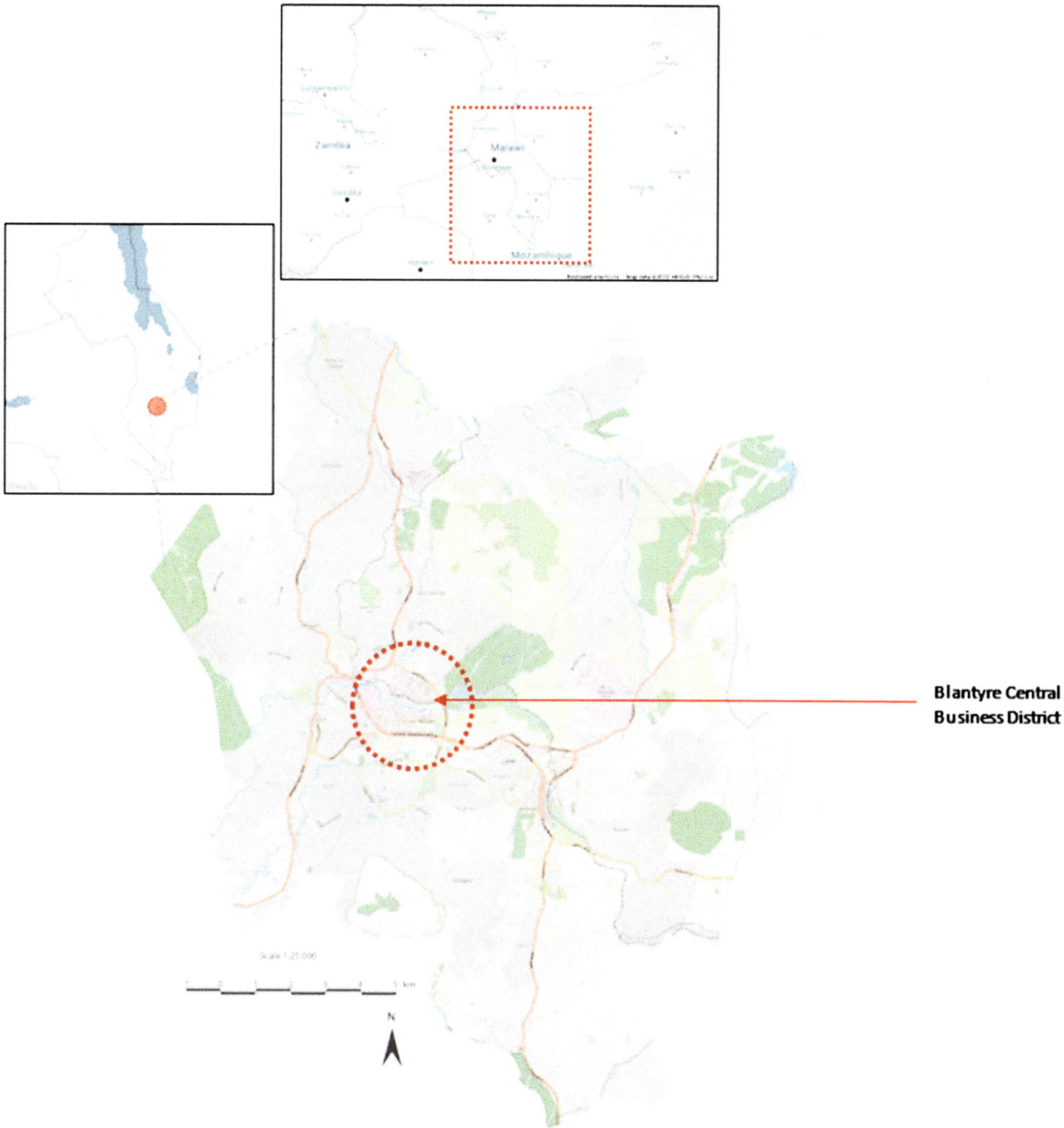

Fig. 7 Location map of Blantyre (by the Author, 2022)

areas of different sizes. While this rate may be considered moderate compared to other countries, the annual urban rate of 5.2% is significantly higher than the national growth rate of 2.8% per year. Projections indicate that the national urbanization level, currently at 15.3%, is anticipated to reach 30% by 2030 and 50% by 2050 (Manda, 2013). Most of the urban population (77%) is in the four cities of Blantyre, Lilongwe, Zomba, and Mzuzu, representing 12% of the total population of Malawi.

Blantyre is the oldest urban center in Malawi. It was founded by Scottish Missionaries in the 1870s and officially declared a planning area in 1897. It is the hub for Malawi's commercial activities, communication, and cooperation activities (UN-Habitat, 2011).

Founded in 1876 as a missionary post for the Church of Scotland, it was named after Blantyre in South Lanarkshire in Scotland, the birthplace of the explorer David Livingstone. The church built the mission while establishing a trading center alongside it, which would later become the home for the Livingstonia Central Africa Trading Company, renamed the African Lakes Cooperation. This commercial activity and its role as a center of colonial trade gave the city of Blantyre its status as Malawi's center of commerce. Colonial rule was abandoned as time passed, leading to the country's independence. Blantyre rapidly grew in population, leading to the high urbanization rates it is currently experiencing and is on track to have one of the highest densities by 2050.

According to UN-Habitat, "the city offers many economic opportunities but lacks the resources to meaningfully implement its strategies and provide the basic social infrastructure and urban services required for economic development. Over 70% of the city population lives in informal settlements, which occupy about 23% of the land in Blantyre" (UN-Habitat, 2011). The poverty rate is 24%, while unemployment stands at 8%.

Being Malawi's commercial and communication hub, Blantyre offers many economic opportunities. According to UN-Habitat, approximately 45% of the population in Blantyre are employed in the private sector, while 12% are employed in the public sector. Additionally, around 36% of residents are self-employed, primarily engaging in informal sector work (UN-Habitat, 2011).

The city is planned into diverse zones resulting from its colonial past and urbanization trends. The most prominent land-use type is residential and is distributed in the following categories: low density, high density, high density permanent, high density traditional, and high density informal. Green areas include agricultural land, grasslands, and rural distribution. In addition to the central business district, the central zones offer institutional, industrial, commercial, railway, civic, and governmental and parking zoning. Finally, the city has bus stations, utilities, private open space, public open space, and cemeteries interwoven into the urban fabric by road networks, railways, dams, and rivers.

2.1 Applying the 15-Minute City Model to Blantyre

The following section will assess Blantyre based on Moreno's four dimensions of density, diversity, proximity, and digitization using the 15-Minute City model. The aim is to understand whether the basic building blocks of this framework exist in the developing city context.

2.1.1 Density

According to the CNU, the maximum density to achieve the 15-Minute City is 5000 people/km sq. for a 5-min walk and 59,570/km^2. for a 15-Minute walk threshold (Duany & Steuteville, 2021). The 2018 Population and Housing Census shows that Blantyre's density is 4.6% (3334/km^2) of a population of 800,264 (NSO-Malawi, 2019). This indicates that Blantyre falls within the optimal range.

Furthermore, the population distribution of Blantyre falls along the socio-economic boundaries of the city. Figure 8 shows the differences in residential density and housing typologies. Typologies range from small to large houses of low to high densities and informal settlements. Most high-density settlements are located close to the city center and amenities. It is also here that most informal settlements in the city evolved.

2.1.2 Diversity

Moreno defines the city's diversity as encompassing six fundamental urban social functions that are vital for sustaining a satisfactory urban lifestyle: *residential living, employment opportunities, commercial activities, healthcare services, educational facilities, and recreational entertainment*. Blantyre has a diverse population in terms of age, sex, income, tribal origin, and socio-economic makeup. However, the social needs of different demographic groups are not being met with the current layout of the city. Below, this paper highlights a few social functions required to sustain a decent urban life in Blantyre. Living:

Data show that the poverty gap in Malawi is at $3.65 a day (World Bank, 2017). According to Oxfam, economic inequality (measured by consumption) in Malawi worsened between the years 2004/05 and 2010/11. The gap between rich and poor is widening, and in just seven years, the gap between the wealthiest 10% of Malawians and the poorest 40% has increased by almost a third (Mussa & Masanjala, 2015).

The Malawi Multi-dimensional Poverty Index (as established by the National Statistical Office and Oxford Poverty & Human Development Initiative (NSO & OPHI)) quantifies various dimensions of poverty, including nutrition, food security, access to drinking water, sanitation, education, school attendance, literacy and schooling, electricity access, environmental conditions, waste management, housing quality, employment, job diversity, asset ownership, unemployment, and prevalence of child labor. Blantyre City measured 0.092, which means that multidimensionally poor people in Blantyre City experience about 9% of the weighted deprivations listed above (NSO-Malawi & OPHI, 2021).

In Malawian urban centers, 14.8% of urban dwellers perceive themselves as poor, and the proportion of households who perceive themselves as richer than their neighbors was 36.6% in 2019 (NSO, 2020). According to the Malawi Wealth Index (National Statistical Office (NSO) and ICF Macro, 2011), the population with the most significant wealth is in urban areas. According to the Malawi Wealth Index, the Gini coefficient of urban areas is 27.7, while that of rural areas is 35.1. Based on this, we can see that urban Malawians spend much of their money trying to find food

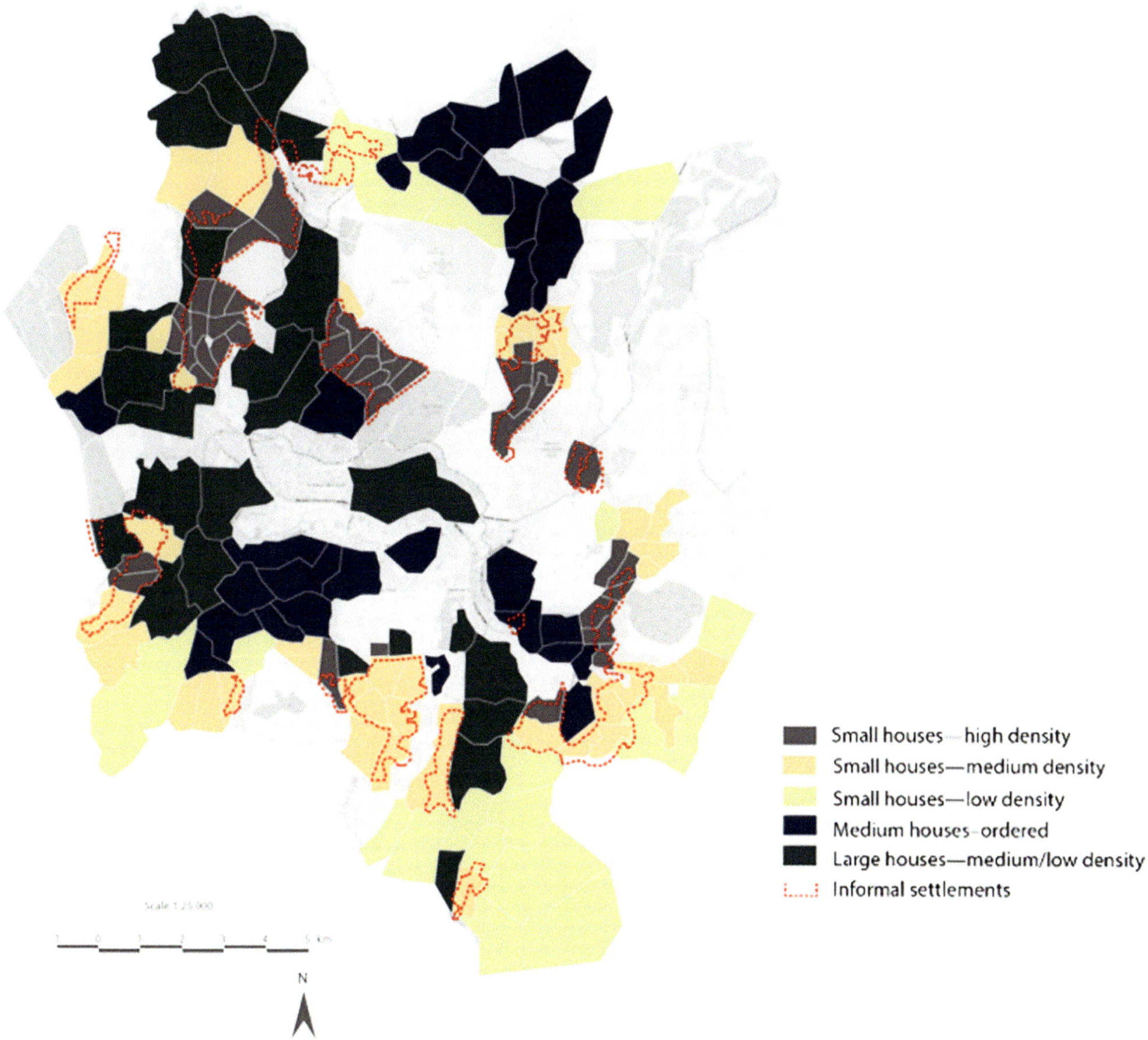

Fig. 8 Map of Blantyre showing residential density by housing typology (by the Author, 2022)

sources to survive (Fig. 9), typically spending their cash on informal markets and street food. In Blantyre, 99% of residents use this as a primary food source, versus only 53% of them using supermarkets. Many residents also use small shops, restaurants, and takeaways locally within the city, and 64% of urban residents in Blantyre grow their food. In addition, there are 17% who live on food remittances, and some share their food with neighbors. This shows the strong culture of sharing between households and how residents support others through community efforts.

Because most Malawians buy or purchase their food from informal sources versus supermarkets, it is worth noting, as shown in Fig. 9, that a minority of them (5%) will go to a supermarket at least once a month. However, 83% will go to informal sources for sustenance and food at least 5 days a week. This shows that informal food production and distribution structures within a city like Blantyre are fundamental when initiating proper strategies for a resilient city (Mvula & Chiweza, 2013).

(a) Working

Regarding working, most urban residents in a developing city like Blantyre are self-employed. The definition of self-employment is mainly related to the informal sector, such as street vendors who sell everyday goods without permits or with only temporary licenses on the side of the streets. This also includes rendering services that would

TABLE 10: Household Food Sources

	Super-market	Small shop/ restau-rant/ take-away	Informal market/ street food	Grow it	Food aid	Food remit-tances	Shared meal with neigh-bours and/or other house-holds	Food provid-ed by neigh-bours and/or other house-holds	Commu-nity food kitchen	Borrow food from others
Windhoek	97	84	76	3	1	5	14	11	0	12
Gaborone	97	56	29	5	6	4	21	22	0	3
Maseru	84	89	49	47	3	14	20	29	1	41
Manzini	90	49	48	10	1	3	9	13	18	18
Maputo	23	78	98	23	1	12	19	10	0	20
Blantyre	53	69	99	64	2	17	23	18	0	11
Lusaka	16	80	100	3	1	13	13	10	0	8
Harare	30	17	98	60	2	19	19	19	3	42
Cape Town	94	75	66	5	3	6	45	34	6	29
Msunduzi	97	40	42	30	5	5	18	21	1	24
Johannesburg	96	80	85	9	2	2	14	13	9	6
Total	79	68	70	22	2	8	21	20	4	21

Multiple responses. N=6,453

Fig. 9 Household food sources (Mvula & Chiweza, 2013)

Fig. 10 Percentage distribution of employment status by residence, Malawi 2018 (NSO-Malawi, 2019)

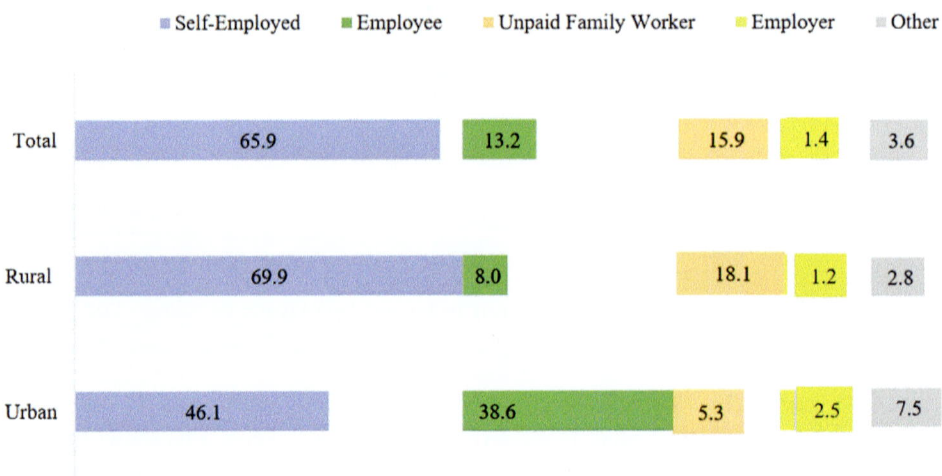

otherwise be inaccessible in formal commercial and market structures. Figure 10 shows that 38.6% of urban dwellers are employed by a legal employer, 5.3% are unpaid family workers, and 2.5% are employers. Types vary between those who work vocational jobs and those who work non-vocational jobs. Some work in factories and have skilled vocations, while others work in offices and gain a significant income based on their educational background and status in the city. Unpaid family workers are typical and are often family members brought in from rural areas to urban areas to serve the needs of families in the capacity of childcare worker, housekeeper, gardener, or cook for remuneration in room and board. These are informal agreements with no set time limit for the family member's service.

According to the 2018 Malawi Population and Housing Census, elementary occupations were the most common job descriptions. These consist of housework or housekeepers; skilled agricultural forestry and fisheries workers; service and sales workers mostly found in stores and businesses; crafts and related trades workers who are significant contributors to tourist industries; and finally, professionals, of whom 2.8% are female and 4.4% are male. Furthermore, 2.9% of the urban population are managers, 9.2% are professionals, 2.9% are technicians and assist associate professionals, 3.1% are clerical support workers, 30.6% are elementary occupations, and 43.3% are service and sales workers (Mvula & Chiweza, 2013).

(b) Education

The literacy rate of Blantyre City is 89.7%. Following the educational trends of the southern region, about 49.4% of citizens have primary education, 24% have secondary education, 22% have no education at all, and 4.1% have a university or other tertiary education (Mvula & Chiweza, 2013).

2.1.3 Digitization

Compared to the developed world, which has ample access to high-speed internet and digital technologies, Malawi has limited access to high-speed internet. In urban households, 85.1% possess cell phones, 46% own a television, 17% pay for TV, and only 4.2% own fixed-phone telephones, as shown in Fig. 11; 61.7% listen to the radio, and 5.6% use the post office. Moreover, only 24% of households own a computer in urban centers. Because the Internet is the main avenue toward digitization, 69.4% of urban households currently use mobile broadband, while only 1.7% use wireless broadband. This is very important because it shows that urban infrastructure for digitization is not yet as established as in the developed world. Instead of using wireless broadband within residences or public spaces, city residents prefer to use dongles to provide wireless hotspots that are much more mobile and portable. The main reason urban residents do not have internet access is that most can not afford it. About 55% of urban residents cannot afford internet, and for 45%, it is not available in their area; very interestingly, 23% do not know how to use it.

Of those using the internet in Malawi's urban centers, 12% percent use it to send or receive email; 74% use it to participate in social networks; 2.6% use it to access chat sites, blogs, newspaper groups, or online discussions; and less than 1% use it to consult Wikimedia online encyclopedias and informational sources. In addition, 9.8% use it for streaming or downloading images, movies, videos, and games, and 0% use the internet to read or download online newspaper magazines or e-books. Due to the popularity and establishment of mobile virtual networks, service providers have taken advantage of this by introducing mobile money and banking, which are very popular and widely used among Malawians in both rural and urban areas. Around 26.5% of urban dwellers used mobile money and banking services. They are dominated by the provider Airtel money, which covers 60.2% of all urban residents in Malawi, followed by

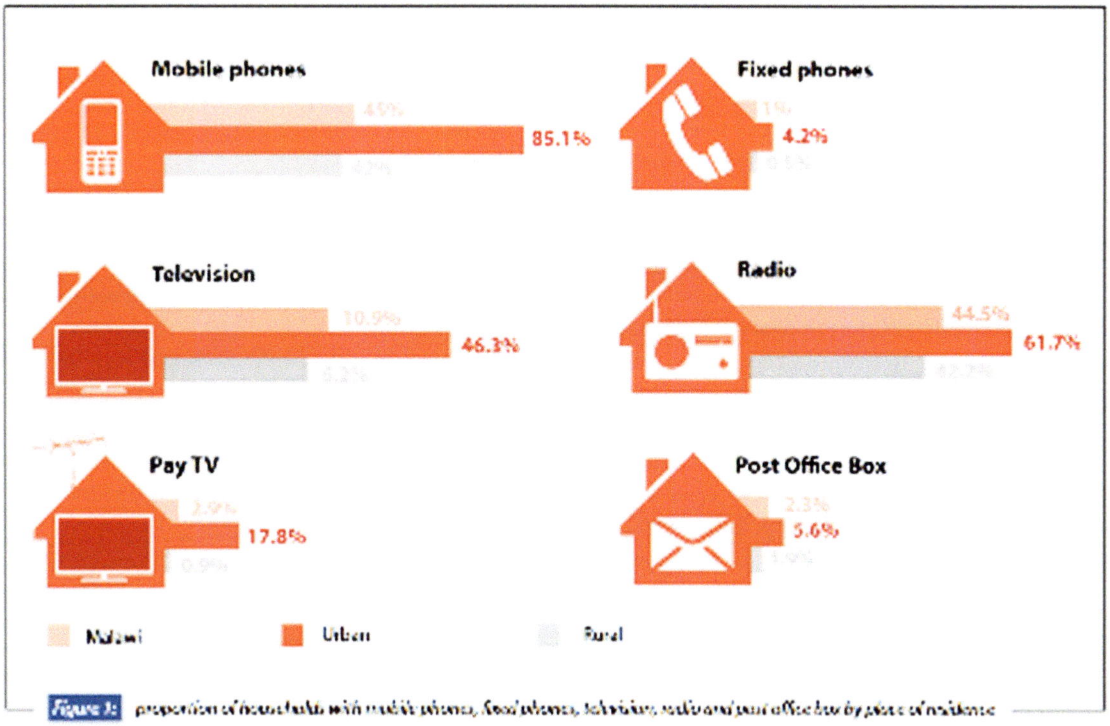

Fig. 11 Proportion of households with mobile phones, fixed phones, television, radio, and post office box by place of residence (Malawi Communications Regulatory Authority, 2015)

TNM Mpamba, which covers 30.1% of coverage in Malawi. This is followed by Mo626, a banking app provided by the National Bank of Malawi. The primary uses for mobile banking apps (known as mobile money top-ups) have been identified as 81% for airtime, 1.9% for salary payments, 3.1% for bill payments, and 13.4% for receiving payments and remittances (Malawi Communications Regulatory Authority, 2015).

Since 14.9% of Blantyre citizens live in poverty and 1.5% are considered ultra-poor (Malawi National Statistical Office, n.d.), access to Information and Communication Technologies (ICT) services is limited or nearly impossible. Although 85% of Malawians have a cellphone or mobile device as their primary access to ICT services, this access is limited to the specific "bundle" they can afford. This means that access to digital connectivity and remote work opportunities is limited (Malawi Communications Regulatory Authority, 2015).

However, this digital network gives rise to innovative digital products such as mobile banking and social network integrations.

2.1.4 Proximity

Proximity is the main focus of Chrono-Urbanism. When tested for connectivity and integration, the road network of Blantyre shows gaps and weaknesses in integration to the main commercial and industrial areas.

DepthMapX software was used to perform angular segment analysis to check if the network was viable and to test the network's connectivity. A DXF file of the network was converted to a segment map and analyzed through the software to test the segment length and angular connectivity attributes. This showed the integration of road network segments at 300, 500, and 1000 m radii. The results revealed the least integrated parts of the network under the "Integration" attribute (segment length–weight).

Following the angular step analysis, a metric angular step depth analysis was performed. The metric angular step analysis runs an evaluation from a specific segment of the network to the other networks. In this analysis variation, the radii tested in the network were specified to be on the level of pedestrians, cyclists, larger cycle networks, and roads (vehicular networks). This resulted in several outcomes, including choice values (how busy the entire road network is) and integration values (how central a road is). The pedestrian choice values show how far a pedestrian travels to get to public transportation (segment) in a large city with an efficient transportation system.

This is important because it shows that it takes the average citizen a long time to cover the distance from where they reside to integrated areas of the city with the most amenities. The step analysis revealed that the network has higher connectivity along main road networks linking the city to others and commercial clusters and nodes. Figure 12 shows that regions in red reflect higher integration into the overall city network, while those in blue show segregation, disjointedness, and the least connectivity to the city network.

When Fig. 12 is overlayed with the Blantyre land-use map, as seen in Fig. 13, the industrial zones are the most integrated and connected. This is because industrial and commercial zones in the current urban design of Blantyre have high proximity to each other and the prominent transportation veins running through and leading out of the city.

This is a legacy of colonial rule, which saw urban plans designed to favor colonialists and foreign interests in the country and not local populations, leading to the formation of informal settlements, as shown in Fig. 14 (McCracken, 1998). Colonial influence has been a pivotal factor in the disjointedness of African cities. Colonialists introduced planning policies that aimed at segregating the indegenous population and systematically assert their "perceived social, cultural and technological superiority over Africans" (Mwathunga & Donaldson, 2018). UN-Habitat (2011) characterized the urban geographies in Africa as: "highly disjointed, dysfunctional and unsustainable urban geographies of inequality and human suffering, with urban areas increasingly composed of small islands of well-being that are spatially and socially segregated from rapidly growing and increasingly impoverished masses —the urban divide. This urban divide was forged over decades of intrusion by foreign governments and corporations, settlers, and a politically repressive postcolonial regime" (Blantyre City Report, 2021). This urban divide also allows us to understand the characteristics of the Blantyre road network. Riley (2012) described it in the following way:

> The metaphor of a disjointed patchwork aptly describes Blantyre, where densely crowded settlements, vast tracts of undeveloped land, traditional villages, mountains, river valleys, commercial and industrial areas, and low-density suburbs all mark distinct kinds of spaces within the municipal boundaries (Riley, 2012).

This legacy of disjointedness aligns with the findings of the step analysis. It also aligns with Gehl's description of developing cities and the consequences of mobility in attempting to achieve better proximity. Traditional pedestrian and cyclist traffic is declining, while rapidly growing motorized traffic is filling cities. Informal transportation systems run city streets and thus drastically reduce opportunities to walk and cycle (Gehl, 2010).

3 Results

3.1 Proximity and Density

Once all the layers of the 15-Minute City parameters outlined by Moreno were translated to the Blantyre context,

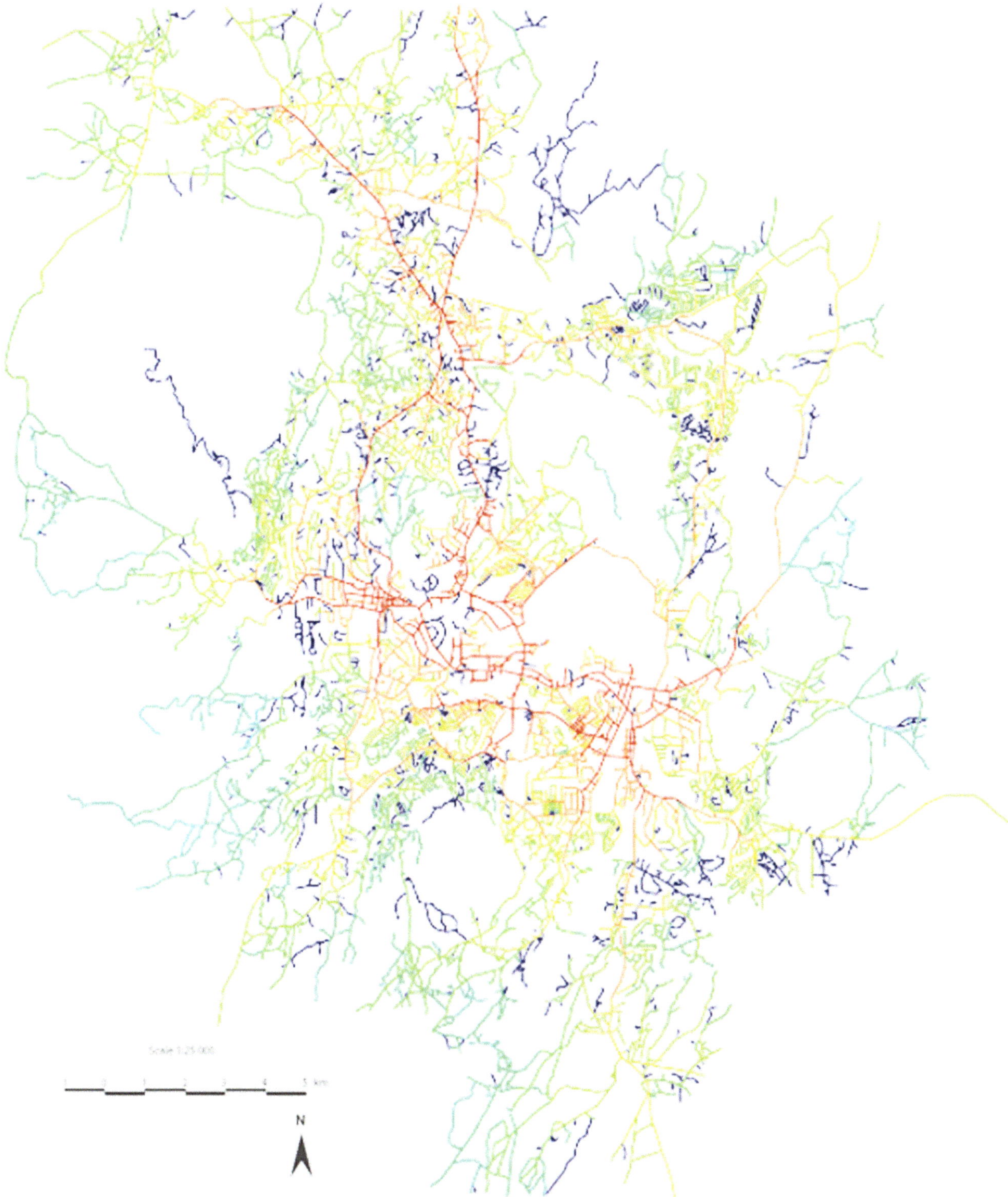

Fig. 12 Segment analysis of Blantyre showing integration and connectivity of City Network (by the Author, 2022)

they were overlayed to create a composite map, as shown in Fig. 15. This map shows the relationship between the locations of varied housing densities, informal settlements, and the road network. The city's well-integrated areas, represented by the red and orange colors, also fall into industrial, commercial, and high-amenity areas such as markets,

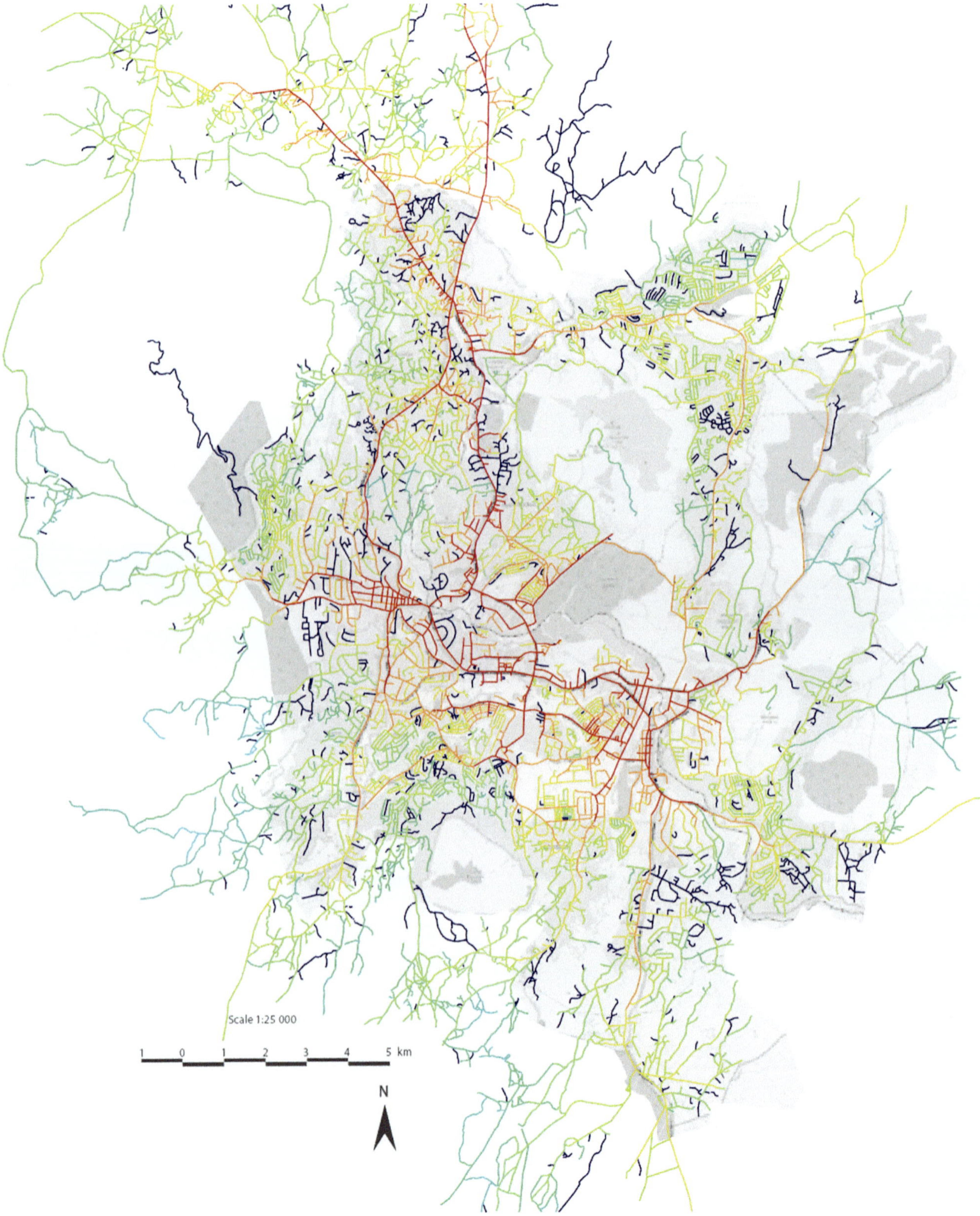

Fig. 13 Blantyre land-use map overlaid with segment analysis shows industrial zones to be the most integrated and connected (by the Author, 2022)

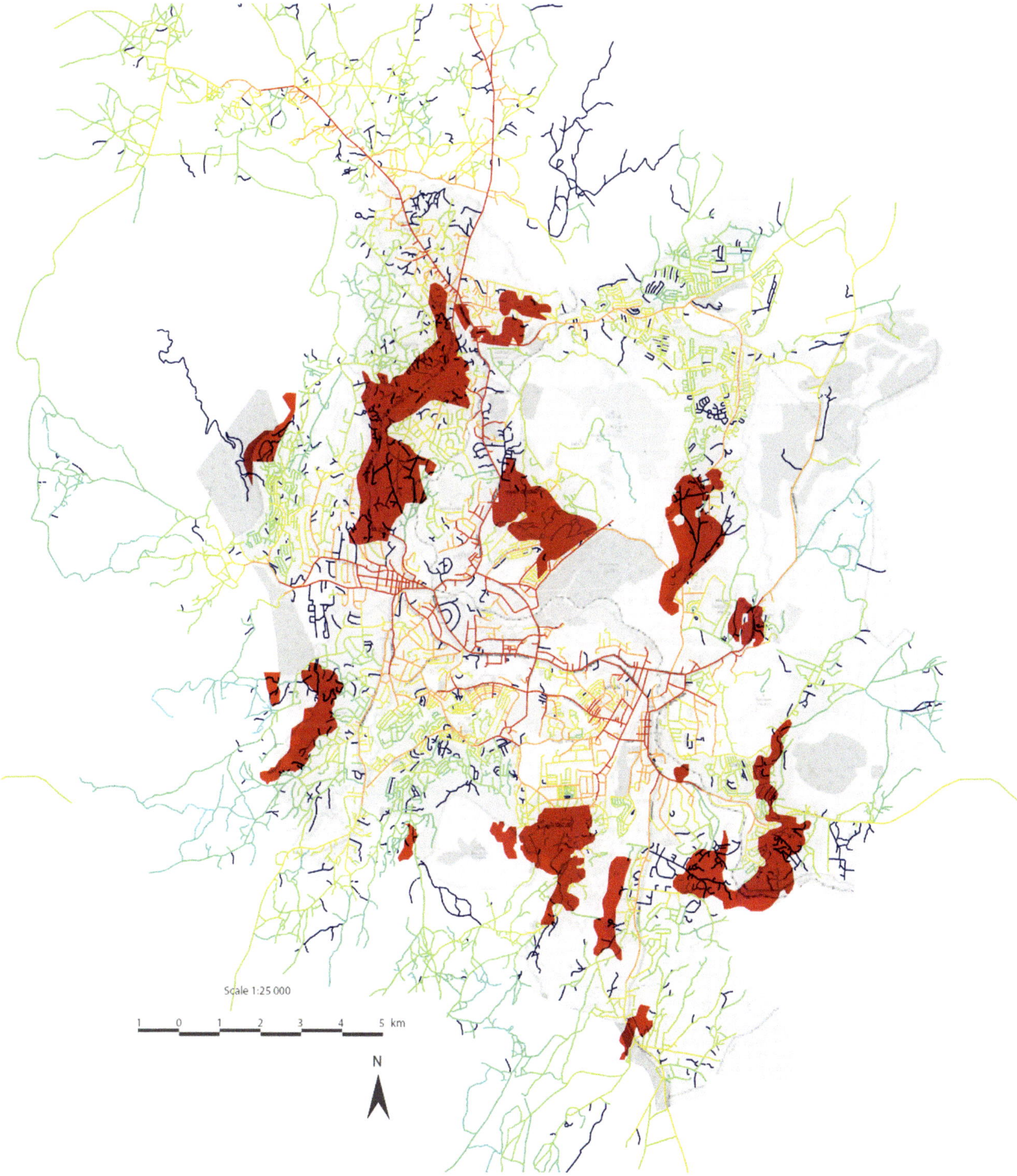

Fig. 14 Blantyre land-use map overlaid with segment analysis and highlighting Blantyre informal settlements (by the Author, 2022)

supermarkets, and shops. It is revealed that the areas with the highest proximity and the most integration to the road networks are larger medium- to low-density residential and informal settlements. The housing typology of large-to-medium low-density housing represents the distribution of higher-income residents.

In contrast, the smaller, high-density, and informal settlements represent the distribution of lower-income residents. The most integrated areas are higher-income, lower-density areas and informal settlements. Historically, informal settlements have developed alongside more affluent neighborhoods in the city due to the urban condition of these areas being closer to city amenities and job opportunities for those of lesser means. We also see that neighborhoods on the city periphery with small medium-to-low-density houses are less connected and integrated into the network and have less proximity to amenities.

After running tests on the transferability of the 15-Minute City, the next step was to see how integral theory could help address the gaps in the case of Blantyre.

3.2 Integral Theory Analysis

Integral Theory analysis allowed a more complete view of all four dimensions of density, proximity, diversity, and Digitization. First, as presented by Fleming and Roberts (2019), the Integral Framework was used to identify the crucial components of a sustainable city, as outlined in Sect. 2.3. These components fall under the quadrants of experience, performance, culture, and systems. The experience quadrant pertains to inner health and psychological well-being and is subjective. The culture quadrant pertains to subjective collective factors such as urban cultural vibrancy, resilience, and social equity. The performance quadrant pertains to outer health, primarily physical, and well-being and is objective and quantifiable. The systems quadrant relates to collective systems that interconnect to make a city function, such as human, active, passive, living, and life-cycle systems (Fig. 16).

Once categorized, the four quadrants were compared to the dimensions of the 15-Minute City (Density, Diversity, Proximity, and Digitization). Then, characteristics of the 15-Minute City dimensions, which aligned with those of the Integral Theory quadrants, were clustered, as shown in Table 2. This led to a modified interpretation of the Integral Framework shown in Fig. 17. The density dimension is interpreted as experiences, diversity with culture, proximity with performance, and digitization with systems.

Density/Experience

This quadrant expresses experience through density and all its social dimensions. Core components of wholesome city living are reflected here, including the closeness of essential services, the reduced timespan of travel, proximity of living to work, commerce, education, entertainment, and basic amenities, walkable streets and bicycle lanes, and the overall ubiquity of city dwellers.

Diversity/Culture

Here diversity is expressed as culture. The human scale is highlighted, and the cultural experience and systems are integrated. Social equity and inclusion are highly emphasized here. This allows communities to build social bonds of trust that favor the disadvantaged. In this quadrant, we also see that the equitable distribution of civic services and services at reduced cost allows for social sustainability. Diversity in culture, people, and experiences is highly encouraged to promote social cohesion and capital.

Proximity/Performance

Proximity is then expressed by performance. This quadrant reconciles the human scale and density through mixed-use neighborhoods. These improve city residents' livability and service delivery, including a sustainable food supply. This quadrant has multiple uses of available spaces and multimodal use of basic infrastructure. We see locally based solutions and the building of healthy landscapes. The city exploits available resources sustainably while simultaneously generating energy.

Digitization/Systems

Finally, in this quadrant, digitization is expressed by systems. Along with existing city systems, digital transactions such as cashless transactions, online shopping, and virtual communications and interactions are encouraged.

Then, using this reconfiguration of the 15-Minute City and Integral Frameworks, an integral analysis was performed on Blantyre to illustrate the complexity and overlapping of quadrants in a developing city context. In performing this analysis, it was clear that the components of the city did not fit in only one but several quadrants simultaneously, as shown in Fig. 18. This was a more accurate depiction of the functioning of a developing city where informality plays a central role.

Table 3 categorizes the prevailing characteristics found in Blantyre based on the modified Integral Framework. The components shown below were what was organized strictly in the various quadrants. However, notable are areas where there were many overlaps, as shown in the next section.

3.3 Integral Theory Analysis Overlaps

Density/Experience

These quadrant links to the other quadrants to varying degrees. Density and experience are integral to socially segregated areas and informal settlements in Blantyre. An informal indigenous economy is represented by informal employment, informal markets with cashless transactions, and street food. These create informal, culturally vibrant spaces with diverse cultures and people and strong cultural ties, and traditions. City dwellers shop in person; however, the cost of goods and services is high for many. Although

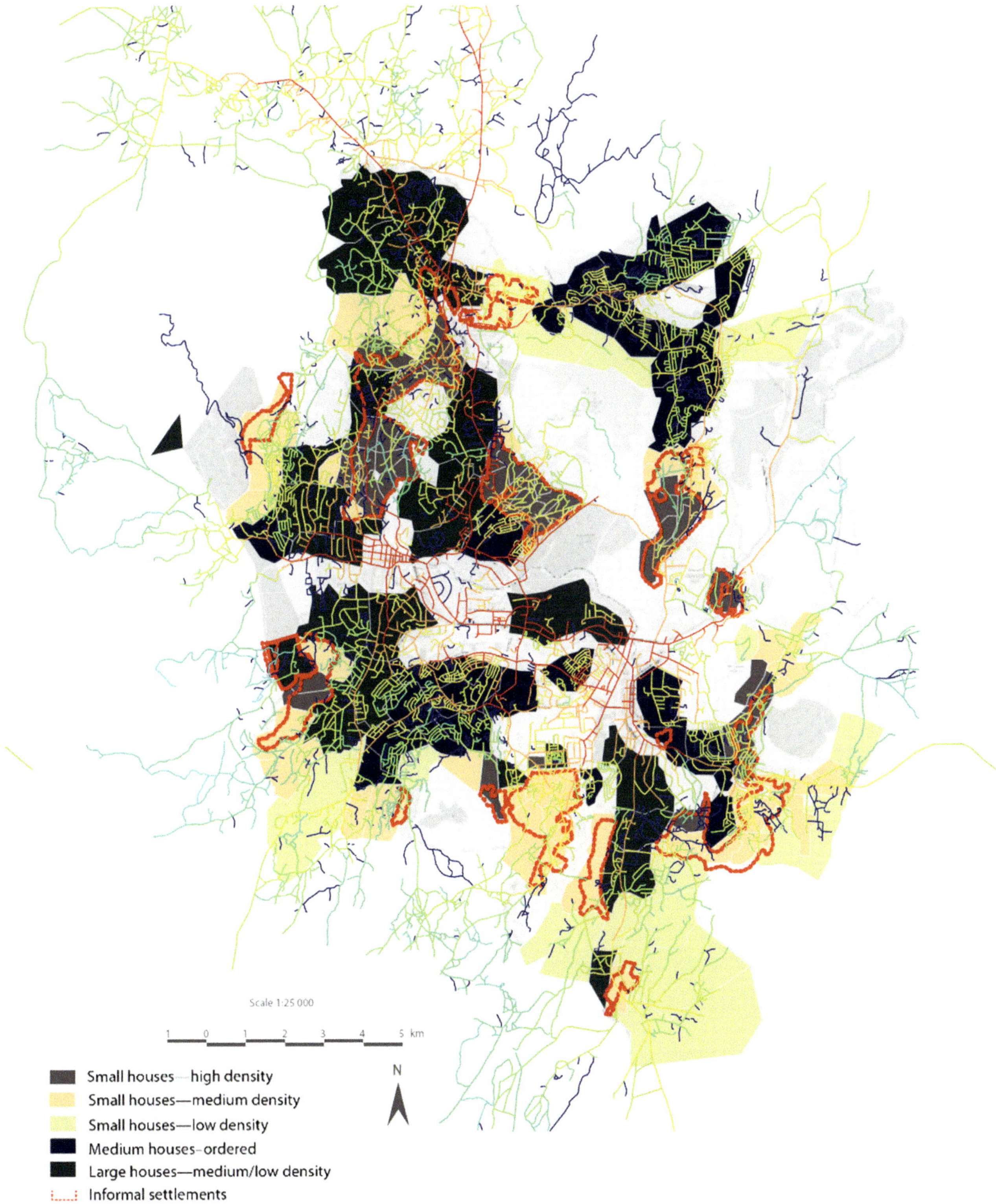

Fig. 15 Composite map showing parameters of proximity and density according to Moreno's 15-Minute city (by the Author, 2022)

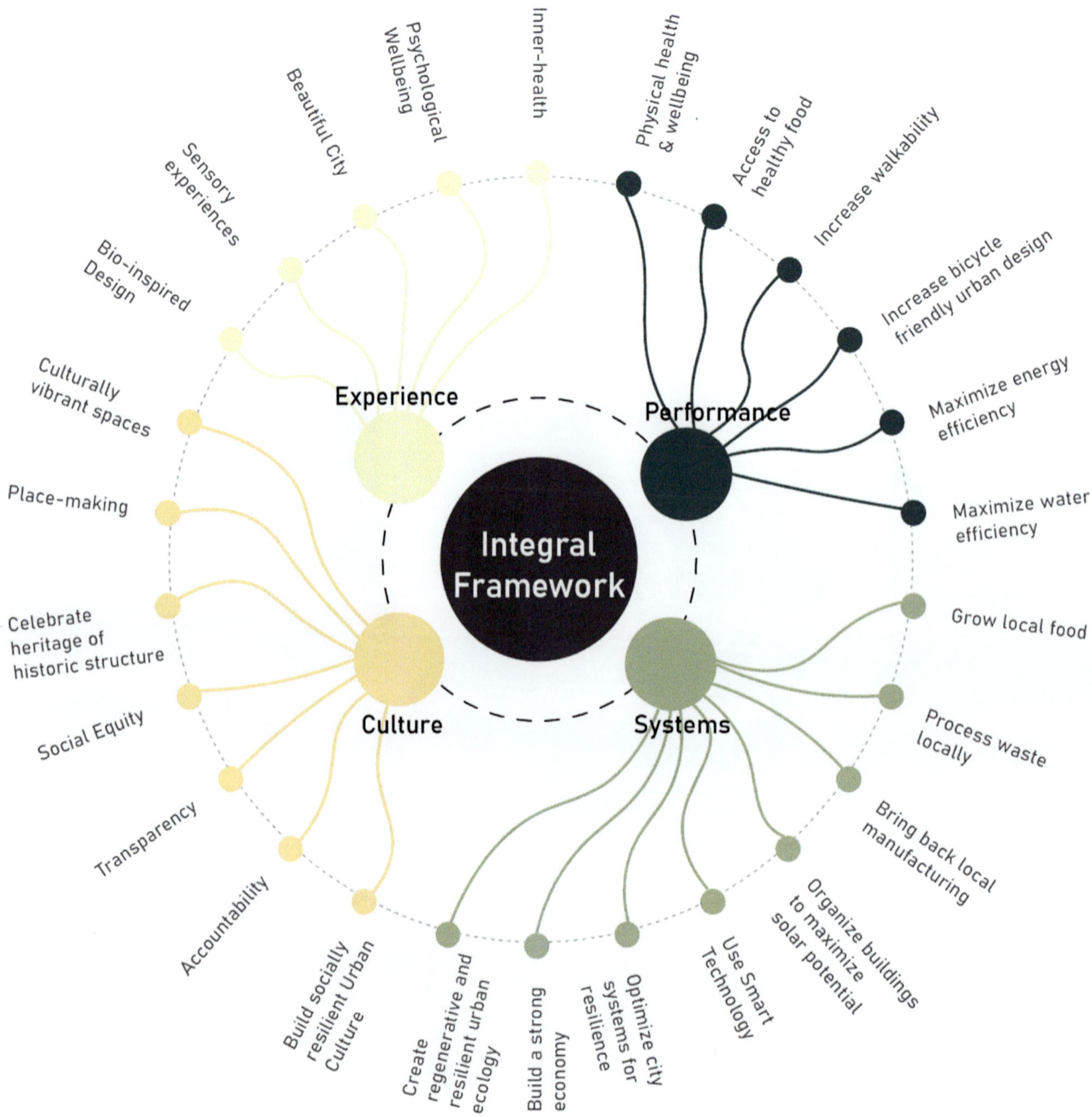

Fig. 16 Integral framework showing four quadrants (by the Author, based on Fleming & Roberts, 2019)

the city is becoming more car-centric, residents primarily walk, cycle, and use public transportation for mobility. For many, there is poor proximity to essential amenities and lack of planned biophilic spaces. Crime and safety are a concern, especially at night, due to unreliable and low electric power capacity.

Diversity/Culture

In addition to the Diversity and Cultural components in Table 3, complex social networks determine in-person communication and interactions and the citizen's relationship to their natural and built environment. Because many residents walk the city, there are many trees and natural corridors. However, these natural corridors are subject to littering and pollution from the consumption of city dwellers and informal urban spaces.

Proximity/Performance

This quadrant also integrates the challenge of the unreliable and low-capacity electric grid. The electricity grid is powered by

Table 2 Integral framework analysis of 15-Minute City (by the Author, 2022)

	Subjective		**Objective**	
	15-Minute City	**Integral framework**	**Integral framework**	**15-Minute City**
	Density	**Experience**	**Performance**	**Proximity**
Individual	• Closeness of essential services • Living • Working • Commerce • Healthcare • Education -Entertainment • Proximity • Walkable streets and bicycle lanes • Ubiquity	• **Work toward inner health and psychological well-being** • Make the city beautiful and engage all the senses • Apply bio-inspired design to parks, roadways, and building exteriors	• **Work toward "outer health" Physical health and well-being** • Increase availability of healthy food • Increase walkability • Increase bicycle-friendly urban design • Maximize energy efficiency • Maximize water efficiency	• Energy generation • Food supply • Multiple uses of available spaces • Improving service delivery • Improving livability • Multimodal uses of basic infrastructure Maximizing public infrastructure • Exploitation of available resources • Mixed-use neighborhoods • Density • Human scale • Locally based solutions Building healthy landscapes
	15-Minute City	**Integral framework**	**Integral framework**	**15-Minute City**
	Diversity	**Culture**	**Systems**	**Digitization**
Collective	• Strengthening of social bonds • Building trust • Social equity • Social inclusion • Favoring the disadvantaged • Equitable distribution of civic services • Services at reduced cost • Diversity • Social sustainability • Creating social capital • Promoting social cohesion	• Create **culturally vibrant urban spaces** and places and celebrate the heritage of historic structures • Embed **social equity** into every decision with transparency and accountability • Build a socially resilient urban culture	• **Life-cycle Systems**: grow local food, bring back local manufacturing, and process waste locally • **Passive Systems**: organize buildings to maximize potential for solar • **Active Systems**: Use Smart City technology to optimize city systems for resilience • **Living Systems**: Create a regenerative and resilient urban ecology • **Human Systems**: Build a strong economy	• Digitization • Online shopping • Cashless transactions • Virtual
		communications and interactions		

hydro-electric power, which is vulnerable to climate change. This has encouraged the high use of fossil fuels and given city dwellers low access to smart city infrastructure. However, city dwellers can access virtual mobile connections and mobile banking networks through services like Airtel Money and TNM Mpamba—Malawi's leading telecommunications providers.

Digitization/Systems

In addition to nature-based systems, informality presents as the most prevalent system in Blantyre. Furthermore, the inadequate wealth distribution echoes the legacy of Malawi's former colonial rule. Access to clean water is especially

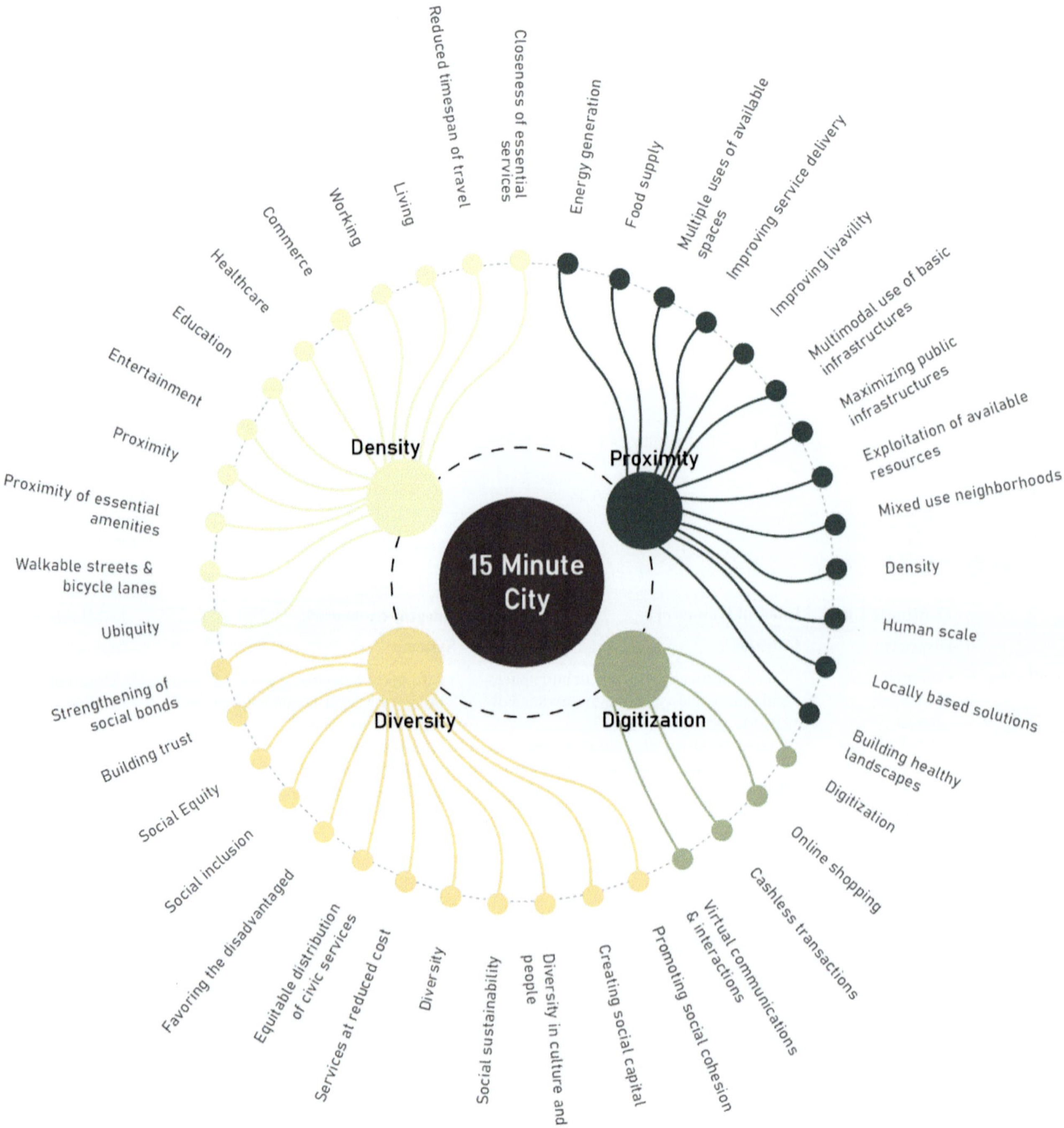

Fig. 17 Modified integral framework (by the Author, 2022)

crucial in informal settlements and for sustainable food production.

3.4 The Integral Developing City Framework

Considering the shortfalls of the 15-Minute City concept in the developing city context, some gaps are bridged by incorporating the modified integral theory framework. This analysis paves the way for the concept of the **Integral Developing City,** developed by the author. This concept delivers a holistic perspective that interweaves the complexities of informality and urban systems of developing cities.

Not only would the Integral Developing City embrace informality in its planning but it would also embody principles of sustainable development, resilience, and people-centric placemaking. The key to its success is that

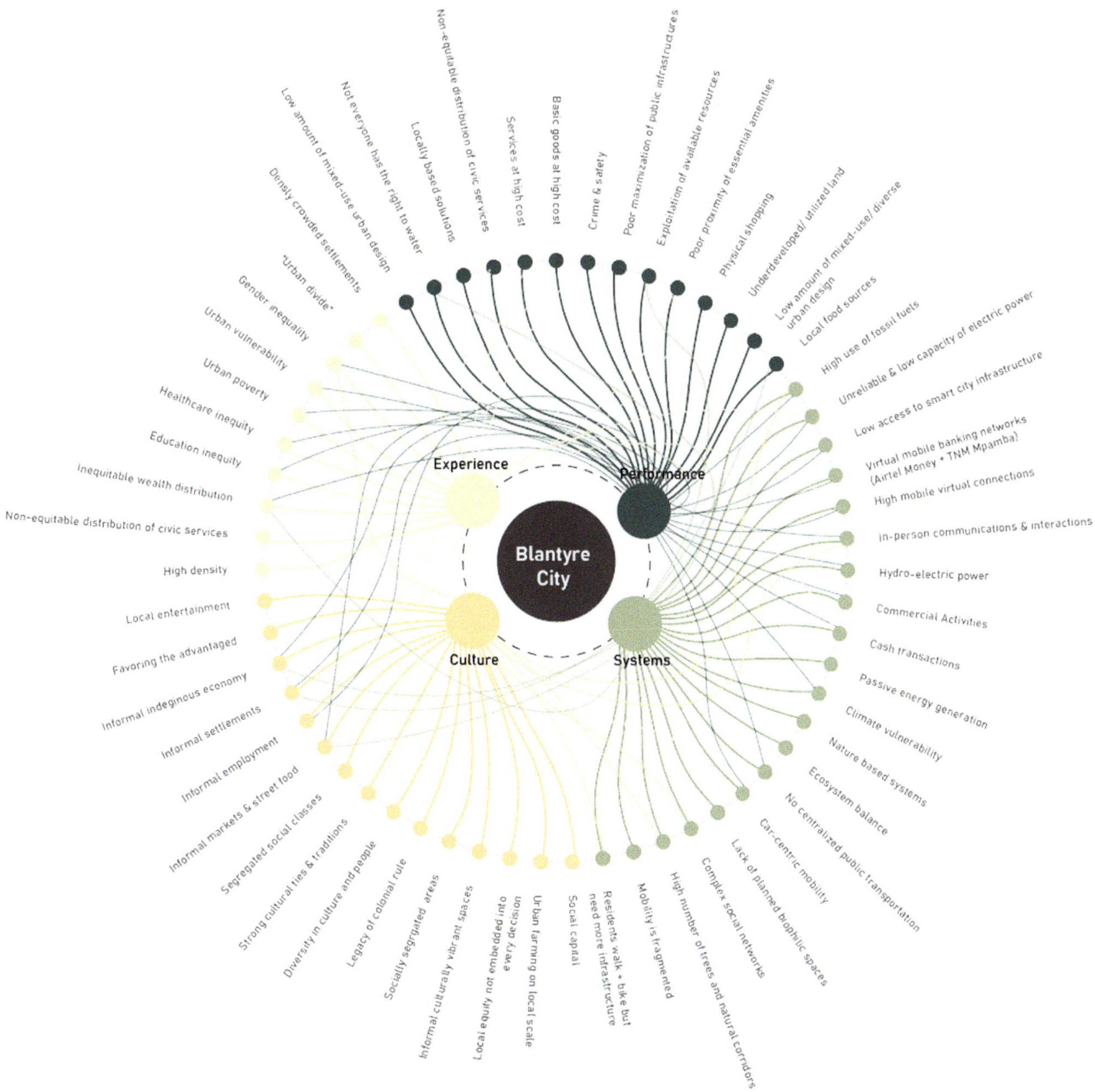

Fig. 18 Integral framework assessment of Blantyre City (by the Author, 2022)

many of the quadrants are interconnected through a systems-thinking model and that no one quadrant is above another, as shown in Fig. 19.

In this concept, we see the creation of spaces that address inner psychological health, physical health, and well-being and include culturally vibrant places. The concept also celebrates the heritage of historic structures while simultaneously celebrating the indigenous economy, high mobile virtual systems, and complex networks. The planned biophilic spaces are prioritized to beautify the city and provide sensory experiences through bio-inspired design and biophilia. Closeness to essential services is paramount while providing walkable streets and open spaces that offer diversity in culture and people and promote social cohesion. Poverty alleviation works with informal markets and high mobile virtual connections to provide connectivity to city dwellers by recreating a hybrid mobility means and reducing fossil fuels through optimizing city systems for resilience and transportation.

Equitable distribution of civic services and advocacy for vulnerable groups in the city work hand-in-hand with diverse commerce, commercial activities, and proximity to essential amenities. Digitization is built upon virtual mobile

Table 3 Integral framework characteristics of Blantyre City (by the Author, 2022)

Density/experience: Densely crowded settlements High density Urban poverty Urban vulnerability Gender inequality Inequitable wealth distribution Urban divide Healthcare inequity Education inequity Non-equitable distribution of civic services and public space	*Proximity/performance*: Small amount of mixed-use urban design Not everyone had access to clean water Locally based solutions Non-equitable distribution of civic services Services at high cost Crime and safety Basic goods at high cost Services at high cost Poor maximization of public infrastructure Exploitation of available resources Poor proximity to essential amenities Shopping in person Underdeveloped/utilized land Small amount of mixed-use/diverse urban design
Diversity/culture: Local entertainment Favoring the advantaged Informal settlements Informal indigenous economy Informal employment Informal markets and street food Strong cultural ties and traditions Diversity in culture and people Socially segregated areas Informal culturally vibrant spaces Legacy of colonial rule Socially segregated areas Local equity is not embedded in every decision Urban farming on a local scale Social capital	*Digitization/systems*: High use of fossil fuels The unreliable and low capacity of electric power Low access to smart city infrastructure Virtual mobile banking networks (Airtel Money + TNM Mpamba High mobile virtual connections In-person communications and interactions Hydro-electric power Commercial activities Cash transactions Passive energy generation Climate vulnerability Nature-based systems Ecosystem balance No centralized public transportation Car-centric mobility The lack of planned biophilic spaces Complex social networks High number of trees and natural corridors Mobility fragmented Residents walk and bike but need more infrastructure

networks and high mobile virtual connections. Finally, mixed-use design encourages informal settlements, markets, and complex social networks. The human scale is emphasized in this model while working to create a sustainable environment in response to climate change resilience. Table 4 presents the Integral Developing City quadrants.

4 Discussion

This research aimed to test the conceptual approach of the 15-Minute City against the framework of Integral Design Theory (Fleming & Roberts, 2019) and explore its transferability to developing cities. The results show that the 15-Minute City model is not transferrable to developing cities. An implication of this finding is that it highlights the need for more appropriate methods and concepts for resiliency of developing cities; concepts such as the Integral Sustainable City can provide the framework needed to make that a reality.

This study presents a novel approach to resilient planning strategies for cities of the global south, especially in the Sub-Saharan African context. It builds on existing urban systems that favor the poor such as informality and socio-economic networks. This is important to the field of urban planning because strategic priorities such as the UN Sustainable Development Goals will require applying an integrated approach to planning rather than the most accessible solutions, which can be limited to only certain development goals. This research adds to the toolkit of urban strategies with an integrated planning approach designed to help developing cities grow sustainably and holistically while encouraging local and indigenous systems. This approach goes a step further in bringing multi-dimensional perspectives to address gaps in urban planning methodologies currently used in Malawi.

Furthermore, the findings of this research highlight that the 15-Minute City framework does not align with the systems in Blantyre. This is due to the following challenges:

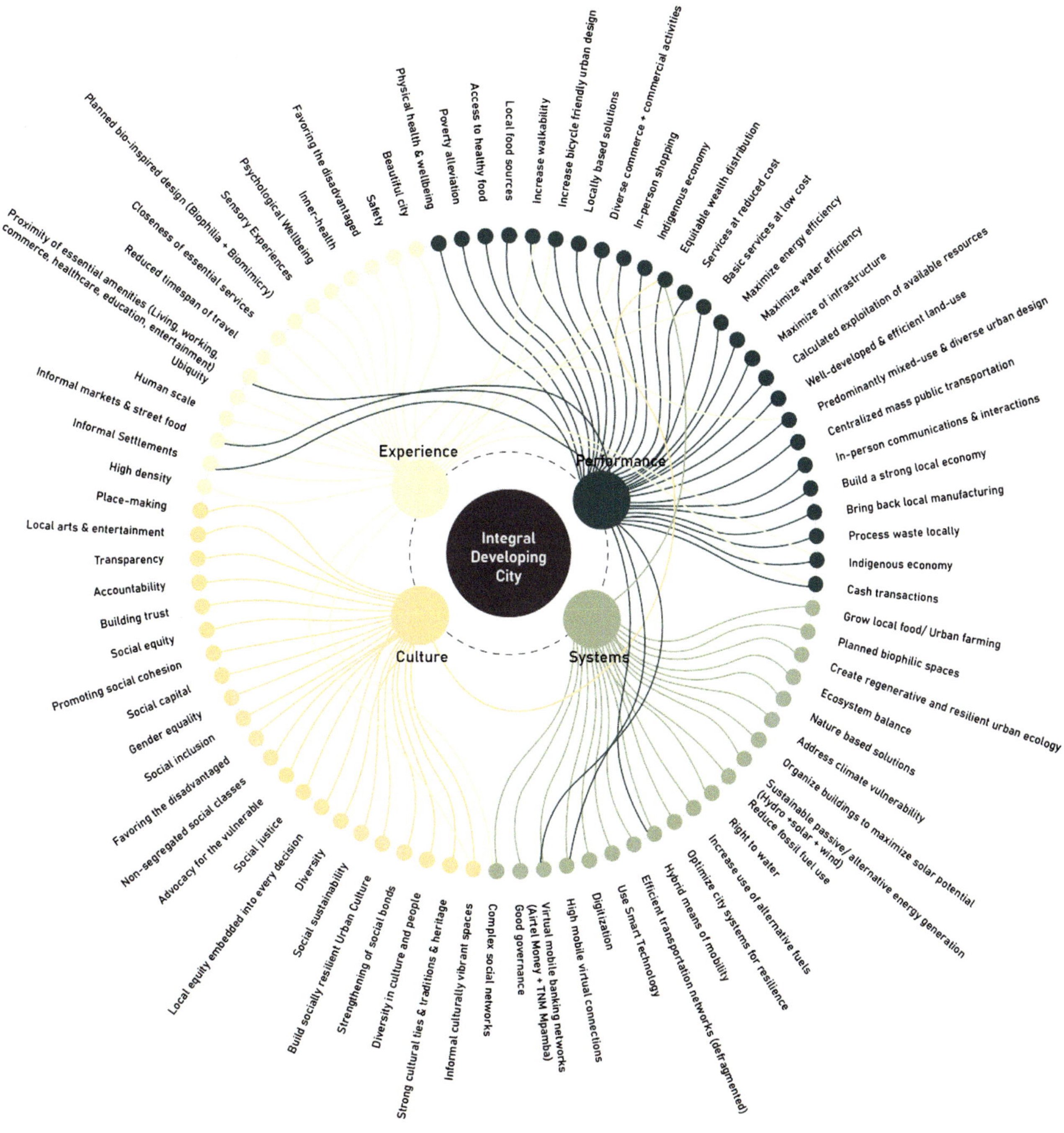

Fig. 19 The integral developing city framework (by the Author, 2022)

4.1 Effects of Colonization

First, the evolution of Blantyre as a city from its colonial origins has created a challenge of inefficient and inequitable distribution of amenities to every neighborhood, separated by the socio-economic standing of Blantyre citizens.

Mobility:

Regarding mobility, there is potential for the development and use of bicycle infrastructure to compensate for the shortfall in access to amenities. However, the current road/bike network is underdeveloped and does not support this.

Table 4 Integral developing city quadrants (by the Author, 2022)

Experience	Culture	Performance	Systems
• Inner health • Psychological Well-being • Beautiful city • Sensory Experiences • Planned bio-inspired design (Biophilia + Biomimicry) • Closeness of essential services • Services at low cost • Basic goods at low cost • Reduced timespan of travel • Proximity of essential amenities (Living, working, commerce, healthcare, education, entertainment) • Walkable streets & bicycle lanes • Ubiquity • Favoring the disadvantaged • Safety • Human scale • Physical shopping • Cash transactions • High density • In-person communications & interactions • Indigenous economy • Informal Settlements • Informal markets & street food • Poverty alleviation • Complex social networks • Mixed-use urban design	• Culturally vibrant spaces • Place-making • Celebrate the heritage of historic structures • Social equity • Transparency • Accountability • Build socially resilient Urban Culture • Strengthening of social bonds • Building trust • Social inclusion • Favoring the disadvantaged • Services at reduced cost • Diversity • Social sustainability • Diversity in culture and people • Promoting social cohesion • Local arts & entertainment • Equitable distribution of civic services • Social capital • Equitable wealth distribution • Non-segregated social classes • Diversity in culture and people • Strong cultural ties & traditions • Complex social networks • Gender equality • Advocacy for the vulnerable • Informal culturally vibrant spaces • Local equity embedded in every decision • Social Justice	• Physical health & well-being • Access to healthy food • Increase walkability • Increase bicycle-friendly urban design • Maximize energy efficiency • Maximize water efficiency • Locally based solutions • Diverse commerce + commercial activities • Basic services at low cost • Hybrid means of mobility • Maximization of infrastructure • Calculated exploitation of available resources • Proximity to essential amenities • In-person shopping • In-person communications & interactions • Equitable wealth distribution • Indigenous economy • Informal Settlements • Informal markets & street food • Poverty alleviation • High mobile virtual connections • Virtual mobile banking networks (Airtel Money + TNM Mpamba) • Well-developed and utilized land • Predominantly mixed-use & diverse urban design • Local food sources • Centralized mass public transportation	• Indigenous economy • High mobile virtual connections • Complex social networks • Good governance • Planned biophilic spaces • Urban farming • Sustainable passive/ alternative energy generation (Hydro + solar + wind) • Right to water • Address climate vulnerability • Efficient transportation networks (defragmented) • Nature-based solutions • Ecosystem balance • Grow local food • Process waste locally • Bring back local manufacturing • Organize buildings to maximize solar potential • Use Smart Technology • Optimize city systems for resilience • Build a strong local economy • Create regenerative and resilient urban ecology • Digitization • Hybrid means of mobility • Reduce fossil fuel use • Increase the use of alternative fuels • Cash transactions • Virtual mobile banking networks (Airtel Money + TNM Mpamba)

Informality:

There is also a big challenge and simultaneous potential for informality in Blantyre, since it is hard to accurately map and record the extent to which the informal and indigenous economy fills in the gaps of proximity to amenities under *proximity, diversity, density,* and *digitization.* Informality displays high potential because it acts as a gap-filler in all neighborhoods and is also the socio-economic glue that serves citizens on all levels of the socio-economic spectrum, fulfilling their basic needs, e.g., “Maokala” (corner stores) and informal street vendors.

4.2 Overcoming These Challenges

Overcoming these challenges will require the addition of an integral framework. In this way, incorporating informality will make the 15-Minute City concept more transferrable, resulting in the **Integral Developing City** as described above (Fig. 19). The Integral Developing City can harness informality and grow its diversity through it.

This can be achieved through strategic planning moves such amenities could be set up tactically for future planning of Blantyre. Hotspots must be identified and the overall supply systems must feed into them. This would then make the proximity network more viable through the strengthening of social networks as the determinants of amenity distribution.

More research is needed to identify the patterns of informal amenity structures in cities like Blantyre and to understand their influence from socio-cultural networks.

4.3 Digitization Needs to Be Developed

Digitization is not on track to work similarly to the 15-Minute City in the developing world because remote work and technological access are luxuries in places like Blantyre. However, there is an opportunity for innovative digital structures to evolve and integrate with the current digital infrastructure available to citizens at various socio-economic levels.

4.4 Informality is an Asset

Informality is very present in developing cities like Blantyre, and while it has been a blight for many governments and city planners, it is a very present opportunity. Research has shown that there are advantages to taking this perspective. A report by Carrizosa et al. presents arguments from planning professionals from sub-saharan African cities describing a specific school of thought called the continuum approach to informality; their continuum shows that informality is not random and chaotic but structured, and that informality is not always a bad thing. Amin Tamale Kiggundu, Head of Department of Architecture and Physical Planning at Makerere University in Uganda, states, “We need to be more imaginative and more inclusive. Informality needs to be thought of as an opportunity for city revenue” (Carrizosa, 2022). Research like this demonstrates that city planners must engage in and learn how to embed informality for future interventions and for the framing of policy regarding informality and sustainable design. There are systems in place, such as the local indigenous economy, food production, urban farming, markets, high-density mixed-use spaces, integrated social systems, and potential for public spaces that promote cultural experiences. These urban spaces create pockets and opportunities for integral and holistic city planning that look at the streetscape and are people-centric. Because there is high social capital, there is a high potential for co-creation opportunities that are more sustainable in the long run. This also allows specific sites and cultures to start to grow and make the city’s diversity more varied and vibrant. In doing so, we begin to use inclusivity, tackle equities throughout the planning process, and increase the quality of life for city dwellers in developing cities. Informality is also an asset because it is a framework or system that is resilient to change and very flexible. We can learn from this change and use flexibility to design more resilient spaces and urban systems.

4.5 Areas of Opportunity

The Integral Developing City framework can be applied to Blantyre in key areas that intersect road network integration and connectivity, high-density distribution of dwelling units, and informal networks (areas highlighted in Fig. 20). These zones can integrate the Integral Developing City principles by implementing small, community-scale interventions using co-creation techniques.

4.6 Challenges and Potentials.

While many opportunities have presented themselves in the Integral Developing City framework, many challenges still remain in developing cities. While the 15-Minute City model is valuable and applicable to developed cities in the global North, there must be a tailored response to developing cities —in this case, to Malawian systems. Figure 20 shows that cultural and social capital is also readily available, even with the opportunities found in informality and informal networks. Digital networks are also available in a localized structure. However, bicycle and walkability infrastructure

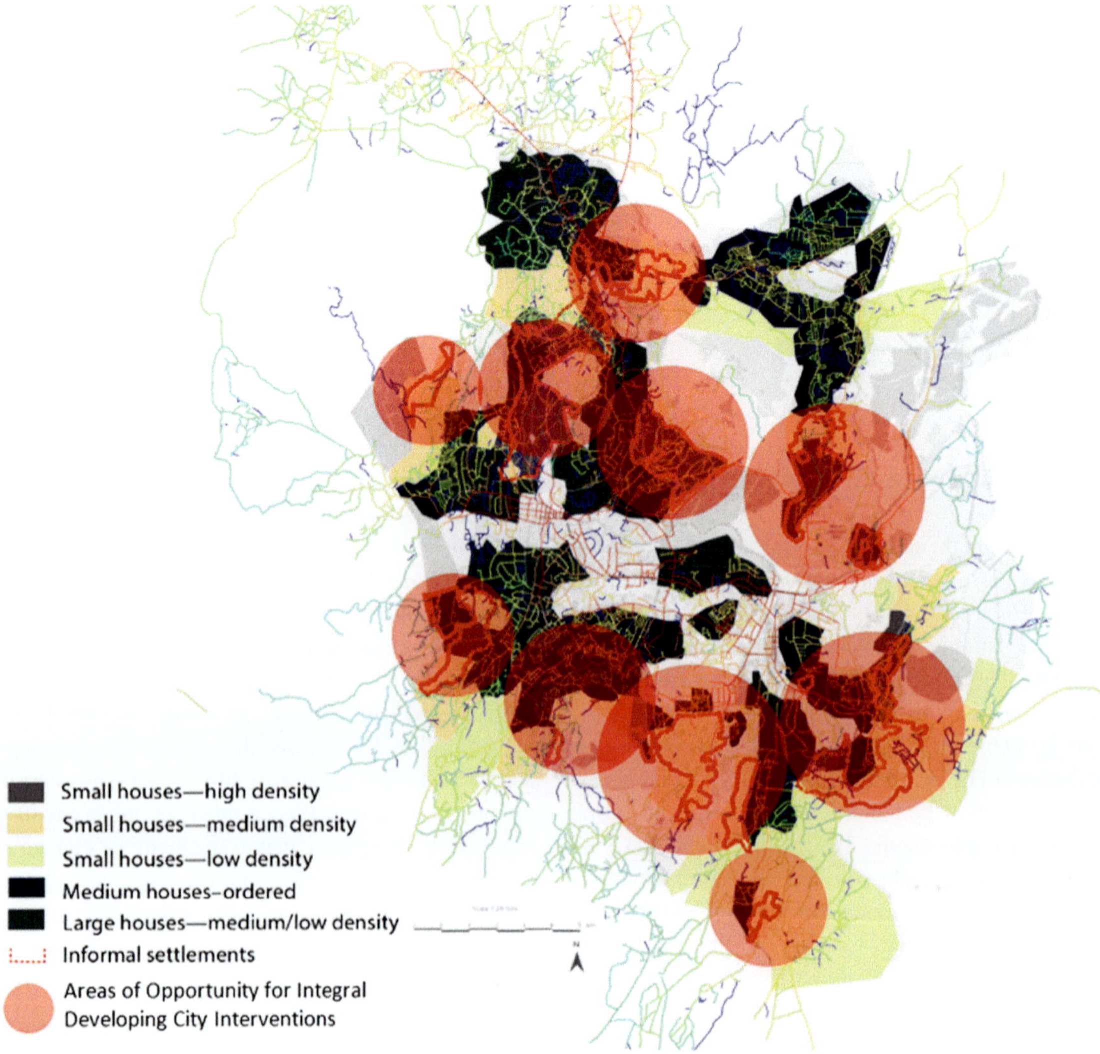

Fig. 20 Composite map showing areas of opportunity for integral developing city interventions (by the Author, 2022)

need to be improved to allow for streetscape access and the right to the city for urban dwellers. Climate change must also be addressed and embedded in planning responses to developing cities. These strategies can be embedded into community resilience responses, such as creating third spaces that strengthen community ties and quality of life. Space must be designed to accommodate informal street life systems such as mall eateries, vendors selling their goods, and informal social interactions. However, this can not happen without a policy review and multi-stakeholder collaboration. Overarching guiding principles must be determined and adhered to for solution-making in these cities. This includes stakeholders from the government, academic institutions, nongovernmental organizations, representatives from the citizenry, representatives from the informal sector, multiple representatives from multiple branches of the government, and city planning officials. Crippling issues such as poor electrification of city grids and the need to increase energy capacity must be addressed and embedded in the Integral Developing City framework. Doing so will not only build more equity in the city but also allow for the development of Smart City technologies and networks to be accessed by all.

This study was limited to non-empirical research. A fuller qualitative analysis could have been performed with a specified case study of a Blantyre neighborhood and interviews to capture a more accurate reading on the social dimension of the Experience and Culture quadrants of the Integral Developing City framework.

5 Conclusion

This study aimed to evaluate the 15-Minute City using the Integral Framework and to assess its transferability to developing cities like Blantyre. The research findings reveal that

Moreno's framework is not transferrable to developing cities as is; however, the Integral Developing City framework is. The results of this research could be extended to inform policy in the sustainable development of resilient and inclusive cities in the global South, in line with the 2030 UN Sustainable Development Goals. In addition, further research is warranted to test the Integral Developing City framework for its strengths and weaknesses for improvement.

References

Allam, Z., Chabaud, D., Gall, C., Pratlong, F., & Moreno, C. (2023). Redefining investable infrastructure in developing nations in a postpandemic era: the case of the 15-Minute City. *Resilient and Sustainable Cities*, 135–146. https://doi.org/10.1016/b978-0-323-91718-6.00001-3

Bibri, S. E., Krogstie, J., & Kärrholm, M. (2020). Compact city planning and development: Emerging practices and strategies for achieving the goals of sustainability. *Developments in the Built Environment*, *4*. https://doi.org/10.1016/j.dibe.2020.100021

Carlo, R., & Muggah, R. (2023). *The pushback against the 15-minute city*. The Boston Globe. https://www.bostonglobe.com/2023/03/06/opinion/pushback-against-15-minute-city/

Carrizosa, M. (2022). *Urban Informality and the Making of African Cities*.

Chamberlain, L. (2022). *The surprising stickiness of the "15-minute city."* World Economic Forum. https://www.weforum.org/agenda/2022/03/15-minute-city-stickiness/

Chirisa, I., Mutambisi, T., Chivenge, M., Mabaso, E., Matamanda, A., & Ncube, R. (2020). The urban penalty of COVID-19 lockdowns across the globe: manifestations and lessons for Anglophone sub-Saharan Africa. *GeoJournal*, *87*. https://doi.org/10.1007/s10708-020-10281-6

Commission on Environment, W. (n.d.). *Report of the world commission on environment and development: our common future towards sustainable development 2. Part II. Common challenges population and human resources* (Vol. 4).

Duany, A., & Steuteville, R. (2021). *Defining the 15-minute city*. CNU Public Square. https://www.cnu.org/publicsquare/2021/02/08/defining-15-minute-city

Duffy, J. D. (2020). *Integral psychiatry: A model for supporting human flourishing-viewpoint. A primer on integral theory and its application to mental health care*. https://doi.org/10.1177/2164956120952733

Fleming, R., & Roberts, S. (2019). *Sustainable design for the built environment* (1st ed.). Routledge.

Gehl, J. (2010). *Cities for people*. Island Press.

Giddings, B., Hopwood, B., & O'brien, G. (2002). Environment, economy and society: fitting them together into sustainable development. *Sustainable Development*, *10*, 187–196. https://doi.org/10.1002/sd.199

Government of Malawi, & Malawi Department of Tourism. (2022). *About Malawi*. Visit Malawi. https://visitmalawi.mw/index.php/about-malawi/

Integral European Conference. (2022). *Integral theory*. https://integraleuropeanconference.com/integral-theory/

Jacobs, J. (1993). *The death and life of great American Cities*. Vintage Books.

Kuepper, J. (2022). *What is a developing country? Developing countries explained*. The Balance. https://www.thebalancemoney.com/what-is-a-developing-country-1978982

Lu, M., & Diab, E. (2023). Understanding the determinants of x-minute city policies: A review of the North American and Australian cities' planning documents. *Journal of Urban Mobility*, *3*(December 2022), 100040. https://doi.org/10.1016/j.urbmob.2022.100040

Lydon, M., & Garcia, A. (2015). *Tactical urbanism*. Island Press.

Mahler, D. G., Lakner, C., Castaneda Aguilar, A. R., & Wu, H. (2020). *The impact of COVID-19 (Coronavirus) on global poverty: Why Sub-Saharan Africa might be the region hardest hit*. Data Blog. https://blogs.worldbank.org/opendata/impact-covid-19-coronavirus-global-poverty-why-sub-saharan-africa-might-be-region-hardest

Malawi Communications Regulatory Authority. (2015). *National survey on access to and usage of ICT services in Malawi*.

Malawi National Statistical Office. (n.d.). *Government of Malawi 2020 Malawi Poverty Report August 2021 National Statistical Office*.

Manda, M. A. Z. (2013). *Malawi Government Ministry of lands and housing situation of urbanisation in Malawi report*.

McCracken, J. (1998). Blantyre transformed: Class conflict and nationalism in urban Malawi. *Journal of African History, 39*, 247–269.

Moreno, C., Allam, Z., Chabaud, D., Gall, C., & Pratlong, F. (2021). Introducing the "15-minute city": Sustainability, resilience and place identity in future post-pandemic cities. *Smart Cities, 4*(1), 93–111. https://doi.org/10.3390/smartcities4010006

Mussa, R., & Masanjala, W. H. (2015). A dangerous divide the state of inequality in Malawi. *Oxfam*, 1–24.

Mvula, P., & Chiweza, A. (2013). The state of food insecurity in Blantyre City. *Malawi. AFSUN Food Security Series, 18*.

Mwathunga, E., & Donaldson, R. (2018). Urban land contestations, challenges and planning strategies in Malawi's main urban centres. *Land Use Policy*, *77*(September 2017), 1–8. https://doi.org/10.1016/j.landusepol.2018.05.025

National Statistical Office (NSO) and ICF Macro. (2011). *Malawi demographic and health survey 2010* (p. 603).

United Nations, Department of Economic and Social Affairs, Population Division. (2019). *World urbanization prospects: The 2018 revision (ST/ESA/SER.A/420)*. United Nations.

Nations, U. (2022). *Least developed countries (LDCs)*. Department of Economic and Social Affairs Economic Analysis. https://www.un.org/development/desa/dpad/least-developed-country-category.html

NSO. (2020). *The third integrated household panel survey (IHPS3)*.

NSO-Malawi. (2019). *Malawi population and housing census report-2018*.

NSO—Malawi. (2021). *2018 Malawi poplulation and housing census—Blantyre City Report*. www.nsomalawi.mw

NSO-Malawi, & OPHI. (2021). *Malawi Multidimensional Poverty Index*.

O'Sullivan, F. (2021). *The downsides of a 15-Minute City—Bloomberg*. BloombergCityLab. https://www.bloomberg.com/news/articles/2021-03-02/the-downsides-of-a-15-minute-city

Ogunbiyi, O. (2022). The disproportionate burden of COVID-19 in Africa. *Elsevier*. https://doi.org/10.1016/B978-0-323-82860-4.00021-5

Racki, J., Patel, P., & Degroot, D. (2014). *Africa 2050: Urbanization*. https://doi.org/10.1177/0974910113511191

Riley, L. (2012). The political economy of urban food security in Blantyre, Malawi. *The Africa Institute Occasional Paper Series*, *1*(2).

Sareen, S., Sikder, S. K., Boza-Kiss, B., Pachauri, S., & Zimm, C. (2021). Deprivations and Inequities in cities viewed through a pandemic lens. *Frontiers in Sustainable Cities, 3*, 645914. www.Frontiersin.Org. https://doi.org/10.3389/frsc.2021.645914

Silveira, S., Hecht, M., Matthaeus, H., Adli, M., Voelkle, M. C., & Singer, T. (2022). Coping with the COVID-19 pandemic: Perceived Changes in psychological vulnerability, resilience and social cohesion before, during and after lockdown. *International Journal of Environmental Research and Public Health*, *19*(6). https://doi.org/10.3390/ijerph19063290

Silver, B. J. (2010). Country classifications. *Forces of Labor*, 204–204. https://doi.org/10.1017/cbo9780511615702.009

Slater, S. J., Christiana, R. W., & Gustat, J. (2020). Recommendations for keeping parks and green space accessible for mental and physical health during COVID-19 and other pandemics. *Preventing Chronic Disease*, *17*(17). https://doi.org/10.5888/PCD17.200204

Steuteville, R. (2021). *The 15-minute neighborhood gets its 15 minutes of fame*. CNU Public Square. https://www.cnu.org/publicsquare/2021/01/25/15-minute-neighborhood-gets-its-15-minutes-fame

The World Bank Group. (2022). *How does the World Bank classify countries?* The World Bank. https://datahelpdesk.worldbank.org/knowledgebase/articles/378834-how-does-the-world-bank-classify-countries

UN-Habitat. (2011). *Regional and technical cooperation division Malawi: Blantyre urban profile*.

UN-Habitat. (2022). *Urbanization in Malawi: Building inclusive & sustainable cities*. https://unhabitat.org/malawi

United Nations Human Settlements Programme (UN-Habitat). (2020). *World cities report 2020: The value of sustainable urbanization*.

World Bank. (2017). *Malawi poverty gap*. https://data.worldbank.org/indicator/SI.POV.LMIC.GP?end=2021&start=2021&view=bar

Assessing the Impact of Urban Regeneration in Districts with Cultural Heritage on Neighbourhood Satisfaction and Social Behaviour: The Case of Heliopolis, Cairo

Ingy M. Rostom, Shaimaa M. Kamel, and Laila M. Khodeir

Abstract

With ever-growing populations in capital cities, governments are becoming more concerned with the economic aspects of urban regeneration and growth. Yet sometimes growth plans fail to create environments that reflect local values; it is then essential to recognise the effect of built cultural heritage on the satisfaction and subjective well-being of the local community. The study explores associations between regeneration of the built environment and corresponding changes in social behaviour as secondary effects of the decline in urban quality of life. The research is based on a case study approach that investigates the impact of urban regeneration in a district with rich cultural heritage, the "Heliopolis district" in Cairo, Egypt, on neighbourhood satisfaction as an indicator of urban livability. The methodology incorporated in this study is divided into two phases. The first phase was concerned with data collection using two tools: first, a site survey to document the changes that have been made to the urban fabric, and second, an online survey to assess neighbourhood satisfaction within four domains: place identity and legibility; transportation and mobility; walkability and safety; activity pattern and social behaviour. Data gathering is then followed by a descriptive analysis of the statistical results to explain frequencies and trends in social behaviour. Findings suggest that cultural heritage, whether tangible (built) or intangible (intellectual), is a key component of subjective well-being and consequently neighbourhood satisfaction. Demographic factors such as age, physical ability, and social status have a strong impact on neighbourhood satisfaction. Place identity, accessibility, safety, and perceived quality of space have all shown positive correlations to neighbourhood satisfaction. It is essential for policymakers to recognise the possible challenges of urban regeneration and the need for creating effective and equitable regeneration policies.

Keywords

Urban regeneration • Cultural heritage • Neighbourhood satisfaction • Social behaviour • Heliopolis

1 Introduction

Urban planning is becoming more and more interested in how to make cities better places to live. The expansion of urban populations globally, because of fast population growth and urbanisation processes, makes urban quality of life more significant for an increasing number of individuals. Consequently, cities adapt their physical qualities to accommodate additional populations.

A deeper comprehension of the relationship between cities' physical environments and their residents' quality of life may promote current and future urban development. There has been a significant increase in understanding of how people's physical surroundings might affect their well-being. Marans and Stimson (2011) offered an introduction to methods for gauging and analysing connections between city living and well-being. Based on their review of the literature, Kent and Thompson (2014) proposed that the built environment can improve people's health and happiness in three ways: by encouraging people to be more physically active, by fostering greater social integration within communities, and by increasing everyone's access to healthy foods. Open, natural, and green areas, as well as urban architecture that promotes social interaction and

I. M. Rostom (✉) · S. M. Kamel · L. M. Khodeir
Department of Architecture, Faculty of Engineering, Ain Shams University, Cairo, Egypt
e-mail: engy.elgazar@eng.asu.edu.eg

S. M. Kamel
e-mail: shaimaa_kamel@eng.asu.edu.eg

L. M. Khodeir
e-mail: laila.mohammed@eng.asu.edu.eg

A. Battisti et al. (eds.), *Greening Our Cities: Sustainable Urbanism for a Greener Future*,
Advances in Science, Technology & Innovation, https://doi.org/10.1007/978-3-031-49495-6_13

safety, were among the key determinants of satisfaction identified by Pfeiffer and Cloutier (2016). On a smaller scale, the neighbourhood-built environment may have an impact on subjective well-being via four routes, including social connections, leisure, physical health, and emotional experiences, according to Mouratidis (2017b). Also, according to Shekhar et al. (2019), involvement and social engagement, accessibility, place identity, and safety are the four factors that influence well-being in communities.

With the growing population and urbanisation on the one hand and the newfound knowledge of the importance of a city's physical setting in determining its citizens' quality of life on the other, the following issue arises: cities have a fundamental choice: whether to accept globalisation based on worldwide models or to place importance on local history and create revitalisation plans that may conserve previous characteristics of the landscape. Many developing countries' efforts to modernise their cities over the past few decades have ignored the need to preserve cultural heritage in the city's core, which has led to a loss of identity among the local population (Steinberg, 2008). This trend not only devalues the city's unique cultural character but also makes it challenging to preserve the city's cultural context, which contributes to the urban core's gradual extinction.

1.1 Problem Definition

This study tackles a very challenging issue when dealing with urban regeneration, and that is how to look at urban areas that are vital to the fabric of the city yet are not deemed deserving of designation as conservation areas. Distinctive populations, historical settings, or street layouts are among some of the morphological and cultural aspects that define these areas. They offer the setting in which the more prominent heritage assets are established, but they should not be viewed as mere settings because value is frequently created by the sum of artefacts and their settings. The study highlights how crucial it is to widen the notion of "built heritage" to include less evident and intangible elements.

1.2 Research Question and Hypothesis

Determinants of neighbourhood satisfaction can vary depending on the cultural context. Each local context has unique traits that may influence not only the environmental characteristics but also how locals experience and perceive them (Yang, 2008). There is a gap in current research regarding studies of cities from countries with different cultures to explain the variance in people's perceptions and responses to spatial and cultural changes with respect to their own local context.

The study draws the hypothesis that a trilateral association links urban context, local cultural heritage, and neighbourhood satisfaction. And hence, the main question addressed through the research is: what are the determinants of neighbourhood satisfaction that are linked to the cultural heritage of the local context? It also contributes to the literature as it explores associations between regeneration of the built environment and corresponding changes in social behaviour as secondary effects of the decline in urban qualities of life.

The paper is organised as follows: Sect. 2 depicts an overview of existing knowledge on the relationship between culture and urban regeneration based on past experiences. And in the second part, determinants of neighbourhood satisfaction are thoroughly investigated and classified. Section 3 presents the methodological background of the study. Section 4 presents the results of the analysis, while Sect. 5 provides a discussion of the findings. In the concluding Sect. 6, a general conclusion is reached, with recommendations for policymakers and future planning discussed in Sect. 7.

2 Literature Review

The literature is divided into two sections; the first section is concerned with the different approaches to identifying cultural heritage. The second section explains neighbourhood satisfaction through definitions, previous research, and relevant domains.

2.1 Heritage by Designation and Appropriation

According to recent studies, the main objective of urban regeneration is to counteract the present fast expansion in urban areas by preserving urban identity, culture, and traditions in addition to other goals, such as raising people's incomes and protecting natural resources (Ramlee et al., 2015). However, existing definitions of built heritage are limited and mostly depend on traditional notions of architectural and historical importance. The preservation of specific buildings and monuments is seldom an issue since it is already explicitly handled by law (Hassler et al., 2002). Similarly, a city or town as a whole might be designated as a conservation area to prohibit any kind of major construction. Therefore, the most important issue is how to deal with parts of cities that are not eligible for designation as conservation areas but are nevertheless crucial to the urban fabric. These pieces of city life often represent a defining characteristic of urban form and culture, such as a very high population concentration, historic significance, or distinctive street layout. However, they should not be dismissed as a mere setting since it is typically the combination of objects and their environment that produces value and not just the objects themselves. The existing concept of built heritage needs to be expanded to include less apparent cases, which can be defined as "heritage by appropriation".

Culture may be designated as heritage or appropriated as heritage, according to Rautenberg (1998). Heritage through designation is the standard method by which places, buildings, and other cultural artefacts are given the honorific name of "heritage" by a panel of experts. Top-down approaches with little public input tend to provide predictable and hardly significant historic designations. On the contrary, most cases of appropriation as a kind of cultural heritage occur as a result of spontaneous public action rather than deliberate lobbying efforts. The practise of "heritage via appropriation" is indicative of increasing cultural liberalisation, in which individuals have a more active role in deciding what is culturally significant. The "black spots" of cultural tourism, such as the death and burial places of modern celebrities, are only one example of the kinds of items and occurrences that have been included into the literature of heritage via appropriation (Tweed & Sutherland, 2007). Consequently, cultural heritage now encompasses a wide range of phenomena. Heritage is increasingly seen as crucial to people's sense of belonging and cultural identity, in part because it is more inclusive.

2.2 Neighbourhood Satisfaction

Different scales of environmental satisfaction, such as the city, the neighbourhood, and the residence (the dwelling unit), have been identified by researchers such as Campbell et al. (1976), yet the neighbourhood scale appears to be the most appropriate scale for studying urban regeneration and growth as well as analysing social and spatial structures, where city-level spatial layouts may be too generalised (Mantey, 2021).

Neighbourhood satisfaction, or basically the residents' degree of contentment with their neighbourhood, assesses how effectively the neighbourhood meets personal or household needs. It reflects the level of congruence between a person's neighbourhood goals and his actual living conditions (Campbell et al., 1976; Cao, 2016). Neighbourhood satisfaction is influenced by both the prospects provided by the built environment and personal assessments of what is significant in life. It is not only a component of a socially sustainable neighbourhood but also a fundamental aspect of quality of life that is translated into life satisfaction. Neighbourhood satisfaction has been proven to be positively correlated with life satisfaction (Cao, 2016) and may represent an important pathway between the urban context and subjective well-being (Campbell et al., 1976; Marans, 2003); Fig. (1).

The framework of determinants of life satisfaction domains proposed by Campbell et al., as seen in Fig. 1, can be used to examine the association between spatial objective qualities and neighbourhood satisfaction (Campbell et al., 1976). Campbell's model suggests that life satisfaction is the combined result of objective (spatial) attributes that act as a stimulus and residents' subjective reactions that represent how these attributes are viewed and appraised, in addition to satisfaction with other life domains that are not necessarily linked to residential factors. The following are some examples of non-residential domains of life satisfaction: health, work, leisure time, and personal life, among others. Since each of the mentioned domains needs substantial research on their effects on life satisfaction, the study is focusing on just one, which is the neighbourhood. The neighbourhood, as previously mentioned, is the most appropriate scale to

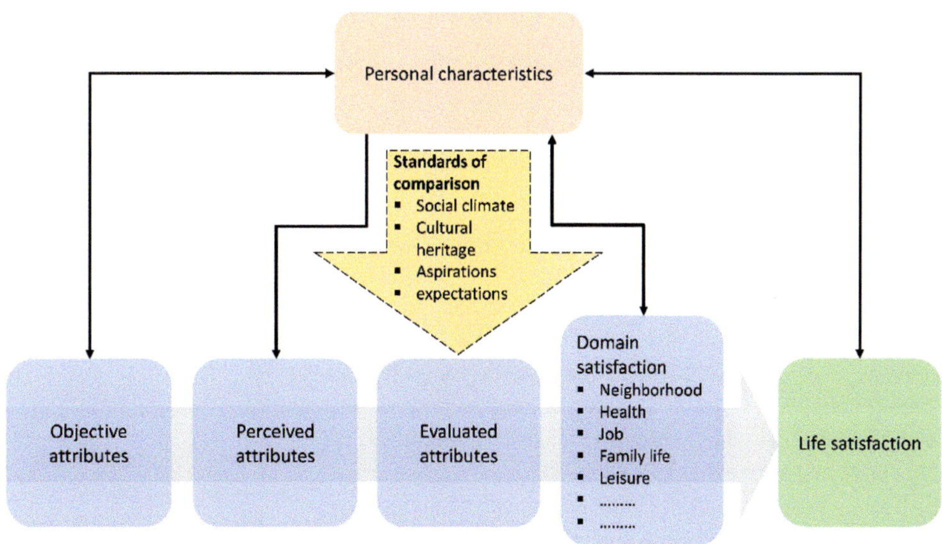

Fig. 1 Relationship model explaining the association between objective environmental attributes and overall life satisfaction based on Campbell's model (Campbell et al., 1976)

measure the effects of spatial morphological changes in a cultural context on the perceived quality of life and satisfaction.

Campbell's approach has been restricted to the relationships between the neighbourhood's objective characteristics and its residents' sensations of contentment. Neighbourhood satisfaction, according to Campbell's model, is the outcome of a chain of reactions, each of which comes after the previous one. The objective (spatial) features come first, representing the stimulus for a reaction (as previously mentioned), then are followed by the cognitive reaction, which reflects how the objective attributes are perceived, and finally, the chain ends with the assessments of these attributes. Objective attributes represent the spatial features of the space and are measured independently, excluding residents' input, by outside observers or by any other method (geospatial data is the most common method of collecting objective and spatial data). The subjective assessment of neighbourhood attributes is a component of two factors: residents' perceptions of the environment's spatial characteristics and (independently) their satisfaction with them. The way a person perceives neighbourhood characteristics depends on but differs from the physical environment; the perceived environment is not always the same as the real environment.

According to Campbell's framework (Error! Reference source not found.), the relation between spatial attributes and subjective perceptions is "indirect". Galster's psychological construct theory of neighbourhood satisfaction explains how personal, household, cultural, and social characteristics can filter or sort the influence of environmental spatial attributes on neighbourhood satisfaction. Residents, in their evaluation of neighbourhood quality, create a "reference" condition for each neighbourhood feature that is based on their personal needs and preferences (Galster & Hesser, 1981). This concept highlights the importance of individual customs and provides insights on neighbourhood satisfaction. Perception and the cognitive map are two ideas that environmental psychologists employ to describe these standards.

A person assesses his or her context in the light of a particular set of perceived needs and goals. From one side, the needs and ambitions that a particular person perceives are a complex of both personal characteristics (such as social status) and cultural standards (which are reflected in the cultural context where a person is located), each dependent on the person. On the other side, cognitive maps are spatial structures in the brain that retain information about how humans conceptualise the built and natural environments. These networks connect a person's recollection of past events with how they currently perceive current events, concepts, and emotions (Young, 1999).

The influence of spatial dimensions on neighbourhood satisfaction, the assessments of neighbourhood qualities, and total neighbourhood satisfaction are all influenced by personal and household characteristics, one of the three key factors determining neighbourhood satisfaction (Mantey, 2021). Nevertheless, past research indicates that subjective evaluation and perception of neighbourhood qualities are more significant drivers of neighbourhood satisfaction than individual or household characteristics (Parkes et al., 2002). It implies that spatial features have a weak and indirect relationship with neighbourhood satisfaction. The same research (Parkes et al., 2002) also revealed that neighbourhood spatial characteristics explained a greater proportion of the variance in neighbourhood satisfaction than did demographic factors. This result confirms that, irrespective of age, gender, or social standard, good neighbourhood design can help boost satisfaction.

2.2.1 Spatial and Socio-cultural Dimensions of Neighbourhood Satisfaction

According to the literature, neighbourhood satisfaction is a component of three factors as seen in (Fig. 2): (1) spatial (objective) dimensions and attributes of the area, such as density, walkability, accessibility, neighbourhood amenities, open public and green spaces, distances to various services and low traffic (Mouratidis, 2017a; Yang, 2008); (2) subjective assessments of the neighbourhood qualities such as perceptions of safety, quietness, social cohesion, place attachment, and perceived aesthetic quality (Buys & Miller, 2012; Hur & Nasar, 2014); and (3) individual/household characteristics (demographic factors) (Mantey, 2021).

Spatial Dimensions of Neighbourhood Satisfaction

There have been a number of attempts to establish a relationship between objectively measured physical characteristics and the general walkability of a city. Imageability, confinement, scale, visibility, diversity, legibility, connectivity, and cohesion were outlined, as well as a few others, as desirable features in urban design (Singh, 2016). Also, a number of studies found a positive relationship between public green spaces and life satisfaction (Ambrey & Fleming, 2013; Shafer et al., 2000). Happiness was studied in ten of the world's most populous cities by Leyden et al., (2011), who concluded that residents are more content when their communities are seen as economically viable, welcoming to families with children, and providing convenient access to public transportation as well as cultural and recreational opportunities.

Three spatial dimensions of neighbourhood satisfaction were selected for study based on the nature of changes that the neighbourhood has gone through. These dimensions are legibility, imageability, and perceived quality of public spaces.

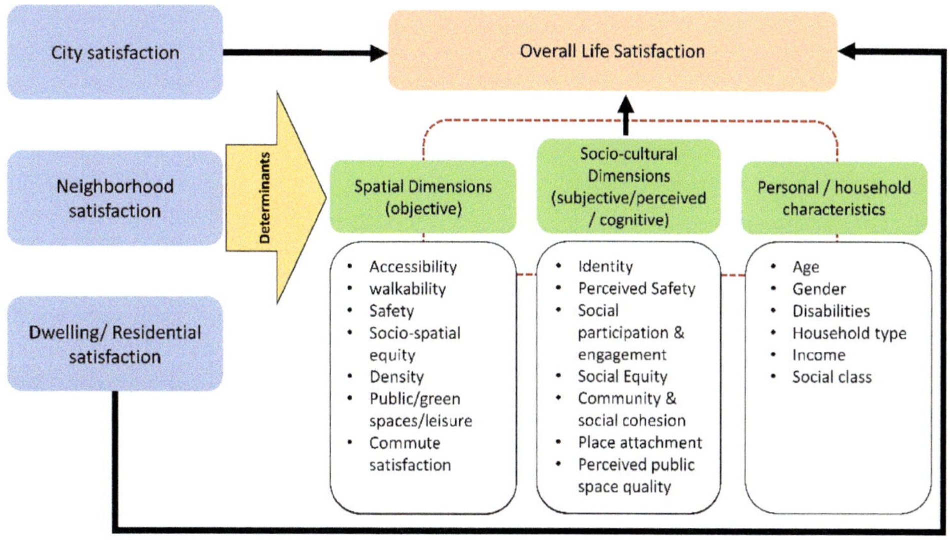

Fig. 2 Factors of neighborhood satisfaction. Conducted by the author

Legibility

"Legibility" is defined by Lynch (1960), who refers to the legibility of the cityscape as "the ease with which its parts can be recognised and can be organised into a coherent pattern". For a city to be considered "legible," its neighbourhoods, landmarks, and streets must all be readily recognisable and be able to be organised into a coherent whole. The environmental image, an individual's overarching mental picture of the external physical world, serves as a crucial connection in the navigational process. The picture serves to understand data and direct behaviour, and it is the result of both current perception and accumulated knowledge. It is possible that being totally lost in a contemporary city is an unusual occurrence for most residents, as maps, street numbers, and directional signs all provide a hand in helping people find their way about. But if disorientation occurs only once, the accompanying worry and even dread will reveal how intrinsically related it is to feelings of balance and well-being. A person's mental and emotional well-being may be greatly improved by developing a positive impression of their immediate surroundings, where balance can be achieved. The pleasant sensation of home is greatest when home is not just familiar but unique, and this is the counterpart of the terror that comes with disorientation. Commonly encountered by newcomers or in new settings, stress and anxiety accompany disorientation in both built and natural surroundings (Cohen et al., 1987), and this may contribute to navigation anxiety. Not only does a recognisable and easily understood setting provide a sense of safety, but it also increases the richness of the human experience. Some of the benefits of a legible environment include increased feelings of well-being, clearer channels of communication, better conceptual structure, and maybe even a deeper appreciation for the everyday experience (Lynch, 1960).

Imageability

The term "imageability" is concerned with the quality of a physical environment to evoke vivid images and broadly shared social meanings among the occupants of the setting. It lies in the details of the built environment, which facilitate creating strong mental images in the mind of the observer (Lynch, 1960). In the research paper "Psychological maps of Paris" (Milgram & Jodelet, 1976), the authors argue that people create nonverbal, spatial ideas about their cities that are totally different from the stereotyped images, as seen by outsiders, which do not reflect their own personal images tied to direct experience. The study argues that the perception of the city is not limited to what actually exists but also to what is highlighted by the community. Maps drawn by city residents are projections of their individual lifestyles and express emotional charges held by residents towards their city.

Perceived Open Space and Aesthetic Quality

Civic officials have lately been thinking about how parks and other outdoor areas impact a city's image. There is an agreement that the finest open areas are found in the best cities (Heckscher & Robinson, 1977). For instance, Rockefeller Centre in New York City has become a landmark because of the Cascade Garden and Plaza, and so has the Fountain Plaza in Portland, which turned the downtown area into a recognisable symbol in the city's core. Cities in these areas are focusing more on enhancing existing urban open spaces or establishing new ones as a means to improve their municipal image. A positive mental picture of a city gives its owner a crucial feeling of security, according to research on mental images of cities by Lynch (1960). The potential richness and intensity of human experience may also be increased by a unique and legible setting. An excellently planned open area with clearly defined, well organised, and often utilised sections improve the user's feeling of place and sense of belonging (Lynch, 1960). As a result, properly planned open space may enhance both the user's perception of a location as well as the image of the city. In conclusion, urban open space is a crucial component in giving a city a strong reputation. A sense of duty or belonging is sparked by functional public areas, which boost civic pride and lower crime rates.

Socio-cultural Dimensions of Neighbourhood Satisfaction

Many studies confirmed the potential for cultural heritage to contribute to the unique character of the town through evoking psychological and emotional impacts on both locals and visitors (Tweed & Sutherland, 2007). The historic centres not only provide the local community with visual and sensuous benefits but also serve as places to recall by preserving historical knowledge. Senior centres' residents in New York City were surveyed by Friedman et al. (2011) to see how their perceptions of their surrounding community affect their satisfaction with life. They determine that walkability has no influence on residents' satisfaction, whereas a sense of perceived safety and community in one's neighbourhood has a positive effect. High environmental, physical, and mental stress have been linked to living in unsafe places, such as those with a high crime rate. This has negative effects on both quality of life and mental well-being (Bazargan, 1994). Research on the quality of life in cities as a whole has shown a correlation between more social equity and resident satisfaction (Ballas, 2013). The built environment in cities may impact physical activity, neighbourhood social cohesiveness, and equal access to nutritious food, according to a research review by Kent and Thompson (2014). Open public areas, natural and green spaces, and urban design fostering social interaction and safety were all cited as contributing to satisfaction in communities by Pfeiffer and Cloutier (2016). According to Shekar et al. (2019), there are four key categories that may be used to classify the main determinants of urban livability: participation and engagement, accessibility, identity, and safety.

Five socio-cultural dimensions of neighbourhood satisfaction were chosen to be studied. These attributes are place attachment, community attachment, social cohesion, place identity and belonging, social equity, and perceived safety.

3 Place Attachment

Place attachment is defined by Stedman (2002) as a cognitive and emotional tie between individuals and their surroundings. Identity is the meaning assigned to oneself and acquired through psychological and social contact with others. Stedman observed that once meanings are established for a certain location, they serve as crucial predictors of attachment, enjoyment, and behaviour. An individual's emotional attachment to a particular location arises through a web of interconnected, symbolic ties to the land and the people who live there (Rubinstein & Parmelee, 1992). Riley (1992) proposed that the average person has an emotional connection to the environment they live in because it serves as a setting for their everyday lives. In the attempts to connect place attachment to age and position in the life cycle, research conducted by Coles (1970) hypothesised that the frequent disruptions experienced by migrating children acted as a sort of chronic trauma, impairing their capacity to establish a stable sense of self. In connection with childhood recollections, adults may demonstrate a simple affinity for a location linked with family love and safety. The location may also be perceived as a live memory, complete with the feelings and emotions that existed at that remote time. This "location" may not exist in the physical world, but it does exist in the heart and thoughts of the individual. Another sort of attachment is the elderly's perception of their life path and their connection to certain places. This feeling of belonging may aid the elderly in continuing to feel competent and independent (Rubinstein & Parmelee, 1992).

3.1 Community Attachment and Social Cohesion

The idea of community attachment is described as residents' pleasure with their living environments, which aids in the formation and maintenance of their identities and ties to their community. The notion of group provides safety and a sense of belonging and is essential for finding one's place in the community. In Coles's research on migrant children (Coles, 1970), it was hypothesised that their continuous, disruptive mobility may operate as chronic trauma, diminishing the brain's capacity to develop a strong sense of self. Community attachment is considered an emotional attachment to the local region; it is the result of people's views, their social position, and the community's attributes. Community connection is most deeply rooted in an individual's social activity. Community identity seems to be constructed from the private significance of life events and the public manifestations of local culture (Hummon, 1992).

3.2 Place Identity and Belonging

According to Proshansky et al. (1983), place identity is a component of self-identification made up of memories, thoughts, values, sentiments, meanings, and concepts of action and experience that are connected to the physical environments that every human being is characterised by. People do more than just experience and "document" their surroundings; their needs and aspirations regarding a place are what define its identity. Because each individual views the same environment through a distinct self-identity, each person has a "different" place identity. Each person absorbs and retains information based on a unique set of values, memories, and experiences. Place belongingness is another name for place identity (Tuan, 1991). Place identity serves as a benchmark against which the present physical environment may be evaluated. Choosing what is familiar and what is unknown is one of the initial tasks. Occasionally, drastic changes in one's physical surroundings might pose a danger to one's identity (Cuba & Hummon, 1993). The elderly view of a location is becoming more significant, and it may become their focal point as they cherish their experiences by storing photographs that aid in bringing back such recollections.

3.3 Social Equity

It is imperative that "accessibility for all" be recognised as a basic citizen right since the usability of urban settings is crucial to enhancing an individual's capacity to actively contribute to her or his own well-being. According to this viewpoint, accessibility may be rethought as an integral part of people-centred and inclusive urban design policies. According to the UN's sustainable development goal no. 10, target 10.2 states that "by 2030, empower and promote the social, economic, and political inclusion of all, irrespective of age, sex, disability, race, ethnicity, origin, religion, or economic or other status" (UN, 2016).

3.4 Perceived Safety

The perception of safety is a significant component that influences the approach and avoidance behaviours of neighbourhood users and residents and may be considered a facilitator or an inhibitor of interaction with the built environment. Perceived safety can be affected by two stimuli: strangers and the loss of identity due to a change in the immediate environment.

Strangers and Defensible Spaces: Defensible space is a concept introduced by architect Oscar Newman (1972) as a route to a safe, productive, and well-kept environment. A "prospective criminal" views an area owned by its people as unsafe for illegal actions since he or she is perceived as an invader who can be quickly discovered, captured, and dealt with (Newman, 1972). As a result, Newman argued for the partitioning of vast areas of public space into smaller sections that might be used and controlled by individuals or organisations. As Newman pointed out, "target hardening," the use of design features like fences, gates, and locks and other components that enable citizens to have both visible and physical control over public places and the neighbourhood environment may provide defensible space and reduce criminal activity. To do this, he identified four elements of physical design that can increase a community's safety: (1) clearly defined boundaries; (2) large windows that give residents a good view of the street; (3) buildings that blend in with their surroundings; and (4) the placement of new neighbourhoods far from areas with a high crime rate. According to Newman's theories of territoriality, access to neighbourhoods, residential streets, communal spaces of apartment buildings, and housing blocks should be restricted.

Loss of Self and Place Identity: Voluntary relocation is a kind of place disturbance. Families that choose to relocate to a new location must adapt to their new surroundings. Even temporary relocation is accompanied by a feeling of loss (Hummon, 1992). In addition to deliberate relocations, involuntary relocations often create huge disturbances in people's homes, thoughts, and emotions. They have been forcibly removed from "home" and cannot return. All their possessions have disappeared along with their memories, and they quickly begin to mourn their losses. People affected by relocation may experience feelings of isolation, fear, and inability to deal with reality. That is one form of perceived safety that is related to the local environment, and that has a direct effect on the subjective assessment of "loss of place identity" and "loss of self-identity".

4 Methodology

There are two main methods of analysis used in "quality of life" research. The first method analyses objective indicators using physically collected secondary data, whereas the second method employs surveys to analyse subjective assessments of domains. Since this study is basically concerned with assessing users' satisfaction with their built environment, it adopts the survey method. It is the method through which people can best assess the quality of their surroundings, and these assessments are typically based on a variety of attitudes and ideas about the environment and the local community. It is not only concerned with the present state but also with people's preferences for their future environments, which can be used as guidelines to improve the design.

4.1 Study Concept

The overall hypothesis of the study is founded on a basic form of Campbell's model (Campbell et al., 1976), which discusses the connection between objective environmental factors and subjective neighbourhood satisfaction. The foundational model has been condensed to the most significant aspects of neighbourhood satisfaction by concentrating on the physical changes that the neighbourhood has undergone throughout the process of urban development. These variables are thought to comprise both the subjective evaluations of many determinants of neighbourhood satisfaction and the objective components of the built environment. "Heliopolis" was chosen as a case study to validate this hypothesis.

4.2 Case Study

The city of Heliopolis was chosen as the case study to assess the satisfaction of the local community with major physical changes made to their urban environment. Reasons for choosing "Heliopolis" include the following:

I. Heliopolis City represents a rich cultural context. It was founded in 1905 by Belgian industrialist Baron Empain, who together with Boghos Nubar headed the Heliopolis Oasis Company (Kamaly, 2013). Designed by Belgian architects "Ernest Jasper" and "Alexandre Marcel", the district features a distinctive architectural style known as Heliopolis architecture, which included Moorish, Persian, Arabic, Islamic, and European elements. It had wide pathways as featured in the European style, vast gardens overlooked by elegant arabesque balconies, and roofs with domes reminiscent of Islamic and Ottoman architecture. "Heliopolis" also features a number of landmarks, such as "The Baron Palace" and "The Heliopolis Palace Hotel", realised by 1910 (Dobrowolska & Dobrowolska, 2006). Heliopolis possesses so much potential in both tangible and intangible cultural heritage.

II. Heliopolis City has undergone a major urban development plan that was put into action starting in 2018. This plan aimed primarily at increasing traffic flow inside the district using a number of design techniques, such as replacing the radial flow of cars with free-flowing straight roads using bridges instead of signalised intersections.

The following section explains the primary urban design scheme of Heliopolis, which was planned in the early twentieth century, to draw a benchmark against which current changes can be compared.

4.2.1 A Historical Background of the Urban Planning Scheme of Heliopolis

The four main neighbourhoods of Heliopolis are "Al-Montazah", "Mansheyet Al-Bakri", "Al-Bostan", and "Al-Nozha" (Fig. 3). Beginning in the early 1900s with the "Al-Montazah" neighbourhood, the growth and planning of Heliopolis had gone through numerous periods, each with its own personality. "Al Montazah" neighbourhood relied heavily on axial design, with many perpendicular axes such as "Al-Ahram" Street, "Al-Baron" Street, and "Haroon Al-Rasheed" Street, and a "garden city" concept for the Al-Korba region (Fig. 4). From the 1930s through the 1960s, radial planning was employed, based on central squares with radial streets linking these squares to construct a large, well-connected network of main streets with residential blocks in between (Fig. 5). In order to preserve the garden city concept of the original design, wide green areas and tram lines were added on the islands of these main roadways. After the nationalisation of the Heliopolis company in the 1960s, the new extensions in the "Al-Nozha" neighbourhood maintained the district's attachment to green spaces by incorporating central gardens between residential blocks (Fig. 6), while allowing floor plans to increase to ten stories per building in contrast to the four-story low-rise buildings of earlier developments (Dobrowolska & Dobrowolska, 2006).

4.3 Data Gathering Tools

The methodology of this study depends on finding the link between objective attributes of the built environment and subjective assessments of the neighbourhood users. To fulfil this goal, two methods were used for data gathering, as follows.

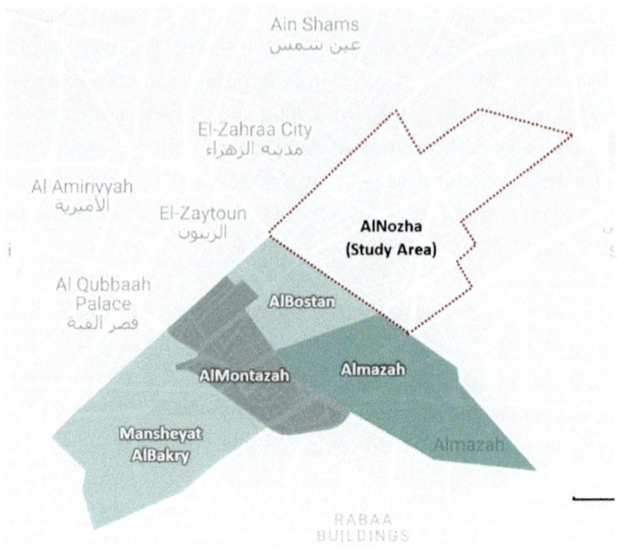

Fig. 3 Main neighbourhoods of Heliopolis. *Source* Google maps (2022)

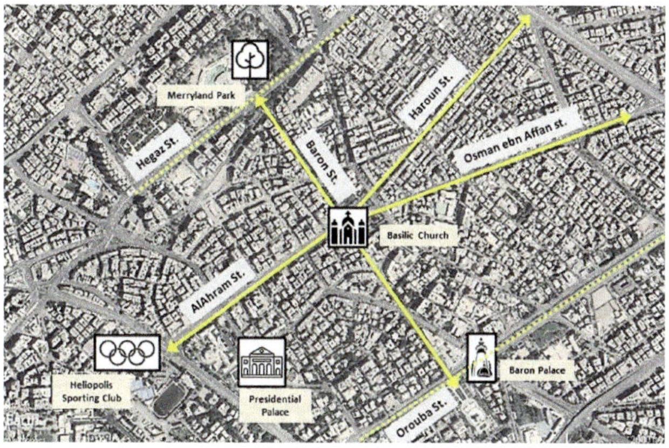

Fig. 4 Urban design of "Al-Montazah" district. *Source* Google Earth (2022)

Fig. 5 Radial street planning of "Al-Bostan" neighbourhood. *Source* Google Earth (2022)

Fig. 6 "Al-Nozha" neighbourhood street layout. *Source* Google Earth (2022)

Fig. 7 "Al-Mahkama" square before and after urban regeneration plan. *Source* Google Earth (2017, 2022)

4.3.1 Study Area Mapping

The first part of the study required collecting objective data reflecting the neighbourhood characteristics in the present after the physical changes that the neighbourhood has gone through since the beginning of the development plan in 2018. A detailed mapping of the major physical changes that had occurred in the "Heliopolis" district after the urban regeneration plan was conducted and categorised into three types of physical change. These changes were the cornerstone upon which the survey questions were designed to assess the users' perceptions and insights. The tools used in this stage were digital mapping using Google Earth and physical mapping and documentation through a site survey.

The changes were grouped into three main categories:

Newly Added Bridges

A number of 11 new bridges were added to Heliopolis to ease the traffic congestion by substituting signalised intersections with free-flowing roads on the street level and flying over bridges across. Most of the squares from the original Heliopolis planning scheme were removed, along with the accompanying green spaces and central islands (Figs. 7 and 8).

Removal of Green Spaces and Street Widening

Street widening came at the expense of central (refuge) island removal or shrinkage in many main streets, such as "Fareed Semeika", "Abd El-Aziz Fahmy", and "Abo-Bakr

Fig. 8 "Al-Hejaz" square before and after urban regeneration plan. *Source* Google Earth (2017, 2022)

Fig. 9 "Heliopolis square" before and after 2018. *Source* Google Earth (2017, 2022)

Al-Sadeek". Using Google Earth pictures of before and after maps of streets lying in "Al-Bostan" and "Al-Nozha" neighbourhoods, the author could estimate an approximate number for the decrease in the square area of the green spaces in the neighbourhood. This number reaches almost 83% loss of the total green spaces located in central islands and squares removed from these neighbourhoods (Figs. 9 and 10).

Addition of New Activities

As an investment in the newly found spaces under the newly constructed bridges, a large number of temporary light-weight structures were built under the bridges and rented to different types of activities (Fig. 11). After surveying the site and taking an inventory, under bridge activities can be mainly divided into three groups:

- Coffee shops and fast-food restaurants,
- Retail shops,
- Parking lots.

4.3.2 Questionnaire

In the second part of the study, an online questionnaire was conducted to assess the subjective evaluations of neighbourhood users. The questionnaire aims at finding the correlation between spatial attributes that have been affected by urban regeneration in the study area, as a district of rich cultural heritage, and neighbourhood satisfaction among residents, visitors, and passers-by. The structure of the questionnaire is based on components of neighbourhood satisfaction as defined in the literature of the study. It provides a measure against previously identified indicators of perceived neighbourhood satisfaction, including those not only associated with perception but also with change and attitudes towards change. A descriptive analysis of the statistical results was used to explain alterations in social behaviour that started to emerge as a consequence of urban morphology and that might become inherent in generations to come.

Questions were designed to assess both spatial (tangible) and socio-cultural (intangible) dimensions. Subjective and perceived neighbourhood characteristics have been evaluated using the questionnaire, in which residents subjectively assessed their satisfaction with the various attributes of their newly revitalised built environment. It consists of 27 questions (indicators) in 4 domains and 7 sub-domains (Table 1), in addition to 6 questions aimed at determining the demographic composition of the respondents.

The questionnaire was conducted online in July–August 2022 with a random sample of Heliopolis residents and

Fig. 10 Decrease in area of green spaces calculated in "Al-Bostan" neighbourhood. *Source* Author based on Google Earth (2017, 2022)

Fig. 11 Under-bridge retail and parking. *Source* Captured by the author (2022)

visitors. To estimate the sample size, the anticipated population in both "Heliopolis" and "Al-Nozha" districts was 365,300 according to the 2017 census (EGY-CAPMAS, 2017). The sample size was estimated based on the equation shown in Fig. 12, and the sample size calculation indicated that 118 respondents should be included to ensure a confidence level of 95% and a margin of error of 9%. A total of 290 responses were collected, of which 127 were residents and 163 visitors.

Limitations of the study: The questionnaire was designed to test both subjective and objective attributes of the urban quality of life, but the scope of this study will only include "subjective attributes" of the urban environment since it investigates the relation between the cultural heritage of the study context and the subjective well-being of the local community.

5 Results

Using the thematic analysis method, results from the questionnaire and open-ended questions were analysed, and seven main attributes were found to be linking built environments with cultural heritage, neighbourhood satisfaction, and subjective well-being. This section will explain the basic literature behind the logic of each of the seven attributes, and it will further link and validate them using statistical values and an analysis of the responses to the questionnaire.

5.1 Socio-cultural Dimensions

This section discusses results related to the socio-cultural characteristics of the urban environment.

5.1.1 Place Attachment

When people were asked through the questionnaire to speak about the ties that link them to Heliopolis and emphasise their sense of place attachment, they spoke of "family gatherings and visits", "childhood memories in clubs and restaurants with friends", "school friends", "church visits", "street strolls with friends and family", and "special foods in their favourite restaurants in Heliopolis". Although the question is highly dependent on the qualities of place, people's top choices for reasons for place attachment always involved "memories and people they love". All open

Table 1 Survey design

Domain	Sub-domain	Indicator	Survey question type	Demographic factors
Built environment	Place identity and legibility	Distinguishing features of the neighbourhood	Multiple choice question	Age, gender, mobility impairment, and user type
		Promoters of place attachment	Multiple choice question	
		Ease of navigation through the neighbourhood (legibility)	Multiple choice question	
		Ability to identify already known places (imageability)	Picture identification	
	Accessibility	Change in accessibility to local services	Multiple choice question	
		Services requiring main-street crossing	Multiple choice question	
	Walkability	Distances walked daily	Multiple choice question	
		Subjective assessment of side walking experience	Likert rating scale	
		Street crossing methods	Multiple choice question	
Travel	Transportation and mobility	Transportation method used	Multiple choice question	
		Sufficiency of public transit	Multiple choice question	
		Proximity to public transit stops	Multiple choice question	
		Car ownership	Multiple choice question	
Living	Activity pattern and behaviour	Change in daily activities	Multiple choice question	
		Coffee shops and restaurants change in visit rate	Multiple choice question	
		Subjective assessment of new introduced activities' effects	Multiple choice question	
		Reasons of changing activity	Multiple choice question	
Subjective well-being	Perceived safety	Subjective assessment of street crossing experience	Likert rating scale	
		Number of experienced car accidents (if any)	Multiple choice question	
		Reasons for neighbourhood insecurities	Multiple choice question	
	Subjective assessment of development plan	Perceptions of newly introduced activities under bridges	Multiple choice question	
		Subjective assessment of image and cognition of the neighbourhood	Likert rating scale	
		Subjective assessment of traffic flow	Multiple choice question	
		User best served by development plan	Multiple choice question	
		Consideration of relocating	Multiple choice question	
		Overall satisfaction	Open-ended question	
		Subjective assessment of neighbourhood priorities for the walking-impaired	Ranking question	

Source Author

$$S = \frac{\frac{Z^2}{e^2} \times p(1-p)}{1 + \left(\frac{Z^2 \times p(1-p)}{e^2 \times N}\right)}.$$

Fig. 12 Sample size calculating equation, where S = sample size, N = population size, e = margin of error (percentage in decimal form), Z = z-score, and p = sample proportion. *Source* Megahed et al. (2020)

responses were fed into a "word cloud" generating program, and the result came as seen in Fig. 13. The largest and hence most frequent words were "family", "memories", and "club" followed by "places", "friends", "childhood", and "school". Places linked with concepts of "family", "love", and "safety" gain special meanings and are forever linked to these places in core memories. They become a part of a person's identity, and hence, his stability and well-being become threatened when these memories are taken away with profound changes to the place he once knew and loved.

Data collected, shown in Fig. 14, regarding residents' responses to what triggers and promotes their sense of place attachment indicates that age and length of residency are positively linked with expressed attachment and possessions or physical attributes of significance that help establish a sense of self-identity within the heart. A 90% of people aged 51–60 and 75% of people aged 60 or older expressed that their walking experience in promenade-type commercial axes, as in Ibrahim Al-Lakkani St.-Roxy, promoted their strongest sense of place attachment through the cognition of images strongly related to the architectural style, commercial activities, and recreational activities that were founded tens of years ago and still exist to this day.

5.1.2 Community Attachment and Social Cohesion

Coming in second and third place, factors promoting place attachment were "meeting others in public spaces" and "neighbourhood social events", where social events have a higher public base among younger residents and visitors

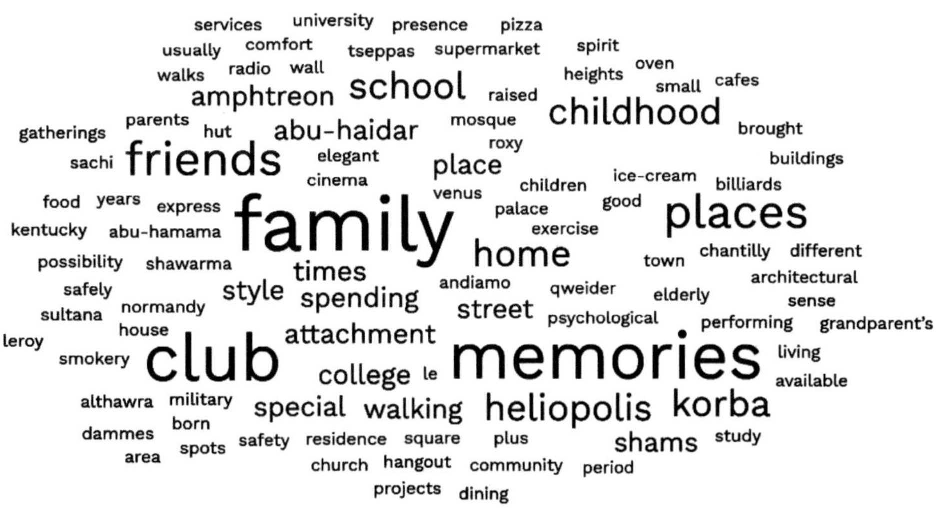

Fig. 13 Word cloud representation of respondents' answers to what promotes their sense of place attachment. *Source* Author

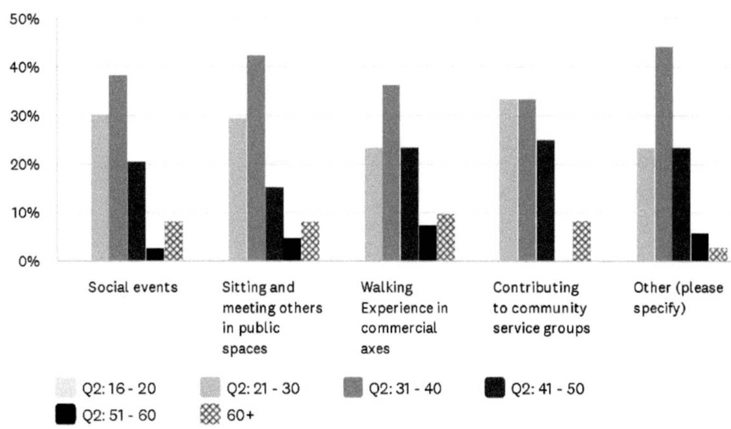

Fig. 14 Graphical representation of respondents' choices of what promotes their sense of place attachment, filtered by age groups. *Source* Author

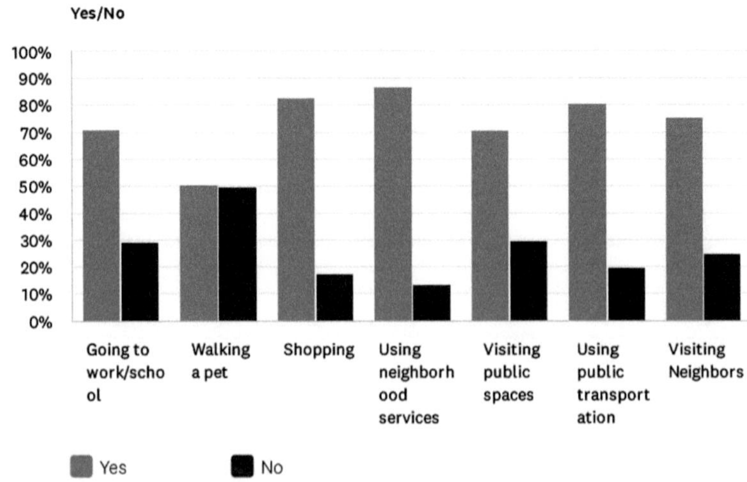

Fig. 15 Activities requiring main street crossing. *Source* Author

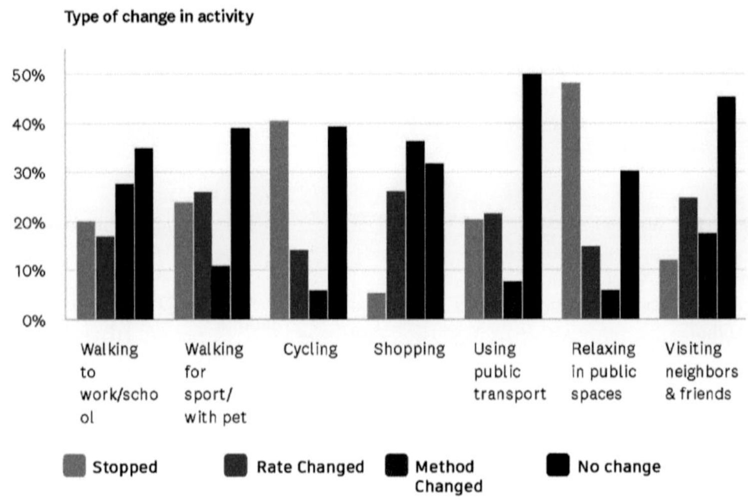

Fig. 16 Statistical analysis of responses to the question "How did the following activities change after the development plan?". *Source* Author

(aged 21–30 and 31–40), as seen in Fig. 14. These two kinds of activities involve mingling with the local community in a familiar setting, creating personalised memories that can only be created in this specific place. It involves feelings of "belonging" and "safety" with a backdrop of the mental images of the famous architectural style of Heliopolis. And in the fourth place is the feeling of attachment gained through contributing to "community service groups", which only gains recognition during times when locals feel thrilled or threatened. This was obvious directly after the "2011 revolution", when Egyptians were thrilled to regain control over their land and wanted to put efforts into revitalising their surroundings, each in his own place. And again, community service groups became active when the "Heliopolis development plan" was announced; they felt threatened of losing their identity with the major changes made to their context. Groupings of people from the same local community can heighten a sense of place attachment, especially when a common goal brings them together.

When respondents were asked to mention the daily activities that required main street crossing (Fig. 15), 70% and 75% mentioned they needed to cross main streets to access public spaces, and visit neighbours and friends living in the neighbourhood, respectively. The question was further developed to ask how these activities have changed due to the previous reason (Fig. 16). Results show that 48.5% of respondents have stopped visiting public/green spaces either due to demolition or inaccessibility after street widening, and 25% have changed their rate of visiting family and friends living in Heliopolis due to inaccessibility, 40% have stopped cycling which was considered one of the social activities grouping cycling communities in Heliopolis on regular tours, and 23% have stopped their daily walking activity due to problem with accessibility or traffic safety (wide road crossing) where walking was also considered one of the social activities where residents get to meet in public spaces for daily walking and jogging (e.g. the side walk surrounding the military academy). It can be concluded that tearing the urban fabric, although a physical action, has resulted in cutting social ties inside the district where the venue to practise these socialising acts and communal events is gone.

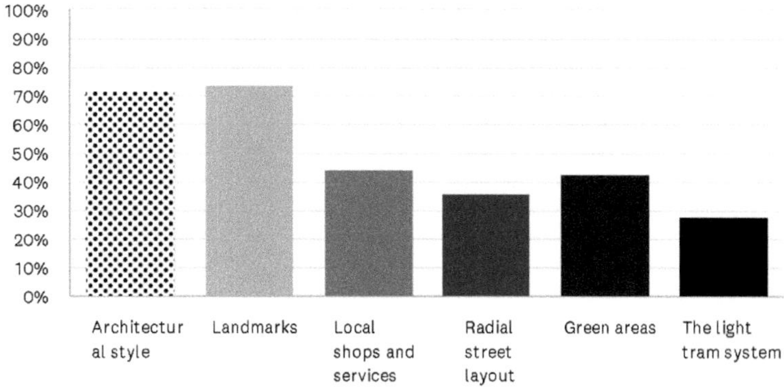

Fig. 17 "Heliopolis" distinguishing features. *Source* Author

5.1.3 Place Identity and Belonging

The first section of the questionnaire was designed to assess the significance of "Heliopolis" identity among residents and visitors, what are the main features that distinguish the neighbourhood from others, how much place identity has been affected by the urban regeneration plan, and how willing are residents to relocate. Data in Fig. 17 represent respondents' choices of the most distinguishing features of "Heliopolis". The data clearly shows that the physical features of the built environment, represented in the architectural style and landmarks, are in the lead of the attributes that are assimilated and recorded by the viewers in their daily lives until they become an inherent part of the place identity. There is a significant gap between the respondents' choice of "architectural style" (71.9%) and "landmarks" (73.8%), as prominent neighbourhood features, and the rest of the neighbourhood's distinguishing features, such as local services (44.29%), green areas (42.86%), street layout (36.19%), and the famous Heliopolis tram (28.10%).

Data extracted from the question, which asked the residents to evaluate the effects of the urban development and regeneration project on different domains of their environment, only came to assert the same results as those extracted from the former question. Again, architectural style and

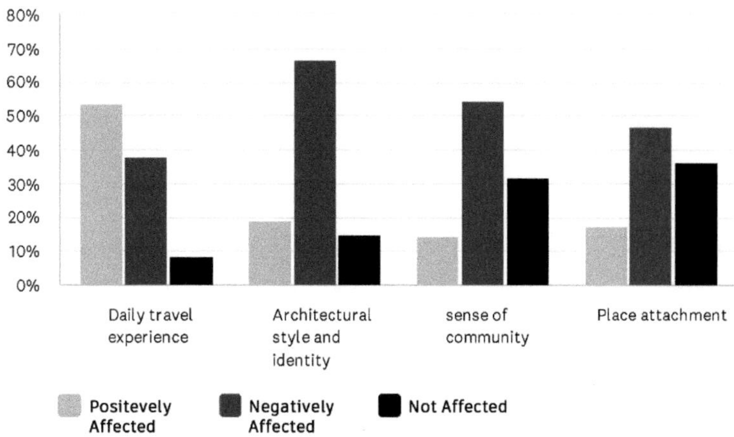

Fig. 18 Residents' evaluation of the effects of the development plan. *Source* Author

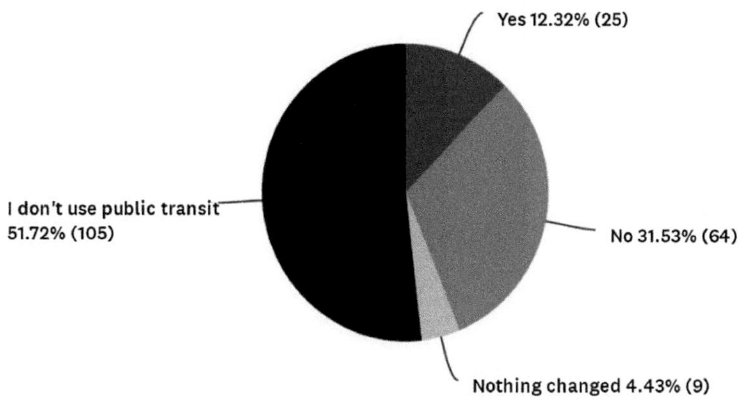

Fig. 19 Responses to sufficiency of public transportation. *Source* Author

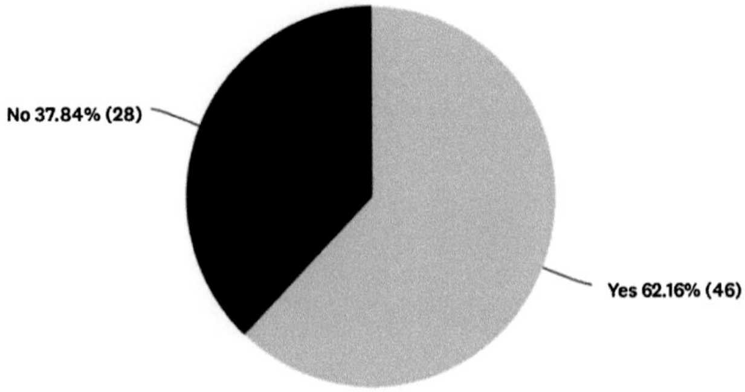

Fig. 20 Non-car owners' considerations for buying a car. *Source* Author

place identity were chosen by the residents as their main concerns when it comes to changes done to their context (Fig. 18).

5.1.4 Social Equity

The survey concludes that 31% of non-car owners think public transportation in "Heliopolis" is insufficient, and this deficiency has gotten worse with the removal of the light tram (Fig. 19). Changes in activity patterns that happened after the development plan, as seen in Fig. 16, due to a reduction in walkability of the district with lower accessibility to public transportation, have resulted in 63% of non-car owners among the questionnaire respondents considering buying a car (Fig. 20).

Through the site mapping study, another challenge of a car-dependent environment was detected, and that is the ability to achieve equity for the disabled. With 25-m-wide streets, pedestrian bridges with no elevator access, and more than 700 m separating pedestrian crossings with no signalised crossing in-between, it can be very challenging for a disabled person (whether physically or visually) to cross the street.

Based on the previous observations, residents were asked who, in their opinion, was best served by the development plan (Fig. 21). Almost 63% of the respondents agreed that the modifications done to Heliopolis to ease traffic congestion and enhance the flow have much better served car drivers who pass through the district on their way to other destinations, while only 16% of the respondents think that the plan is in favour of the residents.

5.1.5 Perceived Safety

Based on the collected data, perceived safety will be explained from two viewpoints: fear of strangers and fear of identity loss.

Strangers and Defensible Spaces

The survey included an open-ended question that asked the respondents to mention their opinions of "newly added activities under bridges". Some of the residents' responses to how they perceive under-bridge activities came as follows:

"تلوث بصري وسمعي وعدم شعور بالأمان لوجود غرباء في الحي لهم انماط سلوكيه غير مرتبطة بطبيعة السكان الأصليين"

"Visual and auditory pollution and insecurity because there are strangers in the neighbourhood with behavioural patterns unrelated to the nature of the indigenous people".

"تجمعات شباب يضايق المارة"

"Gatherings of young people harass passers-by".

Loss of Self and Place Identity

Some of the residents' responses to the question, "If you are facing the problem of insecurity, what is the reason for it?"

"تغير المكان مبالغ فيه"

A 60-year-old resident replied: "The change of place is exaggerated."

Fig. 21 User best-served by the development plan. *Source* Author

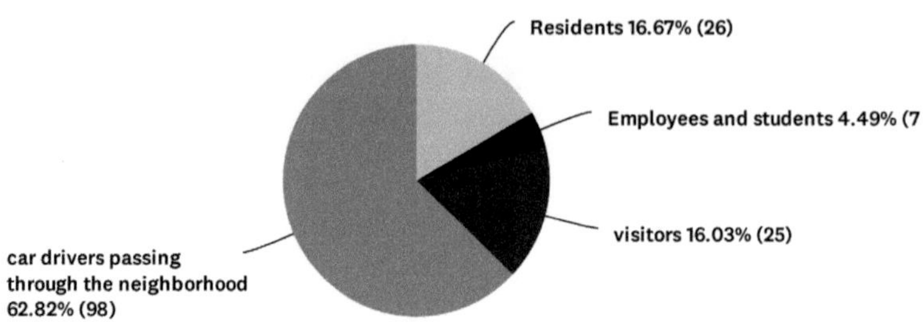

"تغير هوية المكان"

Another 60+ year old replied: "Change of identity of the place."

5.2 Spatial Dimensions

This section discusses results related to the spatial and physical characteristics of the urban environment.

5.2.1 Legibility and Imageability (Logic and Sense)

Imageability is measured through spatial cognition using the place identification technique. Participants in the questionnaire were presented with six images representing different squares from the neighbourhood and asked to identify the place in the pictures and choose the ones representing two specific squares (named in the question form). Correct recognition of pictures shows that the scene is an active part of the participant's representation of the neighbourhood; it contains an element of the neighbourhood that is familiar to the participant and is part of his cognitive mapping of the area of study. The recognition was scored by noting the percentage of participants who were able to correctly identify the square required in the question.

In the first question, respondents were required to identify "Hejaz Square" from three pictures among two other squares, while in the second question, they were required to identify "Triumph square" from three pictures among two other squares (Fig. 22). The questions were designed to

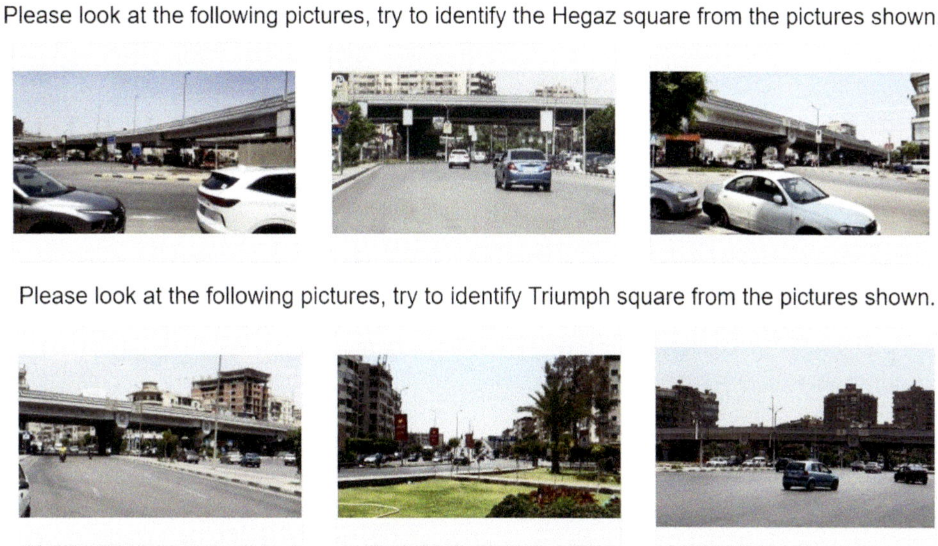

Fig. 22 "AL-Hejaz" and "Triumph" squares identification pictures. *Source* Captured by the author (2022)

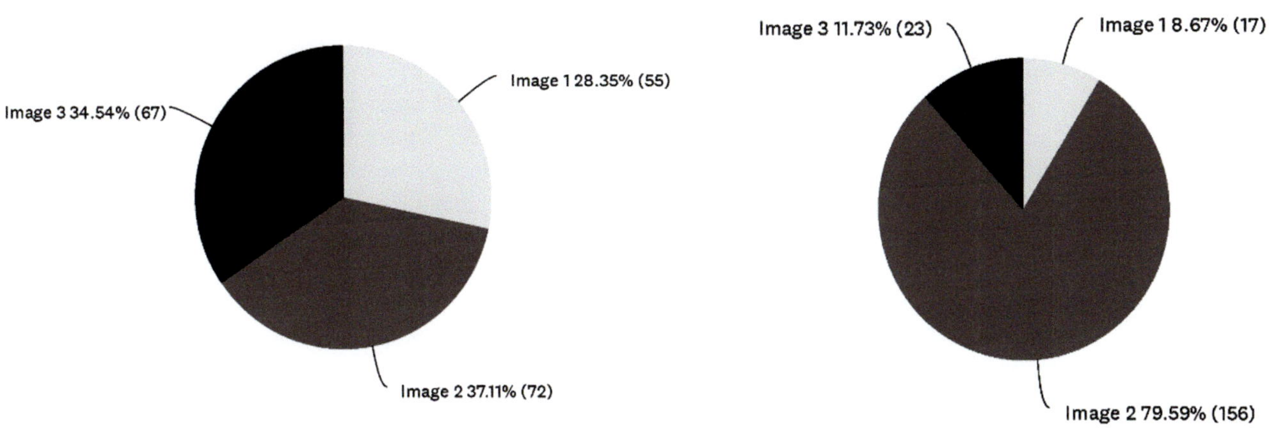

Fig. 23 A statistical representation of the responses to the `picture identification' question. "Al-Hejaz square" to the left and "Triumph square" to the right. *Source* Author

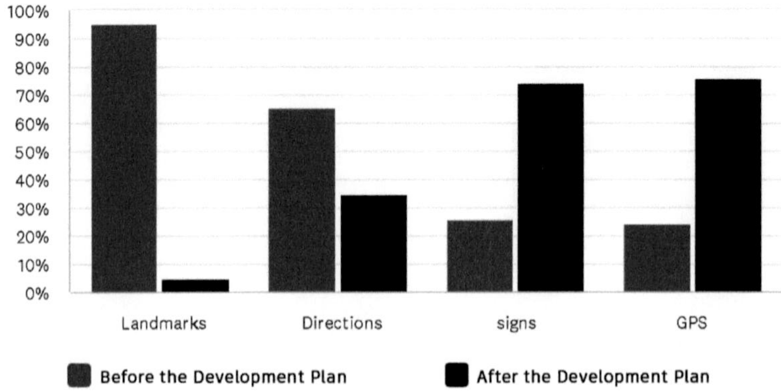

Fig. 24 Responses to the question: "How do you find your way around Heliopolis?". *Source* Author

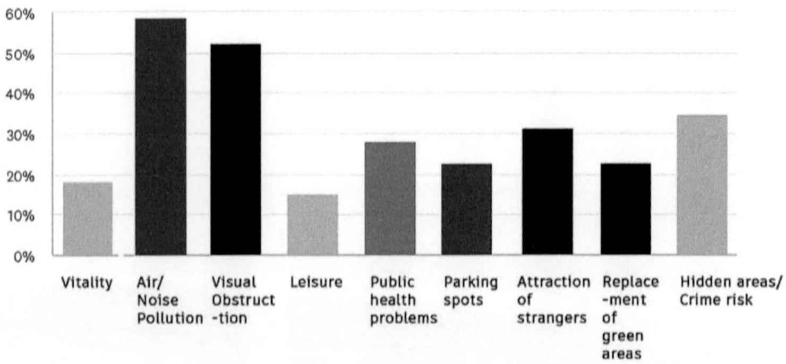

Fig. 25 Residents' responses to their perceptions of under-bridge activities. *Source* Author

assess the importance of the mental image, previously formed in the user's mind, for the recognition of places he used to know. It also assesses how squares that are designed as common squares with a radial pattern and a feature in the centre would be easier to recognise for their pattern (which expresses the legibility of the district) than a square that is covered by a flying-over bridge blocking the line of vision, as seen in the statistical results in Fig. 23. Hence, based on this logic, the two chosen squares were different in their anatomy, where one of them, "Al-Hejaz", is one of the modified squares with a bridge crossing over, while "Triumph square" is not changed and still keeps its characteristics of a radial design and a green space in the centre with a landmark wall.

It is notable that the visual obstruction caused by the construction of mega-structure bridges has caused the squares to lose their identity and, hence, has become difficult to identify. As seen in Fig. 23, while 80% of respondents were able to identify the "Triumph square" among the three pictures, only 28.5% of the respondents were able to identify the "Al-Hejaz" square. The landmarks that once identified each square and gave it its special character are now gone. The bridges have also obstructed the line of vision that had previously linked successive squares, from one landmark to the next, in a narrative fashion.

The cityscape's "legibility" was assessed by asking the residents about their way-finding technique around Heliopolis both before and after the development plan (Fig. 24). The results show that "landmarks" played a vital role in connecting the district together and providing iconic images that guided people on their ways inside Heliopolis. Axial lines connecting the landmarks, as in the original Heliopolis scheme, helped recognise focal points from far distances and gave each axis a distinct image. A 47% of the respondents say they depended on landmarks for wayfinding before the urban development plan, whereas 66% of respondents said they now must use the GPS to navigate their way in Heliopolis after the development plan.

5.2.2 Perceived Open Space and Aesthetic Quality

In the survey, residents were asked to describe their perceptions about the new commercial activities recently added under bridges and assess the effect of these services on their daily lives. As shown in Fig. 25, the second highest response, with a percentage of 51.9%, to how people perceive new alterations to their environment is "visual obstruction". Additionally, the highest third response is that these new structures have developed hidden spaces that are dark, neglected, and could promote higher crime rates.

Another open-ended question was asked to assess the effect of the new commercial activities under the bridges on the neighbourhood residents. Some of the responses came as follows:

Fig. 26 Comparison between squares replaced with bridges and activities, and squares originally planned in the Heliopolis scheme. To the left: "Triumph" square; and to the right: "Al-Mahkama" square. *Source* Captured by the author (2022)

"أصبح الحي الراقي أقرب للعشوائيات"

"The classy neighbourhood has become like a slum".

"مزعج جدا موضوع المحلات أسفل الكباري ويسبب مشاكل مرورية ومنظر غير حضاري."

"The subject of shops under the bridges is very annoying, causes traffic problems, and is an uncivilised view".

6 Discussion

In this section, discussion of the findings will be conducted in light of exploring the contribution of each of the main domains investigated by the survey to neighbourhood satisfaction.

6.1 Exploring the Contribution of the Built Heritage to Neighbourhood Satisfaction

In "Heliopolis", city squares (with all the accompanying green spaces and landmarks) were removed in favour of free-flowing streets and mega-bridges. In a subsequent stage and in an attempt to make use of under-bridge space, new commercial activities were added. These activities were found to be non-congruent and a misfit for the place and community. Mega bridges with their visual obstruction of main axes, which were an inherent part of the "Heliopolis" master plan, in addition to newly added activities that have attracted strangers from other parts of the city to the neighbourhood, have shaken feelings of contentment and security for Heliopolis residents.

Many locals in historically significant areas value environmental preservation for its symbolic and psychological benefits more than the potential financial benefits of neighbourhood redevelopment projects (Firey, 1945). As mentioned in the literature review, distinctiveness and uniqueness are among the leading principles of identity (Twigger-Ross & Uzzell, 1996). Residents were able to best identify themselves within their environment through the most distinctive features represented in the architectural style of the neighbourhood buildings and unique landmarks that do not have a match with any other neighbourhood in Cairo. This was not the case with the other features, such as green areas, which, although they might be an important aesthetic feature of the district, are similarly to be found (maybe even in bigger and better layouts) in other districts across Egypt. Changes made to the image of the neighbourhood in contrast with the cognitive pictures already in residents' minds (Fig. 26) caused feelings of agitation, sorrow, confusion, loss of identity, and loss of pride.

Based on Cole and Hummons theories, social activities and public manifestations of local culture are among the most significant reasons for creating and maintaining community attachment. Since these activities basically take place in public spaces (also known as third places), it is essential to maintain and develop public cultural spaces in order to create a sense of social cohesion and community attachment among the local community. As seen in Fig. 27, all the cross marks (X) represent the removed squares (round-abouts) that were designed as green public spaces that acted as both meeting and leisure spaces for the locals and nodes connecting main axes physically and visually. The dotted lines represent internal streets that were originally designed to have central islands with 6–12 m wide lanes on each side, and to double this width with 12–25 m wide lanes on each side of a 3 m wide central island (Earth, 2022). The widening of internal streets to double their size has resulted in tearing the fabric of the district, forming isolated islands with high-way streets separating them.

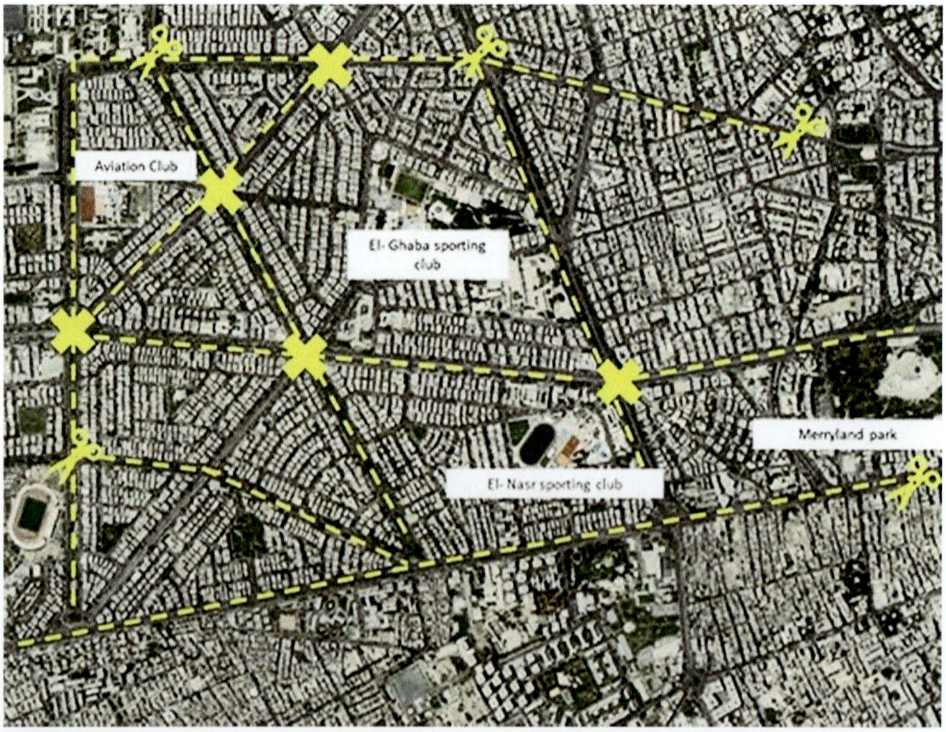

Fig. 27 Heliopolis' tearing of urban fabric. 2022 *Source* Author using Google Earth (2022)

6.2 Exploring the Contribution of Travel to Neighbourhood Satisfaction

Findings show that there is a statistical difference between respondents' perceptions regarding the benefits of the urban regeneration project in facilitating the traffic flow, where around 88 per cent of the survey respondents think that the development project has indeed increased the traffic flow through Heliopolis, while only 12% of the respondents think the project has not improved the traffic. Yet respondents in both areas express similar major concerns about the district's identity, walkability, accessibility, and safety issues, as well as the positive effects that the newly added infrastructure and commercial uses could offer regarding urban revitalisation and development.

Below are some of the respondents' comments regarding their perception of the benefits of easier traffic flow versus deficits in their social and spatial climate:

"خطة التطوير سهلت الحركة المرورية بوضوح لكن ينقصها توفير اماكن للسير والعبور والغاء فكرة الانشطة أسفل الكباري ومحاولة زيادة المساحات الخضراء."

"The development plan clearly facilitated the traffic movement, but it lacks the provision of places for walking and crossing, the abolition of the idea of activities under bridges, and an attempt to increase the green spaces."

"All these developments served motorists as the road became wide, but distorted the landmarks of Heliopolis, which were characterised by beauty and sophistication."

These perceptions have led to the belief that the development project has targeted neighbourhood visitors and passers-by more than it has targeted residents. That is clear from the statistical representation of residents' responses to the question "Who is the user that the development plan has served best?". Almost 63% of the respondents agreed that the modifications done to Heliopolis to ease traffic congestion and enhance the flow have much better served car drivers who pass through the district on their way to other destinations, while only 16% of the respondents think that the plan is in favour of the residents (Fig. 21).

6.3 Exploring the Contribution of Subjective Well-Being to Neighbourhood Satisfaction

Theories of "defensible spaces" explain why residents of "Heliopolis" have expressed their insecurities towards their

"كل هذه التطورات خدمت راكبي السيارات حيث أصبح الطريق واسع، ولكن شوهت معالم مصر الجديدة التي كانت تتميز بالجمال والرقى."

"lost" neighbourhood to "strangers". The newly added activities under bridges have attracted "outsiders" from other districts. The very nature of those activities made them more directed at "visitors" coming to the neighbourhood to attend certain services, such as coffee shops facing the courthouse, where hundreds of daily visitors use these places to meet, communicate, and break from sessions. These shops are placed within the inner fabric of the neighbourhood, where visitors are invading the privacy, order, and quietness of the residential area. They are invading the residents' "defensible spaces". Land uses at the heart of residential blocks are directed towards "outsiders" instead of "residents". Common public spaces between residential blocks are "preached to" and "invaded" by strangers. All of which caused feelings of loss of "territoriality" and "sense of control" over the neighbourhood and consequently the loss of "perceived safety".

These kinds of invasions result in what is known as "psychological stress". Psychological stress happens when perceived environmental demands, such as high levels of noise or congestion, exceed the individual's perceived capacity to cope with those demands (Lazarus, 1966). Social stressors could develop a phenomenon called "aftereffects of stressors", as described by Glass and Singer (Glass & Singer, 1972). Some of the aftereffects of exposure to stressors, as published by the Institute of Medicine (Stelfox, 2006), include:

- Great irritability and aggressiveness,
- Increased errors (e.g. at work),
- Reduced sensitivity to social responsibility,
- Greater vulnerability to disease.

In Heliopolis, residents did not relocate but instead their environment was "replaced", which in the same manner caused feelings of "loss", "detachment", "insecurity" and "confusion". Residents may view physical disruptions to their environment as a potential incubator for negative incidents. Though these changes may not represent a direct threat to the neighbourhood, poor or unusual physical conditions may affect the local community in other ways, including raising perceptions of physical disorder and fear of crime and causing declines in neighbourhood satisfaction (Hur & Nasar, 2014).

An important conclusion derived from this section is that the three main attributes of subjective contentment that have been affected, namely "place identity", "community attachment and social cohesion", and "perceived safety", are highly relatable. This phenomenon can be referred to as "spill-over effects", where certain life domains influence one another even if indirectly (Sirgy, 2012). It is only rational to think that shaking the pillars of "self "and "place" identity in addition to sense of community in a society that has been building up for over 100 years, where current residents might be second or third generations of "Heliopolis" residents, has a major impact on threatening their "perceived safety". Residents have shown more fear and anxiety about losing their cultural roots than they have about losing their lives trying to cross a main street.

6.4 Exploring the Effect of Demographic Characteristics on Neighbourhood Satisfaction

Among the basic values upon which the Heliopolis district was originally designed was the principle and value of "social equity". A neighbourhood where the residences of different social classes were interwoven in a homogenous way where there were no physical barriers. An example of this strategy is the design of the "*Ezbet El Moslemeen*" neighbourhood inside Heliopolis, which was created around the turn of the twentieth century to accommodate the working class, immigrating from rural areas in Egypt to Heliopolis, alongside the residential sections of middle- and upper-middle-class families in an integrated community setting. This was one of the most effective policies utilised at Heliopolis. Accessibility to services and transportation was available for everyone, which comes in deep contrast to an urban regeneration plan that favours car owners over pedestrians and privileges the healthy, young, and capable population over vulnerable groups such as the elderly, disabled, or economically deprived.

One of the methods used to analyse the survey results was finding relations between the means of different variables to find a correlation. In this case, a negative correlation was found between age and satisfaction with the built environment's safety and street amenities (Fig. 28). Although an abnormality appears where residents aged 60 and older show higher satisfaction with safety and street amenities, this could be explained either due to adaptation with long residency to neighbourhood conditions and having a different comparison set of standards or due to fewer street visits and more dependence on services done by others.

Analysis of the results demonstrated that demographic factors such as age, physical ability, social status, and length of stay in the neighbourhood have a strong impact on neighbourhood satisfaction. Figure 29 is a statistical representation of the percentage of residents who are willing to relocate due to dissatisfaction with current neighbourhood conditions. The evidence, supported by literature, shows that residents who are aged 50 or older and who have been living in the same neighbourhood for far longer years are significantly less likely to relocate, while younger adults (aged 31–40 and 41–50) are more likely to relocate by 32–35%.

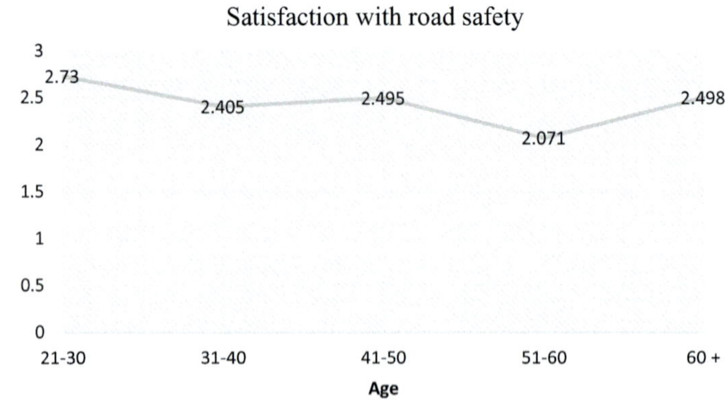

Fig. 28 Statistical representation of "mean values" of the relation between age and satisfaction with road safety. *Source* Author

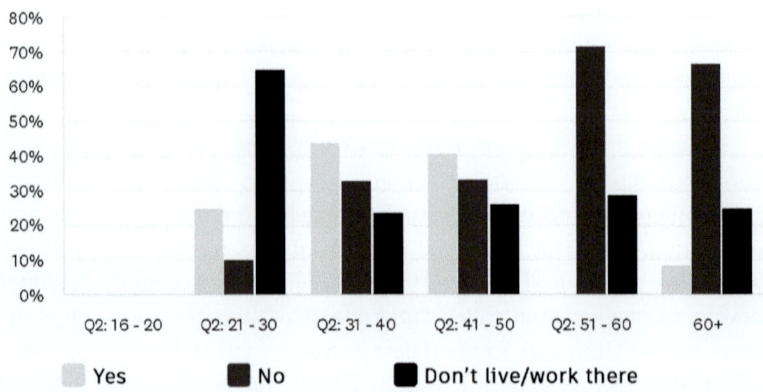

Fig. 29 Residents' susceptibility to relocate, filtered by age. *Source* Author

7 Conclusion

Findings indicate several determinants of neighbourhood satisfaction in culturally rich environments. Perceived safety, place attachment, social cohesion, social equity, and place identity were all positively associated with subjective well-being, which represents an important component of neighbourhood satisfaction. Physical characteristics of the urban environment that are related to the cultural heritage and the city's image, such as legibility and imageability, were also positively associated with neighbourhood satisfaction. This concludes that perceptions of residents carry more weight in explanations of neighbourhood satisfaction than objective neighbourhood characteristics. The connections between individuals and the meanings preserved in

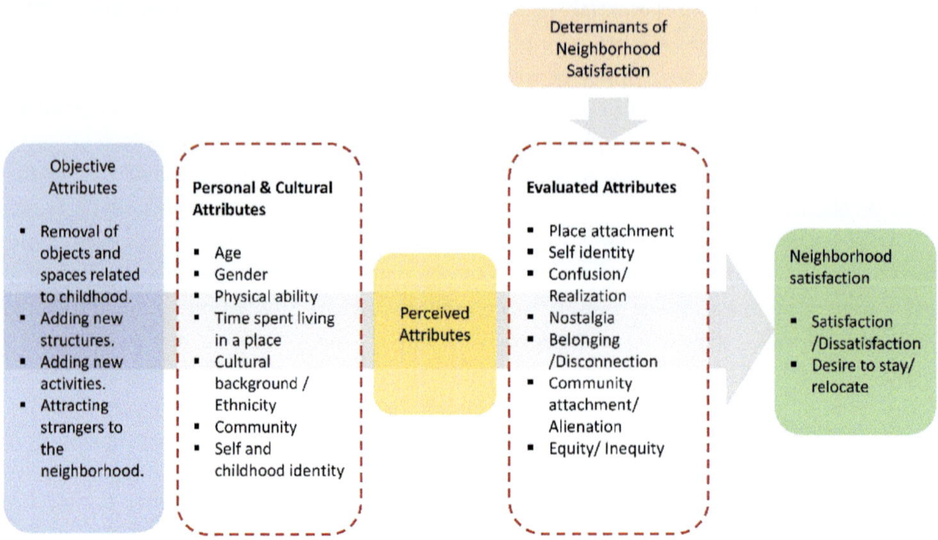

Fig. 30 Concluded diagram of personal and cultural attributes linking the physical environment to neighbourhood satisfaction. Conducted by the author

heritage places are crucial to comprehend. By gaining this knowledge, we will be in a stronger position to devise creative strategies that reap the advantages of both preservation and development.

The study proves that personal, cultural, and social characteristics together with demographic factors act as a lens through which the built environment is perceived and evaluated, as suggested by Campbell et al. (1976) and Galster and Hesser (1981). Age, length of stay in the neighbourhood, social status, and past experience in rich cultural environments were found to positively relate to place attachment, community attachment, and a sense of belonging. Physical disabilities (the elderly and disabled) and low-income have proved to be inversely related with neighbourhood satisfaction in car-dependent environments. And again, age, gender, and physical disabilities have proved to be related to neighbourhood satisfaction in compact, high-density environments due to fear of strangers, loss of identity, and traffic problems. Figure 30 is a translation of the data and literature, based on Campbell's model (Campbell et al., 1976), to explain the chain of reactions that start with changes done to physical objective attributes, filtered through and affected by a number of demographic factors, personal characteristics, cultural, and contextual attributes that eventually makeup a special lens through which each individual (based on his own mixture of the previously mentioned attributes) perceives his environment, evaluates it, and finally is satisfied with it or not.

8 Recommendations

The city has invested in a beautification programme where the bridge passing through "*Al-Mahkama*" square has been fully planted as a compensation for the lost green spaces and to improve the visual quality of the environment. This attempt proves the recognition of the city with the nature of the social climate in "Heliopolis" which was originally planned as a "garden city". To avoid this process of post-project evaluation and modification, building interim street transformation projects gives people opportunities to experience changes before they become permanent, hence cutting the number of modifications that need to be done later. The study recommends pre-development citizen surveys, which can assist local policymakers in planning and optimising the implementation of an urban regeneration project based on the requirements and objectives of the community. Such studies serve as knowledge creation platforms to increase the effectiveness of local policymaking and boost public confidence in it.

References

Ambrey, C., & Fleming, C. (2013). Public greenspace and life satisfaction in urban Australia. *Urban Studies, 51*(6), 1290–1321. https://doi.org/10.1177/0042098013494417

Ballas, D. (2013). What makes a 'happy city'? *Cities, 32*, 39–50. https://doi.org/10.1016/j.cities.2013.04.009

Bazargan, M. (1994). The effects of health, environmental, and socio-psychological variables on fear of crime and its consequences among urban black elderly individuals. *The International Journal of Aging and Human Development, 38*(2), 99–115. https://doi.org/10.2190/fhqy-5t3l-gadm-pud0

Buys, L., & Miller, E. (2012). Residential satisfaction in inner urban higher-density Brisbane, Australia: Role of dwelling design, neighbourhood and neighbours. *Journal of Environmental Planning and Management, 55*(3), 319–338. https://doi.org/10.1080/09640568.2011.597592

Cao, X. (Jason). (2016). How does neighborhood design affect life satisfaction? Evidence from Twin Cities. *Travel Behaviour and Society, 5*, 68–76. https://doi.org/10.1016/j.tbs.2015.07.001

CAPMAS, E. (2017). *Arab Republic of Egypt—Population, Housing and Establishments Census 2017*. https://censusinfo.capmas.gov.eg/Metadata-ar-v4.2/index.php/catalog/1723

Cohen, S., Evans, G. W., Stokols, D., & Krantz, D. S. (1987). Behavior, health and environmental stress. Plenum Press, New York, 1986, 283 pp. *Applied Ergonomics, 18*(3), 253–254. https://doi.org/10.1016/0003-6870(87)90054-8

Coles, R. (1970). *Uprooted children; the early life of migrant farm workers*. University of Pittsburgh Press.

Cuba, L., & Hummon, D. M. (1993). A place to call home: Identification with dwelling, community, and region. *The Sociological Quarterly, 34*(1), 111–131. https://doi.org/10.1111/j.1533-8525.1993.tb00133.x

Dobrowolska, A., & Dobrowolski, J. (2006). *Heliopolis: Rebirth of the city of the Sun*. American University in Cairo Press.

Firey, W. (1945). Sentiment and symbolism as ecological variables. *American Sociological Review, 10*(2), 140–148. https://doi.org/10.2307/2085629

Friedman, D., Parikh, N. S., Giunta, N., Fahs, M. C., & Gallo, W. T. (2011). The influence of neighborhood factors on the quality of life of older adults attending New York City Senior Centers: Results from the Health Indicators Project. *Quality of Life Research, 21*(1), 123–131. https://doi.org/10.1007/s11136-011-9923-6

Galster, G. C., & Hesser, G. W. (1981). Residential satisfaction. *Environment and Behavior, 13*(6), 735–758. https://doi.org/10.1177/0013916581136006

Glass, D. C., & Singer, J. (1972). *Urban stress: Experiments on noise and social stressors*. Academic Press.

Google. (2001). Google Earth VR. Google Earth. Computer software. Retrieved February 23, 2022, from https://earth.google.com/web/search/El+Hegaz+Street,+Al+Matar,+El+Nozha,+Cairo/@30.110727,31.3439432,47.88910522a,899.18426703d,35y,0h,45t,0r/data=CpUBGmsSZQolMHgxNDU4MTU4N2RhYzYxNjViOjB4MmY5ODRjNjgyZDY5Y2M0YxnAyMuaWBw-QCF8rlupDFg_QCoqRWwgSGVnYXogU3RyZWV0LCBBbCBNYXRhciwgRWwgTm96aGEsIENhaXJvGAIgASImCiQJ-zSGPu0aPkARb0ueRoIDPkAZU1y0FL9dP0AhyoDAW5YmP0AoAg

Hassler, U., Algreen-Ussing, G., & Kohler, N. (2002). Cultural heritage and sustainable development in SUIT. *SUIT Position Paper, 3*(9), 1–5.

Heckscher, A., & Robinson, P. C. (1977). *Open spaces: The life of American cities*. Harper & Row.

Hummon, D. M. (1992). Community attachment. *Place Attachment*, 253–278. https://doi.org/10.1007/978-1-4684-8753-4_12

Hur, M., & Nasar, J. L. (2014). Physical upkeep, perceived upkeep, fear of crime and neighborhood satisfaction. *Journal of Environmental Psychology, 38*, 186–194. https://doi.org/10.1016/j.jenvp.2014.02.001

Kamaly, R. (2013, October 2). History of Heliopolis. *Egypt today*. https://www.egypttoday.com/Article/10/562/History-of-Heliopolis

Kent, J. L., & Thompson, S. (2014). The three domains of urban planning for health and well-being. *Journal of Planning Literature, 29*(3), 239–256. https://doi.org/10.1177/0885412214520712

Lazarus, R. S. (1966). *Psychological stress and coping process*. McGraw Hill.

Leyden, K. M., Goldberg, A., & Michelbach, P. (2011). Understanding the pursuit of happiness in ten major cities. *Urban Affairs Review, 47*(6), 861–888. https://doi.org/10.1177/1078087411403120

Lynch, K. (1960). *The image of the city*. The MIT Press. https://mitpress.mit.edu/9780262620017/the-image-of-the-city/

Mantey, D. (2021). Objective and subjective determinants of neighborhood satisfaction in the context of retrofitting suburbs. *Sustainability, 13*(21), 11954. https://doi.org/10.3390/su132111954

Marans, R. W. (2003). Understanding Environmental Quality through quality of life studies: The 2001 Das and its use of subjective and objective indicators. *Landscape and Urban Planning, 65*(1–2), 73–83. https://doi.org/10.1016/s0169-2046(02)00239-6

Marans, R. W., & Stimson, R. (2011). An overview of quality of urban life. *Investigating Quality of Urban Life*, 1–29. https://doi.org/10.1007/978-94-007-1742-8_1

Megahed, G., Elshater, A., & Afifi, S. M. Z. (2020). Competencies urban planning students need to succeed in professional practices. *Archnet-IJAR: International Journal of Architectural Research, 14* (2), 267–287. https://doi.org/10.1108/arch-02-2019-0027

Milgram, S., & Jodelet, D. (1976). La Carte mentale de paris. *PsycEXTRA Dataset*. https://doi.org/10.1037/e668292012-204

Mouratidis, K. (2017a). Is compact city livable? The impact of compact versus sprawled neighbourhoods on neighbourhood satisfaction. *Urban Studies, 55*(11), 2408–2430. https://doi.org/10.1177/0042098017729109

Mouratidis, K. (2017b). Rethinking how built environments influence subjective well-being: A new conceptual framework. *Journal of Urbanism: International Research on Placemaking and Urban Sustainability, 11*(1), 24–40. https://doi.org/10.1080/17549175.2017.1310749

Newman, O. (1972). *Defensible space: Crime prevention through urban design*. The Macmillan Company.

Parkes, A., Kearns, A., & Atkinson, R. (2002). What makes people dissatisfied with their neighbourhoods? *Urban Studies, 39*(13), 2413–2438. https://doi.org/10.1080/0042098022000027031

Pfeiffer, D., & Cloutier, S. (2016). Planning for happy neighborhoods. *Journal of the American Planning Association, 82*(3), 267–279. https://doi.org/10.1080/01944363.2016.1166347

Proshansky, H. M., Fabian, A. K., & Kaminoff, R. (1983). Place-identity: Physical world socialization of the self. *Journal of Environmental Psychology, 3*(1), 57–83. https://doi.org/10.1016/s0272-4944(83)80021-8

Ramlee, M., Omar, D., Yunus, R. M., & Samadi, Z. (2015). Revitalization of urban public spaces: An overview. *Procedia—Social and Behavioral Sciences, 201*, 360–367. https://doi.org/10.1016/j.sbspro.2015.08.187

Rautenberg, M. (1998). L'émergence patrimoniale de l'ethnologie: Entre mémoire et politiques publiques. *Patrimoine et Modernité*, 279–291.

Riley, R. B. (1992). Attachment to the ordinary landscape. *Place Attachment*, 13–35. https://doi.org/10.1007/978-1-4684-8753-4_2

Rubinstein, R. I., & Parmelee, P. A. (1992). Attachment to place and the representation of the life course by the elderly. *Place Attachment*, 139–163. https://doi.org/10.1007/978-1-4684-8753-4_7

Shafer, C. S., Lee, B. K., & Turner, S. (2000). A tale of three greenway trails: User perceptions related to quality of life. *Landscape and Urban Planning, 49*(3–4), 163–178. https://doi.org/10.1016/s0169-2046(00)00057-8

Shekhar, H., Schmidt, A. J., & Wehling, H.-W. (2019). Exploring wellbeing in human settlements—A spatial planning perspective. *Habitat International, 87*, 66–74. https://doi.org/10.1016/j.habitatint.2019.04.007

Singh, R. (2016). Factors affecting walkability of neighborhoods. *Procedia—Social and Behavioral Sciences, 216*, 643–654. https://doi.org/10.1016/j.sbspro.2015.12.048

Consequences of hedonic well-being, life satisfaction, and Eudaimonia. *Social Indicators Research Series*, 45–60. https://doi.org/10.1007/978-94-007-4405-9_3

Stedman, R. C. (2002). Toward a social psychology of place. *Environment and Behavior, 34*(5), 561–581. https://doi.org/10.1177/0013916502034005001

Steinberg, F. (2008). Revitalization of historic inner-city areas in Asia: The potential for urban renewal in Ha Noi, Jakarta, and Manila. *Urban Development Series*. https://www.adb.org/publications/revitalization-historic-inner-city-areas-asia-potential-urban-renewal-ha-noi-jakarta

Stelfox, H. T. (2006). The "to err is human" report and the patient safety literature. *Quality and Safety in Health Care, 15*(3), 174–178. https://doi.org/10.1136/qshc.2006.017947

Tuan, Y.-F. (1991). Language and the making of place: A narrative-descriptive approach. *Annals of the Association of American Geographers, 81*(4), 684–696. https://doi.org/10.1111/j.1467-8306.1991.tb01715.x

Tweed, C., & Sutherland, M. (2007). Built cultural heritage and sustainable urban development. *Landscape and Urban Planning, 83* (1), 62–69. https://doi.org/10.1016/j.landurbplan.2007.05.008

Twigger-Ross, C. L., & Uzzell, D. L. (1996). Place and identity processes. *Journal of Environmental Psychology, 16*(3), 205–220. https://doi.org/10.1006/jevp.1996.0017

UN. (2016). *Goal 10*. United Nations: Department of economic and social affairs. https://sdgs.un.org/goals/goal10

Yang, Y. (2008). A tale of two cities: Physical form and neighborhood satisfaction in Metropolitan Portland and Charlotte. *Journal of the American Planning Association, 74*(3), 307–323. https://doi.org/10.1080/01944360802215546

Young, R. D. (1999). Environmental psychology. In *Encyclopedia of environmental science* (pp. 223–224). Essay, Kluwer Academic Publishers.

Lost Waters of Istanbul: Deciphering Cultural and Natural Landscapes Through Mapping

Kaan Ozgun and Beliz Bayraktar

Abstract

Istanbul has been known as a water city since ancient times, with many conveyance structures built to meet the city's water needs. Its unique landscape and aquifer systems have been part of the city's daily life throughout history, but today they face extinction due to urbanization and extensive groundwater use. Of the 321 creeks, only 174 remain due to these factors. The Lykos Creek, depicted in Christoforo Buondelmonti's engraving, was one of the main creeks that has been lost. The paper examines how natural systems like creeks and springs have shaped Istanbul's urban heritage, specifically the *bostan* and *hagiasma*. Using GIS, we analyzed land use by comparing past and present conditions through historical maps, including the 1845 Istanbul map prepared by Mühendishane-i Berri Hümayun students and the "Carte de Constantinople" map by F. Kauffer and J. B. Lechevalier. Our analysis focuses on the historical peninsula, one of the earliest administrative regions of Ottoman Istanbul. By comparing the historical maps to current urban features, hydrogeological maps, topographical data, and water network systems, we revealed the historical and current relationships of the sites studied. This analysis significantly impacts land use, tourism, and historical heritage development policies.

Keywords

Historical mapping • Water urbanism • Culture-nature • Deep structure • Green–blue infrastructure

K. Ozgun (✉) · B. Bayraktar
Faculty of Architecture and Design, Ozyegin University, Istanbul, Turkey
e-mail: kaan.ozgun@ozyegin.edu.tr

B. Bayraktar
e-mail: beliz.bayraktar@ozu.edu.tr

1 Introduction

With the increasing rate of climate change-induced disasters, it is clear that existing cities are the major contributor to the problem, considering that two people out of three will live in cities by 2050 (Habitat, 2022). Urban planning is a practice that is reliant on political power, and as such, it operates within administrative boundaries. Conventional planning, design, and engineering practices typically adhere to these boundaries when pursuing urban development. However, this can create challenges since nature is a product of millions of years of evolution. The boundaries imposed by urban planning may not be as enduring as those inherent in nature itself, particularly given the rapid changes that have occurred during the Anthropocene era.

Few transdisciplinary studies exist on Istanbul, focusing on the city's prehistory. Studies from natural science disciplines such as geology and geomorphology have their own focus and deal with the prehistorical concerns of Istanbul (See Algan et al., 2001; Şengör & Kındap, 2019). The environmental design and planning disciplines, such as architecture and urban planning, mainly concentrate on the city since human occupation. Meanwhile, the urban morphology subfield researchers generally take an interdisciplinary approach covering the field of architecture, geography, history, philosophy, and archeology (Sarihan, 2021). Similarly, landscape urbanism discourse emerged in the mid of 90s later evolved into ecological urbanism within the landscape architecture discipline, helping to soften this dichotomy between the disciplines of natural and social science and humanities (Steiner, 2011). For example, Anna Spirn, a prolific scholar and theorist in landscape architecture, uses the deep-structure framework not to view nature as an incidental feature neglecting its hidden potential lost over centuries of urban development. Adopting a deep-structure perspective can enable us to better contextualize lost ecological artifacts within the urban condition and re-establish a connection with nature. Spirn (1993) emphasizes that while

A. Battisti et al. (eds.), *Greening Our Cities: Sustainable Urbanism for a Greener Future*,
Advances in Science, Technology & Innovation, https://doi.org/10.1007/978-3-031-49495-6_14

the deep structure may be concealed, it cannot be erased. When surface structures obstruct or obscure the deep structure, it requires additional resources to sustain, including energy, materials, and information. The deep structure will inevitably reassert itself if these resources are no longer applied. As such, understanding and aligning with the deep structure of nature can facilitate sustainable urban planning practices that account for the long-term ecological consequences of development (Spirn, 1993).

A historical city like Istanbul had a rich and intimate relationship with nature, as evidenced by several historical artifacts that still exist today, such as the *hagiasma* and the *bostan* (Yurdakul, 2010). The role of the monumental mosques and complexes that were integrated into the city after the conquest in shaping Istanbul's historical topography and, especially, its silhouette is of great importance (Karakuyu, 2010). Historical maps can provide valuable clues about the relationship between nature and human history in Istanbul. Unfortunately, the current state of these artifacts is insufficient for fully understanding the landscape systems, as the natural factors that initially shaped them have largely disappeared from present-day Istanbul. This has resulted in the loss of Istanbul's deep structure and its impact on the city's natural, artifactual, and sociocultural history.

We aim to gain a deeper understanding of the relationship between nature and urban development in Istanbul throughout history. To achieve this, we have created a chronological matrix map table that represents the critical periods of the city from an ecological and urban perspective. For our case study, we selected the historical peninsula, which witnessed the Byzantine period (330–1453), Ottoman (1453–1923), and the Turkish Republic era (after 1923 to the present time) (Koç & Kubat, 2018). Along with the well-documented historical periods, we have proposed and visualized a predevelopment phase in the matrix template, including topography, surface water network, and groundwater table. By examining the deep structure of Istanbul in this way, we hope to better understand the evolution of the city's ecological and urban features. Our focus is on Lykos Creek and its surrounding landscape, which are the most significant landscape components of the historical peninsula. By analyzing the relationship between water resources and historical artifacts, we aim to raise awareness of Istanbul's deep structure and contribute to its future urban development from an ecological perspective.

2 Background

2.1 Historical and Cultural Importance of Water Systems in Istanbul

In ancient times, Istanbul's rugged terrain underwent massive tectonic movements, resulting in seven hills, plain areas, knobs, and low ridges. For example, the Bosphorus was created by the deposition of four sedimentary units over 26,000 years. This suggests that the Bosphorus was once a freshwater lake from 26,000 to 5300 years ago. Eventually, it transformed into the dual-flow system we see today (Algan et al., 2001). Waterways and springs followed suit, and the Lykos (also known as Bayrampaşa creek) became one of historical Istanbul's most significant water formations. These waterways ultimately shaped natural bays and harbors, which defined the frontiers of the city's historical peninsula (Dinç & Bölen, 2014; Karakuyu, 2010).

Several studies state that the physical space usage of Istanbul over time has shaped its topography significantly. Romans and Ottomans used their knowledge of Constantinople's unique geography when strategically placing important landmarks such as Hagia Sophia or Topkapı Palace, giving them a tactical advantage during times of war (Esin, 1992). Istanbul's unique topography has dramatically influenced its relationship with water. Bono et al. (2001) discuss the relationship between Constantinople's water supply system and its urban development during early medieval times. In 330 AD, when Constantine I founded Constantinople, he built an aqueduct system that connected it with the Belgrade forest to provide clean drinking water for its citizens. Later, emperors such as Constantius II (337–361) and Valens (364–378) added granaries near the Golden Horn harbor district, while Julian (361–363) began construction of another major port at the Kumkapı area along the Marmara Sea coast. These infrastructure developments indicate growth in population over time due to the increased availability of resources like food grains and potable water through these ports and harbors, thus making life easier around this region (Magdalino, 2015). In addition, Sensoy (2015) argues that Istanbul's topography, which includes steep hills and valleys, has made it challenging to manage water resources and has led to water-related problems such as floods and landslides.

Water has always been a crucial issue in Istanbul's history, prompting the construction of various water structures to address these concerns. The Ottoman period, in particular, saw the construction of various water structures, such as aqueducts, bends, pools, water wheels, fountains, *maksemler*, *sebiller*, *şadırvanlar*, baths, and bridges. Islamic culture viewed fountains as an opportunity to perform good deeds, and they were often used to decorate *külliyes* (complexes of buildings) and street corners. Therefore, water structures were practical and had cultural and esthetic significance (Urfalıoğlu, 2015).

2.1.1 Bostans

Directly related to water resources, bostans were the market gardens of Istanbul and can be adapted to modern-day urban agriculture (Kaldjian, 2004). People used to buy fruits and vegetables from the garden next to their homes. Besides parks and meadows of the city, bostans were Istanbul's essential landscape components where the link between humans and nature developed. There were two significant bostans in the historical peninsula called Yenibahçe and Langa. Yenibahçe bostan was located on the city's east side, surrounding Lykos Creek. Also, the bridge of grandmaster Sinan, the architect, is used to cross over Lykos Creek (Dilbaz, 2019). Where Lykos Creek changes its direction from northwest to the southwest, the Yenibahçe bostans start disappearing and then disembogues to the harbor of Theodosius (see Fig. 6). Langa bostans, the market garden that provides the season's earliest vegetables, were situated at the shore, where the Lykos Creek meets the sea and also was cherished for bringing out sizable cucumbers. While bostans were spread throughout Istanbul, they were consistently aggregating around creeks, artesian springs, and active springs from which people drew water (Kaldjian, 2004).

2.1.2 Çukurbostans

Sensoy (2015) highlights Istanbul's vulnerability to water scarcity and pollution due to its reliance on surface water sources such as rivers and reservoirs. In response to drought conditions, the Byzantines constructed open-air and enclosed cisterns for water storage (Ward et al., 2017). These cisterns, which were used for extended periods, helped to enrich the soil, eventually transforming into sunken gardens, or "çukurbostans," where locals planted a variety of fruits and vegetables (Kasapseçkin, 2012). The Hagios Mokios cistern, also known as the Altımermer Çukurbostan, has a long history dating back to the Byzantine era and had a complex relationship with water during the Ottoman period. Studies suggest that the bostan's fertile soil may be attributed to the water harvested and stored at this location (Çoban, 2020). Despite its historical significance, the Altımermer Çukurbostan has been transformed into a park, with its past now largely forgotten.

2.1.3 Spring Waters and Hagiasma

During the Byzantine era, Constantinople was frequently attacked, which caused problems with access to water for extended periods. To address this issue, the city relied on water resources that were brought from distant locations. During these challenging periods, spring waters were particularly crucial since they provided a reliable source of water for people through wells and hagiasmas. Hagiasmas were essential to the city's culture, meaning "Holy waters of Orthodox Greeks." They were sourced from springs, wells, or fountains. People of all faiths, including Muslims, visited hagiasmas to recover from illnesses or make their wishes come true. According to Halit Bayrı's research, there were 126 hagiasmas in 1953 (Bayrı, 1953). Unfortunately, many hagiasmas disappeared due to zoning development and road works. Even for the ones that still exist, their springs are often contaminated or have gone dry (Değer, 2010).

In the historical peninsula, Lykos, alias Bayrampaşa creek, used to divide the valley in the middle, was the only natural water source during the Roman, Byzantine, and Ottoman periods (Eyice, 2010). According to the "Byzantine Constantinople" map by Janin (1964) (see Fig. 6) and other maps illustrated in the method chapter, Lykos Creek consisted of aboveground and underground courses. The visual resources used in this study revealed that the creek's underground system, visible around midway, used to flow aboveground for some periods. In Buondelmonti's Constantinople map (Art, 2004) (see Fig. 1), the Lykos Creek flows under some bridges and reaches Theodosius's harbor. Unfortunately, maps visualizing Byzantine Constantinople do not include specific information about the landscape around Lykos Creek.

The students of Mühendishane-i Berri Hümayun created a map of Istanbul in 1845, during the Ottoman era, which provided valuable information on the city's topography, waterways, settlements, and landscaping (Yılmaz, 2015). While the "Carte de Constantinople" map of 1786, produced by F. Kauffer and J. B., was the first scientific-scale map of Istanbul and gave detailed information about the waterways, the formation around the Lykos Creek was only partially visible (Lechevalier, 1786). Nevertheless, the 1845 map remains the most comprehensive visualization of Istanbul's urban form and landscapes (Yılmaz, 2015).

2.2 Istanbul Throughout the Urbanization

In contrast to the bustling development of twentieth-century Istanbul, the city during the nineteenth century had a significantly lower level of advancement. Despite being a busy city, it lacked a sewer system. It had an unstructured urban fabric of serpentine streets, which often caused neighborhoods to vanish overnight due to fire outbreaks caused by the presence of wooden dwellings. The rapid population increase of Istanbul during this period created many issues, including a scarcity of resources such as clean water and a lack of security, making the city's administration an ongoing challenge (Dilbaz, 2019).

Istanbul was divided into four traditional regions during the Ottoman period. Bilad-Selase included the regions of Galata, Eyüp, and Üsküdar, whereas Dersaadet was the historical peninsula region. Dersaadet consisted of 13 sub-regions called "nahiye." Each nahiye was then subdivided into smaller neighborhoods. The street configurations in the historical peninsula were first applied in 1850 when the irregular organic urban form could no longer fulfill new development needs, and life became unbearable. New roads and sewer systems were constructed alongside transportation

Fig. 1 Engraving by Christoforo Buondelmonti (Art, 2004)

improvements. Many foreign architects and engineers worked on these projects during the city's development (Dilbaz, 2019).

During the Republican era, the government held an international design competition to develop a zoning plan in 1933. Planners such as Henri Prost, Alfred Agache, and Hermann Ehlgötz participated in the competition. Even though Henri Prost's development proposal for the historical peninsula and Beyoglu was omitted, his master plan was accepted in 1937 (Gül, 2009).

Prost envisioned a metropolitan transportation network, including viaducts, bridges, and tunnels. On the other hand, Prost's modernization idea was of such a type that it would severely harm the city's deep structure since he proposed to annihilate Lykos Creek and construct a new boulevard that would be the largest road in the historical peninsula. Prost advocated constructing today's Vatan and Millet streets, which would serve as the historical peninsula's main roads. Vatan Street, 60 m wide, was planned to run along the valley bed where the Lykos Creek flowed. Prost planned this route to pass through a zoo and botanical park that would be constructed in Yenibahçe. Despite Lykos Creek being left to dry up due to Vatan Street (a.k.a Adnan Menderes Boulevard) construction, the creek continued to flow on the construction site until a light rail project was proposed and built underground. The creek stopped flowing and left space for a metro project from Lykos Creek's predevelopment location (Dilbaz, 2019).

3 Method

3.1 Chronological Map Matrix

We created a matrix template (see Fig. 2) to present the morphological periods and their historical maps. The historical peninsula witnessed the Byzantine period (330–1453), Ottoman (1453–1923), and the Turkish republic era (after 1923 to the present time) (Koç & Kubat, 2018). We added a section in the map matrix representing the peninsula's predevelopment phase. To feed this section, we used a groundwater map, a contour map, and a surface water network from a digital elevation model (DEM) using an open-source geographic information system (GIS) software so-called QGIS. By combining these analyses with other historical maps from different periods of human occupation, we aim to identify the lost aspects of the deep structure.

Predevelopment Phase A topographical map showing the surface water network was produced in QGIS to represent the predevelopment phase of the peninsula. A 25 m resolution digital elevation model (DEM) was employed to decipher the predevelopment water network of the studied catchment. Freely available SRTM 30 m resolution DEM data of the United States Geological Survey and Copernicus 25 m resolution DEM data were tested first. To derive the stream network, a slightly better resolution Copernicus raw DEM was selected and prepared in QGIS. The DEM data contains all the surface information, including the DEM's natural and artificial depressions and flaws. The fill sinks algorithm was employed first to prepare a hydrologically corrected DEM, followed by the Strahler order algorithm to identify the surface water network as another feature of the terrain landscape. The Strahler order algorithm in QGIS extracts waterway types and lengths, classifying stream types with numbers. The generated raster indicates a water network order from 1 to 12, representing streams and ephemeral water bodies to rivers. We overlaid this map with an existing aerial map and historical maps in the matrix to ground the truth of the current and lost waterways. The water network becomes visible in our study catchment as fifth and sixth order. Referencing these stream orders, the channel network and drainage basin algorithm was used to generate the catchment boundaries.

A groundwater map extracted from Istanbul Büyükşehir Belediyesi (IBB) earthquake studies (IBB, 2007) was georeferenced with other referential historical maps and current satellite images in QGIS. The map shows the historical peninsula's water depth and groundwater flow direction. The map is essential for understanding Hagios Mokios Cistern's relationship with the underlying groundwater.

Byzantine–Ottoman–Republic Phases Concerning the Byzantine period (see Fig. 2), Boundelmonti's Constantinople map (see Fig. 1), and Raymond Janin's Byzantine map (see Fig. 6) were georeferenced for further analysis. Buoundelmonti's map is the oldest visual representation of Istanbul (Kubilay, 2015). Our study also used the map as the primary resource for recognizing the coexistence of main waterways (especially Lykos) and settlements during the Byzantine period. The Hagios Mokios Cistern and the

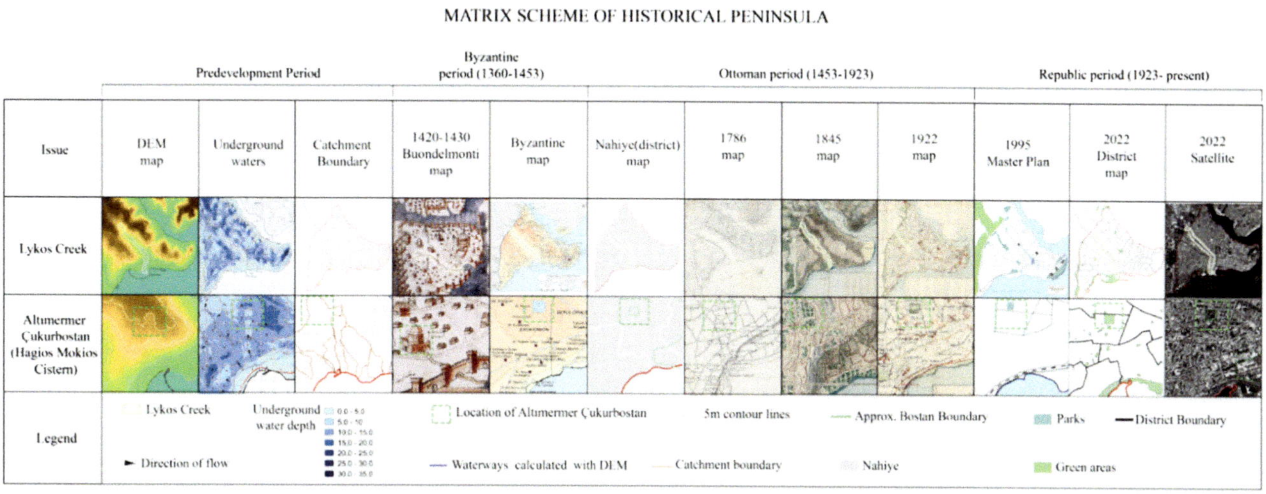

Fig. 2 Chronological map matrix represents four key periods of Constantinople through the use of historical maps and maps generated by the authors

connected waterway were also highlighted on all maps and presented in the matrix.

The path of Lykos Creek was digitized and marked on all maps to observe the landscape changes over time. The 1845 map was superimposed with the satellite map through georeferencing, and the approximate bostan boundaries were digitized to determine how they intervened with the current urban fabric.

Janin's Byzantine Constantinople map, shown in Fig. 6, is a detailed map representing topography, religious buildings, cisterns, waterways connected to them, harbors, roads, and squares. This map was used in the matrix to emphasize the surface and underground waters of Lykos Creek, natural harbors that emerged from the existence of waterways, and the hagios mokios cistern, and its relationship with topography.

Referring to the Ottoman era (see Fig. 2), the first map is the nahiye (district) map; the second is 1786, the third is 1845, and the fourth is the 1922 map. As Istanbul's first scientific scaled map (Kubilay, 2015), we deemed the 1786 map to be the most reliable source for analyzing the lost creeks and their relationship with topography for the purpose of our study. The nahiye map extracted from Orbay's thesis showing the Ottoman-era districts was georeferenced in QGIS and used in the matrix to determine if there is a link between the nahiye boundaries and Lykos Creek (Orbay, 2001). The 1845 map has essential information about former roads, urban fabric, and bostans with its large-scale qualification. In addition to being extremely precise, this map offers the most comprehensive information about Ottoman-era Istanbul, supplemented with Arabic instructions (Yılmaz, 2015). The map served as a method for determining Ottoman-era land use. The 1922 map of Istanbul (Guédik-Pacha, 1922) (see Fig. 2) has been incorporated into our study to track changes in the city's landscape and water relations in the following years. This map is an invaluable resource for understanding the history and development of Istanbul, as it presents a clear and detailed layout of the city's architecture and topography during that era. In fact, after conducting an extensive investigation, we have found that this map is one of the few that represents that particular era clearly, making it a qualified resource for our analysis. By examining the 1922 map, we can gain insight into the city's road networks, districts, and green areas.

The era of the Turkish Republic (see Fig. 2) is depicted through various maps, such as the 1995 master plan, district map, and 2022 satellite map. The 1995 master plan reveals the land-use pattern after the disappearance of Lykos Creek, enabling us to examine the relationship between Vatan Street (formerly Lykos Creek) and the surrounding areas. We utilized the district map to compare the lost water formations with the current district boundaries. Additionally, we overlaid the previous bostan boundaries and Lykos Creek's path onto the satellite map to evaluate the changes in land use over time.

3.2 Land-Use Analysis of Lykos Creek's Surrounding Landscape

Different maps depicting various periods were collected and analyzed to determine land use and landscape changes over time, mainly regarding water. By overlaying these maps, the procedure allowed for comparing current and past land use, using the 1845 map as a reference point. This process is significant, as it identifies bostans' locations and digitizes them on the modern-day urban fabric map. Only Yenibahçe, Langa bostans, and Altımermer Cukurbostanı, related to water in the research area, were digitized on the maps (see Figs. 2, 3, 4 and 5). The ultimate aim is to analyze the lost waters' relationship with the landscape by combining historical maps and current satellite maps supported by data digitization.

Providing sufficient space for waterways is central to delivering green corridors and increasing connectivity across the city as a Waterway restoration strategy (Water, 2013). The Istanbul Water and Sewage Administration (ISKI) and State Department of Water (DSI) have been tasked with protecting and rehabilitating the waterways that flow throughout the city. However, it is important to note that ISKI and DSI's conservation policies are limited to creek beds outside potable water catchments. These policies are not holistic and focus on creating buffer zones to maintain a safe distance between buildings and waterways (Dinc, 2019). From 1981 to 2013, the protection corridor boundary for streams has been 100 m at both banks. Today, the corridor width has been reduced to 10 m for construction purposes for creeks that flow in Bosphorus and 35 m for creeks that flow Golden Horn (Dinc, 2019; Dinç & Bölen, 2014).

To refer to current best practices and represent the predevelopment waterway corridor, Melbourne Water utilizes setbacks of 20–30 m and 50 m, including riparian and vegetated buffer zones on both banks for first, second, third, and fourth-order streams, respectively (Water, 2013). In our study, we analyzed the current land use in the historical peninsula by considering a 100 m buffer zone around Lykos Creek to reveal the potential for future interventions for daylighting. Using QGIS software, we digitized the 2017 Fatih land-use data after georeferencing a scanned land-use map downloaded from IBB (2017). Subsequently, we calculated the land-use percentages within the 100 m corridor boundary (see Fig. 3) to determine a relationship between former and current land use and to interpret if these land uses can be restored.

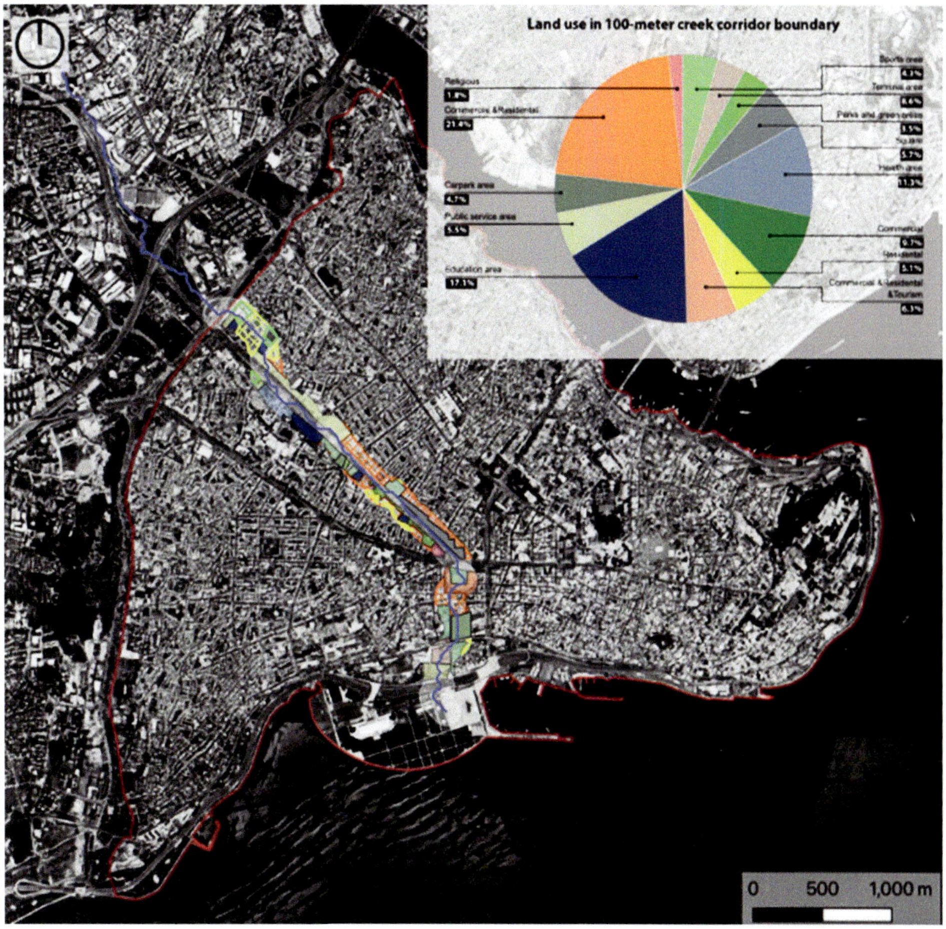

Fig. 3 Land-use analysis conducted within 100 m buffer zone of predevelopment waterway corridor. Inset pie chart shows land-use percentages generated by digitizing IBB's land-use map in QGIS (IBB, 2017) with background satellite photo by Google

4 Findings

4.1 Lykos Creek and the Surrounding Land Use

The land-use pie chart (see Inset drawing in Fig. 3) was created within the 100 m buffer using the data collected from IBB (2017), which shows that commercial, tourism, and, to a large extent, residential zones dominate the buffer area. Although the bostans around Lykos Creek were repurposed as parks in the Fatih master plan of 1995, there is currently no evidence of these parks. The superimposed historical maps informed that the parcels around Lykos Creek previously used for bostans still exist today. The majority of these parcels are currently used for state-owned institutions. Bezmiâlem University, Istanbul police department, Istanbul directorate of tax administration, Fatih central library, and Fatih governorship are some of those institutions partaking within the boundaries of Yenibahçe Bostans. In addition to commercial areas, Yenikapı car park with a capacity of 350 vehicles, Marmaray Yenikapı station, a sizeable coach station, and other government institutions are within the borders of Langa bostans. This may be due to new zoning studies and development plans supported by the "Conservation by expropriation" law to protect natural, historical, and cultural values documented in the 1980 zoning plan reports (Şahin, 2004).

4.2 Hagios Mokios Cistern

Considering the groundwater map to find out the earliest condition of Hogios Mokios Cistern, it can be said that the bountiful groundwater in the vicinity led to the construction of the cistern in the first instance. This also explains the rich soil of Altımermer Cukurbostan, documented by Sahin (2004). Another remarkable fact is that a waterway emerges from the cistern in Buondelmonti's map (Art, 2004) (see Figs. 1 and 2). According to the map, the waterway passes through a gate and flows into the sea. The water flow direction arrows on the groundwater map also prove that this area is a drainage area, in some way connected with the Hagios Moikos Cistern. Perhaps this waterway could be an intermittent creek arising during heavy rain. Today, the only remains of this cultural heritage are the surrounding walls of Fındıkzade Cukurbostan Park.

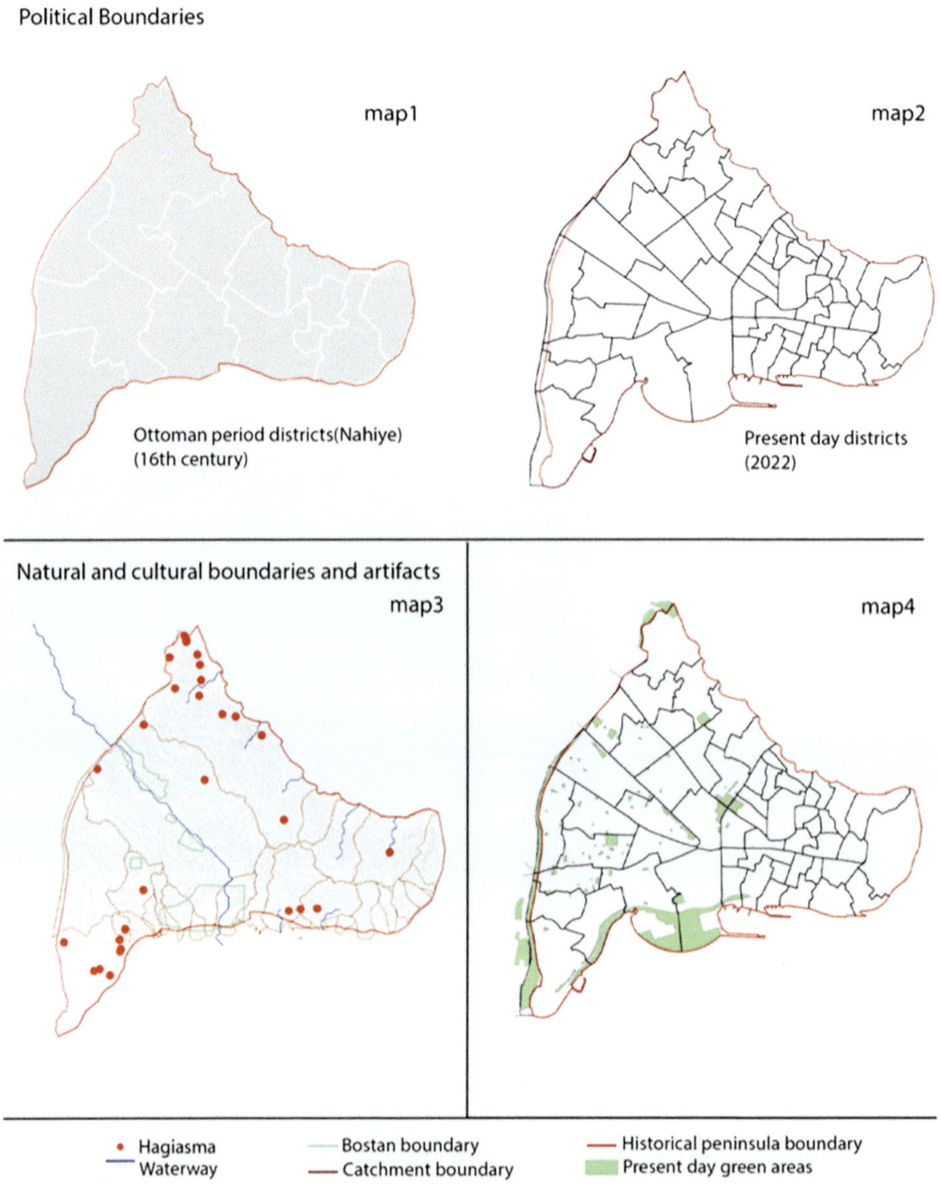

Fig. 4 Diagrams generated in QGIS by the authors shows past and present conditions of political and natural boundaries and artifacts. *Source* Authors

4.3 The Relationship Between Natural and Political Boundaries

A synthesis scheme (see Fig. 5) was created to comprehend the relationship between natural and political boundaries. For the purpose of our study, we define boundaries as a framework as well as a physical edge. Natural boundaries are embedded in the deep structure of nature and can be defined within landscape ecology, geology, and hydrology in practice. When discussing a landscape, we can divide it into different areas with different structures or functions. We call these areas "patches." The boundary framework also examines the movement between these patches: living things, materials, energy, or information. Finally, boundaries act as a point of contact between patches and can be structural or/and functional gradients (Cadenasso et al., 2008). Think of it like a patchwork quilt, with different pieces of fabric sewn together, each with its own pattern or texture, and some areas where the fabrics overlap and gradually change from one pattern to another. These boundaries are important because they help us understand how different landscape parts interact and affect each other. For example, from a hydrological perspective, a catchment, aka watershed or basin, is defined by topographical features called ridges. Water flows on the surface and finds its way into the valley by gravity and within the ridges, a defined boundary by nature. A less permanent natural boundary is the edge between a forest and a field (Cadenasso et al., 2008).

To ensure clarity for our study, we have defined political boundaries as the borders of human social and physical

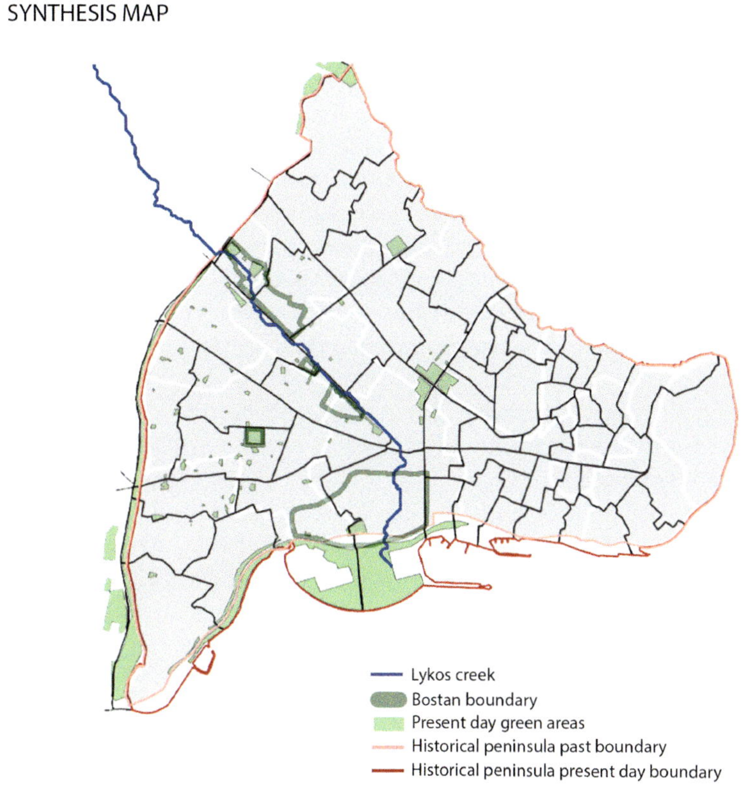

Fig. 5 Synthesis map generated by the authors in QGIS. *Source* Authors

Fig. 6 Cplakidas, CC BY-SA 3.0 <https://creativecommons.org/licenses/by-sa/3.0>, via Wikimedia Commons. Topographical map of Constantinople during the Byzantine period. Main map. *Source* R. Janin, Constantinople Byzantine (1964)

structures, including land ownership, district borders, and neighborhood boundaries. Figure 4 contains Map 1 and Map 2, which display district boundaries as examples of political boundaries (Çoban, 2020). Map 3, also shown in Fig. 4, displays natural and cultural boundaries and artifacts (Bayrı, 1953; Değer, 2010; Kaldjian, 2004). Catchment areas and waterways extracted from DEM represent the natural components, while bostans are considered both natural and cultural artifacts. Additionally, hagiasmas are depicted as cultural artifacts that emerged near water resources. The map displays these components to examine their potential interdependence or influence on one another. Finally, Map 4 in Fig. 4 shows present-day green areas and districts. For comparison, all maps are overlapped to create a synthesis map (see Fig. 5).

In the synthesis map (see Fig. 5), we can see that the political boundaries on the historical peninsula have remained mostly unchanged, with only minor adjustments over time. This indicates that natural and political boundaries have never entirely aligned, and waterways have not significantly impacted the arrangement of urban districts in the area. However, the construction of Vatan and Millet streets has had a noticeable effect on the surrounding boundaries. The urban landscape around Lykos Creek has undergone the most significant changes due to these road constructions. It is possible that the path of Lykos Creek was turned into a boulevard as an easier way to organize the chaotic urban texture. Interestingly, the 1845 map reveals that Lykos Creek was never intended to be a natural design factor in the urban environment. The creek passed through parcels incidentally, and the newly constructed boulevard unknowingly followed its original location without acknowledging its natural origins.

Over the years, the boundaries of the historical peninsula have undergone significant transformations. As depicted in the DEM map (see Fig. 2), natural bays have formed due to waterways flowing into the sea. In the past, the Lykos Creek used to flow into the Harbor of Theodosius (see Fig. 6), which is now filled for the construction of the Yenikapı landfill area. Sadly, the Yenikapı baraza project caused the deterioration of the natural structure of the peninsula. For example, the debris from the sea floor while building the Marmaray project was used to construct this landfill. Even though other ports were not impacted as much, the project still caused harm to the historical peninsula's natural landscape.

5 Discussion and Conclusion

The impact of urbanization on Istanbul's water ecosystem has been devastating, as our study has shown. The historical and ecological loss resulting from the disappearance of water resources and bostans is not limited to the depletion of water as a resource. However, it extends to losing a cultural and historical legacy that links the city with its natural environment. One such example is the lost Lykos Creek, which our research has found to be a significant element in the deep structure of the historical peninsula. Despite its cultural significance, Lykos Creek had little influence on the city's planning decisions and was doomed from the outset. Its extinction and other water resources result from the unintended consequences of Istanbul's chaotic and irregular urban order.

The impact of the drying of Lykos Creek went beyond just its own waters, potentially affecting the spring waters connected to it and putting cultural assets like hagiasmas at risk. It is important to note that Lykos is not the only creek that Istanbul has lost. Historical maps such as "Carte de Constantinople" and the 1845 map reveal that Istanbul was once dotted with many creeks, significantly impacting the city's landscape and urban life.

Projects focused on creek renaturalization and daylighting buried streams are gaining recognition globally as blue–green infrastructure. The benefits of these initiatives are diverse and include reducing overland flooding, replacing culverts with natural creek beds that offer potential landscapes for recreation, restoring lost hydrological and ecological connectivity, improving water quality, regenerating biotic and abiotic environments, beautifying cities, increasing property values, reconnecting nature with people, and creating a sense of belonging and sustainable communities (Pinkham, 2000). Doing a project like this in a complex urban condition with lots of history is daunting. Projects like this exist but require negotiation and consensus with many actors, including politicians, communities, and land owners (Council, 2016). Practically, it is not feasible to reestablish Lykos Creek in its predevelopment bed due to the presence of an underground railway and a boulevard. However, our land-use analysis has revealed that the surrounding land is owned by the state, allowing the municipality to suggest a new path for the creek.

We have shown that the natural boundaries of the peninsula's deep structure and political and cultural boundaries interfere but do not align. Concerning the conflict, the importance of land ownership is highlighted in this study. Land acquisition and the controversial practice of "conservation by expropriation" must address the ongoing conflicts between natural, political, and cultural boundaries. Often, these projects can be tied to more extensive infrastructure plans or meta-master plans to offset land acquisition costs (Pinkham, 2000). For instance, the anticipated earthquake has proposed a citywide urban renewal project in Istanbul. A comprehensive plan could simultaneously demolish existing buildings and reintroduce green–blue infrastructure projects like Lykos Creek. The buildings surrounding the former Lykos Creek (also known as Vatan Street) could be assessed and designated for restoration or demolition while reestablishing the easement for the lost creek.

Figure 6, Raymond Janin's Byzantine map (Janin, 1964), is one of the primary sources for our study and represents the peninsula during its relevant era. We have used it in our findings to emphasize the lost natural features, such as the peninsula boundary and many other harbors and bays indicated in the map, and draw attention to natural, cultural, and political features boundaries. This map, in particular, provides the location of the Byzantine wall, one of the rare artifacts that remains until modern Istanbul. Interestingly, many perceive it as a natural threshold, as the culture dominates nature in the peninsula. The prehistory of the peninsula is almost unrecognizable, emphasizing the significance of studying historical maps and artifacts to understand how the city has evolved.

During the Ottoman period, bostans served as a harmonious blend of nature and culture. These gardens relied on water for their survival. It is fascinating to see how a former Byzantine cistern used for water storage was transformed into a lush garden for food cultivation, now serving as a public park. Sadly, bostans near Lykos Creek were lost with their source. The 1845 map shows the creek flowing through residential areas, and this is supported by the fact that gardeners in Istanbul used to build houses in their gardens for their families. In the past, when there were no proper roads, the less affluent families would visit nearby gardens for fresh air ("Bostan," 1963). The disappearing bostans and the Lykos Creek remind us of the importance of preserving green spaces and water sources in urban areas.

Istanbul has been grappling with frequent droughts for the past four decades, and the most recent one lasted almost two years, posing a looming threat of the worst drought in the future (Akgün, 2021). The water scarcity issue in Istanbul has led to detrimental practices like excessive groundwater extraction. As per Kaldjian's findings (Kaldjian, 2004), certain bostans, once productive gardens, are now being utilized by the municipality as storage facilities for water trucks. The land previously occupied by these bostans has now turned into parking lots, car washing areas, and storage sites. As a result, water trucks extract water from wells that have been sustained for centuries. According to Professor Dr. Hüseyin Öztürk, a geoscience expert, the water from these wells harms human health and could even cause natural calamities like land subsidence (IHA, 2019). These findings indicate that the government primarily makes land-use decisions in the region. Furthermore, the consequences of these actions highlight the importance of implementing sustainable water management practices to secure a healthy future for Istanbul.

In conclusion, the interplay of nature, culture, and politics has shaped Istanbul's historical peninsula over time. Despite the lack of urban planning that considers the natural structure, hagiasmas, bostans, and other water structures in the past indicate a deep connection between Istanbul and its environment. However, this connection has been severely damaged in recent years, leading to many environmental problems, including frequent droughts, water extraction from wells, and the disappearance of once-thriving ecosystems. Our study aims to raise awareness of this issue and highlight the importance of preserving Istanbul's natural structure. Even minor improvements that benefit the aquatic ecosystem, such as restoring lost creeks or promoting green–blue infrastructure, can significantly impact Istanbul's urban life and the well-being of its people. Therefore, we hope our study will inspire future efforts to protect and enhance Istanbul's natural heritage while fostering sustainable urban development.

References

Akgün, İ. (2021). Supply, demand, or prayer. *Eurozine*. Retrieved March 05, 2023, from https://www.eurozine.com/supply-demand-or-prayer/

Algan, O., Çağatay, N., Tchepalyga, A., Ongan, D., Eastoe, C., & Gökaşan, E. (2001). Stratigraphy of the sediment infill in Bosphorus Strait: Water exchange between the Black and Mediterranean Seas during the last glacial Holocene. *Geo-Marine Letters, 20*, 209–218.

Art, M. M. o. (2004). Byzantium: Faith and power (1261–1557) [Exhibition catalogue]. In *New haven*. Yale University Press.

Bayrı, M. H. (1953). İstanbul İlinde Yer Adları X: Ayazmalar, Kaynaklar. *Türk Folklor Araştırmaları Dergisi*, 715–718.

Bono, P., Crow, J., & Bayliss, R. (2001). The water supply of Constantinople: archaeology and hydrogeology of an Early Medieval city. *Environmental Geology, 40*.

Bostan. (1963). In R. E. Koqu (Ed.), *Istanbul Ansiklopedisi (Encyclopedia of Istanbul)* (Vol. 6, pp. 2971–2973). Istanbul: Istanbul Ansiklopedisi ve Nesriyat Kollektif Sirketi, Hüsnü tabiat ve Ercan Matbaasi.

Cadenasso, M., Pickett, S., Groffman, P., Band, L., Brush, G., Galvin, M., Grove, J., Hagar, G., Marshall, V., & McGrath, B. (2008). Exchanges across land-water-scape boundaries in urban systems: Strategies for reducing nitrate pollution. *Annals of the New York Academy of Sciences, 1134*(1), 213–232.

Council, I. C. (2016). *Small Creek transformation project: Community engagement initiatives progress report.* https://www.ipswich.qld.gov.au/__data/assets/pdf_file/0005/94912/Small-Creek-Transformation-Project-Progress-Report.pdf

Çoban, E. Ş. (2020). *1546 ve 1600 tarihli İstanbul vakıfları tahrir defterlerine göre İstanbul'da yeşil alanlar* Lisansüstü Eğitim Enstitüsü. [Thesis].

Değer, M. (2010). *İstanbul'un Şifalı Suları: The Healing Waters of Istanbul*. İstanbul 2010 Avrupa Kültür Başkenti.

Dilbaz, M. (2019). *Kaybolan Tarihin Peşinde*. Timaş Yayınları.

Dinc, H. (2019). The analyses of the physical effect of land use decisions on stream systems and its reflection on urban life: The risk evaluation of flood and overflow in İstanbul. *PLANLAMA-PLANNING, 29*(2).

Dinç, H., & Bölen, F. (2014). The physical structure of streams in Istanbul. *Planning, 24*(2), 107–120. https://doi.org/10.5505/planlama.2014.97269

Esin, U. (1992). İstanbul'un En Eski Buluntu Yerleri ve Kültürleri. *Semavi Eyice Armağanı, İstanbul Yazıları*, 55–77.

Eyice, S. (2010). Tarih Çınarı Eski İstanbul'u Anlattı. *Hürriyet Gazatesi*. https://www.hurriyet.com.tr/gundem/tarih-cinari-eskiistanbulu-anlatti-14144874

Guédik-Pacha. (1922). *Plan d'ensemble de la ville d Constantinople*. Cambridge, Massachusetts, Harvard Map Collection, Harvard University. https://hgl.harvard.edu/catalog/harvard-g7434-i8-1922-s6

Gül, M. (2009). *The emergence of modern Istanbul: Transformation and modernisation of a city*. I.B. Tauris.

Habitat, U. (2022). *Envisioning the future of cities* (World Cities report 2022, Issue. U. N. H. S. P. (UN-Habitat). chrome-extension://efaidnbmnnnibpcajpcglclefindmkaj/https://unhabitat.org/sites/default/files/2022/06/wcr_2022.pdf

IBB, D. v. Z. İ. M. (2007). *İstanbul Mikrobölgeleme Projesi Avrupa Yakası*. https://8luvomezzzsk.merlincdn.net/wp-content/uploads/2020/11/C_MIKROBOLGELEME-AVRUPA-RAPOR.pdf

IBB, Ş. P. M. (2017). *ARAZİ KULLANIM*. https://sehirplanlama.ibb.istanbul/fatih-ilcesi/#arazi

IHA. (2019). İstanbulluları bekleyen büyük tehlike!.. Uzman isim uyardı!.. 'Evlerde kullanımı son derece tehlikeli'. Retrieved December 26, 2019, from https://www.tv100.com/prof-dr-huseyin-ozturkten-istanbuldaki-barajlara-takviye-edilen-kullanilan-kuyu-suyu-uyarisi-haber-478914

Janin, R. (1964). *Constantinople byzantine: développement urbain et répertoire topographique*. Paris, Institut Français d'Etudes Byzantines.

Kaldjian, P. J. (2004). Istanbul's bostans: A millennium of market gardens. *Geographical Review, 94*(3), 284–304.

Karakuyu, M. (2010). İstanbul'un Tarihsel Topoğrafyası ve Literatür Değerlendirmesi. *Türkiye Araştırmaları Literatür Dergisi* (16), 33–60.

Kasapseçkin, M. A. (2012). *İstanbul Tarihi Yarımada'da bulunan yeniden işlevlendirilmiş kapalı sarnıçların malzeme ve yapay aydınlatma yönünden incelenmesi*.

Koç, A., & Kubat, A. (2018). Kent Biçimi Araştırmalarında Karşılaştırmalı Analiz Yöntemleri: İstanbul Tarihi Yarımadası Örneği. II. Kentsel Morfoloji Sempozyumu: "Değişkent"Değişen Kent, Mekân ve Biçim Bildiriler Kitabı.

Kubilay, A. Y. (2015). BATI KAYNAKLI HARITALARDA İSTANBUL [Istanbul in Western Source Maps]. *From A History of Istanbul, 1*, 522–543. https://istanbultarihi.ist/28-bati-kaynakli-haritalardaki-istanbul

Lechevalier, L. e. p. F. K. e. J. B. (1786). *Carte de Constantinople*. https://collections.library.yale.edu/catalog/15502072

Magdalino, P. (2015). Byzantine Constantinople: Topography and settlement. *From A History of Istanbul, 1*, 370. https://istanbultarihi.ist/389-byzantine-constantinople-topography-and-settlement

Orbay, İ. (2001). *Istanbul viewed: the representation of the city in Ottoman maps of the sixteenth and seventeenth centuries* MIT [Thesis]. http://dspace.mit.edu/bitstream/handle/1721.1/8630/49545862-MIT.pdf?sequence=2

Pinkham, R. (2000). *Daylighting: New life for buried streams*. Rocky Mountain Institute.

Sarihan, E. I. (2021). Historical city evaluation in the context of morphological theories (Istanbul, Last Ottoman Period).

Sensoy, F. (2015). The impact of floods on the topography of Istanbul from 1453 to the present. *From A History of Istanbul, 1*, 466. https://istanbultarihi.ist/395-the-impact-of-floods-on-the-topography-of-istanbul-from-1453-to-the-present

Spirn, A. W. (1993). Deep structure: On process, form, and design in the urban landscape. In *City and nature* (Vol. 13, pp. 9–16). Odense University Press.

Steiner, F. (2011). Landscape ecological urbanism: Origins and trajectories. *Landscape and Urban Planning, 100*(4), 333–337. https://doi.org/10.1016/j.landurbplan.2011.01.020

Şahin, İ. (2004). *Türkiye'de kentsel koruma politikalarının gelişimine yardımcı bir yöntem denemesi, örnekleme: İstanbul Tarihi Yarımada* Fen Bilimleri Enstitüsü].

Şengör, A. M. C., & Kındap, T. (2019). The geology and geomorphology of İstanbul. *World Geomorphological Landscapes*.

Urfalıoğlu, N. (2015). İSTANBUL'UN ÇEŞME VE SEBİLLERİ. *Antik Çağ'dan XXI. Yüzyıla Büyük İstanbul Tarihi*. Retrieved March 01, 2023, from https://istanbultarihi.ist/322-istanbulun-cesme-ve-sebilleri

Ward, K. A., Crapper, M., Altuğ, K., & Crow, J. (2017). The byzantine cisterns of Constantinople. *Water Science and Technology: Water Supply, 17*(6), 1499–1506.

Water, M. (2013). Waterway corridors–guidelines for greenfield development areas within the port Phillip and Westernport region.

Yılmaz, C. (2015). Insights from an 1845 map of ISTANBUL. *A History of Istanbul*. Retrieved February 26, 2023, from https://istanbultarihi.ist/399-insights-from-an-1845-map-of-istanbul

Yurdakul, İ. (2010). *Aziz şehre leziz su: Dersaadet, İstanbul su şirketi, 1873–1933* (Vol. 434). Kitabevi.

SRQL Index: An Assessment Tool to Promote Sustainability, Resilience, and Quality of Life in Informal Settlements in Egypt

Salwa A. M. Ahmed, Khaled A. Youssef, Amr Abdallah, and Khaled S. Abdelmagid

Abstract

This paper is part of a research project aiming at developing a smart decision support system (SDSS) to propose proper interventions for upgrading informal settlements in Egypt. The system depends on a multi-criteria assessment tool, namely the SRQL Index, developed to promote sustainability, resilience, and quality of life. The scope of this paper is limited to developing SRQL. A critical review of national and international resources has been conducted to derive 1645 indicators that promote SRQL in informal settlements. These indicators were filtered into 89 relevant indicators, and then were categorized into 4 domains, 10 dimensions, and 27 key-indicators. Using the analytic hierarchy process (AHP), domains, dimensions, key-indicators, and indicators were assigned relative weights, depending on the input of two-round focus group sessions. Representatives of various backgrounds, e.g., academics, planners, local administration, and NGOs, participated in determining the relative weights of domains, dimensions, key-indicators, and indicators, respectively. Focus group sessions revealed that the socio-economic domain has the highest relative weight (43.94%), urban governance (25.11%), the environment (18.26%), and then urban aspects (12.69%). The top relative weight indicators are 'income/capita' (5.09%), and 'population in hazard locations' (4.69%), while the lowest relative weight indicators are 'building heights', 'plot area', and 'vacant lands' (0.04%).

S. A. M. Ahmed (✉) · K. A. Youssef · A. Abdallah · K. S. Abdelmagid
Architectural Engineering, Faculty of Engineering, Assiut University, Assiut, Egypt
e-mail: smegahed@aun.edu.eg

K. A. Youssef
e-mail: khaled.ali.youssef@aun.edu.eg

K. S. Abdelmagid
e-mail: khaled@aun.edu.eg

Keywords

Sustainability • Resilience • Quality of life • Informal settlements • Egypt • AHP

1 Introduction

1.1 Informal Settlements in Egypt: Historical Overview

Informal settlement can be considered the dominant urbanization mode in Egypt; about 60% of the population inhibits these settlements. Informal settlements occupy 60% of greater Cairo, 77% of the urban growth of Bani Mazar, El Minya City, and 87% of the urban growth of Giza City. About 80% of informal settlements are in the form of expansion on privately owned agricultural lands, whereas about 15% of these settlements exist on state-owned lands (Hegazy, 2016).

The history of informal settlements in Egypt began in the late 1950s and sped up during the 1960s. Massive industrialization created employment opportunities around big cities causing massive rural–urban influx to these cities, especially Cairo, the Capital City of Egypt. In the period between 1967 and 1973, financial resources had been directed to fulfill military needs. Consequently, governmental investment in public housing decreased to its minimum and the private sector could not meet the demand, opening the door wide for the informal sector supply. After 1973, informal settlements dominated the affordable housing market and inhibited by low-or-middle income dwellers. In the late 1970s, the majority of Egyptian working in the Gulf countries invested their savings in informal housing areas. Between 1981 and 2011, strict rules against illegal urbanization managed to minimize the appearance of new informal settlements. The 25th of January 2011 Revolution was a turning point for urbanism, pursuing new means of action

A. Battisti et al. (eds.), *Greening Our Cities: Sustainable Urbanism for a Greener Future*, Advances in Science, Technology & Innovation, https://doi.org/10.1007/978-3-031-49495-6_15

and putting pressure on governmental bodies to recognize the rights of informal settlement dwellers (Séjourné, 2009; Khalifa, 2015; Abd Elrahman & El-khateeb, 2016; UN-Habitat, 2020).

1.2 Egyptian Initiatives to Upgrade Informal Settlements

In 1993, the Egyptian government adopted two main strategies to institutionalize the process of upgrading informal settlements, including: (1) developing new laws and regulations, and (2) building a local administrative system. Various institutions have been involved in upgrading informal settlements, e.g., the Ministry of Utilities and Urban Communities, the Ministry of Urban Development and Informal Settlements, and Informal Settlements Development Fund (ISDF), together with governmental bodies and NGOs working at the governorate level.

In 1993, the National Plan for the Development of Informal Settlements focused on drainage networks, potable water, electricity, and road paving. In parallel, local authorities focused on other issues such as street lighting, road widening, and public services. These efforts are argued to be devoted to political rather than development objectives. Lack of development priorities, reduced scope of upgrading these settlements, and limited financial resources negatively affected the achievement of the expected results. In 2000, a participatory approach was proposed as the main policy approach for urban development, redefining the objectives of urban upgrading to: (1) improve environmental, socio-economic, and healthcare related conditions, (2) make sure that the population benefit from upgrading efforts, and (3) encourage citizens to establish income-generating activities. In 2004, governmental programs were launched to restrict the growth of informal settlements, but unfortunately these plans could not achieve the targeted objectives. In 2005, the 'National Program of Social Housing' was established to upgrade existing informal settlements and prevent the emergence of new ones. Due to the lack of control, the expansion of informal settlements increased. In 2008, the Informal Settlements Development Facility (ISDF) was established. ISDF classified these settlements to unplanned areas and unsafe areas (risk to life and health), giving the priority for unsafe areas. Tools and approaches to upgrade informal settlements included: (1) limiting the growth and emergence of informal settlements, and (2) dealing with existing informal settlements (Hegazy, 2016; Abd Elrahman & El-khateeb, 2016; UN-Habitat, 2020). ISDF strategies are divided into three categories, including: short-term strategies for unsafe areas, mid-term strategies for developing unplanned areas, and long-term strategies for preventing the emergence of new informal settlements. ISDF policies include settling dwellers of informal settlements in the same location, if applicable, working with governorates within the framework of decentralization, partnerships with local authorities, the community, and key stakeholders, and ensuring suitable living and economic conditions for dwellers (ISDF, 2022).

Challenges facing upgrade policies and programs can be classified into: (1) regulatory dimension, i.e., strategies for tenure regulation, affordable housing, and supply of urban infrastructure and basic services, (2) institutional dimension, e.g., a strategy for promoting decentralization and institutionalization, (3) financial dimension, i.e., the resources and funds required for upgrade, and (4) participatory dimension, i.e., engagement and participation of the community, stakeholders, and key players in the upgrading process (Hegazy, 2016; Arandel & El Batran, 1997).

1.3 Research Problem

In alignment with Sustainable Development Goals (SDGs), Egypt developed its Vision 2030 to achieve sustainable development. Despite the tenth pillar of the vision, namely urban planning, which aims at improving the quality of life through the management of land and resources (Ministry of Planning & Economic Development, 2016; Neoxero, 2016), infrastructure and public services have been always dominating the intervention in informal settlements; having the priority over promoting environmental, economic, and social sustainability as prime goals (Abd Elrahman & El-khateeb, 2016; Hegazy, 2016).

Through the last four decades, resilience has become a priority in urban planning activities; being a key concern involved with sustainability. Rabid urbanization, climate change, declining economic conditions, severe poverty, and political instability all have had direct and indirect impacts on cities, in general, and on informal settlements particularly (Elsayed et al., 2020; Asian Development Bank, 2022). This promotes resilience to become one of the main pillars when upgrading informal settlements toward rationally reacting to changes.

In another sense, ISDF classifies informal settlements into two broad categories, namely unsafe and undeveloped settlements, proposing relevant policies for upgrade strategies to be executed. A rating/score system that evaluates settlements in each category and afterward prioritize intervention is lacking. Further, methodologies for dealing with these two categories and the tools used are left for planners to develop, case by case. This includes the analysis of the cases studied, the evaluation tools and rating systems prompted, the proper actions proposed, and the priorities of intervention. Moreover, lack of coordination among involved authorities and related bodies, incomprehensive upgrading approaches, and the subjectivity of the upgrading process and the decisions made double the problem.

1.4 Related Work

Sustainability, resilience, and quality of life represent the framework of developing the assessment tool targeted, namely SRQL Index, which is considered the cornerstone of the SSDSS that will be developed in following stage. Hundreds of indicator-based sets, frameworks, and measuring tools have been developed and validated for the sake of promoting urban sustainability, resilience, and quality of life. Scope of application encompasses the national level, the city level, and the single-building level.

Examples of indicator-based sets at the national and city level include, but are not limited to, the 'Urban Sustainability Indicators' developed by the European Foundation for the Improvement of Living and Working Conditions (Mega & Pedersen, 1998), 'Urban Indicators Guidelines' developed by the United Nations Human Settlements Program (Urban Indicators Guidelines, 2004), 'Eurostat Sustainable Development Indicators' (Eurostat, 2022), 'City Resilience Index' (Rockefeller Foundation, 2013), 'Quality of Life Index (QLI)' developed by Carol Estwing Ferrans and Marjorie Powers (Ferrans & Powers, 1985), and 'OECD Better Life Index' (OECD, 2020). At the architectural level, the list of indicator-based sets includes 'Building Research Establishment Environmental Assessment Method (BREEAM)'; a sustainability assessment method launched in 1990, by the Building Research Establishment, and the 'Leadership in Energy and Environmental Design (LEED) Rating System' that gives certification and a framework for healthy, highly efficient, and cost-saving green buildings, among many other indices and rating systems. At the national level, Egypt Vision 2030 identified a set of urban planning indicators to promote sustainability and improve the quality of life, in general, whereas ISDF, and Hayah Karima Initiative (HKI); which means 'decent life' focus more on slums and informal settlements (ISDF, 2022).

Even though international indices and rating systems prove well and serve many purposes in evaluating and assessing urban environments to promote sustainability, resilience, and quality of life, informal settlements require more context-sensitive evaluation and further assessment tools; due to the special nature of and the severe conditions in these settlements. On the other hand, existing local indicators lack comprehensiveness, integration and measurability. Summing up, there is a need for a holistic, integrated, local-sensitive, informal-settlement oriented, measurable, and priority-aware assessment tool capable of promoting urban sustainability, resilience, and quality of life in informal settlements.

1.5 Aim and Objectives

The aim of the research project is to develop a tool capable of objectively assess informal settlements in Egypt. The assessment tool, i.e., SRQL Index, gets its name from the initials of the three domains under which informal settlements in Egypt can be upgraded, namely Sustainability, Resilience, and Quality of Life. This index is the cornerstone of a Smart Spatial Decision Support System (SSDSS) to evaluate informal settlements and further propose proper interventions; according to the priorities resulting from the evaluation process. For the aim to be attained, two main objectives are to be accomplished: (1) identifying assessment criteria; indicators, that to promote sustainability, resilience, and quality of life when upgrading informal settlements in Egypt, and (2) determining the relative weight of indicators that would affect the evaluation of the settlement and the priorities of interventions.

1.6 Scope and Limitations

For the aim of the research project to be attained, an intervention matrix is to be developed; identifying in particularity the most relevant intervention/s based on the evaluation results/score. Further, the SRQL-based SSDSS is to be built, providing decision-makers with a prioritized list of interventions based on the evaluation of the informal settlement studied. At last, a pilot application should be carried out to verify the SRQL list of indicators, the relative weights assigned to these indicators, i.e., SRQL Index, the intervention matrix, and the priorities recommended; i.e., the entire SSDSS. The scope of this paper will be limited to the development of SRQL Index; i.e., the compiled list of indicators, and the relative weight assigned to each of them, while the development of the intervention matrix, the SSDSS, and the pilot application of the system for validation will be the issue of other research papers.

2 Methodology

To develop the assessment tool targeted, potential indicators that promote urban sustainability, resilience, and quality of life were collected and categorized. National and international resources in the last 10 years were reviewed. A matching analysis was applied to remove duplicates. The list of indicators was further fine-tuned being affected by several factors, including, but are not limited to, the significance of the indicator, accessibility to the data required, measurability, priority, and frequency of the indicator in local and international resources. A compiled list of indicators was presented to, and validated by, a focus group of various backgrounds related to the field of the study, e.g., academics, professional planners, local administration, and NGOs representatives, in two sessions (a face-to-face sessions and a virtual meeting). In sessions, focus group members reviewed the compiled list of indicators and the proposed categorization structure. Further, they were asked to assign a weight for each domain, dimension, key-indicator, and indicator via an online survey tool. Results have been analyzed using analytic hierarchy process (AHP) to determine the relative weight of domains, dimensions, key-indicators, and indicators. In a following session, results, i.e., relative weights of indicators, were presented to the focus group members for discussion and feedback. Figure 1 shows the research methodology.

3 Results

3.1 SRQL Indicators

Urban sustainability, resilience, and quality of life-related indicators have been collected out of an extensive critical review of international and national resources, e.g., papers, books, research projects, articles, reports, as well as national and international frameworks, codes, indices, manuals, action plans, and initiatives. Table 1 illustrates a sample of the critical review results carried out on academic research work, whereas Table 2 presents an example of the results of reviewing national and international frameworks.

As a result of the critical review process, a total of 1645 indicators to promote/assess sustainability, resilience, and quality of life, have been collected, i.e., 729 indicators to promote/assess sustainability, 270 indicators to promote/assess resilience, and 646 indicators to promote/assess the quality of life. The initial list (List 1) included numerous duplicate indicators. After removing duplicates, the second list (List 2) included 687 non-duplicate indicators. Then, this list is revised according to the applicability of indicators to informal settlements, and the frequency of indicators in the resources. As a result, the revised list (List 3) included 142 informal settlement-relevant and most frequent indicators. In terms of the measurability of indicators, the potential access to the information required, and the priority of indications, the list of indicators is fine-tuned to finally encompass 89 indicators (List 4).

To be able to apply the analytic hierarchy process (AHP) method[1] for determining the relative weight of indicators; depending on the feedback of the focus group participants, a tree-structure categorization of the indicators has been developed. As a result, the 89 SRQL indicators (List 4) have been categorized into 27 key-indicators, 10 dimensions, and 4 domains, Fig. 2. Both the indicators and the tree structure developed have been presented, discussed, and validated by the focus group, which included 15 academics from 9 universities, 4 professional planners, 10 governmental agency representatives, and 3 NGOs representatives: a total of 32 participants. The hierarchy of SRQL Index is shown in detail in Appendix 1.

3.2 SRQL Index

To determine the relative weight of each SRQL indicator, analytic hierarchy process (AHP) was prompted, as it is considered one of the most widely employed multi-criteria methods to rank different options against each other; based on a set of criteria. Because the total number of comparisons becomes excessive as the number of alternatives or criteria, i.e., indicators in this case, increases (Wedley et al., 1993), SRQL indicators are organized into groups, i.e., categories, following tree-structure hierarchy. Thus, the maximum number of domains, the number of dimensions under each domain, the number of key-indicators under each dimension, or the number of indicators under each key-indicator did not exceed 6. This allowed for a maximum of 15 comparisons in each ranking process.

After discussing and validating the list of indicators and the tree-structure proposed, focus group participants have been requested to compare the domains, the dimensions, the key-indicators, and the indicators pairwise, via an online survey tool. Average relative weights assigned to each domain, dimension, key-indicator, or indicator have been then calculated using AHP method. Relative weights of SRQL domains, dimensions, key-indicators, and indicators are, respectively, shown in Fig. 3a–d, whereas more about the relative weights of these domains, dimensions, key-indicators, and indicators is presented in detail in Appendix B.

4 Discussion

At the SRQL domain level, socio-economic aspects appeared of high priority compared with urban governance, environmental, and urban aspects, with a relative weight of 43.94%; about 1.76 times the average weight of domains

[1] The Analytic Hierarchy Process (AHP), submitted by Saaty, T. L. (1988), is an effective tool for dealing with multi-criteria decision-making. AHP aids to set priorities and make the best decision. By reducing complex decisions to a series of pairwise comparisons, and then synthesizing the results, the AHP helps to capture both subjective and objective aspects of a decision. In addition, AHP integrates an effective technique for monitoring the consistency of the stakeholder's evaluations, thus reducing the bias in the decision-making process.

(25%). This highlights the importance of the social and economic aspects when upgrading informal settlements in Egypt. As per the focus group participants, urban aspects domain, i.e., urban form, housing and population, services and utilities, and transportation and mobility, lags the other three domains, with a relative weight of 12.69%; i.e., 51% of the average weight. This seems consistent with the changing perspective of related authorities concerning the upgrade of informal settlements, and the paradigm shift from focusing on services, utilities, and housing to promoting income-generating activities and improving the social conditions of dwellers. At the SRQL dimension level, results showed the economy to be of central importance, with a relative weight of 24.73%; i.e., almost 2.5 times the average weight of dimensions (10%), to be followed by the social aspects; i.e., demographics of the population, equity, and safety and security, with a relative weight of 19.44%, state agencies and local authority (14.11%), environmental quality (13.73%), and community leadership (11%). Below average dimensions include services and utilities (5.48%), energy (4.53%), housing and population (3.56%), transportation and mobility (2.75%), and urban form (0.90%), respectively. Not surprisingly, the urban form dimension in informal settlements is at the bottom of the list, as more pressing issues should be tackled first.

As for the key-indicators, five aspects come at the top of the list, including: income and expenses (11.59%, i.e., 3.13 times the average weight of key-indicators), safety and security (10.74%), pollution (8.86%), effectiveness (7.86%), and equity and satisfaction (6.99%), respectively; all are above the average weight of key-indicators (from 3.13 to 1.9 times the average). The least average weight key-indicators are physical features (0.17%), public open spaces (0.24%), cultural and recreational services (0.30%), commercial services (0.40%), and size of the area (0.40%), respectively, with about 0.05% to 0.11% of the average relative weight. This reflects the retreat of physical features and open spaces as key-indicators in informal settlements; together with the need for certain types of public services, i.e., cultural, recreational, and commercial activities. Further, the focus group participants thus claim the size of the area to be insignificant.

As for the fourth SRQL level, the relative weight of indicators differs dramatically from 5.09 to 0.17%. This illustrates the significance of the valuable outcomes of the focus group sessions. The top five indicators are income per capita (5.09%), population in hazard locations (4.69%), population living below the national poverty line (4.02%), contamination of drinking water (3.89%), and housing cost/income (3.76%), respectively. Again, this emphasizes the significance of economic conditions as well as the factors that threaten the life of informal settlement dwellers. These top five indicators have from 4.6 to 3.4 times the average relative weight of indicators. On the other hand, several indicators appeared of almost no impact when evaluating informal settlements in Egypt; as per the feedback of focus group participants, such as plot area (0.04%), building heights (0.04%), vacant lands (0,08%), area of residential apartment (0.1%), and private car ownership (0.13%); comprising for about 4–12% of the average relative weight of indicators.

Most importantly, the great disparities in the relative weight among indicators would have a notable impact on the informal settlement evaluation score, as well as the proper intervention that would be selected out of the intervention matrix developed, and the priorities of intervention based on the relative weight of each indicator and the score achieved. Developing a matrix of interventions, that to propose a set of procedures based on the score the informal settlement achieves, and numerically identifying the priorities of interventions are integral to the SSDSS targeted by the end of the research project. This will be the aim of recent future work.

Summing up, the research managed to develop a comprehensive index; compiled list of indicators, as a tool to evaluate and further assess informal settlements in Egypt.

5 Conclusions

In this paper, an index to promote and measure sustainability, resilience, and quality of life, SRQL Index, has been developed. First, a critical review of related resources has been conducted, and 1645 indicators have been collected. After removing duplicates, indicators have been filtered, based on the relevance to informal settlements, frequency in resources, measurability, access to the information required for measurements, and priority, to 89 SRQL indicators. These indicators have been categorized into 4 domains, 10 dimensions, and 27 key indicators. Using analytical hierarchical process (AHP) method, these domains, dimensions, key-indicators, and indicators have been assigned relative weights by 32 focus group participants. Upon developing a scoring system (future work), SRQL Index would be able to assess informal settlements. Based on the score achieved in each indicator, priorities of intervention would be determined. Summing up, the index is considered a cornerstone in the targeted smart spatial decision support system (SSDSS) that would help planners and decision-makers in Egypt upgrade informal settlements objectively.

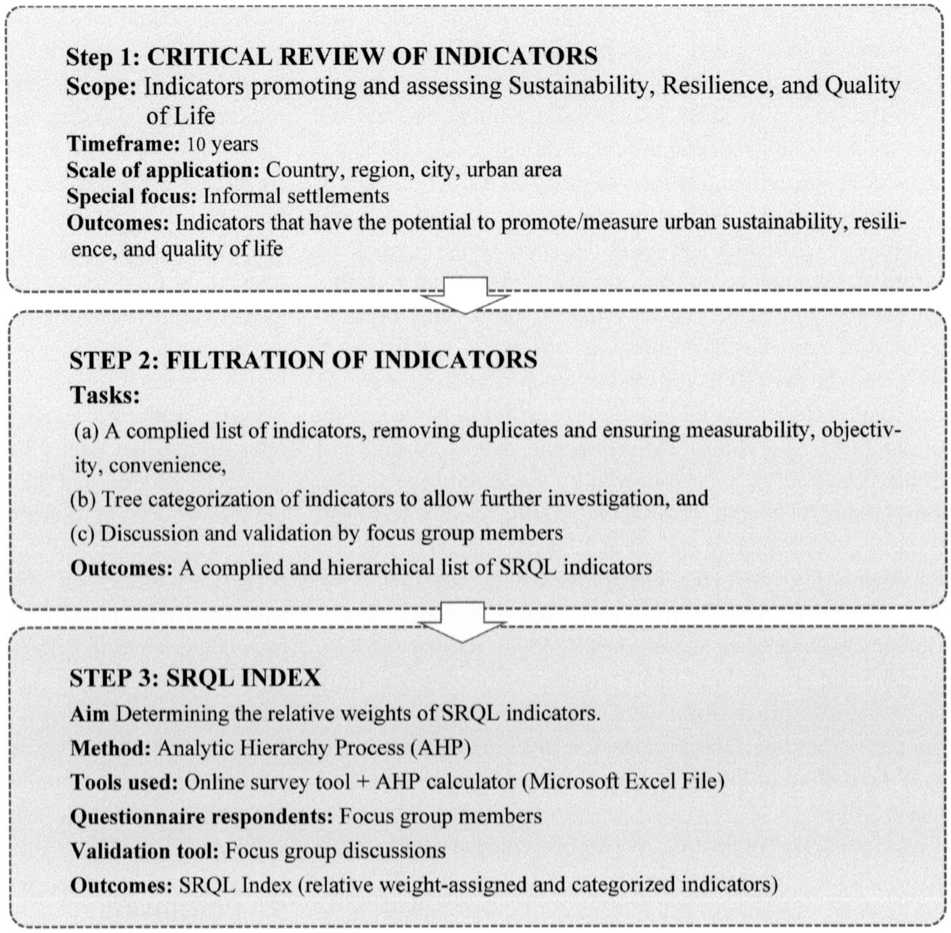

Fig. 1 Research methodology. *Source* The authors

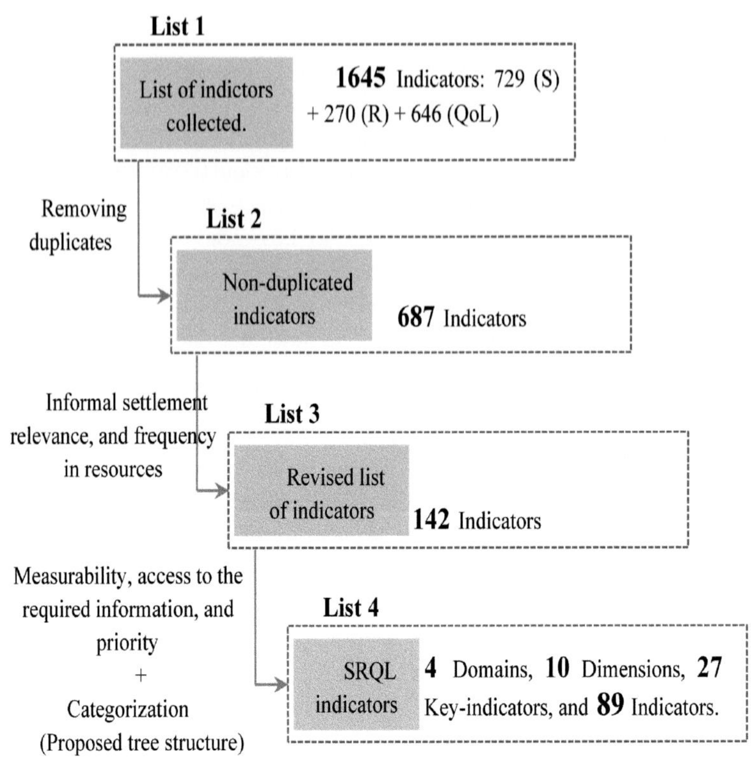

Fig. 2 Filtration and categorization process of SRQL indicators. *Source* The authors

Fig. 3 Relative weights of SRQL domain, dimensions, key-indicators, and indicators. *Source* The authors

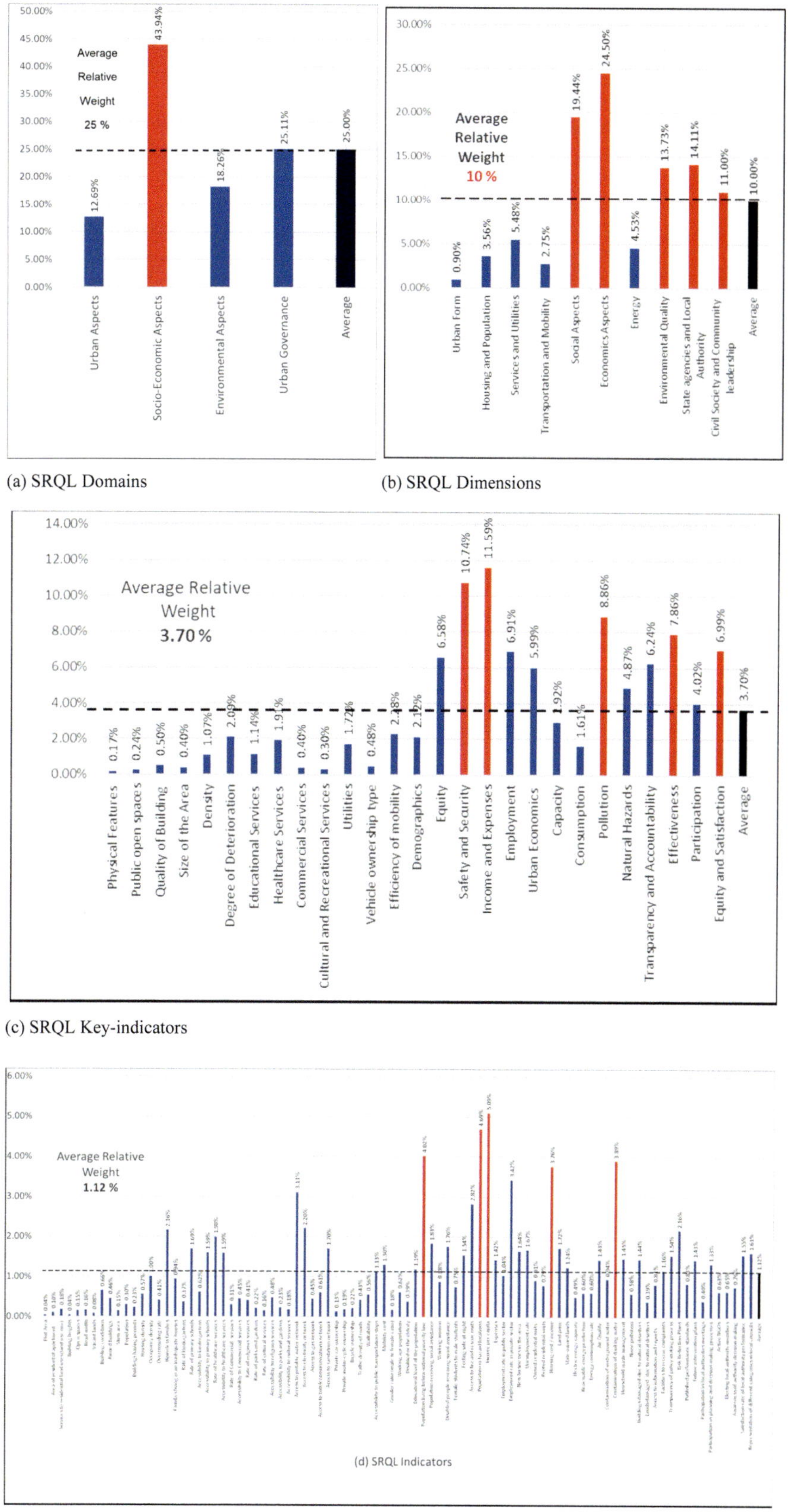

(a) SRQL Domains

(b) SRQL Dimensions

(c) SRQL Key-indicators

(d) SRQL Indicators

Table 1 Review of sustainability (S), resilience (R), and quality of life (QoL) indicators in academia—sample

References	Country/ City	No. of indicators	Analysis method	Remarks/Findings/Results	Scale of application	Domains		
						S	R	QoL
(Superti et al., 2021)	Switzerland	57	Comparative analysis	Conceptual link between economy and sustainability	City	✓	X	X
(Merino-Saum et al., 2020)	Switzerland	67	Multi- and cross-typological systematic analysis	Priority of social issues over environmental aspects Marginal consideration to political aspects	City	✓	X	✓
(Steiniger et al., 2020)	Chile	29	Expert consultation, comparison, and benchmarking	Reacting to urban development challenges of informal settlements in a developing country	Informal settlements	✓	X	X
(González et al., 2018)	Chile	38	Principal component analysis and varimax factor analysis	Multi-level indicators that identify and quantify spatial disparities among urban areas from the perspective of disaster risk reduction	Multiple	X	✓	X
(Leobons et al., 2019)	Brazil	11	Systematic literature review and bibliometric analysis	Resilience indicators used in the academia in the field of transportation	Regions and cities	X	✓	X
(Xu & Xue, 2017)	China	46	Content analysis, expert investigation, and AHP method	The weight of governance, management, and economic development indicators are much higher than other indicators	Public spaces	X	✓	X
(Giri et al., 2021)	Nepal	82	Numerical analysis and Chi-square test	Factors that would help informal settlers adapt to climate change challenges, e.g., infrastructure, health facilities, and social capitals	Informal settlements	X	✓	X
(Amiraslani, 2021)	Tehran, Iran	11	Structured query language (SQL) algorithms	New indicators to assess the QoL, e.g., public complaints, governance capability, green spaces, public transport, public services, and car parking	City	X	X	✓
(Shahraki et al., 2020)	Kabul, Afghanistan	102	Regression and factor analysis	General dissatisfaction with the QoL indicators related to transportation, leisure, governance, and public services	Informal settlements	X	X	✓
(Stańczyk et al., 2012)	Poland	30	Strategy realization analysis	High priority indicators to promote sustainability are strongly related to waste management, sewage system, and number of trees planted annually	Local community	✓	X	X
(Lynch & Mosbah, 2017)	United States	91	Experts	List of strategies to ensure local progress toward achieving sustainability goals in the U.S. and internationally	Wide urban area	✓	X	X
(Scherzer et al., 2019)	Norway	47	Index development (Community resilience index)	Resilience index of the Norwegian municipalities Considerable variations between cities in view of resilience indicators	Community	X	✓	X

Source The authors

Table 2 Review of international and national frameworks to promote sustainability, resilience, and quality of life—sample

Impact	References	Title	Type	No. of indicators	Description/Objective	Domains		
						S	R	QoL
International	(Global platform for sustainable cities, World Bank, 2018)	Urban Sustainability Framework	Urban Sustainability Framework	180	Help cities achieve United Nations sustainable development goal 11—making cities inclusive, safe, and resilient	✓	✓	X
	(UN-Habitat, 2017)	ISO 37120 Sustainable development of communities	Standard code	100	Definitions and methodologies to develop a set of indicators to steer and measure the delivery of city services and quality of life	✓	X	X
	(BRE Global, 2012)	BREEAM Communities Manual	Assessment code	62	A tool to assess, validate, and improve social, environmental, and economic sustainability	✓	X	X
	UN-Habitat, (2005)	Urban Governance Index (UGI)	Advocacy tool (Index)	16	Measuring the accountability, participation, equity, and effectiveness of the local government	✓	✓	X
	(United Nations, 2017)	Sustainable Development Goals (SDGs) for 2030	A universal call (Vision)	248	17 sustainability development goals for the sake of eliminating poverty, promoting natural conservation, and achieving urban sustainability	✓	X	X
	(United Nations, 2017)	Millennium Development Goals (MDGs)	A universal call (Vision)	48	A vision to eliminate poverty, hunger, and diseases	✓	X	X
	(EU., 2016)	Reference framework for sustainable cities (RFSC)	Open access web tool	30	A tool to help cities and urban territories promote and improve their integrated urban development actions	✓	X	X
	(U.S. Green Building Council, 2014)	LEED v4 for Neighborhood Development	Rating system	59	A green certification concept applicable to the neighborhood level, containing a set of measurable standards	✓	X	X
	(JSBC & IBEC, 2001)	CASBEE for Urban Development, Japan	Performance assessment tool	82	A tool to assesses the environmental performance for the aim of promoting sustainability	✓	X	X
	(Arup & Rockefeller Foundation, 2013)	Cities Resilience Index	Advocacy tool (index)	52	A technically robust basis for measuring city resilience	X	✓	X
	(Chan et al., 2005)	CUHK Hong Kong quality of life index	Rating index	23	An index to measure the quality of life in Hong Kong via indicator that to monitor social, political, cultural, economic and environmental conditions	X	X	✓
National	(Ministry of Planning and Economic Development, 2016)	Egypt Vision 2030	Vision & action plan	172	A vision to promote sustainable development for offering a better life to all Egyptians by 2030. It has three main domains, namely economy, society, and environment	✓	X	✓
	(ISDF, 2005)	ISDF Indicators	Organization	N/A	Policies to upgrade informal settlements and ensure social justice and citizen's right to have a proper housing; while at the same time improving living and economic conditions	✓	X	X
	(Hayah Karima Initiative, 2020)	Haya Karima (HK)	National initiative		Urgent intervention to offer a decent life for a portion of Egyptian citizens	✓	X	X

Source The authors

Acknowledgements This paper is based upon work supported by Science, Technology, and Innovation Funding Authority (STDF), project number 45850-phase 1. The title of the project is 'Smart Urban Governance Center (SUGC) for Sustainable, Resilient and Livable Cities'. The research team acknowledges Eng. Islam M. S. Abouel-hamd for his effort in the critical review stage of the research project.

Appendix 1: SRQL Domains, Dimensions, Key-indicators, and Indicators

Domains	Dimensions	Key-indicators	Indicators
Urban aspects	Urban form	Physical features	Plot area
			Area of residential apartment
			Services to residential land uses (land use mix)
			Building heights
		Public open spaces	Open spaces
			Road width
			Vacant lands
		Quality of buildings	Building conditions
			Ruined buildings
	Housing and population	Size of the area	Slum area
			Population
		Density	Buildings having permits
			Housing density
			Occupancy density
		Degree of deterioration	Overcrowding rate
			Homeless families
			Families living in inadequate houses
	Services and utilities	Educational services	Rate of kindergartens
			Rate of primary schools
			Accessibility to kindergartens
			Accessibility to primary schools
		Healthcare services	Rate of healthcare services
			Accessibility to healthcare services
		Commercial services	Rate of commercial services
			Accessibility to commercial services
		Cultural and recreational services	Rate of religious services
			Rate of parks and gardens
			Rate of cultural services
			Accessibility to religious services
			Accessibility to parks and gardens
			Accessibility to cultural services
		Utilities	Access to potable water network
			Access to electricity network
			Access to gas network
			Access to telecommunication networks
			Access to sanitation network
	Transportation and mobility	Vehicle ownership type	Private car ownership
			Private motor-cycle ownership
			Bicycle ownership
		Efficiency of mobility	Traffic density of roads
			Walkability
			Accessibility to public transportation stops
			Mobility cost
Socio-economic aspects	Social aspects	Demographics	Gender ratio (male-to-female ratio)
			Working age population
			Disabled or the elderly
			Educational level of the population
		Equity	Population living below national poverty line
			Population receiving social assistance
			Working women
			Disabled people receiving social assistance
			Female students to male students
		Safety and security	Feeling safe at night
			Access to fire and rescue units
			Population in hazard locations
	Economics aspects	Income and expenses	Income per capita
			Expenses
		Employment	Employment rate in public sector
			Employment rate in private sector
			New businesses in the area
			Unemployment rate
		Urban economics	Owned residential units
			Rented residential units
			Housing cost/income
			Land price
			State-owned lands

(continued)

Domains	Dimensions	Key-indicators	Indicators
Environmental aspects	Energy	Capacity	Electric energy capacity
			Renewable energy production
		Consumption	Energy consumption rate
	Environmental quality	Pollution	Air quality
			Contamination of underground water
			Contamination of drinking water
			Household waste management
			Noise pollution
		Natural hazards	Buildings damaged due to natural disasters
			Lands damaged due to natural disasters
Urban governance	State agencies and local authority	Transparency and accountability	Access to information and reports
			Facilities to receive complaints
			Transparency of policy-making process
		Effectiveness	Risk reduction plans
			Published performance standards
			Future intervention plans
	Civil society and community leadership	Participation	Participation in local authorities meetings
			Participation in planning and decision making processes
			Active NGOs
			Electing local authority members
		Equity and satisfaction	Awareness of authority decision making
			Satisfaction rate of local authority performance
			Representation of different categories in local councils

Appendix 2: Relative Weights of SRQL Domains, Dimensions, Key-Indicators, and Indicators

SRQL Domains

No.	Domain	RW (%)
1	Urban aspects	12.69
2	Socio-economic aspects	43.94
3	Environmental aspects	18.26
4	Urban governance	25.11
Average relative weight		25.0

SRQL Dimensions

No.	Dimension	RW (%)
1	Urban form	0.90
2	Housing and population	3.56
3	Services and utilities	5.48
4	Transportation and mobility	2.75
5	Social aspects	19.44
6	Economics aspects	24.50
7	Energy	4.53
8	Environmental quality	13.73
9	State agencies and local authority	14.11
10	Civil society and community leadership	11.00
Average relative weight		10.0

SRQL Key-Indicators

No.	Key-indicator	RW (%)
1	Physical features	0.17
2	Public open spaces	0.24
3	Quality of buildings	0.50
4	Size of the area	0.40
5	Density	1.07
6	Degree of deterioration	2.09
7	Educational services	1.14
8	Healthcare services	1.91
9	Commercial services	0.40
10	Cultural and recreational services	0.30
11	Utilities	1.72
12	Vehicle ownership type	0.48
13	Efficiency of mobility	2.28
14	Demographics	2.12
15	Equity	6.58
16	Safety and security	10.74
17	Income and expenses	11.59
18	Employment	6.91
19	Urban economics	5.99
20	Capacity	2.92
21	Consumption	1.61
22	Pollution	8.86
23	Natural hazards	4.87
24	Transparency and accountability	6.24
25	Effectiveness	7.86
26	Participation	4.02

(continued)

No.	Key-indicator	RW (%)
27	Equity and satisfaction	6.99
Average relative weight		3.70

SRQL Indicators

No.	Indicator	RW (%)	No.	Indicator	RW (%)
1	Plot area	0.04	46	Disabled or the elderly	0.39
2	Area of residential apartment	0.10	47	Educational level of the population	1.19
3	Services to residential land uses (land use mix)	0.18	48	Population living below national poverty line	4.02
4	Building heights	0.04	49	Population receiving social assistance	1.83
5	Open spaces	0.15	50	Working women	0.88
6	Road width	0.16	51	Disabled people receiving social assistance	1.76
7	Vacant lands	0.08	52	Female students to male students	0.75
8	Building conditions	0.66	53	Feeling safe at night	1.54
9	Ruined buildings	0.46	54	Access to fire and rescue units	2.82
10	Slum area	0.15	55	Population in hazard locations	4.69
11	Population	0.30	56	Income per capita	5.09
12	Buildings having permits	0.23	57	Expenses	1.42
13	Housing density	.057	58	Employment rate in public sector	1.04
14	Occupancy density	1.00	59	Employment rate in private sector	3.42
15	Overcrowding rate	0.41	60	New businesses in the area	1.64
16	Homeless families	2.16	61	Unemployment rate	1.67
17	Families living in inadequate houses	0.94	62	Owned residential units	0.91
18	Rate of kindergartens	0.37	63	Rented residential units	0.79
19	Rate of primary schools	1.69	64	Housing cost/income	3.76
20	Accessibility to kindergartens	0.62	65	Land price	1.72
21	Accessibility to primary schools	1.59	66	State-owned lands	1.24
22	Rate of healthcare services	1.98	67	Electric energy capacity	0.49
23	Accessibility to healthcare services	1.59	68	Renewable energy production	0.60

(continued)

No.	Indicator	RW (%)	No.	Indicator	RW (%)
24	Rate of commercial services	0.31	69	Energy consumption rate	0.60
25	Accessibility to commercial services	0.45	70	Air quality	1.43
26	Rate of religious services	0.41	71	Contamination of underground water	0.94
27	Rate of parks and gardens	0.22	72	Contamination of drinking water	3.89
28	Rate of cultural services	0.16	73	Household waste management	1.45
29	Accessibility to religious services	0.48	74	Noise pollution	0.58
30	Accessibility to parks and gardens	0.23	75	Buildings damaged due to natural disasters	1.44
31	Accessibility to cultural services	0.18	76	Lands damaged due to natural disasters	0.39
32	Access to potable water network	3.11	77	Access to information and reports	0.81
33	Access to electricity network	2.20	78	Facilities to receive complaints	1.16
34	Access to gas network	0.45	79	Transparency of policy-making process	1.54
35	Access to telecommunication networks	0.61	80	Risk reduction plans	2.16
36	Access to sanitation network	1.70	81	Published performance standards	0.83
37	Private car ownership	0.13	82	Future intervention plans	1.43
38	Private motor-cycle ownership	0.19	83	Participation in local authorities meetings	0.40
39	Bicycle ownership	0.22	84	Participation in planning and decision making processes	1.33
40	Traffic density of roads	0.43	85	Active NGOs	0.63
41	Walkability	0.56	86	Electing local authority members	0.65
42	Accessibility to public transportation stops	1.13	87	Awareness of authority decision making	0.76
43	Mobility cost	1.30	88	Satisfaction rate of local authority performance	1.55
44	Gender ratio (male-to-female ratio)	0.18	89	Representation of different categories in local councils	1.61
45	Working age population	0.62	Average relative weight		1.12

References

Abd Elrahman, A., & El-khateeb, S. (2016). Mapping informal areas in Egypt between the past interventions and next urban revolution. *Journal of Urban Research, 21*(1):116–129. https://doi.org/10.21608/jur.2016.89852

Amiraslani, F. (2021) Analysis of quality of life across Tehran districts based on designated indicators and relational database management system. *Urban Governance*, ISSN 2664-3286. https://doi.org/10.1016/j.ugj.2021.09.003

Arandel, C., & El Batran, M. (1997) *The informal housing development process in Egypt*. Centre Nacional de la Recherche Scientifique (CNRS). https://www.ucl.ac.uk/bartlett/development/sites/bartlett/files/migrated-files/WP82_0.pdf

Arup & Rockefeller Foundation. (2013). City Resilience Index. Available at: https://www.cityresilienceindex.org/#/. Accessed October 2021.

Asian Development Bank. (2022). Building resilience of the urban poor in Indonesia. https://www.adb.org/publications/building-resilience-urban-poor-indonesia

BRE Global. (2012). BREEAM Communities technical manual. Available at: https://www.breeam.com/communitiesmanual/. Accessed October 2021.

Chan, Y. K., Kwan, C. C. A., & Shek, T. L. D. (2005). Quality of life in Hong Kong: the Cuhk Hong Kong quality of life index. *Social Indicators Research, 71*, 259–289. https://doi.org/10.1007/1-4020-3602-7_9

EU.(2016). Reference framework for European sustainable cities RFSC. Available at: http://rfsc.eu/. Accessed October 2021.

Eurostat. (2022). Eurostat sustainable development indicators. https://ec.europa.eu/eurostat/web/sdi/indicators

Elsayed, M., El Badrawy, A., & Sheta, S. (2020). An approach towards achieving urban resilience in threatened urban communities. *MEJ-Mansoura Engineering Journal, 44*(4), 1–11. https://doi.org/10.21608/bfemu.2020.94731

Ferrans, C. E., & Powers, M. J. (1985). Quality of life index: Development and psychometric properties. *Advances in Nursing Science, 8*(1), 15–24. https://doi.org/10.1097/00012272-198510000-00005

Giri, M., Bista, G., Singh, P. K., & Pandey, R. (2021). Climate change vulnerability assessment of urban informal settlers in Nepal, a least developed country. *Journal of Cleaner Production, 307*, 127213. https://doi.org/10.1016/j.jclepro.2021.1272

Global Platform for Sustainable Cities, World Bank. (2018). *Urban Sustainability Framework*. (1st ed). Washington, DC: World Bank.

González, D. P., Monsalve, M., Moris, R., & Herrera, C. (2018). Risk and resilience monitor: Development of multiscale and multilevel indicators for disaster risk management for the communes and urban areas of Chile. *Applied Geography, 94*, 262–271. https://doi.org/10.1016/j.apgeog.2018.03.00

Haya Karima Initiative (HKI). (2020). Presidency of the Council of Ministers. https://www.hayakarima.com/about_en.html. Accessed October 2021.

Hegazy, I. (2016). Informal settlement upgrading policies in Egypt: Towards improvement in the upgrading process. *Journal of Urbanism: International Research on Placemaking and Urban Sustainability, 9*(3). https://doi.org/10.1080/17549175.2015.1009845

ISDF. (2005) Informal Settlement Development Fund, Presidency of the Council of Ministers http://www.isdf.gov.eg/AboutGov.aspx?about=6. Accessed October 2021.

ISDF. (2022). Strategies and policies of upgrading informal settlements. Informal Settlements Development Fund. http://www.isdf.gov.eg/AboutGov.aspx?about=6

JSBC—Japan sustainable building consortium and IBEC—Institute for building environment and energy conservation. Comprehensive Assessment System for Built Environment Efficiency (CASBEE). (2001). Available at: https://www.ibec.or.jp/CASBEE/english/. Accessed October 2001.

Khalifa, M. A. (2015). Evolution of informal settlements upgrading strategies in Egypt: From negligence to participatory development. *Ain Shams Engineering Journal, 6*, 1151–1159. https://doi.org/10.1016/j.asej.2015.04.008

Leobons, C. M., Gouvêa Campos, V. B., & de Mello Bandeira, R. A. (2019). Assessing urban transportation systems resilience: A proposal of indicators. *Transportation Research Procedia, 37*, 322–329. https://doi.org/10.1016/j.trpro.2018.12.199

Lynch, A. J., & Mosbah, S. M. (2017). Improving local measures of sustainability: A study of built-environment indicators in the United States. *Cities, 60*, 301–313. https://doi.org/10.1016/j.cities.2016.09.011

Mega, V., & Pedersen, J. (1998). *Urban sustainability indicators*. European Foundation for the Improvement of Living and Working Conditions. https://www.eurofound.europa.eu/sites/default/files/ef_publication/field_ef_document/ef9807en.pdf

Merino-Saum, A., Halla, P., Superti, V., Boesch, A., & Binder, C. R. (2020). Indicators for urban sustainability: Key lessons from a systematic analysis of 67 measurement initiatives. *Ecological Indicators, 119*, 106879. https://doi.org/10.1016/j.ecolind.2020.1068

Ministry of Planning and Economic Development, E. (2016). Egypt Vision 2030. https://arabdevelopmentportal.com/sites/default/files/publication/sds_egypt_vision_2030.pdf

Neoxero, E. (2016). SDS Egypt 2030. Egypt 2030. http://sdsegypt2030.com/?lang=en. Accessed October 2021.

OECD. (2020). Better life index. https://stats.oecd.org/Index.aspx?DataSetCode=BLI

Rockefeller Foundation. (2013). *City resilience index*. The Rockefeller Foundation. https://www.arup.com/perspectives/publications/research/section/city-resilience-index

Saaty, T. L. (1988). *What is the analytic hierarchy process?* (pp. 109–121). Springer Berlin Heidelberg.

Scherzer, S., Lujala, P., & Rød, J. K. (2019). A community resilience index for Norway: An adaptation of the baseline resilience indicators for communities (BRIC). *International Journal of Disaster Risk Reduction, 36*, 101107. https://doi.org/10.1016/j.ijdrr.2019.101107

Séjourné, M. (2009). The history of informal settlements. In: Kipper, Regina, & Fischer, Marion (Eds.) *'Cairo's informal areas between urban challenges and hidden potentials facts. Voices. Visions'*. Participatory Development Programme in Urban Areas (PDP). https://www.citiesalliance.org/sites/default/files/2019-07/CairosInformalAreas_fulltext.pdf

Shahraki, S. Z., Hosseini, A., Sauri, D., & Hussaini, F. (2020). Fringe more than context: Perceived quality of life in informal settlements in a developing country: The case of Kabul, Afghanistan. *Sustainable Cities and Society, 63*, 102494. https://doi.org/10.1016/j.scs.2020.102494

Stańczyk, J., Sleszynski, J., & Grodzińska-Jurczak, M. (2012). Selecting sustainability indicators for local community—case study of Milanówek municipality, Poland. *International Journal of Sustainable Development., 7*, 77–86.

Steiniger, S., Wagemann, E., de la Barrera, F., Molinos-Senante, M., Villegas, R., de la Fuente, H., & Barton, J. R. (2020). Localising urban sustainability indicators: The CEDEUS indicator set, and lessons from an expert-driven process. *Cities, 101*, 102683. https://doi.org/10.1016/j.cities.2020.10268

Superti, V., Merino-Saum, A., Baur, I., & Binder, C. R. (2021). Unraveling how the concept of circularity relates to sustainability: An indicator-based meta-analysis applied at the urban scale. *Journal of Cleaner Production, 315*, 128070. https://doi.org/10.1016/j.jclepro.2021.128070

UN-Habitat. (2020). Towards Arab cities without informal areas. United Nations Human Settlements Programme (UN-Habitat). https://www.isdb.org/sites/default/files/media/documents/2020-06/36%20IsDB%20UN%20Habitat%20Report%20on%20Informal%20Settlements%20in%20the%20Arab%20Region-%20Feb%202020.pdf

UN-Habitat. (2005). Habitat.: global campaign on urban governance: Urban governance index. UN. https://staging.unhabitat.org/content.asp?typeid=19&catid=25&cid=2167. Accessed October 2021.

UN-Habitat. (2017). ISO 37120—oxy.socialoxy.social/wp-content/uploads/2020/06/iso_37120_2014_en... · 2020. 6. 9. · ISO 37120 indicators for City Services and quality of life licensed to director@globalcities.ca—[PDF document]. vdocuments.mx. https://vdocuments.mx/iso-37120-oxy-2020-6-9-iso-37120-indicators-for-city-services-and-quality.html. Accessed October 2021.

United Nations. (2017). Home—United Nations Sustainable Development. https://www.un.org/sustainabledevelopment/. Accessed October 2021.

Urban Indicators Guidelines—UN-Habitat. (2004). United Nations: United Nations Human Settlements Program. https://unhabitat.org/sites/default/files/download-manager-files/Urban%20Indicators.pdf

U.S. Green Building Council. (2014). LEED certification for Neighborhood Development. LEED certification for neighborhood development | U.S. Green Building Council. https://www.usgbc.org/leed/rating-systems/neighborhood-development

Wedley, W. C., Schoner, B., & Tang, T. S. (1993). Starting rules for incomplete comparisons in the analytic hierarchy process Great Britain. *Mathematical and Computer Modelling, 17*(415), 93–100. https://doi.org/10.1016/0895-7177(93)90178-2

Xu, H., & Xue, B. (2017). Key indicators for the resilience of complex urban public spaces. *Journal of Building Engineering, 12*, 306–313. https://doi.org/10.1016/j.jobe.2017.06.018

Analysing Global Approaches for Urban Renewal: A Tool for Urban Conservation, Case Study of India

Kamini Sinha and Ashitha Tharian

Abstract

The purpose of the paper is to provide insight into the theoretical as well as practical processes of urban renewal. The study helps to understand the term urban renewal, urban revitalisation, urban regeneration, urban redevelopment, urban rehabilitation, and urban transformation through literature study. This study also analyses the various past and present approaches towards urban renewal in a pursuit to unlock the potential of future urban renewal projects. Today, as the world progresses on the path of urbanisation, the scope of urban renewal is very high. The paper tries to evaluate the current status of urban renewal and understand its future possibilities. It aims to analyse the global approaches for urban renewal, in order to understand its evolution towards becoming more holistic and inclusive, catering to social, economic, environmental, and cultural needs of contemporary areas. The objective is to review some of the approaches taken by leading countries in major urban renewal projects carried out through time. It is also to understand the various terms used by the countries for such interventions through extensive literature study. It is to glance at the Indian context to look into urban renewal schemes and practices within India, critically analyse newer approaches like participatory urban renewal and also scrutinises the future of urban renewal. This has been done by analysing Nizamuddin urban renewal project, India, where urban renewal has also been used as a tool for urban conservation. The paper concludes on a note on the future of urban renewal towards a positive impact. Urban renewal is the future of contemporary cities. Planned urban extension along with urban renewal is a reality that will shape the urban future. Next generation of urban renewal projects will have to combine considerations of economic growth and social equity, along with environment and cultural concerns, in order to make the whole exercise more relevant and impactful.

Keywords

Urban renewal • Community participation • Revitalisation • Regeneration • Urban transformation

1 Introduction

The essential requirement of any society is a safe, secure, and healthy condition where the population can carry out its daily life happily and progressively. However, it has been seen that the settlements developed various issues regarding hygiene and infrastructure over time. Administrations and societies, in general, have been trying to deal with the problem in multiple ways. These may be, redefining areas, beautifying them or upgrading their infrastructures. The approach varies in different nations, which has yielded results of all types. With the fast pace of development and urbanisation, the process of deterioration in areas followed by efforts for their upgradation is continuously going on. A study of these efforts shows the evolution process of the urban operation, which is referred to as urban renewal. This research tries to understand this process of evolution by analysing the processes.

1.1 Methodology

The research has been dealt with in five segments. The first segment reviews the various definitions proposed for the urban processes targeting the upliftment of existing areas. The second segment studies the different historical approaches to urban renewal worldwide to understand its evolution

K. Sinha (✉)
National Institute of Technology Patna, Patna, India
e-mail: kamini@nitp.ac.in

A. Tharian
IIT Bombay, Mumbai, India

A. Battisti et al. (eds.), *Greening Our Cities: Sustainable Urbanism for a Greener Future*,
Advances in Science, Technology & Innovation, https://doi.org/10.1007/978-3-031-49495-6_16

over the years. The third segment is a glance at the Indian context of urban renewal. It looks into urban renewal schemes and practices within India. The fourth segment critically looks into one of the newer approaches to urban renewal: participatory urban renewal. The final part scrutinises the future of urban renewal with an example of Nizamuddin urban renewal. The paper concludes on a note on the future of urban renewal towards a positive impact.

1.2 Part 1: Analysing Definitions from Various Sources

Various definitions for the act of urban renewal have been developed in the past fifty years by practitioners and academicians. The definition has changed over the years in response to changes in understanding and approaches taken towards the same. Specific terms, similar in meaning, have been introduced at various points in history, denoting a shift in perception towards the subject.

The terms urban revitalisation or urban regeneration can often mean the same as urban renewal and are used interchangeably, but a deeper inspection can unravel the differences between them. The term urban revitalisation has often been used to describe the urban renewal projects in western countries after the 1970s, emphasising neighbourhood preservation and housing rehabilitation. Whereas urban regeneration is the term used in European countries starting in the mid-1950s after conservation and rehabilitation got acceptance as a part of urban renewal in the European region (Broudhoax, 1994). So, urban renewal can be defined as an umbrella term that includes, but is not limited to, urban redevelopment, rehabilitation, and transformation. The term urban redevelopment is used when the renewal programme involves the replacement of old buildings and upgradation of roads and services roads with new ones. Rehabilitation is returning something to good condition, as defined by the Cambridge dictionary.

Moreover, the term urban rehabilitation can be used to describe a derelict area being restored to good condition as it once was. Finally, urban transformation is used when the renewal programme brings about a complete change in the appearance or character of an area. Table 16.1 explains better the similarity or differences in the terms.

The analysis of the matrix helps conclude that urban renewal is a tool that can be used for improvement of dilapidating neighbourhoods in cities. The dilapidation could be a result of the development and changes of urban processes and socioeconomic structuring. The improvements can be in terms of providing a better way for the social psychological needs, physical needs like better accessibilities and transportation systems, economic growth, suitable changes in land use and activity patterns, and augmentation of facilities and services for improving quality of life. The process may include both demolition and refurbishing. Demolition can be carried out in areas where improvement in living and sanitary conditions is not possible because of both settlement design and status of existing structures. In all other cases, refurbishing and reuse can be considered. When demolition and rebuilding is the major action for achieving targeted changes, the process is termed redevelopment. Urban revitalisation can be understood as a management strategy for facilitating economic, social, environmental, cultural, and historical redevelopment of a problematic, deprived, and derelict area. The emphasis is on restrengthening and revitalising the area via reintegration with the urban system. The tool of urban renewal, redevelopment, or management strategy of urban revitalisation may be used for rehabilitating a blighted area and improving its quality. If during such processes there is a complete change in the appearance or character of the area, it is termed urban transformation. It is achieved by implementing comprehensive and integrated approaches.

1.3 Part 2: Historical Approaches Towards Urban Renewal

Study and analysis of historical approaches better explain the evolution of the processes of urban renewal. Many historians and urban planners analysed past urban renewal projects and classified them into eras and generations. However, it is interesting to understand the need for renewal projects and the significant steps different countries have taken in this direction. Movements like the Park Movement and City Beautiful in America of the late nineteenth century explain the initial urban renewal projects, which started in response to the environmental degradation owing to urbanisation and industrialisation.

By the 1930s, public works programs shifted attention to the clearance of slums and the construction of low-income housing in their place. This was followed by the Housing Act of 1949. One such example is the Williamsburg houses of Brooklyn, New York, built-in 1936–1938 under the Housing Division of the Public Works Administration (PWA) when twelve congested slum blocks were cleared for 1622 low-wage bracket families (Fig. 16.1).

The urban renewal programs faced sharp criticism shortly, from 1949 to1954, as the application led to the destruction of homes and neighbourhoods, the displacement of small businesses, and demolished blighted areas making room for luxury housing. The period had more houses destroyed than what was rebuilt. The conditions were worse for those previously bad and got better for those previously good enough (Broudhoax, 1994). Similarly, for the same reasons such as the displacements of

Table 16.1 Matrix of definitions

	References	Urban renewal	Urban revitalisation	Urban regeneration	Urban redevelopment	Urban rehabilitation	Urban transformation
1	Cambridge dictionary	Building improvement and replacement in cities, including neighbourhoods	Revitalisation: the process of making something grow, develop, or become successful again		Redevelopment: the act or process of changing an area of a town by replacing old buildings, roads, etc., with new	Rehabilitation: the process of returning something to a good condition	Transformation: a complete change to improve the appearance or character of something
2	Adams (1953)	Tools for a holistic environment meeting social, psychological, and physical needs					
3	Kulshrestha (2006)	Revitalising decayed areas through improved accessibility, facilities, land use, and economic activities					
4	Spandou et al., (2010)		A multifaceted process for (re) developing problem areas through strategic urban managements				
5	Erbas and Erbil (2013)	Responds to deteriorated urban spaces, adapting to evolving functions and socioeconomic structures, by restructuring through demolition and redevelopment to improve living conditions and sanitation	Addressing of impacts that affect deterioration and strengthening integration to urban system		Revitalising urban textures with degraded economic and structural qualities through new planning design	Rearranging wrecked buildings, preserving original quality, and reusing old urban texture with partial renewal	Reintegrating outdated, worn-out regions into the city through comprehensive strategies for economic, social, physical, and environmental improvement
6	Yi et al., (2017)	a response to shifting expectations that abandoned and derelict urban buildings be refurbished and reused, rather than demolished					
7	Kulukshetra (2018)	Urban renewal process of rebuilding areas of the cities that have become obsolete and abandoned or are in a state of considerable decay		Integrating physical, economic, and social reconstruction to address dynamics, diversify economy, and combat environmental degradation	Applicable to dilapidated areas, with crumbling structures, declared unfit for human habitation or any other function (commercial, industrial, etc.) that needs to be demolished and rebuilt	Is a process of improving the quality of a blighted area by one or more of the following actions: partial rebuilding, repair, remodelling, renovating, conserving, providing educational, health care, and recreational facilities or augmenting utilities and services	

Fig. 16.1 Williamsburg Houses of Brooklyn, New York. *Source* United States Public Works Administration, and United States Committee on Architectural Surveys. Williamsburg houses, Brooklyn, New York City. New York, 1939. Photograph https://www.loc.gov/item/2004672462/

Fig. 16.2 Betances Houses block, New York. From the New York City Planning Commission's Plan for New York City, published 1969, https://urbanomnibus.net/2016/10/model-cities-redux/

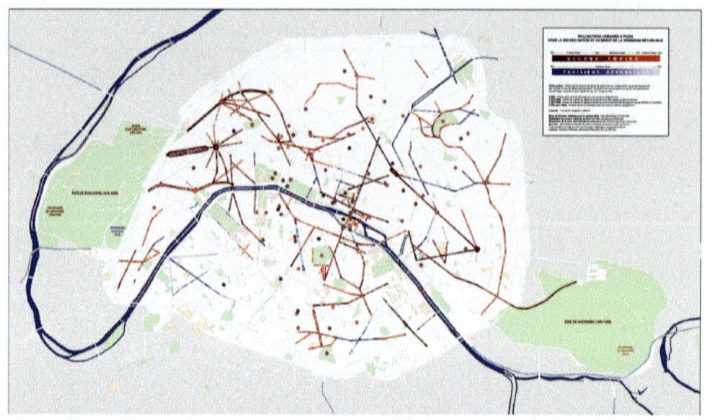

Fig. 16.3 New streets designed by Haussmann, Paris. *Source* By Dimitri Destugues—Own work, CC BY-SA 3.0, https://commons.wikimedia.org/w/index.php?curid=15714684

businesses and neighbourhoods, the interstate highway proposals were also receiving backlashes. After a period of criticism, the immediate action was to make urban renewal programs profitable, making them more attractive to private investors.

Eventually, by the 1960s, it was understood that physical renewal could not ultimately improve urban conditions. The Model Cities program of 1966 was introduced to pay greater attention to social renewal though for various reasons the program had to be shut down by 1974. Also, when the national historic preservation Act (1966) became effective in America, the federally funded projects included urban renewal programs that mainly changed the older and historic business districts and residential neighbourhoods. This technique was also used to construct interstate highways in cities. Betances Houses block released between 1970 and 1974 in Mott Haven, New York, showcased diverse approaches in design, finance, and management, blending row housing, rehabilitation, and mid-rise development for low-income communities (Fig. 16.2). By the 1970s, "Urban Revitalisation" was introduced to emphasise neighbourhood preservation and housing rehabilitation.

The history of urban renewal in Europe started in the mid-nineteenth century aiming to improve the unsanitary conditions in working-class neighbourhoods by carrying out a slum clearance program. One of history's first large-scale urban renewal projects is accredited to Paris by Haussmann under Napoleon III between 1853 and 1870 (Fig. 16.3). It resulted in the demolition of mediaeval overcrowded and unhealthy neighbourhoods as defined by officials of that time. In addition, the plan included the creation of wide avenues, new parks, and squares, along with annexing the suburbs of Paris. It also included the construction of sewers, fountains, and aqueducts within the city.

However, by the 1920s, the focus shifted to the renewal of war-damaged cities and towns, and this was the most extensive urban renewal process in history. An example would be the West German Social Housing Reconstruction program. Moreover, this loss due to war also increased people's consciousness about the historical continuum in historic urban scenes. Thus, by 1954, conservation and rehabilitation both became a part of urban renewal in the European region, and this was termed urban regeneration.

Comparing the above with the Asian country Hong Kong, urban renewal initiatives started with large-scale slum clearance in 1954, followed by a disastrous fire in one squatter area. It was the largest and most congested squatter area in Hong Kong. The squatters were cleared from all dangerous slums/sites and given temporary housing. By 1972, redevelopment of these temporary estates had begun. Temporary housing was converted to new blocks, and school and community facilities were introduced. By 1974, the in-situ development of inner-city districts was underway, while public–private partnership was introduced for carrying out the development in 1987 (Hong Kong Housing Authority, 2004).

Thus, we see that the evolution of policies regarding urban renewal (Fig. 16.4) followed a pattern that gradually evolved from a demolition and reconstruction approach to a softer and more socially oriented approach, which concentrated on the renovation of existing structures (Broudhoax, 1994).

A well-known and cited article (Naomi, 1997) analysed the history of urban renewal in western countries. The research identified three generations of deliberate interventions in urban areas in distress. It has been summarised in Table 16.2. The first generation is the bulldozer approach, which focuses on clearing slums. The second generation emphasised social goals and benefited their people, not their places. The third generation of urban renewal highlighted the public–private partnership concept, which benefited the places but not the people. As per Naomi, there is still a call for a fourth generation that combines economic growth and social equity considerations in urban renewal projects.

1.4 Part 3: Glance at the Indian Context of Urban Renewal

Urban renewal in India has a long history, with early examples seen in the rejuvenation of Old Delhi after its invasion in 1739 and the revitalisation of Mysore city in the 1870s. The urban renewal efforts in Old Delhi after British control in 1803 aimed to showcase British dominance by demolishing buildings near Redfort, introducing a tram, widening Chandini Chowk (a market square), and converting “Bheegum ka bagh” (a garden of historic importance, built during Mughal rule) into Delhi Railway Station Road. Post-independence (after 1947), the first generation of urban renewal programs focused on slum clearance. However, these efforts had limited impact. With economic reforms in 1991, urban renewal gained prominence, and the focus shifted to the redevelopment of slums in cities (Kulukshetra, 2018). Government initiatives like JNNURM and AMRUT were launched in 2005 and 2015, respectively, to allocate funds and regenerate cities. Delhi, in particular, has comprehensive plans for state initiatives, seen in the Master Plan Delhi 2021. It identifies four types of urban renewal areas as given below:

Fig. 16.4 History of urban renewal (*source* drawn by authors)

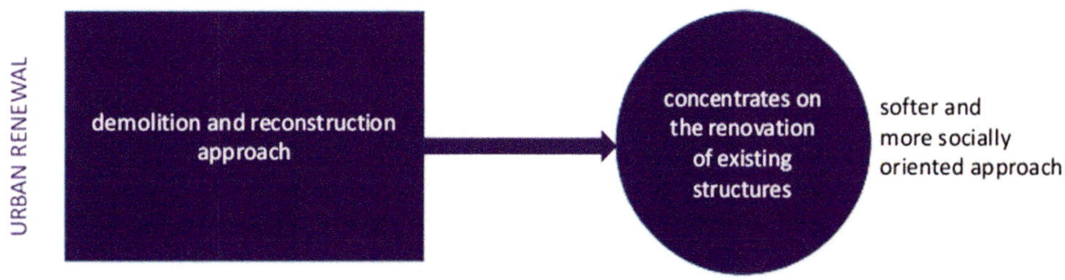

Table 16.2 Generations of urban renewal [*source* adapted from (Naomi, 1997)]

Categorised generation	Urban renewal approach highlight	Description	Benefits
First generation	Bulldozer approach	Engaged in physical solutions, eliminated slum areas, and constructed modern roads and towers in their place	–
Second generation	Project renewal	Emphasised social goals and worked for existing residents in existing housing units	Benefitted people but not the places
Third generation	Public–private partnership (PPP)	Economic considerations at the top of each municipal/urban action and developed the concept of PPP	Benefits places but not the people
Fourth generation	Combines benefits of third and fourth generation		

First Point	Particular areas compromising the traditional inner-city area
Second Point	Unplanned/unauthorised areas, which include urban villages; unauthorised colonies, and slums and clusters
Third Point	Planned blighted areas such as group housing, DDA flats, and industrial areas.
Fourth Point	Heritage areas located within the city.

Other examples of urban renewal in India include the habitat improvement project in Indore, heritage conservation in cities like Kolkata, Mysore, and Pune, and the redevelopment of railway stations by the Ministry of Railways (Verma, 2000; Jawaid & Khan, 2020). The Pradhan Mantri Awas Yojana—Urban (PMAY-U) and the Smart City Mission, both initiated in 2015, also play important roles in urban renewal efforts.

Overall, urban renewal in India has a rich history, spanning from early revitalisation efforts in Mysore and Old Delhi to current initiatives focused on heritage conservation, and infrastructure improvement, reflecting a dedication to transforming urban areas and enhancing the quality of life.

1.5 Part 4: Newer Approaches in Urban Renewal: Participatory Urban Renewal

The concept of participatory urban renewal focuses on including the persons directly involved from the beginning of the planning process. It has interwoven social, political, and capital interests (Kos & Potocknic, 2005). However, with participatory urban renewal, the needs of each project are unique, because of varying geographical location, cultural setting, economic framework, and political background. Thus, the methodology for each project/proposal would be an experimental approach ranging from public debates to public polls, and what works for one case may not work for the other.

A participatory approach does not guarantee a successful urban renewal programme. Implementing agency need to be cautious of "speculative urban development" which occurs when an individual or a group of individuals steer the urban renewal policies towards what they hope would benefit the society. However, it fails to do so. This was observed in the urban regeneration project taken up in Barcelona and urban redevelopment project conducted in Seoul (Križnik, 2018). The main features of both the projects are summarised in Table 16.3. It is to be noted that both places are characterised by strong social connections and place attachments, as well as a distinct local culture and collective identity.

Though both programs created new jobs, provided new housing and public space, and improved social amenities and infrastructure, they could not be equally beneficial for all. As a result, it negatively affected the perception of transformation among some residents. Thus, both projects had negative feedback during the implementation stage (Križnik, 2018). This can be attributed to a lack of proper citizen participation and social cohesion, the two primary dimensions of localised social sustainability. Therefore, both programs can be concluded as casualties of speculative urban development.

Specific measures as mentioned below can be taken to avoid speculative urban development and to channel the participatory approaches towards a successful urban renewal program (Kos & Potocknic, 2005).

First Point:	Primarily inhabitants need to be motivated.
Second Point:	Gaining and strengthening confidence among the inhabitants
Third Point:	Build trust between the various parties involved. (at least 80% of the directly involved inhabitants)
Fourth Point	Create an atmosphere for constructive discussion (chances are that initially the atmosphere will be tense and argumentative)
Fifth Point	Organise legal aid for solving individual problems and clearing doubts regarding communal property-legal difficulties
Sixth Point	Communication is to be always maintained between principal movers of renewal (Fig. 16.5).

Table 16.3 Comparison of urban redevelopment: Seoul and Urban Regeneration: Barcelona

Urban regeneration: Barcelona	Urban redevelopment: Seoul
Comprehensive long term: this plan integrated new development with the existing social and urban fabric	Short-term plan: the concerned locality was demolished, and a new development occurred in its place
The local government prepared the plan The local government implemented it in partnership with the private sector	Plan was prepared by local government, and residents were also involved Implementation was carried out by private corporations and supported by local government

2 Part 5: Analysing the Future of Urban Renewal with an Example: Nizamuddin Urban Renewal Project, India

The literature review has provided insights into various approaches towards urban renewal taken globally, the past approaches taken in India, and the challenges of the newer approaches. It can be concluded that the future urban renewal projects should steer away from bulldozer approach, at the same time it should make sure it is beneficial to both its people and the place. It can be drawn out that urban renewal has to address four of the critical issues, namely urban identity, environment concern, social concern, and cultural concern (Broudhoax, 1994). The newer approach in urban renewal is participatory urban renewal, and all efforts need to be put to control speculative urban development.

Analysis of one of the major urban renewal projects of India in recent years, namely "Nizamuddin urban renewal", initiated in 2007, showed that its three key objectives are: built heritage conservation, development of a city park, and socio-economic development of the residents of the area. It was a non-profit public–private partnership between Agha Khan Trust and multiple government agencies. The Agha Khan Trust for culture took care of the cultural aspects of the project area. Government agencies like Archaeological Survey of India, Central Public Works Department, and the Municipal Corporation of Delhi were the agents of action from the conception up to the implementation (Chauhan, 2020). This project laid focus on the UNESCO World Heritage Site Humayun's Tomb along with the Hazrat Nizamuddin Basti and Sunder Nursery. The Nizamuddin Basti, is an ancient area in Delhi, is centred around the dargah of Hazrat Nizamuddin Auliya, a Sufi saint known for promoting universal peace and harmony among people of different faiths. This region has had a profound cultural influence for centuries, extending across Central Asia. However, inappropriate developments in recent decades had tarnished its image as a ghetto, turning it into a neglected area associated with crime. Nizamuddin area has a dense ensemble of mediaeval period Islamic buildings. The local community here is very vibrant and has a continuous presence of 700 years.

Along with the Humayun's Tomb, The Nizamuddin urban renewal project conserves numerous built structures in the heart of Delhi, including mausoleum of Abdur Rahim Khan, Rahim kavya sankalan, Sabs Burj, Nila Gumbad, Jamaat Khana mosque, Tilangani's mausoleum, Dargah canopy, and Jahanara's gravestone. It makes use of a traditional craft-based conservation approach, restoring architectural and artistic values with skilled craftsmen and scientific evidence. It integrates local heritage and traditional crafts. These conserved mediaeval period Islamic structures act as a tourist destination, and here, the local community is involved in sharing Nizamuddin's tangible and intangible heritage with visitors and school children. Trained youth from Hazrat Nizamuddin basti serve as heritage guides, offering a unique experience and showcasing their living heritage through heritage walks, street theatre, and dastangoi performances. These guides, often students, have the opportunity to support their education or pursue a career as full-time guides. Apart from this, the Agha Khan Trust for culture celebrates Nizamuddin, by conducting various capacity building workshops, and awareness programmes to commemorate various historic figures/cultures/cuisine/music in connection to the area. Heritage awareness programmes also include celebrating special days and festivals to reach larger and often new groups. The Sundar nursery, the next focus point of the project, was originally established during the planning and construction of Imperial Delhi in the early twentieth century; it served as a place for growing and testing trees and plants for the new capital city. Today, the nursery is home to a variety of flourishing trees, some of which are rarely found elsewhere in Delhi. Additionally, it holds archaeological significance, with scattered remains of Mughal structures and nationally protected monuments.

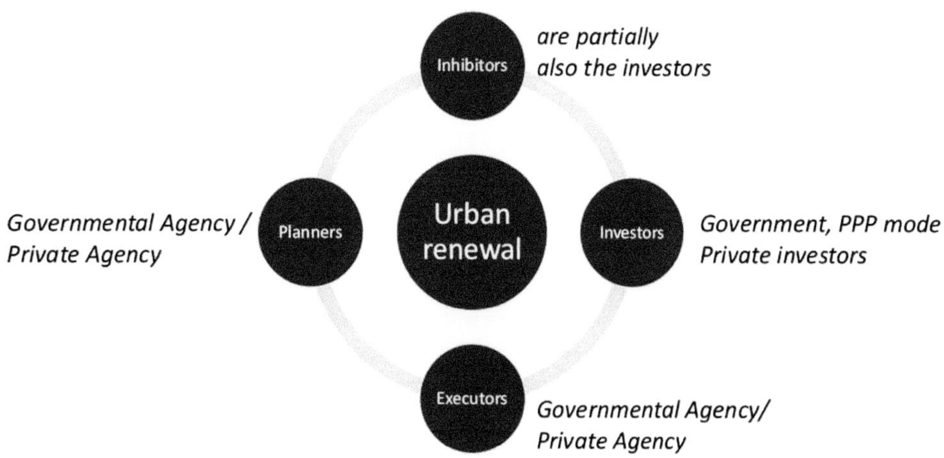

Fig. 16.5 Principal movers of any Urban Renewal Program. *Source* drawn by authors

Table 16.4 Nizamuddin urban renewal initiative analysis

Critical issues	Nizamuddin urban renewal initiative
Urban identity	The restoration of the monument precinct
Environmental concerns	Creating a Microcosm of Delhi's Original Landscape and restoring Sunder Nursery's significance in Delhi's natural history. They support a diverse ecosystem providing a rich bio-habitat on site
Social concerns	The program has, through heritage walks, street theatre, dastangoi performances—all led by youth from Hazrat Nizamuddin Basti, informed thousands of school children about the historical significance of the Nizamuddin area, its cultural and architectural heritage. The programme also aimed for education and wellbeing of children of the Nizamuddin area
Cultural concerns	Traditions at the Dargah, the famed cuisine, music, and handicraft traditions

The project also aimed to take up the concerns of improving quality of life, education, health and sanitation conditions, upliftment of infrastructures, open spaces, housing, and street conditions along with heritage and environmental concerns (Agha Khan Development Network, 2021). It is a model project addressing culture-based development of India's historic city centres. It directly contributes to seven out of 17 United Nations Sustainable Development Goals, namely goals 1, 3, 5, 10, 11, and 17 (no poverty, good health and wellbeing, quality education, gender equality, reduced inequalities, sustainable cities and communities, and partnership for the goals).

Table 16.4 summarises the analysis of the above initiative explaining how it addresses each of the four critical issues of urban identity, environment concern, social concern, and cultural concerns. It has been found to satisfactorily address each of the four critical issues.

Analysing the project, based on parameters for participatory urban renewal in Table 16.5, it has been found that all measures to avoid speculative urban development have been taken into consideration.

The project's unique approach of combining heritage conservation with urban renewal, environmental, and socioeconomic development has been recognised globally

Table 16.5 Nizamuddin urban renewal initiative analysis-participatory approach

No.	Measures to avoid speculative urban design	Case of Nizamuddin urban renewal project
1	Primarily inhabitants need to be motivated	Two years prior to the commencement of the actual projects, the discussions were initiated with community leaders on measures required to enhance the quality of life for residents of Hazrat Nizamuddin Basti. Similarly, discussions were carried out with all stakeholders on the methods to be employed for conservation needs of the project. The project started with a programme for the youth in the area to identify and map Nizamuddin's heritage and to understand the cultural legacy that they inherit. These served as motivations to get the community involved from the onset of the project
2	Gaining and strengthening confidence among the inhabitant	Including the community leaders into the initial discussions and ensuring that those discussions and agreements were captured in the Memorandum of Understanding (MoU) signed before the commencement of the program helped gain confidence among the inhabitants. The confidence was further strengthened by the implementation of various initiatives aimed at improving the quality of life for resident communities through simultaneous action towards providing education, health, sanitation, and vocational training facilities
3	Build trust between the various parties involved	The Agha Khan Trust had negotiated partnerships between diverse stakeholders, including artists' factions, residents, the local administration, and government agencies in charge of conservation. This was made possible through numerous meetings and discussions and by working in close partnership with all the stakeholders
4	Create an atmosphere for constructive discussion	Capturing the correct target audience for any discussion paves way for creating an atmosphere for constructive discussion. For example, by including health workers into a discussion with the community members, aided in the identification of major reasons for the high malnutrition among children in the area

(continued)

Table 16.5 (continued)

No.	Measures to avoid speculative urban design	Case of Nizamuddin urban renewal project
5	Organise legal aid for solving individual problems and clearing doubts regarding communal property-legal difficulties	Archaeological Survey of India (ASI) had taken up the clearing of more than 150 illegal encroachments and constructions from the project area. There were long drawn legal battles and the High court of Delhi ruled in favour of the ASI, stating that the land in question belonged to the ASI and the illegal encroachment on it should be removed (THG Publishing Pvt. Ltd., 2010). Many consultations and workshops were organised to better educate people on the legal aspects of conservation along with its ecological aspects (Agha Khan Development Network, 2021)
6	Communication is to be always maintained	The Nizamuddin Urban Renewal Initiative has pioneered a successful prototype for urban revitalisation by encouraging a collaborative public–private partnership model. Through effective communication and cooperation, this initiative paved the way for sustainable urban transformation

and has attracted some of the prestigious awards like the London-based world tourism award, Award for Excellence at the 2020 UNESCO Asia–Pacific Awards for Cultural Heritage Conservation, UNESCO's the Special Recognition for Sustainable Development, in 2020, and Outlook Group's Award for Indian Responsible Tourism Awards.

3 Conclusion

Globally, over 50% of the population lives in urban areas today. Planned urban extension along with urban renewal is a reality that will shape the urban future. Understanding how the evolution of various approaches to urban renewal shifted from a bulldozer approach to a softer and socially oriented one, can help evaluate any ongoing urban renewal projects to know where they stand. Nizamuddin urban renewal project can be taken as an example of the fourth generation of urban renewal projects which combines the considerations of economic growth and social equity that is an extension of Naomi's three generations of urban renewal. Nizamuddin urban renewal project can also be a model for participatory approach of urban renewal where all the six measures to avoid speculative urban design are adhered to, along with addressing the four critical issues of urban identity, environment concern, social concern, and cultural concerns. This also shows that urban renewal can be a tool for urban conservation.

Next generation of urban renewal projects needs to be humane and provide social justice for everyone involved by combining the benefits of its second and third generations. Urban renewal is a multifaceted process, and a thorough comprehension of the past approaches will help steer it in a direction, where a more holistic and integrated planning for the contemporary cities can be achieved. Understanding Global Trends holistically can aid its local application.

References

Project Annual Report

Agha Khan Development Network. (2021, February 2). *Empowering community: Nizamuddin Urban Renewal Initiative*. Retrieved from Nizamuddin Urban Renewal Initiative: https://www.nizamuddinrenewal.org/cultural-revival/

Research Paper

Adams, F. J. (1953). The future of cities and urban redevelopment. *American Journal of Public Health and the Nations Health, 43*(7), 930–930. https://doi.org/10.2105/ajph.43.7.930-a

Broudhoax, A.-M. (1994). *Neighbourhoodregeneration in beijing: An overview of projects implemented in the inner city since 1990*. McGill University. https://www.mcgill.ca/mchg/student/neighborhood/chapter1

Chauhan, E. (2020). Challenges to community participation in heritage tourism development: Case studies of Shahjahanabad and Nizamuddin basti in New Delhi, India. *WIT Transactions on Ecology and the Environment, 248*. https://doi.org/10.2495/ST200191

Erbas, A., & Erbil, T. (2013). Neoliberal challenges and practices of urban regeneration projects in Istanbul. *Corp. At, 4*, 20–23.

Hong Kong Housing Authority. (2004, May 28). *Memories of home, 50 years of public housing*. Retrieved from housing authority: https://www.housingauthority.gov.hk/en/aboutus/events/50yrsexhibition/highlights.html

Jawaid, M. F., & Khan, A. R. (2020). The smart city mission in India and prospects of improvement in the urban environment. *IOP Conference Series: Materials Science and Engineering, 955*(1). https://doi.org/10.1088/1757-899X/955/1/012001

Kos, D., & Potočnik, A. J. (2005). Participatory urban renewal. *Urbani Izziv, 16*(2), 141–146. http://www.jstor.org/stable/24920707

Križnik, B. (2018). Transformation of deprived urban areas and social sustainability: A Comparative Study of urban regeneration and urban redevelopment in Barcelona and Seoul. *Urbani Izziv, 29*(1), 83–95. https://doi.org/10.5379/urbani-iziv-en-2018-29-01-003

Kulshrestha, S. K. (2006). *Dictionary of urban and regional planning*. In S. K. Kulshrestha (Ed.). Kalpaz Publications.

Kulukshetra, S. K. (2018). *Urban renewal in India: Theory*. Sage Publications India Pvt Ltd.

Naomi, C. (1997). Neighbourhood regeneration: The state of the art. *Journal of Planning Education and Research, 17*, 131–144.

Spandou, M., Garcia, C., & Macário, R. (2010). Urban revitalization and transport: Local factors and driving forces from a stakeholders view. In *CITTA 3rd annual conference on planning research: bringing city form back into planning*, January 2018, 1–22.

THG Publishing Pvt. Ltd. (2010, May 24). *Encroachments cleared near Humayun's Tomb*. Retrieved from The Hindu: https://www.thehindu.com/news/cities/Delhi/Encroachments-cleared-near-Humayuns-Tomb/article16303108.ece

Verma, D. G. (2000). Indore's habitat improvement project: Success or failure? *Habitat International, 24*(1), 91–117. https://doi.org/10.1016/s0197-3975(99)00031-4

Yi, Z., Liu, G., Lang, W., Shrestha, A., & Martek, I. (2017). Strategic approaches to sustainable urban renewal in developing countries: A case study of Shenzhen, China. *Sustainability, 9*(8). https://doi.org/10.3390/su9081460

Urbanization and Homicide in Modern Cities

Miranda Rira and Julian Çota

Abstract

The twenty first century and all that is happening in the local and global arena show us that everything is changing, everything is being transformed in political, urban, historical, and cultural content. Due to globalization, technological innovation, and rapid urbanization, the morphology of cities is also changing like other institutions of society including family, state, health, environment, etc. Apart from the buildings and construction, we should also take into consideration other issues of human capital, such as deviance and crime within urban life. In this paper, we will see if there is a correlation between homicide in the family and the pressure that the modern city offers. This hypothesis is true, especially in a culture where honor is a strong cultural syndrome, traditional norms are strong, and the gender gap in victimization is quite big. We build our theory on a qualitative approach where forty-five interviews were made in different Albanian jails with inmates who have killed someone in the family. Is it urban or rural deviance? The results of the interviews will be included for analysis in this paper.

Keywords

Urbanization • Homicide • Urban crime • Honor • Traditional cultures

M. Rira (✉)
Department of Behavior and Health Sciences, Barleti University, Tirana, Albania
e-mail: m.rira@umb.edu.al

J. Çota
Department of Social Work and Civic Education, University of Elbasan, Elbasan, Albania
e-mail: julian.cota@uniel.edu.al

1 Introduction

In our modern era, cities and urban environments are rapidly and constantly changing. Apart from the buildings and construction of the cities, which are easily seen, there are also other sides of human capital, part of urban life such as mortality rates, inequality, minority issues, deviance and crime, poverty and urban violence, also known as the dark part of the cities. In his study on urban violence and crime, made in three countries, namely Brazil, El Salvador, and South Africa, Ferreira (2020), asserts that there was high inequality, high mortality rates, a high index of violence toward minorities, blacks, and other minorities living in the same urban areas. Crime is a social phenomenon that is present at all times and in all societies. There are no communities or cities that are immune of deviance, violence, and criminal affairs.

When we talk about urban life, there is so much to talk about, because city life is thriving, the labor markets are significantly changing, the socioeconomic structure is changing, and the morphology of the cities is constantly changing. Cities are living systems, transformed, and experienced by the people (Castells, 1983). Cities are not mere objects with neither life nor spirit. The material and non-material aspects of city life are intensively changing. Thus, there are two main ways how we approach city life, the cultural and the structuralist approaches. In this paper, we are going to talk about the culturalist approach (Flanagan, 2010) and see how social phenomena such as crime, deviance, and different criminal affairs are intertwined with people who live in urban areas. Marcos Ferreira reminds us that ongoing rates of violence and high rates of crimes are by no means urban issues (Ferreira, 2020). Other authors such as Park and Burgess (2019) agree that the city is not just buildings and streets, rather they saw the city as the agora of social organization and human behavior. Cities are engines of growth and innovation (ibid). Some of the questions that have engaged urban sociologists are as follows: how do

A. Battisti et al. (eds.), *Greening Our Cities: Sustainable Urbanism for a Greener Future*, Advances in Science, Technology & Innovation, https://doi.org/10.1007/978-3-031-49495-6_17

economic functions, political and cultural institutions, combine themselves in a process of urbanization? Manuel Castells in his twelve years of study raised the question of the framework of the urban site. How do class, sex, race, ethnic origins, and cultural traditions contribute to and intervene in urban scene (Castells, 1983)? Most of the researchers have done so from a dichotomy point of view by portraying macro-sociological analyses such as society and crime; city and crime, culture and crime; crime and poverty, deviance, and social class.

2 The Theoretical Approach

The framework of this paper goes back to Chicago School sociologists, who analyzed the city and its social relations, by evaluating the city as a social laboratory. If we see at the time and the place during which the Chicago school developed itself, it was during the first part of the twentieth century, from the 1920-s until the 1950-s, they were also known as the end of the Great Depression years, periods of great growth and change. The Chicago school sociologists were fascinated with the complexities of the urban community and the cultural patterns of urban life. According to them, the city is not an artifact or a residual arrangement. The city itself embodies the real nature of human life. It is an expression of human life in general and the tradition, custom, economic, industrial, and ecological factors (Park & Burgess, 2019). The city of Chicago and other American cities were going through intense industrial changes mainly because of the shift from the rural, agrarian community to the vast, heterogeneous, and urban metropolis (Lutters & Ackerman, 1996). Apart from the development of the structure of the city, some of the topics that were fascinating in the study of city life were race relations, immigrant communities, the ghettos, deviance, high rates of poverty, and criminal affairs. The essential philosophy of the Chicago School sociologists was that social problems such as suicide, mental illness, poverty, inequality, homicide, deviance, and violence were mainly and basically urban issues.

Their ecological studies estimated that different levels of deviance were linked with different zones of the city. It was especially in the disorganized social communities where crime and criminal affairs were lavished. The poorest and the most populated neighborhoods of the city, usually form "the transit zones", where unknown people come and go. In such turmoil and disorganized social environment, people become violent and deviant (Jensen, 2009). City is seen as a very mobile entity, where people come in and go out. Thus, the mobility itself becomes the spirit and the essence of urban metabolism (Park & Burgess, 2019). The sociologists of the Chicago School stressed the idea that it is neither the individual himself who is born a criminal nor other economic factors that affect him. Furthermore, it is the environment itself that plays a key role in the process of socialization. The individual is a product of the environment where he is born and grows up. The study of crime and deviant behavior has been studied by different anthropologists, sociologists, criminologists, and social psychologists. Apart from Park et al. (1967), another social theory that explains the causes of criminal behavior is the social disorganization theory (Sutherland, 1942). Sociological theories of anomie (Agnew, 1992; Dicristina, 2005; Durkheim, 1897; Kim & Pridemore, 2005; Merton, 1938; Messner & Rosenfeld, 2007; Rose, 1966; Stack, 1983) highlight the state of *anomie* as a state of human existence and the prerequisite for deviant actions and criminal affairs. The lack of laws in society creates different types of anomie, such as institutional anomie (Merton, 1938) where people cannot fulfill their goals in society because of the malfunction of the institutions. There are others who stress the cultural background of different communities. A cultural conflict is a specific aspect of social disorganization. Also, the morphology and the intensity of crime can change from one place to another, from one city to another, and from one time to another.

Criminality in urban areas is a dynamic activity. Sometimes it is even harder to identify the records of crimes because what is known as deviant in a country, it is not perceived the same in another country. Similarly, in his paper Sellin (1938) argued that it is important to think of crime as a cultural conflict. Sometimes during social cohesion, there can be a disconnection between cultural norms, hence creating a culture conflict between people whose reactions are violent responses toward others. Other authors such as Shaw and McKay (1969) believed that violence and deviance were typical for certain zones of the city.

Homicide and urban crime is not a new topic in the sociology of deviance and criminology. Different studies were made in the previous century by sociologists and criminologists to find a relationship between crime and socioeconomic conditions in rural and urban areas (Bernburg, 2002; Hirschi, 1969; Hopkins, 1975; Kim & Pridemore, 2005; Lee, 2009). Some of them agree that the causes of crime include the individual itself, some put the focus on society, and others on the environment, the stress of urban life, and the anomic state where people feel lost and they feel that they belong to no one. This is exactly where urban ecologists and sociologists of deviance meet because they stress the impact that outside factors have on the individual, especially the environment where they live and belong. For Manuel Castells, the city and the environment are at the heart of the politics, because politics shapes the city and the economy of the people. In that framework, urban and ecological dimensions are two fundamental axes of social organization (Castells, 1978). There is also a wide range of research that has found that poverty or economic reasons are a major source of homicide (Ferracuti & Wolfgang, 1962; Polanyi, 2001).

As urban ecologists agree, the city is not merely an artifact, but an organism (Wirth, 1956). In that sense, it is constantly changing and constantly growing. It is very interesting to see the city as a living organism, one that grows and changes constantly, even when we do not recognize it. As in every living community, old and new parts, cohesion and coercion, poverty and wealth, ghettos and towers, and love and violence co-habit with each other in daily urban life. Apart from the good parts, there are also dark sides of the city life. When we try to analyze cities and urban crimes, the rates and the records of criminal affairs, it is of crucial importance to also understand what is conceived as deviance and what is social control, under which the communities in the city co-habit together and groups of small and millions of people such as those in megacities co-exist and co-live together. Cities are indeed marked by the physical traces of dead and passive layered morphologies (Jensen, 2009). As Racine (2004) puts it, violence and cities are intertwined. It is almost impossible to find a city with no records of criminality or offenders.

Having the background of these urban sociologists, we will argue that there is a correlation between homicide in the family, mental health issues, and the pressure that the modern city offers. We build our theory on a qualitative approach trying to see the correlation between crime and the city. This hypothesis is true, especially in our country in Albania, a subculture where honor is a strong cultural syndrome, where traditional norms are strong, and where the gender gap in victimization is quite big.

3 What Are Deviance and Social Control?

Anthony Giddens wrote that "….no one is as normal as he thinks he is" (Giddens, 1996). Crime and deviance are two concepts that are intertwined with each other. For Franzese (2009), "the study of deviance is the study of differences". There are perceptions of deviance such as that of Clinard (2006) who wrote that: "deviant behavior is every behavior that is against the moral and ethical norms of a given society". Many sociologists think that homicide is the extreme facet of a conflict. Orru (1985) explained deviance by referring the ancient Greek notion "*anomias*" which describes the lack of law, the sinful, and wicked human nature. When people were running away from God, they were breaking the law and being deviant. The *Homo Sapiens* history, Greek mythology, folklore, and beyond, are full of cases of homicides within city life and even family members. There is evidence in evolutionary psychology that both sexes have been victims of spousal homicide. Most family homicides are spousal homicides fueled by male sexual dominance and male sexual jealousy (Wilson & Daly, 1993). Since all societies are made by humans, men and women, and their human emotions such as anger, fear, rage, and hate are the same, spousal homicide is present almost in every society. In some societies, where the rule of law is stronger and where gender roles are clearly identified, the homicide of intimate partners is less present.

We cannot ignore the culture and the history of a city, when we study deviance and criminal affairs in a given society (Rira, 2012a). Because cultural values, cultural norms, and symbols are crucial elements in understanding society (Durkheim, 1897). We would like to pinpoint the Albanian family which in our view provides a specific case study because of a strong culture of honor that derives from its five-century customary traditional laws, known as the *Kanun*. All that is known about the Balkan societies is that in some of them, there is still a strongly patriarchal society. The long-run wars between these countries with each other have developed a rigid sense of their ethnic belongings and nationalism. Later, the process of urbanization, industrialization, and surely the process of globalization transformed these societies into more complex and diverse arena of the public sphere. In "high context" cultures like Japanese, Middle Eastern, East European, and Asian South East cultures, the concept of culture serves as a barometer for social behavior (Cohen et al., 1996). Some Dutch sociologists employ the concept of *impulsive violence* (Spierenburg, 1994) which was quite common in Middle Ages. Although we believe that impulsive reactions are human reactions and they can happen during all times. As Daly and Wilson (1982) put it, the victims and offenders are usually closely acquainted and connected.

Albanian society, as other parts of the Balkans, in an ethnic-based society, from its ancestors Illyrians toward modern up to date Albanians. Through long wars and fights with its neighbors, it has a history filled with legends and heroes who die for their country and their national hymn. Their existence and their cultural identity is well prescribed in their customary traditional laws that are bequeathed from a generation to another. They are the unseen pillars which form the social control. Hence, the cultural framework is crucial when we study violence and social coercion of a given society. Modern society is, however, deeply suspicious of ritual (Durkheim, 2006).

The correlation between crime or homicide and anomie has also been emphasized in the second half of the twentieth century, especially by American sociologists and criminologists. Agnew (1992) with his strain theory of crime and delinquency focused on negative relationships with others. The individuals experience strain/stress or pressure because of the disjunction between their aspirations/expectations and their actual achievement. A culture of honor exists always, where a culture of shame exists, for the honor/shame syndrome is inseparable. The cultural anthropologists who have studied Mediterranean societies and East Asian cultures agree on the existence of a cultural syndrome that develops on the concepts of honor and shame (Cohen & Nisbett, 1994). In many cultural systems, honor is associated with men while shame is associated with women (Weidman, 2003). In the Albanian society, honor of women is linked with their sexual purity and honor killings of women happen because of honor (Rira, 2012b). "Even in the 1990s during the Balkan crises, the honor/shame complex was invoked to explain the ancient hatreds that supposedly accounted for the political and ethnic cleansing conflicts of the region" (Weidman, 2003). The American sociologist Marvin Wolfgang in his study of 588 homicides in the city of Philadelphia in 1958 found out that jealousy was the third most common reason of the domestic killings (Wolfgang, 1958).

Is honor so important for criminologists, anthropologists, and sociologists in order to understand the victimization of women or homicide? Thousands of women are killed for family "honor" each year all around the globe, which means that crime has not disappeared from modern life. It has not disappeared because it is deeply rooted in the mentality of people in patriarchal societies. The culture of honor produces honor killings, because honor is a strong moral code especially in traditional societies. In such mindsets, people are ready to die in order to put honor in place.

4 Empirical and Analytical Analyses of Crimes in the Family in Albania, 2016–2020

Crimes of violence cover a wide range of acts of violence, such as killing, rape, theft, homicide, and assault. The study of crime in Albania and in other countries as well is sometimes hindered by a lack of confidence in the completeness and accuracy of the criminal record. Usually, there is a gap in the number of crimes that happen and those that are reported at police stations. This happens because many crimes remain unrecorded or unreported or there is always a "dark area" of such crimes that is of the order of 80% (Herbert, 2004). The National Crime Survey in United States has confirmed that the overall rate of crime is larger than the reported crime index (Giddens, 1996). Especially the domestic violence is usually perceived as a private issue and often victims do not go to declare it at the police station. Thus, many crimes in the city and beyond it remain in the dark zone, they are not recorded, the victims remain victims, crimes remain crimes, and they are never brought to light. Suicide and mental health issues are other "dark zones" of urban life and modernity. If we look at it from a macro-sociological framework, suicide is one of the main causes of death of people. Hence, more than 800,000 people die globally every year by assassinating themselves or through suicide (Bilsen, 2018). During the year 2020, around 1.5 million people committed suicide, which means that every 20s, a record of suicide is happening (ibid). This is happening in urban and rural areas. In such framework, suicide can be seen as a core problem of public health that risks the new generations and their future.

If we compare Albania with other countries in Europe, we find out that Albania has a low rate of the number of suicide. According to Eurostat, Albania had in 2017, 7.5 suicides per 100,000 inhabitants, while Serbia for example our neighbor had 14.5 suicides per 100,000 inhabitants. Other countries in Europe, such as Lithuania, Slovenia, and Hungary, have higher rates of suicide, with almost 17 suicides per 100,000 inhabitants (World Health Organization, 2012).

The rate of suicide can change from one place to another, according to the psycho-social factors, according to the role of traditions, values, and religious beliefs (Diekstra, 1996). Other studies show that suicide acts happen in urban and in rural areas as well, there is no correlation between them. Durkheim asserted that many times suicide coexists with homicide (Durkheim, 2006). This has been a true statement through the years since many people who commit suicide murder someone else before they commit a suicide, especially when the victim is a family member or an intimate sexual partner.

In this part of the paper, we will look at the number of murders, and killings in the family that happened in Albania, referring to a 5-year period, from 2017 to 2021. This will give us an idea, about the dependent and independent variables, such as where murders happen more, in the urban areas or in the countryside. All the statistics are taken from the Department of Statistics, General Directorate of State Police, in Albania.

According to Table 1, the total number of evidence shows that the recorded criminal offenses are almost the same in the years 2019 and 2021, 2017, and 2018. There are urban and rural crimes as well, and thus, homicide happens in every place where humans live and interact with each other. Rural violence and its effects can be equally devastating, but in terms of scale and the existence of an urban effect, it is a problem associated with the environment of the city. Let us look at the figures during the last five years in Albania, what is the prevalence of criminal offenses referring rural or urban areas; in the study, there are taken into consideration 12 cities of Albania, which actually represent the 12 districts of it, with a population of 2.8 million inhabitants.

During 2016 year, there were

- 41 homicides in the city and
- 46 homicides in the country.

Table 1 Total criminal offences 2017–2022

Recorded criminal offences					
Criminal offences	2017	2018	2019	2020	2021
Intentional homicide	50	51	58	52	61
Attempted murder	129	114	93	114	101
Intentional crimes against health	2650	2493	2449	2521	2176
Sexual crime	113	110	155	146	136
Theft	4351	4532	4290	3909	3508
Violent robbery	169	108	85	88	85
Armed robbery	13	22	13	14	11
Theft through abuse of office	118	70	76	46	67
Fraud	1127	1040	959	876	975
Counterfeit	85	82	84	86	67
Domestic violence	1412	1374	1341	1505	1631
Drug crimes	2070	1907	1989	2086	3078
The violation of road traffic rules	5316	5713	5652	5300	5854
Other	16,714	16,852	17,966	15,993	17,425
Total	34,317	34,468	35,210	32,736	35,175

Source General Directorate of State Police

The cities that have the highest rate of homicide are Tirana, which is the capital of Albania and has the biggest population in Albania with almost 1 million inhabitants followed by Vlora, Elbasan, and Shkodra. As Fig. 1 shows, Tirana has the highest homicide rate due to its population and also because it is a metropolitan city with a high mobility of people coming in and out, especially due to internal migration of Albanians who come to Tirana to find a job and live a better life. People who leave their quiet countryside places and come to new communities in Tirana find themselves lost, and they go through the state of *anomie*, which is a prerequisite for crime and criminal behavior.

During the year 2017, there were

- 41 homicides in the city and 61 homicides in the country.

As Fig. 2 shows, the number of homicides in 2017 is almost the same with the number of homicides in 2016.

- During the year 2018, as Fig. 3 shows, there were 24 homicides in the city and 64 homicides in the country.
- During the year 2019, there were 39 homicides in the city and 50 homicides in the country.
- During 2020, there were 54 homicides in the city and 71 homicides in the country.

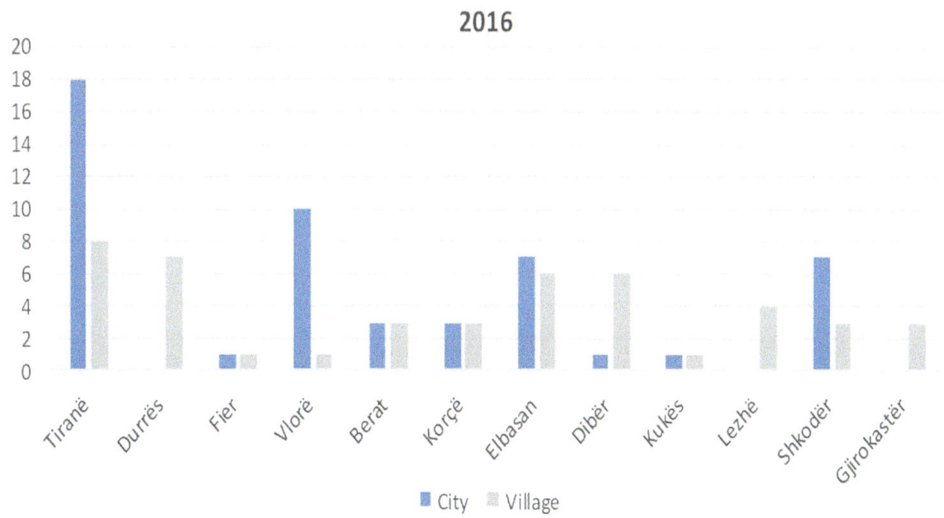

Fig. 1 Homicide in 2016, urban and rural areas. *Source* General Directorate of State Police

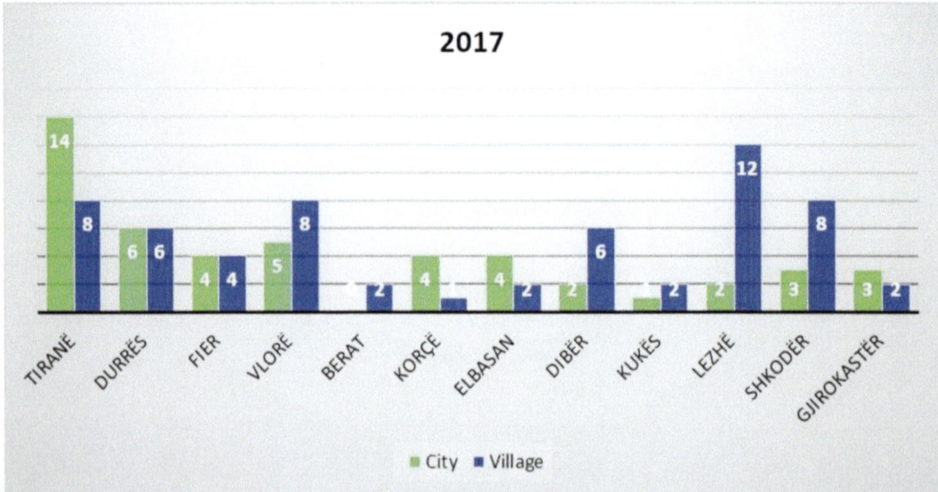

Fig. 2 Homicides in 2017, urban and rural areas. *Source* General Directorate of State Police

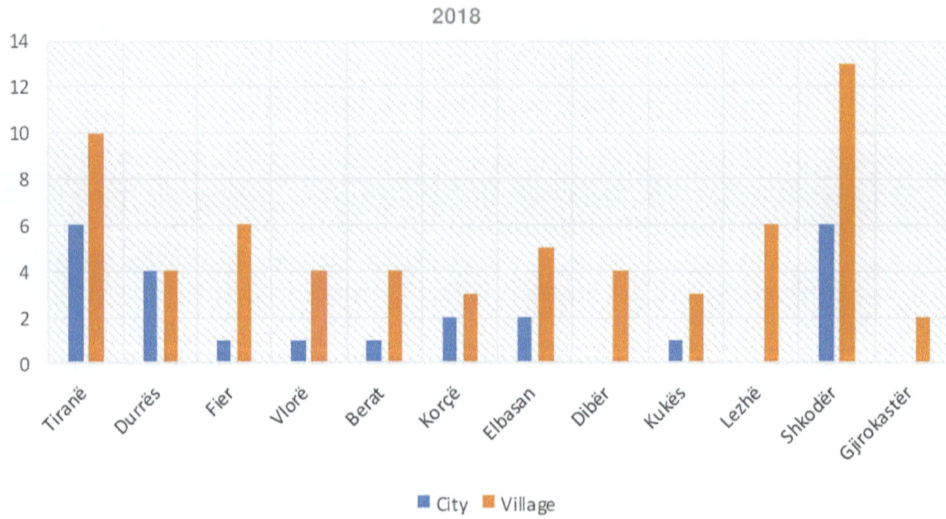

Fig. 3 Homicides in 2018 in rural and urban areas. *Source* General Directorate of State Police

As Figs. 4 and 5 show, the 12 cities (which actually represent 12 districts of Albania) taken into consideration during a 5-year period show differences in the number of the murders from a city to another city and from rural to urban areas. Furthermore, it is the percentage of crimes or murders committed in these areas compared with the population of Albania in these 5 years that gives us a better understanding of the problem. As it is shown in Fig. 6, there are more homicides in the rural areas than in the urban ones. That is a questioning point of all the theories that look at the city as "the perfect environment" for criminal affairs. Theorists who perceive the city as a social, dynamic, and living organism, bustling with life should not always pay tribute to the city as the habitat for perpetuators and deviant behavior.

The qualitative approach to criminal offenses

As we have mentioned in the abstract of this paper, we will try to see the correlation between urban life or urban environments and criminal affairs. Hence, forty-five interviews were made in different Albanian jails with inmates who have killed someone in the family. Is it urban or rural deviance? Do most crimes happen in the city or in the countryside? The inmates who were interviewed and were part of this study were suffering punishment in high-security jails in Peqin, in the "Ali Demi" jail known as jail 325, and the "Mine Peza jail known as jail 302, in Tirana. Every interview lasted from 20 to 147 min each inmate.

- The inmates (women and men) were asked about the cause for killing their husband/wife/or sexual partner that they were living with. Almost 80% of all the interviewers blame the society or the environment they live in, which imposes on them new rules that they cannot cope with. Some of them agree that they have little or no impact in their society. Their aspirations become less limited (Kim & Pridemore, 2005), and their social norms are unclear.

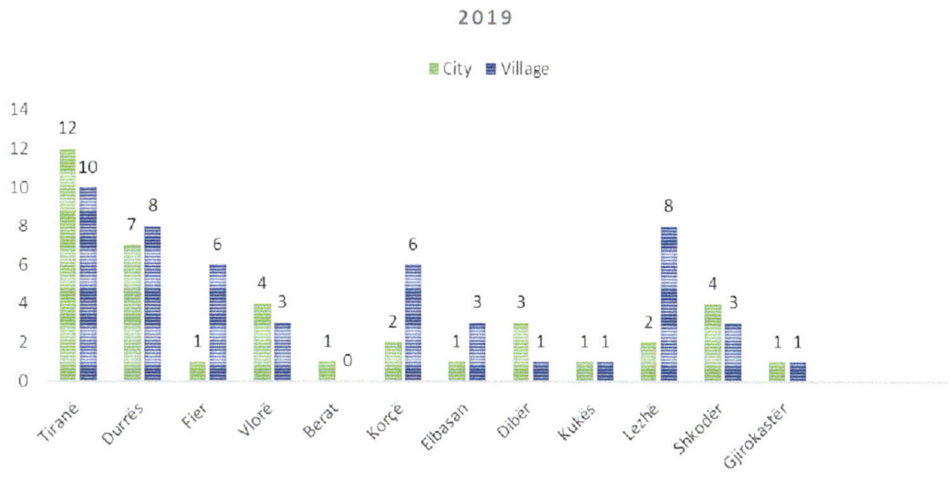

Fig. 4 Homicides in 2019 in urban and rural areas. *Source* General Directorate of State Police

Fig. 5 Number of homicides in rural and urban areas. *Source* General Directorate of State Police

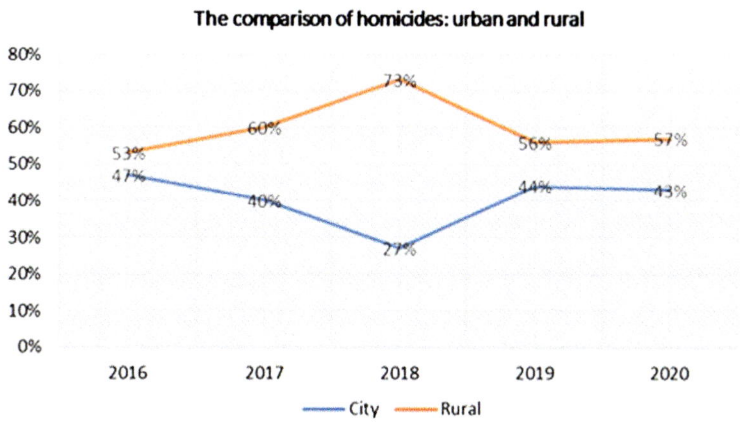

Fig. 6 Comparison of homicides: urban and rural. *Source* General Directorate of State Police

- Regarding the question of "how much city life impacts them", 27 of them thought that it is not important since the problems were inside their walls, thus explaining that domestic violence is something private that does not have to do with outside factors.
- 40 inmates out of 45 agreed that they had severe conflicts and acts of physical violence from their partners for long periods of time, before the homicide happened.
- The inmates (men) who had killed their female partners or wives said that did not repent of their acts of violence or homicide. The inmates (female) partners who had killed their male sexual partners or their husbands said that they felt very sorry for what had happened. They would never do the same thing again.
- 22 of the interviewers men and women (out of 45) when asked if urban life impacted them negatively said that

they felt a lot of pressure, because there was high inequality in living conditions, especially in big cities, where they felt lost and lonely.

If we have a look at all the interviews, we will see that it is hard to find a correlation between the victims and urban crime. In the case of the women, all the interviewers said that their economic state was horrible, and they lived in very poor neighborhoods. Some of them accused the state, the community, and the municipality, of not implying social policies in helping poor families. Unemployed men were more aggressive toward their wives and other family members. The long-lasting domestic violence (be it in the city or in the village), in most cases, ended up with homicide, suicide, familicide, and other criminal offenses.

5 Summary

Industrial city life or modern city arrangements cannot be understood if we do not look at them as living entities, that are constantly changing and being developed. Hence, the ties of this paper go back to the Chicago School sociologists who were amazed at the development of the cities, thus being real ecological laboratories, daily changing. They tried to see at the cities not just as blocks of buildings and streets but as units full of spirit and living organisms. They wrote of magic and mentality, of myth and intellectuality, of social values and social norms, which create the social organization of a given community (Park & Burgess, 2019). They focused their analyses and discussions mostly on negative issues of the cities, such as poverty, deviance, race relationships, and the ghetto. Building on that, we have tried to see one of the dark sides of city life which is crime and homicide in the city, by taking into consideration the Albanian family.

Albanian society is a transitional society, for almost three decades now. In those countries where the rule of law is weak, where the state institutions fail to reconstruct *the collective conscience* and genuine public interest, there is always a risk of growing violence in society and high prevalence of criminal affairs. This can particularly be seen in Albania, where different criminal groups have engaged in the drug industry, homicide, robbery, etc. Although, there is no clear evidence that links city size with crime or violence. There is rural and urban violence, and its effects can be equally devastating, but in terms of scale, crime is an urban problem (Herbert, 2004). Some other paradigms build on the idea that the nexus of urban crime are fueled by economic and political reasons (Reid, 2003). There are sociologists who think that crimes of violence, murder, assault, and rape are mostly urban crimes, while robbery, theft, and property crime are basically rural crimes (Giddens, 1996). Also, other theorists believe that the majority of crimes and conflicts committed in rural areas are primarily acts of social and political protest. Even crimes of burglary, horse-stealing, crop-raiding are not rural crimes, since there is an element of protest about them (Shakesheff, 2003). The World Bank (2011) in a study of 50 countries found out that there is no clear evidence between city size and violence. However, a positive correlation between the annual pace of urbanization and city homicide rates was found (Nowak, 2012). This is not always true. Cities like Mumbai grow rapidly, but they have a low homicide rate (ibid). According to Anthony Giddens, there are more reported murders each year in Detroit with a population of just over 1.5 million inhabitants than in the whole of Britain, with a population of 55 million (Giddens, 1996). The increase in the homicide rate in Albania corresponds with rapid social, economic, and political changes. There are evidences that this premise is true for other Eastern European countries like Russia that have gone through this transit period as well (Pridemore & Kim, 2007). Such changes generate major social disruption, as hundreds of people move from traditional, well-organized rural areas to urban, unknown new communities.

Western urban society has never been homogenous and equal. There is research done by different sociologists such as Bauman (2006), who proclaimed the idea that we live in a liquid world and human beings bond in a liquid reality. Bauman coined the term *"liquid"* meaning that due to the change of the character of the work and labor, society is unable to keep any constant shape (2006:139), that is why it is consistently changing. People and societies are no longer mere objects; furthermore, they are always changing, with people moving from a place to another. Societies are constantly changing, and so are their living habitats. Indeed, the image of the city in our culture has often been a dark one that of a place of fear and anger (Park et al., 1967). Cities are often perceived to be indwelled by dangerous groups of people (Remy & Voye, 1981) or as organisms under forces that are uncontrolled (Wirth, 1956). Thus, when we think about city life, we unconsciously and quite biasedly, think about violence and deviance. Violence is not usually constructed in our minds with the rural environment which most of the time we perceive as a place of nature and relax. There are other authors who do not see city or urban areas as natural geographical arrangements that promote crime and murder, rather they emphasize the grey and dark area of unrecorded/unreported criminal activities (Herbert, 2004). Furthermore, the concept of Margit Mayer, who building on Castells' urban social movement, tells us that cultural identity, local politics, and collective consumption are structured in urban life (Mayer, 2006). It is very essential to realize that there is no place, economic or political system, no human society that is immune of crime and deviance.

In this paper, we tried to see if there is a correlation between homicide in the family and urban–rural homicide.

From all the people that were interviewed, who were inmates in five different jails in Albania, they argue that they experience issues related to criminal activity, because of the society itself, with its rules and regulations, not exactly the city, as a living habitat. In the case of Albania, during the last three decades, because of the transition of the political system there have been high inequality rates and massive demographic movements of people from countryside toward the cities. These uncontrolled movements of people and the lack of social policies created a big gap between the poor and the rich, especially in the urban zones. Pridemore and Kim (2007) agree that societies that go through transitional years are more prone to crime and corruption. In one hand there is a pressing need for money, and on the other they try to cope with the new capitalist trade market rules. In such turmoil, people are more prone to violent attitudes and deviant behavior (Rira, 2012a).

According to the statistics taken from the General Directory of the Police in Albania, there is a growing number of criminal affairs that happen within the family. In 2005, there were 94 cases, and in 2015, this number goes up to 3670 cases. Also, in 2013, there were only 23 women killed, while in 2016, the number goes up to 32 and in 2021 the number is 47 women killed, all victims of domestic violence. During January–September 2017, approximately 3000 cases, victims of domestic violence and murder, were reported in police stations in Albania. The numbers are growing from one year to another, because of the rapid social and economic changes in Albanian society. This state is not sustainable for many people, and they have to fight a lot in order to cope with life. Hence, people go through an anomic state, which is a prerequisite for criminal acts and deviance. Societies in transition are characterized by immense changes that destabilize the social structure of society, and there is a growth in the number of crimes.

Crimes and deviant behavior happen in urban areas and in rural communities, in reach neighborhoods, in ghettos, in rich family members, and in poor relatives. The trade market and the capitalist culture provoke great tension for people who want to achieve economic success. Meanwhile, other institutions of society such as family, education, religion, and so forth go through the institutional anomic state, because their role in the life of the individual is quite small. Such a situation creates more freedom on one hand and more crimes on the other. Thus, people go through an individual anomic and institutional anomic state (Merton, 1938). That is when they are more prone to crime, no matter if they live in the countryside or in the urban areas.

The interviews made with inmates who have committed murders or homicides in three different jails in Albania showed us that there is no clear cut between urban crimes and rural crimes. Perpetuators and offenders come from rural and urban communities, men and women are both victims of rural or urban violence demonstrated in suicide, robbery, homicide, filicide, forcible rape, or murder. The empiric data of homicides during a five year period, for all 12 districts of Albania, showed us that the number of homicides in the rural areas is higher that the number of homicides committed in urban areas. Thus, the mindset and the theories that crime and deviance are mostly urban issues are not always true. Although, the city settlement can be more stressful for some people whose emotional and cultural ties with other people become weak. In such conditions, they experience feelings of despair, lost, *anomie,* and not belonging to anyone or anywhere. These significant human conditions are usually prerequisites for criminal affairs and aggressive behavior.

References

Agnew, R. (1992). Foundation for a general strain theory of crime and delinquency. *Criminology, 30.*

Bauman, Z. (2006). *Liquid modernity*. Polity Press.

Bernburg, J. G. (2002). Anomie, social change and crime. A theoretical examination of institutional-anomie theory. *The British Journal of Criminology, 42.*

Bilsen, J. (2018). Suicide and youth, risk factors. *Front Psychiatry, 9.*

Castells, M. (1983). *The city and the grassroots. A cross-cultural theory of urban social movements.* The University of California Press.

Castells, M. (1978). *City, class and power.* The Macmillan Press Ltd.

Clinard, M. (2006). *Criminal behavior systems: A typology*. Wildeman.

Cohen, D., & Nisbett, R. E. (1994). Self-protection and the culture of honor: Explaining southern violence. *Journal of Personality and Social Psychology, 20.*

Cohen, D., Nisbett, R., Bowdle, B., & Schwarz, N. (1996). Insult, aggression, and southern culture of honor: An 'experimental ethnography'. *Journal of Personality and Social Psychology, 70.*

Daly, M., & Wilson, M. (1982). *Homicide and Kinship. American anthropologist, New series* (Vol. 84(2)). Blackwell Publishing on behalf of the American Anthropological Association.

Diekstra, R. F. W. (1996). The epidemiology of suicide and par suicide. *Archives of Suicide Research,* 1–29.

Dicristina, B. (2005). Durkheim's latent theory of gender and homicide. *British Journal of Criminology, 46.* Published by Oxford University Press on behalf of the Centre for Crime and Justice Studies.

Durkheim, E. (1897). In S. Lukes (Ed.), *The division of labor in society*. Free Press.

Durkheim, E. (2006). *(First published in 1897) On suicide*. Penguin Classics Books.

Ferracuti, F., & Wolfgang, M. (1962). A socio-psychological study of subculture of violence. *British Criminology, 7*(3), 37.

Ferreira, M. (2020). Urban violence and crime. In *The Palgrave encyclopedia of peace and conflict studies.*

Flanagan, W. G. (2010). *Urban sociology, images and structures.* Rowman and Littlefield Publishers.

Franzese, R. J. (2009). *The sociology of deviance, differences, tradition and stigma* (2nd ed., p. 21). Charles Thomas.

Giddens, A. (1996). *Introduction to sociology* (2nd ed.). W. W. Norton & Company, Inc.

Herbert, D. (2004). Urban violence. *Journal of Geographica Helvetica, 3*, 208–216.

Hirschi, T. (1969). *Causes of delinquency*. University of California Press.

Hopkins, A. (1975). On the sociology of the criminal law. *Social Problems, 22*. Univeristy of California Press.
Jensen, O. B. (2009). Flows of meaning, cultures of movements—Urban mobility as meaningful everyday life practice. *Mobilities, 4* (1), 139–158.
Kim, S., & Pridemore, W. A. (2005). Poverty, socioeconomic change, institutional anomy and homicide. *Social Science Quaterly, 86*.
Lee, M. (2009). Concentrated poverty, race and homicide. *The Sociological Quarterly, 41*.
Lutters, G., & Ackerman, S. (1996). *An introduction to the Chicago school of sociology*. Interval Research Proprietary.
Mayer, M. (2006). Manuel Castells' the city and the grassroots. *International Journal of Urban and Regional Research, 30*.
Merton, R. K. (1938). *Social structure and anomie*. In S. H. Traub, & C. B. Little (Eds.), *Theories of deviance* (4th ed.). Peacock.
Messner, S., & Rosenfeld, R. (2007). *Crime and the American dream*. Belmont.
Nowak, M. (2012). Urban armed violence. *Research Notes, 23*.
Orru, M. (1985). *Anomie and social theory in ancient Greece*. Cambridge University Press.
Park, R. E., & Burgess, E. W. (2019). *The city. With a new foreword of Robert J. Sampson*. University of Chicago Press.
Park, R. E., Burgess, E. W., & McKenzie, R. D. (1967). *The city. Suggestions for investigation of human behavior in the urban environment*. University of Chicago Press.
Polanyi, K. (2001). *The great transformation: The political and economic origins of our time*. Beacon Press.
Pridemore, W. A., & Kim, S. (2007). Socio economic change and homicide in a transitional society. *The Sociological Quaterly, 48*.
Racine, J. B. (2004). Urban violence: A challenge for geographers? Introduction to the special issue. *Geographica Helvetica*, 59.
Reid, L. (2003). *Crime in the city. A political and economic analyses of urban crime*. LFB, Scholarly Publishing.
Remy, J., & Voye, L. (1981). *Ville, ordre et violence, formes spatiales et transaction sociale*. Presses Universitaires de France.
Rira, M. (2012a). Deviance and crime within the family in Albania. *International Researcher, 1*(3).
Rira, M. (2012b). Anomie and honor in the Albanian society. *Mediterranean Journal of Social Sciences, 3*(3).
Rose, G. (1966). Anomie and deviation—A conceptual framework for empirical studies. *The British Journal of Sociology, 17*(1).
Sellin, Th. (1938). Culture, conflict and crime. *American Journal of Sociology, 44*(1), 97–103. Published by the University of Chicago Press.
Shakesheff, T. (2003). *Rural conflict, crime and protest, Herefordshire 1800–1860*. The Boydell Press.
Shaw, C. R., & McKay, H. (1969). *Juvenile delinquency and urban areas*. The University of Chicago Press.
Sutherland, E. H. (1942). Development of the theory. In K. Schuessler (Ed.), *Edwin H. Sutherland on analyzing crime*. University of Chicago Press.
Spierenburg, P. (1994). Homicide trends and cultural meanings: Amsterdam, 1431–1816. *Journal of Social History, 27*.
Stack, S. (1983). Homicide and property crime, the relationships to Anomie. *Aggressive Behaviour, 9*.
Weidman, A. (2003). Beyond honor and shame: Performing gender in the mediterranean. *Anthropological Quarterly, 7*.
Wolfgang, M. (1958). *Patterns in criminal homicide*. The University of Pennsylvania.
Wilson, M., & Daly, M. (1993). Spousal homicide risk and estrangement. In *Violence and victims* (Vol. 08). Springer Publishing Company.
Wirth, L. (1956). *The ghetto*. The University of Chicago Press.
World Health Organization. (2012). *Global Strategy for Dengue Prevention and Control*. Geneva, Switzerland. www.who.int
World Bank. (2011). *Violence in the city. Understanding and supporting community. Responses to urban violence*. Washington DC. elibrary.**worldbank**.org.

Designing of Urban Regeneration

Design New Educational Building Energy-Efficient Shape with TS-825 Housing Standard Integrated Simulation

Ashraf Najmaldeen Hussein and Mustafa Acaroglu

Abstract

In the investigation, we try to design the newest energy-efficient educational buildings for Selçuk University Technology faculty by simulating the energy consumption in the building using Autodesk Revit BIM programs based on actual weather data of the site of the existing building to design near-zero energy educational buildings. To reach the goal of this paper, we suggested two new compact buildings enveloped (square shape with an aspect ratio of 1:1 and another basic circle cylindrical shape) to maximum reduce heat transfer from outside walls to the environment at the same volume of existing buildings of technology faculty. The second step is to select the optimum condition of envelopment constructions parameters. Select the energy-efficient envelope element from TS- 825 used in housing building codes by the Turkey government at the same high and level as the selected building model. As the result of simulations in the square envelop model saving of total air-condition load (19.304)% and about (20.76)% in heating load consumptions compared with the existing technical faculty building, but in the cylindrical envelope building model, the energy loads saved approximately equal to cubic model calculation about (18.7458)% in total air-conditioning load and (19.9)% of heating loads compared to existing building loads. At last, results of iterations by envelope construction elements (ground, walls, floor, windows, and doors) concerning TS-825 code hierarchical additional saving ratio to each other plus saving in the shape of model about (65.59) and (61.1)% for total air-condition load but the heating load saving was calculated at about (78.209) and (70.4)% in cubic and cylindrical educational building models at same floor area of technology faculty existing building.

Keywords

Educational buildings • Energy efficiency • Near-zero • Sustainable • Revit • Envelopment • Livable environment

1 Introduction

On a global scale, the issue of energy is one of the most important problems, and scientific research shows that the energy resources taken up in parallel with the rapid increase in energy consumption will be exhausted shortly. 40% of the world's energy is consumed in buildings, and it is a widely accepted reality that measures and changes to be taken in buildings will create significant energy consumption reduction. For the first time in Turkey, the scope of the Energy Efficiency Law dated 09.06.2008; in the regulation on Increasing Energy Efficiency in Transportation, which was published in the Official Gazette and entered into force on 02.05.2019, covering the years 2014–2018 (Erdede & Bektaş, 2018).

In this study, to reduce energy consumption in existing buildings on the Selçuk University campus, a new energy-efficient building design model with a similar volume and purpose was considered in the Faculty of Technology location. Firstly, at the Selçuk University Faculty of Technology, the best building envelope elements and objects selected in the energy direction suggested using two types of the same volume, rectangular 1:1 aspect ratio, and cylindrical round. It is aimed to model a new technology faculty that will be self-sufficient in terms of sufficient energy and energy consumption. The second examines the climate of

A. N. Hussein (✉) · M. Acaroglu
Technical Faculty Mechanical Engineering Department Energy Division Kampüs Konya, Selçuk University, Konya, Turkey
e-mail: falseAshrafnajmaldden@gmail.com

M. Acaroglu
e-mail: acaroglu@selcuk.edu.tr

A. Battisti et al. (eds.), *Greening Our Cities: Sustainable Urbanism for a Greener Future*,
Advances in Science, Technology & Innovation, https://doi.org/10.1007/978-3-031-49495-6_18

building regions and their location with wind breeze and solar radiation due to their location around it, the shape of the building envelope element (ground, walls, floor, windows, and doors), construction materials, and insulations. In Turkey, most published papers and investigations by researchers about energy efficiency parameters affect educational buildings.

When the simulated the green building in the Huazhong University building for the day. In the Wuhan city climate, where the building they bought consisted of 17 zooms on the ground floor and 23 zoos on the other floors, they saw a 4% decrease in total energy consumption after simulating four energy yield parameters/insulation, meat blood, roof heat transfer, and solar energy gain. (Li et al., 2021).

The first school building of the same type project in Konya as an example and made an instrumental energy analysis of passive designs in the building through the Autodesk Revit program. As a result of their study, they stated that the windows were cheerful in the case of a transition between the two plasters in the south direction8, but the wall insulation used was harmful. Finally, they recommended that the issues related to building height, heating, cooling, and ventilation be dwelt too much (Tulukcu & Murat, 2021).

Calculated the relationship between gardens and green areas around the faculty of architecture education building on the campus of Çukurova University, the city of Adana with a Mediterranean climate, which is the sixth largest city in Turkey, by calculating the ENVI-met landscape design simulation of the predicted average vote index (PMV) factor and proposing a new landscape design on campus. After taking six different sizes (0.5 * 0.5, 1 * 1, 2 * 2, 4 * 4, 5 * 5, 10 * 10) m between 5:00 and 14:00 on February 5 and August 16, they saw an average increase of 0.5 in cold hours but a decrease of 1.75 in hot hours in all trials. At the same time, two new Çukurova University Faculty of Architecture in the North, South, and Southwest have proposed landscape design around the education building (Altunkasa & Uslu, 2020).

Yaşar University calculated the total energy consumption and daylight ratio, respectively, through the Groos-shopper environment and Rhino-platform simulation program using four different angle moving shading elements with 60% window wall proportion in a south-facing classroom. Using horizontal shutters, vertical shutters, and a protruding shading device on the southern façade, the length of the first two elements varies between 0.1 and 0.3 m and the angle of v (− 90° to 90°) and the horizontal shutters whose first and second slices are restricted after the trials. As a result of the trials, the first horizontal shutters used showed maximum daylight gain and maximum total energy consumption as 77% and 5382.29 kWh/year with a length of 0.298 and an angle of − 5°. Still, the minimum daylight gain and total energy consumption were 66% and 51,422.5 kWh/year for 0.3 m and case 88 (Keskinel, 2020).

At the Istanbul Yeditepe University Faculty of Engineering campus building as an example, and after evaluating LEED, BREEAM, and SBTools certificates, they prepared the documents of the parameters with Edraw MAX varta and had them newly created. If the sample base building of the certificate created was evaluated with the LEED certificate, it was awarded the highest 110 points, but with BREEM and SBTools, it was awarded 150–270 points. Based on the three certificates they used, they designed a new certificate called YUCEC and classified them into four scoring levels (green between 40 and 50 points, platinum between 51 and 70 points, and gold between 71 and 90 points) (Gelisen & Guzelkokar, 2019).

Green building as an existing M-block education building at Bahçeşehir University and gaining LEED certification. Their research has considered the application and advantages of climate change, cost, and evaluation in green buildings, performance, and green building characteristics in existing buildings. As a result, exemplary education returns to green and sustainable building return investments, landscaping light reflective stones coverings on the roof terrace interior operations in winter reduce the temperature from 22 to 20 ° C and use of sanitary ship-owners and self-cleaning paints in the building using self-cleaning paints, it is aimed to be recognized by the LEED Certificate of the education building of Istanbul Bahçeşehir University by leveling the realization of garbage cleaning and insulation in the building in a programmatic manner (Gelisen & Guzelkokar, 2019).

Conducted a test with a Relux simulation Varta to evaluate the BREEAM certificate criteria for the natural lighting performance and building parameters according to the climate in Izmir and London City. In his study, two rooms with a volume of (3 * 5.4 * 2.7) m were used as references, and A-Model window area was used as (3 * 15) m but in B-model (1.5 * 1.5) m. After simulating the transformation of the year in two climates and the equinox days, the BREEAM certification proved that it is necessary to use 300 lx artificial lighting per year for 2000 h annually because the problem covers 60% lower lighting in university and higher education fathers. BREEAM does not take window width in natural lighting evaluation specifications; this parameter is essential (Sönmez, 2019).

Conducted an energy analysis according to the climates of five different regional cities of Turkey (Denizli, Erzurum, Konya, Samsun, and Urfa) by taking education building learning from the Ministry of National Education Type first school building project building in Konya using Revit program. As a result of his analysis, he saw that the electrical energy system died in the cities of Urfa with the highest and the lowest Samsun. Still, the highest Konya and the lowest if carbon emissions were released, and compared the use of alternative energy systems, and calculated that the highest energy survey among the five climates was achieved in the city of Konya when the wind turbine building was not suitable for the land where it was located. Finally, she calculated that the internal temperature level consumes electricity that produces 0.2–0.5% every time it increases by 1° (Tulukcu, 2019).

2 Methodology

Selçuk University (S.U.) Technology Faculty Model Demission's and Characteristic We obtained the building measurement and carnivores of the technology faculty from two sources. The first source was from the university's construction works and technical department office. We learned the seating area of the building, the shape of the internal volumes of the building, and the dimensions of the floors of the building by obtaining the original of the project. In addition to the wall, base, roof, window, and container objects used during construction.

The second source of information collection on the learned building is the Google Earth online Internet program shown in (Fig. 18.1) for the 3D shape of the building, direction, and height of the windows sitting.

2.1 S.U. Technology Faculty New Models Design Proposals.

To S.U. Technology faculty shown in (Fig. 18.2) has proposed two new building forms in the same floor area and with the same façade design distribution and considering the heat transfer in terms of heat transfer (having the lowest external wall surface areas).

2.1.1 S.U. Faculty of Technology Square Shape Proposal

Regarding energy efficiency, the construction work with the lowest external drums with a 1:1 aspect ratio within the same area was initially proposed to construct the original building facades (Fig. 18.3), shaped like the original building.

Fig. 18.1 S.U Faculty of Technology Building and location (Selcuk university 2022), (earth, 2022)

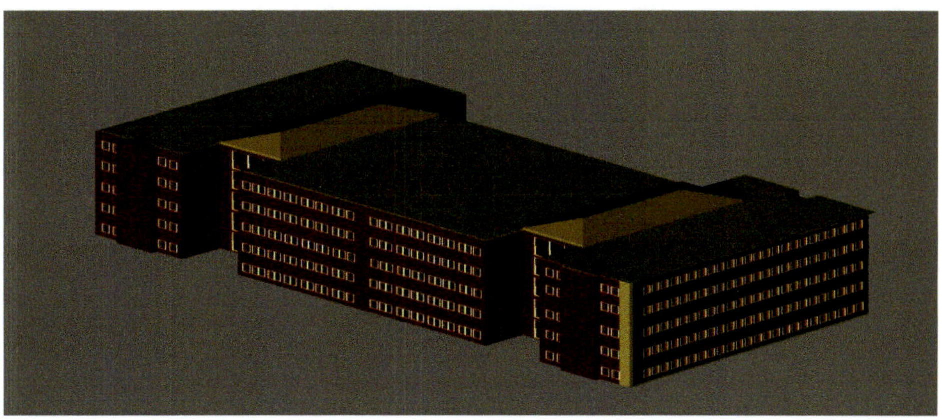

Fig. 18.2 S.U. Technology Faculty Draws 3D Sketch Autodesk by Revit. (Self-drawing by Autodesk Revit 2020) (Self, 2022)

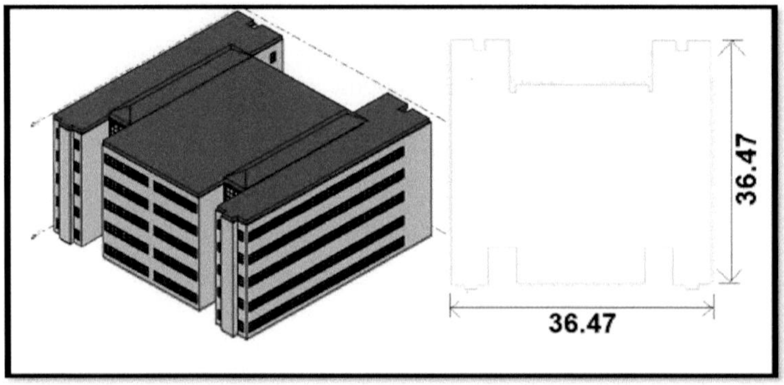

Fig. 18.3 S.U. Technology Faculty New Square 1:1 Aspect Ratio Envelope Proposal. (Self-drawing by Autodesk Revit 2020) (Self, 2022)

2.1.2 S.U. Faculty of Technology Cylindrical Shape Proposal

The second envelope sample at the same floor area with the lowest surface floor area ratio (SFAR) factor than a square shape. This is seen in the same proportional way as the surface façade measurements of the original building, which was proposed for the cylindrical envelope technology faculty building (Fig. 18.4).

3 Samples Models Used in Buildings Envelope Optimization

TS-825 for residential buildings insulation energy-efficiency standard that complies with the proposed new educational building samples example guide. It was that a dwelling building model same floors would be built in the 3rd-degree day zone of Turkey (including the city of Konya), and it was decided to reach the most efficient building envelope by experimenting with different EPS and XPS insulated base, wall, roof samples, and various primary building materials.

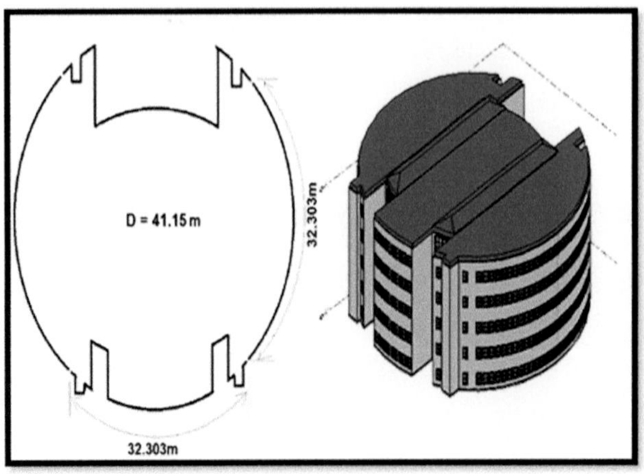

Fig. 18.4 S. U. Technology Faculty New Cylindrical Envelope Proposal. (Self-drawing by Autodesk Revit 2020) (Self, 2022)

3.1 Ground

In the newly proposed square and cylindrical building designs, energy increase was compared with the materials used in the original building by using samples whose components were determined as following (Figs. 18.5 and 18.6) and Revit family arrangement in Fig. 18.7.

3.2 Floor

In Turkey, the new roof models, components in the third-day temperature geography in the TS-825 guideline, have been made three iterations using the new roof sample with different wall layers using floor 1 and floor 2 sample models in a square and cylindrical envelope.

3.3 Windows

In the proposed new building envelope, experiments were carried out by preparing various clear windows (double and triple layers of 4 mm thickness of glasses) that are the source of visual communication to the external environment of the building.

3.4 Doors

In the old and proposed new buildings, which had glazed and glassless, two different types of doors of two building façade used.

On the southern façade, two (4 * 4) m self-opening and closing (or revolving doors) doors were used for personal/student uses and maximum provide loss of air-conditioning energies.

Two wooden doors at (1 * 2) m area were used on the northern facades for services or genuine emergencies.

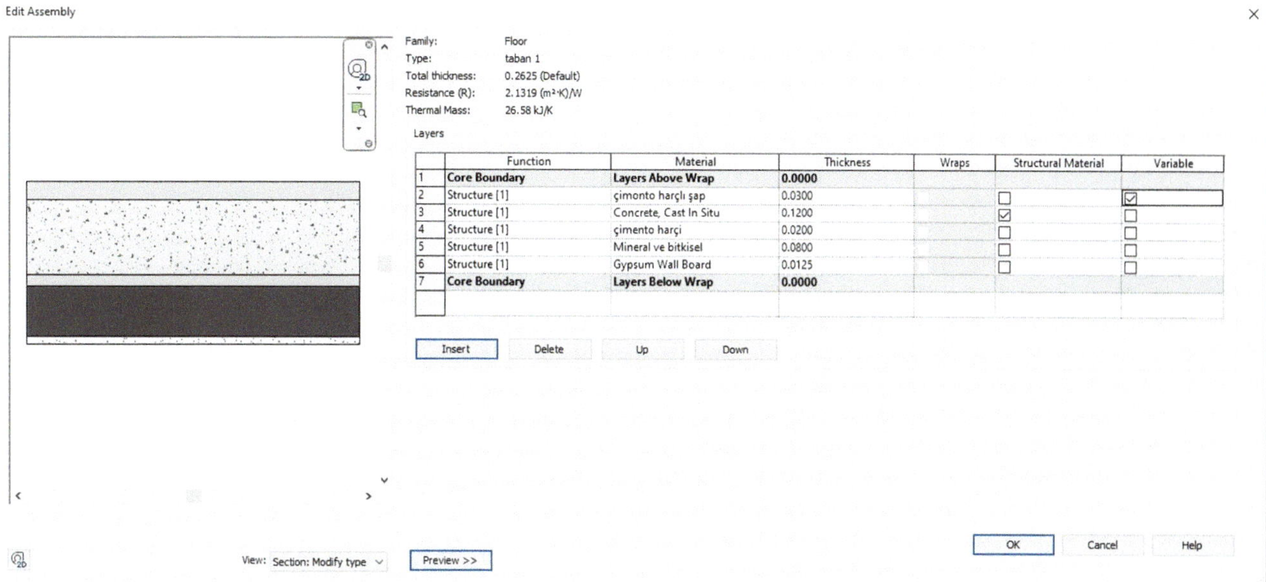

Fig. 18.5 Floor 1 samples model input to the Revit Program Families. (Self-drawing by Autodesk Revit 2020) (Self, 2022)

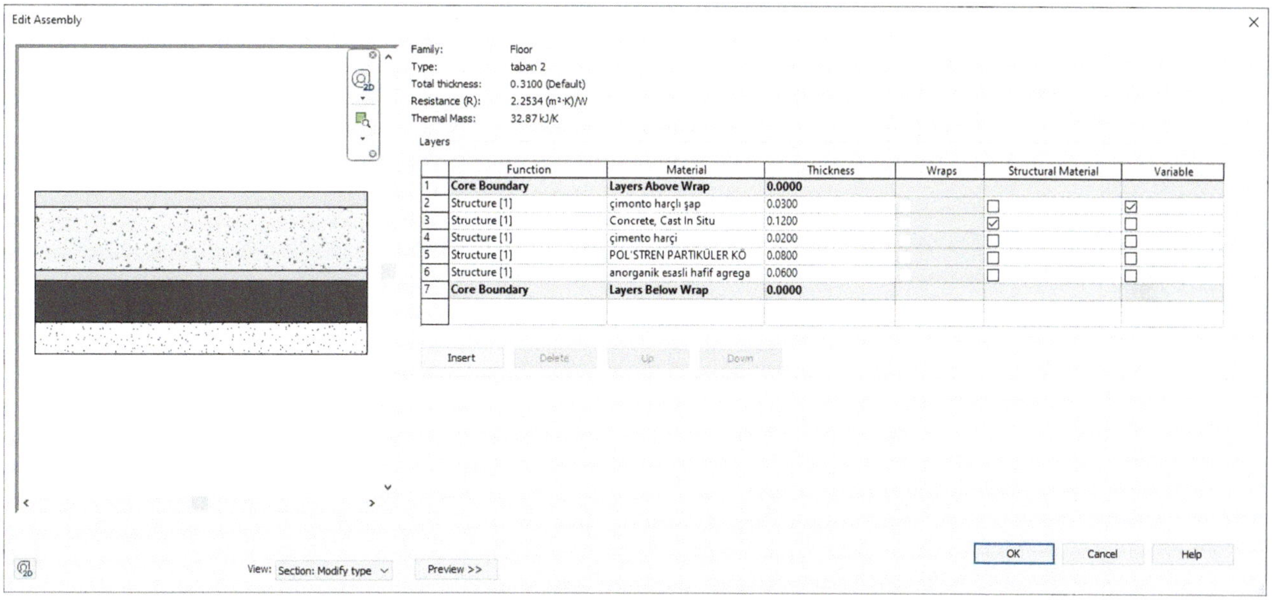

Fig. 18.6 Floor 2 samples model input to the Revit Program Families. (Self-drawing by Autodesk Revit 2020) (Self, 2022)

4 Result and Conclusion

4.1 Results of Proposal Building Envelope Shapes Analysis

The result of the analysis of the newest building envelope shapes and compared amount air- conditioning load energies (Table 18.1) with the old technology faculty loads. In the square building when saving 20.76% of the heating load and 16.714% of the cooling load, the total air-conditioning load saving reached to 19.304%.

However, when we compare the cylinder building model with the original building, the maximum heating load-saving ratio is 19.9%. The cooling load-saving ratio is 16.714%, and the total load-saving ratio nearest to the square building model saving ratio is 18.7458%.

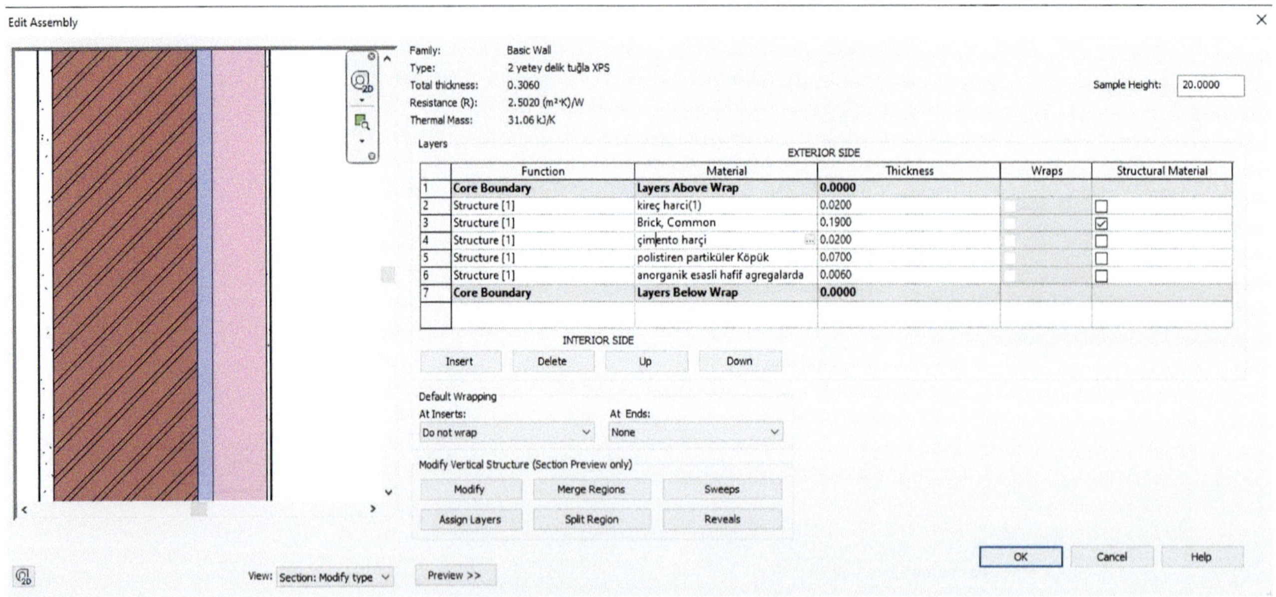

Fig. 18.7 Sample of hollow brick walls layer thickness. (Self-drawing by Autodesk Revit 2020) (Self, 2022)

Table 18.1 Old and proposal buildings maximum heating, cooling, and total air-conditioning loads

	Max. heating (W)	Max. cooling (W)	Total air-conditioning (W)
Original	407,059	228,886	635,945
Square	322,553	190,631	513,184
Cylindrical	326,054	190,629	516,683

4.2 Results of Optimization of Floor Models Simulations

The results in the simulation of the floor model samples (floor 1 and floor 2) in square and cylindrical buildings the amount of energy consumption shown in Table 18.2.

In the square shape envelope, when selecting floor 1 models obtained 26.594% energy saving in maximum heating loads and 22.612% in maximum cooling loading because of this saving ratio, the total air-conditioning energy saving ratio was calculated as 25.161%.

At the otherwise when used floor 2 sample was in the same building shapes, the maximum heating load-saving ratio increased to 35.623%, the rate of the maximum cooling load-saving ratio decreased to 14.524%, and the total maximum energy consumption for floor 2 sample in the building increased to 28.029%.

In the cylindrical shape, the load-saving ratio in the floor 1 model increased to 23.099, 22.995, and 23.062% in heating, cooling, and total air-conditioning load energies. But the saving ratio in the floor 2 sample models increased to 25.58%, 26.322%, and 25.847%, respectively. Results of optimization of wall models simulations When the new wall basic construction materials used (hollow brick and reinforced concrete) in the square building envelope and compared with the original building envelope shapes and materials (Fig. 18.8), obtained energy saving ratio of 35,495% in heating load, 24,678% in cooling load, and 31,602% in total air-conditioning load at the case of hollow brick used with floor 1 models. With the floor sample model,

Table 18.2 Amount of air-conditioning loads on floor 1 and floor 2 for news building envelope

	Max. heating (W)	Max. cooling (W)	Total air-conditioning (W)
Square			
Floor 1	298,804	177,130	475,934
Floor 2	262,051	195,642	457,693
Cylindric			
Floor 1	313,031	176,253	489,284
Floor 2	302,933	168,639	471,572

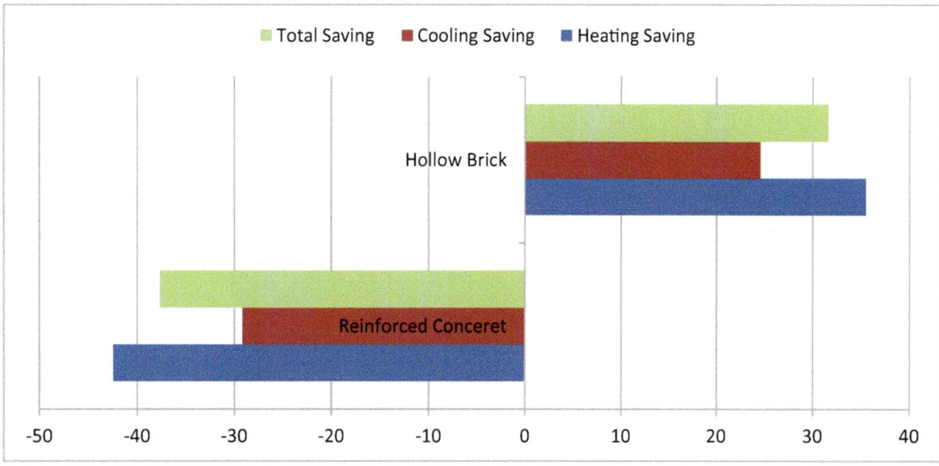

Fig. 18.8 Hollow brick and reinforced concrete saving increase/decrease in square envelopment (*source* Authors)

hollow brick contributed approximately 31.521% in the heating load savings and 24.570% in the cooling load, and then, the total air-conditioning savings ratio is equal to (31.580%). In the cylindrical envelope, when used hollow brick walls with floor 1 models, it achieved 33.492% savings ratio in heating load, 28.601% ratio in cooling load, and 31.732% savings ratio in total air-conditioning load energy. But, when using the same walls model with floor 2, the total savings ratio decreased to 29.503%, and maximum heating and cooling loads saving contributed at 31.705%–25.585%, respectively.

Secondly, use of reinforced concrete primary building materials on the walls, a total energy consumption of rare increases ratio − 37.667% with the floor 1 and 2 in the square envelope building shape. And the total air-conditioning energy consumption rate with reinforced concrete increases between (− 61.184 and 81.839)% in the cylindrical model envelope building show in Fig. 18.9.

At the second step of walls when used EPS and XPS insulation were with hollow brick models and analyses of 16 iterations between four faces in each square and cylindrical building envelope. We calculated that using insulation on all facades case in the square black building and not using insulation on east/west wall cases would have a 5% effect on the total energy consumption of the building. On the other hand, when the east or west façade does not use insulation alone (the other three walls are insulated) it has been energy consumption saving ratio of the building can 37.375% in total air-conditioning consumption savings, and the savings ratio rate increases by 42% when all walls are insulated. In all the energy analysis experiments, it has been observed that the heating loads are much higher than the cooling load's energy in the Turkey Konya city climate. In actual conditions, we see that cooling air-conditioning systems are not used in the buildings in Konya because the shading and natural or mechanical ventilation ventilators are sufficient to keep the building cool in human conditions for hot days.

4.3 Results of Optimization of New Roof Model Simulations

In analyzing the new roof sample model, a square building, both floor 1 and floor 2, for four walls facades use insulation. The total air-conditioning energy saving is increased by approximately 61%, including the wall, floor, and building shape. A maximum of 2% achieves energy savings. It has been willing to save 70% in heating energy by incorporating

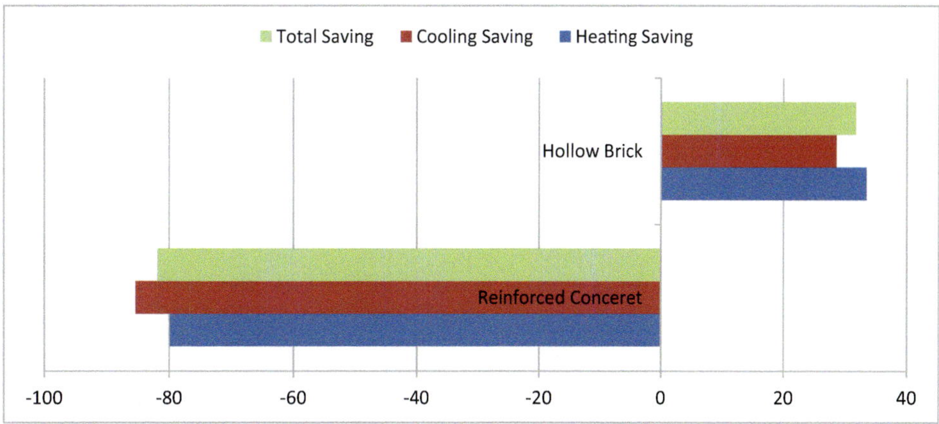

Fig. 18.9 Hollow brick and reinforced concrete saving increase\decrease in cylindrical envelope (*source* Authors)

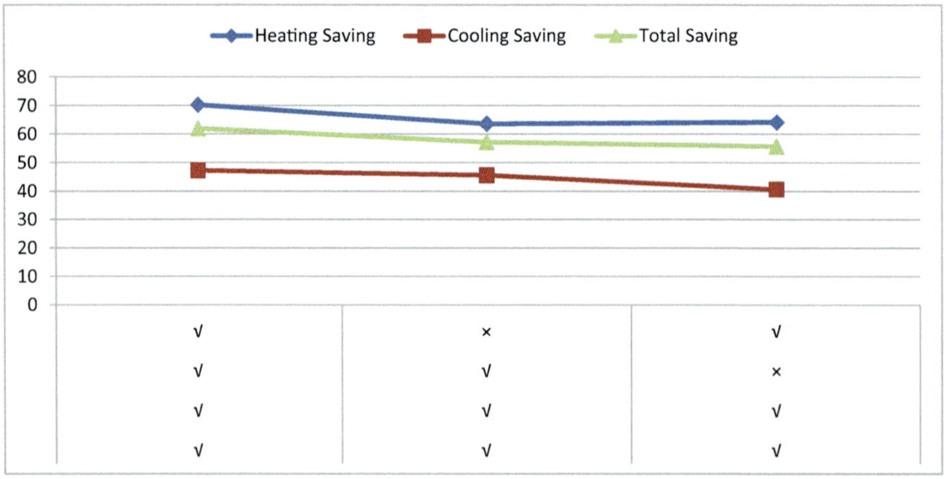

Fig. 18.10 Saving ratio for a new roof model floor 1 in square buildings (*source* Authors)

the new roof and the combined wall (Fig. 18.10), floor, and shape savings in the square shell.

Even though the same simulations in a cylindrical envelope, we see that the maximum total air-conditioning savings decreased from 61 to 58%. The savings rate was calculated at the maximum heating from 70 to 64%. At the same time, it was observed that the cylindrical building at the total walls with insulation or east or west un-insulated energy saving increased by approximately 2%. In comparison, it increased with floor 1 and approximately 5% with floor 2. Results of Optimization of Windows Glasses Layers Simulations 5 0.1 Double Glazed Windows As a result of the analysis observed that the maximum savings in the double-glazed windows used in the square building model were about (57.533–52.748)% of the total air-condition load and 'its equal of − 4.11% dissipated of maximums optimization of all envelopes opaque part in the total energy of the building. The reason for this increase in energy consumption is that when double glasses are used in the building, it provides a 5.333% savings ratio in heating energy, but on the other hand, -a 20.763% consumption ratio increases in building cooling loads.

In square models with floor 1 in the building without calculated cooling load capacity or without using cooling air-conditioning systems (Table 18.3), the total energy savings in the heating load can be achieved at 75.05% of air-conditioning loads. With floor 2 the maximum savings increased from 57.553 to 59.894%, with a decrease of − 2.992%, relative to floor 1 models, resulting in the total heating saving ratio increasing to 76.009%.

When using double glasses with floor 1 in a cylindrical building, the total cooling load energy consumption (Table 18.4) increases from (31.093) to (37.783)%, and the total air-conditioning savings ratio between (57.533 and 52.748)% decreases. And in the floor 2 models, the energy consumption ratio is increased by about 2.916% when compared with square building envelopes.

When triple glass is used in the square building, it saves 2.832% more energy than the total envelope elements show in Table 18.5. The total air-conditioning energy saving is about (64,822–58,140)%. In the heating load simulations, the maximum savings are calculated by 75.05% in the double-glazing uses, but in triple glazed, the savings increase to 77.249% in the same floor.

Table 18.3 Air-conditioning loads for double glassed windows using in square building envelope

	South	East	West	Max. heating (W)	Max. cooling (W)	Total air-conditioning (W)
Floor 1						
√	√	√	√	101,562	168,375	269,937
√	√	√	×	127,769	157,718	285,487
√	√	×	√	129,939	170,557	300,496
Floor 2						
√	√	√	√	97,656	157,393	255,049
√	√	√	×	124,584	156,963	281,547
√	√	×	√	126,702	169,820	296,522

Table 18.4 Air-conditioning loads for double glassed windows in cylindrical building envelope

North	South	East	West	Max. heating (W)	Max. cooling (W)	Total air-conditioning (W)
Floor 1						
√	√	√	√	132,762	131,744	264,506
√	√	√	×	142,280	142,570	284,850
√	√	×	√	141,293	142,406	283,699
Floor 2						
√	√	√	√	138,557	135,038	273,595
√	√	√	×	157,727	136,670	294,397
√	√	×	√	155,288	143,843	299,131

Table 18.5 Air-conditioning loads for triple glassed windows using in square building envelope

North	South	East	West	Max. heating (W)	Max. cooling (W)	Total air-conditioning (W)
Floor 1						
√	√	√	√	92,609	131,106	223,715
√	√	√	×	125,255	105,936	231,191
√	√	×	√	120,987	145,222	266,209
Floor 2						
√	√	√	√	88,704	130,123	218,827
√	√	√	×	120,035	131,957	251,992
√	√	×	√	117,749	144,485	262,234

Table 18.6 Air-conditioning loads for triple glassed windows in cylindrical building envelope

North	South	East	West	Max. heating (W)	Max. cooling (W)	Total air-conditioning (W)
Floor 1						
√	√	√	√	120,155	126,966	247,121
√	√	√	×	145,723	138,150	283,873
√	√	×	√	143,620	138,566	282,186
Floor 2						
√	√	√	√	120,279	127,011	247,290
√	√	√	×	147,041	139,502	286,543
√	√	×	√	144,823	139,478	284,301

Finally, the minor savings in using triple glass in the square building envelope (Table 18.6) were calculated as 0.73% in the un-insulated eastern facades.

5 Summary

Proposal square and cylindrical envelope shape more energy efficient than existing long rectangular shapes of technology faculty. Square building envelope total air-conditioning energy saving about 19.304%, and the new cylindrical building envelope about 18.7458% compared with an existing educational building. Using the TS-825 housing energy standard for Turkey proves efficiency in educational building at hierarchical addition simulation of building envelope optimizations at a total saving ratio of 65.59% and 61.6% for square and cylindrical building envelopes, respectively.

References

Altunkasa, C., & Uslu, C. (2020). Use of outdoor microclimate simulation maps for a planting design to improve thermal comfort. *Sustainable Cities and Society, 57*, 102137.

Earth, G. (2022). *Selcuk university teknology faculty, Konya, Turkey.* Konya. https://www.google.com/maps/uv?pb=!1s0x14d08d15e76415

db%3A0x57050886b797f0ee!3m1!7e115!4s%2Fmaps%2Fplace%2Fselcuk%2Buniversity%2Bteknolgy%2Bfakulty%2F%4038.028333%2C32.5096882%2C3a%2C75y%2C315.54h%2C90t%2Fdata%3D*213m4*211e1*213m2*211sGKOT5jUVi8tyRGiGBT_YGg*212e0*214m2*213m1*211s0x14d08d15e76415db%3A0x57050886b797f0ee%3Fsa%3DX%26ved%3D2ahUKEwjY8PWyhqCBAxVUcfEDHVVmCRcQpx96BAg6EAA!5sselcuk%20university%20technology%20faculty%20-%20%D8%A8%D8%AD%D8%AB%20Google%E2%80%8F!15sCgIgAQ&imagekey=!1e2!2sGKOT5jUVi8tyRGiGBT_YGg&hl=ar&sa=X&ved=2ahUKEwjY8PWyhqCBAxVUcfEDHVVmCRcQpx96BAhMEAw

Erdede, S. B., & Bektaş, S. (2018). Türkiye için yeşil bina sertifika sistemi gerekliliği. In *2nd international symposium on innovative approaches in scientific studies*, November.

Gelisen, G., & Guzelkokar, O. (2019). Mevcut yapıların sürdürülebilir yeşil binalara dönüştürülmesi. *Ulusal Çevre Bilimleri Araştırma Dergisi, 2*(2), 76–90.

Keskinel, Y. (2020). *Performance of movable façade panels in terms of dynamic daylight metrics for a classroom* Izmir. Institute of Technology.

Li, Q., Zhang, L., Zhang, L., & Wu, X. (2021). Optimizing energy efficiency and thermal comfort in building green retrofit. *Energy, 237*, 121509.

Selcuk University, B. C. U. (2022). *The data and dimesion of Teknology College in Selcuk University*.

Self, D. I. (2022). *Drawing of Selcuk university teknology faculty by autodesk revit 2020* (901–49876657) [Drawing an simulation]. https://accounts.autodesk.com/logon?resume=%2Fas%2FwPOrLvQTMT%2Fresume%2Fas%2Fauthorization.ping&spentity=null#username

Sönmez, İ. (2019). *Testing room and window design parameters for daylight performance according to BREEAM assessment criteria: Cases of London and İzmir*. Izmir Institute of Technology.

Tulukcu, D. A. (2019). *Tip ilköğretim okulu projesinin enerji etkinlik performansının iklimsel veri değişkenleri üzerinden sorgulanması*. Konya Teknik Üniversitesi

Tulukcu, D. A., & Murat, O. (2021). Örnek Bir Binanın Isıl ve Çevresel Performansının Autodesk Revit Simülasyon Programı ile Analizi. *Avrupa Bilim ve Teknoloji Dergisi, 23*, 197–206.

Examining the Impact of the Built Environment on Walkability and Physical Activity Among the Disadvantaged Population

Ingy M. Rostom, Shaimaa M. Kamel, and Laila M. Khodeir

Abstract

The term "walkability" was coined to explain the effect that the built environment has on people's propensity to walk about and be physically active. Walkability has been used to evaluate the physical aspects of walking facilities by comparing them to pedestrian perception. The evaluation result can then be used to directly address the concerns of pedestrians. Despite recent advances in research, the walkability notion still doesn't account for the requirements of disadvantaged pedestrians. The gaps in research have been realized through anomalies in research findings which suggest deficiencies that failed to accommodate for the needs and perceptions of all pedestrians. This paper is a systematic structural review of the research on the walking behaviors, physical activity, and perceptions of the built environment for several disadvantaged groups, namely the elderly, the physically challenged, low-income, and racial minorities. A systematic search of the literature was conducted using Scopus search engine. Combinations of the keywords; built environment, walkability, age, disabilities, and income/cultural factors were used. A qualitative approach was used to analyze all evidence. Three disadvantaged population groups were identified as meeting the inclusion criteria. Identifying the factors that influence neighborhood satisfaction for the disadvantaged will assist with planning and design of urban developments from a holistic view, enhancing quality and appeal for all social groups to help ensure a lower resident turnover rate. This review helps to identify groups that are marginalized in urban design and assessment methods as a step to design more holistic urban environments that include everyone.

I. M. Rostom (✉) · S. M. Kamel · L. M. Khodeir
Department of Architecture, Faculty of Engineering, Ain Shams University, Cairo, Egypt
e-mail: engy.elgazar@eng.asu.edu.eg

S. M. Kamel
e-mail: shaimaa_kamel@eng.asu.edu.eg

Keywords

Built environment • Walkability • Physical activity • Disadvantaged population • Neighborhood satisfaction

1 Introduction

An environment that is "walkable" is one that welcomes and facilitates pedestrian traffic. It is now generally accepted that the ability to walk to one's destination is a crucial factor in accomplishing health and environmental objectives. According to the sustainability literature, walkability can lessen the carbon impact, lower traffic, and reduce fuel use (Ewing et al., 2008), all of which are connected to the aim of an "Active Community" (Doyle et al., 2006).

Important built environment features have been connected to mobility, physical activity, and related health impacts such as decreased levels of obesity, diabetes, and heart disease (Hardman & Stensel, 2009; UN, 2015). This body of work fails to account for the possibility that marginalized groups' perceptions of the impact of built environments on walkability and activity patterns differ from the norm. Simple and conventional walkability measurements like "walking speed" or "accessibility" will have radically different implications for a "normal" pedestrian versus a pedestrian in a wheelchair or the elderly strolling with a cane.

Studies have shown that the built environment (BE) is a major factor in determining whether a neighborhood is walkable, but other studies have shown that an individual's unique traits also have a part in how they perceive their environment and how happy they are, which in turn determines their propensity to engage in physical activity. Attempts to quantify walkability by isolating the effects of individual environmental factors may result in misleading conclusions.

What people think about the built environment's overall quality has a significant impact on how walkable a place is.

A. Battisti et al. (eds.), *Greening Our Cities: Sustainable Urbanism for a Greener Future*,
Advances in Science, Technology & Innovation, https://doi.org/10.1007/978-3-031-49495-6_19

Given the gaps in data on pedestrian characteristics, local authorities often struggle to meet the needs of their constituents by building and maintaining adequate pedestrian infrastructure. The purpose of this review is to determine which aspects of walkability are most important for senior pedestrians, the physically impaired, and the culturally and economically disadvantaged populations. By identifying these features, future urban planning may be made to better accommodate and please a wide variety of pedestrian populations.

2 Disadvantaged Population Groups

In this section, different groups of the disadvantaged population are identified; stratified by reasons of disadvantage, and reviews of research and experiments on their specific cases are discussed. Later, in the following section, these reviews are further analyzed and discussed.

2.1 Disadvantaged by Age

The United Nations (UN, 2015), the World Health Organization (WHO, 2002), and other such organizations advocate for the "age-friendly city," "livable city," and other similar concepts to improve the quality of life of the elderly by trying to regulate their living environment in a way that permits them to stay active and independent. One of the main tenets of the "livable cities" concept is making sure that people of all ages, including seniors, have easy access to the resources they need to live independently and with dignity in a community environment. All people, but notably the elderly, benefit greatly from physical environments that are accommodating to their ages. Isolation, melancholy, poor health, and mobility concerns are magnified in elderly people who live in unsafe environments or in areas with numerous physical impediments (Aksoy & Korkmaz-Yaylagul, 2019).

The World Health Organization (WHO, 2002) has established three requirements for "active aging:" involvement, health, and safety. Because of this pattern on a global scale, urban planners are shifting their attention to these areas. The needs of an aging society must be factored into any urban planning paradigm that aims to be inclusive. As the population ages, falls pose an increasing threat in terms of the number of people hurt and the money spent on their care. Environmental dangers like slippery or irregular walking surfaces, or an absence of handrails, greatly increase the likelihood of a fall. Concerns about personal safety have a direct impact on seniors' sense of happiness and, by extension, their appreciation for where they live. The world's aging population and the trend toward urbanization must be factored into a planning model for cities if the seniors are to be able to maintain their level of social engagement and productivity into old life (Bookman, 2008).

When thinking about the elderly's capacity to walk safely, walking speed is an important but often overlooked element. Crosswalks can be helpful for pedestrians who need to get across busy streets or highways. It is important to understand how the built environment interacts with individuals of different ages and abilities because crossing distances or crossing timings computed using the average walking pace among healthy individuals may not be suitable for other demographic groups (Lachapelle & Cloutier, 2017).

In a study done in Winnipeg, Canada (Arango & Montufar, 2008), researchers compared normal and crossing walking speeds of senior pedestrians with and without assistive equipment to see whether the use of such devices slowed their walking pace. Regular walking speed was defined as the pace at which individuals move about their daily lives unconstrained by the need to cross a signalized junction, whereas crossing walking speed was defined as the rate at which pedestrians move across that particular intersection. Figures 1a and b depict the overall ranges of regular and crossing walking speed of senior pedestrians using mobility aids at signalized junctions. U.S. and Canadian MUTCDs (Manual on Uniform Traffic Control Devices for Streets and Highways) (FHWA, 2003) indicate a design threshold of 1.2 m/s, although less than 5% of this population achieves that speed.

The study suggests that:

- Senior pedestrians who rely on mobility aids walk at an average of 0.78 m per second (m/s) at signalized intersections and 0.75 m/s (m/s) along the road segment.
- Over 95% of senior pedestrians who rely on mobility aids would be left out of the planning process if the proposed design value of 1.2 m/s is adopted, as suggested in the most recent Canadian and US MUTCDs [10]. With the proposed change to the US MUTCD, the walking speed is reduced to 0.91 m/s, reducing the value to roughly 85%.

2.2 Disadvantaged by Physical and Visual Challenges

It is possible for people to suffer stress on both a physical and mental level in extremely demanding and restrictive circumstances because they aren't adapted to them or aren't in harmony with them. Individuals with disabilities are sometimes disregarded owing to a lack of data on their road walking behavioral patterns, despite the fact that it is vital to construct walking facilities to fit the needs of all walkers.

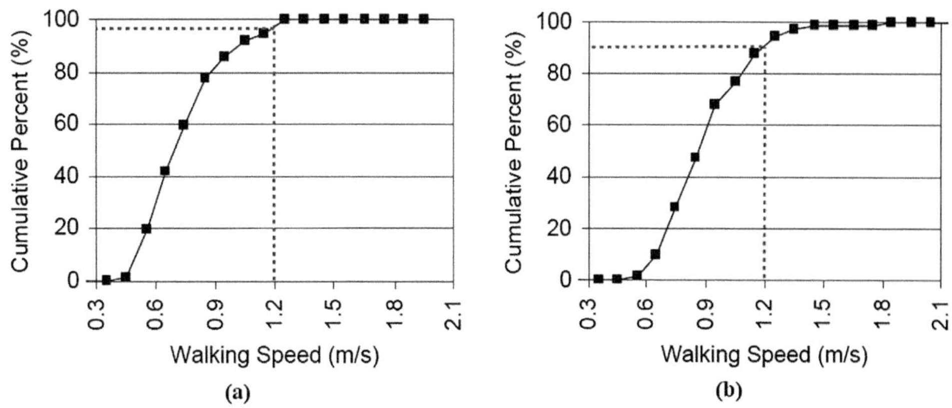

Fig. 1 Walking speeds of senior pedestrians who use assistive walking equipment for mobility at signalized intersections: **a** regular speed and **b** crossing speed (1 ft/s = 0.305 m/s). *Source* Arango and Montufar (2008)

Only a modest number of research have examined the movement patterns of people with disabilities. The street-crossing habits of people with impairments were studied by Pecchini and Giuliani (2015). In another study, thirty people with visual impairments were put through a walking speed test by Cook et al. (1999). And in order to characterize how individuals behave when they see someone in a wheelchair in a corridor of varied lengths, Miyazaki et al. (2003) performed a series of studies involving 30 people and one person in a wheelchair. Daamen and Hoogendoorn (2010) conducted an experiment in the Netherlands to evaluate the capability of entrances for pedestrians of varying ages and abilities.

Research literature study reveals that the vast majority of studies have concentrated on the evacuation behavior of people with impairments, while just a few publications have addressed their capacity to deal with constructed surroundings in crowded settings. Utah State University (USU) conducted many large-scale controlled pedestrian behavior tests with impaired participants to fill this need (Sharifi et al., 2017). The purpose of the research was to compare the reported and actual pedestrian walking behaviors of persons with impairments in different situations (e.g., straight, oblique, narrow, and steep). Researchers in this study compared the movement patterns of adults with and without physical limitations.

The results showed that people with disabilities moved substantially more slowly than non-disabled peers in all types of walking environments. This demonstrates how design ideas based on a person's typical walking speed may ignore the needs of fragile walkers. More importantly, the statistics show that the intricacy of walking facilities impacted people's pace. Space limitations and intricate turning procedures can significantly reduce walking speed, as all subjects experienced in bottleneck and right-angle facilities.

Clark-Carter et al. (1986) also observed that the intricacy of the built environment significantly influenced the walking rate of those with impaired vision, and these results were found to be consistent with their findings. In order to accommodate the needs of several demographic subsets, complex-geometry walking infrastructures may need to be expanded.

Similarly, Frazila conducted a study into the physical aspects of walkability from the perspective of the visually impaired (Frazila & Zukhruf, 2018). Perspectives of blind pedestrians were used as a basis for the study since they are grounded in the real world. Therefore, the findings from the physical features might be used to evaluate the necessary walking facilities for the visually impaired. The research opens with a detailed discussion of the unique characteristics of blind people's mobility. The characteristics have to do with how much time may be spent walking, how people act while strolling in groups, and what kinds of transportation are available. Figure 2 shows how those who are blind and those who are not have different estimates of how long it will take them to complete a walk. Most blind pedestrians can only walk for around 10 min, while non-blind pedestrians have an average walking duration of 20 min. Further, sighted pedestrians estimate they can cover the maximum distance in 80 min, but blind pedestrians estimate they would take around half as long. In order to go about, many opted for the public transit system. Meanwhile, despite its drawbacks as a public transit option, walking trips have only increased in popularity.

Next, the study summarizes the necessary walkability features, which are derived from the basic amenities for visually impaired pedestrians. The results indicate that:

- Some of the most important infrastructure features are associated with safety and security.
- Walking aids for the visually impaired, such as tactile pavement and braille information, are essential.
- The blind pedestrian is not concerned with the full width of the sidewalk, but rather with the effective width, and the greater this width, the more satisfied the blind pedestrian is likely to be.

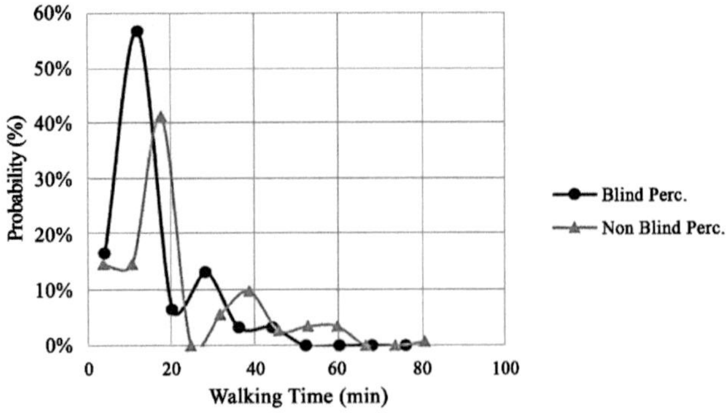

Fig. 2 Possible walking time perception of blind and non-blind pedestrians. *Source* Frazila & Zukhruf (2018)

- The convenience of the environment for blind pedestrians was shown to be correlated with the condition of the tactile pavement, perhaps as a result of the blind pedestrians' habit of using the tactile pavement as a navigational aid.
- The study focused primarily on pedestrian crossings since most traffic signals in the study region lacked the necessary sound equipment. A positive correlation was shown between the availability of zebra crossings and people's reports of how simple it was to use them.
- Blind commuters will face new difficulties as a result of changes to the sidewalk layout. There is a negative association between the number of intermediates, which causes the walker to worry about the shift in sidewalk level, and the perception of blind pedestrians.

In conclusion, unsuitable walking facilities contribute to the problem of dependent living, as well as feelings of loneliness and worry, and hence they are two of the most pressing issues facing individuals with disabilities today.

2.3 Disadvantaged by Economic and Cultural Factors

Although there is some evidence that socioeconomic disadvantages, such as low income and cultural minority status, are related to lower levels of walkability, only a few studies have been conducted to recognize this link. Income has a role in how much and how often people travel, and how their households choose to get about. Families struggling financially are less likely to own vehicles and more likely to use modes of transportation other than private autos. It is difficult to pinpoint the specific factors that influence low-income people's mobility.

While more people than ever before want to live in walkable urban communities, those who are already marginalized confront significant obstacles to doing so. They are also at risk of being uprooted due to prejudice in planning procedures, marginalization in infrastructure and basic amenities, and physical exclusion. Ethnic minorities and those of lower socioeconomic status are disproportionately affected by lifestyle-related disorders such as heart disease and weight-related disorders (Hardman & Stensel, 2009). Unfortunately, low-income and racial minorities have greater rates of pedestrian injury and death than more wealthy populations, despite widespread promotion of the advantages of walking. In the United States, males of color had a mortality risk for pedestrians that is two times that of white males (CDC, 2013).

Comprehensive study of the associations between built environments and walkability, segregated by socioeconomic contexts (Adkins et al., 2017), reveals that the built environment impacts walkability and physical activity less for marginalized groups than for more privileged populations. Residents in supportive contexts achieved higher walkability and physical activity rates than those in non-supportive contexts. Figure 3 shows how the environment influences the amount of walking done in both unsupportive and supportive environments. The built environment effects of physical exercise and recreational walking are less than those of walking for transportation (Sundquist et al., 2011), but the pattern of a larger environmental effect for privileged groups remains consistent.

The likelihood that low-income people will walk or be physically active may be higher than expected in built environments that are not optimized for walking due to two factors. To start, two major issues limit the choices available to vulnerable populations: Fewer people owning cars which means more reliance on mobility and active transportation (Lachapelle et al., 2015); and the difficulty of moving to a more walkable neighborhood, specially where there is a high demand causing higher market prices (Tremoulet et al., 2016). Secondly, certain low-income neighborhoods may make up for some of environmental flaws with more engaging social environments (Clark & Scott, 2013).

Based on the results of a study associating rates of mobility to emission exposure in Vancouver, Canada (Marshall et al., 2009), it appears that less advantaged and more walkable areas of town have documented some of the highest levels of nitrogen monoxide exposure which is an air pollutant generated by car engine combustion. Walking on busy roads puts pedestrians in harm's way, both because of their close interaction with cars and because their breathing rates are higher, making them more susceptible to the negative health effects of the pollutants they inhale (Bigazzi & Figliozzi, 2014). Disparities in infrastructure and environmental conditions have been linked to the already mentioned safety and health problems. The quality and accessibility to amenities decline in poor neighborhoods, according to a number of case studies (Lowe, 2016).

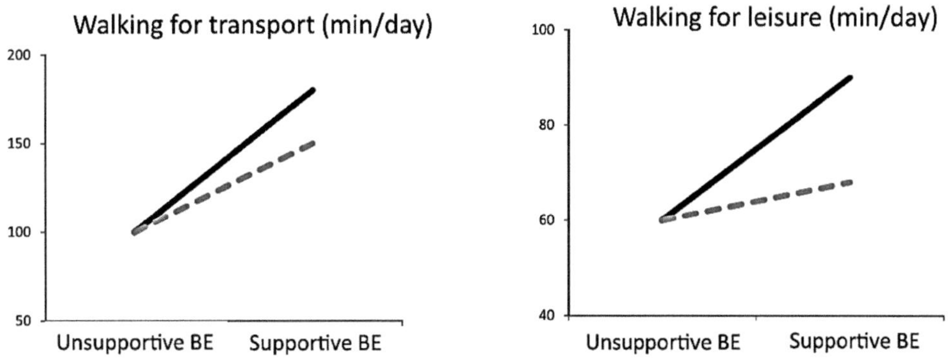

Fig. 3 Illustration of built environment effects on walkability; the walking rates in unsupportive and supportive environments. *Source* Adkins et al. (2017)

3 Discussion and Conclusion

This study is part of a research project concerned with investigating the correlation between the built environment and the quality of urban life for its users in the district of "Heliopolis" located in Cairo-Egypt. The study is concerned with "walkability" as one of the most important determinants of the quality of urban life and hence neighborhood satisfaction. "Walkability" becomes an even more pressing daily challenge for the disadvantaged populations, namely the elderly, the physically challenged, low-income, and racial minorities. Low walkability of the neighborhood can severely affect rates of mobility and physical activity, which eventually causes impacts to the health of its residents as well as environmental issues such as higher levels of air pollution (Ewing et al., 2008).

The results reveal that "safety" is the most important and widely cited factor in encouraging walkability and physical activity among underprivileged populations. This feature of the built environment is now more important than the environment's appearance or its physical characteristics. Urban programs that aim to enhance pedestrian health should prioritize increasing people's sense of safety and their propensity to use urban areas in order to create a welcoming and inclusive environment for all residents. Policies impacting these characteristics of the built environment should consider things like lighting, design approach, visibility, accessibility, camera surveillance, specialized equipment (like sound equipment for the visually impaired), and so on. This doesn't mean that accidents or crimes won't happen, but these rules might pave the way for better environmental control and lessen people's anxiety about being hurt or violated.

When running the search for the body of literature using the "Scopus" search engine, limiting the findings to fields of engineering and environmental science, the results show that 80% of the literature is dedicated to studying the effect of the built environment on walkability and physical activity of "the elderly", while only 20% of the literature on the same subject is dedicated to "the physically challenged" and "economically or culturally challenged" put together. Since Europe is at the forefront of the globe in terms of nations with slowing population growth and an aging population, it is apprehensible that "the elderly" is an important subject of European research. Yet it is essential for low- and middle-income countries to address issues of "safety", "health", and "accessibility" among their populations by conducting and applying scientific research to solving problems related to economic and cultural issues.

All pedestrians, regardless of age, physical ability, socioeconomic status, or cultural background, should be considered while designing the built environment. Therefore, a meta-narrative analysis strategy is recommended for evaluating the built environment to support future urban development plans. Taking a multidisciplinary perspective like this helps clarify the many ways in which the environment is defined and related to different forms of behavioral change. By grouping similar indicators into broader themes or categories, designers may overcome the restrictions imposed by the diverse nature of measurement in this field.

If walkability is defined only in terms of physical features, this may overlook important social, individual/household attributes, and other non-built environmental aspects that might be a barrier for underprivileged groups, such as fear of crime and a lack of social support. The information acquired through a comprehensive literature analysis highlighted the fact that there is a void in urban policies and evaluation frameworks where a minority of disadvantaged groups who do not adhere to conventional metrics are not included in the design or renovation of existing locations. Planners need to consider community concerns about redevelopment and relocation, as well as the fact that planned built environment improvements may mainly benefit the already advantaged.

References

Adkins, A., Makarewicz, C., Scanze, M., Ingram, M., & Luhr, G. (2017). Contextualizing walkability: Do relationships between built environments and walking vary by socioeconomic context? *Journal of the American Planning Association, 83*(3), 296–314. https://doi.org/10.1080/01944363.2017.1322527

Aksoy, E., & Korkmaz-Yaylagul, N. (2019). Assessing liveable cities for older people in an urban district in Turkey using the analytical hierarchy process. *Urban Planning, 4*(2), 83–95. https://doi.org/10.17645/up.v4i2.1943

Arango, J., & Montufar, J. (2008). Walking speed of older pedestrians who use canes or walkers for mobility. *Transportation Research Record: Journal of the Transportation Research Board, 2073*(1), 79–85. https://doi.org/10.3141/2073-09

Bigazzi, A. Y., & Figliozzi, M. A. (2014). Review of urban Bicyclists' intake and uptake of traffic-related air pollution. *Transport Reviews, 34*(2), 221–245. https://doi.org/10.1080/01441647.2014.897772

Bookman, A. (2008). Innovative models of aging in place: Transforming our communities for an aging population. *Community, Work & Family, 11*(4), 419–438. https://doi.org/10.1080/13668800802362334

CDC, M. (2013). *Motor vehicle traffic-related pedestrian deaths—United States, 2001–2010*. Morbidity and Mortality Weekly Report. https://www.cdc.gov/mmwr/pdf/wk/mm6215.pdf

Clark, A. F., & Scott, D. M. (2013). Does the social environment influence active travel? An investigation of walking in Hamilton, Canada. *Journal of Transport Geography, 31*, 278–285. https://doi.org/10.1016/j.jtrangeo.2013.06.005

Clark-Carter, D. D., Heyes, A. D., & Howarth, C. I. (1986). The efficiency and walking speed of visually impaired people. *Ergonomics, 29*(6), 779–789. https://doi.org/10.1080/00140138608968314

Cook, G. K., Wfiot, M. S., Webber, G. M., & Bright, K. T. (1999). Emergency lighting and wayfinding provision systems for visually impaired people: Phase II of a study. *Lighting Research and Technology, 31*(2), 43–48. https://doi.org/10.1177/096032719903100202

Daamen, W., & Hoogendoorn, S. P. (2010). Emergency door capacity: Influence of door width, population composition and stress level. *Fire Technology, 48*(1), 55–71. https://doi.org/10.1007/s10694-010-0202-9

Doyle, S., Kelly-Schwartz, A., Schlossberg, M., & Stockard, J. (2006). Active community environments and health: The relationship of walkable and safe communities to individual health. *Journal of the American Planning Association, 72*(1), 19–31. https://doi.org/10.1080/01944360608976721

Ewing, R., Bartholomew, K., Winkelman, S., & Anderson, G. (2008). Urban development and climate change. *Journal of Urbanism: International Research on Placemaking and Urban Sustainability, 1*(3), 201–216. https://doi.org/10.1080/17549170802529316

FHWA, U. S. department of transportation. (2003). *Manual on uniform traffic control devices for streets and highways*. MUTCD. https://mutcd.fhwa.dot.gov/index.htm

Frazila, R. B., & Zukhruf, F. (2018). Exploring physical attributes of walkability from perspective of blind pedestrians. *MATEC Web of Conferences, 147*, 02002. https://doi.org/10.1051/matecconf/201814702002

Hardman, A. E., & Stensel, D. J. (2009). *Physical activity and health: The evidence explained* (2nd ed.). Routledge.

Lachapelle, U., & Cloutier, M.-S. (2017). On the complexity of finishing a crossing on time: Elderly pedestrians, timing and cycling infrastructure. *Transportation Research Part a: Policy and Practice, 96*, 54–63. https://doi.org/10.1016/j.tra.2016.12.005

Lachapelle, U., Frank, L. D., Sallis, J. F., Saelens, B. E., & Conway, T. L. (2015). Active transportation by transit-dependent and choice riders and potential displacement of leisure physical activity. *Journal of Planning Education and Research, 36*(2), 225–238. https://doi.org/10.1177/0739456x15616253

Lowe, K. (2016). Environmental justice and pedestrianism: Sidewalk continuity, race, and poverty in New Orleans, Louisiana. *Transportation Research Record: Journal of the Transportation Research Board, 2598*(1), 119–123. https://doi.org/10.3141/2598-14

Marshall, J. D., Brauer, M., & Frank, L. D. (2009). Healthy neighborhoods: Walkability and air pollution. *Environmental Health Perspectives, 117*(11), 1752–1759. https://doi.org/10.1289/ehp.0900595

Miyazaki, K., Katsuhara, M., Matsukura, H., & Kiriya, N. (2003). Evacuation experiments for wheelchair user and pedestrian group. *The Journal of Japan Institute of Navigation, 108*, 129–138. https://doi.org/10.9749/jin.108.129

Pecchini, D., & Giuliani, F. (2015). Street-crossing behavior of people with disabilities. *Journal of Transportation Engineering, 141*(10). https://doi.org/10.1061/(ASCE)te.1943-5436.0000782

Sharifi, M. S., Christensen, K., Chen, A., Stuart, D., Kim, Y. S., & Chen, Y. (2017). A large-scale controlled experiment on pedestrian walking behavior involving individuals with disabilities. *Travel Behaviour and Society, 8*, 14–25. https://doi.org/10.1016/j.tbs.2017.03.003

Sundquist, K., Eriksson, U., Kawakami, N., Skog, L., Ohlsson, H., & Arvidsson, D. (2011). Neighborhood walkability, physical activity, and walking behavior: The Swedish neighborhood and physical activity (SNAP) study. *Social Science & Medicine, 72*(8), 1266–1273. https://doi.org/10.1016/j.socscimed.2011.03.004

Tremoulet, A., Dann, R. J., & Adkins, A. (2016). Moving to location affordability? Housing choice vouchers and residential relocation in the Portland, Oregon, region. *Housing Policy Debate, 26*(4–5), 692–713. https://doi.org/10.1080/10511482.2016.1150314

United Nations, P. Division. (2015). *World population ageing 2015*. https://www.un.org/development/desa/pd/content/world-population-ageing-2015

World Health Organization. (2002). Active ageing: a policy framework. *World Health Organization*. https://apps.who.int/iris/handle/10665/67215

The Urban Regeneration of Detention Spaces: The Case Study of the Bolzano Prison

Claudia Battaino and Filippo Marconi

Abstract

Urban regeneration requires a new look, attentive to social phenomena and their implications, even in relation to issues apparently distant from the community, such as prisons. There is a great opportunity in Italy for repurpose and redevelopment, linked to the disposal of old prison properties or prisons no longer able to ensure the necessary spaces and conditions to guarantee humanity and the dignity of the sentence. These buildings, often of architectural value, due to their size and location in the urban fabric can play an important role in the revitalization of historic centers. Another pressing issue is prison overcrowding. In many Italian territories, processes are underway to remove prisons from inhabited centers: the choice of the greatest possible distance, however, risks generating new forms of marginalization, discrimination and spatial and social isolation. The aim of the paper is to reflect on prison space and prison design as a space compatible with human and environmental health, in the idea of a new "empowering" prison model, which guarantees the prisoner a new centrality. The imprisoned person must be able to become an active subject of the fulfillment of the sentence so that it can become an opportunity for redemption, capable of guaranteeing conditions that conform to humanity and dignity and above all relations with the social, economic and environmental contexts of reference. The architectural project, as a form of knowledge, has the dutiful task of contributing to the change of course, dictating the rules and modes of coexistence of everyday environments. The question of the architectural quality of the prison is a complex issue that concerns both the search for innovative spatial models and the urgent need for adaptation and redevelopment of the existing heritage, as well as the forecasting of new interventions in the territory for urban regeneration. It is within this framework that the Bolzano case study is placed, whose prison turns out to be a structure absolutely unsuitable for a detention that respects human rights. The main problems associated with this reality are due to the scarcity of human and financial resources, but above all to the inefficiency and inadequacy of the spaces. The contribution analyzes the scenarios of the progressive decommissioning of the current prison, starting from the municipality's forecast of a new construction outside the historical center. This solution will make it possible to study innovative spaces, with a focus on the single dimension of the prisoner.

Keywords

Prison • Urban-social regeneration • Architectural research • Space • Hypogeum

1 Introduction

Over the course of a century, two founding paradigms of detention spaces, central surveillance and isolation of the individual, have shifted due to cultural changes.

In 1979, with his design for the De Koepel prison, Rem Koolhaas highlighted the spontaneous adaptation of an existing structure, the panopticon, outlining a complex programme for collective facilities and surfaces where prisoners could move freely, restoring higher forms of social relations. The prison, an expression of the social condenser theorized in the manifesto Delirious New York (Koolhaas, 1978) and of another project that remained on paper, the Parc de la Villette, transforms the heterotopia, the space-*outil* described by Foucault (1976) for the surveillance of a closed

C. Battaino (✉)
DICAM, University of Trento, Trento, Italy
e-mail: claudia.battaino@unitn.it

F. Marconi
Architectural Engineering, University of Trento, Trento, Italy

A. Battisti et al. (eds.), *Greening Our Cities: Sustainable Urbanism for a Greener Future*,
Advances in Science, Technology & Innovation, https://doi.org/10.1007/978-3-031-49495-6_20

community, into an *outil* programme to resolve congestion, open to the co-presence and conflict of many needs, more flexible and habitable than the Panopticon.

The effects of the prison and its spaces on human psychology suggest the fundamental role of architecture and architectural design on the mental health of the prisoner. The safety and life of inmates in prison are strongly influenced by the architectural features of prison spaces, their variety and breadth, the amount of air and light. Prisoners' survival seems to be affected by the lack of spatial and acoustical privacy, and this can cause physical and mental deterioration (Moore, 1981).

Prison today is felt to be a complex place, with a high social and human content, which must guarantee spaces for multiple activities, ensuring the needs for a dignified human life.

Deteriorating conditions and insufficient space have led to a progressive dismissal of the prisons located in urban centers, in favor of new buildings in the peripheral areas. However, the decommissioning of the old structures has not always generated new opportunities for the redevelopment of cities, leaving buildings and spaces abandoned, in increasing degradation.

Dealing with urban regeneration implies the need for a careful look at social phenomena and their implications, even regarding issues apparently distant from the community, such as prisons. The bond that binds historic centers and prison buildings is profound. The detention compartments, as well as the military structures, have contributed greatly to the development and growth of the cities. Although the relocation of these structures to the outskirts of the city may seem like a ghettoization operation, it is a strategy that should be supported by the municipalities. The places dedicated to punishment, in fact, can contribute to the rehabilitation of prisoners only through an innovation of their spaces, in quantitative and qualitative terms. This would also open up new possibilities, to give back to the community buildings no longer capable of fulfilling the function of detention, strategic areas because they are located in the heart of the city.

The first four metropolitan cities (Rome, Milan, Naples and Palermo) have penal institutions inserted within the city center: Regina Coeli, San Vittore, Poggio Reale and the Ucciardone. Buildings full of history, which due to the increase in the prison population, the scarcity of spaces and state resources end up being absolutely unsuitable for the purposes of detention.

Can urban redevelopment start from prisons? How to solve the problems of detention and enhance some of the largest and most central real estate complexes in the cities, transforming the urban space in favor of the citizen?

The reflection starts from the penitentiary structures which, due to their dimensional and historical–architectural characteristics, have urban and/or territorial relevance, the prison contexts that highlight unsustainable operating costs, which are inefficient and inadequate, the urban sectors are also often forced to needlessly endure serious and burdensome hardships.

2 Prisons and Cities in Italy

Italy sees the birth of a national prison system about a century later than most European countries. Only after the royal decree of October 9, 1861 n. 255 a national body is established in charge of managing, organizing and maintaining the entire system.

Tracing an excursus on the Italian situation, with a focus on the main moments of development and evolution of the penitentiary system in Italy, starting from the aftermath of the Second World War until today, is useful for understanding how the prison structures have influenced the growth of the center's urban areas and how the decisions made over time have affected the well-being of the prison population (Ferrari, 2015).

At the end of the Second World War the structures present on the Italian territory were almost totally in poor condition. The prison system was neglected, and the living conditions within the institutions were inhumane. Until 1949 the Fascist code remained in force, responsible for violent revolts in all the main Italian institutions after the end of the regime. The first reform was passed with law no. 460 of 1949, which initiated a period of renewal and typological research. Between this period and the end of the 1970s, 65 new penitentiary complexes were built, among the best known those designed by architects belonging to the Roman School, including Mario Ridolfi, Sergio Lenci and Pasquale Carbonara (Petrella, 2002).

The year 1975 is a significant step in the creation of innovative programs and projects related to prison architecture. Law no. 354 or the "rules on the penitentiary system and on the execution of privative and limitative measures of freedom", to adapt the penitentiary system to the principles enshrined in the Constitution, or those expressed by article 27 paragraph 3, according to which "the penalties cannot consist in treatments contrary to the sense of humanity and must aim at the re-education of the condemned". However, the spirit of change dictated by the law soon clashed with emergencies related to public order, which swept Italy in the seventies and eighties. This had serious consequences. The subsequent prison constructions, rather than going toward the reformulation of "re-educational prison open to the city", moved away from the guidelines issued by the law, favoring "compact" typological solutions (Fig. 1), that were configurations organized by blocks developed in height, based exclusively on the security and control of detainees (Dé Rossi, 2016).

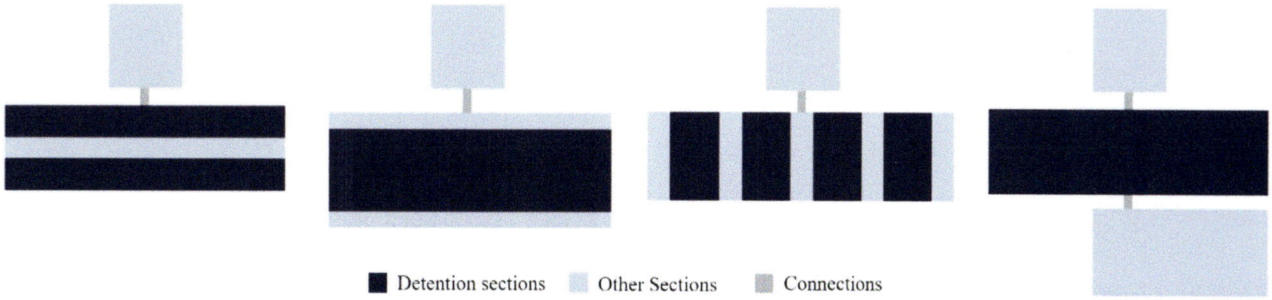

Fig. 1 "Compact" typological solutions of prisons. © Marconi, F

Very often, the penitentiary structures have been placed in peripheral sites, difficult to reach, choosing, in extreme cases, completely marginal sites, such as, for example, islands, also places of great ecological-naturalistic interest and inaccessibility. This also resulted in a progressive and substantial reduction of outdoor spaces, as if to deny the outdoor activities as a punishment, in favor of the activities carried out inside the prison. All of these choices, location, typology and distribution, along with the fact that the use of prefabricated reinforced concrete panels is preferred in buildings, have made any future expansion or change difficult.

With the end of the crisis, in the nineties, the compact, high-security typology was gradually abandoned, in order to favor the birth of new typological schemes, considering the prison also as a place in which to combine spaces for work, education, social and recreational activities, alongside with the guidelines of the aforementioned 1975 reform. This change of course favors the progressive development of the "telegraph pole" typological model (Fig. 2), a system that provides long central corridors, through which are distributed the different bodies that contain a plurality of functions and spaces.

We had to wait a long time, until the year 2000 for the issuance of a new "regulation containing rules on the penitentiary system and on the deprivative and restrictive measures of liberty" (Presidential Decree June 30, 2000). It has the merit of introducing some innovations in terms of the design of prisons. In particular, the indications mainly concern architectural and plant engineering aspects, as well as functional, in order to guarantee the prisoner an acceptable standard of living. The new rules define the spaces dedicated to housing, work, leisure and study and introduce a fundamental theme, the relationship between prisoners and family members. However, not even this regulation will serve to mark a real turning point, due to the existing overcrowding in prisons (Fig. 3) it is not possible to achieve the goal of contact, dynamic and relational spaces (Astarita & Bonatelli, 2006). The prison architectures built in this period propose a cross typological scheme, already extensively explored in the past and now outdated.

Moving on to the current consistency of Italian prisons, it can be seen that the building structures are extremely varied, about 50% of them being built before 1945, they are located in urban centers, often presenting architectural and historical features of value.

From the data available from the Ministry of Justice (www.giustizia.it), it emerges that 8.5% of prisons in Italy were built before 1800, very often the result of the readjustment of ancient castles, convents and historic buildings. An example of this is the prison house in Venice Giudecca, in an ancient twelfth century monastery; the district house of Pozzuoli, a convent complex of the fifteenth century; the Eboli prison house, inside a medieval castle; the prison house of Castelfranco Emilia, whose construction was commissioned by Pope Urban VIII in 1626 to defend the Papal State; the Volterra prison house, in the Medici Fortress built in 1474; the district house of Sciacca, formerly an old

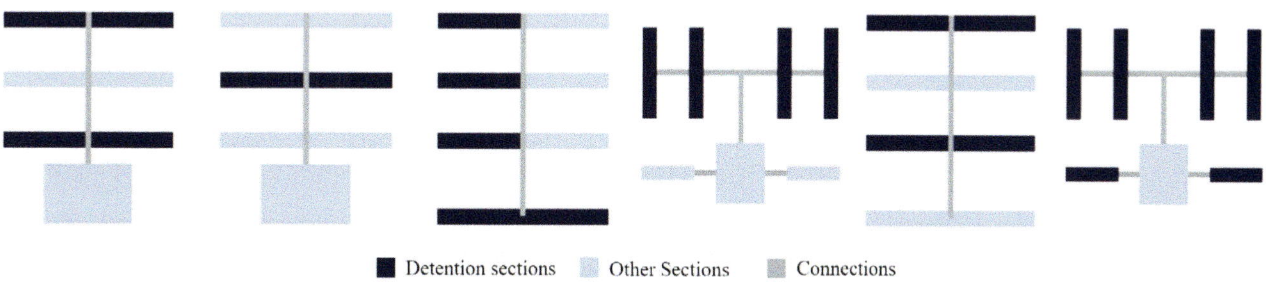

Fig. 2 "Telegraph pole" typological models of prisons. © Marconi, F

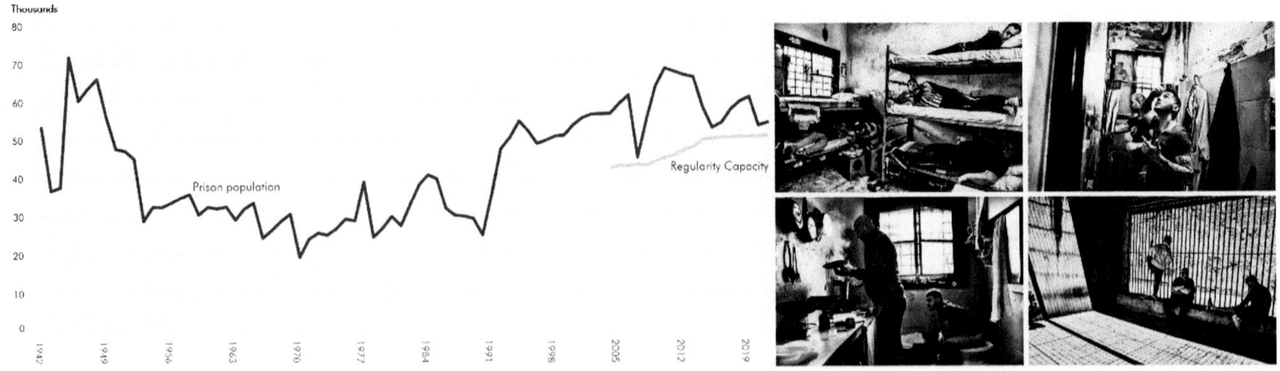

Fig. 3 Overcrowding of Italian prisons. Photos © 2004–2022 Bispuri, V

convent of 1200 and many other structures. If new and more efficient structures were built, all these prisons, which no longer have all the necessary spaces, but occupy buildings of historical and cultural interest, are an urban resource of great cultural interest for the community and the tourist system, if reopened to the public.

9% of the prisons were built between 1800 and 1900, rooted in the historical fabric of urban centers, currently no longer suitable due to an objective lack of space. Some examples are: the Milan San Vittore prison, the prison in Civitavecchia, built in 1864 at the behest of Pope Pius IX, one of the first prisons in the United Kingdom of Italy; the Palermo Ucciardone prison house, designed in the nineteenth century; the Regina Coeli prison in Rome, a convent converted into a penitentiary. A careful reflection on the possibilities of recycling and reuse must be made on this historical heritage.

25% of the prisons were built in the first half of the 1900s; these are buildings inserted in the urban context, born as penal institutions, of requires a technical–economic evaluation, to be carried out on a case-by-case basis, to identify sustainable interventions for the readjustment of the existing or replacement with new buildings and the public reuse of old complexes.

50% of the structures were built between 1950 and 2000. These are outdated prisons, far from urban centers and difficult to reach, which have no relationship with the contexts in which they are located. Amongst them: the Milano Opera prison; the prison of Turin; the Sollicciano Prison of Florence. These structures, while not lacking in space, are, however, affected by the need for renovation works.

Only 5% of prisons were therefore designed and built after 2000. Although these prisons are recent, they are far from what are considered to be the international models (Bernheimer & O'Brien, 2017) for the atonement of the sentence. They are prisons totally disconnected from the city and with strong critical connections with the public sphere. Comparing the Italian situation with the European one, it is clear that the architecture of the prison, in countries such as Austria and Norway, is much more advanced.

Consequently, the data reported above make us understand, in their crudeness, the overall state of the Italian prison system, which presents highly critical elements, linked on the one hand to the age and inadequacy of the buildings and areas, on the other to overcrowding.

To this is added a consideration that concerns all those buildings that until a few years ago covered the role of prisons in the cities and which currently appear to be abandoned and forgotten. It is complex to estimate how many cubic meters or square meters the assets made up of public buildings intended for disused penitentiaries amount to. It can be estimated, by comparison and on the basis of some research regarding vacant or dysfunctional spaces, that they are many. All this immense heritage represents a problem in terms of urban regeneration. Finally, we must not forget the entire islands or parts of them, real ecological paradises, destined in the past to penitentiaries or penal colonies, naturalistic and architectural–cultural wonders, now abandoned for which recovery is urgently needed. Examples include the former Procida prison, inside a medieval fortification of Terra Murata; the former prison of the island of Santo Stefano, Ventonene the former prison of Asinara, the former prison of Santarcangelo di Romagna located in an ancient polygonal tower of the fifth century and many other cases.

3 Possible Scenarios

The decommissioning of old prisons in urban centers involves in many cases a modification of the areas and of the city itself, which must be carefully studied and understood on a case-by-case basis. A first element to take into consideration is the historical memory of the prison buildings: The prisons that are today located within the urban fabric have an important social significance for the place in which they are inserted (Di Franco & Bozzuto, 2020). However, this significant heritage, layered through history, has several critical issues such as lack of space, old buildings, and accessibility.

Most of the historic buildings, still used for detention, with some exceptions such as prison of S. Vittore and Milan (Biagi, 2012) are functional only for security. Among the main problems, in addition to the dated age of the structures, there is the lack of spaces necessary for inmates, in particular outdoor spaces. The overcrowding of prisons and the incorrect organization of spaces can determine, in addition to the onset of various pathologies, also the consolidation of a profound social hatred.

Prison, as an ideal model, should therefore take into account that a quality architectural project of the entire complex is a sine qua non for the civil rehabilitation of the individual (Fransson & Giofré, 2018).

The improvement of the prison`s infrastructural system requires more spacious and characterized environments, sequences of open spaces, spaces of relationship with the city and adequately served by connections, so that a link can be established between the community and inmates (Santangelo, 2017). Prisons perform a public function of primary importance; they are the place where punishment can become an opportunity for redemption, where conditions are in line with humanity and dignity and a relationship with the reference social, economic and environmental contexts (Hoffs, 2014).

In many territories, processes of relocation of prison facilities are underway. In the search for suitable places for new prison sites, urban plans generally identify these areas in the vicinity of the neighborhoods planned in the 1900s for economic–social housing: the perspective of "the greatest possible distance" (Marcetti, 2009). This outplacement philosophy risk generating new forms of discrimination and strong social isolation, both for inmates and those working in prison, especially if the prison project is not an inclusive and innovative process.

The urban planning choices regarding the potential sites where to place the new prison structures in the area cannot be solved simply by removing the prison from the city, identifying a free area. Its main role is that the proposition of reasoned choices for the sustainable transformation of the city by means of rational architectures, in the face of an in-depth data collection that includes historical, social, economic analyzes, surveys and complex and interdisciplinary assessments. For the new prisons, there is generally no search for contextualization and the design of "hinge" spaces with respect to the surrounding area, which can contribute to resolving even the unsolved areas of the contemporary city.

By drawing up some useful guidelines for the innovative design of the prison with attenuated custody, some key points are identified: It must be located in a portion of the city suitable for the well-being of users and staff; it must guarantee easy accessibility to main roads and rapid access to essential services (courts, hospitals and police stations); must ensure adequate security, etc. A careful study on the possible impact on the city from multiple points of view, environmental, urban and social, must guide the choices. The localization process can be carried out according to different scales, in terms of territorial scale, evaluating the relationships between the identified area and the existing social services system and on a local scale, taking into account the environmental quality, climate, light and configuration of the area, considering its size and relations with the surrounding area. The connections with the coordination and support structures that contribute to the functioning of the institutional machine must also be analyzed in order to understand the connections with the main road networks (roads, railways, motorways, etc.). In addition to the analysis of the structures inherent to the criminal circuit, the possible connections with additional indispensable services must be identified, including productive, welfare and social services (Vessella, 2016).

Finally, as an architectural and urban project, the relationship with the existing landscape, residential, agricultural or industrial and the criticalities and potentialities of the chosen place must be identified, paying attention to the design of the connections. The correct design of a penitentiary structure in the territorial context is very important for interpreting the cultural climate and the idea that citizens have of the penitentiary function.

Above all, the prison must be rethought, as was done in the past for historical prisons, as a constitutive element of the contemporary city and its landscape, which must relate, in terms of scale and figure, with the surrounding spaces and realities. Taking up the considerations of Ludvig HIlberseimer, in the großstadt architektur 1926 schools, hospitals, airports, courts, etc., as well as prisons, constitute the architecture of the city and should be part of a large, rational and shared civil project.

Therefore, research and solutions must be carried out in which the new location and the decommissioning and consequent reuse of the old buildings are included in the same regeneration project of the city. This can only be done through an innovative integrated planning process, which addresses, at the same time, the issue of the reuse of former prisons and the design of new realities. This need now becomes an opportunity to outline a systemic methodological process, applied to urban planning. Hence, the need for strategic planning in the penitentiary sector which must have, among other objectives, the integration between transport networks and the system of public functions that can be activated in the area.

There is a great opportunity in Italy regarding the disposal, reuse and transformation of properties which, due to

Fig. 4 Perugia Network. Project of the redevelopment of the former prison (2014). © Battaino, C

their size and location in the urban fabric, can play an important role in urban regeneration. This is the case, for example, of the former prison complexes of Perugia, located in the historic center, closed places and full of memories, abandoned in 2006. After years of debates, research and design experiments (Fig. 4), which have contributed to highlight the need to redevelop them as "cultural factories" at the service of citizens (Battaino, 2015), currently, thanks to the Pnrr, the National Recovery and Resilience Plan, the male and female sections and the Paradiso pavilion will be subject to redevelopment and restoration and transformed into a citadel judicial.

Another example is Padua. Located in the center of the city, a former castle from the 1300 s, built by incorporating the old city walls had lost its identity during the two long centuries of prison life, as it had lost it in the memory and cognitive maps of the Paduans, becoming the pale shadow of what once was. The enhancement of contemporary art and design in a large cultural space is currently being perfected.

4 The Project as Research. The Bolzano Case Study

Starting from experiences that confirm the appropriateness of a multidisciplinary and multi-scalar design, adopting the method of research by design, the project investigates the possibilities and limits of architectural composition as a tool for defining a complex system, capable of interpreting, through its configuration, the transformation of urban space and the relationship with it. Reasoning on imaginative logics, in search of a more harmonious relationship between construction, environment and human life, the project—starting from the traditional tools concerning the typology of educational, working and socialization spaces—aims to deepen the possible changes, form and sense of prison, in the relationship between architecture-city-landscape, within the new system of ecological relations. The study of architectural spaces is based on a careful analysis of international case studies and a comparison of Italian and European regulations in terms of prison buildings. The project of relationships is taken as the backbone of research and teaching.

What makes the prison space distant from everyday life is its alienating dimension, its constitution as a single enclosed space. The innovative devices of the architectural-urban project therefore aim to re-interpret some of the traditional elements of prison construction. First of all, the theme of the "fence" and the "wall", as an opportunity to imagine a "multi-dimensional" place in which, precisely the limit, is transformed into a relational device with the city. Another fundamental concept is the binomial nature-artifice, the relationship with the earth and the earth line, which unfolds in the imaginative projection of new sections for the territory—through architecture as a complex system, with high ecological performance (Telesca, 2019).

The current Bolzano Prison (Fig. 5) is located in an Austro-Hungarian building dating back to the late 1800s, placed in the historic center. The prison consists of 33 2–5-8-seater cells which are regularly doubled due to overcrowding. It is divided into three; to the north and south the prison is built against other private residential buildings. Structurally, the building is affected by its age; in addition to being ancient, it is also poorly maintained. Many sections of the prison are in the most critical situations. The building, not large in size, is divided into two sections; each of them includes overnight rooms that accommodate from three to eight inmates; each cell is equipped with a sink and a toilet, but not a shower. As for the common areas, there is only one room on the first floor, which is very small. There are two gyms, located one on the first and one on the second floor in a bad state of conservation, and inmates can attend them four at a time, and consequently, the spaces present are not enough for everyone. There is a library and a space dedicated to school and training, but there is no place dedicated to processing. As for the outdoor spaces, there is a courtyard, which is absolutely narrow to ensure the necessary needs.

Currently there are 107 inmates, compared to a maximum capacity of 88 places (www.giustizia.it, data updated to 05/09/2022). As can be seen, the prison is overcrowded, unsuitable for its purpose, degraded and lacking in basic spaces. The serious situation is not due to a carelessness of the prison system but, as in other cases, to the age, location and size of the building.

Prisons have been the subject of a heated debate concerning how to intervene to remedy a problem that is still unsolved. The positions on the field see, on the one hand, the irremediable obsolescence of the old prison, with the

Fig. 5 Building of the prison of Bolzano (2021). © Marconi, F

provision of transfer to a new suitable structure, on the other hand the realization of works for the adaptation and safety of the building.

Considering the layout of the building and its conditions in contact with other buildings, it is not possible to think of extensions for prison use, and also in the interiors, the possibilities for improvement are very limited.

Having assessed the situation, the 2010 National Penitentiary Intervention Plan therefore envisaged the construction of a new medium-security prison, to be located in the southern outskirts of Bolzano, with a capacity of 220 inmates.

During a meeting in 2022 between the current president of the province of Bolzano and the government's Minister of Justice, it emerged that the current prison is at the bottom of the rankings in terms of quality standards, so state intervention to support the construction of a new facility is urgently needed.

The area identified by the municipality for the new prison is located on a site adjacent to the Bolzano airport (Fig. 6). This decision is aligned with the design criteria of the guidelines of the DAP, Department of Penitentiary Administration which suggest to place the newly built penitentiary complexes outside the residential perimeter of the city, preferring peripheral areas that are accessible by infrastructures and means of transport (Pedrotti & Rottensteiner, 2014). The identified area is located in the industrial area, which extends for an area of approximately 42,000 square meters.

Like a new ecological infrastructure, between the airport, the residential fabric and the countryside, through the section and the excavation, the prison is conceived as an open and productive landscape, green and terraced, which includes multiple spaces en plain air and, where the social needs are combined with the cultivation of vineyards and other fruit trees (Fig. 7).

The project, with the aim of increasing the quantity and quality of services and public space, plays a role of "mending" and "filter" between different spaces, between detention and freedom. The prison is imagined as a hybrid, hypogeous and vital place, designed without external walls for welcoming visitors and for the well-being of prisoners and workers, reducing the perceptual and social distance "between what is inside and what is outside"(Marconi, 2022). The new prison outlines three macro strategies for the detention spaces: working with the ground line and ground modeling, for a prison without walls toward the city; the configuration of the urban and typological layout and the design of the open spaces, to control the large size and define spaces suited to the multiple needs, individual and common, welcoming; the energy configuration of the buildings to ensure environmental comfort and the well-being of the users.

The most important part of a prison is certainly the cell. The project consists of single residential units, both to guarantee privacy and autonomy and for the mental health of the prisoner himself. The convivial aspect is instead implemented and thought out in the design of the common parts. This is intended to encourage conviviality and sociability during daytime activities and then to guarantee an intimate space during the night. Great importance is also given to the design of green spaces, recognizing the fundamental importance of the presence of agricultural spaces and sports

Fig. 6 Area identified by the municipality for the new Bolzano's prison. © Marconi, F

Fig. 7 Design of the new Prison of Bolzano. © Marconi, F

facilities. The interior design of the residential cells is based on two fundamental design principles: the spaciousness of common spaces, such as corridors, conceived not only as a connecting element but also as a relational space; the presence of light thanks to the use of glass-cement as a separating element that guarantees the filtering of natural light (Fig. 8).

For the city of Bolzano, which is pursuing the construction of the new penitentiary structure, an opportunity not to be missed is obviously also that of redeveloping the historic building in the city, to accommodate new public spaces for the community.

5 Conclusions

The possible scenarios regarding the design of the detention spaces are wide. The former prisons, due to their large size or strategic location within the historical fabric, are undoubtedly an important opportunity for urban regeneration. The specificity of the individual cases requires different mappings, surveys and reflections depending on the individual situation as the panorama of Italian prison construction is very varied. The outcome of the applied research on Bolzano shows how rethinking the prison is decisive, not only to restore human dignity to detention, but also as an opportunity to strengthen, improve and enhance contemporary cities, both in the historic centers and in the suburbs, through equipment and quality public spaces. Finally, sustainable decommissioning construction reuse processes need to be explored through interdisciplinary and punctual methodological paths. The goal of a "new" penitentiary building plan, which will allow Italy to adapt to the European models of regeneration, giving the cities, community spaces and adequate public services. Themes such as energy, the environment and the comfort of architectural spaces are fundamental to a sustainable, ethical and rehabilitating prison. The effects of these issues on inmates' mental health mark the urgency of new and definitive structural interventions on Italian prisons, which cannot be postponed any longer.

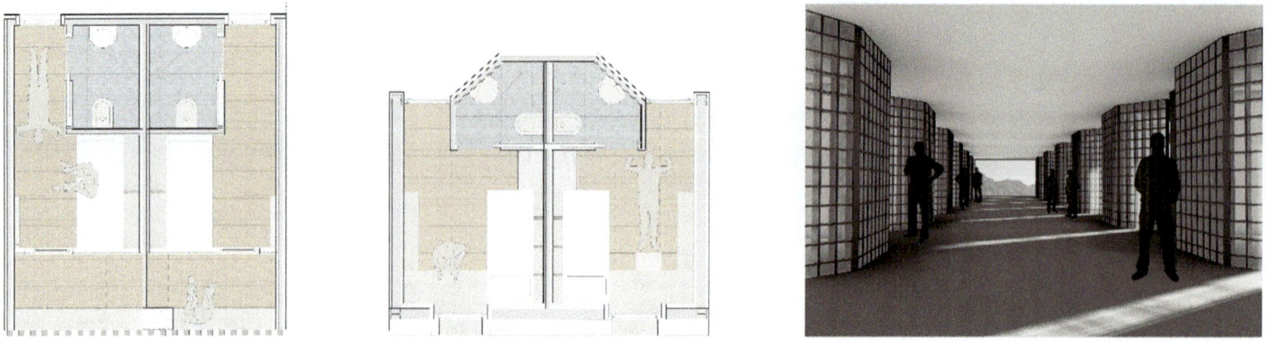

Fig. 8 Internal residential design of the project. © Marconi, F

References

Astarita, L., & Bonatelli, P. (2006). *Dentro ogni carcere. Antigone nei 208 istituti di pena italiani. 4° Rapporto sulle condizioni di detenzione*. Carrocci.

Battaino, C. (2015) Perugia network. In Belardi, P., Menchetelli, V. *Kultur Fabrik Perugia. Ricercare, rilevare, rigenerare l'ex carcere maschile di Perugia*. Il formichiere, 80–88.

Bernheimer, L., & O'Brien, R. (2017). *Wellbeing in prison design. A guide*. Matter Architecture.

Biagi, M. (2012). *Carcere, città e architettura. Progetti per il carcere di San vittore a Milano 2004–2009*. Maggioli Editore.

Dé Rossi, A. (2016). *Non solo carcere. Norme, storia e architettura dei modelli penitenziari*. Mursia Editore.

Di Franco, A., & Bozzuto, P. (2020). *Lo spazio di relazione nel carcere. Una riflessione progettuale, a partire dai casi milanesi*. LetteraVentidue.

Ferrari, L. (2015). *No prison. Ovvero il fallimento del carcere*. Rubbettino.

Foucault, M. (1976). *Surveiller et punir. Naissance de la prison*. Editions Gallimard.

Fransson, E. Giofré F. (2018). *Prison architecture and humans*. Datapage India.

Hoffs, B. A. M. (2014). *Begin Again. New prison design which supports the process of rehabilitation*. www.research.tue.nl

Koolhaas, R. (1978). *Delirious new york*. The Monacelli Press.

Marconi, F. (2022). *Lo spazio della detenzione. Un nuovo carcere per la città di Bolzano*. Student thesis: Master 2021–2022, supervisors Battaino C., De Jaco M. University of Trento.

Marcetti, C. (2009). *Gli spazi della pena e l'architettura del caercere. L'edilizia che non c'è*. http://www.antoniocasella.eu/

Moore, E. O. (1981). *A prison environment's effect on health care service demands*. Baywood Publishing Co.

Pedrotti, A., Rottensteiner, M. (2014) *Dentro le mura fuori dal carcere. Una ricerca sul nuovo carcere della Provincia di Bolzano*. Caritas.

Petrella, P. (2002). *Le architetture della pena*. Luciano Editore.

Santangelo, M. (2017). *In prigione, architettura e tempo della detenzione*. LetteraVentidue.

Telesca, D.T. (2019). *Carcere e rieducazione. Da istituto penale a istituto culturale*. Aras Edizioni.

Vessella, L. (2016). *L'architettura del carcere a custodia attenuata*. FrancoAngeli.

Assessing and Improving Walkability and Cycling in Urban Areas: A Case Study in Kuwait

Mohamed M. Salem, Fatmah A. Abul, Fatemah A. Altarrad, and Aseel D. Alrashidi

Abstract

Historically, street design in Kuwait has been focused on accommodating automobiles. Not only in Kuwait but also in other Gulf Cooperation Council (GCC) nations, this has been a common design approach. Unfortunately, the vehicle-oriented approach to urban street design contributes to issues such as increased traffic congestion and travel delays, decreased safety, deteriorating public health, and increased environmental impacts. Moreover, it does not encourage the use of alternative modes of transportation, such as public transportation, cycling, or walking. Due to COVID-19, the significance of implementing road retrofits in Kuwait has recently been recognized. As the number of pedestrians and cyclists increased dramatically, people became acutely aware of the poor condition of the city's streets and sidewalks and the need for a retrofit. The sidewalks were not designed to accommodate such a large number of pedestrians and cyclists. This has been a dilemma in Kuwait for many years. This research will therefore employ a case study approach to investigate walkability and cycling in two local neighborhoods in Kuwait (namely Mishrif and Salem Sabah Al-Salem Al-Sabah Coastal Street) with contrasting characteristics and will propose urban solutions to the problems. It will discuss the procedures taken to retrofit the selected streets and present the possibility of simple, low-cost adjustments that can alter the level of service provided on the streets of Kuwait.

Keywords

Walkability • Cycling • Street retrofit • Complete street • COVID-19

1 Background

Walking and cycling can play a major part in the movement of people within a city environment. These nonmotorized modes contribute to better health, convenient local journeys, and help the environment by reducing emissions and energy consumption. To develop areas and facilities, the population will use them without resorting to the private car. A complete streets package needs to be implemented alongside developments. Through careful planning and implementation of facilities, schemes, and ideas, the outside environment can become a pleasant place where people enjoy being, feel safe, and pedestrian mobility is encouraged. Planting, shading, and water strategies will all contribute to reducing the radiant temperature and maximizing cooling breezes to extend the time that people can spend outdoors.

Complete streets can be defined as the streets which are designed and operated to enable safe access for all users (i.e., pedestrians, bicyclists, motorists, and transit riders). People of all ages and abilities must be able to safely move along and across a complete street.

In recent years, there has been growing interest in promoting walkability and cycling in cities around the world, including developing countries. However, the effectiveness of specific design interventions for promoting walkability and cycling in developing countries is not well understood. In this research, we examine walkability and cycling in developing countries with a focus on Kuwait.

A review of the literature reveals that there is a growing body of research on walkability and cycling in developing countries, but there is still a significant research gap in

M. M. Salem (✉)
Industrial Practice, Australian University, Mishrif, P.O. Box 1411 13015 Safat, Kuwait
e-mail: msalem@au.edu.kw

F. A. Abul · F. A. Altarrad · A. D. Alrashidi
Australian University, Mishrif, P.O. Box 1411 13015 Safat, Kuwait

A. Battisti et al. (eds.), *Greening Our Cities: Sustainable Urbanism for a Greener Future*, Advances in Science, Technology & Innovation, https://doi.org/10.1007/978-3-031-49495-6_21

understanding the effectiveness of specific design interventions in these contexts.

In Qatar, a developing nation, Shaaban (2019) evaluated the walkability of sidewalks and corridors. The study employs a mixed-methods approach, incorporating a walkability audit tool, surveys, and geographic information system (GIS) analysis to evaluate the pedestrian infrastructure and walkability in three distinct Qatari neighborhoods. The study reveals that Qatar's pedestrian infrastructure is inadequate and in need of improvement. The auditing tool for walkability identified several problem areas, including the quality of the sidewalk surface, the availability of crosswalks, and traffic-calming measures. The survey data also revealed that residents of the three surveyed neighborhoods felt the pedestrian infrastructure was inadequate and not conducive to walking. The study concludes that improving pedestrian infrastructure and walkability in Qatar necessitates a multifaceted strategy involving improvements to sidewalk design, crosswalks, and traffic-calming measures. The study also emphasizes the significance of involving residents in the design and planning of pedestrian infrastructure, as well as the need for additional research on walkability in developing nations.

Similarly, Butler (2015) evaluated the walkability of neighborhoods in New Zealand using criteria such as pedestrian network density, sidewalk coverage, the walkability index, connectivity, traffic-calming measures, and pedestrian amenities. The study discovered that neighborhoods with a higher walkability index score had higher rates of walking and cycling, and that traffic-calming measures such as speed humps and chicanes were effective at reducing traffic speeds and volumes, which could encourage walking and cycling. These results suggest that similar design interventions could promote walkability and cycling in Kuwait and other developing nations.

Other studies have also evaluated the effectiveness of specific design interventions for promoting walkability and cycling in developing countries. However, while these studies provide valuable insights into the effectiveness of specific design interventions for promoting walkability and cycling in developing countries, there is still a need for more research on how these interventions might need to be adapted to the specific context of these countries. For example, developing countries may have limited resources for infrastructure development and maintenance, which may impact the feasibility and effectiveness of certain design interventions. Developing countries may also have different cultural attitudes and norms toward walking and cycling, which may influence the demand for and use of pedestrian and cycling infrastructure.

Kuwait lacks complete streets, which hinders walking and cycling for its residents. Poorly designed sidewalks, obstructions, encroachments, and inadequate maintenance are some reasons that contribute to this issue (Atef Elhamy Kamel, 2013). The typical street design in Kuwait prioritizes neither the safety nor the welfare of pedestrians and cyclists. Consequently, walking and cycling are challenging in Kuwait, especially given the harsh climate and the population's dependence on private vehicles. This lack of active mobility has contributed to weight gain and obesity in Kuwait, as is the case with other Arab nations (Weiderpass et al., 2019).

With recent urbanization and modernization, population density in Kuwait has increased, highlighting the need for sustainable urban development. The United Nations predict a fifty percent increase in the global population by the year 2100, further emphasizing the urgency of sustainable urban development in Kuwait (Gomes, 2020).

To make Kuwait more walkable, policymakers are now promoting sustainable design and considering urban retrofits and redesigns to make the city culturally adaptable and climate appropriate. A strategy that incorporates community design, infrastructure availability, pricing, and quality is necessary to increase walking and cycling in urban environments. In 2017, Kuwait introduced the 4KMP Master Plan for 2040, envisioning future neighborhoods that incorporate advanced technologies. To enhance mobility, the plan employed GIS-based spatial network analysis to create comprehensive and convenient transportation networks that prioritize walking and cycling (Objective #2 in the 4KMP, known as accessible connections) (Kuwait Municipality, 2018). The Master Plan proposes the redevelopment of over 200 neighborhoods throughout the country to improve travel. However, without implementation plans, it is unclear how Kuwait will transform these neighborhoods.

The COVID-19 pandemic has highlighted the need for street retrofits in Kuwait. The sudden increase in pedestrians and cyclists, especially during curfew (see Fig. 21.1), drew widespread attention to the poor condition of the streets and sidewalks, emphasizing the need for a retrofit. The sidewalks were inadequate to accommodate the surge in the number of pedestrians and cyclists, a problem that has persisted in Kuwait for years.

1.1 Importance of Walkability and Cycling in Urban Streets

Most urban nations today promote walking and cycling for health reasons. Several studies indicate that increasing the walkability of streets can contribute to better public health, increase social capital production, reduce obesity rates at the societal level, and increase the value of residential properties (Tiemann et al., 2012). For instance, a study conducted in the Netherlands found that people who cycled to work had a 40% lower risk of dying than those who drove or took public

Fig. 21.1 Cyclists during curfew time (April 2021). *Source* Authors photos

transportation (de Hartog et al., 2010). Another study in the United States found that people who lived in walkable neighborhoods were more physically active and had lower rates of obesity and diabetes (Frank et al., 2008). Research has also shown that walkable neighborhoods are associated with increased social interaction and stronger community ties. For instance, a study in the United States found that residents of walkable neighborhoods had more social contacts and were more likely to trust their neighbors than those in less walkable neighborhoods (Ewing & Cervero, 2010).

Promoting walking and cycling in urban areas can lead to numerous benefits, including improved public health, increased social capital, reduced obesity rates, increased property values, and environmental benefits (Forsyth & Krizek, 2010; Shaaban, 2019).

1.2 Street Evaluation and Retrofit for Walkability and Cycling

To make the streets of an urban city more walkable, Shaaban (2019) presents several walkability indicators. These indicators can evaluate the walkability of the streets in Kuwait and other urban nations. Included among the indicators are obstructions, continuity, physical condition, encroachments, cleanliness, slope, shade trees, and buffer zone. Objects such as utility poles, signposts, fire hydrants, and tree trunks constitute obstructions. These objects should be kept to a minimum because they narrow the sidewalk. Additionally, sidewalks must be continuous, which would benefit the elderly and the disabled. The surface of the sidewalks must be firm, nonslip, devoid of cracks, and comfortable for walking. To avoid obstructing the sidewalk, encroachments such as trash cans, bushes, and advertising signs must be kept to a minimum. To provide a pleasant experience for pedestrians, sidewalks should be clean. The slopes must be clear to improve wheelchair accessibility. To shield pedestrians from the sun's heat, sidewalks must be equipped with shade trees. For the safety and comfort of pedestrians, sidewalks should include buffer zones.

The planning of bicycle-friendly infrastructures in metropolitan areas must be incorporated by urban designers. Bicycle facilities, such as bike trails and multi-use paths, should be provided foremost. The report by Delaware Valley Regional Planning Commission (2011) and the work by Butler (2015) present the essential principles for the creation of bike trails. Therefore, the bike trails must supplement the off-road paths rather than serve as a substitute. There should be a few crossings and independent routes. These paths should also connect to the primary transportation system.

Additionally, there are guidelines for designing bike trails. The clearance and width should be approximately 3 m, and the path should be two-way and shared. To facilitate the speed of the cyclists, the grade of the path should not exceed five percent. To optimize the management of storm water, the trail drainage slopes should be in a single direction. Along the trail, there should be signs and pavement markings. These signs function as warnings, information, distance markers, and others (DVRPC, 2011; Butler, 2015).

The bike lanes are components of the roads. Through the use of signage, striping, and pavement markings, they are distinguished. These lanes must be one-way, on the right side of the roadway, and located between the parking and travel lanes. These lanes have a minimum width of 1.2 m and lack a gutter and curb. When a street is one-way, bike lanes must be placed on the left side of the road. This location can reduce street conflicts (DVRPC, 2011).

2 Research Aims and Methodology

This study aims to assess and improve walkability and cycling in urban areas, with a focus on a case study in Kuwait. The study will investigate walkability and cycling in two neighborhoods and streets in Kuwait (specifically Mishrif and Salem Sabah Al-Salem Al-Sabah Coastal Street) to provide a comprehensive understanding of the current state of walkability and cycling in the case study area and to identify potential design interventions for improvement.

2.1 Research Design

The research design will comprise three main phases:

2.1.1 Preliminary Data Collection

In this phase, the research team will preliminarily review existing literature on walkability and cycling in urban areas, with a focus on developing countries and Kuwait. This review will help inform the development of evaluation criteria and design interventions for the case study area.

2.1.2 Walkability and Cycling Assessment

In this phase, the research team will conduct a comprehensive assessment of walkability and cycling in the case study area. The assessment will be conducted using a range of evaluation criteria, such as sidewalk coverage, street connectivity, safety, and comfort for pedestrians and cyclists, as suggested by Shaaban (2019), and Butler (2015).

2.1.3 Design Intervention Evaluation and Implementation

In this phase, the research team will evaluate the effectiveness of potential design interventions for improving walkability and cycling in the case study area. The design interventions will be developed based on the results of the walkability and cycling assessment and will include infrastructure improvements, such as new sidewalks, bike lanes, and traffic-calming measures. The research team will use the relevant codes and standards for this purpose.

After evaluating the streets, the research team will propose retrofit solutions for the selected streets to demonstrate the possibility of low-cost, easily implementable modifications that can affect the level of service provided on Kuwait's streets.

3 Data Collection and Analysis

In order to achieve the study aim, the following data collection and analysis strategy was originated (Fig. 21.2).

3.1 Selected Sites Information

The following two sites were selected for evaluation: The sites were selected to ensure that it covers residential, commercial, and mixed-use sites:

3.1.1 Residential (Mishref)

In this study, Mishref has been selected to have five of its roads evaluated. Mishref was one area most appropriate to this study as it attracted cyclists and pedestrians, because of the points of interest available nearby, its aesthetic sights, and having its road length equal to 15.42 km for a full course (Fig. 21.3).

3.1.2 Commercial/Mixed-Use: Salem Sabah Al-Salem Al-Sabah Coastal Street

Salem Sabah Al-Salem Al-Sabah Coastal Street is in the south of the State of Kuwait and extends to many coastal areas, for instance, Mahboula, Abu Halifa, Manqaf, and Fahaheel (Fig. 21.4).

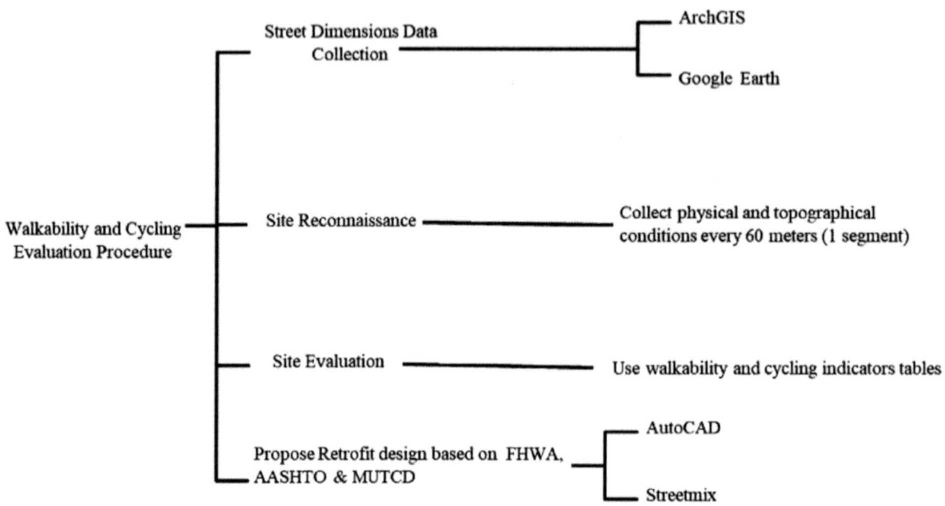

Fig. 21.2 Data collection and analysis. *Source* Authors, developed for this study

Fig. 21.3 Mishref block 3 roads. *Source* Google Earth, 2021

Fig. 21.4 Salem Sabah Al-Salem Al-Sabah Coastal Street. *Source* Google Earth, 2021

Salem Al-Sabah Coastal Street length back and forth is approximately 17.850 km. The street contains mixed types of properties, such as Abu Halifa port for the export of oil, commercial and residential buildings, resorts, hotels, shopping malls, stores, restaurant complexes, mosques, schools, and hospitals.

4 Results

4.1 Data Collection and Site Reconnaissance

From ArcGIS and Google Earth, the road and sidewalk dimensions were acquired. In Mishref, the road is a 4-lane

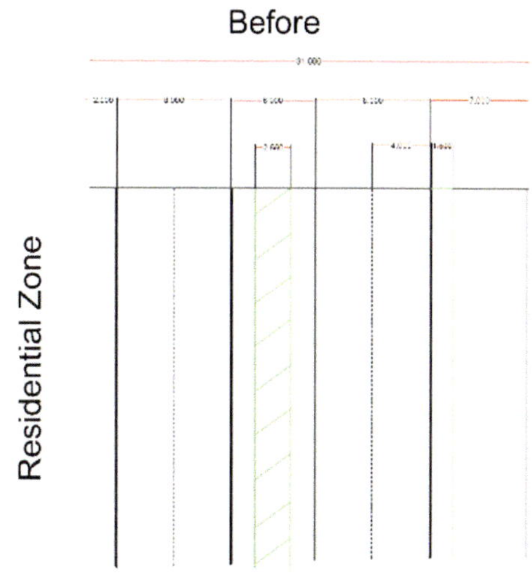

Fig. 21.5 General street design (before). *Source* Authors measurements

urban locally divided street, with each lane being 4 m wide. The road is divided by a 6 m wide median, including the 0.125 m curb on each of its sides. The median has a 2.55 m planter zone in the middle, and two empty zones that can be utilized as sidewalks on each side. One of these zones is 1.7 m wide, while the other is 1.75 m wide. The sidewalk on side 1 (shoulder) is only 2 m wide, which is enough for a pedestrian zone only. However, a long part of side 1 is on the opposite side of the residential zone. This sidewalk is mostly empty, with only trees on it and a cleared zone for pedestrians to use that is only 1.5–2 m wide. However, the actual width of the sidewalk is 7 m and can be redesigned efficiently to use the space.

Each lane on Salem Al-Sabah Coastal Street is 3.67 m wide, and there are six lanes. In addition, the width of the sidewalk is variable, ranging from 3 m to dimensions as small as 2.5 and 2 m. In addition, there is a 7-m-wide sidewalk in front of Abu Halifa Port for the export of oil.

The island, which is the middle sidewalk between two streets, has a width of 6.5 m along the road. In addition, there is an intersection between the areas of Manqaf and Abu Halifa, with a radius of 45 m for the pavement edge corner and 6 m for the channelizing island.

The project team conducted a site reconnaissance exercise for the study areas to determine and comprehend its physical and topographical conditions, which included the following activities: recording the physical characteristics of streets in terms of cross-sections, identifying and recording the condition of existing sidewalks, assessing pedestrian routes and identifying pedestrian crossing locations, recording existing land uses and right of-way, evaluating existing pavement condition, and documenting existing land uses and right-of-way.

4.2 Indicator Tables

Three comprehensive evaluation tables designed by Shaaban (2019) and Butler (2015) were utilized to systematically identify design issues. This was a crucial step in discovering the major design flows of the selected streets.

The total length of roads for the two areas, namely Mishref and Salem Sabah Al-Salem Al-Sabah Coastal Street, is 33.27 km. Moreover, each road was divided into segments, and each segment is 60 m long, totaling 555 segments.

4.3 Walkability (Sidewalk) Indicators

Following Shaaban (2019), indicators of walkability have been measured (see Table 21.1). This evaluation determines which streets need to be retrofitted or modified to accommodate pedestrians.

The sidewalks on Mishref and Sabah Al-Salem streets received a score of 2.95 because they were clean, had a good slope, and had few obstructions, as shown in Table 21.1. There are one to two obstructions on the sidewalk for every 60 m. The sidewalks are continuous, but there are no buffer zones adjacent to them.

4.4 Corridor Indicator Table

4.4.1 Mishref

According to Table 21.2, Mishref's course lacks crosswalks, pedestrian signals, benches, and security patrols. The proportion of drivers who yielded to pedestrians was between two and five, and the same holds true for drivers who speed. In addition, because of the low volume of traffic, most drivers do not obey traffic signals, and the road has few traffic-calming measures.

On side 1, there are few parking sunshades to provide protection from the sun and rain. There is only one bus stop along the entire length of the roadway, and there are hardly any curb ramps. This side's illumination comes from houses or sports facilities, which is sufficient. Side 2 has an abundance of trees that provide protection from the sun and rain. However, this side of the road lacks transit ceases and curb ramps. This side's sidewalk poles provide excellent illumination.

4.4.2 Salem Sabah Al-Salem Al-Sabah Costal Street

Salem Al-Sabah Coastal Street comprises six routes, three routes from the Fahaheel area to the Mahboula area, and three routes from the Mahboula area to the Fahaheel area. In addition, the street has two sides of the corridor.

Side 1: This side contains a few trees. In addition, side one lacked pedestrian crossings, bridges, and underpasses to safeguard pedestrians. No traffic signs or curb ramps for pedestrians with mobility devices were present. In addition, there was a lack of seating and lighting along the street, which would increase the discomfort and safety of pedestrians and make them invisible to motorists. Heavy traffic but few trucks are present. Also, the lack of traffic-calming devices such as speed bumps along the street. Numerous accidents are caused by drivers' disregard for pedestrians, but some drivers obey the traffic lights. Few police patrols and cameras were available to ensure the safety and security of pedestrians and motorists (Table 21.2).

Side 2: This side has good lighting along the road, which increases the comfort and safety of pedestrians, and many trees shield pedestrians from direct sunlight and heavy rain but may obstruct drivers' vision if trees are not trimmed monthly. There was no traffic sign and no curb ramps for mobility-aided, blind, and visually impaired pedestrians. This complicates the movement between the sidewalk and the road. Moreover, there are no chairs or benches. Heavy traffic but few trucks are present. Also, the lack of traffic-calming devices such as speed bumps along the street.

Table 21.1 Results of sidewalk indicator

Indicator											
Mishref	Side 1*	3.9**	3.7	1.9	2.7	4	2	4	2.1	1	2.8
	Side 2*	3.9	3.7	4	3	4	2	4	2.6	1	3.1
	Overall	3.9	3.7	2.95	2.85	4	2	4	2.35	1	2.95
Salem Sabah Al Salem Al Sabah Costal Street	Side 1	3.93	3.74	3.96	2.85	4	1.26	4	1.86	1	3
	Side 2	3	3.81	3.95	2.92	4	1.04	4	2.34	1	2.89
	Overall	3.48	3.84	4	2.92	4	1.15	4	2.18	1	2.95

*Side 1: shoulders of the street, and side 2 is the island (median of the road)

**Scores are from 1 to 4, where 1 is worst, and 4 is best

Table 21.2 Results of corridor indicator

Indicator																				
Mishref	Side 1 Side 2 Overall	R/C *	1** 1 1	1 1 1	1 1 1	1.4 1 1.2	1 1 1	2.1 2.6 2.4	1 1 1	1 1 1	2 3 2.5	2 2 2	1 1 1	1 1 1	2 2 2	3 3 3	3 3 3	1.1 1.1 1.1	1.6 1.7 1.65
Salem Sabah Al-Salem Al-Sabah Costal Street	Side 1 Side 2 Overall	R/C	1.03 1.03 1.03	1 1 1	1 1 1	1 1 1	1 1 1	1.62 1.90 1.78	1.02 1.02 1.01	1.06 1 1.03	1 3 2	1.96 1.96 1.96	1.96 1.96 1.96	2.92 2.92 2.92	2 2 2	2.95 2.95 2.95	2 2 2	1 1 1	1.50 1.63 1.57

*R denotes residential land use; C denotes commercial land use
**Scores are from 1 to 3, where 1 is worst, and 3 is best

4.5 Bicycle Lane Indicator

4.5.1 Mishref

As per Table 21.3, there are no bicycle lanes in Mishref, but the road is highly visible. The roads are convenient for cyclists and have safe intersections. Cyclists are treated with contempt by motorists, who also drive at excessive speeds. In addition, cyclists have no way to navigate around corners and roundabouts.

4.5.2 Salem Sabah Al-Salem Al-Sabah Costal Street

There are no bicycle lanes on the first and second sides of the street, and the street is unsuitable for cyclists because of its uneven surface. Moreover, the road is lined with numerous commercial complexes, commercial and residential buildings, schools, hospitals, and stores, resulting in heavy traffic congestion. On the street, vehicle speeds between medium and high posed a danger to the safety and security of cyclists. Furthermore, motorists disregard the law and cyclists. There is contention between vehicles and cyclists at roundabouts and turns because cyclists lack a designated lane to shield them from danger. Additionally, there is sufficient lighting and dense trees that provide shade and shelter from the sun and rain, but trees that obstruct drivers' vision because they are not trimmed monthly (Table 21.3).

4.6 Retrofit Maps

After identifying existing design issues through site visits, preliminary design options were explored based on design criteria extracted from various sources, including the Federal Highway Administration (FHWA), the American Association of State Highway and Transportation Officials (AASHTO), and the Manual on Uniform Traffic Control Devices (MUTCD). Hence, and based on the data collected, analyzed, and the international codes and standards reviewed, and best practice design criteria identified, retrofit designs were developed where layout plans, concept profile for the street's typical street cross-sections, right-of-way limits and requirements, typical pavement structure, design criteria were considered.

AutoCAD software package and Streetmix (https://streetmix.net/) were used for presenting the retrofitted designs.

4.6.1 Mishref

The retrofitting process began with the export of the current roads data into AutoCAD. Henceforth, the development process commenced by determining the optimal placement of bike lanes for the safety and comfort of cyclists of all ages. It is common knowledge that bike lanes are placed on roads according to various codes and standards. Based on AASHTO specifications, a 1.5 m wide bike lane was

Table 21.3 Results of bicycle indicator

Indicator												
Mishref	Side 1 Side 2 Overall	1* 1 1	3 3 3	4 4 4	2.7 2.8 2.75	4 4.6 4.3	2 2 2	2 2 2	3 3 3	1 1 1	1 1 1	2.4 2.4 2.4
Salem Sabah Al Salem Al Sabah Costal Street	Side 1 Side 2 Overall	1 1 1	4 4 4	2 2 2	2.82 2.89 2.79	3.70 4.63 4.20	2 2 2	2.07 2.07 2.07	2.60 2.36 2.48	1 1 1	1 1 1	2.22 2.29 2.25

*Scores are from 1 to 5, where 1 is worst, and 5 is best

selected for this design (DVRPC, 2011). The road in Mishref has a 7-m sidewalk that is underutilized. It was determined to be an ideal location for the 1.5 m wide bike lane, as the bike lanes of both directions in this design will be 3 m wide and adjacent to one another. Because the median and sidewalk adjacent to the residential area could not accommodate a bike-only lane, the bike lanes of both directions were placed next to one another for a complete course. Figures 21.5 and 21.6 depict the road before and after adjustments were made.

After the bicycle lane was positioned strategically, the pedestrian zone had to be designated. The pedestrians could utilize both the median and the residential zone sidewalk. However, the median is not specifically designated for pedestrians, as it can be utilized by both pedestrians and cyclists as a shared lane. Because of the U-turns, it would be considered a short course, but it would be ideal for children and adolescents. The median is 6 m wide with a 2.55-m planter and light pole zone in the center, a 1.7-m-wide sidewalk on the right and a 1.75-m-wide sidewalk on the left. These dimensions are ideal for pedestrians, as the minimum pedestrian zone width according to AASHTO is 1.525 m (DVRPC, 2011), and they remain unmodified as this study seeks efficient and low-cost adjustments.

By providing crosswalks along the road, cyclists can travel from their bike lanes on the far right to the medians, and pedestrians can travel from the median to the sidewalk next to their homes. The crosswalks specified in this design are wider than the MUTCD minimum width requirement of 1.8 m. This was done to increase pedestrian and cyclist visibility, safety, and well-being (Fig. 21.7). Crosswalks are strategically positioned before U-turns where there is a high likelihood of existing speed humps. The speed bumps must be placed between 5 and 10 m from the crosswalk to slow down and alert speeding drivers. In addition, pedestrian crossing signs must be placed at this distance before the crosswalk. Alongside crosswalks, curb ramps are provided to facilitate pedestrian and bicycle access, particularly for those with special needs.

Several segments of the sidewalk on the far left of Fig. 21.6 are in deteriorating physical condition and are lined with parked cars. To address this issue, a portion of the 7-m-wide sidewalk was sacrificed to create on-street parking on the far-right side. According to the Kuwait Highway Design Manual (Kuwait Municipality, n.d.), the parking lot's width without an angle is 2.6 m. This can be used for residences with a limited amount of land available for car parking. After the cars are removed, the sidewalk must be renovated so that it is suitable and comfortable for pedestrians.

Mishref has a 2.55-m-wide planter zone in the center of the median. It has lampposts, trees in the median, and traffic

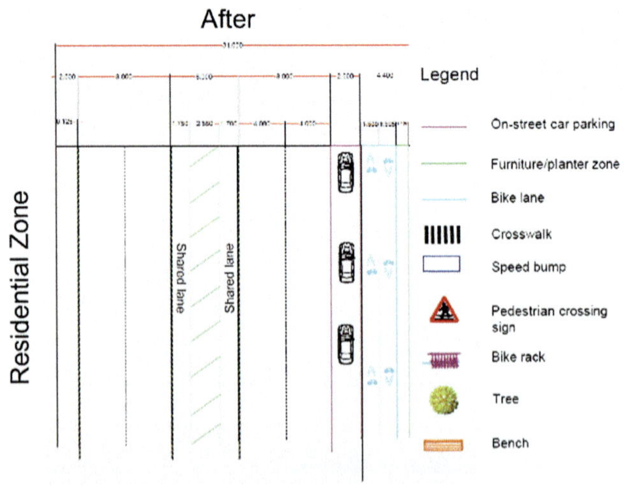

Fig. 21.6 General street design (after). *Source* Authors measurements

signs. However, in this design, another planter zone with a width of 1.125 m was added. It should have a few amenities for pedestrians and cyclists, including trees, benches, bike racks, and light poles.

This would therefore summarize the general design concept for Mishref's intended retrofit. As depicted in Figs. 21.8 and 21.9, the design was subsequently applied to multiple roadway segments.

Figure 21.8 demonstrates how a bicycle lane would function around a roundabout. This is accomplished by placing a crosswalk beneath the roundabout to connect the lane to the opposite side for a safer crossing. However, speed bumps are not available at that location; therefore, they should be placed to protect cyclists from vehicles exiting the freeway. This concept would be implemented at every roundabout on the Mishref course.

Figure 21.9 depicts a problem in which the bike lane course is interrupted by the road and continues on the opposite side. A crosswalk was placed before a U-turn to connect the lanes and ensure the presence of speed bumps. By connecting the bike lanes in this manner, it is guaranteed that the lanes will pass by the bus stop for the convenience of bike riders.

4.6.2 Salem Sabah Al-Salem Al-Sabah Coastal Street

After the dimensions and measurements of the road and the side and center sidewalks have been determined using Google Earth and ArcGIS, the next step is to enter the identified data into AutoCAD.

First, development measures have facilitated the construction of cycling paths for amateur cyclists. It is common knowledge that bicycle lanes are placed on the road, but after collecting data on the streets, it was discovered that most motorists lack sufficient traffic awareness regarding cyclists'

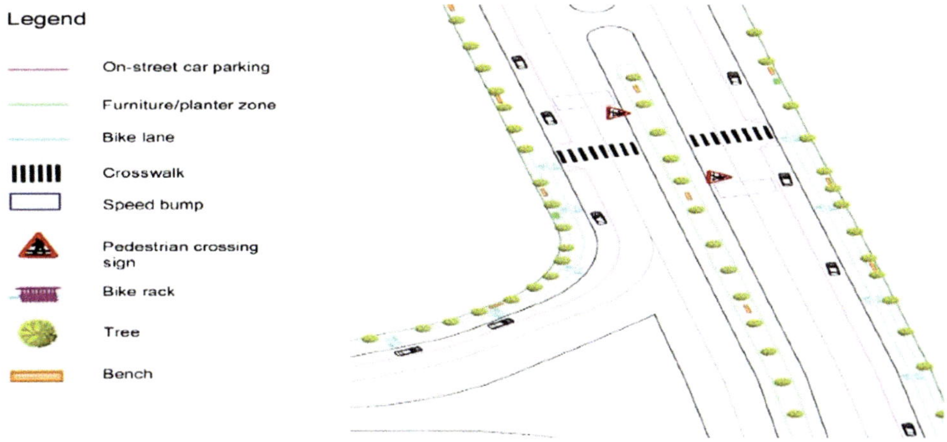

Fig. 21.7 Sample segment of Mishref's road. (*source* Authors)

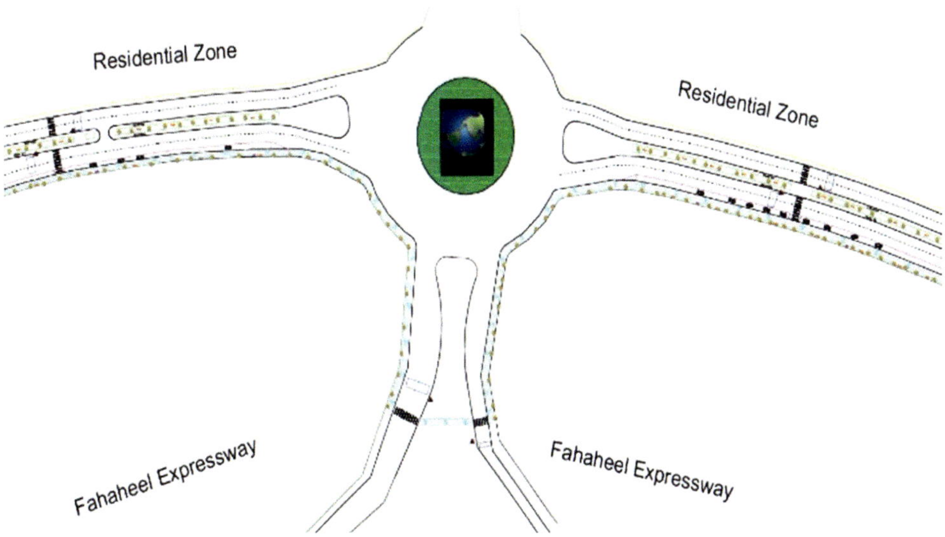

Fig. 21.8 Mishref's first roundabout into the area (*source* Authors)

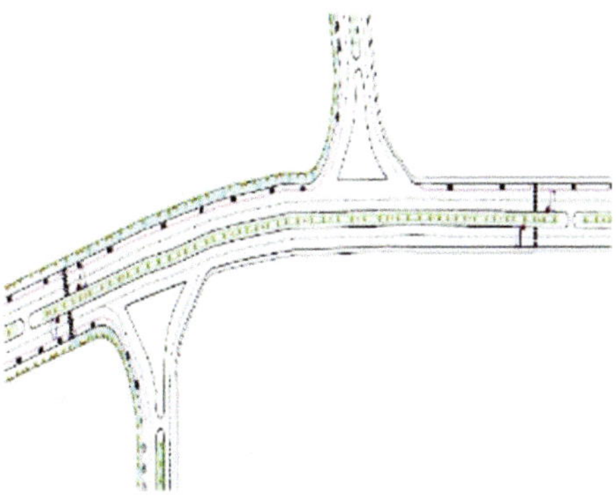

Fig. 21.9 Sample segment of Mishref's road comprising U-turns (*source* Authors)

safety and do not adhere to the regulations. To ensure the safety of cyclists and drivers, special bicycle lanes will be installed on the sidewalk. AASHTO and DVRPC (2011) specifies that the minimum width of the bicycle lane on roads with a curb or adjacent to parking is 1.5 m. It turns out that Salem Al-Sabah Street contains a usable, 6.5-m-wide central berth on which a bicycle path is placed.

Figure 21.10 depicts an intersection between the Mangaf neighborhood and Abu Halifa, where the bicycle path begins. In this design, the bicycle path will be 1.5 m wide for each lane in both directions, with a total width of 3 m.

Second, the Ministry of Public Works performed numerous repairs, maintenance, and expansions on the road, resulting in a significant reduction of shoulder space in both directions. Therefore, the sidewalks must be widened so that pedestrians can use them. In addition, before beginning the expansion of the sidewalks, the Ministry of Municipality and private offices of residential buildings owned by citizens on Salem Al-Sabah Street confirmed empty spaces used by some building owners as car parks are owned by the State of Kuwait and not private property, allowing the sidewalks to be redesigned without obstructions.

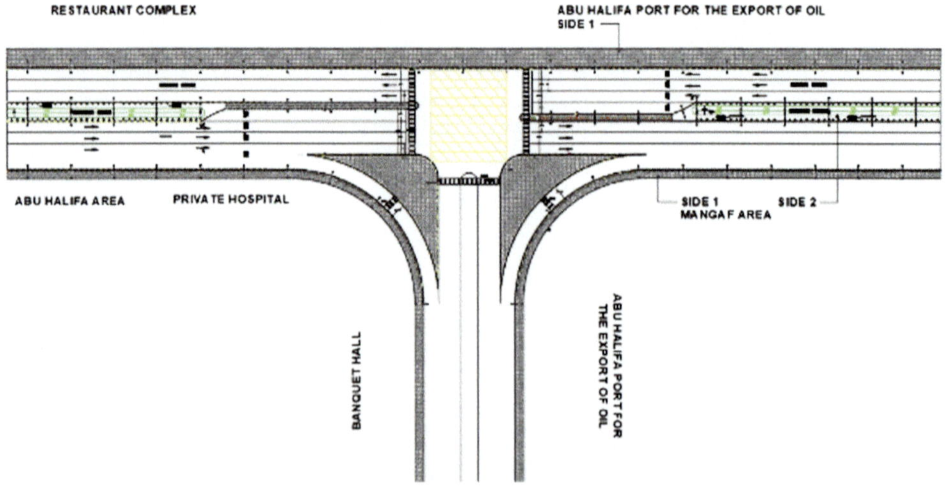

Fig. 21.10 Intersection between the Mangaf neighborhood and Abu Halifa (*source* Authors)

As depicted in Fig. 21.11, there is no private sidewalk for pedestrians, and some residential buildings have used the sidewalk as car parks, although each residential complex has a private basement for parking. Consequently, a 5-m-wide sidewalk that is divided into several sections was constructed, with a width of 5 m (Fig. 21.12).

Third, as depicted in Fig. 21.13, the width of the sidewalk on the right is 2.5 m, and it has been extended by 5 m to the right, bringing the total width to 7.5 m. According to Kuwait Public Transport Company, a portion of the sidewalk with a width of 4.5 m and a length of 90 m has been cut to accommodate multiple buses stopping at the station, noting that the station's length previously measured 31.5 m (Figs. 21.12 and 21.14). In addition, after the reduction, the pavement is now three meters wide and divided into several sections along the length of the road.

Fourthly, along the road, pedestrian walkways have been installed to reduce the number of accidents involving pedestrians being run over. Cyclists can also use the pedestrian walkways to access Salem Al-Sabah coastal street's vital services, such as Abu Halifa Port for oil export, commercial and investment buildings, resorts, hotels, shopping centers, stores, restaurant complexes, mosques, schools, and hospitals. Additionally, corridors are strategically placed before corners and at intersections (Fig. 21.15). The width of the pedestrian path is 2 m, whereas the

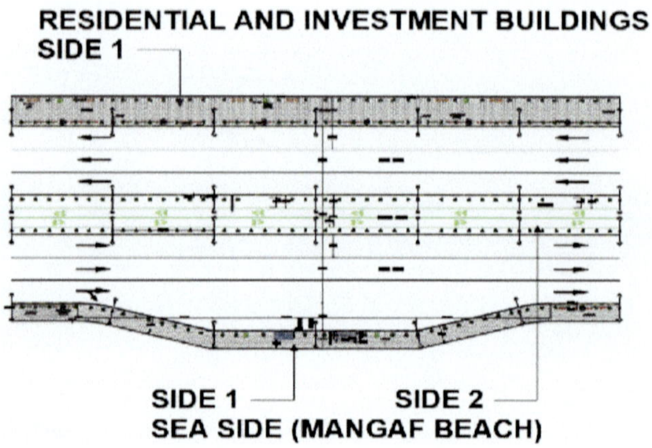

Fig. 21.12 Retrofit sidewalks and bus station (*source* Authors retrofit design)

Fig. 21.13 Sidewalk and bus station. *Source* Google Maps (2021)

Fig. 21.11 No private sidewalk for pedestrians. *Source* Google Maps (2021)

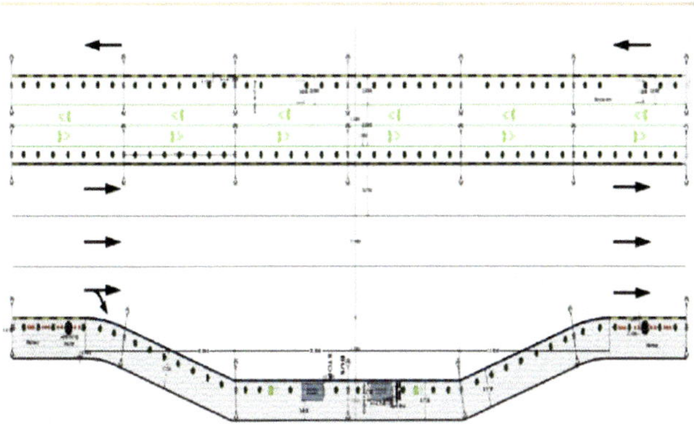

Fig. 21.14 Retrofit sidewalk and bus station (*source* Authors retrofit design)

MUTCD specifies that the minimum width for pedestrian paths is 1.8 m. Also, approximately 15 m before the pedestrian crossing site, warning signals must be installed, and at the pedestrian crossing site, a manual traffic light must be installed to facilitate the access of pedestrians, cyclists, and people with special needs to ensure the safety of all.

5 Retrofit Impact and Discussion

5.1 Mishref

After the redesign, it is necessary to determine how the retrofit affected the performance of the road. Table 21.4 displays sample segments whose scores have been drastically altered as a result of the retrofit by comparing their current and improved scores. All scores increased, indicating that walkability and cycling have improved. The selected criteria in the table are those that required the most attention, as they represent the majority of what is severely lacking on the road to achieve a better outcome.

Consequently, the installation of bike lanes on the sidewalk would prioritize its use for cyclists and enhance their comfort and safety. The crosswalk and pedestrian signs would increase the safety of pedestrians and cyclists. The crosswalk would be visible from a distance to give drivers time to slow down, and the curb ramp would ensure that cyclists and individuals with special needs have a smooth transition from the sidewalk to the road. Before the crosswalk is a speed hump as a traffic-calming measure to slow down drivers. Seats and benches on the sidewalk and median are necessary for the comfort of pedestrians. The sidewalk would be cleared of parked vehicles, allowing pedestrian access. Therefore, its physical condition will improve when the friction between car tires and sidewalk tiles ceases to exist. In addition, the furniture/planter zone contains shade trees, benches, light poles, and bike racks. However, they may also include drinking fountains, surveillance cameras, and artwork.

The road's improved performance would encourage pedestrians and cyclists to utilize it. Figures 21.16 and 21.17 depict the road's general layout before and after the retrofit.

Fig. 21.15 Corridors provided before turns and at intersections (*source* Authors)

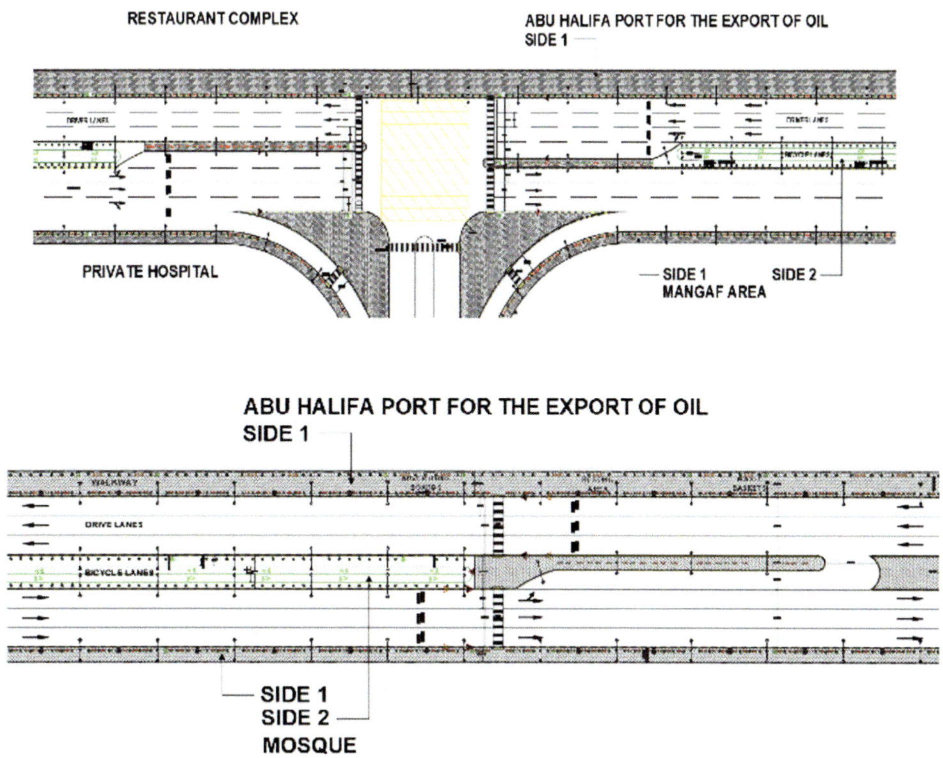

Table 21.4 Scoring improvement in Mishref (sample segments)

Segment	Current	Retrofit	Impact on walkability	Impact on cycling	Score (before)	Score (after)
96–108	Bicycle lane = 1 Directness and Efficiency = 2.7 Crosswalk = 1 Crossing sight distance = 1	Bicycle lane = 2 Directness and Efficiency = 2.7 Crosswalk = 3 Crossing sight distance = 3	Increase ↑	Increase ↑	16.5	28.7
	Pedestrian signal availability = 1 Curb ramp = 1.2 Seats and benches = 1 Traffic-calming measures = 1.2 Encroachments = 3 Physical condition = 3.4	Pedestrian signal availability = 2 Curb ramp = 4 Seats and benches = 3 Traffic-calming measures = 2 Encroachments = 3 Physical condition = 4				

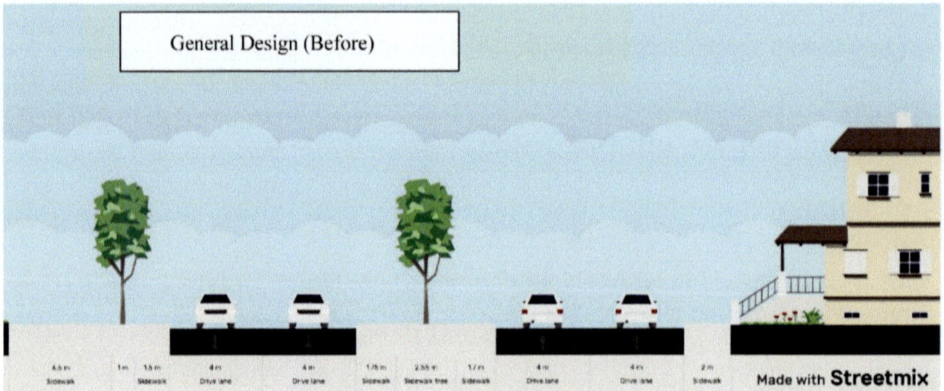

Fig. 21.16 Mishref general design (before) (Streetmix, 2022)

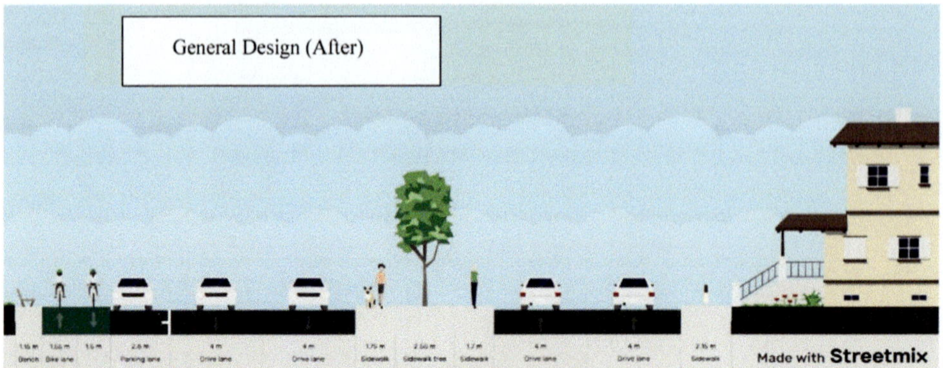

Fig. 21.17 Mishref general design (after) (Streetmix, 2022)

5.2 Salem Sabah Al-Salem Al-Sabah Coastal Street

After the redesign of Salem Al-Sabah Street using AutoCAD has been completed, it is necessary to determine the effect of the retrofit on the road and sidewalk performance. Table 21.5 illustrates sample segments and the major components that have undergone significant modification as a result of retrofitting by comparing current and modified results. All scores increased, which indicates that walking and cycling have improved. Moreover, the selected criteria in the table are those that require the most attention, as they

Table 21.5 Scoring improvement in Salem Al-Sabah Costal Street (sample segments)

Segment	Current	Retrofit	Impact on walkability	Impact on cycling	Score (before)	Score (after)
2280–2400	Bicycle lane = 1 Crosswalk = 1 Pedestrian signal availability = 1 Curb ramp = 1 Obstructions = 3.5 Seats and benches = 1 Shelters from rain and Sun = 1.5 Street lighting = 2 Shade trees = 2 Width = 1 Traffic-calming measures = 1	Bicycle lane = 3 Crosswalk = 3 Pedestrian signal availability = 3 Curb ramp = 4 Obstructions = 4 Seats and benches = 3 Shelters from rain and Sun = 3 Street lighting = 3 Shade trees = 3 Width = 3 Traffic-calming measures = 2	Increase ↑	Increase ↑	16	34

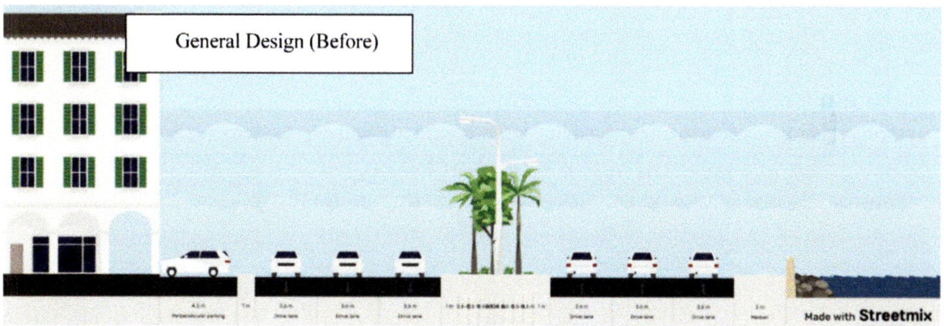

Fig. 21.18 Salem Sabah Al-Salem Al -Sabah Costal Street (Before). *Source* Streetmix, 2022

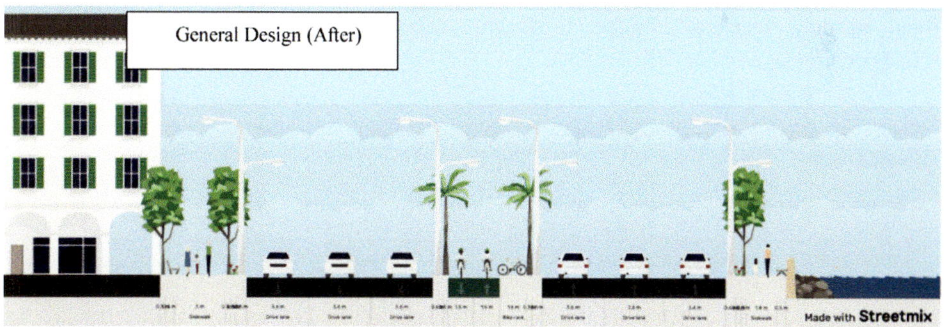

Fig. 21.19 Salem Sabah Al-Salem Al-Sabah Costal Street (After). *Source* Streetmix, 2022

represent the majority of what the path is sorely lacking in its pursuit of a better outcome.

The lighting is provided to enhance the quality of life in the community by making residents feel safe at night and enhancing their visibility. Landscapes are an ideal source of beauty, comfort, energy, and relaxation for pedestrians, so trees are planted to reduce environmental pollution, provide shade from the sun, and filter rainwater. Figures 21.18 and 21.19 depict the road's general appearance before and after the retrofit.

6 Conclusion

This study stems from an effort to better understand, answer and add to the literature questioning how do design codes and standards affect "livability" and what role do they play in determining future urban quality in a community.

As a case study, this research examined walkability and cycling in two local neighborhoods in Kuwait. In addition, it was decided to use simple tools for collecting, assessing, and analyzing road data to identify deficiencies and problems, as well as what is required to improve it, to meet the aim of improving the experience of pedestrians and cyclists (i.e., livability) in various regions of the State of Kuwait.

For information on how to evaluate roads, the literature was consulted. For data collection and redesign of existing streets, codes and standards as well as multiple applications, such as Google Earth, ArcGIS, AutoCAD, and Street mix, have been utilized. The study proposed an improved urban design for the selected streets, which can be applied in several sites and different settings.

Mishref and Salem Sabah Al-Salem Street would benefit from bike lanes, crosswalks, an improved furniture/planting zone, and pedestrian signs. This would improve roadside safety, usability, visual appeal, and service quality. Besides being practical and cost-effective, the suggested solutions can be implemented. The proposed solutions are assumed to be cost-effective because they do not require the construction of new bridges, tunnels, overpasses, or underpasses.

The methodology used in this research provides a comprehensive approach for assessing and improving walkability and cycling in urban areas, with a focus on a case study in Kuwait. It allowed for a comprehensive understanding of the current state of walkability and cycling in the case study area, and the potential effectiveness of design interventions for improvement.

While the study has potential limitations, the findings from this research will be valuable for local authorities and community groups in Kuwait and other developing countries, as they seek to promote sustainable and healthy transportation options in urban areas.

References

Atef Elhamy Kamel, M. (2013). Encouraging walkability in GCC cities: smart urban solutions. *Smart and Sustainable Built Environment, 2*(3), 288–310. https://doi.org/10.1108/SASBE-03-2013-0015

Butler, A. (2015). *Multiple criteria bicycle route assessment.* Retrieved from http://essay.utwente.nl/84015/1/butler.pdf

de Hartog, J., Boogaard, H., Nijland, H., & Hoek, G. (2010). Do the health benefits of cycling outweigh the risks? *Environmental Health Perspectives, 118*(8), 1109–1116. https://doi.org/10.1289/ehp.0901747

Delaware Valley Regional Planning Commission. (2011). Pedestrian and bicycle friendly policies, practices, and ordinances (Issue November). Retrieved from http://www.dvrpc.org/reports/11019.pdf

Ewing, R., & Cervero, R. (2010). Travel and the built environment. *Journal of the American Planning Association. American Planning Association,* 76(3), 265–294. https://doi.org/10.1080/01944361003766766

Forsyth, A., & Krizek, K. J. (2010). Promoting walking and bicycling: Assessing the evidence to assist planners. *Built Environment, 36*(4), 429–46. http://www.jstor.org/stable/23289968

Frank, L. D., Andresen, M. A., & Schmid, T. L. (2008). Obesity relationships with community design, physical activity, and time spent in cars. *American Journal of Preventive Medicine, 27*(2), 87–96. https://doi.org/10.1016/j.amepre.2004.04.011

Gomes, A. (2020). *Urban planning and its legacy in Kuwait.* Retrieved from https://blogs.lse.ac.uk/mec/2020/04/23/urban-planningand-its-legacy-in-kuwait/

Kuwait Municipality, (n.d.) *Kuwait highway design manual.* Kuwait.

Kuwait Municipality. (2018). 4th Kuwait master plan 2040.

Shaaban, K. (2019). Assessing sidewalk and corridor walkability in developing countries. *Sustainability, 11*(14). https://doi.org/10.3390/su11143865

Tiemann, T., Scott, A., & Atkins., K. (Eds.). (2012). Sidewalks, streets and walkability. *Spaces and Flows: An International Journal of Urban and Extra-urban Studies.*

Weiderpass, E., Botteri, E., Longenecker, J. C., Alkandari, A., Al-Wotayan, R., Al Duwairi, Q., & Tuomilehto, J. (2019). The prevalence of overweight and obesity in an adult Kuwaiti population in 2014. *Frontiers in Endocrinology, 10*, 449. https://doi.org/10.3389/fendo.2019.00449

Deficiencies Within CE and LCA Models: Agrowaste Fibre and Conventional Insulation Materials

Farres Yasser, Anton Ianakiev, Marisela Mendoza, and Hynda Aoun Klalib

Abstract

Energy consumption is rising in buildings due to constantly growing populations and the rapid growth of urban building construction projects. The poor thermal management of buildings adversely affects the economy, environment, and living standards. These issues entwine with the general need to mitigate greenhouse gas emissions. Hence, many researchers focus on energy efficiency in buildings through active and passive techniques using sustainable materials. Such passive techniques include agrowaste fibre insulation panels, which have received significant attention due to their environmental benefits, cost efficiency, and energy efficiency potential through satisfactory/good thermal insulation levels. However, further in-depth studies are required to understand how agrowaste fibre insulation performs within broader contexts. This chapter scrutinises the current literature on the Circular Economy (CE) and Life Cycle Assessment (LCA) of the most promising agrowaste materials, as well as the current conventional foam and plastic-based insulation. The previous studies conducted juxta positioned, fragmented, incomplete, and unrelated CE and LCA studies of each type of insulation panel rather than purposefully and holistically analysing the overall performance of conventional and agrowaste insulation from cradle to cradle or cradle to grave. Often only one or none of the methods utilised in the manufacturing, processing, disposal, and reuse of insulation phases are considered for judging which insulation type is better for the environment and energy efficiency. It is usually the thermal insulation value and cost of the raw materials used for production that determines which is chosen. This chapter has concluded that this methodology of juxta positioning LCA studies and simulation results presents an incomplete argument. An argument–which cannot be utilised—states that agrowaste fibres are or are not as preferable, from an environmental or thermally insulation perspective, as conventional insulation materials.

Keywords

Life-cycle assessment • Circular-Economy • Agrowaste • Insulation panels • Conventional materials • Energy consumption

1 Introduction

Commercial buildings alone consume 20% of the total energy produced in North America, as well as greenhouse gas emissions. Over 40% of that energy demand and emissions are attributable to their Heating, Ventilation, and Air Conditioning (HVAC) systems (Bisset, 2007; Collins et al., 2010; Laustsen, 2008). Approximately, 27% of all UK carbon emissions are attributed to carbon emissions from residential buildings (Department for Environment Food and Rural Affairs (DEFRA), 2006). Heating and cooling are not solely responsible for these energy consumption levels since infiltration can lead to an increase of three times the required kilowatt-hours per m^2 (kWh/m^2) depending on the infiltration rate. This is particularly the case in `Georgian and Victorian homes where the standard infiltration rate is 0.8–1 air change per hour (ACH), where the best practice is 0.5 ach. Having an infiltration rate of 0.5 ACH may lead to the consumption of 17 kWh/m^2 instead of consumption

F. Yasser (✉) · A. Ianakiev · M. Mendoza · H. A. Klalib
Design and the Built Environment (ADBE), Nottingham Trent University, Maudsley Building, Nottingham, UK
e-mail: farres.yasser@ntu.ac.uk

A. Ianakiev
e-mail: anton.ianakiev@ntu.ac.uk

M. Mendoza
e-mail: marisela.mendoza@ntu.ac.uk

H. A. Klalib
e-mail: hynda.klalib@ntu.ac.uk

A. Battisti et al. (eds.), *Greening Our Cities: Sustainable Urbanism for a Greener Future*,
Advances in Science, Technology & Innovation, https://doi.org/10.1007/978-3-031-49495-6_22

levels reaching 44 kWh/m^2 (Mahadevan, 2006). Nevertheless, insulation still plays a fundamental role in the grand scheme of things as its production, use and then end-of-life disposal or reuse cause the most harm to the environment.

The present rate of housing stock renewal in the United Kingdom is less than 1% (Environment Change Institute, 2005). Given the barriers to technological development, high confidence in new products (such as agrofibre waste insulation) is needed before the innovative solutions can reach the market and go through rigorous testing. Currently, agrowaste fibre insulation is not portraying as a potential to the large group of stakeholders concerned with housing stock renewal (Collins et al., 2010). Presenting complete LCA and CE models that compare conventional insulation materials and agrowaste insulation may increase the consumer demand for novel and more sustainable housing, possibly increasing the turnover rate from conventional to agrowaste insulation. Currently, the situation is quite the opposite. Incomplete models of agrowaste insulation are compared to the fully holistic studies conducted for conventional materials. This gives the impression to the stake holder that agrowaste fibres are a 'blue-sky' initiative at best and not a possible substitute for conventional materials.

While 29% of carbon emissions from the United Kingdom are attributed to building heating due to the lack of building stock renewal and renovation, the insulation used in these buildings also contributes to the degradation of the environment. As such, thermal insulation is seminal to achieving adequate levels of energy efficiency in buildings through cooling and heating operations. Commercial materials such as plastic foams, glass foams, mineral wools (in the form of rollable mats), and rock wools among others, are usually used as thermal insulators. Their in-production emissions make them a threat to the environment. In many cases, they may even carry toxic gases and particles after disposal, threatening human health. Overall commercial and conventional synthetic materials do not propose a solution to the problem, but only a feeble alternative (Liang & Ho, 2007).

Companies focus on how to allocate energy and decarbonisation. They do not focus on the fact that the insulation materials themselves can become carbon sinks and can be more economical and environmentally conscious due to them being derived from waste, specifically agricultural waste. This provides favourable end-of-life processes to the environment, such as reuse, composting or facilitated and efficient recycling. This is contrary to foams such as EPS which are carcinogenic, cause leaching the nutrients of the soil and damage marine ecosystems for centuries before they can degrade. (Ellen MacArthur Foundation, 2016; Pearlmutter et al., 2020).

Agro waste-based insulation is the most comprehensive alternative to conventional synthetic materials considering their promising energy efficiency, thermal insulation, health benefits, and environmental safety. There are of course other natural-based solutions such as mycelium. However, such solutions are not as widely available as agrowaste fibres. They also require specialised manufacturing processes or binders to be functional. Hempcrete is an example of an advantageous agrowaste material. It mixes agrofibres and lime (Bedlivá & Isaacs, 2014). However, this is a relatively controversial product due to the ethical concerns that surround it. It was used as a hybrid for construction and insulation. Cork is also instrumental and environmental, though being a by-product of cork oak mainly located in Portugal (Quercus suber L.) (Abenojar et al., 2014) and harvested only every 9–12 years. Cellulose Fibre Insulation (CFI), whether loose fill or wet spray, has proven its usability with its thermal insulation value of 0.04 W/mK and a density of 50 kg/m^3. Despite its positive environmental impact, from an LCA standard (ISO 14040) perspective, the majority of the time, it is not a sustainable solution. This is because it contains several additives, which leaves no option but to incinerate the CFI at the end of its life. Recycling can only occur where there are no contaminants from the in-use phase or additives from the paper industry (Lopez Hurtado et al., 2016). As a result of the amount of methane that is released from CFI in landfills, it has reportedly triple the global warming impact factor (Schmidt et al., 2004). As such, this study aims to understand if the current LCA standards obtained for conventional and natural materials can properly represent each of their sustainable capabilities of appropriately or not. This would help provide an equal opportunity and consumer demand for developing agrowaste fibres into fully functional, tested, and purchasable products.

Sustainable and renewable fibres that are available in abundant quantities have proven their thermal insulation performance levels. They have been used in experiments that involved producing particle boards in many regions. These fibres include but are not limited to coconut husk (Van Dam et al., 2004; Viswanathan et al., 2000), cotton (Alma et al., 2005), bagasse (Widyorini et al., 2005), wood (Kawasaki & Kawai, 2006), oil palm (Khalil et al., 2007), coffee husk and hulls (Bekalo & Reinhardt, 2010), and waste tea leaves (Shi et al., 2006). They provide a multitude of triumphs not only by providing material from renewable recourses but also low thermal conductivity and provide ongoing economic growth from an agricultural process waste product (Tangjuank, 2011a). Though they provide promising performance capabilities in their energy efficiency during the construction at the hypothetical level and in terms of their use and end-of-life process, there have been very few studies on their capacity within the CE and LCA. This causes a shallow understanding of their true potential as possible staple insulation material alternatives for the construction industry.

2 Promising Materials Lacking Sufficient CE and LCA Data

Intensive research has been undertaken for various types of agrowaste fibres in different regions of the globe. They show great potential for thermal conductivity, and acoustic and thermal resistance properties. Many studies have concentrated mainly on their thermal insulation properties. However, they have not been thoroughly studied through the LCA evaluation to be appropriately considered for mass production, official commercial manufacturing processes, or validation by independent quality assurance or environmental impact assessment bodies. A review of the various unique characteristics of these agrowaste fibres provides an insight into their potential and why they should be further investigated for a holistic approach to their utilisation.

Among these agrowaste, fibrous materials are pineapple leaves (Tangjuank, 2011a), sunflower composites with plaster or epoxy as a binder (Binici et al., 2014), straw bale (Ashour et al., 2010; Carfrae et al., 2009; Chougan et al., 2020; Palladino et al., 2021), seeds and bagasse. The thermal insulation properties of a pineapple can be utilised in various forms, including panels made from shredded and dried pineapple leaves fused through natural rubber latex. The results showed that panel densities of 178 and 232 kg/m^3 had thermal conductivity levels that ranged between 0.035 and 0.043 (W/mK) and fire resistance of 1.35 min according to the American Society for Testing and Materials ASTM D 635–98 fire resistance standard (American Society for Testing & Materials Standards, 2003; Tangjuank, 2011b). Other panels use the hot-pressing method to bond pineapple leaf fibres and natural rubber; a 338 kg/m^3 dense panel showed a thermal conductivity of 0.057 W/mK (Kumfu & Jintakosol, 2012). Please see Table 1 for a list of some of the aforementioned agrowaste insulation products, their thermal performance and densities.

Sunflower waste appears mainly in Europe, representing approximately 60% of its global waste. It is used to form particleboards, and fibreboards manufactured using ground sunflower piths (Lopez Hurtado et al., 2016) with various grain densities, sizes, and diameters. Also, a de-oiled cake was produced during the sunflower refinery processes (Evon et al., 2014). A thermal conductivity value of 0.0885 W/mK was apparent, though when using the stalks of sunflowers combined with textile waste, the resulting thermal conductivity performance was 0.0728 W/mK (Sañudo et al., 2016). Which is not as ideal as the 0.03–0.045 range that conventional foam insulation provide.

Strawbale is a widely available wheat industry by-product that has a thermal conductivity of 0.067 W/mK and sample density 60 kg/m^3 (Goodhew & Griffiths, 2005). Some studies also gave insights into how the straw fibres' orientation can provide extreme improvements in thermal conductivity levels using the same density, such as when the straws are placed perpendicular to the panel's direction of heat flow (Pruteanu, 2010). Through case studies in residential homes, a Lithuanian study proved that using straw bale insulation panels can provide day-heating reductions by about 60%, translating into approximately 2000 kWh of solar gains during the year's warmer seasons (Milutiene et al., 2012).

Reeds are usually found in eastern Europe and combined with iron or nylon wires to form insulation tiles. However, they are not widely used as panels for wall or roof insulation, internally or externally, despite them providing thermal insulation levels which range between 0.045 and 0.056 W/mK, with density levels varying from 130 to 190 kg/m^3 (Moncada et al., 2013; Reet, 2015).

With these materials being burnt locally or disposed of in landfills due to a lack of efficient incinerations, they introduce a great hazard to the environment if left untreated. However, they provide great potential economic and environmental benefits as guaranteed by-products of agriculture since they are produced in great amounts. Since 2010, these numbers range from 1.5 to 8 × 1010 kg, and in some cases 1.5–3 × 1012 kg. These are some of the best-performing agrowaste fibres for thermal conductivity. However, they have not been tested thoroughly to truly understand their performance in the circular economy (CE). Thus how they

Table 1 A list of the thermal conductivity and product density specifications of the mentioned natural insulation product solutions. Adapted by the authors

Insulation material type	Product density (kg/m^2)	Thermal conductivity (W/mK}	References
Cellulose fibre	50	0.04	Lopez Hurtado et al. (2016)
Pineapple leaves	178–232	0.035–0.045	Tangjuank (2011a, 2011b)
Sunflower, sunflower stalks and cotton textile waste	200	0.1642	Binici et al., (2014)
Sunflower, rape straw, sunflower bark, sunflower pith and a mix of sunflower pith and bark	235–714	0.055–0.156	Jones et al., (2020)
Straw bales	1 5	0.155	Palladino et al., (2021)

can either increase or decrease energy consumption, and global emissions impact their life cycle Assessments is yet to be discovered.

As mentioned earlier, hemp has proven very efficient as an insulation material, however, the main reason why hemp has ethical issues surrounding it, is how the general public misperceives it as a psychedelic. It is also not as abundant as wheat straw in the United Kingdom. The crucial difference between hemp and marijuana is regarding the psychoactive component: THC (tetrahydrocannabinol). Marijuana has a THC content that's usually higher than 5% whilst hemp has 0.3% or less. Hemp is used medically as a mood-enhancer, anti-oxidant, neuroprotective, anti-inflammatory agent and relaxant with little to no side effects. Marijuana on the other hand, causes appetite stimulation, euphoria states, and drowsiness and a risk of paranoia. It is worth noting that hemp is legalised in more than 50 states in the USA whereas marijuana's legality is much more limited (Malone, 2021).

To understand the negative effects of misunderstanding or confusing hemp with marijuana, research by Hota et al. (2023) provided insights into an institute named Alpha which works closely with rural farmers and female artisans regarding the cultivation and processing of hemp. They have the challenge of investing into the communities that are pre-existing, which reduces the initiation of new communities, and to support farmers in the production of hemp and promotion and information regarding hemp and its products via campaigns and also online and offline mediums. However, since the general public lack a basic understanding regarding the difference between both hemp and marijuana, two challenges arise; whether to conform to the social norm of the communities where hemp is seen as unethical thus jeopardising the livelihood of many families or communities, or to actively engage in the industry and what is seen as an "unethical activity" which can then further stigmatise the families. Therefore it is essential to increase the level of awareness regarding the differences between hemp and marijuana but also to use crop straws that do not face this controversy and possibly make them the larger portion of a composite panel. The success of crop straws could be an indirect method for farmers to find an alternative crop as well (Hota et al., 2023).

3 The Circular Economy and Agrowaste Fibre Insulation

Though CE is a novel paradigm that attempts to end the lack of compatibility and harmony between economic and environmental growth (Hart et al., 2019), it is slowly progressing into a more detailed and elaborate model with various diagrams, suggested approaches, and methodologies for implementation. Pomponi and Moncaster (2017) suggest that the circular economy is the initial understanding of a closed-loop management system commonly referred to as "the product life factor" (Karami et al., 2015). On the other hand, Blanco et al. (2021), refer to Lyle's regenerative design, which expresses the importance of a collaborative ecosystem to prolong a product's life and the sustainability for the environment through the flow of resources, energy, and services. For instance, the Ellen MacArthur Foundation and McKinsey Centre for Business and Environment (EMF model) have set a trifold of concepts that stress CE. This is further illustrated in Fig. 1 (ARUP, 2016).

1. Early design-out of waste. A prescribed cycle is assigned before product manufacturing, disassembly, and refurbishment to minimize or eliminate waste (Ghisellini et al., 2018).
2. Design for reuse. The butterfly diagram of the EMF (Ellen MacArthur Foundation) considers two life cycles: of technical and biological materials (Geldermans, 2016). Multiple reuse and recycling apply to biological materials several times before decomposition, unlike technical materials for example plastics or metals, which require reuse due to their toxicity (Esposito et al., 2018).
3. Renewable resources. Renewable energy as a byproduct reduces resource dependency (Esposito et al., 2018; Jones & Comfort, 2018).

The ReSOLVE framework stands for regeneration, sharing, optimising, looping, virtualising, and exchanging materials and energy systems, a vital output of the Ellen MacArthur Foundation's research. These six elements are meant to be applied to products, buildings, neighbourhoods, cities, regions, and entire economies. It can help current and future projects incorporating circular economy principles to reach zero waste since very few of these projects or institutions have yet demonstrated exemplary levels of sustained reuse. Regarding material development, the ReSOLVE framework can regenerate the use of agrowaste materials in processes such as anaerobic digestion, composting or bio-refining to provide energy and products whilst diverting waste from waste landfills. Sharing agrowaste from rural areas and crop produce would reduce the need for initial capital, materials, logistics, design, and planning costs. Designing for disassembly in the early process of agrowaste materials increases product lifespan and the standardisation of components across multiple applications. Optimising these systems for long-term use would consider the natural properties of agrowaste and how it can be altered or merged with novel technologies. This would make the products suitable for indoor and outdoor use for long periods and allow for new functions after previous ones have expired in the looping

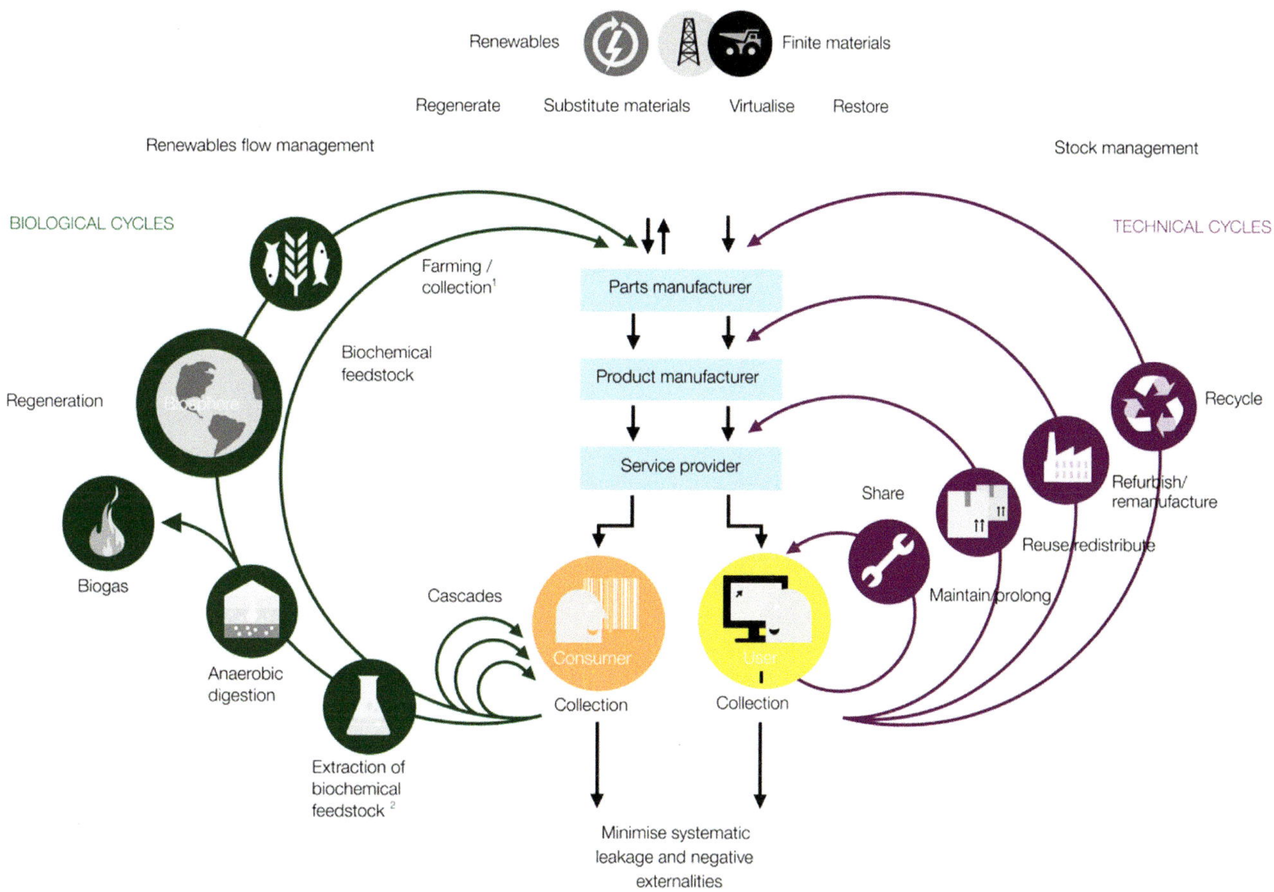

Fig 1 Circular economy model for renewables and finite materials. (ARUP, 2016)

process. Virtualising the supply and demand of waste of agricultural products would save time and money for users that would otherwise be spent on physical interventions.

Using natural materials from agrowaste as part of a buildings' main built components, such as insulated building blocks or insulation panels, can influence the current status quo of heavily processed and hard-to-reuse materials, to using materials which utilise recycled building components and biological, sustainable and renewable materials. This would surface through agrowaste-based building components replacing current product-centric delivery models instead with service-centric ones. This could influence manufacturers to keep their products as operable as possible for the longest possible time. It also allows for the optimised reuse and recycling of the products due to the manufacturers' advantage of having complete knowledge of their products.

3.1 Conventional Synthetic Materials Within the Circular Economy

The depletion of non-renewable resources leads to severe ecological and social impacts, and long-lasting footprints are left by the heavy use of raw virgin resources. To move towards a more sustainable economic system, the CE has recently become an oft-discussed approach. CE can move towards overcoming the current linearity of product lifecycles, where the value of resources, materials and products is maintained and circulated (through recycling activities) in the economy for as long as is feasible (ARUP, 2016). For instance, the UK can reduce expenditure in primary resources by 60bn GBP by 2030 (Jones et al., 2020) by applying circular economy strategies and a £3-5bn reduction to London's GDP by 2036 (LWARB, 2017). This is essential to the European Union's efforts to develop a sustainable, resource-efficient, low-carbon and competitive economy. In this context, insulation from non-renewable sources is employed in the CE system to mitigate their harms and reap as much of their productivity, energy efficiency, and environmental benefits as possible.

The conventional synthetic materials incorporated within the CE model to produce insulation are mats (blankets) manufactured from recycled plastic bottles using thermal bonded nonwoven technology. This uses mainly polyester and polyethene terephthalate (PET). They are then carded while mixed with varying proportions, followed by thermoplastic fibres.

The results are products with different properties depending on the fibre proportions, temperature and pressure at which the fibre panel was exposed (Vasile et al., 2019).

Simultaneously, ecological solutions that consist of materials that enable carbon sequestration and require small amounts of energy for production, are becoming progressively more popular from a building construction stand point. These include agrowaste materials, which are cost-efficient, abundant in quantity and somewhat competing for low thermal conductivity levels compared to conventional insulation.

An example would be semi-rigid boards made with a core material that is an agrowaste material (straw), then coupled, reinforced, and finished with fibre glass (a synthetic fibre). The core of these boards is comprised mainly of agricultural waste from the Triticale family. These are made by compressing raw dry vegetable material, precisely long straws, until they reach a density of 125 kg/m^3; then fibreglass is applied to both sides to tighten the ensemble and allow for the application of plaster. Polyester fibre yarn is then used to fasten the boards (Goodhew & Griffiths, 2005; Pruteanu, 2010). (See Fig. 2).

It would seem troubling to consider conventional thermal insulation materials such as plastics and various synthetic fibres to be reused and adopted within the circular economy framework since they cause hazards in every step of their manufacturing process. They cause severe and long-lasting ecological footprints and social impacts due to their hefty consumption of virgin resources (Ritzén & Sandström, 2017). The extent of damages caused by conventional thermal insulation are not wholly apparent as also reach the molecular level. Rochman et al. (2013) and Ishibashi et al. (2004) demonstrated that plastic waste and many other additives could deeply and widely infiltrate ecosystems when fragmented into smaller pieces of plastic of micrometre-size, which caused the incorporation of these pollutants into the food chain. This further impacts the health of organisms through unusual and unprecedented materials.

(a) (b) (c) (d)

Fig 2 Examples of conventional and agrowaste fibre insulations **a** Bulk cellulose within testing frame **b** Straw board **c** Recycled plastics board **d** wool mat on top of low-quality virgin wool (Vasile et al., 2019)

3.2 Agrowaste Fibres Within the Circular Economy

As a solution to the previously mentioned hazards of using conventional materials in building insulation, recent studies increasingly document efforts towards utilising agricultural waste in the circular economy model. Gellert (2010), for instance, exhibited that insulating raw natural materials have, at under 10%, gained a modest but steadily increasing-share of the European market since the 1990s. Gellert (2010) also stressed that conformity with regulatory requirements, including European standards like the European Technical Approvals (ETA), is now part of the consumer acceptance of an insulation product. This standard however is usually more focused on the specification of the material rather than the LCA and CE evaluation. The following are animal and plant-based examples of tested products that focus on thermal conductivity properties (Vasile et al., 2019). First, cellulose, a by-product made of cellulose acetate from tobacco and finely shredded thin paper, is utilised as a bulk insulation product, for example, for various types of cavities. Second, low-quality wool (80% mass) is formed into blankets or rolls of insulation from undemanding local sheep breeds. These are resilient to harsh weather and illness yet yield lesser long and thick regulated fibres without the required creases and elasticity suitable for the textile industry, but make them a possible circular economy agrowaste insulation candidate.

Overall, the literature suggests that it is achievable to attain useful insulation materials from agrowaste materials and industrial by-products. Material density was a major factor in the thermal conductivity performance of the materials concerning different temperature ranges. Some agrowaste materials decreased in thermal conductivity levels with density increase, whilst others increased with increases in material density, as demonstrated by Vasile et al., (2019) (See Fig. 3).

For instance, the wool-based mat (density = 45 kg/m^3) had a conductivity level of 0.0335 W/mK. In contrast, the wool-based mat (density = 65 kg/m^3) had a lower thermal conductivity of 0.0315 W/mk. This shows a decrease in conductivity levels with the increase of density contrary to the other three materials, plastic waste mat, cellulose, and straw boards, which had higher density and conductivity levels in the order they are listed. Also, during most instances, materials increased in thermal conductivity with the increase in convection heat, except for cellulose and

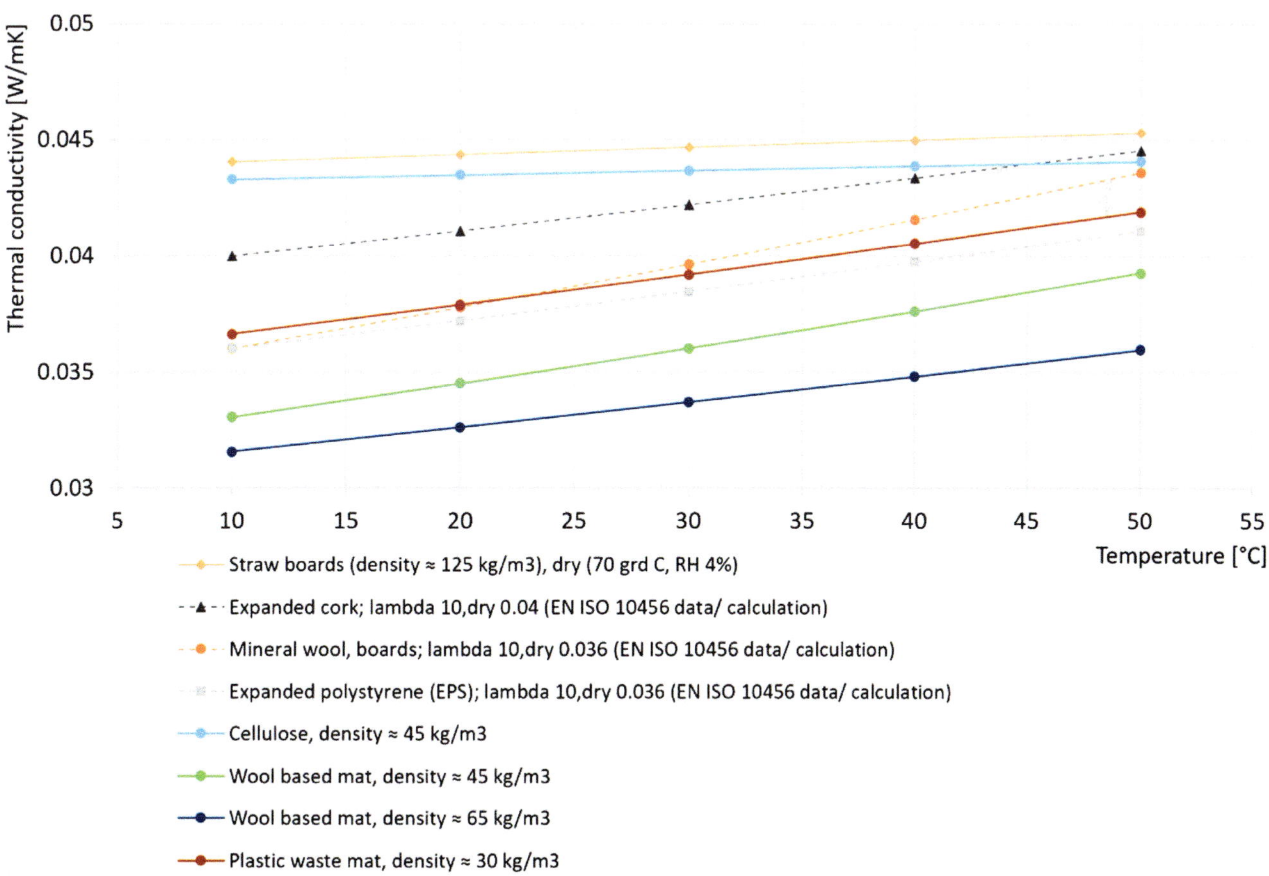

Fig 3 Comparison between tested agrowaste and conventional waste materials results (solid lines) and existing insulation materials (dashed lines) (Vasile et al., 2019)

strawboard, which had an increase of only 0.001 W/mK, as opposed to an increase of 0.06 W/mK in the other materials (see Fig. 3) (Vasile et al., 2019).

The thermal conductivity of each of the four products relevant to a circular economy model, the two plastic-based materials (straw/fibreglass and nylon, recycled plastics) and the two agro-based materials (cellulose and low-quality wool), was closely analysed. As demonstrated in Fig. 3, the wool-based panels performed much better than the conventional expanded polyurethane foam (EPS) boards with a lower thermal conductivity level in the lower 30 kg/m^3 and higher 65 kg/m^3 densities at 0.0315 W/mK and 0.035 W/mK, respectively. Although the Cellulose and the straw boards showed higher thermal conductivity levels (0.043 W/mK and 0.044 W/mK, respectively) than the conventional EPS foam and the competing recycled plastics board at 0.036–0.037 W/mK, the values still provide a satisfactory substitute to using plastics. Further testing and intermixing with other agrowaste fabrics, such as wool, could lower thermal conductivity levels without turning to plastics to enact a circular economy.

3.3 Global Circular Economy Potentials of Argowaste Fibres

The potential impact of utilising insulation panels made from agricultural waste on the global circular economy is significant. By repurposing agricultural waste for product creation, it would reduce the consumption of natural resources, enhance economic benefits through the use of existing materials, and save costs associated with procuring materials from external sources. In the next decade, this approach has the potential to save approximately $1 trillion in the United States alone. In contrast to the current linear materials economy where only one job is created per 10,000 tons of waste, whilst a circular economy-based construction industry could generate nearly 20 jobs per 10,000 tons of recycled waste (Esposito et al., 2017). However, this calculation primarily considers conventional waste materials and does not fully explore the possibilities of a circular economy centred around agricultural waste, which has unique properties and potentials. The goal of the circular economy is to decouple production and growth from reliance on natural resources, while simultaneously utilising agricultural waste to positively impact the environment. This approach encourages collaboration and inclusivity, harnessing waste from agriculture without depleting non-renewable resources. It stands in contrast to the linear economy of the industrial revolution, which fails to recognise the potential value of ecological collaboration and waste fibre utilisation.

In hot and arid regions, large-scale circular models are particularly effective compared to linear models. Megacities in these regions experience increased agrowaste production, as well as the heat island effect and pollution. These cities typically have agricultural lands surrounding them, providing a continuous supply of agricultural goods. When combined with the need to reduce cooling energy costs, these diverse waste materials can drive the success of circular models. For example, in Dubai, the 'Nabasco® 8010' exterior cladding tile was developed using local agrowaste materials, including reeds, grass, sanitary paper, textiles that are recycled, treated waste from one non potable water, and bioplastic resins or polyester which is biobased. Upon the application of the tile in an experimental apartment and reduced the interior temperature by up to 12 °C. Its design, which features trilateral surfaces with act as continuous pockets for airflow and overshadowing, facilitated the dissipation of exterior heat (Pujadas-Gispert et al., 2020). This example illustrates the practical application of a predominantly waste material insulation panel that also serves as an innovative facade insulation system external within the hot and arid region construction industries. The tile design had all four main stages of the circular economy cycle in mind. These were: collection, recycling, design, and reuse or retransformation as well as practical use and transportation. The tiles were container-friendly, easy to install and remove from various surfaces, and suitable for installation on building corners (Fig. 4) (Pujadas-Gispert et al., 2020).

The inception of waste insulation tiles from agrowaste can occur at the micro-level CE. This is primarily due to the lack of initiation by bodies of legislation and industrial hubs or platforms to adopt and encourage the development of circular economy strategies since it requires a considerable initial investment to implement. Therefore, through initiating the movement to micro-level CE, the issue of having a shortage of CE professionals and investment into research and development of CE strategies for insulation panes at the level of government can be bypassed. The relationship with smaller businesses and end-consumers would be the main channels for the CE model growth (Fig. 5) (Al Hosni et al., 2020; Ferronato et al., 2019a, 2019b).

A 'Circular Economy Package' was developed in Europe with a set of policies and regulations which focus on waste management, overcoming the use of landfills, and ensure the resources needed (European Commission, 2017). The CE package was developed into 9R CE principles (Fig. 6) (Kirchherr et al., 2017).

Short-term CE initiatives were introduced in France, Germany, Belgium, and the Netherlands, who have already established a long-term strategy expected to reduce 50% of raw materials consumption by 2030 (Van Buren et al.,

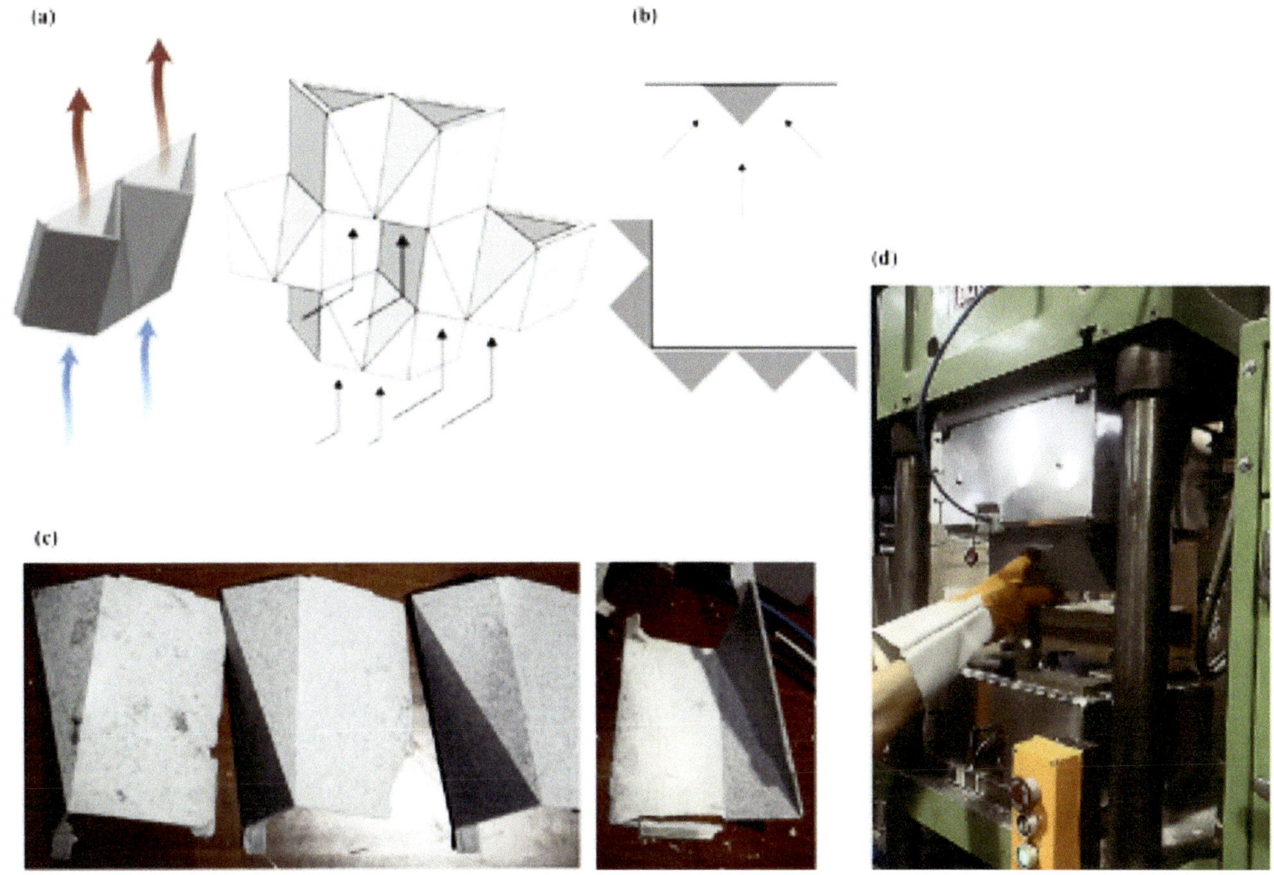

Fig 4 Production process of exterior cladding tile 'Nabasco® 8010' made from bio-based waste materials. **a** Airflows through the tile cavities dispersing air's temperature. **b** Smooth finish at apartment corners **c** End product from bio-based waste materials (VIRTUe, 2018). **d** Easy manufacturing process using mold pressuring system (VIRTUe, 2019). Adapted by Authors

2016). India, Japan, and China have also introduced legislation for the circularity of materials (Braw, 2020), whilst very few projects have done so in the Middle East (Al Hosni et al., 2020).

As proven by the previous analysis of agrowaste and conventional materials according to the EMF model, it is evident that agrowaste materials need to be studied further. In theory they provide superior cradle-to-gate (C2GA) and cradle-to-cradle (C2C) processes which incorporate both economic growth and environmental sustainability. Though proven successful in many variants of thermal insulation and within the CE model, conventional materials waste still produce high emissions. They require a lot of energy and other recourses to produce, let alone recycle or reuse. At the outset though, both material groups seem to have their prospects regarding how and why they should be incorporated within the CE model. Therefore, a deeper look into their actual energy efficiency performance through the LCA model is needed to understand the intricate details and differences between incorporating conventional material wastes and agrowaste into the production of thermal insulation solutions.

4 The Life Cycle Assessment Model

Life Cycle Assessment (LCA) is a study that evaluates the impacts of products and services on the surrounding environment and also human health, during their lifetime. Better known as cradle-to-grave (CTGR), this evaluation starts from the point of raw material extraction to the discarding of the exhaust product. On the other hand, the cradle-to-gate (C2GA) approach is utilised where the collection of the product is terminated prior to its transportation to the end-users. An authentic LCA analysis is performed per the ISO standards 14,040 and 14,044 (Hao et al., 2016; Hawes et al., 1993). The assessment requires a specific definition of the boundaries within the system since any environmental burdens which are not taken into account within the assessment are then ignored. Normalisation within the LCA analysis is conducted concerning a specified functional unit. In the case of thermal insulation materials, this is typically the mass of material required to acquire a 1 m^2 K/W thermal resistance. Of course, this does not consider the different

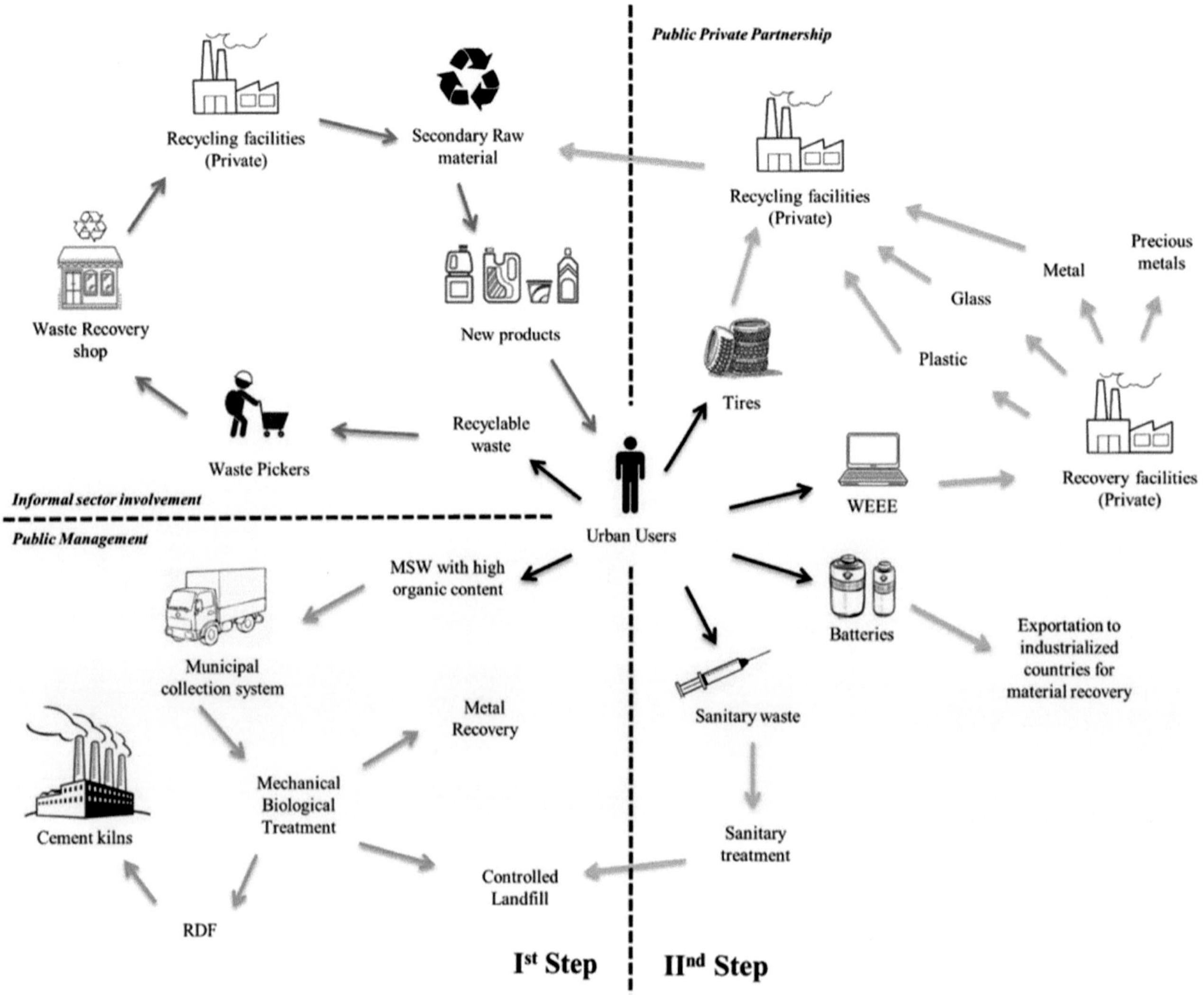

Fig 5 Three-scale module of the circular economy (Micro, top left; Meso, bottom left; Macro, right) to implement where it is still in its novel stages (Al Hosni et al., 2020; Ferronato et al., 2019)

surface area values of each insulation product tested since these data are not always present. According to the European Commission (2019), to compare the various insulation materials, a functional unit of $1 m^2$ for an insulation panel with a thickness that gives a thermal resistance $R = 1\ m^2$ K/W permits the comparison of insulation materials.

The LCA study uses two well-defined measurement indicators to evaluate the potential of services and products, which are Cumulative Energy Demand (CED) and Global Warming Potential (GWP) (ISO, 2007). The former is measured in MJ per functional unit and is the principal energy consumed during the entire life cycle. The second is defined by the Intergovernmental Panel on Climate Change (IPCC) as a method that evaluates the greenhouse gas emissions of the functional unit during the life cycle of the measured material. The indicator is evaluated on a tri-faceted timescale of 20, 50, and 100 years and in kilograms of equivalent CO_2 per functional unit. Examples of applications of LCA in real-life case studies looking at various energy provision strategies within sustainable building operations are low to zero carbon emission gases (Salkuyeh & II, 2015), wind turbines (Khare et al., 2013), and photovoltaics (Mekhilef et al., 2012). However successful these energy recourses may prove to be within the LCA, a deeper study of how agrowaste materials can help better harness these energy systems in the form of insulation to properly utilise them rather than waste the produced energy is needed. The LCA of agrowaste fibre insulation products will dictate whether they can or cannot be produced on a large scale due to price, use of non-renewable fuel resources, or environmental favourability (Asdrubali et al., 2015).

LCA is usually required to present a cradle-to-grave or cradle-to-gate evaluation. The primary evaluation accounts for the entire life cycle of the product or service, from the

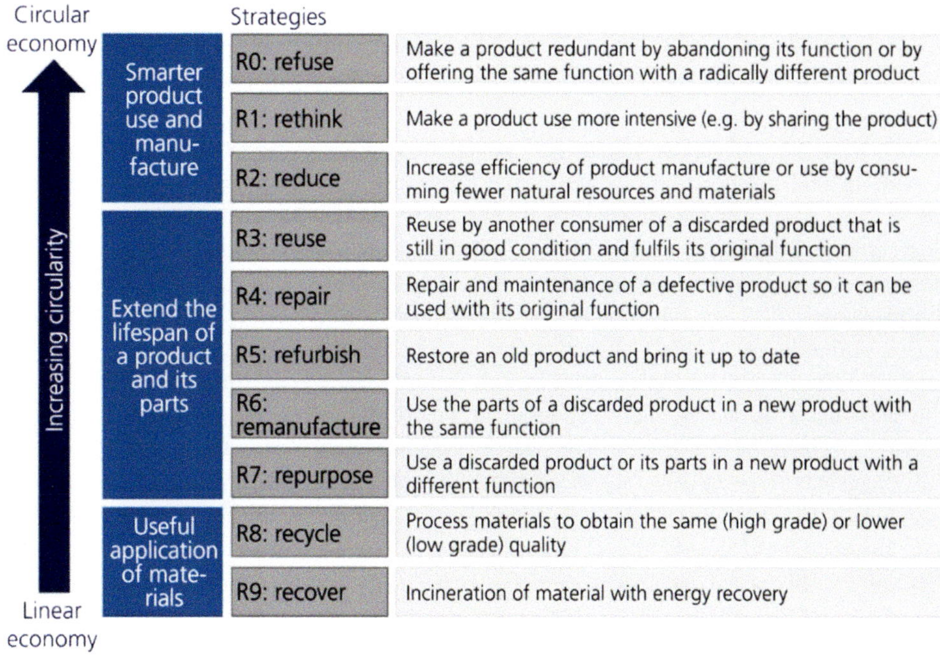

Fig 6 **9R** Circular Economy Principles model featuring smooth transition from a linear economy to a circular economy (Kirchherr et al., 2017)

extraction of the raw materials to disposal. The secondary analysis considers phases up to the transportation to the consumers, disregarding the use and the disposal phases.

The environmental impacts are normalised to a functional unit, f.u.—the mass in kg of material that is needed to have a thermal resistance value equal to 1 m^2 K/W for a 1 m^2 panel. This is important since in some cases, higher GWP and energy efficiency values are exhibited via hydrocarbon-based insulation materials than through inorganic or cellulosic alternatives, when reported in product weight. However, when they are compared on an FU basis the cellulosic-based materials may portray higher energy efficiency values (Hill et al., 2018).

Based on this, most of the LCA studies collected, focusing on conventional or agrowaste insulation, do not qualify for cradle-to-grave evaluations, let alone cradle-to-cradle analysis. According to (Silvestre et al., 2016), the cradle-to-cradle process typically consists of the A1 to A5 processes. However, this is often calculated with various estimations without an accurate account of what the process entails, as suggested by the researchers who carry out the assessment themselves (Casas-Ledón et al., 2020).

4.1 Agrowaste Fibres Within the LCA Model

Few studies closely evaluate the environmental impacts of agrowaste fibres using the LCA approach. These research papers mainly cause this lack of data, often primarily focusing on the various performance properties of the insulation prototype (2011b; Nguyen et al., 2018; Platt et al., 2019; Tangjuank, 2011a). Further analyses should be performed regarding fire resistance and resistance to water vapour, but more importantly, evaluations of their effect on the environment and energy consumption. Such parameters tend to be available and tested for commercial materials only, therefore further studies are required for unconventional materials such as agrowaste. The various properties of agrowaste fibres can be enhanced through the use of additives or systems (i.e., vapour barriers). However, in this case, the additional environmental impact that is then produced should be assessed. In sum, issues still remain that require solutions before the widespread use of unconventional materials. Some properties need further investigation, including durability, water vapour diffusion, fire resistance and fungal resistance.

An essential requirement influencing these insulators' sustainability is their components' availability since using local materials reduces economic and environmental impacts. The production of insulation products made from natural materials should not interfere negatively with the planting and harvesting of food crops. It is important to use residues and by-products from the agricultural sector. Use of unconventional materials can lead to a reduction in the utilisation of non-renewable or oil-based resources (Asdrubali et al., 2015).

LCA analysis methods must primarily consider and include the energy and emissions utilised and discharged from cultivation to transportation, production, operation, and discarding processes. However, most research initiatives concentrated solely or mainly on the operating period in reference (Sierra-Pérez et al., 2016). A prevalent issue with agrowaste insulation analysis is chemical analysis. Moreover, the elemental composition of the agrowaste is analysed without any further explanation as to how these analyses can affect how we, in turn, manufacture, produce and reuse them, which would thus make them catalysts for energy efficiency, environmental preservation, and economic prosperity (Liu et al., 2017). For instance, after careful analysis of 136 high-quality papers, the number of papers dedicated to the three main analysis types for agrowaste insulation are 94% experimental, 12% simulations, and 6% theoretical. These focus primarily on methods of manufacturing (pre-treating and moulding) and testing (properties) of the agrowaste fibres at hand. However, these experiments mostly ignore large-scale manufacturing processes that use technological advancements in processing, advanced simulation software, and theoretical calculation that can share the true capabilities of agrowaste thermal insulation (Liu et al., 2017).

4.2 Incomplete LCA Analysis Case Study: Cork Insulation

ICB, or Insulation Cork Board, is a thermal insulation board made from the bark of trees which are found mainly in Portugal. It is considered a bio-based, agrowaste, and environmentally sustainable insulation material. It is considered an agrowaste material since it is a by-product of trees. It is also available in other regions, so the study of its LCA does not apply only to a specific location, making it lose credibility if applied theoretically to another country or region. ICB's central studied technical characteristics are thermal performance, density, available thicknesses, and its 'Unformité Européene' marking (Silvestre et al., 2016). Figure 7 presents a complete process of ICB's production stage (Cradle-to-gate stage).

This falls within one of the other three main stages presented in Table 2 (Silvestre et al., 2016), which any material should be subjected to for proper and complete analysis of the material's CED and GWP. However, in this instance, cork is only evaluated at the cradle-to-gate stage (production and construction) or Cycle A (Fig. 7). Therefore, regardless of the promising results portrayed, the study is regarded as incomplete and inconclusive to the preferability between conventional and agrowaste materials for a holistic approach towards energy efficiency and the environment (Silvestre et al., 2016). The existing frailties in the B, C and D cycles, which represent the application, discarding and reuse stages, demonstrate the limited reliability of assessing the construction phase (cradle to gate). It needs to be analysed as a whole system, incorporating the cradle-to-grave loop and then closing the loop to conclude the cradle-to-cradle loop.

The expected service life of ICB is 50 years, void of decay in thermal performance when applied internally or within cavities of buildings. The National Civil Engineering

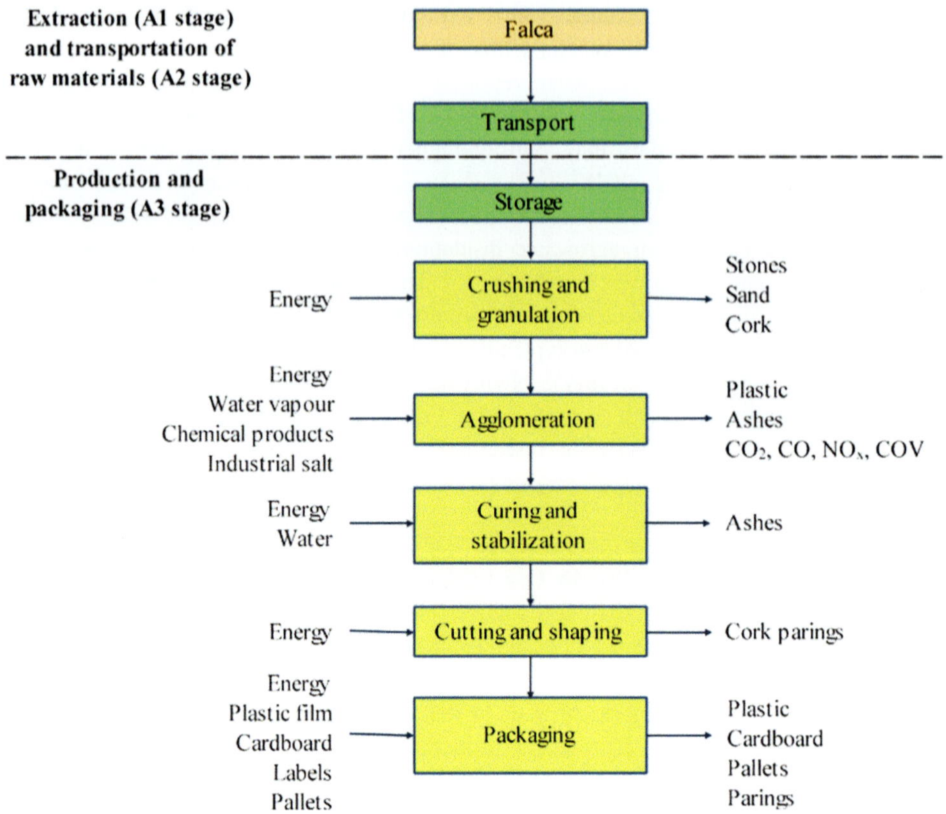

Fig. 7 First phase of the 3 LCA stages for Insulation Cork Board (Silvestre et al., 2016)

Table 2 The 3 main stages and 5 substages of the Life Cycles assessment for building materials according to European Standards (Silvestre et al., 2016)

	LCA boundaries	Life cycle stages/LCA information modules	Life cycle stage designation and description
Cradle to grave	Cradle to gate	Product stage (Al-A3)	A1-raw material extraction and processing, processing of secondary material input: A2-transport lo the manufacturer A3-manufacturing
	Cradle to grave	Construction process stage (A4-A5)	A4-transpor! to the building site AS-installation in the building
		Use stage-information module related to the building fabric (Bl-BS) Use stage-information modules related to the operation of the building (B6-B7)	Bl-use or application of the installed product; B2-maintenance; B3-repair; B4-replacement B5-refurbishment B6-operational energy use B7-operational water use
		End -of-life stage {Cl-C4)	C1-de-construction, demolition; C2-transport to waste processing; C3-waste processing for reuse, recovery and/or recycling (3R); C4-disposal
		Benefits and loads beyond the system boundary (D)	D-reuse, recovery and/or recycling (3R) potential

Laboratory, or the Laboratório Nacional de Engenharia Civil (LNEC), tested samples of ICB that had been removed from a building that was undergoing refurbishment and with an age of five decades. The tiles had a thermal conductivity of 0.0392 W/mK, lower than their initial value. However, in the case of external cladding, data is not clear as to whether the ICB performed as well, better, or less efficiently than when it was first applied. Ideally, in this case, stochastic data from service life prediction (SLP) studies should be utilised in C2C within the C2 LCA of ICB external cladding to understand the unpredicted durability or complex pathologic process within the Operational Energy use Cycle (Silvestre et al., 2015).

The GWP and the loads of ICB at the stages C1 to C4 and D are dependent upon the building after demolition. Current studies show that the end-of-life process of granulation provides a granulated product of ICB, making them a part of a seemingly complete LCA. This is because samples were in satisfactory condition upon demolition of the building and samples were cleared of adhesive and demolition stones, and then, they underwent a granulation process. However, this logic cannot be sustained within a true and accurate LCA. Since once the ICB is within the ETICS (External Thermal Insulation Composite System) or ITICS (Internal Thermal Insulation Composite System) system, it is not possible for it to be physically separated from the stages of the LCA that are remaining. Consideration must be given to the fact that they have been mixed with construction waste as undifferentiated construction and demolition waste. Also, they have undergone processes to separate them from other substances for reuse according to the Construction and Demolition Waste Code (CDW) 17 09 04—mixed construction and demolition waste (Commission, 2000). Many components need to be calculated, such as:

- Discarded product transportation for waste processing (C2, to a recycling site)
- Transportation of waste (C2, to final disposal)
- Waste processing (C3, assimilation of waste segments from demolition, waste processing of materials calculated for reuse, recycling, and energy recovery)
- Waste disposal (C4, 3R potential of CDW and other waste flow) needs to be calculated (See Table 2).

The aforementioned environmental impacts that arise from demolition (sub-stage C1) were not considered since they were similar for all the alternatives under assessment. Therefore, the Ecoinvent database was used to model each waste stream produced and its disposal. The total GWP of ICB recycling at end-of-life can amount to savings of 4% non-renewable energy during the boards' production representing 5% of the cradle-to-gate GWP. However, when all transportation costs are considered, ICB units' transportation to the endpoint or construction site can represent between 1% Acidification (AP) and 71% Eutrophication (EP) in the cradle-to-gate environmental impacts. This applies to internal or external ICB units. This demonstrates the importance of a holistic approach towards the tested agrowaste materials utilised in experiments for thermal insulation.

The current Insulation Cork Board (ICB) analysis is the most accurately analysed agrowaste material from an LCA standpoint. However, it still lacks various components of the assessment model to be considered complete. The current incomplete analysation of ICB, used three separate sources for the analysis: an Environmental Product Declaration (EPD), previously published in 2015 in the Portuguese EPD Program; an extensive study that combined two works from recognised database inventories on the materials' performance (Silvestre et al., 2016), and a series of environmental impacts studies based on waste operators in Portugal (Coelho & de Brito, 2011). It is clear from the data, that although this is the most seemingly comprehensive and illustrative set of studies for an agrowaste insulation product, it is far from providing conclusive and insightful perceptions. Despite its popularity, corkboard studies cannot fully fulfil how energy efficient system and environmental preservation can be achieved.

4.3 Incomplete LCA Analysis Case Study: Cellulose Fibre Insulation

The second most popular biobased thermal insulation material is CFI (cellulose fibre insulation). It is utilised to provide the necessary specific value of thermal resistance, be it 1 kg of material or 1 m^2K/W, so 1.280 kg of material in the case of loose-fill CFI. The LCA standard (ISO14040) investigates cellulose fibre insulation, composed mainly from ground paper fibres treated with inorganic additives. These provide fire retardancy and resistance to mould. It has a consistency likened to cotton wool. Usually, newspapers are the source of cellulose fibres. Mechanical pulp or chemical pulp is incorporated. Mechanical pulp includes organic additives including china clay, kaolins, or cationic starch to enhance properties including strength, paper moisture retention and opacity. The inks are either inorganic carbon or chromatic pigments (Lopez Hurtado et al., 2016).

In this assessment, a study considered that CFI, when considered LCA for reuse to produce loose-fill CFI, entered the system with a pre-LCA history of being incorporated into the newsprint industry, representing almost 90% of the overall energy consumption of CFI's LCA. This realisation was only possible since the study considered the previous history of CFI before studying how it fits into the LCA model. This includes newsprint production, manufacturing of reusable CFI, addition of additives, as well as the installation, the use, and the disposal for its inventory computation. As a result, we can conclude that CFI is even more energy-consuming than stone wool. This is contrary to most conclusions from other research papers, which do not account for the rigorous newspaper print process that cellulose fibres undergo before reusing or recycling in the LCA.

Other methods of incorporating cellulose fibres within the building construction industry, may triple the GWP. One example is using part of the cellulose fibres (usually around 20%) reclaimed from demolition for energy in incineration or disposing of some of the CI in landfills. This results in methane and exhaust released during landfilling and incineration (AS, 2016).

In contrast, studies by Takano et al. (2014) on Finland's computational tools for selecting suitable insulation materials for construction projects revealed that using non-newsprint CFI instead of conventional rock wool could decrease emissions from greenhouse gasses emissions and the embodied energy of building envelopes by up to 15%. Similarly, it was found that by refurbishing a multi-story residential building using the criteria of BBR 2012 (Boverket's building regulations) and Passivhaus 2012 energy-efficiency standards, energy minimisations decreased by 7%, and CO_2 emissions during material production decreased by almost 10% (Tettey et al., 2014). See Table 3 by Lopez Hurtado et al., (2016). Clearly, agro-based waste installation not only improves the thermal properties of a building but it also reduces its impact ecologically. Interestingly, these incentivising factors in choosing an agro-based waste for insulation are only possible when the initial hypothesis considers the CFI's complete analysis based on its pre-LCA history (Lopez Hurtado et al., 2016).

Table 3 The Non-newsprint cellulose in the Table outperforms the other conventional materials used in all of thermal conductivity, energy consumption and environmental impact according to the BBR standards (Lopez Hurtado et al., 2016)

	Building product density (kg/m^3)	Thermal conductivity (W/mK)	Primary energy ($MJ\text{-}kg/m^3$)	Global warming Potential ($kg\ CO_2\text{-}Eq/kg$)	Water demand (l/kg)
EPS foam slab	30	0.0375	105.486	7.336	192.729
Rock wool	60	0.04	26.393	1.511	32.384
Polvurethane rigid foam	30	0.032	103.782	6.788	350.982
Cork slab	150	0.049	10.487	0.807	30.337
Cellulose fibre	50	0.04	51.517	1.831	20.789
Wood wool	180	0.07	20.267	0.124	2.763

5 Additional Advantageous Material Properties

Agrowaste has many attractive characteristics. Two of the most helpful in terms of their contribution to energy efficiency and their preservation of the environment are their natural porosity and ability to amalgamate with natural binders. Porous materials are particularly interesting for energy conservation and storage within material sciences (Li et al., 2012). Their solid matrix and interconnected pores can be utilised for various energy systems like insulation, thermal energy saving, adsorption, evaporation, and geothermal systems (Vafai, 2015). They provide both low heat conductivity and structural stability. In recent experiments, the use of porous agrowaste-based materials on the surface of a roof demonstrated the cooling effects caused by passive water evaporation. Upon the exchange of heat between the upper surface of the roof and the atmosphere, water in liquid form diffuses into the internal substrates of the surface then evaporates, generating a cooling effect. This makes porous material evaporation cooling a recognised and efficient technique for roof cooling (Rashidi et al., 2018). Other non-hygroscopic materials, including cellulose insulation were tested by Peuhkuri et al. (2008), concluding that the temperature gradient in porous material drives humidity from warmer sides toward the colder ones allowing the humidity to escape the building rather than be trapped in it.

Agrowaste materials have been bound with non-environmental binders, which led to numerous issues in their preparation and methods of decommissioning or disposal. Therefore, rare but satisfactory experiments were conducted with natural agents that eliminate the use of inorganic toxic binders and resins (Jiang et al., 2013; Shi et al., 2013; Wei et al., 2015; Zhou et al., 2010). An experiment was conducted to achieve energy and resource conservation, whereby wheat straw was used as the aggregate, and a geopolymer was utilised as a natural mineral binder, surfactant and foaming agent. The thermal conductivity of the wheat straw insulation panel decreased with the increase in hydrogen peroxide (H2O2) content, with a sharp decline at the H2O2/Solid content of 0.0062. This is primarily related to the pore size and distribution, which increases the foaming degree of the agrowaste panel, leading to larger pores and lowering the thermal conductivity (Fig. 8) (Liu et al., 2019).

Such a product would pose two issues, the first being that it is almost inseparable from agrowaste fibre. Since it is bound with it or that it is separable, yet requires vast amounts of resources and time to separate it. In this case, incineration or disposal in landfills would be the most practical option for manufacturers. The second is that using a geopolymer rather than a resin or binder creates the third complication for end-of-life usage of agrowaste materials or environmentally conscious disposal such as composting. Including foam-based geopolymers, conventional industry binders or resins with agro waste fibres does not contribute to the end-of-life stage in any way. Thus, any product that includes those composite solidification processes cannot be deemed favourable from an LCA or CE perspective.

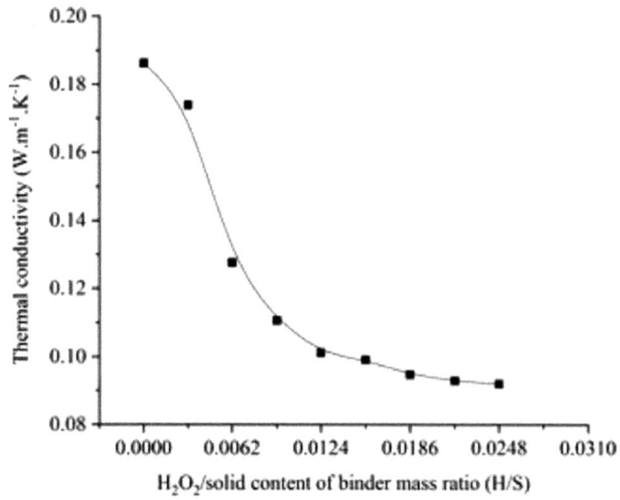

Fig. 8 Wheat straw-based insulation panel decreases in thermal conductivity as the H_2O_2 to solid ratio content increases (Liu et al., 2019)

6 Non-Objective LCA Comparison Between Synthetic and Agrowaste Fibres

A group of agrowaste fibres (Kenaf, Waste Wool, and Paper mixed with Recycled Textiles) were compared to conventional synthetic fibres (PET, textiles, XPS, and Rock wool) according to the available LCA data regarding each material. Both the CED and GWP were compared to evaluate the impacts of each material and whether or not unconventional agrowaste fibres should be utilised further in the insulation industry (Asdrubali et al., 2015). Kenaf and sheep wool was compared to extruded and expanded polystyrene and rock wool. Since little information on the agrowaste materials was available as they were at the early stages of pilot testing at best and not complete enough for full commercialisation, the embodied energy of the materials was compared based on the most synonymous data. This provided non-objective and inaccurate data for comparison. One study provided the cradle-to-gate approach (C2GA) on kenaf and recycled textile and paper. Others only gave a partial cradle-to-grave approach (CTGR) of sheep wool, recycled PET, and recycled textile marked in bold in Table 4 (Asdrubali et al., 2015).

This makes the comparison even more complicated and unsystematic. The software used to calculate the LCA of PET bottles was SimaPro. It has an approach oriented towards environmental product declaration, with impact categories used to classify and characterise the life cycle impact assessment. The software meets the ISO 14040/44 (Lee & Inaba, 2004) standards to fulfil the cradle-to-gate life cycle inventory for the mass of product of thermal resistance $R = 1(m^2\ K/W)$ (Intini & Kühtz, 2011). The software SimaPro was also used for extruded expanded polystyrene and rock wool to determine the Energy Demand (CED), IPCC (International Panel on Climate Change) and Eco-Indicator 99 were chosen (Asdrubali, 2009). For other materials such as kenaf fibres, the guarded hot plate and hot box apparatus methods were used to gain information, as well as previous studies which had already presented the LCA information regarding the materials used for insulation (Ardente et al., 2008; Asdrubali, 2006). The recycled textiles were tested by the insulation company Maiano, but no further details were provided, only the end thermal insulation values (Buratti et al., 2016; Maiano, 2021).

This trend comparing the embodied energy of the materials based on the most synonymous data can be observed in many research papers. Moreover, they do not mention the stages that the embodied energy value encompasses. In Table 7, several studies were revised to understand which LCA stages they fulfilled when presenting the embodied energy values, they presented for both the conventional and agrowaste materials. Some of these values were based on assumptions of future changes to the type of transportation used, that is, less consumption of fossil fuels, and not on the current transportation systems. Please refer to Table 2 as a guide for the life cycle stages designation mentioned in Table 7 (Casas-Ledón et al., 2020; Intini & Kühtz, 2011; Papadopoulos & Giama, 2007; Ricciardi et al., 2014; Schiavoni et al., 2016; Sierra-Pérez et al., 2016; Silvestre et al., 2016).

The results prove that using data from other independent studies as evidence is flawed. For instance, all the materials used in Table 4 are characterised as good in terms of thermal conductivity, except for recycled textile and paper, which are based on other sources. These sources contain incomplete LCA studies, which also state that the fibres used from paper incorporated into the materials used in Table 4 have extreme energy consumption values (267.7 MJ/functional unit) and global warming potential (14.38 $kgCO_2eq$/functional unit). Whereas the original paper featuring Table 2 (Silvestre et al., 2016) states much lower levels of energy consumption and GWP. This creates a dilemma that eliminates presenting the objective reality of a particular material as it has not been tested from CR2GR, but instead a subjective choice of whether a material will be presented as environmentally conscious or taxing according to different sources.

These results do not account for the fact that the original research papers did not include the embodied energy of the PET during the following stages:

Table 4 The Table demonstrates the inconsistent data between agrowaste fibre and conventional synthetic fibres used for insulation. It provides a Notes column with a key noting the various flaws of information amongst the materials. Adapted by the authors (Asdrubali et al., 2015)

Material	Functional unit weight (kg)	Thermal conductivity (W/m*K)	Energy consumption (MJea per functional unit)	Global warming potential ($kgCO_2eg$ per functional unit)	Approach and System Boundary	References
Recycled PET	1.065	0.0355	83.723	1.783	CTGA, Italy	Intini and Kühtz (2011)
Recycled PET (commercialised)	1.48	0.037	21.056a	3.117	CTGR, NA	Maiano (2021)
Reccled textile (commercialised)	1.79	0.0358	17.57a	1.545	CTGR, NA	Maiano (2021)
Recycled textile and paper	14.7	0.034–0.039	267.7	14.68	CTGA, Italy	Ricciardi et al., (2014)
Extruded expanded polystyrene	1.75	0.035	127.31	13.22	CTGA, European	Asdrubali (2006), Asdrubali (2009)
Kenaf	1.52	0.038	59.37	3.17	CTGA, Italy (400 km)	Ardente et al., (2008)
Rock wool	1.2	0.04	53.09	2.77	CTGA, European	Asdrubali, (2006), Asdrubali (2009)
Sheep wool	0.76	0.038	17.119	1.457	CIGR, NA	Maiano (2021)

Table 5 A summarised account of the papers reviewed

Material	Density Kg/m^3	Thermal conductivity	Completed an LCA assessment of the binder and construction type selection	Inconsistent/ Patched or mended LCA calculations	Promising thermal conductivity results which compete with conventional materials	Insulation material properties' tests have been performed	Natural Insulation properties outperform or are similar conventional materials	Further investigation of properties needed	Further development to manufacturing, resin selection or materials preparation is needed	References
Cellulose fibre	50	0.04		x	X	x	x	x	x	Lopez Hurtado et al., (2016)
Pineapple leaves	178–232	0.035–	0.045		X	x		x		Tangjuank (2011a, 2011b)
Sunflower, Sunflower stalks and cotton textile waste	200	0.1642			X	x		x	x	Binici et al., (2014)
Mycelium composite construction	235–714	0.055– 0.156	x	x	X	x	x			Jones et al., (2020)
Straw bale from top table	0.15	0.115		x				x		Palladino et al., (2021)
Insulation Cork Boards (ICB)	110	0.0392	x	x	X				x	Silvestre et al., (2016)
Nabasco 8010 Tile	N/A. the tile works to decrease heat gain through providing an air gap on the facade, but not strictly act as an insulation material	N/A		x		x		x	x	Pujadas-Gispert et al., (2020)
Cellulose board Straw board Plastic waste Wool based mat Wool based matt dense	45 125 30 45 65	0.0433 0.0437 0.0357 0.033			X			x		Vasile et al., (2019)

Table 6 A detailed account of the papers reviewed

Insulation material Type	Product density (kg/m^2)	Thermal conductivity (W/mK)	Advantageous common findings	Disadvantageous common findings	References
Cellulose fibre	50	0.04	CFI is an innovative, eco-friendly insulation material that has similar thermal insulation characteristics to its non-renewable counterparts Properties were predicted based on other studies. These provided very precise insights and details but did not test the specific sample at hand. Moisture properties, Air infiltration Openings, Fire properties and Fungal development were presented	The material performance needs further research The LCA model is performed using six different studies, these do not complement each other Further manufacturing and installation processes development are needed for optimisation. This is challenging due to the large variety of newsprint paper suppliers each with different compositions as well as the different manufacturing methods More environmentally friendly fire and antifungal retardants need to be added to replace the current mineral additives. These would also tackle the effects of tackifiers in the wet-spray method, of which its effects on CFI are unknown	Lopez Hurtado et al., (2016)
Pineapple leaves	178–232	0.035–0.045	The thermal properties of pineapple leaf boards varied between 0.035 and 0.043 W/mK, similar to fiber glass wool (0.039 W/mK), and polystyrene foam (0.041 W/mK). Though it was slightly higher than that of polyurethane	Many of the parameters required for an insulation material's overall success have not been tested except for fire resistance, moisture content, and water absorption. These cannot provide a holistic view of the material's performance. These include thermal insulation, durability, cost, ease of application and fire resistance The fire resistance according to the ASTM D 635- 98 standard proved successful at (1.34 min) at 1:3 ratios of particle: binder	Tangjuank (2011a, 2011b)
Sunflower, Sunflower stalks & cotton textile waste	200	0.1642	Under certain pressures, epoxy escapes, causing more airgaps in boards. The thermal conductivity level from these samples pass the TS 805 EN 60155 standard which requires a thermal conductivity of less than 0.1, with ranges of 0.07 to 0.09 W/mK	Ureaformaldehyde (a nonbiodegradable resin) was used to form the organic insulation panel. Thus, further research is needed to understand the usable resins that would not hinder the composability or	(Binici et al., 2014)

(continued)

Table 6 (continued)

Insulation material Type	Product density (kg/m^2)	Thermal conductivity (W/mK)	Advantageous common findings	Disadvantageous common findings	References
			The findings demonstrate that sunflower stalks, the cotton waste, can produce insulation materials that are organic in nature and that can compete commercially with non-organic materials when further developed All of the following properties have ben tested: Compressive and flexural strength,, water absorption and unit weight values, ultrasonic sound penetration velocity coefficient	biodegradability of the insulation panel Properties such as compressive and flexural strength, water absorption and unit weight values, ultrasonic sound penetration and velocity coefficient were tested. However crucial properties such as fire resistance, specific heat etc. were not thus requiring further testing	
Mycelium composite	235–714	0.055–0.156	Mycelium composites are particularly well suited for thermal and acoustic insulation applications, exhibiting similar or lower thermal conductivities than commercial thermal insulation materials and 70–75% acoustic absorption outperforming traditional ceiling tiles, polyurethane foams and plywood They also exhibit better fire re- action and fire safety properties than traditional construction materials, such as extruded polystyrene insulation and particleboard and good termite resistance utilising natural termiticides The growing trends in the research and commercialisation of mycelium composite materials and their useful material properties makes them an effective, cheap and environmentally sustainable technology emerging with the potential to significantly contribute to the future of green construction A life cycle analysis was performed by an external study which utilised a software with extensive datasets to report each material and process with great detail.	The typically foam-like mechanical properties, high moisture uptake and many gaps in material property documentation currently limit the application and usage of mycelium materials. Thus further research and development of these materials is necessary, in order to understand the usage as non- or semi-structural supplements to traditional construction mate- rials for insulation, panelling and furnishings The LCA process is still lacking tangibility which can only be achieved through an onsite LCA process	Jones et al., (2020)

(continued)

Table 6 (continued)

Insulation material Type	Product density (kg/m^2)	Thermal conductivity (W/mK)	Advantageous common findings	Disadvantageous common findings	References
Straw bale from top table	100–110	0.115	The new innovative wall-sample with DW straw bales (wall 1, with natural materials) significantly reduced heat losses through the building's envelope, by 5–10%, with a decrease of 10–20% in global total primary energy demand, compared with traditional envelopes	An LCA using the Edilclima software was established for the electrical energy consumption alone which confirmed better energy thermal conductivity No onsite variable conditioned tests have been conducted on the material, rather references to various standards to within energy performance simulation software to test the materials performance. This gives an inaccurate and non-holistic view of the insulation material. The heat losses through the investigated walls provided 2 different results when compared according to calculation procedure (EN 52016 and UNI TS 11300. With the former showing an increase of 10–12% and 26–36% for heating and cooling loads. The UNI TS 11300 showed an increase of loads by up to 75% higher than the EN 52016 standard when ventilation, heat gains, and solar radiation were all considered This was due to the energy need of the building being much lower in the hourly-energy-calculation procedure (EN 52016), resulting in a lower primary energy need, leading to HVAC increases of 10–12% overall, of which 6–7% for heating and over than 90% for cooling)	Palladino et al., (2021)
Insulation Cork Boards (ICB)	110 (kg/m^3)	0.0392	Considering only the A1–A3 stages of the life cycle, ICB has a low contribution to several impact categories when compared to expanded and extruded polystyrene, mineral wool, and polyurethane. These include consumption of primary non-renewable energy and Global	ICB though underperforms in consumption of primary energy, renewable and water consumption CMsay in full?L only considers CO_2 fossil emissions and does not consider CO_2 capture during tree growth or biogenic CO_2 emissions of cork oak tree growth	Silvestre et al., (2016)

(continued)

Table 6 (continued)

Insulation material Type	Product density (kg/m^2)	Thermal conductivity (W/mK)	Advantageous common findings	Disadvantageous common findings	References
			Warming Potential over a time span of 100 years Depending on the criteria of the specifier, and on the sources of electric energy, ICB may be a sustainable choice for building insulation from an environmental point of view The LCA results presented here are considered scientifically sound since they were achieved through biogenic CO_2-accounting and disseminating of the results, particularly in the scientific community involved in the development of insulation materials	"falca" or cork powder bought externally for the boiler. The manufacturing process thus needs development to mitigate its CO_2 usage Values such as the GWP figures for the "product stage" (A1–A3) do not consider the CO and CO_2 emissions from the boiler and from the autoclave, and thus are lacking These emissions are biogenic and are presently not considered by the most used EIAM Environmental Impact Assessment Method Two methodologies were used to calculate the LCA of the ICB. 1. Biogenic CO_2 emissions. 2. CML Despite these conclusions, it is clear that there is a lot of patchwork at hand, and that there is more reliance on which part the specifier would like to focus on to determine the circular economy and LCA of ICB rather than have a true holistic number that represents the board	
Nabasco 8010 tile	N/A	N/A	The facade made of bio-based materials, which are sourced from renewable and recyclable resources, required less control of indoor ambient temperatures, and had a low environmental impact from the cradle-to- cradle materials life-cycle perspective The panel consists of raw vegetable matter that is resistant to moisture, pressure, pests, fungal growth, and bacteria The tiles were made of a specific bio-based material (Nabasco? 8010) made of waste (sanitary paper, grass, reeds, recycled textiles, drinking-water treatment waste, and bio-based polyester resin); calcium carbonate ($CaCO_3$), a waste material	Additional improvements are needed for standardising types of screws and nails required, facilitate quality construction performance and assembly speed The project used a "vapor-proof film" for water- repellency/proofing and damp-proofing. These were not taken as part of the cradle-to-cradle calculations of the panel since they were used to enhance the panels' performance on the structures' façade	VIRTUe, () Pujadas-Gispert et al., (2020)

(continued)

Table 6 (continued)

Insulation material Type	Product density (kg/m^2)	Thermal conductivity (W/mK)	Advantageous common findings	Disadvantageous common findings	References
			from drinking water companies; and a bio-based polyester resin (50% waste glycol from biodiesel production and 50% animal waste and cooking fats) It is locally produced thus reducing transport emissions The tile is fire resistant and qualified as a class BeS1 d0 material in accordance with EN 13501e1 It has slightly lower mechanical and density properties than conventional glass It requires low maintenance throughout its operational life and can be crushed and reused as a basic raw material in the same process at the end of its useful life		
Cellulose board, straw board, plastic waste, wool based mat, wool based matt dense	45,125,30,45,65 kg/m3	0.0433 0.0437 0.0357 0.033	In this paper, the thermal conductivity of products made from waste such as cellulose, straw, recycled plastics and low-quality wool were investigated, as they are a few of the materials relevant for a circular economy. The study focused on thermal insulating and phono-absorption properties of the tested products, concluded that it was possible to obtain useful construction materials from recycled materials or using industrial by-products Conversely here, more air gaps gave higher conductivity, whilst more dense panels with less gaps gave lower conductivity Thermal characteristics of the analysed products greatly depend on density, in respect to thermal conductivity and its values in a temperature range Higher densities usually translated into lower and stable conductivity through various temperature variations	Further research is needed to facilitate a widespread use of these materials in construction technologies. This includes the characterisation of specific heat (as crucial for the evaluation of thermal dynamic properties and for dynamic simulations) and in the investigation of fire resistance, water vapor diffusion and humidity influence on thermal and acoustic characteristics, fungal resistance and behaviour in the event of moist conditions	Vasile et al., (2019) Palladino et al., (2021)

Table 7 Comparative analysis of typical thermal insulation materials and their LCA deficiencies

Materials	ρ(kg/m3)	λ (W/mK)	Embodied Energy (MJ/f.u)	Lifecycle Stage Designation	References
Expanded polyurethane 30	30	0.03	125	A1-A5	Ricciardi et al., (2014)
Expanded polystyrene	20	0.04	135	A1-A5	Ricciardi et al., (2014)
Stone wool	25	0.064	50	A1-A5	Papadopoulos and Giama (2007)
Cork Insulation				A1-A5	Sierra-Pérez et al., (2016), Silvestre et al., (2016)
Eucalyptus bark fibers	25	0.045	16.2	A1-A3	Casas-Ledón et al., (2020)
Recycled PET Polyethylene terephthalate	40	0.037	21.1	A1-A4	Intini and Kühtz (2011) Schiavoni et al., (2016)

- From inception before they were reused.
- The use-phase data which is retrieved once the panels were installed in buildings for testing.
- The cradle-to-cradle breakdown of each process (See bottom Key and Notes column in Table 4).

This makes the study at hand, whether the accumulative study or the originals, futile and impractical regarding the accurate data needed to decide further action. Tables 5 and 6 provide both a summarised and detailed account of the similarities and differences between the papers studied for this review.

This further clarifies the deficiencies when comparing the mentioned studies in reaching an objective method of judging the level of sustainability for each material. It is worth mentioning that the studies have more in common in terms of product specifications. However, less so in terms of LCA studies. This is because some of them do not have LCA studies performed altogether or have very limited studies conducted.

7 Conclusions

This research provides a comprehensive review of research on agrowaste fibres used in insulation panels and their thermal conductivity. Agrowaste fibres have shown equivalent thermal conductivity to synthetic materials and offer additional benefits such as acoustic absorption, compressive strength, fire retardancy, and the possibility of binding with natural binders. These factors make them suitable for mass production and can contribute to economic growth and energy preservation. Agrowaste fibres also offer a solution to the environmental hazards caused by landfilling mass amounts of unused agrowaste. However, the studies highlighted the need for a cradle-to-cradle approach to fully understand the potential of agrowaste materials as insulation. The paper emphasises the importance of conducting more life cycle assessment (LCA) models evaluations to understand the practicality and performance of agrowaste fibres compared to conventional materials. The transportation phase of the agrowaste materials was identified as a significant factor in their overall environmental impact. The paper also points out flaws in comparing synthetic materials to agrowaste materials, such as not considering the previous service life of synthetic materials and the lack of clarity regarding the disposal phase. Agrowaste fibres can offer advantages in terms of their disposal as they can be incorporated into various applications such as textiles, fertilisers, animal feed, and construction materials, depending on their separation and binding properties.

Conflicts of Interest The authors declare no conflict of interest.

Data Availability Statement All data, models, and codes generated or used during the study appear in the submitted article.

References

Abenojar, J., Barbosa, A. Q., Ballesteros, Y., Del Real, J. C., Da Silva, L. F. M., & Martínez, M. A. (2014). Effect of surface treatments on natural cork: Surface energy, adhesion, and acoustic insulation. *Wood Science and Technology, 48*(1), 207–224. https://doi.org/10.1007/s00226-013-0599-7

Al Hosni, I. S., Amoudi, O., & Callaghan, N. (2020). An exploratory study on challenges of circular economy in the built environment in Oman. *Proceedings of Institution of Civil Engineers: Management, Procurement and Law, 173*(3), 104–113. https://doi.org/10.1680/jmapl.19.00034

Alma, M. H., Kalaycıoğlu, H., Bektaş, I., & Tutus, A. (2005). Properties of cotton carpel-based particleboards. *Industrial Crops and Products, 22*(2), 141–149.

American Society for Testing and Materials Standards. (2003). Rate of burning and/or extent and time of burning of plastics in a horizontal position, *1*. 1–7.

Ardente, F., Beccali, M., Cellura, M., & Mistretta, M. (2008). Building energy performance: A LCA case study of kenaf-fibres insulation board. *Energy and Buildings, 40*(1), 1–10. https://doi.org/10.1016/j.enbuild.2006.12.009

ARUP. (2016). *The circular economy in the built environment* (pp. 1–93). Callifornia Academy of Sciences.

AS, A. S. (2016). Environmental product declaration. *Equilibrium, 350*, 140.

Asdrubali, F., D'Alessandro, F., & Schiavoni, S. (2015). A review of unconventional sustainable building insulation materials. *Sustainable Materials and Technologies, 4*(2015), 1–17. https://doi.org/10.1016/j.susmat.2015.05.002

Asdrubali, F. (2006). Survey on the acoustical properties of new sustainable materials for noise control. In *EURONOISE 2006—The 6th European conference on noise control: Advanced solutions for noise control*, 1–10.

Asdrubali, F. (2009). The role of Life Cycle Assessment (LCA) in the design of sustainable buildings: Thermal and sound insulating materials. In *8th European conference on noise control 2009, EURONOISE 2009—Proceedings of the Institute of* Acoustics, 31 (PART 3).

Ashour, T., Wieland, H., Georg, H., Bockisch, F. J., & Wu, W. (2010). The influence of natural reinforcement fibres on insulation values of earth plaster for straw bale buildings. *Materials and Design, 31*(10), 4676–4685. https://doi.org/10.1016/j.matdes.2010.05.026

Bedlivá, H., & Isaacs, N. (2014). Hempcrete—An environmentally friendly material? *Advanced Materials Research, 1041*, 83–86. https://doi.org/10.4028/www.scientific.net/AMR.1041.83

Bekalo, S. A., & Reinhardt, H.-W. (2010). Fibers of coffee husk and hulls for the production of particleboard. *Materials and Structures, 43*(8), 1049–1060.

Binici, H., Eken, M., Dolaz, M., Aksogan, O., & Kara, M. (2014). An environmentally friendly thermal insulation material from sunflower stalk, textile waste and stubble fibres. *Construction and Building Materials, 51*, 24–33. https://doi.org/10.1016/J.CONBUILDMAT.2013.10.038

Bisset, R. (2007). Buildings can play a key role in combating climate change. *Bulletin on Energy Efficiency, 7*(4).

Blanco, E., Zari, M. P., Raskin, K., & Clergeau, P. (2021). Urban ecosystem-level biomimicry and regenerative design: Linking ecosystem functioning and urban built environments. *Sustainability (switzerland), 13*(1), 1–12. https://doi.org/10.3390/su13010404

Braw, E. (2020). Five countries moving ahead of the pack on circular economy legislation. In the Guardian. https://www.theguardian.com/sustainable-business/2014/oct/29/countries-eu-circular-economy-legislation-denmark-sweden-scotland

Buratti, C., Belloni, E., Lascaro, E., Lopez, G. A., & Ricciardi, P. (2016). Sustainable panels with recycled materials for building applications: Environmental and acoustic characterisation. *Energy Procedia, 101*(September), 972–979. https://doi.org/10.1016/j.egypro.2016.11.123

Carfrae, J., Wilde, P., Goodhew, S., Walker, P., & Littlewood, J. (2009). Detailing the effective use of rain-screens to protect walls made from straw bales in combination with hygroscopic, breathable finishes.

Casas-Ledón, Y., Daza Salgado, K., Cea, J., Arteaga-Pérez, L. E., & Fuentealba, C. (2020). Life cycle assessment of innovative insulation panels based on eucalyptus bark fibers. *Journal of Cleaner Production, 249*. https://doi.org/10.1016/j.jclepro.2019.119356

Chougan, M., Ghaffar, S. H., Al-Kheetan, M. J., & Gecevicius, M. (2020). Wheat straw pre-treatments using eco-friendly strategies for enhancing the tensile properties of bio-based polylactic acid composites. *Industrial Crops and Products*, 155(July). https://doi.org/10.1016/j.indcrop.2020.112836

Coelho, A., & de Brito, J. (2011). Economic analysis of conventional versus selective demolition—A case study. *Resources, Conservation and Recycling, 55*(3), 382–392. https://doi.org/10.1016/j.resconrec.2010.11.003

Collins, L., Natarajan, S., & Levermore, G. (2010). Climate change and future energy consumption in UK housing stock. *Building Services Engineering Research and Technology, 31*(1), 75–90. https://doi.org/10.1177/0143624409354972

Commission, E. (2017). On the implementation of the circular economy action plan. *Com, 2017*(33), 1–21.

Commission, E. (2000). EC. Commission decision of 3 May 2000 replacing decision 94/3/EC establishing a list of wastes pursuant to article 1(a) of council directive 75/442/EEC on waste and council decision 94/904/ec establishing a list of hazardous waste pursuant to article 1(4) o.

Department for Environment Food and Rural Affairs (DEFRA). (2006). Climate change: The UK programme 2006, *193*. https://www.gov.uk/government/publications/climate-change-the-uk-programme-2006

Ellen MacArthur Foundation. (2016). *The new plastics economy rethinking the future of plastics* (p 120). Ellen MacArthur Foundation.

Environment Change Institute. (2005). *40% House: The energy performance of buildings directive (EPBD): Improving the energy efficiency of the existing housing stock*. Oxford University.

Esposito, M., Tse, T., & Soufani, K. (2017). Is the circular economy a new fast-expanding market? *Thunderbird International Business Review, 59*(1), 9–14. https://doi.org/10.1002/tie.21764

Esposito, M., Tse, T., & Soufani, K. (2018). Introducing a circular economy: New thinking with new managerial and policy implications. *California Management Review, 60*(3), 5–19.

European Commission. (2019). Product Environmental Footprint Category Rules (PEFCRs) for thermal insulation. December, 1–141. https://ec.europa.eu/environment/eussd/smgp/pdf/Thermal Insulationfinal-Oct2019.pdf

Evon, P., Vandenbossche, V., Pontalier, P.-Y., & Rigal, L. (2014). New thermal insulation fiberboards from cake generated during biorefinery of sunflower whole plant in a twin-screw extruder. *Industrial Crops and Products, 52*, 354–362.

Ferronato, N., Rada, E. C., Gorritty Portillo, M. A., Cioca, L. I., Ragazzi, M., & Torretta, V. (2019). Introduction of the circular economy within developing regions: A comparative analysis of advantages and opportunities for waste valorisation. *Journal of Environmental Management, 230*(September 2018), 366–378. https://doi.org/10.1016/j.jenvman.2018.09.095

Ferronato, N., Rada, E. C., Gorritty Portillo, M. A., Cioca, L. I., Ragazzi, M., & Torretta, V. (2019). Introduction of the circular economy within developing regions: A comparative analysis of advantages and opportunities for waste valorization. *Journal of Environmental Management, 230*(September 2018), 366–378. https://doi.org/10.1016/j.jenvman.2018.09.095

Geldermans, R. J. (2016). Design for change and circularity–accommodating circular material & product flows in construction. *Energy Procedia, 96*, 301–311.

Gellert, R. (2010). Natural fibre and fibre composite materials for insulation in buildings. *Materials for Energy Efficiency and Thermal Comfort in Buildings*, 229–256. https://doi.org/10.1533/9781845699277.2.229

Ghisellini, P., Ripa, M., & Ulgiati, S. (2018). Exploring environmental and economic costs and benefits of a circular economy approach to the construction and demolition sector. A literature review. *Journal of Cleaner Production, 178*, 618–643.

Goodhew, S., & Griffiths, R. (2005). Sustainable earth walls to meet the building regulations. *Energy and Buildings, 37*(5), 451–459.

Hao, J.-H., Chen, Q., & Hu, K. (2016). Porosity distribution optimization of insulation materials by the variational method. *International Journal of Heat and Mass Transfer, 92*, 1–7. https://doi.org/10.1016/j.ijheatmasstransfer.2015.08.076

Hart, J., Adams, K., Giesekam, J., Tingley, D. D., & Pomponi, F. (2019). Barriers and drivers in a circular economy: The case of the built environment. *Procedia Cirp, 80*, 619–624.

Hawes, D. W., Feldman, D., & Banu, D. (1993). Latent heat storage in building materials. *Energy and Buildings, 20*(1), 77–86. https://doi.org/10.1016/0378-7788(93)90040-2

Hill, C., Norton, A., & Dibdiakova, J. (2018). A comparison of the environmental impacts of different categories of insulation

materials. *Energy and Buildings, 162*, 12–20. https://doi.org/10.1016/j.enbuild.2017.12.009
Hota, P. K., Bhatt, B., & Qureshi, I. (2023). Institutional work to navigate ethical dilemmas: Evidence from a social enterprise. *Journal of Business Venturing, 38*(1), 106269. https://doi.org/10.1016/j.jbusvent.2022.106269
Intini, F., & Kühtz, S. (2011). Recycling in buildings: An LCA case study of a thermal insulation panel made of polyester fiber, recycled from post-consumer PET bottles. *The International Journal of Life Cycle Assessment, 16*(4), 306–315.
Ishibashi, H., Matsumura, N., Hirano, M., Matsuoka, M., Shiratsuchi, H., Ishibashi, Y., Takao, Y., & Arizono, K. (2004). Effects of triclosan on the early life stages and reproduction of medaka Oryzias latipes and induction of hepatic vitellogenin. *Aquatic Toxicology, 67*(2), 167–179.
ISO. (2007). *6946: 2007-Building components and building elements-Thermal resistance and thermal transmittance-Calculation method*. International Organization for Standardization ISO.
Jiang, D., Cui, S., Song, X., & Zhang, J. (2013). Analysis of micro-morphology and heat-insulating property of leaf concrete. *Construction and Building Materials, 49*, 663–671. https://doi.org/10.1016/j.conbuildmat.2013.08.055
Jones, P., & Comfort, D. (2018). The construction industry and the circular economy. *International Journal of Management Cases, 20* (1), 4–15.
Jones, M., Mautner, A., Luenco, S., Bismarck, A., & John, S. (2020). Engineered mycelium composite construction materials from fungal biorefineries: A critical review. *Materials and Design, 187*, 108397. https://doi.org/10.1016/j.matdes.2019.108397
Karami, P., Al-Ayish, N., & Gudmundsson, K. (2015). A comparative study of the environmental impact of Swedish residential buildings with vacuum insulation panels. *Energy and Buildings, 109*, 183–194. https://doi.org/10.1016/j.enbuild.2015.10.031
Kawasaki, T., & Kawai, S. (2006). Thermal insulation properties of wood-based sandwich panel for use as structural insulated walls and floors. *Journal of Wood Science, 52*(1), 75–83.
Khalil, H. P. S. A., Hanida, S., Kang, C. W., & Fuaad, N. A. N. (2007). Agro-hybrid composite: The effects on mechanical and physical properties of oil palm fiber (EFB)/glass hybrid reinforced polyester composites. *Journal of Reinforced Plastics and Composites, 26*(2), 203–218.
Khare, V., Nema, S., & Baredar, P. (2013). Status of solar wind renewable energy in India. *Renewable and Sustainable Energy Reviews, 27*, 1–10.
Kirchherr, J., Reike, D., & Hekkert, M. (2017). Conceptualizing the circular economy: An analysis of 114 definitions. *Resources, Conservation and Recycling, 127*, 221–232.
Kumfu, S., & Jintakosol, T. (2012). Thermal insulation produced from pineapple leaf fiber and natural rubber latex. *Advanced Materials Research, 506*, 453–456.
Laustsen, J. (2008). Energy efficiency requirements in building codes, energy efficiency policies for new buildings. *IEA Information Paper*.
Lee, K. -M., & Inaba, A. (2004). Life cycle assessment: Best practices of international organization for standardization (ISO) 14040 series. *Committee on Trade and Investment*, February, 99. http://publications.apec.org/publication-detail.php?pub_id=453
Li, Y., Fu, Z.-Y., & Su, B.-L. (2012). Hierarchically structured porous materials for energy conversion and storage. *Advanced Functional Materials, 22*(22), 4634–4667.
Liang, H.-H., & Ho, M.-C. (2007). Toxicity characteristics of commercially manufactured insulation materials for building applications in Taiwan. *Construction and Building Materials, 21*(6), 1254–1261.
Liu, L., Zou, S., Li, H., Deng, L., Bai, C., Zhang, X., Wang, S., & Li, N. (2019). Experimental physical properties of an eco-friendly bio-insulation material based on wheat straw for buildings. *Energy and Buildings, 201*, 19–36. https://doi.org/10.1016/j.enbuild.2019.07.037
Liu, L. F., Li, H. Q., Lazzaretto, A., Manente, G., Tong, C. Y., Liu, Q. Bin, & Li, N. P. (2017). The development history and prospects of biomass-based insulation materials for buildings. *Renewable and Sustainable Energy Reviews, 69*(December 2016), 912–932. https://doi.org/10.1016/j.rser.2016.11.140
Lopez Hurtado, P., Rouilly, A., Vandenbossche, V., & Raynaud, C. (2016). A review on the properties of cellulose fibre insulation. *Building and Environment, 96*, 170–177. https://doi.org/10.1016/j.buildenv.2015.09.031
LWARB. (2017). London's circular economy route map. 1–60. https://www.lwarb.gov.uk/wp-content/uploads/2015/04/LWARB-London's-CE-route-map_16.6.17a_singlepages_sml.pdf
Mahadevan, P. (2006). Domestic retrofitting strategies in the UK: Effectiveness versus affordability. In *PLEA 2006—23rd international conference on passive and low energy architecture, conference proceedings*, September, 6–8.
Maiano. (2021). Maiano construction catalog technical data. 1–2.
Malone, T. (2021). CBD, marijuana and hemp: What is the difference among these cannabis products, and which are legal? *The Conversation*, 1–5.
Mekhilef, S., Saidur, R., & Kamalisarvestani, M. (2012). Effect of dust, humidity and air velocity on efficiency of photovoltaic cells. *Renewable and Sustainable Energy Reviews, 16*(5), 2920–2925.
Milutiene, E., Staniškis, J. K., Kručius, A., Auguliene, V., & Ardickas, D. (2012). Increase in buildings sustainability by using renewable materials and energy. *Clean Technologies and Environmental Policy, 14*(6), 1075–1084.
Moncada, L. G. G., Asdrubali, F., & Rotili, A. (2013). Influence of new factors on global energy prospects in the medium term: comparison among the 2010, 2011 and 2012 editions of the IEA's World Energy Outlook reports. *Economics and Policy of Energy and the Environment*.
Nguyen, D. M., Grillet, A. C., Bui, Q. B., Diep, T. M. H., & Woloszyn, M. (2018). Building bio-insulation materials based on bamboo powder and bio-binders. *Construction and Building Materials, 186*, 686–698. https://doi.org/10.1016/j.conbuildmat.2018.07.153
Palladino, D., Scrucca, F., Calabrese, N., Barberio, G., & Ingrao, C. (2021). Durum-wheat straw bales for thermal insulation of buildings : Findings from a comparative energy analysis of a set of wall-composition samples on the building scale.
Papadopoulos, A. M., & Giama, E. (2007). Environmental performance evaluation of thermal insulation materials and its impact on the building. *Building and Environment, 42*(5), 2178–2187. https://doi.org/10.1016/J.BUILDENV.2006.04.012
Pearlmutter, D., Theochari, D., Nehls, T., Pinho, P., Piro, P., Korolova, A., Papaefthimiou, S., Mateo, M. C. G., Calheiros, C., Zluwa, I., Pitha, U., Schosseler, P., Florentin, Y., Ouannou, S., Gal, E., Aicher, A., Arnold, K., Igondová, E., & Pucher, B. (2020). Enhancing the circular economy with nature-based solutions in the built urban environment: Green building materials, systems and sites. *Blue-Green Systems, 2*(1), 46–72. https://doi.org/10.2166/bgs.2019.928
Peuhkuri, R., Rode, C., & Hansen, K. K. (2008). Non-isothermal moisture transport through insulation materials. *Building and Environment, 43*(5), 811–822.
Platt, S. L., Walker, P., Maskell, D., & Laborel-Préneron, A. (2019). Re-baling straw for better insulation. *3rd International Conference on Bio-Based Building Materials, 37*(2), 6. https://purehost.bath.ac.uk/ws/portalfiles/potal/189875835/RE_BALING_STRAW_FOR_BETTER_INSULATION.pdf%0A. https://researchportal.bath.ac.uk/en/publications/re-baling-straw-for-better-insulation

Pomponi, F., & Moncaster, A. (2017). Circular economy for the built environment: A research framework. *Journal of Cleaner Production, 143*, 710–718. https://doi.org/10.1016/j.jclepro.2016.12.055

Pruteanu, M. (2010). Investigations regarding the thermal conductivity of straw. *Buletinul Institutului Politehnic Din Lasi. Sectia Constructii, Arhitectura, 56*(3), 9.

Pujadas-Gispert, E., Alsailani, M., van Dijk (Koen), K. C. A., Rozema (Annine), A. D. K., ten Hoope (Puck), J. P., Korevaar (Carmen), C. C., & Moonen (Faas), S. P. G. (2020). Design, construction, and thermal performance evaluation of an innovative bio-based ventilated façade. *Frontiers of Architectural Research, 9*(3), 681–696. https://doi.org/10.1016/j.foar.2020.02.003

Rashidi, S., Esfahani, J. A., & Karimi, N. (2018). Porous materials in building energy technologies—A review of the applications, modelling and experiments. *Renewable and Sustainable Energy Reviews, 91*(September 2017), 229–247. https://doi.org/10.1016/j.rser.2018.03.092

Hiss Reet. (2015). Hiss Reet. ecological insulation qualities, areas of application : Hiss Reet Platte—ecological insulation. Retrieved from https://archive.org/details/pdfy-E1piiczDcv6tXMxk/page/n1/mode/2up

Ricciardi, P., Belloni, E., & Cotana, F. (2014). Innovative panels with recycled materials: Thermal and acoustic performance and life cycle assessment. *Applied Energy, 134*, 150–162. https://doi.org/10.1016/j.apenergy.2014.07.112

Ritzén, S., & Sandström, G. Ö. (2017). Barriers to the circular economy–integration of perspectives and domains. *Procedia Cirp, 64*, 7–12.

Rochman, C. M., Hoh, E., Kurobe, T., & Teh, S. J. (2013). Ingested plastic transfers hazardous chemicals to fish and induces hepatic stress. *Scientific Reports, 3*, 3263.

Salkuyeh, Y. K., & II, T. A. A. (2015). Co-production of olefins, fuels, and electricity from conventional pipeline gas and shale gas with near-zero CO_2 emissions. Part I: process development and technical performance. *Energies, 8*(5), 3739–3761.

Sañudo, R., Dell'Olio, L., Casado, J. A., Carrascal, I. A., & Diego, S. (2016). Track transitions in railways: A review. *Construction and Building Materials, 112*, 140–157.

Schiavoni, S., D'Alessandro, F., Bianchi, F., & Asdrubali, F. (2016). Insulation materials for the building sector: A review and comparative analysis. *Renewable and Sustainable Energy Reviews, 62*, 988–1011. https://doi.org/10.1016/j.rser.2016.05.045

Schmidt, A. C., Jensen, A. A., Clausen, A. U., Kamstrup, O., & Postlethwaite, D. (2004). A comparative life cycle assessment of building insulation products made of stone wool, paper wool and flax: Part 1: Background, goal and scope, life cycle inventory, impact assessment and interpretation. *International Journal of Life Cycle Assessment, 9*(1), 53–66. https://doi.org/10.1007/BF02978536

Shi, J.-S., Li, J., Fan, Y., & Ma, H. (2006). Preparation and properties of waste tea leaves particleboard. *Forestry Studies in China, 8*(1), 41.

Shi, Jianjun, Lu, L., Guo, W., Zhang, J., & Cao, Y. (2013). Heat insulation performance, mechanics and hydrophobic modification of cellulose–SiO_2 composite aerogels. *Carbohydrate Polymers, 98*(1), 282–289. https://doi.org/10.1016/j.carbpol.2013.05.082

Sierra-Pérez, J., Boschmonart-Rives, J., Dias, A. C., & Gabarrell, X. (2016). Environmental implications of the use of agglomerated cork as thermal insulation in buildings. *Journal of Cleaner Production, 126*, 97–107. https://doi.org/10.1016/j.jclepro.2016.02.146

Silvestre, J. D., Silva, A., & de Brito, J. (2015). Uncertainty modelling of service life and environmental performance to reduce risk in building design decisions. *Journal of Civil Engineering and Management, 21*(3), 308–322.

Silvestre, J. D., Pargana, N., De Brito, J., Pinheiro, M. D., & Durão, V. (2016). Insulation cork boards-environmental life cycle assessment of an organic construction material. *Materials, 9*(5), 1–16. https://doi.org/10.3390/ma9050394

Takano, A., Hughes, M., & Winter, S. (2014). A multidisciplinary approach to sustainable building material selection: A case study in a Finnish context. *Building and Environment, 82*, 526–535.

Tangjuank, S. (2011a). Thermal insulation and physical properties of particleboards from pineapple leaves. *International Journal of Physical Sciences, 6*(19), 4528–4532. https://doi.org/10.5897/IJPS11.1057

Tangjuank, S. (2011b). Thermal insulation and physical properties of particleboards from pineapple leaves. *International Journal of Physical Sciences, 6*, 4528–4532.

Tettey, U. Y. A., Dodoo, A., & Gustavsson, L. (2014). Effects of different insulation materials on primary energy and CO_2 emission of a multi-storey residential building. *Energy and Buildings, 82*, 369–377.

Vafai, K. (2015). *Handbook of porous media.* Crc Press.

Van Buren, N., Demmers, M., der Heijden, R., & Witlox, F. (2016). Towards a circular economy: The role of Dutch logistics industries and governments. *Sustainability, 8*(7), 647.

Van Dam, J. E. G., van den Oever, M. J. A., & Keijsers, E. R. P. (2004). Production process for high density high performance binderless boards from whole coconut husk. *Industrial Crops and Products, 20*(1), 97–101.

Vasile, V., Petcu, C., Meiţą, V., & Zaharia, M. C. (2019). Innovative thermal insulation products for a circular economy. *IOP Conference Series: Earth and Environmental Science, 290*(1). https://doi.org/10.1088/1755-1315/290/1/012037

VIRTUe. (2018). Solar decathlon middle east 2018. Student team VIRTUe. Jury report. Available online at: https://www.solardecathlonme.com/.

VIRTUe. (2019). Student Team VIRTUe. Eindhoven University of Technology. Available online at: https://teamvirtue.nl/

Viswanathan, R., Gothandapani, L., & Kailappan, R. (2000). Water absorption and swelling characteristics of coir pith particle board. *Bioresource Technology, 71*(1), 93–94.

Wei, K., Lv, C., Chen, M., Zhou, X., Dai, Z., & Shen, D. (2015). Development and performance evaluation of a new thermal insulation material from rice straw using high frequency hot-pressing. *Energy and Buildings, 87*, 116–122. https://doi.org/10.1016/j.enbuild.2014.11.026

Widyorini, R., Xu, J., Umemura, K., & Kawai, S. (2005). Manufacture and properties of binderless particleboard from bagasse I: Effects of raw material type, storage methods, and manufacturing process. *Journal of Wood Science, 51*(6), 648–654.

Zhou, X., Zheng, F., Li, H., & Lu, C. (2010). An environment-friendly thermal insulation material from cotton stalk fibers. *Energy and Buildings, 42*(7), 1070–1074. https://doi.org/10.1016/j.enbuild.2010.01.020

Development of Urban Regeneration

Enriching the "Communities-Cultural Heritage" Relationship to Ensure Effective Culture-Based Urban Development

Nour Zreika, Daniele Fanzini, and Elena Vai

Abstract

As urban populations continue to grow, cities are compelled to tackle new challenges of a social, economic, and environmental nature. Therefore, it is necessary to limit what has been hindering the sustainable development of cities to overcome them and ensure their future prosperity. Since cities also depend on local and regional material and immaterial cultural resources, this contribution will consider culture-based urban development as a means to promote citizen participation. In the last decade, cultural heritage has become a prominent tool for social and economic improvement. It has also made its way into various European and international policies where cultural heritage is placed at the center of sustainable development. Consequently, interest in cultural heritage can help cities develop by attracting investments and creating job opportunities, as well as by embracing cultural identity, sense of belonging, and social cohesion. However promising culture-based strategies may seem, ensuring their successful implementation for the development of urban contexts is challenging and requires further exploration. For this reason, community participation is considered a promising approach. Local communities can themselves point out the difficulties facing their cities and take part in their problem-solving processes. Nonetheless, collaborating with local communities is problematic due to the digital gap within communities and the struggle to promote cooperation between non-experts with policymakers and other experts. Thus, the objective of this contribution is to enhance the relationship between "communities" and their "cultural heritage" through the exploitation of enabling digital tools as the means for enriching participatory approaches. As affirmation, this contribution will demonstrate the methods and processes through which new forms of community participation are possible and will showcase projects and good practices where these strategies have been effectively realized.

Keyword

Community engagement • Cultural heritage • Digital tools • Cities • Culture-based development

1 Background Statement

Today, cities are home to more than half of the world's population, and by 2030, that number is expected to rise. Given that cities are engines of economic growth and social transformation, the migration of people from rural to urban areas is unavoidable. Despite its opportunities, urbanization gave rise to social, economic, and environmental challenges. It has also become a substantial burden for cities and a barrier to their sustainable development (UNHABITAT, 1996). Thus, urban growth has led to a decline in life quality, lack of urban services, poverty, unemployment, loss of natural resources, and deterioration of environment. Moreover, the COVID-19 pandemic revealed further weaknesses in cities' existing structures; most cities were unprepared for the social and economic impacts caused by the pandemic.

Culture-based urban development is considered a solution to the problems of urban areas, and it is significantly recognized by the European Union for aiding both the present and future populations of cities. Thus, this notion has become predominant in local and regional conventions,

N. Zreika (✉) · D. Fanzini
Department of Architecture, Built Environment and Construction Engineering at Politecnico di Milano, Milan, Italy
e-mail: nour.zreika@polimi.it

D. Fanzini
e-mail: daniele.fanzini@polimi.it

E. Vai
Department of Architecture, Alma Mater Studiorum—Università di Bologna, Bologna, Italy
e-mail: elena.vai@unibo.it

A. Battisti et al. (eds.), *Greening Our Cities: Sustainable Urbanism for a Greener Future*,
Advances in Science, Technology & Innovation, https://doi.org/10.1007/978-3-031-49495-6_23

declarations, documents, and policies. In order to gain social, and economic benefits, culture-based initiatives rely on local and regional cultural resources that are present in cities. Here, the "transdisciplinary" role of cultural heritage, which is capable of delivering advantages at a cultural, social, and economic level, is highlighted (Della Torre & Canziani, 2007). Beginning with the Faro Convention, increased awareness of cultural heritage has made its way to the center of European and international policies (Council of Europe, 2005). The 2013 Hangzhou Declaration, for example, highlights the crucial role of culture in achieving sustainable development (UNESCO, 2013). Thus, from an economic standpoint, cultural heritage is capable of attracting investments, promoting tourism, enhancing life quality, and boosting employment. From a social perspective, it can embrace social cohesion, promote community participation, and strengthen identity, memory, and sense of belonging (Ferrilli, et al., 2017).

Although it might appear that there exists a firm understanding of culture-based strategies, this is not entirely the case. In fact, these strategies are of a complex nature and ensuring their successful implementation for the development of urban context is challenging and requires further exploration. Consequently, this contribution considers community participation as a promising approach. Thus, the success of culture-based urban development relies on the collaboration and involvement of local communities that can themselves point out the difficulties facing their cities and take part in their problem-solving processes.

2 The "Communities-Cultural Heritage" Relationship

Community participation, as highlighted in several national and international policies, is considered a vigorous and inclusive approach that results in long-term benefits and successful transformation (Li et al., 2020). An important factor to consider for this approach is the "communities-cultural heritage" relationship; it is evident that local communities have a strong bond with their heritage and that they generally work to preserve it. Thus, individuals and communities are expected to actively participate by contributing to the formation, implementation, and assessment of projects and policies related to cultural heritage (Rifkin & Kangere, 2001). Active participation is capable of influencing local communities at cultural, social, and economic levels. According to UNESCO, the active participation of local communities in cultural heritage is a means of reaching sustainable development. This has allowed UNESCO to assess how well cultural heritage can encourage local communities to participate. (UNESCO, 2019).

Therefore, adopting a bottom-up approach is encouraged while paving the way for greater community participation. In contrast to a top-down approach, a bottom-up model grants local communities leading roles in the decision-making process and offers governments and authorities supporting roles as consultants and facilitators of these communities. For this reason, it is insinuated that the bottom-up approach can ensure effective culture-based urban development (Nared & Bole, 2020). Established by the European Union, the LEADER approach (Liaison Entre Action de Développement de l'Économie Rurale—Links between Actions for the Development of the Rural Economy) is an example of the growing popularity of bottom-up approaches in local development. Since LEADER considers local communities as territorial experts, it fosters the participation of local actors and makes use of their intrinsic knowledge and skills. Thus, active participation of these local actors is highly recommended for the conceptual and practical development of "local strategies," "decision-making," and "resource allocation" (European Commission, 2021).

Nonetheless, collaborating with local communities is problematic, and with the COVID-19 pandemic, it has become even more so. In this section, the challenges of participatory methods will be categorized as the following: power, management, financing, implementation, and sustainability (see Fig. 23.1). The concept of power is a major challenge in participatory approaches; administrative and governmental bodies are notably resistant to transferring their power to the people. Thus, this shift from a top-down to a bottom-up approach is accompanied by the administrative and governmental bodies' reluctance to involve local communities in decision-making processes (Nasrolahi et al., 2022). Moreover, the failure of authorities to decentralize the handling of participatory projects has given rise to a number of management-related problems. In some cases, participatory approaches may be difficult to finance since they are considered cost- and time-consuming. Finding the most suitable methods of implementation is another challenge for participatory approaches. Thus, researchers are compelled to look for the most appropriate methods and tools to ensure the best outcomes of involving local communities in problem-solving and decision-making processes (Nasrolahi et al., 2022). Since most participatory approaches have short-term benefits, local communities experience short-lived impacts, and as a consequence, they lose interest in contributing to these projects. Thus, sustainability is an obstacle for community participation that must be predictable and plannable (Nasrolahi et al., 2022). Another recent challenge facing local participation is the COVID-19 pandemic, which oversaw the cease of all forms of involvement. However, the pandemic is an incident that has led to a reconsideration of ways in which to approach local

POWER	MANAGEMENT	FINANCING	IMPLEMENTATION	SUSTAINABILTY	PANDEMIC
The resistance of the administrative and governmental bodies to the transfer of power to the people.	Failure of authorities to decentralize the handling of participatory projects	Participatory approaches are considered cost - and time - consuming.	Finding the most suitable methods to implement participatory approaches can be challenging.	Ensuring a long-term benefit is an obstacle for community participation.	The COVID pandemic has led to a disruption of community participation.

Fig. 23.1 Challenges of participatory methods. *Source* (Author)

participation, not only to deal with the current crisis, but also to be prepared for future emergencies. Here, communities have been crucial in ensuring the continuity of participation by encouraging innovative and developed approaches.

3 A New Approach to Community Participation

Since this contribution considers that enhancing the relationship between "communities" and their "cultural heritage" could lead to cultural, social, and economic benefits, it will suggest significant tools for enriching community participation. The past few years have witnessed the emergence of co-design and participation processes as appropriate methods through which sustainable urban development can be ensured. Thus, the co-design approach considers actively involving both experts and citizens in the design and decision-making processes, therefore, addressing the challenges of the urban context.

Even though methods and materials for co-design and participation processes have already been developed, they must be reconsidered. Since the COVID-19 pandemic forced us to rethink the ways in which communities can engage and interact, digital tools were seen as promising for co-design and participation processes. Unlike conventional means of participation, digital technologies are capable of reaching and engaging a larger platform of citizens and becoming a reliable source for obtaining and sharing information (Stelzle et al., 2017).

Based on the results of the "*Stakeholders' Survey on a European Collaborative Cloud for Cultural Heritage*," conducted by the European Union, it is realized that the cultural heritage sector is aware of the inevitability of a digital transition. In order to smoothen this transition, stakeholders agree that a digital infrastructure can allow for greater collaboration and support within the sector (European Commission, 2022). Thus, the integration of digital tools can take place throughout the design process as well as during the intervention phase (refer to Fig. 23.2). During the *design phase*, citizens must be involved in the *data collection* and the *decision-making* processes. Here, online platforms, social media, and mobile applications can be used to conduct campaigns, surveys, and questionnaires. The installation of digital technologies can also be used to gather data; these technologies can either take the form of interactive tools such as interactive screens (iPads, tablets, etc.) or digital polling devices such as audio devices with buttons (YES/NO buttons, rating devices, etc.). Another participatory method is that of *information mapping,* where citizens rely on mapping tools to communicate and share their experiences, feedback, critique, and debate. As for the *intervention phase*, citizens must be able to experience their cities in a new and improved manner. A popular method involves the distribution of QR codes around urban areas. Once scanned, these QR codes can provide citizens with significant data and valuable knowledge about the area. This is also accompanied by an augmented reality (AR) or a virtual reality (VR) experience, where a previously constructed 3D model of the urban area is used for the personalized exploration (using mobile devices, smartphones, tablets, head-mounted displays, glasses, etc.) (De Lange & De Waal, 2019).

4 The Living Labs Experience

This contribution will consider the Living Labs concept as a good practice where the "communities-cultural heritage" strategies can be effectively realized for culture-based urban development. Living Labs have witnessed an expansion in Europe by embracing a new trend for user involvement in open innovation. Thus, Living Labs create a real-life environment where innovative solutions can be examined, assessed, and produced.[1] This notion involves several actors, such as users, researchers, firms, public partners, and stakeholders, within the research and innovation processes in order to create new and improved products and services (Bergvall-Kåreborn et al., 2009). Although there are several implications in literature for what Living Labs are, this contribution will limit itself to two definitions. A Living Lab can be considered a model through which "*stakeholders*

[1] European Network of Living Labs (ENoLL). *What are Living Labs.* Retrieved from European Network of Living Labs: https://enoll.org/about-us/what-are-living-labs/ (accessed on 27 March 2023).

PHASES OF COMMUNITY PARTICIPATION	DESIGN PHASE			INTERVENTION PHASE
METHODS	Data Collection	Decision-making	Information Mapping	Phygital experience of Urban Context
MATERIALS	Online platforms, social media, and mobile applications. Interactive tools such as interactive screens (iPads, tablets, etc.). Digital polling devices such as audio devices with buttons (YES/NO buttons, rating devices, etc.).		Mapping tools to communicate and share experiences, feedback, critique, and debate.	The distribution of QR codes around urban areas. The use of mobile devices, smartphones, tablets, head-mounted displays, and glasses for an augmented reality (AR) or a virtual reality (VR) experience.

Fig. 23.2 Digital tools used to Enrich community participation. *Source* (Author)

form public–private-people partnerships (4Ps), all collaborating for creation, prototyping, validating, and testing of new technologies, services, products and systems in real-life contexts" (ENoLL, 2016). As defined by Bergvall-Kåreborn et al. (2009), a Living Lab is "*a user-centric innovation milieu built on every-day practice and research, with an approach that facilitates user influence in open and distributed innovation processes engaging all relevant partners in real-life contexts, aiming to create sustainable values*" (Bergvall-Kåreborn et al., 2009).

Today, the application of the Living Labs concepts is witnessed across various disciplines, such as culture and creativity, social innovation, social inclusion, education, and smart cities and regions (Bergvall-Kåreborn et al., 2009). Social innovation is considered as one of the building blocks of Living Labs; it has been successful in cultivating the power of citizen participation to overcome the urban challenges of today's cities.

In order to be effective, a Living Lab must adhere to the following five principles: openness, impact, realism, value, and sustainability. Openness encourages innovation to design and produce new products, services, and markets; influence necessitates the recognition of the "users" role in the process; realism requires Living Labs to incorporate real-life environments for their activities; value considers economic, business, and consumer/user dimensions for Living Labs; and sustainability concerns the Living Labs viability and continuity in their urban areas (Bergvall-Kåreborn et al., 2009).

As seen in Fig. 23.3, the key components of Living Labs include ICT and infrastructure, management, partners and users, research, and approach. With the use of ICT and infrastructure, Living Labs were able to create new and improved methods for participation and co-creation for stakeholders. Management of Living Labs is facilitated by researchers, consultants, and companies; these actors are responsible for planning, organizing, and managing their projects and activities. Partners and users play a crucial role in the acquisition and transmission of knowledge and skills. Through the intricate processes of learning, analysis, and interpretation, research can result in both conceptual and operational contributions to Living Labs. In order to identify the best methods and materials, the approach of Living Labs relies on the collection of best practices (Bergvall-Kåreborn et al., 2009).

5 Focus on the ROCK Project

This contribution will refer to the experience of ROCK (Regeneration and Optimization of Cultural Heritage in Creative and Knowledge Cities), a project funded by Horizon 2020 (2017–2020), to validate the effectiveness of culture-based urban development, the value of participation and co-design processes, as well as the usefulness of digital tools.

According to the ROCK project, culture-based initiatives are capable of stimulating the successful regeneration and adaptive reuse of historic city centers. By placing cultural heritage at the center of urban development, this project encouraged the transformation of historic city centers into real-life laboratories where the challenges and opportunities of cultural heritage and the city were explored (Leoni et al., 2020). Thus, it tackled the themes of conservation and valorization of cultural heritage within the urban context and led to the revival of public spaces, the adaptive reuse of abandoned buildings, the renewal of existing resources, and the development of new services within the historic city centers.

Another factor to consider owing to the ROCK project is the link between citizens and their cultural heritage. In this project, citizens played a crucial role in participation and

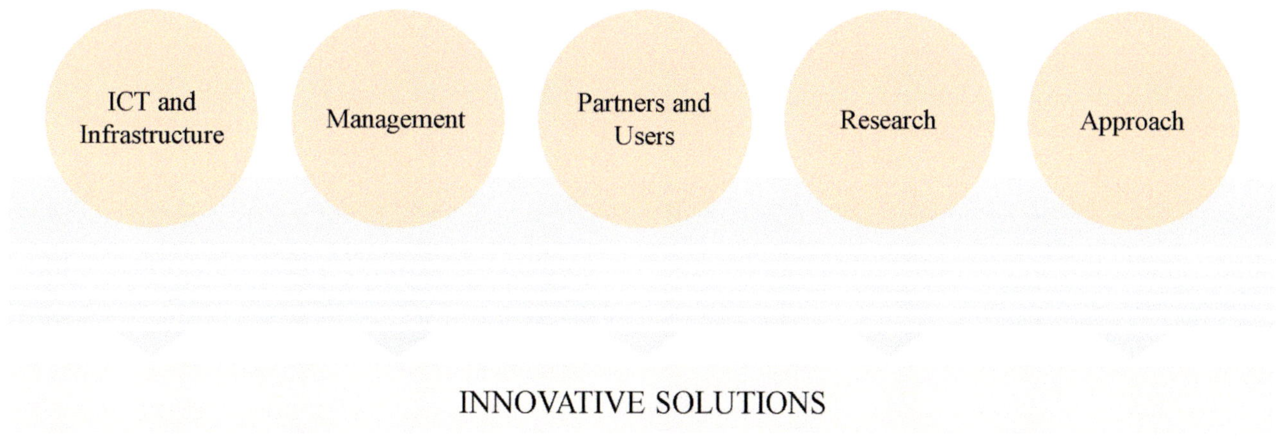

Fig. 23.3 Diagram illustrating the key components of living labs. *Source* (Author)

co-design processes. By combining non-expert information with that of expert knowledge, the success of these processes could be ensured. As a result, the ROCK project promoted the participation of a wide range of actors in order to strengthen the "communities-cultural heritage" relationship and ensure the viability of cultural heritage in cities. It collaborated with municipalities, universities, research centers, associations, networks, and technology providers from various European countries to envision an inclusive, knowledgeable, and creative model of a city (Leoni et al., 2020).

Moreover, the ROCK project has enabled cities to harness the power of cultural heritage, community participation, and technology through the use of digital tools. With the development of a digital infrastructure, the ROCK project was able to enhance the collaboration between various actors as well as create a platform where the testing, prototyping, upgrading, and sharing of skills and knowledge can be possible (Leoni et al., 2020).

5.1 The Cases of the Three Living Labs

This part will look into the efforts of Skopje, Lisbon, and Bologna, the three cities of the ROCK project that took from the good practices of the cities of Turin, Eindhoven, Lyon, Vilnius, Liverpool, Cluj-Napoca, and Athens, and replicated their inventive, collaborative, and sustainable efforts (Leoni et al., 2020). Hence, the experiences of each city will be extracted with respect to the three factors mentioned above: the effectiveness of culture-based urban development, the value of participation and co-design processes, and the usefulness of digital tools.

The Skopje Living Lab (SkULLab)

Based on the cultural and historical significance of the Old Bazaar, the Skopje Living Lab (SkULLab) explored the possibility of deriving social and economic impacts from this heritage site, as well as transforming the urban environment to which it belongs. By making use of the Old Bazaar's cultural and commercial significance, the SkULLab proposed the introduction of contemporary crafts to the traditional place. This action allowed the Old Bazaar (as seen in Fig. 23.4) to acquire a new creative, social, and economic role based on artisanship rather than a touristic role based on visitor consumption (ROCK, n.d.). Thus, the SkULLab functions as a HUB where new and improved models for the urban environment could be developed. The SkULLab provided both conceptual and operational strategies for the culture-based redevelopment of the Old Bazaar.

Within its efforts, the SkULLab invited the local community of the Old Bazaar to participate in finding creative and innovative ways to support artisanship while meeting the contemporary needs of the urban environment. The SkULLab suggests, assesses, and implements a variety of activities to support this creative industry. It also ensured the effective revitalization of the Old Bazaar by supporting the area's continued improvement and establishing a network of stakeholders and an infrastructure for data collection (ROCK, n.d.).

In order to support its initiatives, SkULLab realized a digital platform that supported "open knowledge" and "open data" for collecting information on land use and gathering information on the urban context. ICT and digital tools were also used to create new products and services that can complement traditional artisanship. The proposed cultural and social activities of SkULLab were promoted through ICT and digital tools. These technologies were also used to allow various actors, such as researchers, students, artisans, citizens, organizations, municipalities, and authorities, to take part in the co-design process of the Old Bazaar (Siljanoska, 2020).

The Lisbon Living Lab (LLL)

The Lisbon Living Lab recognized the tangible and intangible, natural and cultural heritage of the former rural and

Fig. 23.4 Old Bazaar, Skopje, Macedonia. *Source* (Masevski, n.d.)

Fig. 23.5 Working session of lisbon living lab. *Source* (ROCK, n.d.)

industrial district of Marvila. Within its culture-based urban strategy, citizens were treated as both the producers and consumers of cultural heritage; not only could citizens participate in workshops, hackathons, and events (see Fig. 23.5), but they could also curate tours, activities, and exhibitions to make the cultural heritage of Marvila accessible to all. From a social perspective, the LLL ensured social cohesion, strengthened a sense of belonging, and promoted intercultural and intergenerational dialogue. From an economic standpoint, it stimulated the creative sector by recruiting new creative talents. From a cultural perspective, the LLL raised awareness among local communities about the problems facing their cultural heritage and encouraged the adoption of best practices for the conservation and valorization of this heritage (ROCK, n.d.).

The participation and co-design approaches adopted for the Lisbon Living Lab (LLL) made it possible to recognize the social, economic, and environmental values of the Marvilla district and find ways to strengthen them to ensure sustainable urban development. Thus, the LLL invited local communities and stakeholders to get involved in developing temporary, creative, and innovative projects for its four key themes of "urban voids," "ecological networks," "territories of continuity," and "local communities vs. gentrification" (ROCK, n.d.).

Moreover, the LLL utilized digital tools for the development of these participation and co-design processes. Digital tools allowed citizens to recognize, research, and promote the cultural heritage in the Marvila district. An online database was used to gather citizens' experiences with their cultural heritage and exhibit these stories through multimedia tools and applications. Through "participatory mapping," "geo-referencing," and "cartography," the collected data was transformed into a new and improved mapping of the Marvila district. Moreover, citizens' experiences were transported into the digital world in the form of video games that illustrate the past, present, and future of the district (Houpert, 2020).

The ROCK Living Lab in Bologna (U-Lab)

The ROCK Living Lab in Bologna (U-Lab) is aimed at transforming the historic city center on the grounds of culture-based urban development. It focused on finding innovative ways for boosting the reputability and livability of the university district of Via Zamboni by designing a set of actions for the city's public spaces, which include the Municipal Theatre, Piazza Verdi, Piazza Scaravilli, and the porticoes. According to U-Lab, these public spaces have become nodes for the city, and interconnections between them result in a system where actions could occur (Boeri, et al., 2019).

Involving local communities in the co-designing activities played a major part in the U-Lab project. Thus, it is through participatory processes that solutions to problems in the university district of Via Zamboni could be found (ROCK, n.d.). As a first step, U-Lab sought to identify a list of potential stakeholders for the process by reaching out to citizens, business owners, universities, organizations, administrations, etc. It later focused its efforts on identifying actions to address the topics of accessibility, sustainability, and new collaborations for these selected open spaces (Boeri et al., 2019). Afterward, by involving these stakeholders in workshops, U-Lab was able to foster urban regeneration, enhancement, and adaptive reuse, as well as create new services and experimental activities. As a result of temporary transformation, the U-Lab was able to make use of urban spaces (see Fig. 23.6) to test, evaluate, validate, and implement new features and functions for the city that are more in line with the demands and needs of its citizens (Djalali et al., 2019).

Additionally, the U-Lab project promoted the use of digital tools within its co-design practices as an attempt to increase the transmission of knowledge and involvement between its various actors (ROCK, n.d.). Thus, digital tools in the form of sensors, applications, and platforms were used to support and enhance the successful transformation and adaptive reuse of the city's public spaces. These tools enabled the collection of data to identify the present usage, climate, and environmental conditions, as well as map the experiences of the users within the public space. Moreover,

Fig. 23.6 Citizens interacting in the Public Space, U-Lab, Bologna, Italy. *Source* (ROCK, n.d.)

the U-Lab made use of digital technologies to increase the accessibility of cultural heritage and improve the lives of disabled inhabitants (ZONA-U, 2019).

5.2 Discussion

The experiences of the ROCK project have shown that all the three abovementioned factors (the effectiveness of culture-based urban development, the value of participation and co-design processes, and the usefulness of digital tools) are intertwined. Hence, participatory and co-design approaches are continuously supported for what they can offer in designing new and improved methods and processes for culture-based urban development. Based on these approaches, the "communities-cultural heritage" relationship is strengthened, and the viability of these cities and their cultural heritage is perceived by enhancing the conditions of urban spaces and increasing the accessibility and usability of cultural heritage (Gianfrate et al., 2018). Thus, urban spaces become laboratories where the collection of data, practices, and relationships is converted into "accessible-to-all" knowledge and testing places for creative and innovative processes leading to the transformation of the urban context (Djalali et al., 2019).

The collaboration of multiple stakeholders allows us to perceive an innovative and experimental model for the city, where its usability, capability, and accessibility are reformed. This also applies to cultural heritage, where, for instance, collaborations between the cultural and technology sectors have shifted the conservation and valorization processes away from traditional spheres toward more progressive ones, based on new fruition models (Djalali et al., 2019).

As for the three Living Labs of the ROCK project, there is a need to improve the interaction between experts and non-experts, where this difficulty in dialogue can reduce the effectiveness of participatory and co-design processes. The urge to create an inclusive design approach is clearly expressed in the document *"Technologies and Tools for Better Access to Cultural Heritage."* This approach must be capable of improving the livability and accessibility of cities without having to adapt or modify the physical space.[2] Thus, this contribution will try to address the aforementioned need by referring to the model of the "mutant city," which is constantly changing in form and function in response to the needs and conditions of its citizens. Due to its mutating character, design and technology are seen as the enabling tools for its urban transformation (Celaschi et al., 2020). Relying on digital tools not only resolves the challenge of communication between actors but also overcomes the limits of time and cost of real-time interventions by providing cultural data access to all and offering a virtual experience similar to that of the real world. These tools can also allow designers to relate what citizens experience in both the digital realm and the physical world. This contribution proposes the adoption of a "phygital" (physical and digital) Living Lab where citizens, artists, and creatives can develop innovative projects for themselves and their cities by integrating art, culture, and new technologies.

Thus, as shown in Fig. 23.7, the proposed digital-based methodology will consist of the following three phases: knowledgeability, capability, and accessibility. Knowledgeability includes activities such as the selection and investigation of the urban context, collecting data on its cultural, historical, and environmental aspects from experts and non-experts' knowledge, and sharing this information through data visualizations and digital twins. Capability consists of gathering experiences of users from both physical and digital interactions. This phase also includes the setting up of co-design activities and participatory actions in order to discuss challenges and solutions for the urban context. Accessibility involves the envisioning of the design solutions through an immersive experience, where these solutions can be staged, tested, evaluated, and approved by a wide range of users before being realized in the real world.

6 Conclusion

Culture-based urban development can be recognized as a sustainable strategy, model, and process for reviving and regenerating urban areas, thanks to the participation and engagement of communities. Therefore, cultural heritage

[2] Regeneration and Optimization of Cultural heritage in creative and Knowledge cities (ROCK). *Technologies and Tools for Better Access to Cultural Heritage.* Retrieved from ROCK Project: https://rockproject.eu/documents-list (accessed on 28 March 2023).

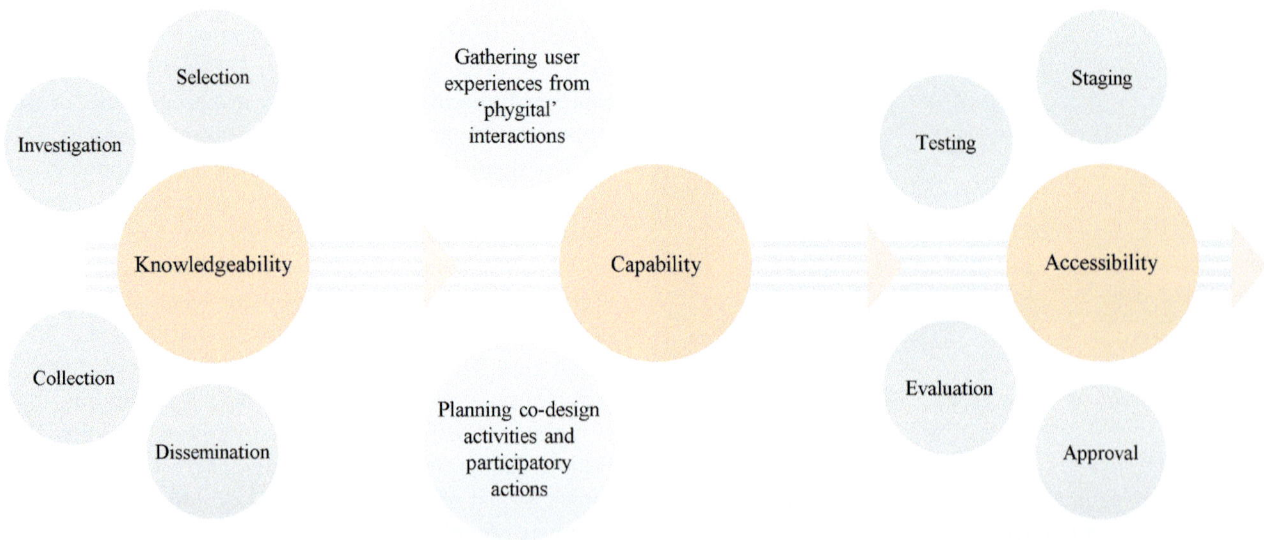

Fig. 23.7 Diagram showing the proposed digital-based methodology. *Source* (Author)

must be considered a source of economic and social enhancement rather than a mere cultural asset. There is great prosperity in adopting participation and co-design approaches for achieving long-term benefits and successful transformation. However, it is vital to design engagement processes based on local identities to enhance the relationship between "communities" and their "cultural heritage" and enrich participatory approaches. New models of the fruition of cultural heritage also need to be co-designed, involving all actors (institutions, researchers, professionals, stakeholders, and citizens), together with innovative policies and digital strategies. These new models can be supported and enabled through the adoption of ICT and digital tools to guarantee effective citizen participation and co-design activities. This in turn could improve the accessibility of a variety of target groups, from the elderly, who are the keepers of cultural memory, to the youth, who are the innovators of digital processes capable of exploiting cultural heritage and the creators of future memories. To conclude, culture-based urban development fits the model of the contemporary knowledge city, where material and immaterial values live alongside concepts such as beautiful and sustainable, as desired by the New European Bauhaus[3] initiative.

[3] The New European Bauhaus is an initiative that combines innovative and transdisciplinary strategies in response to the European Green Deal. It facilitates social transformation by making EU funding available for projects that value sustainability, aesthetics, and inclusion. Retrieved from https://new-european-bauhaus.europa.eu/index_en (accessed on 27 March 2023).

References

Bergvall-Kåreborn, B., Eriksson, C. I., Ståhlbröst, A., & Svensson, J. (2009). A milieu for innovation: defining living labs. In *ISPIM Innovation Symposium*. Retrieved from https://www.diva-portal.org/smash/record.jsf?pid=diva2%3A1004774&dswid=7951

Boeri, A., Longo, D., Gianfrate, V., & Roversi, R. (2019). Cultural heritage-led initiatives for urban regeneration. Pilot implementation actions in Bologna Public spaces. *Diverse City*, 463–472. Retrieved from http://www.cittacreative.eu/wp-content/uploads/2019/06/boeri_DIVERSEcity-2019.pdf

Celaschi, F., Fanzini, D., & Formia, E. (2020). Enabling technologies for continuous and interdependent design. In M. Lauria, E. Mussinelli, & F. Tucci (Eds.), *Producing projects* (pp. 487–493). Maggioli Editore. Retrieved from https://re.public.polimi.it/retrieve/e0c31c10-7bc1-4599-e053-1705fe0aef77/_e-book_AAVV_Producing-Project.pdf

Council of Europe. (2005). Council of Europe framework convention on the value of cultural Heritage for society. In *The Faro convention*. Council of Europe. Retrieved from https://www.coe.int/en/web/conventions/full-list/-/conventions/rms/0900001680083746

De Lange, M., & De Waal, M. (Eds.). (2019). *The hackable city: Digital media and collaborative city-making in the network society*. Springer. Retrieved from https://doi.org/10.1007/978-981-13-2694-3

Della Torre, S., & Canziani, A. (2007). Comprehensive plans for a culture-driven local development: Emergence as a tool for understanding social impacts of projects on built cultural Heritage. In G. Minati, M. Abram, & E. Pessa (Eds.), *Processes of emergence of systems and systemic properties: Towards a general theory of emergence* (pp. 79–90). World Scientific.

Djalali, A., Gianfrate, V., Longo, D., & Vai, E. (2019). An advanced design approach to support urban transformations through multi-stakeholder collaborations. *7th International Forum of Design as a Process, 12*(2).

ENoLL. (2016). *Introducing ENoLL and its living lab community* (1 ed.). In A. Garcia Robles, T. Hivikoski, D. Schuurman, & L. Stokes (Eds.) European Network of Living Labs. Retrieved from https://issuu.com/enoll/docs/enoll-print

European Commission. (2021, June 29). *LEADER/CLLD explained*. Retrieved from The European Network for Rural Development

(ENRD): https://enrd.ec.europa.eu/leader-clld/leader-toolkit/leaderclld-explained_en

European Commission. (2022). *Stakeholders' survey on a European collaborative cloud for cultural Heritage: Report on the online survey results.* Directorate-General for Research and Innovation. Publications Office of the European Union. Retrieved from https://data.europa.eu/doi/10.2777/691331

Ferrilli, G., Sacco, P., Blessi, G. T., & Forbici, S. (2017). Power to the people: When culture works as a social catalyst in urban regeneration processes (and when it does not). *European Planning Studies, 25*(2), 241–258. https://doi.org/10.1080/09654313.2016.1259397

Gianfrate, V., Boeri, A., Celaschi, F., Longo , D., & Vai, E. (2018). Design e Tecnologia Applicate al Contesto Urbano: Le Città come Laboratorio di Conoscenza e Innovazione. *MD Journal.* Retrieved from https://cris.unibo.it/handle/11585/731570

Houpert, C. (2020, February 3). Lisbon highlights the Story of Marvila Residents. *ROCK Project News.* Retrieved from https://rockproject.eu/news-details/221

Leoni, G., Boeri, A., Longo, D., Boulanger, S., Gianfrate, V., Massari, M., & Roversi, R. (2020). Introduction. In G. Leoni, & C. Hein (Eds.), *Cultural Heritage leading urban futures: Actions and Innovations from ROCK Project.* TU Delft Open Faculty of Architecture and the Built Environement, Delft University of Technology. Retrieved from https://books.bk.tudelft.nl/press/catalog/view/787/898/881-4

Li, J., Krishnamurthy, S., Pereira Roders, A., & van Wesemael, P. (2020, February). State-of-the-practice: Assessing Community Participation within Chinese Cultural World Heritage Properties. *Habitat International, 96.* https://doi.org/10.1016/j.habitatint.2019.102107

Masevski, T. (n.d.). *"New souvenirs for Skopje old bazaar". ROCK. https://rockproject.eu/news-details/267*

Nared, J., & Bole, D. (2020). *Participatory research on Heritage and culture-based development: A perspective from South-East Europe.* In J. Nared, & D. Bole (Eds.) Springer Open. https://doi.org/10.1007/978-3-030-28014-7

Nasrolahi, A., Messina, V., & Gena, C. (2022). A model and tool for community engagement case study: Community engagement in the Bisotun World Heritage Site. *AVICH Workshop.* Rome. Retrieved from https://arxiv.org/abs/2207.02353

Rifkin, S. B., & Kangere, M. (2001). What is participation? In S. Hartley (Ed.), *CBR: a participatory strategy in Africa.* University College London. Retrieved from https://www.researchgate.net/publication/311607512_What_is_Participation#:~:text=community%20participation%20in%20three%20distinct,policies%20and%20activities

ROCK. (n.d.). *ROCK Factsheet 1: The use of living labs in culture-led urban regeneration.* Retrieved March 23, 2023, from https://rockproject.eu/documents-list

Siljanoska, J. (2020). Urban living labs for sensitive city cultural Heritage regeneration. *Keeping up with Technologies to Act Responsively in Urban Environment, International Academic Conference on Places and Technologies, 7,* 165–172. https://doi.org/10.18485/arh_pt.2020.7.ch19

Stelzle, B., Jannack, A., & Noennig, J. R. (2017). Co-design and Co-decision: Decision making on collaborative design platforms. In *International conference on knowledge based and intelligent information and engineering systems*, vol. 112, (pp. 2435–2444). Procedia Computer Science. Retrieved from https://www.sciencedirect.com/science/article/pii/S1877050917314394

UNESCO. (2013). Culture: Key to sustainable development. *The Hangzhou Declaration: Placing Culture at the Heart of Sustainable Development Policies.* UNESCO. Retrieved from http://www.unesco.org/new/fileadmin/MULTIMEDIA/HQ/CLT/images/FinalHangzhouDeclaration20130517.pdf

UNESCO. (2019). *Culture 2030 indicators.* United Nations Educational, Scientific and Cultural Organization. UNESCO. Retrieved from https://whc.unesco.org/en/culture2030indicators

UNHABITAT. (1996). *An Urbanizing world: Global report on human settlements 1996.* Oxford University Press. Retrieved from https://mirror.unhabitat.org/downloads/docs/GRHS.1996.0.pdf

ZONA-U. (2019, February). *Il progetto ROCK a Bologna.* Retrieved February 24, 2023, from Zona-U: https://bologna.rockproject.eu/il-progetto-rock-a-bologna/

From Temporary to Permanent: Strategy for Developing Public Spaces—Case of Bengaluru

Sushmita Mudhol, Ann Jacob, Monica Kashkari, and Manjula Vinjamuri

Abstract

Traditionally, planning and design have been limited to planners and engineers, leaving the voices of the public directly impacted by the design unheard. Cities today are undergoing unprecedented transformations, necessitating a rethinking of the approach to planning, and designing public spaces, and therefore have adopted unique practices such as participatory planning and tactical urbanism. The former is a method of designing liveable cities that is driven by active community participation where solutions are built with the help of local knowledge from the community, expert knowledge from professionals, and support from public authorities resulting in stronger outcomes whereas the latter is a temporary intervention using low-cost materials to catalyze long-term change. In Bengaluru, several placemaking projects were taken up using a data-driven approach, guided by tactical urbanism, and grounded in community participation to create better public places. Data collected through methods such as perception surveys, activity mapping, and interviews helped to strengthen the rationale for developing the interventions, tactical interventions helped in making crucial design decisions guided by temporary intervention and its impact, and public engagement helped create a functional design proposal through citizen interactions and build support for the project. The process was tested in three real-time projects; the first project was a "slow street project" which aimed to prioritize the safety of cyclists and pedestrians by introducing traffic calming measures; the second project was a street redevelopment project where pedestrian safety and the walking experience was prioritized through a series of measures including road dieting, and the third project revitalized an underutilized pocket in a densely populated residential area into an active community space. In all these cases, the importance and benefit of data-driven and participatory planning in shaping public space projects were gauged and, in a way, it also demonstrated that public support and stewardship demonstrate an alternate and emerging strategy for public placemaking. The outcome of these tactical urbanism projects led to formalizing a framework for undertaking tactical urbanism projects for public spaces using the three-prong strategy of test, observe and measure, and learn and modify. This research paper is an effort of the Directorate of Urban Land Transport, Government of Karnataka, to document and demonstrate the strategies used to develop and test public space designs on the ground driven by data, public engagement, and tactical urbanism, i.e., interventions that are short term, low cost, and scalable to catalyze long-term change, in a citizen-led format.

Keywords

Participatory planning • Tactical urbanism • Data-driven planning • Public spaces • Community engagement

1 Introduction

Public Space is a space that functions as a holistic integrator, actively promoting social inclusion, public health, and social interaction with built form and nature (Rudd & Mutai, 2019). When considering the social component of planning, it is paramount to provide an environment where individuals

S. Mudhol · A. Jacob · M. Kashkari (✉)
Directorate of Urban Land Transport, Bengaluru, India
e-mail: monica.kashkari@dultbangalore.in

S. Mudhol
e-mail: sushmita.mudhol@dultbangalore.in

A. Jacob
e-mail: ann.jacob@dultbangalore.in

M. Vinjamuri
Government of Karnataka, Bengaluru, India
e-mail: commissioner@dultbangalore.in

A. Battisti et al. (eds.), *Greening Our Cities: Sustainable Urbanism for a Greener Future*,
Advances in Science, Technology & Innovation, https://doi.org/10.1007/978-3-031-49495-6_24

and communities feel a sense of belonging, more so for the people that depend on public infrastructure and public space for recreation and socialization (Metropolis, 2022).

The form and geometry of public space in cities are often a consequence of the act of spatially designing a city but the end-users are seldom involved in deciding the nature of such public space or its design. Since the creation of public space directly impacts local residents and their quality of life, it is imperative to assess the need and function of the space and design mindfully to ensure a positive impact. Any proposals planned and implemented from the top-down approach that is not backed by contextual understanding are inevitably perceived by the citizens who are directly impacted, as attempts to deprive them of their own space and feeling like a stranger in their own neighborhoods (Filip, 2022). Hence, with the evolving changes across the world, cities can and must relook at the process of development of public spaces with a reinvigorated urban planning and design approach for the sake of shared prosperity and harmonious development (Garau, 2016).

Through the use of participatory planning, cities around the world are witnessing a paradigm shift in the way public spaces are built. Participatory planning is successful in bringing together the local knowledge of communities and the technical know-how of professionals leading to stronger outcomes. The opportunity to participate in civic life has been identified as a core human need, essential to the psychological health of individuals and communities. The community, when involved in the initial stages of planning and design, helps to develop a clear understanding of the design process, and its benefits to the community, to foster a feeling of ownership, commitment, and responsibility for the plan (Cilliers & Timmermans, 2014). It further builds trust between the organization and the community bridging the gap between the planner and the beneficiary. Although a challenge, it is important and necessary to broaden the influence of citizens on interventions small and big for the future of the cities as it is for the citizens themselves (Filip, 2022).

Planning public spaces also requires a systematic approach in order to identify and comprehend the challenges, and demonstrate the impact of the project to the citizens to foster trust. Data-driven planning and design is one such approach that helps cities make informed decisions using insights gathered from strategically collected data. Three broad categories of spatial data and analysis that help to spatially visualize the data for informed decision-making are (1) inventory assessments (existing spaces and types); (2) spatial patterns, land use, urban form, and climatic data; and (3) user activity data (Soukup et al., 2020). The New York City Plaza Program, an initiative of the New York City Department of Transport (NYC DOT) which aims to create neighborhood plazas by transforming unused street spaces into vibrant public spaces, is one such unique public-placemaking project where data played a vital role to establish the needs and success of the project. As part of the implementation, Gehl Studio and J. Max Bond Centre (JMBC) conducted extensive surveys of the NYC plazas varying in scale, context, and economic standing by undertaking desktop research to capture data on demographics and economic profile of the neighborhood, observational surveys to map and quantify activities of the people using the space, intercept surveys to gather perception of users on the plaza (Gehl & Center, 2015). The data analysis aided in the understanding of the user demographics of these new public spaces, measure the levels of user satisfaction, map the transformation in user activity, and the amount of time spent in the plaza thereby helping establish the impact of the project and assess its success.

In the recent past, another unique and transformative approach called "Tactical Urbanism" has been widely used to facilitate temporary testing of design interventions, observe and measure their impact, and use the learnings to help make informed design decisions for permanent implementation. This practice turns the city into a testing ground for new ideas and enables diverse, experimental forms of economic, social, and artistic life that are usually repressed by the fixities of urban form and its management (Stevens & Dovey, 2022). Through the practice of tactical urbanism, city planners and designers can test new concepts,and observe the impact in the local context to determine if the changes are acceptable to the masses, before making essential financial or political commitments (Lydon & Garcia, 2012). In the public realm, tactical urbanism practices have a multi-faceted use, such as testing a cycle lane on a street by pop-up cycle lanes, testing bus priority lanes on a street by pop-up bus-only lanes, turning a parking lot into a parklet, creating pedestrian zones on a wide street, transforming underutilized pockets of land into quality public spaces, adding vibrancy to a place by art interventions on a large blank wall, etc. (Lydon et al., 2015). Several of these international practices are a testimony to how tactical urbanism, a new reform in design practice helps establish a more innovative and resilient future for planning and designing public spaces in the city.

Cities around the world have created their own blueprint for developing quality public spaces. In some Indian cities, tactical urbanism has been employed to help make informed design decisions. For example, in Mumbai, the HP Junction was transformed and redesigned overnight using simple tools like traffic cones and paint to fix the road for safe pedestrian movement. The trial was deemed to be a major success which led to a 30% reduction in pedestrian crossing distance and reduced pedestrian-vehicular conflicts. The findings from the intervention were incorporated into the design for permanent implementation (Tak, 2017). Similar tactical trials were conducted to test the junction improvement projects in Bhubaneswar with a special focus on safety and access from a child's perspective. Children and adults were involved in the workshop to ideate design strategies to improve safety at Rupali Square junction. The ideas were drawn over existing maps and ideas so developed helped develop a holistic design that catered to the concerns of all user groups. The intervention was further tested for a month and random selection surveys were conducted to measure the impact of the project. Tactical trials can be conducted for different types of projects like streets, improving access to facilities, creating pop-up parks and plazas, etc. but most cases of tactical implementation in Indian cities mainly include street and junction improvement such as along market streets like the Big Bazaar street trials in Coimbatore, Pondy Bazaar Street in Chennai (ITDP, 2020) and M.G Road in Ranchi, along major street intersections like H P Junction and Mithchowki in Mumbai and Bus Stand road in Rohtak, and school zones in Udaipur. All of these projects aimed at improving pedestrian safety by creating walkable footpath spaces and safe crossings, reducing vehicular conflicts and improving turning radius, and introducing traffic calming measures using materials like paint, traffic cones, barricades, ropes, signages, potted plants, etc., and focused extensively on citizen engagement and participation (GIZ GmbH, 2020).

This paper focuses on the development of public spaces in the city of Bengaluru which is the capital city of a south Indian state called Karnataka. It is one of the fastest-growing cities in the country and is branded as the "Silicon Valley of India." With the growing population and rapid urbanization, the city is under developmental and infrastructural pressures to provide the services and facilities required by the citizens. This developmental pressure also impacts the quality of public spaces in the city. Public space typology in Bengaluru can be broadly classified into streets, parks and gardens, lakes and markets, and religious spaces (Vagale, 2004). Most of these public spaces lack inclusive designs that cater to people of all abilities and ages, as the process of creating or developing public spaces seldom involved analysis of user groups for whom the space is planned, sensitivity to the needs of various user groups, and more importantly involving them during the planning and design process. Most streets, among all public spaces, are considered to be the most essential as they are engines of economic activities, social hubs, and platforms for civic engagement (Deore & Lathia, 2019) are not walkable and are designed mainly to cater to the vehicular needs over that of pedestrians. As better public space planning, design, and management have emerged as major global issues that require attention. Sustainable Development Goals (8 and 11) and the New Urban Agenda have also emphasized the significance of an inclusive and sustainable economy and safe, accessible, and quality public spaces for all (United Nations, 2015). Hence, there is a pressing need to rethink and evolve new forms of public spaces, which can cater to the needs of people with all abilities for social interaction and urban life. Therefore, this paper is an effort to showcase and discuss the strategies adopted in the city of Bengaluru, by the Directorate of Urban Land Transport (DULT), a state-government organization with a mandate to plan, design, and fund sustainable mobility projects across the state of Karnataka. These strategies were developed through learnings from various implemented cases across the world and through years of practice and experience. The chapters share the key strategies and initiatives, case studies on public-placemaking projects implemented in Bengaluru in Richmond town, Wilson Garden and Shantinagar neighborhoods and their learnings (Fig. 1).

2 Temporary to Permanent—The Initiatives and Strategies

Public space development is driven by a rationale that is grounded in various strategies. These strategies must essentially aim to explore the application of a "process-oriented

Fig. 1 Location of projects in Bengaluru, Karnataka, India (1. Alexandria Street, Richmond town 2. Siddaiah Road, Wilson Garden 3. Putaani Park, KSRTC Quarters, Shantinagar)

approach" that pays more attention to the processual nature of public spaces that can further enrich public-space research (Spierings & Melik, 2020). Public spaces should ideally be adaptable to varied user groups and accommodate all temporal activities. As history tells us, though society may be attached to a particular public space (for a short while or a longer period), the meanings can alter with time. Either way, public space's representation, design, and function must be accepted by the public to ensure its utilization (Hatuka, 2020).

The Directorate of Urban Land Transport (DULT) undertook several placemaking projects across the city and through those experiences developed strategies to ensure a holistic approach to planning and designing public spaces. The three key strategies are Data Collection and Analysis, Public Engagement, and Tactical Urbanism. The paper details the use of these strategies on three unique projects in the city. Although the development of public spaces involved all three strategies, each project highlights one strategy in particular based on project type and need.

2.1 Data Collection and Analysis

Data-driven planning and design provide authorities with real-time insights into the functionality of the space, user personas, issues, and problems, both behavioral and structural, perception of the space, and aspirations of the people. These revelations aid in the improvement of design during the design development process and the making of practical choices for long-term sustainability.

Data can be collected from secondary sources like reports, city development plans, etc., and through primary sources like surveys, interviews, etc. Pre- and post-implementation studies are crucial to understanding and assessing the impact of the design. Table 1 lists the approaches employed for developing designs for public spaces.

Hence, the strategy of data collection and analysis facilitates design development by addressing existing functions and the needs of different user groups. As a proof of concept, it also aids in evaluating the effects and successes or failures.

2.2 Public Engagement

Public engagement is a process through which citizens and stakeholders come together to engage in thoughtful discussion, express points of view, and co-create solutions for a project or a proposal. It entails proactively communicating information to stakeholders or persons who could be directly impacted by a project and obtaining their opinions, suggestions, and viewpoints.

It is crucial to recognize the benefits of citizen engagement and provide opportunities that create meaningful public involvement. Early and continuous public involvement ensures that decisions reflect public needs and interests, consider diverse viewpoints and values, and are made in collaboration and consensus with all stakeholders. This process builds mutual understanding and trust between government entities and the people they serve (University of Delaware, 2012).

Table 1 Methods of data collection

Sl. No.	Method	Brief
1	Intercept surveys	For qualitative information from diverse user personas so as to gauge the perception of the space, track sentiment, and understand dwell times or reasons for visiting the space
2	Stationary activity mapping	For information on the functionality of the existing space, by observing, mapping, and tracking different activities, user personas, and types of interactions
3	Perception surveys	To collect impressions or perceptions about a space, program, or issue by active users of the space and the community
4	Spot speed surveys	To determine the speed distribution of a vehicular movement at a specific location. The data gathered in spot speed studies are used to determine vehicle speed
5	Parking and vending surveys	To assess existing parking and vending activities, type of activities, and duration
6	People moving count	To understand the business of the space, accessibility of the space with different modes, and movement pattern of different user personas
7	Activity and suggestion boards	To engage with the community/participants and seek suggestions in interesting ways where one can write their views, paste stickers or post-its, etc. The activity can help gather insights on the project while on site and build support on the ground

Hence, the strategy of public engagement is crucial for the success of the project. The methods could be conventional like public meetings, focus group discussions, workshops, task forces, etc., or technology-driven like dissemination of information or seeking feedback through social media and new technology. The integration of both traditional and technological engagement approaches can produce beneficial results.

2.3 Tactical Urbanism

Tactical urbanism is a city and/or citizen-led approach that uses short-term, low-cost, and scalable interventions to catalyze long-term change. Tactical urbanism Interventions are suited to test and demonstrate change in our physical environments by using temporary and inexpensive materials to further modify the designs tested based on field observations (Street Plans Collaborative, 2016).

The execution of tactical urbanism has various phases.

- **Identifying the Need for a Tactical Urbanism Intervention**: Tactical Urbanism is a helpful approach to test technically sound but contested ideas that lack wide local support, refinement of spatial design proposals based on use patterns on the ground, bring behavioral changes among user groups before permanent infrastructure is built or to rally community support.
- **Establishing a Dedicated Team**: A team to undertake planning and design activities, and ensure smooth implementation of the project and on-site observations.
- **Identifying and Meeting the Stakeholders**: A list of possible stakeholders that includes community stakeholders and government stakeholders needs to be identified and kept informed of progress to ensure their support during the event.
- **Conceptual Plan and Costing**: A preliminary design needs to be in place with key aspects to be tested and identified. At this stage, a decision needs to be made regarding materials to be used to translate the proposal to the ground. Based on the materials identified and the quantity required, costing can be derived.
- **Permissions and Approvals**: To execute tactical trials, permissions need to be sought from relevant government agencies. Therefore, letters seeking permission and assistance need to be sent with details of the project intent, location, time, and specifics of any assistance, if required.
- **Site readiness for Implementation**: A site reconnaissance is essential to assess the readiness of the site and based on the site analysis, relevant agencies need to be contacted to prepare the site for implementation.
- **Communication Plan**: Communication strategies must incorporate both digital and traditional modes. Social media content, flyers, and posters in regional languages need to be part of the implementation and planning process.
- **Data Collection:** Based on the data to be collected before, during, and after the event, survey forms need to be prepared. Photo and video documentation is to be carried out before, during, and after the event from the same angles to create powerful visuals. Spot-by-spot feedback is to be captured during and after the event using the drawings.

Hence, this strategy is an effective method to demonstrate the feasibility of a solution before a more permanent infrastructure or solution is taken up for implementation. It can help in sensitizing the public, enlighten the public on the impacts of proposed changes, and garner the acceptance of stakeholders and political support.

3 Turning Spaces into Places—Case Studies

Planning and design of urban streets are often seen as a desk job with technocrats taking decisions but are rarely practiced as a process where the local community has a say in the outcomes. DULT has always relied on stakeholder consultations to shape grounded proposals, but a participatory approach aims to extend planning and design beyond meeting rooms to the ground where the community is an active participant in problem-solving together with the government. Through the several tactical interventions DULT carried out, to test, learn, and implement has become a mainstream approach to practice at DULT. DULT's experience in conducting tactical interventions has been documented in this paper to provide insights into how the team planned the project, the roadblocks faced, and lessons gained during execution.

3.1 Slow Street, Alexandria Street, Richmond Town, Bengaluru

A "Slow Street" is a street redesign project undertaken by DULT to make neighborhood streets safer for pedestrians and cyclists through design interventions. Slow street involves the implementation of traffic calming measures to reduce the speed of motorized vehicles and improve road safety for pedestrians and cyclists (Kulkarni et al., 2021). To reduce fatalities, the maximum permissible speed of a motor car near an educational institute is defined to be 25km/h as per the road safety question bank on the website of the Ministry of Road Transport and Highway (MoRTH), and the same was defined to be the threshold for vehicular speeds on a Slow Street. The concept of Slow Street was tested on Alexandria Street through the strategy of tactical intervention.

Alexandria Street is a 500 m neighborhood spine, that passes through a residential area housing two major schools which attract students from across the city. The street is also a local destination for all commercial needs of the neighborhood in the Richmond Town area of Bengaluru. The street has high pedestrian use during pockets of time and is widely used by people of all age groups. As the street is the main linkage connecting the schools and shopping centers with the cycle lanes constructed on Rhenius Street and GS Thimmaiah Road, it was chosen for tactical intervention. This project set the ball rolling for undertaking tactical urbanism as a proof of concept of the designs developed.

The strategy of data collection and analysis was at the forefront of this project to validate success, the need, and design strategies for the project.

The methodology followed in testing the Slow Street concept involved conducting a reconnaissance survey to understand the area and the challenges relating to the safety of pedestrians and cyclists, identifying a resident welfare association (RWA), and conducting public consultation with community representatives from various age groups (Fig. 2) to understand the issues and pain points for pedestrians, mapping neighborhood-level activity zones like schools, parks, shopping centers, and co-create location-specific solutions to resolve the identified issues such as speeding of vehicles near school zones, lack of safe crossings, unregulated parking hampering pedestrian movement, etc. Following the public consultation designs were developed based on the inputs received from the local RWA and residents of the street, these designs were tested through low-cost tactical interventions with community participation. Finally, an impact assessment was conducted through extensive data collection and analysis. The strategy of public engagement facilitated garnering deeper insights about the movement pattern in the neighborhood and building ground support for the initiative.

The project collected and analyzed the following metric to measure the impact:

- Qualitative-Through speed analysis using spot speed survey and parking compliance and mapping of parking violation spots across (parking violation is when cars are parked in non-designated locations on street), and
- Qualitative-Through opinion surveys and interviews with residents and visitors on the street.

To achieve the aim of Slow Street, the street was temporarily converted into a one-way street with certain tactical interventions (Figs. 3 and 4), including the bulbing out and kerb extensions as well as regularizing the carriageway widths and making it one lane at some locations were implemented. The junctions were designed for one-way movement and accordingly, the kerb extensions were marked with white-colored thermoplastic and painted with colorful hexagonal patterns to slow down the traffic. Pedestrian crossings were marked with thermoplastic paint and a placemaking intervention was taken up around a tree to act as a pause point on the street. To assess if these design interventions were effective, a spot speed survey was designed. The survey was designed to measure the speed of a vehicle on a particular stretch and understand if vehicles were moving below the threshold defined for a Slow Street, i.e., 25 km/h.

The analysis showed that post the tactical intervention at the site, the average speeds of vehicles along the six stretches were recorded to be below the threshold defined for Slow Street, i.e, 25 km/h (Table 2).

Fig. 2 Public consultation on Alexandria Street. (Source DULT Bangalore)

Fig. 3 "Slow Street" painting on the street. (*Source* DULT Bangalore)

Fig. 4 Public engagement during tactical trials. (*Source* DULT Bangalore)

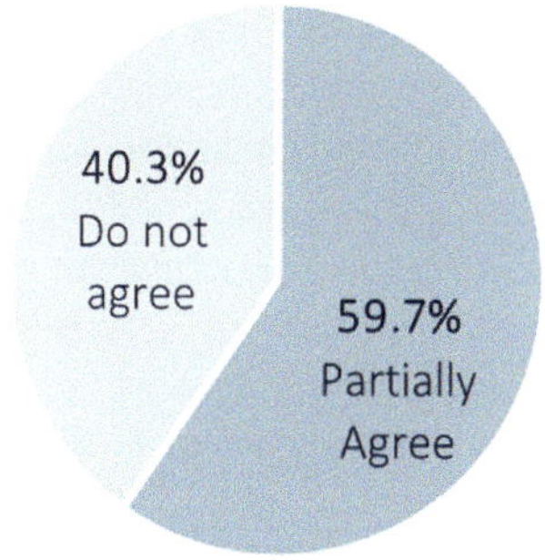

Fig. 5 Percentage of people who agreed to walk more often in the neighborhood (Kulkarni et al., 2021)

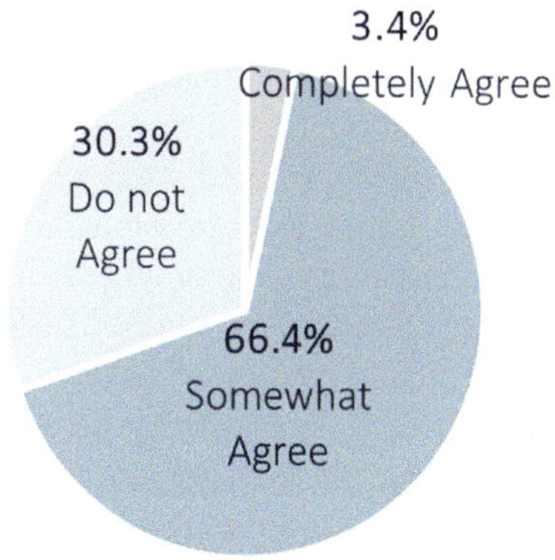

Fig. 6 Percentage of people who felt safer walking on street post-intervention (Kulkarni et al., 2021)

The qualitative surveys captured the satisfaction levels of the people and their perception of safety. These surveys were conducted physically through one-on-one interviews with pedestrians using the street. The survey results showed that 59.7% partially agreed to walk more often within the neighborhood (Fig. 5) and about 69.8% found the street to be safer than before (Fig. 6).

The detailed methodology and survey analysis are presented in the paper titled "Evaluating slow street concept in a residential neighborhood: pilot at Alexandria Street, Bengaluru" (Kulkarni et al., 2021).

The analysis of the survey also threw light on the lack of active communication to create awareness of the intent and objective of the tactical trials as some residents visiting the street were unaware of the purpose of the painted street spaces and road markings. It was evident that there was a need for signboards to explain the notion of slow streets to the public to ensure the dissemination of relevant information for greater impact.

Post-tactical trials the designs were reworked to incorporate the various observations made during the trials. The street was reverted to have two-way movement as it was observed that Alexandria Street was sandwiched between two existing one-way streets and hence this street was the only looped connection available for circulation. It was also observed that the street naturally behaved as a one-way street during school hours due to pick-up/drop-off activity and hence it was proposed to explore a time-bound one-way street accommodating local conditions. The junctions were redesigned to now address the two-way movement. Locations of pedestrian crossings were revised to incorporate the actual movement pattern observed during the trails. The placemaking proposal around the tree was successful and hence was incorporated into the final design. A stakeholder consultation was held to

Table 2 Spot speeds at six (6) stretches along Alexandria Street

	Weekday		Weekend		Average Speed	
	Toward Baldwin	Toward Richmond park	Toward Baldwin	Toward Richmond Park	Toward Baldwin	Toward Richmond Park
LOC 1	21.2	16.6	18.6	19.4	19.9	18.0
LOC 2	20.9	22.4	21.9	20.4	21.4	21.4
LOC 3	13.9	13.8	18.7	13.4	16.3	13.6
LOC 4	13.9	14.8	21.1	15.2	17.5	15.0
LOC 5	18.7	22.0	14.6	23.8	16.6	22.9
LOC 6	18.4	20.0	20.9	19.5	19.6	19.7
Average speed on the street					**18.6**	**18.4**

Note From "Evaluating slow street concept in residential neighborhood: pilot at Alexandria Street, Bengaluru" (Kulkarni et al., 2021)

discuss the final designs following which the project is now taken up for permanent implementation.

The learnings and findings from the tactical trials and surveys analysis facilitated the revision of designs to aptly cater to the needs of the community for permanent interventions.

3.2 Pedestrian Priority Street, Wilson Garden, Bengaluru

A "Pedestrian Priority Street" is a street redesign project undertaken by DULT to improve walkability in the neighborhood by prioritizing the needs of a pedestrian. The concept of Pedestrian Priority Stree was tested on Siddaiah Road in Wilson Garden through the strategy of tactical intervention. Wilson Garden is a planned residential neighborhood with diverse open spaces, located in close proximity to a Shantinagar Bus Station, a major transit hub. Siddaiah Road is the main spine of the neighborhood which caters to the bus movement and is lined with commercial establishments giving it a unique vibrant character. The high through traffic, discontinuous footpaths, unregulated parking, and hard institutional edges made the street unsafe and unsuitable for walking. The project was therefore taken up to redesign the street into a safe, accessible, and complete street facilitating both vehicular and pedestrian movement. The design interventions developed for the street were first tested through tactical trials and hence, in this project, the strategy of tactical trials played a major role in shaping design decisions.

The purpose of the tactical intervention, i.e., interventions that are low cost and temporary, is to test a proposal quickly for its effectiveness, learn from the outcome, and incorporate learnings in the final proposal to be adjusted as needed. The strategy of tactical urbanism was applied in several intervention areas such as pedestrian and bicycle mobility, resolving conflicts between mobility and liveability, improving transit experience, placemaking, and wayfinding. These interventions are not only aimed at solving urban problems for a short period but serve as an example to test and demonstrate solutions to addressing these problems.

The local community and government stakeholders were involved extensively through public engagement programs (Figs. 7 and 8). Several surveys (Figs. 9 and 10) were conducted before the implementation to understand the perception of the people, grade their walking experience on the street, and their aspirations for a pedestrian-friendly street. The data showed (Fig. 11) that about 36% of people using the street expressed having a poor walking experience and the street lacked continuous footpaths, safe pedestrian crossings, and designated parking or vending spots. About 54% of the respondents shared that they walked on the road instead of the footpath due to poor footpath conditions (Fig. 12). Through the survey, suggestions were received from the respondents to provide walkable and continuous footpaths, safe crossings, improve junction geometry, and provide designated parking and vending bays. These

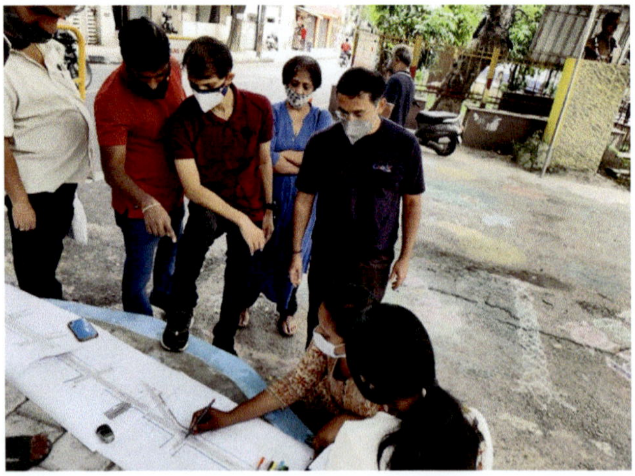

Fig. 7 Public consultation in Wilson Garden. (*Source* DULT Bangalore)

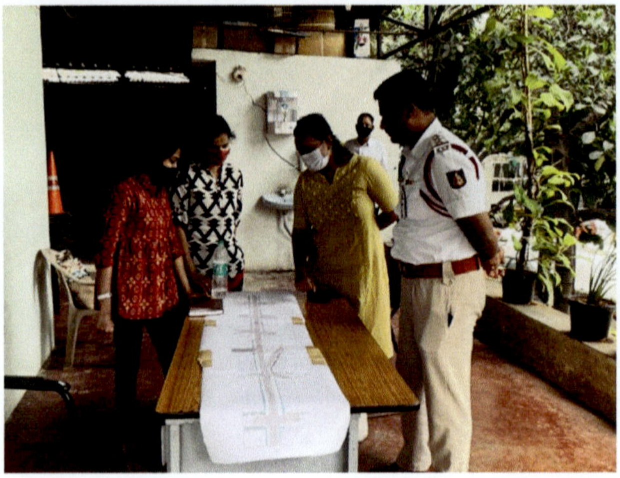

Fig. 8 Stakeholder consultations with traffic police. (*Source* DULT Bangalore)

Fig. 9 Perception surveys with cyclists. (*Source* DULT Bangalore)

Fig. 10 Perception surveys with pedestrians. (*Source* DULT Bangalore)

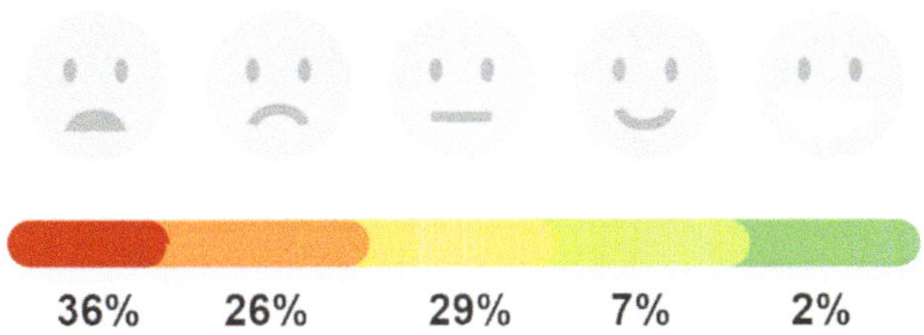

Fig. 11 Rating of walking experience. (*Source* Surveys conducted by DULT Bangalore)

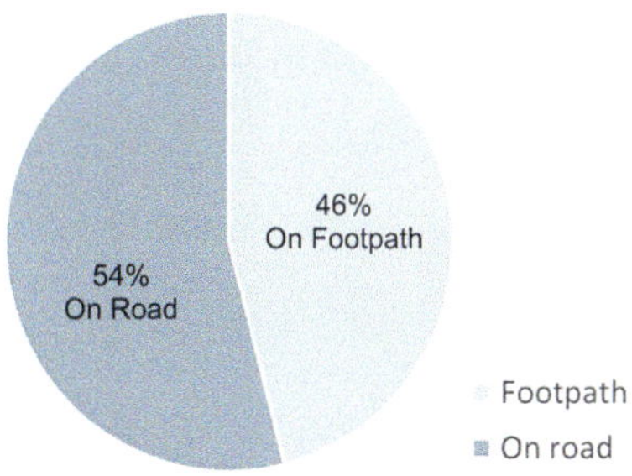

Fig. 12 Percentage of people walking on road and footpath. (*Source* Surveys conducted by DULT Bangalore)

suggestions were incorporated into the design and tested on the ground through tactical trials.

The trials were planned to be held for a week to turn the stretch into a pedestrian-friendly space with zebra crossings, reclaim a dead space for pause and play, and test the uniform carriageway. The conceptual designs were at various stages discussed with officials from the Municipal Corporation, traffic police, elected representatives, and most importantly the citizens of the community.

The tactical urbanism project tested a uniform carriageway of 7.5 m to ensure a minimum footpath width of 2 m, which over the course of implementation was reduced and tested for 7 m and 6.5 m. The strategy of tactical urbanism helped test the concept on the ground, to observe and assess the impact of change and arrive at the most strategic outcome for the design. During the trials, the junction geometry was also tested to ensure safe pedestrian opportunities at the crossroad, designating spaces kerbside for parking bays and street vending activities and reclaiming street space by blocking off the road to create a pocket for pause and play (Figs. 13 and 14).

A local representative from the community resident welfare association helped mobilize the residents of the neighborhood to actively participate during the tactical trials. The pre-event outreach and engagement resulted in overwhelming participation by citizens both young and old (Figs. 15, 16 and 17). The community was involved in translating every design element to the ground. The trial run helped bring out a sense of ownership and credibility in the local community about an infrastructure project in their backyard. The tactical trials required a lot of physical and mental presence to successfully pull off the test run. The process helped garner valuable insights about site conditions and what works for the community and neighborhood of this scale and character.

The team at DULT also conducted perception surveys during, and after the tactical trials to understand various pain points on the street, the issues of a pedestrian, the impact of the trial run, and suggestions to improve the design. Some of the key issues that were highlighted through the survey were narrow and uneven footpaths, uncontrolled on-street parking, vending obstructing pedestrian movement, open drains/dumping, and speeding of vehicles. The design was revised to incorporate the learnings from the tactical trials which included retaining of 6.5 m carriageway, universally accessible footpaths with access ramps and safe crossings, dedicated parking and vending bays, lighting fixtures, and reverting the pop-up public plaza to allow vehicular movement and turning small residual pockets into seating spaces.

The design so developed hence became the one that was widely accepted by the community and government stakeholders. Through the process, the idea of tactical urbanism also gained momentum and helped sensitize other government officials on the need for testing, observing, modifying, and then implementing the design to garner wider societal support and acceptance.

The detailed designs have been developed incorporating all the learnings from data analysis, public engagement, and tactical trials. The project has now been taken up for permanent implementation.

3.3 Putaani Park (Pocket Parks), KSRTC Quarters, Shantinagar, Bengaluru

A "Putaani Park" which in the language of Kannada means "tiny parks" is a public-placemaking project which aimed to create pocket parks in underutilized setback spaces between

Fig. 13 Before implementing pop-up plaza on Siddaiah Road. (*Source* DULT Bangalore)

Fig. 14 After implementing pop-up plaza on Siddaiah Road. (*Source* DULT Bangalore)

residential buildings. The space identified were underutilized pockets of government land in densely populated and built parts of Shantinagar Ward with no access to parks or playgrounds. The goal of Putaani Park was to increase access to public spaces in disadvantaged or underserved areas by creating micro-public spaces, retrofitting existing parks, or using underutilized pockets of land to provide Infant, Toddler, and their Caregiver (ITC) focused amenities and play equipment.

Surveys were conducted to understand user behavior, their needs, and their perception of the space. The analysis showed that 80% of respondents felt that the area lacked quality play areas for children (Fig. 18). About 80% of caregivers responded that children usually played on the streets outside their homes and 20% inside their homes (Fig. 19). Some of the key issues that were highlighted through the survey were the lack of quality play areas in the vicinity, underutilized pockets being used for open dumping and parking, and children playing on the street since the caretakers cannot accompany them to other parks in the neighborhood. The parents expressed that the nearest park is located on the main road, and they find it unsafe for children to venture out by themselves and therefore are usually asked to play in front of their houses. The peculiar observations on site were that older children were often observed playing on the street, and the toddlers were seen playing and seated with their caregivers in front of their homes and the space so identified was being used as a waste dumping site by the residents.

Therefore, it was deduced that the creation of pocket parks would not only need infrastructural improvements but also a certain behavioral change from the residents to

Fig. 15 Community members partaking in creating pedestrian crossings. (*Source* DULT Bangalore)

Fig. 17 Young members of the community partaking in creating pop-up plaza. (*Source* DULT Bangalore)

Fig. 16 Community members partaking in creating pop-up plaza. (*Source* DULT Bangalore)

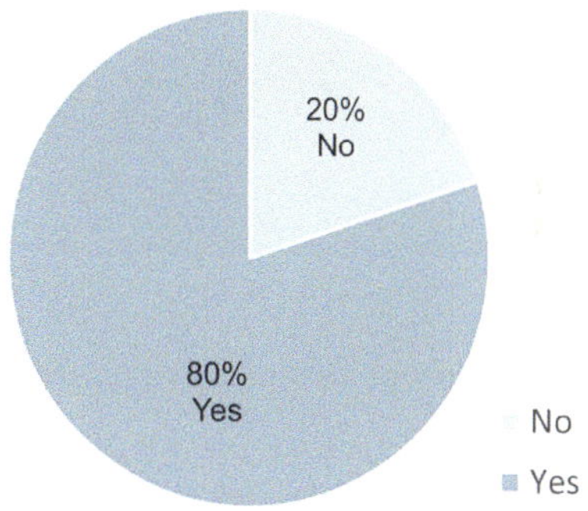

Fig. 18 Percentage distribution of people feeling the lack of quality play area. (*Source* Surveys conducted by DULT Bangalore)

actively engage and take ownership in maintaining the proposed pocket park. Hence, the strategy of public engagement was of utmost importance among the three strategies. The concept of Putaani Park was tested through tactical urbanism to build ground support, create awareness, gauge the impact of the project, and understand the needs of the community (Figs. 20 and 21).

The initiative of tactical urbanism involved setting up simple play opportunities at the location using tires, creating activities with balancing beams placed across tires to encourage the child to safely master balance and gross motor skills, creating an obstacle course with a series of tires placed in various orientations to encourage careful treading, crawling, and balancing, a running art wall in the pocket park allows children to draw their vision for their area, making attractive patterns on the road to creating a slow street stretch, setting up seating for accompanying caretakers, setting of a design board with proposed plans and feedback wall to gather insights into issues toddlers and their caretakers, build momentum, support, and ownership for the project (Figs. 22 and 23). The community engagement program included the creation of a toddler play area, a drawing competition for children to envision their ideal playground, and a discussion with caregivers on the need for quality public spaces and the rationale behind the design.

Residents gathered during the event were briefed about the project and proposed design elements. The community members requested for shaded seating to be provided for caretakers to watch over children playing in the park and also interact with other parents. The caretakers were elated by the idea of putaani parks, they stated that the development of the park would curb the issue of open dumping and provide quality public space for everyone in the neighborhood.

The design proposal was further developed with the insights from the community engagement program and the

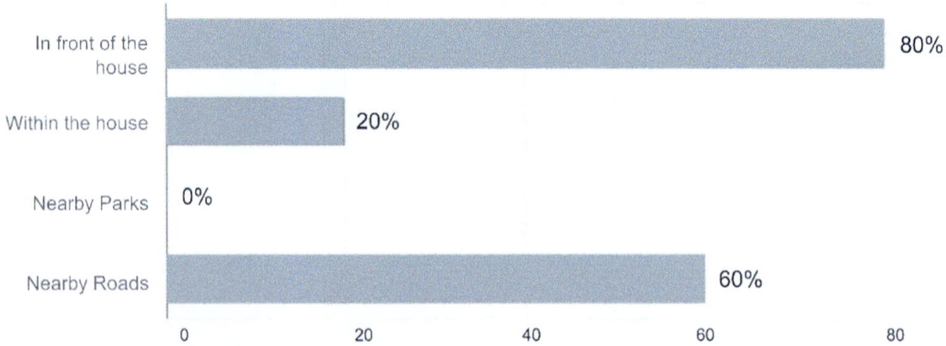

Fig. 19 Percentage distribution of choice of a play area. (*Source* Surveys conducted by DULT Bangalore)

Fig. 20 Designs exhibited for feedback. (*Source* DULT Bangalore)

activity helped the residents realize the value of quality public spaces, internalize the need to maintain cleanliness, and build a sense of ownership of the space. The project has now been taken up for permanent implementation and is under construction.

All three projects are past the design development stage and are now taken up for permanent implementation. The success and impact of the projects will be evaluated after the completion of the execution of the project through post-implementation assessment studies. However, the strategies adopted have helped garner acceptance for the project and build support from elected representatives, government departments and stakeholders, and the community. The three-pronged strategy aided data-driven decision-making for developing holistic designs and securing support from all significant stakeholders to undertake the implementation of the project.

4 Observations and Learnings

This section discusses the diversity in approach and observations from the three case studies. Although all three strategies of data collection, public engagement, and tactical urbanism were used, in each project one of the three strategies held greater significance in shaping the outcomes and influencing the design.

Fig. 21 Drawing competition held for children as part of public engagement program. (*Source* DULT Bangalore)

Fig. 22 Before implementing pop-up park. (*Source* DULT Bangalore)

Fig. 23 After implementing pop-up park. (*Source* DULT Bangalore)

The paper examines three different types of projects, the first examined the creation of a slow street on a neighborhood road, the second examined the creation of a pedestrian priority street on a major arterial road in Wilson Garden, and the third examined the creation of parklets from an unutilized plot of land in a dense low-income neighborhood. In the first project, the strategy of data collection and analysis was crucial to help establish the impact of slow street interventions through spot speed surveys and analysis of the perception of residents and street users on the impact of the initiative. The designs were revised to incorporate the various observations made during the tactical trials. Stakeholder consultations were held to discuss and finalize the designs for permanent implementation.

The final designs for the second project were greatly influenced by the tactical urbanism approach. Throughout the tactical trials, the generated designs underwent a cycle of testing, observation, and modification. The trials helped shape critical design decisions during the revision stage. Tactical urbanism, as a process, also helped sensitize other government officials on the need for testing critical designs on the ground to garner wider societal support and to showcase and sensitize the citizens on the change the project would bring to the neighborhood and build ground support. The design so developed hence became the one that was widely accepted by the community and government stakeholders.

In the third project, the strategy of public engagement was crucial because the identified unutilized pocket of land was being used for dumping garbage and it was important to induce behavioral changes in the community to establish the need for a safe, clean, and engaging public space, build stewardship for the project and help in maintenance and upkeep of the space. The public engagement program helped showcase the value of quality public spaces to the residents, nudge them into internalizing the need to maintain cleanliness, and build a sense of ownership of the space.

The process of planning, designing, testing, and implementing public space designs in Bengaluru helped establish that a "one size fits all" approach is not ideal in the creation of public spaces as each character, nature of space and requirement would vary throughout the city with the scale and typology of the project. This thought is also applicable to the methodology DULT has used in creating public spaces. The strategies developed have time and again proven to be apt and facilitated establishing a positive impact and hence have been adopted as a mainstream practice at DULT.

5 Conclusion

A key part of developing and building inclusive and functional public spaces lies in comprehensive planning and design-this is one of the greatest challenges faced by cities today and the solutions are generally not "one-size fits all" as different spaces would require different methodologies. Developing, managing, and enhancing a public space required a "top-down" and "bottom-up" approach where the former being the leadership essential for the transformation of public spaces at a larger city scale and the latter for organizing, mobilizing, and ideating designs at the neighborhood level. Hence, the beginning of any design or planning process for developing any public space should be to identify and interact with the community, the people who can provide historical perspective, insights into the functioning of the area, and understanding of the need based on local context. This strategy of public engagement from the outset of the project helps a sense of ownership in the project and ensures its success. The complexity of designing public spaces also creates peculiar challenges and hence the strategy of tactical urbanism helps to test the designs and build on them based on their success. The data collected through surveys, interviews, observations, etc. helps to further validate design decisions and measure the impact. Therefore, the strategy of data collection and analysis also plays a major role in ensuring the success of a planned public space.

We thank our colleagues from the Directorate of Urban Land Transport who provided support and expertise during the execution of the different public space projects that greatly assisted the research. We also thank officials from Bruhat Bengaluru Mahanagara Palika (BBMP), Karnataka State Regional Transport Corporation (KSRTC), Bengaluru Traffic Police, and representatives from resident welfare associations for assisting us during the projects and providing their valuable feedback for design development.

References

Cilliers, E. J., & Timmermans, W. (2014). The importance of creative participatory planning in the public place-making process. *Environment and Planning B Planning and Design, 41*, 413–429. https://doi.org/10.1068/b39098

Deore, P., & Lathia, S. (2019). Streets as public spaces: Lessons from street Vending in Ahmedabad, India. *Public Space in the New Urban Agenda: Research into Implementation, 4*(2), 138–153.

Filip, A. J. (2022). Scale compromise: Three challenges of large-scale but citizen-driven urban planning. *Sustainability, 14*(23).

Garau, P. (2016). *Global public space toolkit: From global principles to local policies and practice.* United Nations Human Settlements Programme (UN-Habitat).

Gehl Studio & J. Max Bond Center. (2015). *Public Life and Urban Justice in NYC's Plazas.* Gehl Studio & J. Max Bond Center.

GIZ GmbH. (2020). *Tactical urbanism guidebook.* Deutsche Gesellschaft für Internationale Zusammenarbeit GmbH.

Hatuka, T. (2020). Alternating narratives. In V. Mehta, & D. Palazzo, *Companion to Public Space* (p. 564). Routledge.

ITDP (2020). *From pilot to permanent : How to scale tactical urbanism using lessons from the global south.* ITDP.

Kulkarni, S., Jacob, A., Kashkari, M., & Vinjamuri, M. (2021). Evaluating slow street concept in residential neighbourhood: Pilot at Alexandria Street, Bengaluru. *International Conference on 'Future is Urban': Livability, Resilience and Resource Conservation.* Ahmedabad.

Lydon, M., & Garcia, A. (2012). *Tactical urbanism 2: short-term action, long-term change.*

Lydon, M., Garcia, A., & Duany, A. (2015). *Tactical urbanism: Short-term action for long-term change.* Island Press.

Metropolis. (2022). *Public space strategies for a sustainable metropolitan future-A collection of best practices.* Metropolis.

Rudd, A., & Mutai, J. (2019). *City-wide public space strategies: A compendium of inspiring practices.* United Nations Human Settlements Programme (UN-Habitat).

Soukup, T., Anderson, C., & Kaw, J. K. (2020, February). Evidence-Based Planning and Design. In J. K. Kaw, S. Wahba, & H. Lee, *The hidden wealth of cities: creating, financing, and managing public spaces.* (pp. 94–108). World Bank Group.

Spierings, B., & Melik, R. V. (2020). Researching public space. In V. Mehta, & D. Palazzo, *Companion to public space* (pp. 16–26). Routledge.

Stevens, Q., & Dovey, K. (2022). *Temporary and tactical urbanism: (Re)assembling urban space.* Routledge.

Street Plans Collaborative. (2016). *Tactical urbanist's guide to materials and design—Version 1.0.* California: Street Plans Collaborative.

Tak, R. (2017, September 04). *How tactical urbanism can improve road safety.* Retrieved from WRI India: https://wri-india.org/blog/how-tactical-urbanism-can-improve-road-safety

United Nations. (2015). *Transforming our world: the 2030 Agenda for sustainable development.* UN General Assembly.

University of Delaware. (2012). *Complete communities toolbox.* Retrieved from Complete communities: https://www.completecommunitiesde.org/planning/inclusive-and-active/public-engagement/

Vagale, U. (2004). *Bangalore-future trends in public open space usage. Case study: Mahatma Gandhi Road, Bangalore.* Virginia Polytechnic Institute.

Lessons from Coastal Slums of Global South Toward Flood Resilience

Anubhav Goyal and Maria Matos Silva

Abstract

Urban slums are among the areas most vulnerable to the impacts of climate extremes. At the same time hosts a major urban population in developing countries. More than a billion people in developing countries live in slums. Coastal and river basin slums are particularly vulnerable because they are located in low-lying marshy swamps and have limited options for prevention and absorption, far from the way developed countries do. The challenge put forward has forced slum dwellers to adopt vernacular measures. The objective of the paper is to identify and present these measures as a framework or a compendium of vernacular flood adaptation and mitigation actions. The excess of water which has been generally seen as a threat is progressively changing into also being seen as an opportunity. Cities from developing countries such as Jakarta, Dhaka, and Mumbai set an example while showcasing urban slum adaptation through measures particularly pertinent to public space retrofits. The methodology to identify the measures involves observations and interpretation from assessed cities. A framework of vernacular measures is created and presented from coastal urban slums within each studied city. Results provide a theoretical framework of vernacular measures, conceived for coastal urban slums in developing countries that complement state-of-the-art measures from the developed countries. It is therefore highlighted the significant role of urban slums of the global south in the search for local adaptation building for flood resilience.

Keywords

Adaptation • Developing countries • Flood resilience • Slums • Vernacular measures

1 Introduction

Urban slums have grown rapidly over the past few decades in developing nations, from mega-cities in Latin America to India, China, the Philippines, and Indonesia in the global south. Over 1 billion people now live in slums globally, with three regions accounting for 80% of that total: Eastern and South-Eastern Asia (370 million), Sub-Saharan Africa (238 million), and Central and Southern Asia (227 million) (UN Habitat, 2019). In light of the experienced hazards, cities are among the most vulnerable areas to the impacts of climate extremes. Cities are particularly vulnerable since they are the centers of the world's constructed assests, economic activities, and population (Silva, 2016; Leichenko, 2011). Urban slums in mega-cities of developing countries form a major part of the urban fabric. Coastal urban slums, in particular, are more vulnerable to flooding as they occupy the neglected low-lying marshy mangrove swamps unsuitable for formal development. The slums in mega-cities of developing countries have limited capacity to prevent and absorb in difference with adaptive measures of the developed north. Poor urban dwellers are often forced to live in contested areas that are highly exposed to natural hazards (Bott & Braun, 2019) with floods continued to affect the largest number of people, 35.4 million people in 2018 (UNISDR, 2019). Around 110 million people currently live on land below the high tide line. Based on the sea level projections for 2050, land currently home to 300 million people will fall below the elevation of an average annual coastal flood. Sea level rise and extreme events of hurricanes, cyclones, and precipitation present significant risks based on the topography of the city. Figure 1a–c highlights the impact of sea

A. Goyal (✉) · M. M. Silva
CIAUD, Research Centre for Architecture, Urbanism and Design, Lisbon School of Architecture, Universidade de Lisboa, Lisbon, Portugal
e-mail: urbtext@gmail.com

M. M. Silva
e-mail: mmatossilva@isa.ulisboa.pt

A. Battisti et al. (eds.), *Greening Our Cities: Sustainable Urbanism for a Greener Future*, Advances in Science, Technology & Innovation, https://doi.org/10.1007/978-3-031-49495-6_25

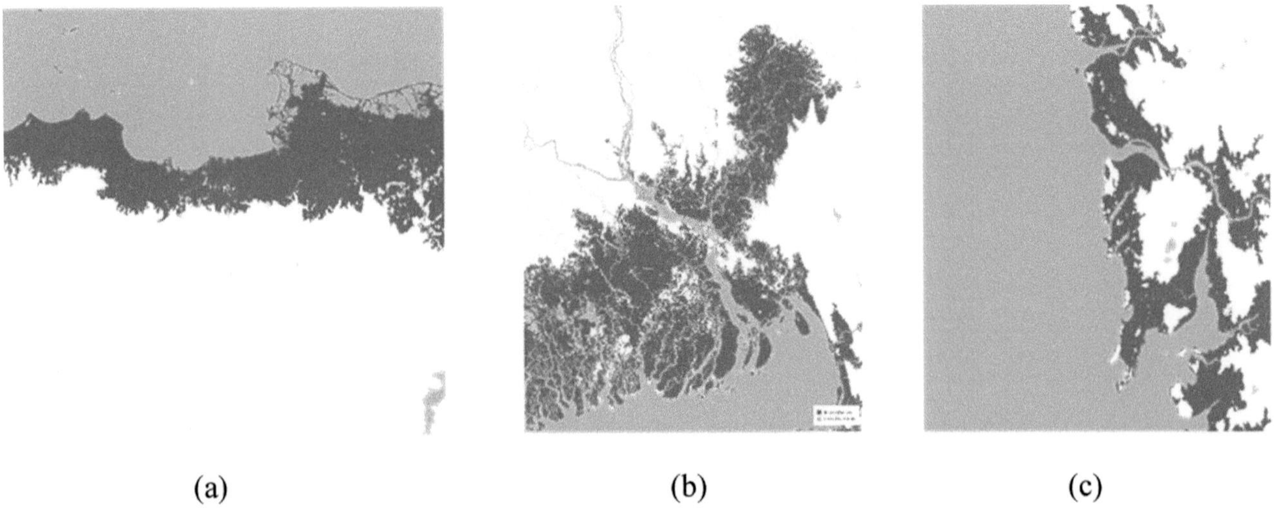

Fig. 1 Land projected to be below annual flood level in 2050 **a** Jakarta, **b** Dhaka, **c** Mumbai (Climate central coastal risk screening tool. Retrieved April 20, 2021, from https://www.climatecentral.org

level rise in a moderate scenario as proposed by Kopp for the climate central modeling tool. By 2100, land now home to 200 million people could sit permanently below the high tide line (Climate Central, 2019).

The 2003 World Development Report notes the pronounced challenges the underprivileged confront when a calamity strikes (World Bank, 2003). Because they have limited ability to prevent and absorb the effects of natural disasters, developing nations are particularly vulnerable. The likelihood of dying in a natural disaster is four times higher in low-income countries than it is in high-income countries. Poor people and poor communities are frequent victims of natural disasters, partly because priced out of the more disaster-proof areas and dwell in crowded, makeshift houses. Poor families are especially hard hit as injuries, disabilities, and loss of life directly affect their most important asset and their labor. Disasters disrupt social assistance programs as well as devastate the natural, physical, and social assets of poor households (Dodman & Satterthwaite, 2010). Slum dwellers have been facing climate injustice with insufficient infrastructure and climate resilience, economic and social inequality burdened with failed policies of redevelopment and rehabilitation. Furthermore, findings in the 'climate change 2014: Impacts, Adaptation and vulnerability,' from the IPCC Working Group II contribution to the Fifth Assessment Report, indicate a high degree of confidence that, as a result of sea level rise, coastal systems, and low-lying areas will increasingly experience submergence, flooding, and erosion throughout the twenty-first century and beyond (Pachauri et al., 2014). The failure of formal climate change mitigation and adaptation is stressed as the major risk of highest concern for the next 10 years with great potential damage (WEF, 2016).

The lack of government's response to address flood risk and vulnerability for slums and recurring floods has forced the urban poor to adopt vernacular measures that learn from past climate extreme events in order to adapt to the future. With the capacity to spontaneously adapt to long-term mean climate conditions, people and societies have dealt with climate, climate variability, and extremes throughout history with varying degrees of success (Pachauri et al., 2014). However, when dealing with extremes and year-to-year variations in climate, the same cannot be said (Burton et al., 2001). Because people may react differently to the same bad occurrence, developing more pertinent, focused, and effective emergency management techniques and actions can be challenging (Adger et al., 2005; Guan & Chen, 2014). In these contexts, per se, evidences the need to critically examine the responses of urban slum dwellers that provide for flood risk reduction.

The preparation for addressing the risk of flooding in the future largely depends on the previous and present adaptation measures. This study aims to identify, characterize, and present a compendium or a framework of embedded local measures of flood adaptation that the coastal urban slum dwellers adopt to survive in harmony with the vagaries of water from the cases, including observation and interpretations using common sense based on practical experiences. In scope of the paper, the local vernacular measures are characterized as measures that are self-adopted at individual and community levels without formal or government interventions and support. Such measures that are inherited in the slums, which are sometimes overlooked and absent from the government approaches of addressing floods, are effective and the consequence of constant flood adaptation by the slum dwellers. To deal with the intensity of the monsoons, they employ simple and low-cost measures that are specific to the place and community (Goyal, 2021).

2 Methods

This manuscript identifies, categorizes, and presents local vernacular flood adaptation and mitigation measures that urban slum dwellers adopt based on the lessons and experiences from past events, while acknowledging the risk of flooding and the lack of government interventions in building flood risk resilience for coastal urban slums.

The methodology applied for the development of this paper has an exploratory and investigative character that encompasses a literature review and empirical observations and assessment. The cities of Jakarta, Dhaka, and Mumbai, that host a major proportion of the urban population in coastal slums witnessing recurring floods over the years, are assessed for identification of the local flood adaptation and mitigation measures that the coastal slum dwellers adopt. The identification of the measures is done through case study analysis approach by direct observations from the sites through interpretation from collected images, literature review of the formal policies, action plans and previously published research aiming to present a framework of local vernacular flood adaptation measures that not just signifies the adaptation capacity of slum dwellers but also contributes to research toward a lexicon of flood adaptation for coastal urban slums. Identified measures are presented in response to the reasons of flooding and categorized into local vernacular measures, that are self-adopted by the slum dwellers without external government support and formal or government measures. Together, they are not meant to offer an exhaustive collection but rather a significant framework of solutions that endorse further reliable research and decision-making.

In order to assess the pertinence of the identified measures, cross-sections were developed from Dharavi, a coastal urban slum in Mumbai, to showcase the employment of the measures that operate at different sites; sea, river, and street fronts, in different times; dry and wet seasons, of a flood event in a coastal urban slum.

3 Literature Review

The cities of Jakarta, Dhaka, and Mumbai are assessed as case examples for identification, interpretation, and categorization of flood adaptation and mitigation measures with respect to the causes of recurring flood events, acknowledging higher proportions of urban slum population that embrace experiences over the years. Figure 2 highlights the locations of the assessed cities; Jakarta in Indonesia, Dhaka in Bangladesh, and Mumbai in India, from the developing countries of the global south. The assessed cities host major urban population in coastal urban slums that engrain vernacular endeavors of addressing flood.

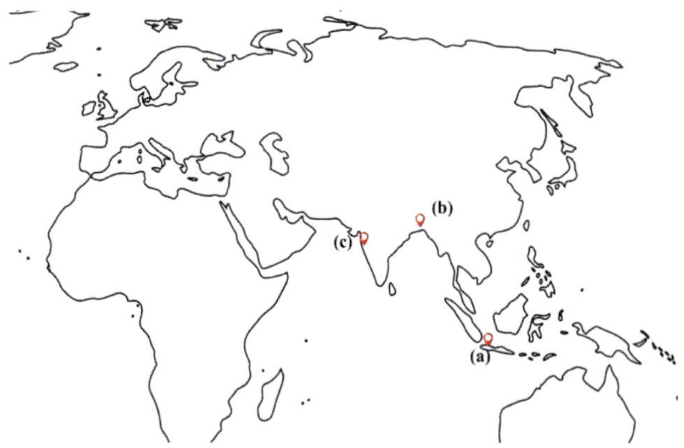

Fig. 2 Location of assessed cities in global south **a** Jakarta, **b** Dhaka and **c** Mumbai (Authors)

3.1 Jakarta

One of the coastal mega-cities with the highest population growth, Jakarta, the capital of Indonesia, is plagued by unusually high tides, sea level rise, and land subsidence caused by excessive groundwater extraction to meet urbanization needs. According to Ng et al. (2012), the pace of land subsidence is as high as 20–28 cm per year, and it has accumulated to about 5 m (Cao et al., 2021). Forty percent of the city's area, mostly in the north, already lies below sea level (The World Bank, 2011). The potential flood extent is predicted to grow by 110.5 km^2 between 2000 and 2050. 88% of this rise is a result of land subsidence (Takagi et al., 2016).

In Jakarta, planning for adaptation to urban flooding goes back to colonial times and has seen various approaches and phases, recently triggered by catastrophic flooding events in 2007 and 2013 (Herbeck & Flitner, 2019). National Capital Integrated Coastal Development Plan in Jakarta, reinforces the existing sea walls, integrated development of the coastline of Jakarta Bay, and construction of offshore sea wall and other reclamation activities as protection structures for flood adaptation. The Jakarta Urgent Flood Mitigation Project proposes to clear the urban waterways and retention ponds by relocating people dwelling in informal settlements. Globally, the focus of land-based and formal adaptation has shifted toward sea and land-based protection with more flexible approach, but still lacks in providing immediate flood risk reduction from recurring floods for dwellers in urban slums. The slum dwellers resort to local adaptation such as raising floors and constructing barriers of bricks and

Fig. 3 **a** Sea wall in northern Jakarta, with house outside the sea wall already submerged in water, **b** construction of wall as barrier to block the inflow of water (Pleasance, 2018), **c** construction of sea wall to keep the residents dry in Jakarta (Pleasance, 2018)

Fig. 4 **a** Traditional architecture for the common man, featuring the Kampong House (Yilun, 2020), **b** floating Kampung Apung in Jakarta

wood to prevent flood water from entering the houses, etc. (Fig. 3a–c).

The construction of *kampong*[1] houses in the *bugis*[2] style of vernacular architecture is an example of flood adaptation since primitive times (Fig. 4a–c). Such vernacular structures comprise knowledge distilled over years and passed on to contemporary adaptation strategies. There is a wealth of knowledge in local vernacular practices within vernacular communities that provide for effective adaptation and lessons. Kampongs often precisely exhibit these features already in their strong culture of self-reliance, economic resilience, habitability (ensuring physical safety), and a robust sense of community (Cairns, 2019).

The World Bank Report on urban challenges in a changing climate in Jakarta (The World Bank, 2011) lists a number of infrastructural projects and substantiates the mismatch between the ongoing adaptation measures and the magnitude of potential disasters. The report highlights the various adaptation programs and actions at the community level. The establishment of the connection between the adaptation actions at individual and community levels with the spatial plans and policies is essential. The urban poor is highly resourceful and also adaptive to climate change. The identified local measures along with the causes of flooding are presented in Table 1.

3.2 Dhaka

Bangladesh is a low-lying delta, with about one-fifth of its area just above 1 m above mean sea level (Saroar & Routray, 2010). About 80% of Bangladesh's territory is often affected by floods (NAPA, 2005). Here, there are four different types of floods which could happen: rain floods, monsoon floods, coastal floods, and flash floods brought by overflowing of rivers. The major flood events and causes of floods are listed in Table 1. Dhaka lies in a

[1] Kampong is used to refer urban slums in Indonesia. The term applies to traditional informal settlements, especially of the indigenous people.
[2] Bugis style house in the form of a stilt traditional house conceal uncovered science and traditional knowledge. This house is built with wood, bamboo, and leaves or reeds. The wood is used for the primary structure; wood and bamboo used for secondary structure; leaves or reeds are used for covering as building envelopes as wall and roof. Such ancient architectural design, created with old technology in a specific place, reflects sage understanding of how to minimize building damage from natural calamities and prevent human injuries.

Table 1 The framework of flood adaptation and mitigation measures for coastal urban slums

Coastal cities	Major flood events	Reasons of flooding	Measures to address flood pertinent to coastal urban slums	(a)	(b)	(c)	(d)
Jakarta	1996 1999 2002 2007 2013 2014 2020 2021	Meteorological Reasons					
		Tidal surge and seawater intrusion through water channels and drains	Concrete dykes [Ex. proposed offshore ring-shaped dike: Giant seawall]	×			×
			Construction of embankments made of wood or brick (Community physical adaptation)	×		×	
			Construction of sluice and tide gates	×			×
			Mangrove planting	×	×		×
		Extreme precipitation	Construction of pump stations	×			×
			Raising floors and adding more levels to the building [Ex. in Kamal Muara] often in the Bugis style	×		×	
			Constructing barriers using bricks and wood	×		×	
			Relocation to upper floors [Ex. in Kampung Melayu]	×		×	
		Topographic Reasons					
		Composite delta of several rivers Main river: Ciliwung River	Large-scale reclamation of land from the sea	×	×		×
			East and west flood canal construction as two new major drainage canals for Jakarta	×	×		×
		Hydrological Reasons					
		Land subsidence > 5 m: Due to excessive extraction of underground water and rapid, unchecked urbanization	Elevating streets	×		×	×
			Construction of retention walls	×		×	×
			Concrete dykes [Ex. proposed offshore ring-shaped dike: Giant seawall]	×			×
			Raising floors and adding more levels to the building	×		×	
			Relocation to higher sites	×		×	
		Extensive use of concrete leading to reduction of permeable surface and increment in land subsidence	Normalizing the waterways and retention ponds by relocating informal dwellers	×	×		×
			Re-development of waterfronts and construction of green public spaces	×	×		×
		Sea water rise	Concrete dykes [Ex. proposed offshore ring-shaped dike: Giant seawall]	×			×
			Relocation to higher sites	×		×	
		Anthropogenic Reasons					
		Extensive reclamation of river catchment for slums	Normalizing the waterways and retention ponds by relocating informal dwellers	×	×		×
		Increase in built-up areas leading to a reduction in permeable areas	Re-development of waterfronts and construction of green public spaces	×	×		×
			Small-scale environmental clean-ups like clearing smaller drains	×	×	×	
			Community-based cleaning (social adaptation)	×		×	
		Solid waste disposable in river channels leading to reduced carrying capacity	Small-scale community-based environmental clean-ups like clearing smaller drains	×	×	×	
		Inadequate drainage system	Community dredging of drains and water lines	×	×	×	
			Small-scale environmental clean-ups like clearing smaller drains	×	×	×	
			East and west flood canal construction as two new major drainage canals for Jakarta	×	×		×
		Blockage of canals due to informal settlements	Normalizing the waterways and retention ponds by relocating informal dwellers	×	×		×

(continued)

Table 1 (continued)

Coastal cities	Major flood events	Reasons of flooding	Measures to address flood pertinent to coastal urban slums	(a)	(b)	(c)	(d)
Dhaka	1954 1974 1984 1987 1988 1998 2004 2009 2016 2017	Meteorological Reasons					
		Cyclones, tidal surge	Construction of embankments	×		×	×
			Reforestation in coastal areas [Ex. Green belt project]	×	×		×
			Building dykes, pump stations, and floodgates	×			×
			Installing community-level warning systems for early preparedness	×		×	×
			Building retaining walls	×		×	×
			Dismantling and relocation of dwelling structures at higher grounds	×		×	
			Temporary relocation into floating boats [Ex. Solar-powered boats are used for temporary living, floating schools]	×		×	
		Extreme precipitation	Construction of flood shelters	×		×	
			Construction of drainage channels	×		×	
			Construction of barriers across the doors	×		×	
			Construction of higher plinths and higher storage facilities	×		×	
			Building dykes, pump stations, and floodgates	×			×
			Laying sandbags and plastic sheets on slopes	×		×	
			Construction of bamboo platforms and 'macha'—bamboo structure for temporary relocation	×		×	
			Dismantling and relocation of dwelling structures at higher grounds	×		×	
			Temporary relocation into floating boats [Ex. Solar-powered boats are used for temporary living, floating schools]	×		×	
			Building stilts inside the room of the individual dwelling	×		×	
		Topographic Reasons					
		Composite delta of several rivers: the Ganges, the Brahmaputra, and the Meghna	Mangrove protection in Sundarbans	×	×		×
			Reforestation in coastal areas [Ex. Green belt project]	×	×		×
		Low-elevation coastal zone	Construction of embankments	×		×	×
			Construction of higher plinths and higher storage facilities	×		×	
			Building on stilts	×		×	
			Building dykes, pump stations, and floodgates	×			×
			Building retaining walls	×		×	×
			Construction of hanging community toilets	×		×	
		Hydrological Reasons					
		Rising sea levels	Construction of embankments	×		×	×
			Building on stilts	×		×	
			Building retaining walls	×		×	×
			Dismantling and relocation of dwelling structures at higher grounds	×		×	
		River bank erosion	Construction of embankments	×		×	×
			Mangrove protection in Sundarbans	×	×		×
			Reforestation in coastal areas [Ex. Green belt project]	×	×		×
		Flash floods due to overflowing of rivers	Construction of flood shelters	×		×	
			Construction of barriers across the doors	×		×	
			Construction of higher plinths and higher storage facilities	×		×	
			Building on stilts	×		×	
			Building dykes, pump stations, and floodgates	×			×
			Building retaining walls	×		×	×
			Construction of bamboo platforms and 'macha'—bamboo structure for temporary relocation	×		×	
			Temporary relocation into floating boats [Ex. Solar-powered boats are used for temporary living, floating schools]	×		×	
			Building stilts inside the room of the individual dwelling	×		×	

(continued)

Table 1 (continued)

Coastal cities	Major flood events	Reasons of flooding	Measures to address flood pertinent to coastal urban slums	(a)	(b)	(c)	(d)
		Anthropogenic Reasons					
		Drainage congestion	Cleaning drains and canals	×	×	×	×
		Encroachments of the floodplains	Dismantling and relocation of dwelling structures at higher grounds	×		×	
			Reforestation in coastal areas [Ex. Green belt project]	×	×		×
		Inadequate pumping facilities	Building dykes, pump stations, and floodgates	×			×
			Floating agriculture and adopting salinity-tolerant crop species	×	×	×	
			Creation of floating gardens	×	×	×	
		Disposal of waste from tannery industries into water bodies	Cleaning drains and canals	×	×	×	×
Mumbai	2005	Meteorological Reasons					
76 events between 2000 and 2008		Monsoon heavy rains	Plastic sheet to cover windows and over the roofs	×		×	
			Raising of plinths in individual dwelling units	×		×	
			Temporary relocation to upper floors in individual dwelling	×		×	
			Temporary relocation to higher public spaces like community grounds, squares, etc	×		×	
			Cleaning and clearing of bottlenecks in the Mithi River	×	×		×
			Removal of encroachments and establishment of recreational parks as buffer zones	×	×		×
			Construction of river embankments	×		×	×
			Cleaning of community drains	×		×	
			Revival/ restoration of traditional water bodies (Ex. wells and water ponds, blue–green infrastructure action plan)	×	×		×
			Building temporary barriers	×		×	
			Dewatering pumps to drain floodwater	×		×	×
			Constructing barriers across the door fronts (Ex. Sandbag walls)	×		×	
			Construction of flood defenses and dikes	×			×
		Cyclones, tidal surge	Plastic sheet to cover windows and over the roofs	×		×	
			Temporary relocation to higher public spaces like community grounds, squares etc	×		×	
			Building temporary barriers	×		×	
			Removal of encroachments and establishment of recreational parks as buffer zones	×	×		×
			Shoreline stabilization and protection (Ex. concrete tripods)	×			×
			Construction of flood defenses and dikes	×			×
		Topographic Reasons					
		Reclamation and alteration to the original seven islands	Restoration of natural drains	×			×
			Revival restoration of traditional water bodies (Ex. wells and water ponds, blue–green infrastructure action plan)	×	×		×
		Flat gradients and mud flats cause excessive siltation	Cleaning and clearing of bottlenecks in the Mithi River	×	×		×
			Dredging of rivers	×		×	×
		Hydrological Reasons					
		Rising sea levels	Shoreline stabilization and protection (Ex. concrete tripods)	×			×
			Construction of flood defenses and dikes	×		×	×
			Removal of encroachments and establishment of recreational parks as buffer zones	×	×		×

(continued)

Table 1 (continued)

Coastal cities	Major flood events	Reasons of flooding	Measures to address flood pertinent to coastal urban slums	(a)	(b)	(c)	(d)
		Anthropogenic Reasons					
		Extensive reclamation Faulty zoning regulations	Cleaning and clearing of bottlenecks in the Mithi River	×	×		×
			Removal of encroachments and establishment of recreational parks as buffer zones	×	×		×
			Community-based fishing	×	×	×	
			Diversifying the livelihoods (Ex. aquaculture) and creating fish farms	×	×	×	
			Revival/ restoration of traditional water bodies (Ex. wells and water ponds, blue–green infrastructure action plan)	×	×		×
		Inadequate rainwater drainage system Faltering drainage systems	Restoration of natural drains	×			×
			Cleaning of community drains	×		×	
			Dewatering pumps to drain floodwater	×		×	×
		Encroachments over Mumbai's natural drain	Restoration of natural drains	×			×
		Lack of waste management resulting in clogged drains	Cleaning and clearing of bottlenecks in the Mithi River	×	×		×
		Reduction in river catchment areas/Depletion of flood plain due to encroachments	Removal of encroachments and establishment of recreational parks as buffer zones	×	×		×
		Bandra–Worli sea link Change of path of natural rivers (Mithi River due to airport project)	Cleaning and clearing of bottlenecks in the Mithi River	×	×		×
			Restoration of natural drains	×			×
		Lack of open grounds in slum neighborhoods to allow water seepage	Water diversion for laundry grounds (Ex. Dhobi Ghat)	×		×	
			Removal of encroachments and establishment of recreational parks as buffer zones	×	×		×
			Revival/restoration of traditional water bodies (Ex. wells and water ponds, blue–green infrastructure action plan)	×	×		×
		Destruction of mangroves and marshes due to mushrooming slums	Removal of encroachments and establishment of recreational parks as buffer zones	×	×		×
		Deforestation of forests in Mumbai					
		Destruction of wetlands due to reclamation of land for infrastructural projects	Revival/restoration of traditional water bodies (Ex. wells and water ponds, blue–green infrastructure action plan)	×	×		×
		Loss of holding ponds					

low-elevation coastal zone with 35% urban population residing in slums (Herbeck & Flitner, 2019). Urbanization and rapid increase of urban population, which put additional strain on the city's aging infrastructure, aggravate the climate risks for Dhaka (Jabeen & Johnson, 2013). It is surrounded by six rivers that overflow in monsoons. Urban poor are not well-served by the policies and institutional frameworks for adaptation planning and urban development. The two main policy attempts, the Bangladesh Climate Change Strategy and Action Plan (BCCSAP) from 2008 and the National Adaptation Programme of Action (NAPA) from 2005, both fall short in addressing urban challenges, especially the vulnerability of the urban poor (Jabeen & Johnson, 2013). The ability to adapt for such group of population is closely tied to the quality and location of their home (Shakil & Hasan, 2018).

Coping with extreme climatic events like floods is not new to the people of Bangladesh. The government of Bangladesh has constructed over 200 flood shelters and about 3931 km long coastal embankments to protect coastal land from inundation by tidal waves and storm surges (NAPA, 2005). Also, the green belt project is implemented with community participation approach involving reforestation to reduce the impacts of cyclones in coastal regions. Floating

(a) (b)

Fig. 5 **a** Raising of plinth in Korail, Dhaka (Maples, 2020). **b** Adaptation by slum dwellers on waterfront in Korail (Waterstudio, 2012)

(a) (b)

Fig. 6 **a** Relocation or shifting of mosque, moved periodically in north Dhaka, **b** floating garden in Gaibandha District, Dhaka allows inhabitants to grow food on flooded land (Bendiksen, 2011)

agriculture in monsoon season and the development of salinity-tolerant crop species are recent adaptation measures being implemented in coastal regions. In Korail, one of the biggest slum in Dhaka, dwellers adopt mainly impact-minimizing or spontaneous rather than planned or preventive measures (Shakil & Hasan, 2018). Such actions involve the construction of barriers and higher plinths etc. as illustrated in Fig. 5. Korail sets an example, where residents on the waterfront are able to adapt the existing dwelling to address the environmental extremes.

Other measures include converting the water-logged fields into ponds for salt-tolerant shrimps and crabs. Dwellers construct floating platforms and bamboo macha for temporary relocation during the event of flooding. Floating boats are used as temporary schools. Hyacinth plants are used to create floating gardens, where squash, okra, and other food crops are planted as illustrated in Fig. 6. The slums in Dhaka also demonstrate examples of floating community spaces, such as raised platforms and hanging toilets, as adaptive space to live in harmony with water.

3.3 Mumbai

Mumbai, which was formerly an archipelago of seven marshy islands off the coast of India, exhibits a strong affinity toward the sea. The foundation of the island city is the continuous reclamation and transformation of urban and natural landscapes shaped by colonial exploitation, political objectives, and global competitiveness. Mumbai, which is struggling with growing development, frequently experiences flooding. The most frequent risk is flooding, and as a result of climate change, this risk is increased due to heavy precipitation and sea level rise. According to the database of flooding incidents between 2000 and 2008, the island city experienced 76 flooding incidents annually on average, 57 instances in the eastern suburbs and 129 flooding instances in the western suburbs. The monsoon season, which lasts from June to September, brings roughly 220 cm of rain to Mumbai annually. Since 2004, the average annual rainfall has been more than 240 cm (Patankar, 2015). The heaviest rain event experienced was in 2005. Inadequate and blocked water routes and drains, the conversion of permeable space into asphalt, and the construction of low regions that result from reclamation between the original distinct islands are the main causes of floods. The lack of green and open surfaces in the city resulted in a low surface infiltration capacity. Many creeks have been filled in or paved over to accommodate urban expansion. The most adverse catastrophe occurs when heavy rains and high tides coincide.

More than 50% (54% in 2001) of Mumbai's urban population lives in slum settlements (Census of India, 2011) which are primarily found in low-lying, flood-prone marshy areas. With recurrent flood experiences, dwellers in the

(a) (b)

Fig. 7 **a** Flood adaptation at individual scale with construction of barriers, raising the furniture, and draining out the water after flooding in Mumbai (Charuchandra, 2020). **b** Clay embankment built to create fish farms in mangrove forest trapping waters during high tide (Kirtane, 2010)

urban slum embrace significant knowledge and embed coping measures. Local vernacular measures at the community and individual levels are crucial for reducing the risk and damage from floods in the absence of institutional or formal interventions. For the urban poor, living in peace with the water and regularly adapting with their limited abilities is a crucial aspect of daily life. The most common protection measure used prior to the anticipated precipitation at the individual scale in Mumbai's slum is raising the plinths and covering the roof and openings with plastic sheets. Other local adaptation measures include building barriers, lifting furniture, and migrating to higher ground (Fig. 7a, b). To reduce waterlogging, community-level measures include cleaning neighborhood roadways, drains, and other areas.

4 Results and Discussion

Urban slums, as also in the assessed cities, are mostly without any legal address and conceived as illegal and thus neglected part of the urban fabric. They lack the necessary formal or government interventions to improve their ability to address and prevent the climate hazards. The lack of infrastructure and services for prevention and preparedness causes the size of climate-induced hazards to expand. Table 1 illustrates the identified multiple reasons of flooding from past flood events, within the assessed cities and corresponding applied measures of flood adaptation and mitigation, with focus upon local vernacular measures adopted by coastal slum dwellers. The common causes of flooding in assessed cities of Jakarta, Dhaka, and Mumbai are rapid urbanization leading to expansion in built-up areas and reduction of green permeable space, reclamation of river catchment for slums, and solid waste disposal in waterways and river channels. Urban poor in the slums of these cities adopt local vernacular measures and prepare to deal with recurring floods without assistance from government interventions. These local adaptation measures differ in sites within the dense slum fabric, in the time of day or the year (in seasons), before, during, or after the flood event. The common measures of constructing flood barriers, raising plinths, and making adjustments to existing structures for facilitating quick out-flow of water, exemplify the fluidity of the built environment to address flooding without relying on the grey infrastructures and relocations. The vernacular measures of addressing floods in coastal urban slums identified from three cities certainly can represent many other similar coastal urban slums of mega-cities of the developing south.

The identified local flood adaptation and mitigation measures (framed in Table 1) that are inherited in the urban slum fabric, mostly unnoticed and missing out in the formal approaches of flood risk reduction, function at different sites (sea, river, and street fronts) and in different times of the day or seasons (dry and wet seasons), before, after and during the flood event in a coastal urban slum. This occupation of measures is illustrated in Fig. 8, in cross-sectional drawings derived from the Dharavi slum in Mumbai. It is realized that the present locally self-built form and even the architectural characteristics of individual houses such as building on stilts, laying plastic sheets, barriers, etc. are the result of consistent flood adaptation, over the years, in the continuous endeavor of living in harmony with water and survival. With the limited ability and formal resources, slum dwellers with learned experiences in accordance with the local topography, existing infrastructure, materials, etc. develop indigenous solutions that complement the adaptation and mitigation measures developed and adopted by the developed countries for addressing and designing future flood resilient cities.

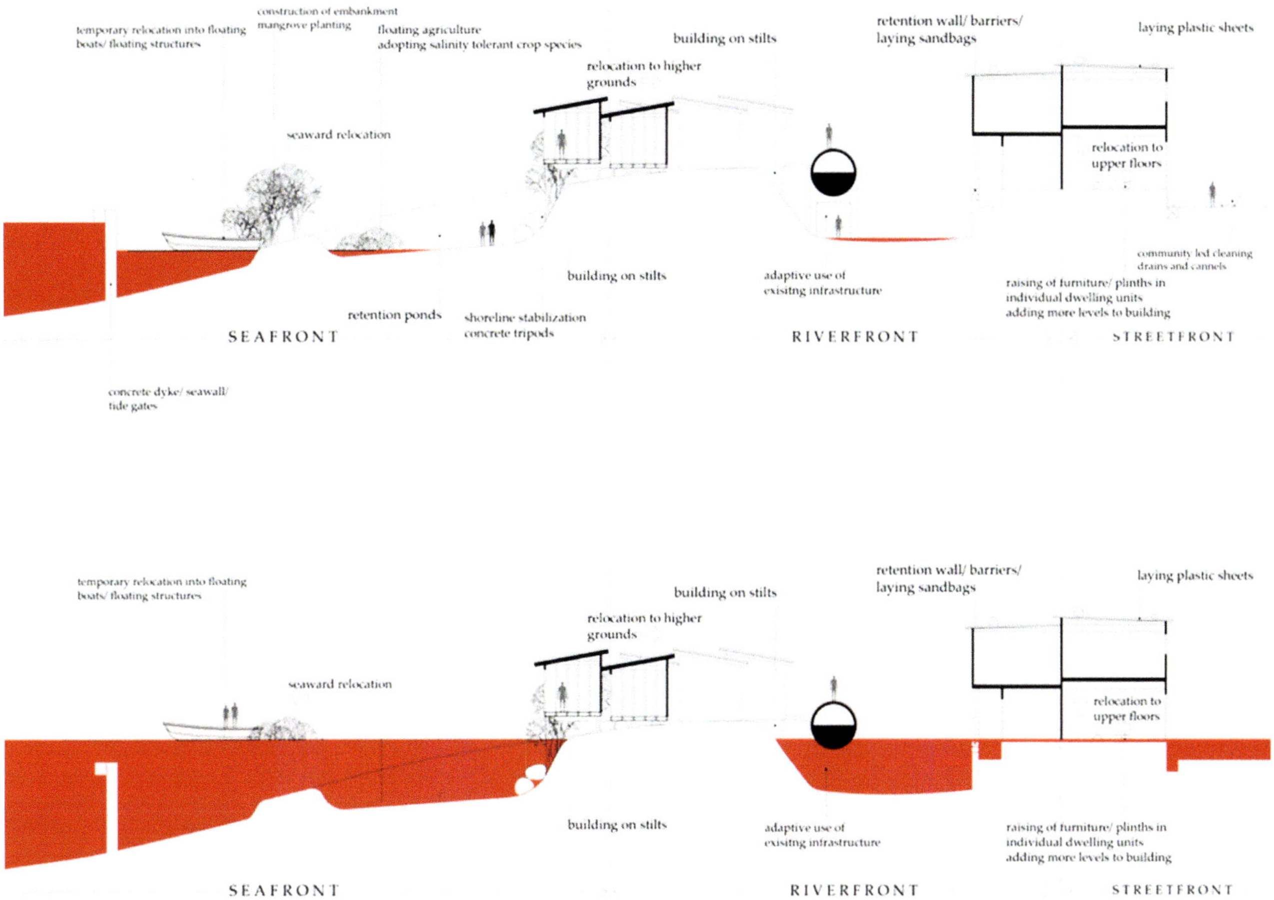

Fig. 8 Cross-section derived from Dharavi slum in Mumbai illustrating local flood adaptation and mitigation measures adopted by coastal slum dwellers. Red: vagaries of water in dry and wet seasons (Authors)

5 Conclusions

The research into the identification of reasons of flooding and the corresponding local adaptation and mitigation measures from the selected cities of developing countries in Asia presents that the causes and the adopted measures have similar characteristics, given that slums comprised similar topographic locations of marshy low-lying riverside sites. The research also presents the adaptive capacity of the slum dwellers, in the absence of adequate formal or government interventions, mainly depends on their locally self-adopted measures learned over the years with experiences. It is realized that local adaptation and mitigation besides playing a significant role in addressing the contemporary flood risk, also will provide for designing and rendering future coastal urban slums flood resilient with involving creative employment of knowledge and physical space that exist in the dense urban slum fabric, rather than merely depending upon external government measures.

Recurring urban flood events are expected to increase. The assessment into the identification of measures for addressing the floods by the slum dwellers, within the scope of this research, presents that government action plans lack in involving local community-led vernacular measures and in safeguarding the local identity and characteristics of existing coastal urban slums. The risk paradigm shift from traditional to more integrated approaches with the involvement of local community-led vernacular measures along with those at risk, the slum dwellers, and covering the vernacular physical infrastructure through public space interventions in flood risk management process are required and proposed to create flood risk resilience for coastal urban slums. It is important to examine and capture slum dweller's responses toward flooding events, especially in cities in developing countries with major proportions of the urban population residing in urban slums.

The formulation of the framework of local flood adaptation and mitigation measures and illustration of the same measures in an urban slum fabric through cross-sectional drawings provide for the initial grounding of preparing a design toolkit, or a handbook, for rendering future coastal urban slums flood resilient.

In conclusion, urban slums need to be conceived as a solution and require critical examination. This involves learning from the remarkable treasure of skills of everyday living that is embedded in the informal space of urban slums of developing countries. Informality allows and provides for greater ease to adapt to the risk of flooding for the urban poor. Flooding in coastal urban slums can be addressed from the creative combination of adaptation and mitigation measures as learned from coping endeavors, allowing for an inclusive community-based approach to climate adaptation and mitigation.

Author Contributions Both authors conceived of the presented idea. The first author developed the theory and wrote the manuscript with support from the second author. Both authors provided critical feedback, discussed the results, and contributed to the final manuscript.

Acknowledgements

The authors would like to acknowledge the support provided by CIAUD, Research Centre for Architecture, Urbanism and Design, at the Lisbon School of Architecture, Universidade de Lisboa in the development of this research paper. This work is financed by national funds through FCT—Fundação para a Ciência e a Tecnologia, I.P., under the Strategic Project with the references UIDB/04008/2020 and UIDP/04008/2020.

We would like to thank the anonymous reviewers for the feedback that assisted toward development of this paper.

Appendix

See Figs. 1, 2, 3, 4, 5, 6, 7 and 8; Table 1.

References

Adger, W. N., Hughes, T. P., Folke, C., Carpenter, S. R., & Rockstrom, J. (2005). Social-ecological resilience to coastal disasters. *Science, 309*(5737), 1036–1039.

Bendiksen, J. (Photographer). (2011, May). The coming storm. National Geographic. Retrieved from https://www.nationalgeographic.com/magazine/article/bangladesh

Bott, L.-M., & Braun, B. (2019). How do households respond to coastal hazards? A framework for accommodating strategies using the example of Semarang Bay, Indonesia. *International Journal of Disaster Risk Reduction, 37*, 101177. https://doi.org/10.1016/j.ijdrr.2019.101177

Burton, I., Challenger, B., Huq, S., Klein, R. J. T., & Yohe, G. (2001). Adaptation to climate change in the context of sustainable development and equity. In *Climate change 2001: Impacts, adaptation, and vulnerability: Contribution of Working Group II to the Third Assessment Report of the Intergovernmental Panel on Climate Change* (pp. 879–906). essay, Cambridge University Press.

Cairns, S. (2019, March 29). What slums can teach us about building the cities of the future. World Economic Forum, Cities and Urbanization. Retrieved from https://www.weforum.org/agenda/2019/03/why-slums-could-provide-the-housing-inspiration-of-the-future/

Cao, A., Esteban, M., Valenzuela, V. P. B., Onuki, M., Takagi, H., Thao, N. D., & Tsuchiya, N. (2021). Future of Asian deltaic megacities under sea level rise and land subsidence: Current adaptation pathways for Tokyo, Jakarta, Manila, and Ho Chi Minh City. *Current Opinion in Environmental Sustainability, 50*, 87–97. https://doi.org/10.1016/j.cosust.2021.02.010

Census of India. (2011). *District census handbook Mumbai, Maharashtra.* Directorate of Census Operations Maharashtra, Series 28 Part XII-B.

Charuchandra, S. (2020, October). To Weather Climate Change, Mumbai Must Embrace Its Wetness. Environment, Urban, The Wire. Retrieved from https://science.thewire.in/environment/mumbai-climate-change-urban-flooding-cyclones-water-pollution-wetness-open-spaces/

Climate Central. (2019). *Flooded future: Global vulnerability to sea level rise worse than previously understood.* https://www.climatecentral.org/news/report-flooded-future-global-vulnerability-to-sea-level-rise-worse-than-previously-understood

Dodman, D., & Satterthwaite, D. (2010). Adapting cities to climate change: Understanding and addressing the development challenges. In Earthscan (Ed.) (p. 206). Earthscan, London.

Goyal, A. (2021). Strategies and approaches pertinent to public space for addressing floods. In *A: Seminario Internacional de Investigación en Urbanismo. "XIII Seminario Internacional de Investigación en Urbanismo, Barcelona-Bogotá.* Universitat Politècnica de Catalunya, Grup de Recerca en Urbanisme, Universitat Politècnica de Catalunya. https://doi.org/10.5821/siiu.10153

Guan, X., & Chen, C. (2014). Using social media data to understand and assess disasters. *Natural Hazards, 74*(2), 837–850.

Herbeck, J., & Flitner, M. (2019). Infrastructuring coastal futures: Key trajectories in Southeast Asian megacities. *DIE ERDE—Journal of the Geographical Society of Berlin, 150*(3), 118–130. https://doi.org/10.12854/erde-2019-451

Jabeen, H., & Johnson, C. (2013). Perceptions of climate variability and coping strategies in informal settlements in Dhaka, Bangladesh. In H. Joffe, T. Rossetto, & J. Adams (Eds.) *Cities at risk. Advances in natural and technological hazards research* (Vol. 33). Springer. https://doi.org/10.1007/978-94-007-6184-1_9

Kirtane, G. (2010). Making the sewer a river again: Why Mumbai must reclaim its Mithi. Observer Research Foundation Mumbai. Retrived from https://admin.indiawaterportal.org/sites/default/files/iwp2/why_mumbai_must_reclaim_its_mithi_gautam_kirtane_orf_2011.pdf

Leichenko, R. (2011). Climate change and urban resilience. *Current Opinion in Environmental Sustainability, 3*(3), 164–168. https://doi.org/10.1016/j.cosust.2010.12.014

Maples, H. (2020). Agency and Climate Resilience in Korail, Dhaka, Bangladesh. Critical Perspective on International Planning. Retrieved from https://sites.utexas.edu/internationalplanning/case-studies/agency-and-climate-resilience-in-bangladesh/

NAPA. (2005). National Adaptation Programme of Action. Ministry of Environment and Forest, Government of the People's Republic of Bangladesh, Final report. http://nda.erd.gov.bd/files/1/Publications/CC%20Policy%20Documents/NAPA2005.pdf

Ng, A. H., Ge, L., Li, X., Abidin, H. Z., Andreas, H., & Zhang, K. (2012). Mapping land subsidence in Jakarta, Indonesia using persistent scatterer interferometry (PSI) technique with ALOS PALSAR. *International Journal of Applied Earth Observation and Geoinformation, 18*, 232–242.

Pachauri, R. K., Allen, M. R., Barros, V. R., Broome, J., Cramer, W., Christ, R., Church, J. A., Clarke, L., Dahe, Q., Dasgupta, P., Dubash, N. K., Edenhofer, O., Elgizouli, I., Field, C. B., Forster, P., Friedlingstein, P., Fuglestvedt, J., Gomez-Echeverri, L., Hallegatte, S., Hegerl, G., Howden, M., Jiang, K., Cisneroz, B. J., Kattsov, V., Lee, H., Mach, K. J., Marotzke, J., Mastrandrea, M. D., Meyer, L., Minx, J., Mulugetta, Y., O'Brien, K., Oppenheimer, M., Pereira, J. J., Pichs-Madruga, R., Plattner, G. K., Pörtner, H. O., Power, S. B., Preston, B., Ravindranath, N. H., Reisinger, A., Riahi, K., Rusticucci, M., Scholes, R., Seyboth, K., Sokona, Y., Stavins, R., Stocker, T. F., Tschakert, P., van Vuuren, D., & van Ypserle, J. P. (2014). *Climate Change 2014: Synthesis Report. Contribution of Working Groups I, II and III to the Fifth Assessment Report of the Intergovernmental Panel on Climate Change*. In R. Pachauri, & L. Meyer (Eds.). IPCC. ISBN: 978-92-9169-143-2. hdl:10013/epic.45156.d001.

Patankar, A. (2015). *The exposure, vulnerability, and ability to respond of poor households to recurrent floods in Mumbai*. The World Bank. Retrieved from http://documents.worldbank.org/curated/en/418931467991989631/pdf/WPS7481.pdf

Pleasance, C. (2018, April 5). The real-life Atlantis: Jakarta is sinking by up to 25cm a year as experts warn these is just a decade left to save it. MailOnline. Retrieved from https://www.dailymail.co.uk/news/article-5581499/Jakarta-sinking-25cm-year-experts-warn-just-decade-left-save-it.html

Saroar, M., & Routray, J. K. (2010). Adaptation *in situ* or retreat? A multivariate approach to explore the factors that guide the Peoples' preference against the impacts of sea level rise in coastal Bangladesh. *Local Environment, 15*(7), 663–686. https://doi.org/10.1080/13549839.2010.498813

Shakil, S. H., & Hasan, M. H. (2018). An assessment of effective approaches for adapting to the impacts of climate change on urban poor communities to make the urban area more resilient: Perspective from Bangladesh. In *54th ISOCARP Congress*.

Silva, M. M. (2016). *Public space design for flooding: Facing the challenges presented by climate change adaptation* [Doctoral dissertation, Universitat de Barcelona]. Universitat de Barcelona. Facultat de Belles Arts. http://hdl.handle.net/10803/400001

Takagi, H., Esteban, M., Mikami, T., & Fujii, D. (2016). Projection of coastal floods in 2050 Jakarta. *Urban Climate, 17*, 135–145. https://doi.org/10.1016/j.uclim.2016.05.003

The World Bank. (2011). *Jakarta: Urban challenges in a changing climate, mayors' task force on climate change, disaster risk and the urban poor, Jakarta.* http://documents.worldbank.org/curated/en/132781468039870805/pdf/650180WP0Box360ange0Jakarta0English.pdf

UN Habitat. (2019). *The sustainable development goals report* (p. 44). Routledge. https://unhabitat.org/sites/default/files/documents/2019-07/sdg_report_2019.pdf

UNISDR. (2019). *2018: Extreme weather events affected 60 million people*. CRED. https://www.undrr.org/news/2018extreme-weather-events-affected-60m-people

World Bank. (2003). *World Development Report 2003: Sustainable Development in a Dynamic World—Transforming Institutions, Growth, and Quality of Life.* World Bank. © World Bank. https://openknowledge.worldbank.org/handle/10986/5985. License: CC BY 3.0 IGO.

World Economic Forum. (2016). These are the world's five biggest slums. Oct 19, 2016. Accessed March 2023. https://www.weforum.org/agenda/2016/10/these-are-the-worlds-five-biggest-slums/

Yilun. (2020, May 29). Vernacular architecture: Traditional architecture for the common man, featuring the Kampong House. Singapore Architecture and Urban Planning. Retrived from https://sgarchi.wordpress.com/2020/05/29/vernacular-and-everyday-architecture/

Public Involvement in Sustainable Development: Sustainable Mobility Accords Programme (Case Study of Bengaluru, India)

Manju George, Sanmay Bohidar, and Manjula Vinjamuri

Abstract

Public involvement is widely recognised as an important element in planning and decision making pertaining to developmental works with public investment. People or communities, for whom any project or work is meant for, are best suited to identify what are their primary requirements and which solutions work well for them. The research paper covers a unique programme "Sustainable Mobility Accords (SuMA)" initiated by a government organisation, namely, Directorate of Urban Land Transport (DULT) partnering with communities at neighbourhood level. The programme is being implemented in the city of Bengaluru in India and focuses on creating sustainable neighbourhoods with major thrust on promotion of non-motorised transport (NMT) including walking and cycling and public transport (PT) for better health, environment, liveability, and ease of commuting. Since Bengaluru is one of the most congested cities fraught with vehicular pollution issues, the government has undertaken several measures including SuMA. In order to get a in depth understanding of the unmet requirements and key issues of the neighbourhood and their potential solutions, SuMA programme is designed with deep involvement of respective communities even ensuring representation of women, children, elderly people, and persons with disabilities. The involvement spans from baseline data collection (covering Household Survey, Focus Group Discussions, and Infrastructure Audits) through issues identification, preparation of proposals, prioritisation by means of consultations, and designing to implementation stage. The support to community partners includes capacity building, funding assistance to improve NMT and PT infrastructure and support in showing measurable impacts on ground. The main research aim is to evaluate the effectiveness of SuMA programme's implementation from public engagement perspective and to underline the insights and learnings for reassessing how and where such public engagements can be further improved to derive additional benefits.

Keywords

Public involvement • Sustainable neighbourhood • NMT and PT • Capacity building • Impact assessment • Insights and learnings

1 Introduction

The GUIDEMAPS handbook from the EU CIVITAS sustainable mobility project asserts that: "There is a growing belief that communities would support transport schemes more readily if they were more actively involved in designing them. They would better understand the need for the project and perhaps be more willing to accept compromises, and they would be able to suggest ways in which the proposals could be better adapted to meet their local needs. In short, they would 'own' the scheme, instead of regarding it as having been 'imposed' on them from above" (Kelly et al., 2005).

The International Association of Public Participation has listed five types of engagement at different stages of development and implementation of the sustainable mobility options. These include informing, consulting, involvement, collaboration, and empowerment for the success of any project. These stages could be used in isolation or as a continuous process; however, at every stage community

M. George (✉) · M. Vinjamuri
Directorate of Urban Land Transport, Bengaluru, India
e-mail: manju.george@dultbangalore.in

M. Vinjamuri
e-mail: commissioner@dultbangalore.in

S. Bohidar · M. Vinjamuri
Government of Karnataka, Bengaluru, India
e-mail: sanmay@dultbangalore.in

A. Battisti et al. (eds.), *Greening Our Cities: Sustainable Urbanism for a Greener Future*,
Advances in Science, Technology & Innovation, https://doi.org/10.1007/978-3-031-49495-6_26

engagement plays a major role in making appropriate and suitable decisions (Alnabelseya et al., 2021).

Various studies have supported the observation that public engagement in transport planning has become fundamental to finding solutions that are both well accepted and technically sound. In the case of the cycle motorway, Aalborg, Denmark, the planners who were interviewed agreed that public involvement was also important to avoid costly and divisive conflicts further down the road. In the paper, one of the planners highlighted that if people are involved in the beginning of the planning process, it would help in minimising negative feedback and better utilisation of resources (time and money) (Driscoll & Vladimir, 2012).

The authors of one study related to a slum redevelopment project in the city of Srirangapatna, India have described how they implemented the design thinking process over a period of one year to involve slum dwellers in the re-design of their own neighbourhood, and how the process led to greater ownership of the design so developed, and an increased willingness in the community to work with local political authorities for development of the selected design" (Kumar et al., 2016).

The Directorate of Urban Land Transport (DULT) is a government organisation set up in Karnataka, a state in India, with a mandate to promote sustainable mobility in urban areas. DULT has been following a robust public engagement strategy while planning various projects. One of its flagship programmes focusing on the promotion of cycling is an open street event called Cycle Day. Cycle Day started in October 2013 in Bengaluru as a Government-Community initiative and is the longest running open street event in India. DULT has engaged with multiple neighbourhoods through more than 550 Cycle Days. Cycle Day is a unique event in which the entire planning and organising of the event is done by the Resident Welfare Associations or other civil society organisations while DULT coordinates with other government organisations like police, local municipal body to facilitate permissions to block the street, etc. Over time, building and leveraging on the relationships built during the Cycle Day, DULT has started associating with communities to implement interventions like walk/Cycle to School, building Cycle Parking facilities in schools, etc. Such small and relatively inexpensive interventions boosted the confidence of the people to use non-motorised options for short commutes.

Extending the Cycle Day initiative to a deeper engagement through SuMA: Cycle Day served as a catalyst to generate the required momentum and the much-needed awareness for sustainable transport in the city. Having realised that "Planning for people with people" is a successful strategy for promoting sustainable mobility, the DULT launched the "Sustainable Mobility Accords (SuMA) in 2020 to further bolster public engagement and to provide a structured framework for planning and implementing projects focused on improving pedestrian, cycling and public transport infrastructure at the neighbourhood level.

2 Objectives of SuMA

The objectives of the programme are:

- To engage communities at the neighbourhood level to further spread awareness on benefits of sustainable modes of transport like walking, cycling, and public transport.
- To support the community to formulate specific goals and action plans for transforming itself into a sustainable neighbourhood by providing necessary support in the form of capacity building, design of interventions, grants for execution, etc.

The expected outcomes of the SuMA initiative include capacity building of neighbourhoods to advocate for sustainable mobility practices/interventions and more neighbourhoods with better non-motorised transport (NMT) and public transport (PT) infrastructure that will promote walking, cycling, and the use of PT. An advisory body, SuMA Task Force, has been constituted with representation from DULT, other relevant government agencies, academia, and civil society to guide the planning and implementation of SuMA.

3 Methodological Approach

The methodological approach followed during the implementation of SuMA encompasses various stages including selection of neighbourhoods, creation of baseline data, development of proposals and design and implementation of the proposals for the first SuMA Cohort. With major focus on public involvement, due emphasis is given to data and data-driven decision making so that the quantitative and qualitative aspects of mobility issues are properly inferred and understood.

3.1 Selection of Neighbourhood

For facilitating involvement of community, applications were invited from interested communities (represented by registered Resident Welfare Associations (RWAs)/Societies/Trusts/Section 8 Companies). The applications were processed based on evaluation criteria that mainly centred on involvement in developmental/awareness-creation activities for their respective community, in general, and sustainable mobility areas, in particular. Community partners were

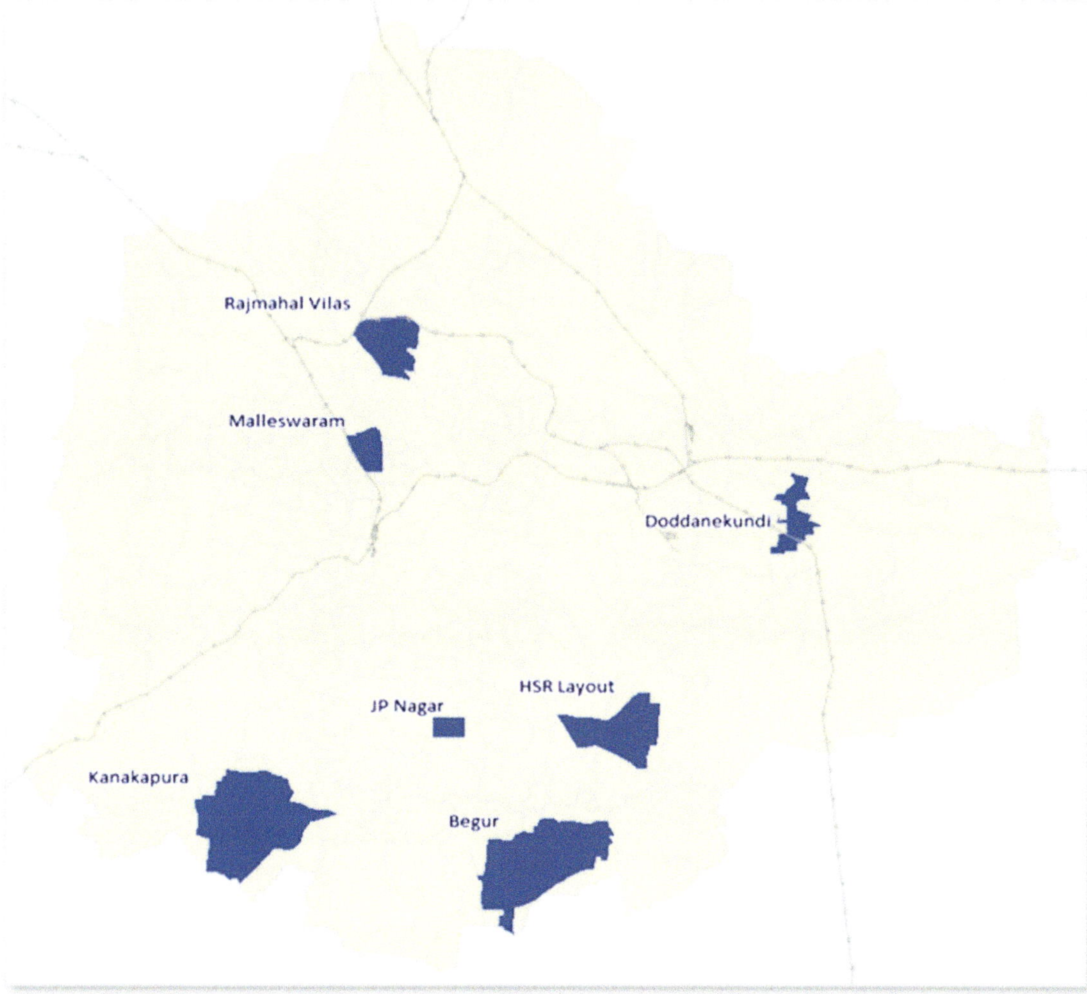

Fig. 1 Seven SuMA neighbourhoods selected as part of first cohort on the city map. (*Source* DULT)

selected by the SuMA Task Force through a two-stage evaluation process with due consideration of their experience in terms of community engagement, basis for selection of neighbourhood and vision for neighbourhood. Figure 1 graphically represents the location of the seven neighbourhoods that are dispersed across the city on the city map. The selection process was followed by formalising the association with the signing of an agreement that mainly covers the roles and responsibilities of the associated stakeholders and other terms and conditions.

The initial phase of work also involved

- Demarcation of neighbourhood boundary in the context of SuMA with active involvement of the community, considering the geographical distribution of residential pockets, existing administrative boundaries, road network, socio-economic aspects of the residents, etc. The geospatial map of the boundary was created together with other relevant base map layers.
- Informing concerned elected representatives and officials of urban local body and departments (through communications and meetings) on the programme and about the selected community partner and seeking their support and cooperation for the initiative.

3.2 Capacity Building (CB)

As articulated in the objective of the programme, capacity building is a major component of the programme and was undertaken through workshops, training programmes, and demonstrations managed by DULT team with the assistance of selected subject matter experts. These sessions spanned from the beginning of programme through implementation phase and included capacity building for conducting household survey and focus group discussions; developing conceptual proposals after analysis of qualitative and quantitative data collected; conducting community consultation workshops; implementing tactical interventions etc. Core members of the community partner and their identified volunteers were the main participants of CB sessions, most of which happened jointly for all the selected neighbourhoods. The technical/training materials and supporting videos were also shared with participants for their further reference.

3.3 Data Collection and Analysis

For making informed decisions on identification and prioritisation of interventions/proposals supported with suitable assessment of key issues, multiple levels of data collection and analysis were envisaged in SuMA. The primary ones were household surveys, focus group discussions, and infrastructure audit (complemented with additional site visits).

3.3.1 Preparatory Activities for Data Collection

The preparatory activities comprised designing of data collection formats, deciding minimum sample, guidelines for household survey, and the design of questionnaires for focus group discussions. Care was taken to ensure even distribution of sample, inclusivity, and heterogeneity mainly with respect to gender parity, representations of all age groups/abilities and different income categories. These were finalised after intense deliberations with the community partner and with the guidance of the advisory body.

3.3.2 Household Data Collection/Survey

The household data collection focused on capturing baseline information like demographic, socio-economic, travel diary, mobility-related challenges etc., covering an evenly distributed sample in terms of age, gender, socio-economic status, abilities, and geographical area. Household survey was conducted by the volunteers of the community partner through an app-based data collection format for around 300 households to capture travel patterns of the household, mode choices, factors influencing the mode choices, concerns, and issues, if any with mobility-related infrastructure. A total of 5882 individuals from 1982 households were surveyed across seven SuMA neighbourhoods.

Kobo Toolbox was used for designing the format and administering app-based survey for which extensive training was provided for the volunteers/surveyors. The survey format was also translated in the local language, Kannada, for ease of understanding and communication with those respondents who were comfortable with Kannada. To inform the community partner regarding the status of data collection and the qualitative aspects of data, a dashboard was created that facilitated a simplified and quick way for providing regular updates. Data validation was a vital component in the process, particularly for household survey and infrastructure audit. The main measures taken included checking/validation of data during data collection stage to identify common errors, frequent sharing of data gaps with surveyors/volunteers and conducting additional training to explain about the errors and required corrective actions (with supporting documentations/illustrations).

3.3.3 Focus Group Discussions (FGDs)

Focus group discussions were conducted to understand the needs and priorities of the underprivileged and marginalised segments of the community. FGDs were conducted with children, women, senior citizens, people with disabilities, people who come for work to the neighbourhood, and people from the lower income strata for all the identified neighbourhoods. The insights drawn from the discussions were compiled, analysed, and documented to complement the findings from other means of data collection. The FGDs were particularly helpful in capturing the voices, pain areas, and suggestions of the marginalised segments.

Figure 2 depicts the aggregated number of participants under each group for the seven neighbourhoods. For example, the total number of participants in the FGDs conducted with senior citizens in all neighbourhoods was 69.

Figures 3 and 4 show the photographs relating to FGDs conducted with people from lower socio-economic strata and people with disabilities respectively.

3.3.4 Infrastructure Audit

Local Area Mobility Parameters (LAMP) audit was done in the neighbourhoods to assess the presence and quality of infrastructure for sustainable mobility. LAMP is a set of indices which help in assessing the mobility-related infrastructure in the neighbourhood based on the Level of Service (LOS) scale of 1 to 4 with 1 being good and 4 being poor. The audits were conducted through transport planning/urban planning students who were extensively trained and guided by the DULT. The data collection for LAMP parameters was done through the ArcGIS collector app. The details of parameters investigated are graphically displayed in Fig. 5. Some sample outputs of LAMP audit are depicted in Fig. 6.

3.3.5 Data Compilation and Analysis

The data collected through the above was compiled, analysed, and documented in the form of reports by the DULT, separately for household survey, FGD, and LAMP audit. These reports were shared with the community partners for their feedbacks and concurrence and finalised based on their inputs. Analytical outcomes and inferences were represented through graphs and tables for easy reading.

The following illustrations—Table 1, Figs. 7, 8, and 9 show a few representative inferences of household data analysis and LAMP audit pertaining to one of the neighbourhoods.

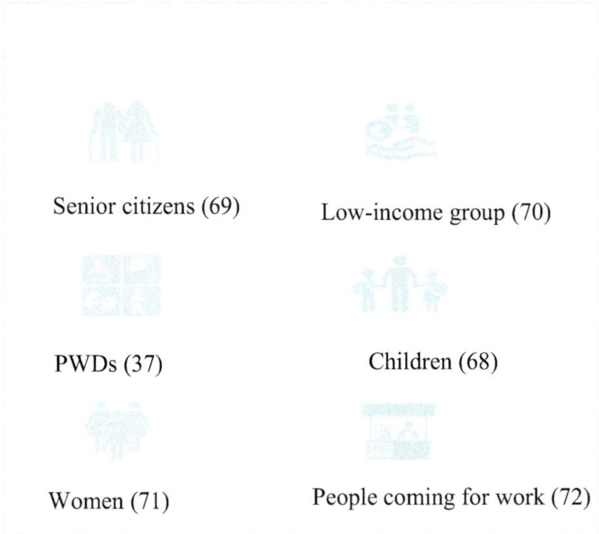

Fig. 2 Illustration showing 6 groups for which FGDs were conducted. (*Source* Focus Group Discussion Report, DULT)

Fig. 3 FGD participants of people from lower socio-economic strata. (*Source* Focus group Discussion, DULT)

The reports of the household survey, FGDs, and LAMP audit of all the SuMA neighbourhoods can be accessed at—https://dult.karnataka.gov.in/68/sustainable-mobility-accords-(suma)/en

3.4 Proposal Development and Design

3.4.1 Shortlisting of Potential Interventions/Proposals

A list of potential interventions was prepared by the community partner and the DULT after extensive discussions on findings of household data analysis, insights of FGDs, and LAMP audit outcomes. Site visits were carried out to selected pockets. Consultations were also carried out with political representatives, local officials, and with institutions/organisations working for the cause of inclusivity and universal accessibility.

Fig. 4 FGD participants of people with disabilities. (*Source* Focus Group Discussion, DULT)

Fig. 5 Details of the parameters assessed through LAMP audit. (*Source* LAMP Audit, DULT)

"For one of the neighbourhoods, lack of a universally accessible mobility infrastructure was a major issue. The neighbourhood had two institutions working for the welfare

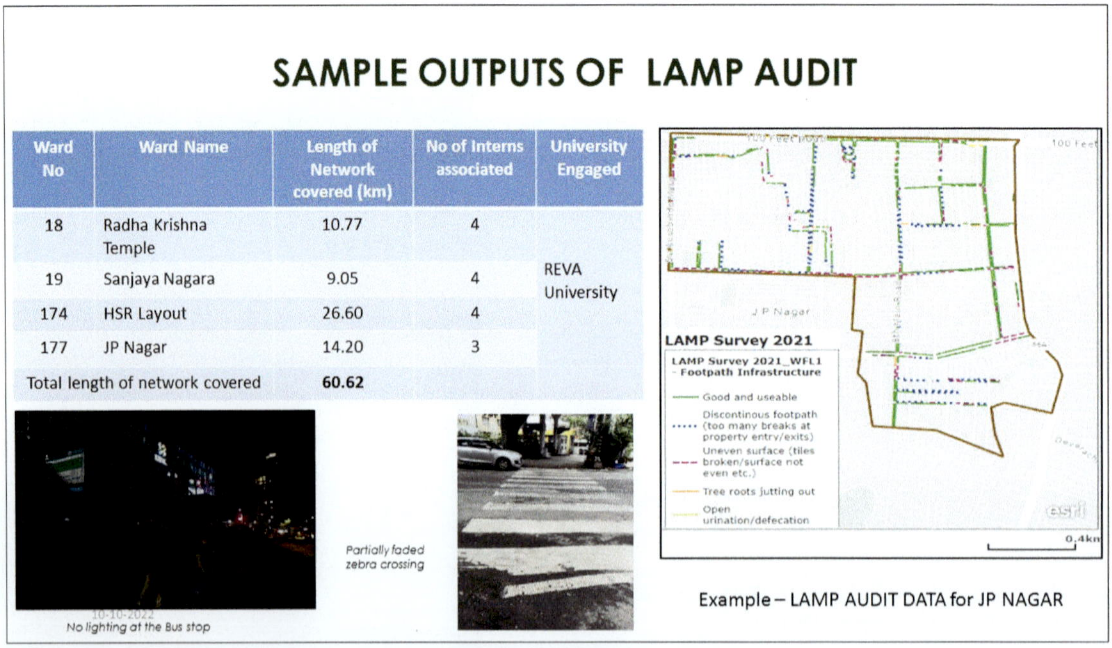

Fig. 6 Examples of the outputs derived for a neighbourhood. (*Source* LAMP Audit, DULT)

Table 1 Means of transportation for one of the neighbourhoods based on purpose of trips

Mode	Work (%)	Education (%)	Recreation/shopping (%)	Other local trips (%)
Walk	19.80	21.40	43.20	19.80
Cycle	1.90	6.30	4.80	2.80
Public Transport	13.40	41.70	5.00	11.10
Autorickshaw/Taxi/Cab	9.30	12.70	5.10	9.80
Personalised-2 Wheelers	19.80	9.90	15.00	17.30
Personalised-4 Wheelers	35.80	8.00	26.90	39.20

(*Source* Household Survey, SuMA—DULT)

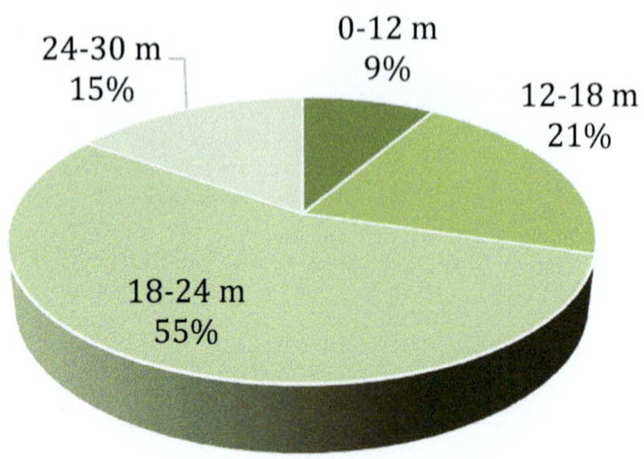

Fig. 7 Size of roads-right of way width. (Source: Household Survey, SuMA-DULT)

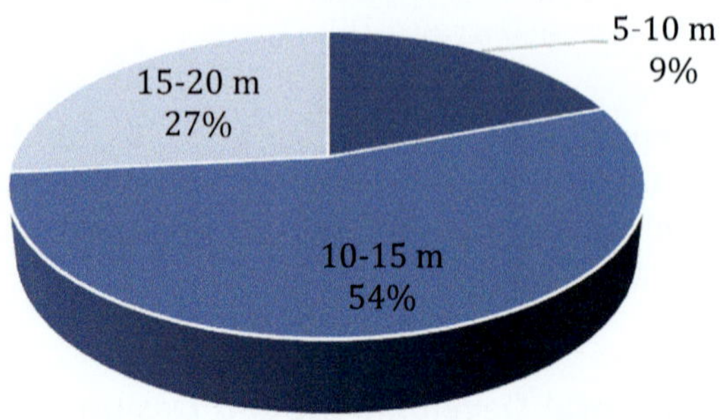

Fig. 8 Size of roads-carriageway width. (Source: Household Survey, SuMA-DULT)

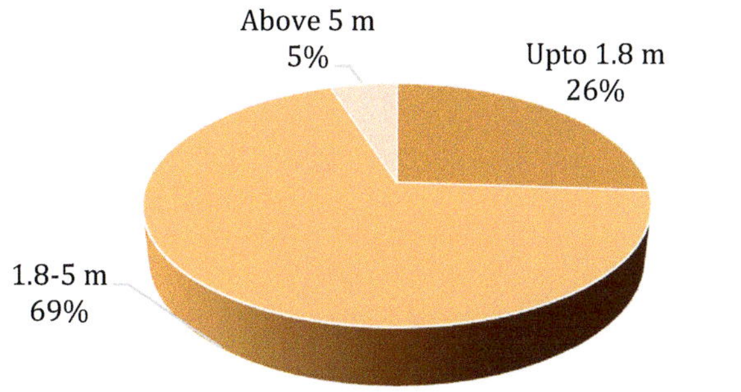

Fig. 9 Size of footpaths-footpath width. (Source: Household Survey, SuMA-DULT)

and empowerment of People with Disabilities (PWD). In order to benefit from their expertise and experience, the core teams of SuMA held consultations with the officials of these institutions. In addition, the teams conducted a ***tag-along survey*** *along the neighbourhood roads/pedestrian access involving the PWDs to understand and get a feel of common (mobility related) pain areas. The survey and subsequent discussions brought out many insights regarding the nature and extent of problems/ their possible solutions and helped immensely in developing appropriate proposals focused on inclusivity".*

3.4.2 Community Consultation Workshop

Community consultation workshop was conducted physically at a convenient location within the neighbourhood mainly to familiarise the large community about the programme and the proposals, and to seek their views/preferences.

Before conducting the workshop, conceptual designs were prepared for the draft (shortlisted) proposals by the DULT with the assistance of interns specialising in Urban Design/Urban Planning or the embedded specialist agency (in two neighbourhoods, DULT engaged a specialist agency to support the community partner). Posters covering the highlights/main features of SuMA, and the draft proposals were displayed in the community consultation workshop for easy understanding of the public for a period of two to three days. The workshop was organised with the active participation of the community partner and volunteers from the neighbourhood, beginning with a launch attended by elected representatives and government officials associated with the neighbourhood.

During the workshop, proposals were explained to the participants and their views, suggestions, and preferences were sought through a short feedback form. (Figs. 10 and 11). The information gathered by means of feedback forms was analysed and the proposals were ranked as per the preferences of the larger community. In consultation with community partners, the top ranked proposals were further examined and refined.

Fig. 10 Shortlisted proposals being briefed to the residents. (*Source* SuMA Exhibition, DULT)

Fig. 11 Shortlisted proposals being briefed to the officials. (*Source* SuMA Exhibition, DULT)

The proposals along with block cost estimate, proposed community engagement plan, phasing of implementation, and post-implementation aspects including maintenance were presented by the community partner to the SuMA Task Force. An interactive session was scheduled after the presentation to get comments and suggestions of the SuMA Task Force on the proposals and other associated aspects.

3.4.3 Pre-Implementation Activities

Based on the suggestions of the SuMA Task Force, the proposals were finetuned and detailed designs were prepared along with final cost estimates in discussion with the community partner. The implementing agency and other concerned departments/stakeholders were informed about the outcomes of the advisory body meeting and the proposals through formal communications by the DULT. Subsequently, the teams of DULT and community partner held

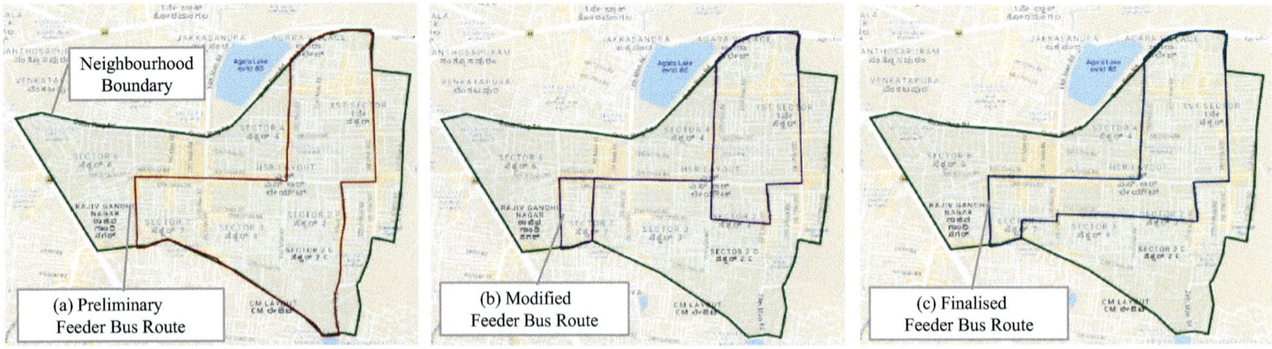

Fig. 12 Representative illustration showing how feeder bus route proposal evolved from conceptual stage to final stage

multiple consultations and site visits along with the stakeholders in relation to various aspects of implementation. The designs and estimates were further refined, if required, and finalised. The Fig. 12 illustrate how the route for a feeder bus project[1] evolved from conceptual stage (Fig. 12a and b) to final bus route (Fig. 12c) through consultative process and based on the contributions/inputs from community partner and key departmental stakeholders. Elected representatives were also apprised on the finalised proposals and the implementation plan.

In accordance with the roles and responsibilities agreed upon and technical details finalised, agreements were formalised with the community partners and the implementing agency covering major milestones, timelines, grant disbursal plan, etc., for the implementation phase. Work related clearances were also requested and obtained from associated agencies wherever necessary.

Fig. 13 Bicycle Lane installed at a neighbourhood. (*Source* DULT)

3.5 Implementation

An amount of US $60,000 has been earmarked by the DULT for every neighbourhood. About 3 % of it could be used for preparatory activities like data collection, conduct of FGDs, and community consultation workshops, etc. Funds were released to the community partners for the preparatory activities. Fund for implementation is released either to the community partner in case the community partner indicates willingness to execute the works itself or to the appropriate implementing agency of the Government, in consultation with the community partner. The work execution commences in line with the finalised designs, activities schedule, and agreement. Photographs related to cycling infrastructure implementation for one of the neighbourhoods are shown in Figs. 13 and 14.

In addition to the core implementation work, various allied and supporting activities as listed below were also undertaken for ensuring smooth implementation.

- Community partner organising community engagement activities/activation events in association with DULT to sensitise and engage with the residents of the neighbourhood on aspects pertaining to SuMA and sustainable practices (representative photographs are given in Fig. 15a, b). These included campaigns to promote certain projects, Cycle Days, etc.
- Developing templates for progress reporting, work monitoring, etc. While the community partners are mandated to take regular assessment of the work, DULT technical team undertakes site inspection at least on a fortnightly basis to facilitate proper supervision and monitoring of the work with respect to schedule and quality. The progress status and site inspection details are reported

[1] The feeder bus proposal was developed for one of the neighbourhoods in consideration of the need for better public transport connectivity within the neighbourhood which is relatively larger in terms of geographical area covering about 5 Sq. Km. as well as for improved linkage with the existing bus routes around the locality (as inferred from household data analysis and insights of FGD).

Fig. 14 Data matters—bicycle counters installed at a neighbourhood. (*Source* DULT)

through monthly status reports and fortnightly reports by community partner and DULT team respectively. The documentation also entails uploading the before and after implementation photographs and videos periodically.

- Holding sensitisation and coordination meetings with officials of relevant stakeholders depending on the projects' requirements.
- DULT facilitating evaluation and review of implementation by the advisory body and seeking their suggestions.
- Dissemination of pertinent information on SuMA activities to the media for transparency and creating further awareness (applicable to all stages of the programme).

3.6 Post Implementation

The SuMA programme also involves various activities that need to be taken upon completion of the project implementation. These include (a) maintenance of infrastructure and assets being developed under SuMA funding and (b) impact assessment of completed projects to evaluate how the interventions have benefited the neighbourhoods and the residents on various aspects, particularly in respect of sustainability/sustainable mobility. The above post-implementation activities will be undertaken with the participation and support of the associated communities which in turn will help in ensuring their continuation of support and advocacy for promoting sustainable transport.

4 Evaluation of Effectiveness

As depicted in the below illustration (Fig. 16), community engagement/public involvement is an integral part and important component in various stages of SuMA programme. While the programme implementation is under various stages of progress, the experience hitherto reveals that wider participation of community has helped in deriving multiple benefits.

Community engagement/public involvement is observed to be significant and beneficial in the following aspects:

- Involvement of a local community partner helped in obtaining responses from residents/households during data collection process during the pandemic.

a b

Fig. 15 **a** Once a defunct conservancy lane, currently being used as a community space (*Source*: DULT), **b** Residents from the neighbourhood utilise the space for social gatherings (*Source* DULT)

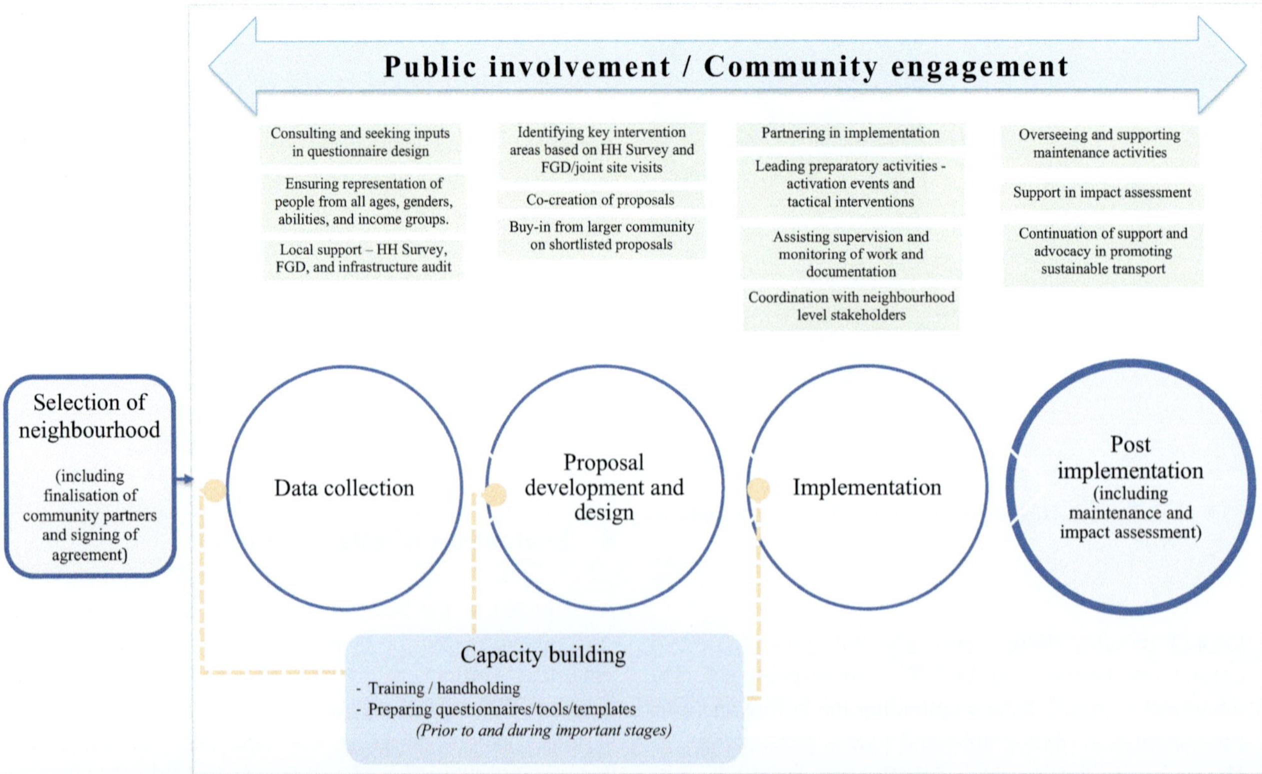

Fig. 16 Illustration showing public/community engagement at different stages of the programme

- The consultative process at various stages led to the formulation of appropriate proposals, based on local concerns and preferences.
- As an organisation, DULT has been able to reach out to larger community through the community partner for activation/community engagement events and tactical interventions.
- Involvement of students from transport/urban planning/urban design disciplines in SuMA as a part of SuMA internship to conduct infrastructure audits or prepare conceptual plans in consultation with community provided the students with a valuable exposure to public engagement and an experience of working with community.
- Involvement of community partners in project implementation/supervision and monitoring of work has helped in ensuring quality of work.

5 Major Insights and Learnings

The major insights and learnings captured so far from the implementation of the SuMA programme are given below alongside the pertinent stage of work.

5.1 Selection Process

Since community partner plays a crucial role in respect of the involvement of the people, its accessibility and compatibility with different sections/groups of the selected neighbourhood needs to be good. In case the community partner has a limited reach due to various reasons (for example, most of its members are professionals unable to spend required time), the government partner must step up to reach out to similar organisations in the neighbourhood for consultations on proposals to get a better buy-in from the community.

5.2 Data Collection

Community partners/their volunteers may require certain trainings multiple times particularly for those activities in which they do not have core competence. It is also observed that the actual availability of all the trained volunteers for a given work (for which they are trained) remains uncertain. This fact needs to be understood and accounted for while scheduling activities.

5.3 Community Consultations

Involvement of people of all ages, gender, and abilities can bring holistic perspective.

5.4 Design Development

A professional agency must be engaged to give support to the community partner at various stages of design development. The DULT involved an agency with urban planners and designers in two of the neighbourhoods while DULT itself provided the required services in the rest. Active involvement of the government stakeholders like the DULT in various activities taken up under SuMA in the neighbourhood is vital in maintaining community interest and in ensuring adherence to schedule.

5.5 Implementation

Formats for monitoring and reporting must be simple, and not elaborate to prevent community partner from losing interest. Its members have full-time jobs and cannot be saddled with too much paperwork. The DULT resolved this issue by designating a core team from its staff for each neighbourhood so that they could fulfil the tasks relating to the documentation required as per government norms.

6 Conclusions

The main research aim of the paper is to evaluate the effectiveness of SuMA programme's implementation from public engagement perspective and to underline the insights and learnings for reassessing how and where such public engagements can be further improved to derive additional benefits. Since the SuMA programme is still under implementation, the assessment of impact will be taken up after the completion of the projects. The key benefits associated with public involvement/community engagement as observed during various activities undertaken so far are summarised as follows. These are eliciting responses for data collection process with relative ease during the pandemic, formulation of proposals based on local concerns and preferences, instilling a sense of ownership during project development, finalisation, and implementation, ensuring better quality of work with the involvement of community in monitoring and supervision. In addition, it facilitated DULT in reaching out to the larger community through community partners to bring out greater consensus from the community. It also helped transport and urban planning students in getting exposure to public engagement and experience of working with communities.

In respect of insights and learnings, the experience showed that selection of community partner who has accessibility and compatibility with different sections/groups of the neighbourhood and making the involvement inclusive with people of all ages, genders, economic strata, and abilities participating in the process are significant in the success of programme. In view of the uncertainty over the availability of all trained volunteers and the fact that they do not have core competency on some activities, community partners may require multiple CB sessions and simplified explanations supported with graphics. Similarly, the monitoring and reporting formats need to be kept simple and short to facilitate them to provide pertinent information without spending much time and losing interest. For citizen-government initiatives such as SuMA, defining the roles and responsibilities of key stakeholders in an appropriate way (considering their major strengths) becomes important. Community is better equipped in conducting questionnaire/interaction-based data collection, identifying main problem areas based on their spatial context, and proposing interventions that could work well in their neighbourhood. On the other hand, government organisations having expertise in planning, designing, and implementation are more suited for activities like data analysis, generating analytics, design and cost estimation, design of templates/data collection formats/developing tools and dashboards, formal communication with government departments/elected representatives, and implementation, which need to be complemented with timely involvement and inputs of the community/public. At various stages of design development, a professional agency must be engaged to give support to the community partner.

Therefore, careful design of any similar programme with the right mix and proper sequencing of activities keeping community/public involvement as the main focus, will bring more success and benefits in respect of sustainability. The insights and learnings of the ongoing SuMA programme will help in assessing how and where such public engagements can be further improved to derive additional benefits.

Acknowledgements We thank the community partners of SuMA Cohort 2020–2021 and our colleagues from the Directorate of Urban Land Transport for actively engaging and contributing for the programme that greatly helped us in fine-tuning the public involvement strategies as part of the research. We also thank the officials from Bruhat Bengaluru Mahanagara Palike (BBMP), Bangalore Metro Rail Corporation Limited (BMRCL), Bengaluru Traffic Police (BTP), and Bengaluru Metropolitan Transport Corporation (BMTC) for providing their valuable insights and active engagement during the programme development and implementation.

References

Alnabelseya, B., Beaudry, M., Hotte, M., & Tebinka, R. (2021). *Public engagement in sustainable mobility projects (Volunteer project report to the mobility management committee of the transportation).* Transportation Association of Canada (TAC). Retrieved from https://trid.trb.org/view/1887365: http://library.tac-atc.ca/publications/CA6ARH1272021P72.pdf

Driscoll, P. A., & Vladimir, M. (2012). Targeted public participation in sustainable mobility planning: A case study of Aalborg, Denmark. *AESOP Congress 2012.* Retrieved from Researchgate.net: https://www.researchgate.net/profile/Vladimir-Mrkajic/publication/287202107_Targeted_Public_Participation_in_Sustainable_Mobility_Planning_A_Case_Study_of_Aalborg_Denmark/links/5672e08d08ae04d9b099b0e1/Targeted-Public-Participation-in-Sustainable-Mobility

Kelly, J. J., Barta, P., Hossinger, F., Witte, R., & Christian, A. (2005). *Successful transport decision-making: A project Management and stakeholder engagement handbook.* Retrieved from www.civitas.eu: https://civitas.eu/sites/default/files/guidemapshandbook_web.pdf

Kumar, A., Lodha, D., Mahalingam Ashwin, A., Prasad, V., & Sahasranaman, A. (2016). Using 'design thinking' to enhance urban re-development: A case study from India. *Engineering Project Organization Journal, 6*(2–4), 1–11.

Digitalization and Smart Lab Concept

Sokol Shurdhi and Saimir Shtylla

Abstract

Introduction of smart labs in academic environments is a big step towards innovative and efficient learning spaces in order to develop and test ideas for a better future, including those for future smart cities. The diffusion of such facilities depends on a country's degree of digitalization and technological progress. During the last three years, there have been registered many successful examples worldwide, preluding for their larger diffusion in future. In countries like Albania, after the important initiatives within the framework of the national digital agenda 2015–2020 and the more ambitious ones expected in the latest (2022–26 digital agenda), introduction of smart labs in academic institutions may result very advantageous. On this basis, the paper aims to describe the approach to adopt, and the eventual benefits that may result from the diffusion of smart labs in Albania. Before analysing the technical aspects, a literature review will be conducted a better comprehension of their importance and impact, as well the differences with other innovation spaces. In addition, it will be made a comparison of recent smart labs examples with a special focus on the recent the smart lab project in the Barleti University in Tirana.

Keywords

IoT • Smart lab • Albania • Technology • Educational buildings

S. Shurdhi (✉) · S. Shtylla
Faculty of Applied Sciences and Creative Industries, Barleti University, Tirana, Albania
e-mail: s.shurdhi@umb.edu.al

S. Shtylla
e-mail: s.shtylla@umb.edu.al

1 Introduction

Nowadays, there is an ongoing process of digital transformation affecting various fields aiming the creation of interconnected networks. The introduction and combination of advanced technology systems is important independently of the final decisions and stakeholders involved. Although, technologies such as the Internet of Things (IoT), big data, additive manufacturing (AM), artificial intelligence (AI) and robotic technologies are very important, but Cyber-Physical Systems (CPS) are more considering the safety they provide. Provision of appropriate learning environments for education and training for the implementation of such sophisticated systems (Knight et al., 2020) is an important contribution for the future "smart" cities, too. At the urban scale, considering the complexity of the processes and having as main goal the improvement of life quality for their inhabitants, the process of becoming "smart" requires facilities to test innovations. FabLabs, Makerspaces, Living Labs and Smart Labs are well-known for their openness towards innovations' test, despite their definitions and purposes are different. In academic institutions, smart laboratories or as commonly known as smart labs represent an advanced model of performing research and didactic activities, due to the affordability and high degree of customization of the IoT devices controlling and monitoring the equipment.

In Albania, innovation spaces have known a large diffusion the recent years but there have not been many initiatives for the introduction of smart labs in academic institutions. The smart lab project at the Barleti University in Tirana represents a new starting point in terms of the vision, approach and expected outcomes that such facilities in the country should offer as spaces where education and research activities occur. This paper contains four parts. The first part makes an overview on the dimensions of the "smart" city concept and its importance in the light of the important climate and urban challenges ahead. The second part

A. Battisti et al. (eds.), *Greening Our Cities: Sustainable Urbanism for a Greener Future*, Advances in Science, Technology & Innovation, https://doi.org/10.1007/978-3-031-49495-6_27

analyses the smart lab concept and its comparison with similar facilities. The third part discusses about Albania's approach in the digitalization process and introduction of facilities for future smart environments. Finally, the fourth part focuses in the implementation of the smart lab in Albania describing approach and last the conclusions.

2 The Smart City. A Necessity

The need for smart and efficient cities is unneglectable in the light of the current critical issues and future climate goals ahead as well. According the New Urban Agenda, in the light of the current urbanization trends, world's population will double by 2050, meaning that cities' role as important economic social and cultural poles will increase (United Nations, 2017). Simultaneously, important challenges arise in the environmental point of view due to the growing demand for facilities, infrastructure, food security, etc. Moreover United Nation (2019) expects in 2030, a change in terms of size due to urban population growth resulting with more medium-sized, large-sized and megacities based on the number of inhabitants as illustrated in Fig. 1.

The 2030 Agenda for sustainable development[1] with its 17 goals and 169 targets, considers cities important contributors in at least three goals. Obviously, planning activity paves the way for cities as more inclusive, safe, resilient and sustainable places (Goal 11) and preventing risks related with climate change (Goal 13), too. In terms of energy (Goal nr. 7), research and technology are important to address the necessary interventions and economic efforts towards higher energy efficiency rates and higher reliance on renewable energy sources for affordable, reliable and modern energy services. Responsibility of cities increases further in the energetic point of view in the light of the two important COP26 global goals, such as the achievement of the global net zero by mid-century and keep 1.5 degrees within reach.[2] The well-known affirmation "Think globally and act locally" recall us the importance cities have to manage themselves better and efficiently in order to properly address policies and actions.

Usually, the "smart" definition refers to cities that make use and implement Information and Communication Technologies (ICT) to properly manage their most common issues. Various authors have different points of view on its definition, fields of action and necessary steps to implement it. Besides the embracement of Information Technology (IT) in infrastructure and even on human bodies to help cities to properly address the various social, economic and environmental problems, Figueiredo et al. (2020) consider the definition "smart" as an indicatory higher degree of safety, convenience, healthiness and more efficient forms of living. Facilitating diversity and enabling better conditions through optimization makes this term utopic according to Schürer (2020) and will not bring major inclusivity and better conditions in cities. In the same wavelength is also (Callander, 2020) when identifying two important dimensions that this synonymous of modernity of the urban space incorporates. Indeed, there is a permanent risk of shifting from a "wired city" (first dimension) exchanging advanced information and communication into a capitalist utopia (second dimension) due to unequal access to digital technologies. "Smart" is not synonym with "wise" for Mehta et al. (2020) of the social risks (unemployment, marginalization, segregation) and problematics due to climate change.

Although, the risk of unsuccess is present, their number will grow worldwide, even at national scale for some countries. Again Callander (2020) expects a competition between existing and new cities to become "smart", a phenomenon that will affect especially the so-called global cities resulting with more than 26 smart cities or 4.6 billion people (58% of the world's population) by 2025. That utopia of a decade ago for Karvonen et al. (2018) has become a national goal to pursue in the United Kingdom (UK) and United States of America (USA) considering that one-third of UK cities with over 100,000 inhabitants and two-thirds of US cities are investing in some forms of smart technology. The intelligent machines for complex processes (Hnilica, 2020) require a set of measures that (James et al., 2021) considers important such as: (i) Human interaction, for equal access to technology and protection users' privacy; (ii) Institutions, acting as coordinator of the technological development and setting strong cooperation with private and academic institutions; (iii) Data, for their importance in the decision-making process and leading towards the improvement and increase of efficiency; (iv) Technology, as necessary infrastructure for a safe and efficient collection, storage and protection of various data; and (v) Climate and energy, for the control and efficient management of energy grid and of the Distributed Energy Resources (DER) available within the smart city environment.

Transitioning into a smart city requires a "case by case" approach (Vihanninjoki & Lehtinen, 2020) as long as experiential factors allow to identify the most suitable urban environments allowing people to produce and enjoy genuine well-being. Different planning approaches in relation to the (urban or rural) context, facilities such as living laboratories for Hirvonen-Kantola et al. (2015) help to better comprehend opportunities and advantages related to smart cities. Consequently, testing technology in order to comprehend its effectiveness becomes prior, meaning that the smart city needs to be a laboratory above all considering that

[1] For further information: https://sdgs.un.org.

[2] For further information: https://ukcop26.org/cop26-goals/.

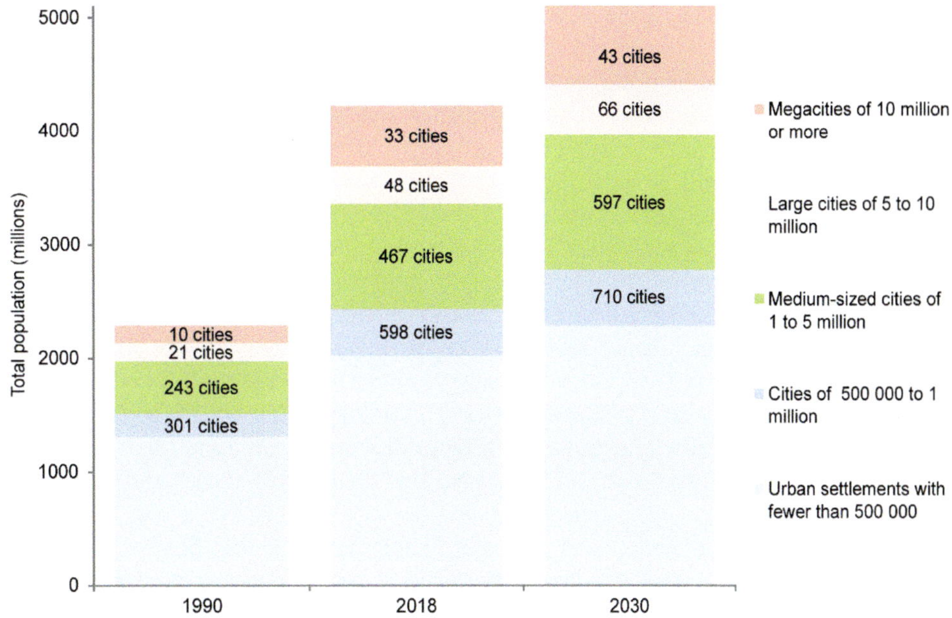

Fig. 1 Population and number of cities of the world, by size class of urban settlement. *Source* United Nation (2019)

technology evolves. Taking gradual steps by constantly documenting them according to Bolici and Mora (2017) should include a great openness towards knowledge and research, as resulted in the case of the city of Amsterdam.

3 Innovating and Testing. The Smart Lab

As mentioned before, testing technologies is important but in smart cities it does not occur exclusively by researchers in laboratories. Indeed, when citizens take part at various degrees during the testing phases, their feedback is very helpful to comprehend the need for any eventual improvement. These facilities or laboratories commonly referred as "labs" have various definitions and (almost) the same meaning though they differ in purpose and users involved. FabLabs, Makerspaces, Living Labs, Smart Labs and Innovation districts are the well-known. In terms of the area to accommodate such spaces, all but the last one need to be of limited area or located within the building.

The Fablabs, abbreviation of Fabrication Laboratories, emerged for the first time in the end of 1990s in the Massachusetts Institute of Technology (MIT) with the purpose to make the production of objects accessible to ordinary users (Le Roux, 2015), by combing art, technology and hands-on "Maker" approaches (Haldrup et al., 2017). Although they are "born for sustainability" (Galuppo et al., 2019), they are extremely important especially in design education (Škec et al., 2020). In this collaborative environment (Wong & Partridge, 2016), members to access tools and workspaces (Van Holm, 2015a). Quite often referred as "hackerspaces" and other names, Van Holm (2015b) highlights differences in philosophy, tools used, and access. Indeed, the term "Makerspace" introduced in 2015 refers to workshops involving sharing of tools, whereas the others focus in electronics and computers (Hackerspaces) or with digital fabrication (Fablabs). Unlike the last one, Makerspaces and Hackerspaces are more "public", while FabLabs are in academic instructions and slightly more business oriented.

Another well-known term "Hackathon", which is the combination of the terms "hacking" and "marathon" where small groups of people produce a working software prototype in a limited amount of time (Komssi et al., 2015). Living Labs (LL) refers to an experimental environment that allows technology to be applied and tested in real life. As "co-producers", (End) users' knowledge is part of policy-process (Nesti, 2017) and cities become a test bed for the so-called smart-from-the-start cities (Halegoua, 2020). For Cosgrave et al. (2013), along with Innovation districts ID, they represent the most efficient option to test and evaluate smart technologies. If the first ones (as a bottom-up concept) enhance innovation, inclusion, usefulness and usability of ICT and its applications in society in large-scale projects, the second ones (as a top-down concept) tend to cluster in large, skilled, economically diverse, well-connected urban environments. They also differentiate in terms of stakeholders involved with the LL involving public, private and academic sectors, whereas ID involve the Government, Industry, University/research institutes, Industry and research Labs, Start-ups/ SMEs Landlords and developers.

Introduced for the first time in 1966 (Lesk et al., 1997), smart labs are common in academic institutions where there is a high demand to efficiently perform labs tasks. The benefits emerging from the integration of IoT devices go beyond didactic activities as they improve teachers' and students' productivity (as main stakeholders). In fact, they also include indoor environmental quality, energy and safety aspects. Thanks to the use of IoT devices in such spaces, students will be taught to be "smarter" (Kassab et al., 2018) and didactic activity will be more inclusive, especially for students with disabilities. Also, considered "remote" labs by Lyalina et al. (2012), because unlike traditional face-to-face labs it enables a new way of "remote" learning from mobile (M-learning) or other electronic devices (E-learning) offering context-aware or location-based services (Deepak et al., 2019). Research methods will drastically change at a point to define them as "paperless" laboratories (Frey, 2004) by providing visual, centralized, and efficient management for teaching environments (Yan et al., 2013). Yin and Hu (2019) analysed its importance, benefits and future development.

Unlike traditional (computer) laboratories including only Internet connection and network security, the "smart" ones allow the use and IOT technology, wireless connectivity and remote management even from mobile devices (independently from the platform). Considering the importance of the data, a high-security system provides the necessary safety where only administrators of the academic institution can access. In terms of benefits, buildings' managers, teachers and students (as main users) are the most advantaged. Indeed, it is possible to manage safety and solve problematics of various nature thanks to the real-time control even remotely. On the other hand, teachers can easily manage their didactic process due to the auxiliary of the electronic devices, whereas students can comprehend the effectiveness of such facilities and increase their productivity too.

Figure 2 illustrates the general concept of a modern smart lab in terms of management of the didactic duties and important aspects such as indoor quality, energetic performance and security.

The operation of this typology of laboratory includes a constant adaption between the attendees' presence and devices' operation in relation to indoor quality factors such as natural lighting, illumination, air temperature and humidity that properly address the operation of illumination terminals' system and Heating, Ventilation and Air Condition HVAC units installed. Beside allowing the conduction of didactic activity under optimal indoor conditions, each IOT device and sensors "captures" all the data, afterwards transferred and stored to the server unit located in the building or university's campus. For the didactic activity, the teacher's workstation connected with the server, which performs the control and monitoring of the various aspects related with it. Through the specific electronic devices, remote control and monitoring can occur even outside the campus, reducing the need for human (physical) presence in the surveillance and monitoring of the laboratory. Teachers and students can access remotely to didactic content or attend online (laboratory) classes thanks to the cameras installed for this purpose. Besides their innovative architecture, smart lab examples in academic institutions have registered positive results on users' satisfaction, energy performance and general management.

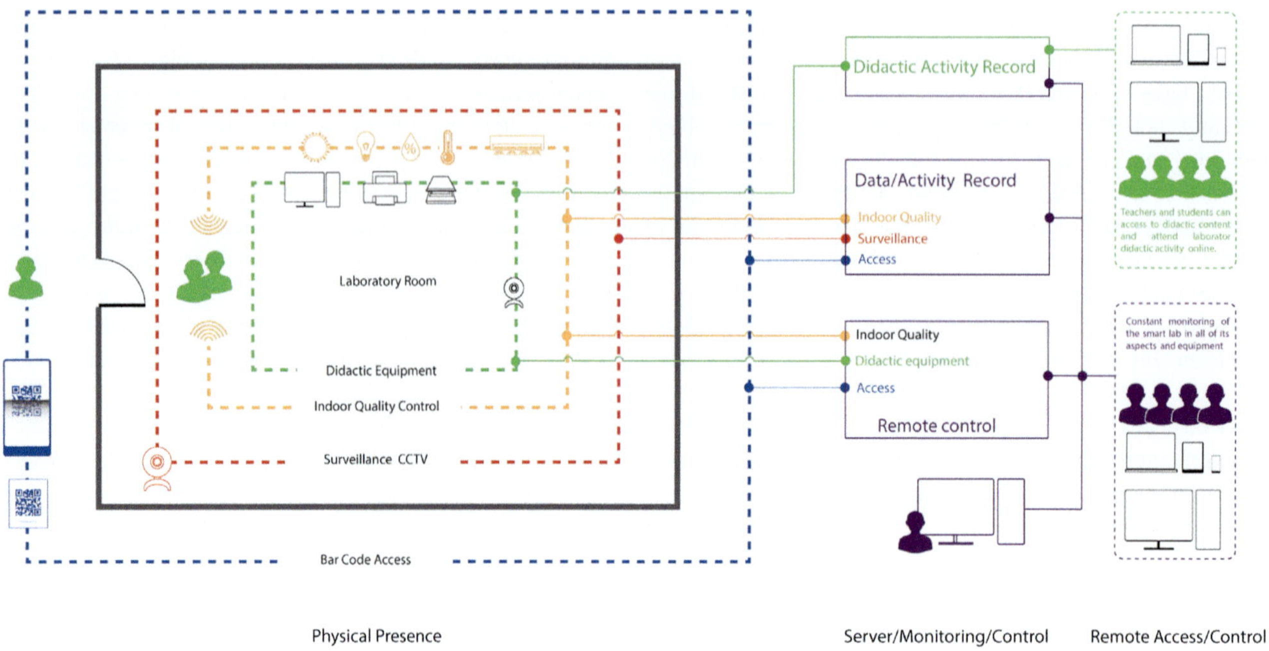

Fig. 2 Concept scheme of a modern smart lab. *Source* The authors

Table 1 Smart labs experiences: description and outcomes

Case study	Description and outcomes
Lei et al. (2013)	Installed a CPS to regulate indoor parameters in relation to the number of users present in the room. 1.9% consume of energy in comparison with the previous non-adaptive system and 96.91% of reliability
Kadar (2016)	Developed for higher efficiency and productivity, minimized costs and readiness in extreme situations. Positive feedback from 14 out of 16 teachers and high curiosity from students for the technology implemented
Poongothai et al. (2018)	Acquisition of energy consumption data from the installed devices involved in the control of the indoor environment and other functionalities based of System on Chip (SoC)
Khriji et al. (2020)	Integration of various devices to monitor indoor conditions in the academic laboratory, in particular temperature and air humidity. Storage of collected data and easy access by users
Vishnoi et al. (2020)	Facial recognition access in laboratory with Raspberry Pi and Amazon Web Services AWS cloud, also including speaker. Lighter and faster operation in comparison to PC-based face recognition system
Banagar and Khattar (2020)	Remote control of Lighting, Temperature and Air Humidity parameters. Reduction of energy consumption by 30% in an hour
Knight et al. (2020)	Voice-input access to the laboratory. Beside the monitoring of equipment and indoor parameters, there were text interaction, camera and dashboards. Positive feedback from users
Li et al. (2020)	Large data processing thanks to the Cloud computing, as well monitoring of the laboratory environment and equipment
Sulistiyowati et al. (2020)	Control of indoor conditions by detecting users' presence and the indoor temperature. Temperature detection accuracy of 97%

Table 1 listed and explained briefly in a chronological order in terms of operations and outcomes achieved, some of the smart labs' cases analysed. The examples of Kadar (2016) and Knight et al. (2020), users' satisfaction was very high. In terms of innovation in the cases of Vishnoi et al. (2020) and Knight et al. (2020) introduced face and voice-based access of users respectively in the laboratory, whereas on data processing, the cases of Lei et al. (2013), Poongothai et al. (2018) and Li et al. (2020) represented the most interesting innovations. For a better and tangible comprehension of the impact, smart labs have in comparison from the traditional ones, and the results of Lei et al. (2013) and Banagar and Khattar (2020) are very interesting in terms of energy savings. In the first case, the implementation of CPS requires only 1.9% of energy consume in comparison with a non-automated system in a laboratory of 150 users, whereas in second case, energy savings grew by 30%. Besides the documented benefits, the examples mentioned above confirm that smart labs are becoming a growing trend in academic institutions because they do not limit in the better management and increase of stakeholders' productivity, but they contribute a lot in the energetic point of view regardless their size.

4 Introducing Smart Labs in Albania

Even though there are no smart cities in Albania, important initiatives have recently been taken from this perspective. In 2015, the launch of the platform e-Albania[3] established a new approach to provide services to citizens and business owners and starting from May 2022, all the public services offered by the central government and public administrations across the county are online. Relying in the principles of the Open Government Partnership Initiative, this platform is one the outcomes of the cross-cutting strategy "Digital Agenda of Albania 2015–2020", whose main mission is the consolidation of digital infrastructure in the national territory with 3 strategic priorities, 15 objectives and over 80 major products. In addition, it includes higher quality online public services and greater governance transparency (Council of Ministers, 2022).

Education and research are also among the top priorities. Included in the strategic priority nr. 2,[4] the national education system will become 100% digitalized with the creation of e-education spaces, a model that is more than welcomed from teachers according to the study conducted by Noti (2014) despite the need for specific formative courses. For research activities, the expected incubators for ICT and development of business support, scientific research and innovation in the ICT field services will lead to the blossom of new businesses and jobs in these fields. The ElementIT[5] project is an important step towards the introduction smart labs in elementary and lower-secondary schools in the country, expecting the set-up of 100 ICT smart labs (with 1600 computer devices) and introduction of ICT curricula from 1st to 3rd grade in elementary schools, although existing school buildings in the country have serious classroom-capacity issues constraining the split of the

[3] For further information: https://e-albania.al/Pages/eAlbania.aspx.

[4] Policy for the development of electronic communications in health, education, etc.

[5] For further information: https://aadf.org.

didactic activity in the morning and for some in the afternoon (Shtylla, 2019). The country's second cross-cutting strategy "Digital Agenda of Albania 2022–2026[6]" represents a further confirmation of the country's ambition leading towards the digitalization of didactic activity in public schools. Unlike the previous version, it states that "smart" laboratories must be present in every pre-university institution within 2023. Although it does not include academic institutions, it aims the introduction of e-learning approach to help students to better enhance knowledge.

The introduction of smart labs in Albanian academic institutions could have many benefits. Starting from the national level, the introduction of a new digital agenda based on a larger diffusion of such facilities in the territory, will result with a major increase of awareness from normal citizens and direct stakeholders (teachers and students) of the productivity, safety they guarantee. It will lead to a new learning approach combining theory and practice, also allowing students to acknowledge the importance of smart devices in practice. Moreover, it will stimulate them to research on issues related to energy efficiency and cybersecurity. At the local level, although they do not permit the test of large-scale solutions, they can be fundamental to implement a new e-learning approach as well. In fact, past experience with online classes during the covid-19 pandemic is another reminder of the importance of remotely accessing teaching materials to help students better understand lessons.

Last but not least, academic institutions have to be the first promoters of the introduction of such facilities because they are the main playground for future e-learning platforms. Stimulating creativity is important, but the management of didactic activity and their energetic performance are as well. There will be more benefits from partnerships with private investors, by exchanging ideas and promoting research activities. The introduction of various makerspaces in the country can be a good starting point as described before but the diffusion of smart labs may trigger faster the attraction to "smart" technologies.

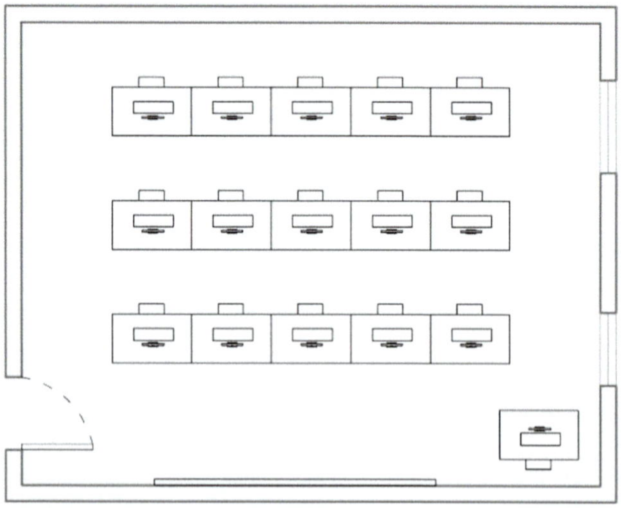

Fig. 3 Layout of the Barleti University smart lab. *Source* Barleti University

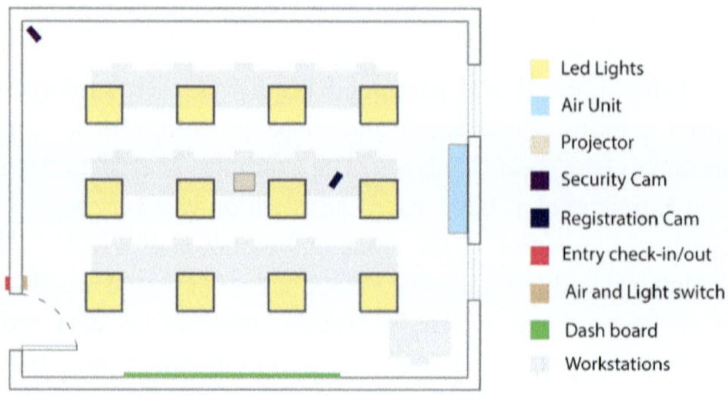

Fig. 4 Components of the Barleti University smart lab. *Source* Barleti University

4.1 Case Study: Barleti University Smart Lab

The Barleti University smart lab will be in the campus of the Faculty of Applied Research and Creative Industries will open in the end of 2022 and it is very early to comprehend its impact. This laboratory of thirty-five square metres will host 16 workstations. As illustrated in Figs. 3 and 4, this lab will be monitored constantly no matter there will be or not any didactic activity going on. The check-in/out interface positioned at the entrance, will detect the presence of the "attendees" in the laboratory. Afterwards, the lighting and air unit switches will programme their operation in relation to the users present in the lab. They will set parameters on illumination, air humidity and temperature through the specific sensors described in Table 2.

The LED lights will operate in relation to the amount of natural light exposure in the lab. It will be possible to record the didactic activity and eventually broadcast it online for those non present in the lab. All the data related with the didactic activity, surveillance, equipment and indoor quality will be stored in the dedicated server in the campus. As usual, there will be developed an application for the remote control of the laboratory and another one for students/teachers to access to the didactic content.

Even if energy consumes will be constantly under control, so as to make this lab even more "green", two photovoltaic (PV) panels will be provided to make it "carbon free" considering the advantages deriving from the country's grid (Belussi et al., 2021). Once fully operational, its

[6] For further information: https://akshi.gov.al/wp-content/uploads/2022/06/vendim-2022-06-01-370.pdf.

Table 2 Expected sensors and their functionalities in the smart lab

TE connectivity	Campbell scientific (CSI)	STEINEL	SPLUSS
Humidity sensor HS1101LF	Solar irradiance sensor (pyranometer) SR20-T2-L	Presence sensor (photoelectric) IES3360 COM1	CO_2 sensor AERASGARD RCO_2-W
Mass airflow sensor LMM-H03			
Temperature sensor LMM-H03			

performance will be detailed in future studies. As a result of its performance, other faculties of the University of Barleti and academic institutions in the country will open similar ones, thus further contributing to the national digital agenda. In the light of the examples described in Table 1, this facility is expected to guarantee a better management of the didactic activities carried out from the departments of the Faculty of Applied Research and Creative Industries. Moreover, there is expected low consume of energy as in the cases of Lei et al. (2013) and Banagar and Khattar (2020).

5 Conclusions

The introduction of smart labs in Albania may result very advantageous in the light of the country's attempt to open labs for the training and qualification of future technicians to promote a new model of e-learning and an efficient management of such facilities from its stakeholders. The recent labs represent an important stimulus towards the digitalization of the country but a new digital agenda to stimulate the diffusion of smart labs in academic institutions is more than urgent because it can pave the way towards more ambitious goals such as the transition towards smart academic institutions. Anyways, the implementation of this model should be triggered above all from academic institutions as long as they are places where innovation occurs and their positive impact will increase productivity of the stakeholders involved and in particular safety as foreign experience suggests. On this basis, the synergic cooperation of various stakeholders is essential to scale up this approach across the country to address major challenges. The smart lab in the Barleti University, as the first attempt to introduce such facilities in the academic institutions in Albania, has paved the way for detailed studies and assessments in terms of energetic performance and management of didactic activity that will be detailed in future.

References

Banagar, A. N., & Khattar, R. (2020). IoT based smart laboratory system. *International Journal of Engineering Research & Technology (IJERT), 9*(1), 315–318. https://doi.org/10.17577/IJERTV9IS010178

Belussi, L., Danza, L., Dobjani, E., Ghellere, M., Meroni, I., Salamone, F., Shtylla, A., & Shtylla, S. (2021). Achieving near zero energy building in Albania: An approach for the retrofit of a public-school building. *E3S Web of Conferences, 312*, 02005. https://doi.org/10.1051/e3sconf/202131202005

Bolici, R., & Mora, L. (2017). How to become a smart city: Learning from Amsterdam. In A. Bisello, D. Vettorato, R. Stephens, & P. Elisei (Eds.), *Smart and sustainable planning for cities and regions, green energy and technology* (pp. 251–266). Springer. https://doi.org/10.1007/978-3-319-44899-2_15

Callander, A. (2020). The trouble with the capitalist utopia. A totalizing scheme of subsumption and planetary urbanization. In S. M. Figueiredo, S. Krishnamurthy, & T. Schroeder (Eds.), *Architecture and the smart city* (pp. 56–67). Routledge. https://doi.org/10.4324/9780429324468

Cosgrave, E., Arbuthnot, K., & Tryfonas, T. (2013). Living labs, innovation districts and information marketplaces: A systems approach for smart cities. *Procedia Computer Science*, 668–677. https://doi.org/10.1016/j.procs.2013.01.070

Council of Ministers. (2022, September 20). akshi. Retrieved from akshi.gov.al: https://akshi.gov.al/wp-content/uploads/2018/03/Digital_Agenda_Strategy_2015_-_2020.pdf

Deepak, A., Rahul, P., Ravi, S., & Sachin, T. (2019). Smart laboratory. *International Journal of Trend in Scientific Research and Development (IJTSRD), 3*(3), 504–509. https://doi.org/10.31142/ijtsrd22840

Figueiredo, S. M., Krishnamurthy, S., & Schroeder, T. (2020). Introduction. Our new brave world. In S. M. Figueiredo, S. Krishnamurthy, & T. Schroeder (Eds.), *Architecture and the smart city*. Routledge. https://doi.org/10.4324/9780429324468

Frey, J. G. (2004). Dark lab or smart lab: The challenges for 21st century laboratory software. *Organic Process Research & Development, 8*(6), 1024–1035. https://doi.org/10.1021/op049895g

Galuppo, L., Kajamaa, A., Ivaldi, S., & Scaratti, G. (2019). Translating sustainability into action: A management challenge in FabLabs. *Sustainability, 11*(6), 1676. https://doi.org/10.3390/su11061676

Haldrup, M., Hobye, M., & Padfield, N. (2017). The bizarre bazaar: FabLabs as hybrid hubs. *CoDesign*, 1–16. https://doi.org/10.1080/15710882.2017.1378684

Halegoua, G. R. (2020). *Smart Cities*. MIT Press.

Hirvonen-Kantola, S., Ahokangas, P., Iivari, M., Heikkilä, M., & Hentilä, H.-L. (2015). Urban development practices as anticipatory action learning: Case arctic smart city living laboratory. *Procedia Economics and Finance, 21*, 337–345. https://doi.org/10.1016/S2212-5671(15)00185-9

Hnilica, S. (2020). The metaphor of the city as a thinking machine. A complicated relationship and its backstory. In S. M. Figueiredo, S. Krishnamurthy, & T. Schroeder (Eds.), *Architecture and the smart city* (pp. 68–84). Routledge. https://doi.org/10.4324/9780429324468

James, P., Astoria, R., Castor, T., Hudspeth, C., Olstinske, D., & Ward, J. (2021). Smart cities: Fundamental concepts. In J. C. Augusto (Ed.), *Handbook of smart cities*. Springer. https://doi.org/10.1007/978-3-030-69698-6_2

Kadar, M. (2016). Smart learning environment for the development of Smart City applications. In *IEEE 8th International Conference on Intelligent Systems (IS)* (pp. 59–64). IEEE. https://doi.org/10.1109/IS.2016.7737500

Karvonen, A., Cugurullo, F., & Caprotti, F. (2018). Introduction: Situating smart cities. In A. Karvonen, F. Cugurullo, & F. Caprotti (Eds.), *Inside smart cities: Place, politics and urban innovation* (pp. 1–12). Routledge. https://doi.org/10.4324/9781351166201

Kassab, M., DeFranco, J., & Voas, J. (2018). Smarter education. *IT Professional, 20*(5), 20–24. https://doi.org/10.1109/MITP.2018.053891333

Khriji, S., El Houssaini, D., Barioul, R., Rehman, T., & Kanoun, O. (2020). Smart-Lab: Design and implementation of an IoT-based laboratory platform. In *IEEE 6th World Forum on Internet of Things (WF-IoT)* (pp. 1–5). https://doi.org/10.1109/wf-iot48130.2020.9221143

Knight, N. J., Kanza, S., Cruickshank, D., Brocklesby, W. S., & Frey, J. G. (2020). Talk2Lab: The smart lab of the future. *IEEE Internet of Things Journal, 7*(9), 8631–8640. https://doi.org/10.1109/JIOT.2020.2995323

Komssi, M., Pichlis, D., Raatikainen, M., Kindstrom, K., & Jarvinen, J. (2015). What are hackathons for? *IEEE Software, 32*(5), 60–67. https://doi.org/10.1109/MS.2014.78

Le Roux, S. (2015). The intangible economy: FabLabs "individualised production of objects". A stage in liberating the function of innovation. *Journal of Innovation Economics, 17*(2), 99–116. https://doi.org/10.3917/jie.017.0099

Lei, C.-U., Liang, H.-N., & Man, K. L. (2013). Building a smart laboratory environment at a university via a cyber-physical system. In *2013 IEEE International Conference on Teaching, Assessment and Learning for Engineering (TALE)* (pp. 243–247). IEEE. https://doi.org/10.1109/tale.2013.6654439

Lesk, M., Harman, D., Fox, E. A., Wu, H., & Buckley, C. (1997). The SMART lab report. *ACM SIGIR Forum, 31*(1), 2–22. https://doi.org/10.1145/263868.263870

Li, S., Gao, X., Wang, W., & Zhang, X. (2020). Design of smart laboratory management system based on cloud computing and internet of things technology. *Journal of Physics Conference Series, 1549*, 022107. https://doi.org/10.1088/1742-6596/1549/2/022107

Lyalina, Y., Langmann, R., & Krisilov, V. (2012). Smart lab concept for different training modes as an extension of the remote lab. In *9th International Conference on Remote Engineering and Virtual Instrumentation (REV)* (pp. 1–4). IEEE. https://doi.org/10.1109/rev.2012.6293125

Mehta, G., Malu, S., & Mathew, M. (2020). Five strategies of socially smart cities. In S. M. Figueiredo, S. Krishnamurthy, & T. Schroeder (Eds.), *Architecture and the smart city* (pp. 165–185). Routledge. https://doi.org/10.4324/9780429324468

Nesti, G. (2017). Living labs: A new tool for co-production? In A. Bisello, D. Vettorato, R. Stephens, & P. Elisei (Eds.), *Smart and sustainable planning for cities and regions, green energy and technology* (p. 267). Springer. https://doi.org/10.1007/978-3-319-44899-2_16

Noti, K. (2014). Curricular objectives and the challenge of the digital school in Albania. *Journal of Educational and Social Research*. https://doi.org/10.5901/jesr.2014.v4n1p463

Poongothai, M., Subramanian, P. M., & Rajeswari, A. (2018). Design and implementation of IoT based smart laboratory. In *5th International Conference on Industrial Engineering and Applications (ICIEA)* (pp. 169–173). https://doi.org/10.1109/iea.2018.8387090

Schürer, O. (2020). The answer is "smart"—But what was the question? About some properties of utopian conceptualization. In S. M. Figueiredo, S. Krishnamurthy, & T. Schroeder (Eds.), *Architecture and the smart city* (pp. 43–55). Routledge. https://doi.org/10.4324/9780429324468

Shtylla, S. (2019). Contrast between energy efficiency policy and renovation interventions in existing school buildings in Albania. In *Difficulties of Turning School Buildings into nZEB. TDW 2019* (pp. 223–228). Polis University.

Škec, S., Özkil, A. G., & McMahon, C. (2020). Fablabs, makerspaces, and design spaces. *Artificial Intelligence for Engineering Design Analysis and Manufacturing, 34*(3), 298–299. https://doi.org/10.1017/S0890060420000426

Sulistiyowati, I., Sugiarto, A. R., & Jamaaluddin, J. (2020). Smart laboratory based on Internet of Things in the Faculty of Electrical Engineering, University of Muhammadiyah Sidoarjo. *IOP Conference Series Materials Science and Engineering, 874*. https://doi.org/10.1088/1757-899x/874/1/012007

United Nation. (2019). World urbanization prospects. In *The 2018 Revision*. United Nations. https://doi.org/10.18356/b9e995fe-en

United Nations. (2017). *New urban agenda*. United Nations.

Van Holm, E. J. (2015a). Makerspaces and contributions to entrepreneurship. *Procedia—Social and Behavioral Sciences, 195*, 24–31. https://doi.org/10.1016/j.sbspro.2015.06.167

Van Holm, E. J. (2015b). What are makerspaces, hackerspaces, and fab labs? *SSRN Electronic Journal*. https://doi.org/10.2139/ssrn.2548211

Vihanninjoki, V., & Lehtinen, S. (2020). Moving in the metropolis. Smart city solutions and the urban everyday experience. In S. M. Figueiredo, S. Krishnamurthy, & T. Schroeder (Eds.), *Architecture and the smart city* (pp. 210–220). Routledge. https://doi.org/10.4324/9780429324468

Vishnoi, M., Saxena, K., Anas, M., Singh, S., & Agarwal, N. (2020). Smart lab with automated access. *International Journal of Science and Management Studies (IJSMS), 3*(3), 78–85. https://doi.org/10.51386/25815946/ijsms-v3i3p108

Wong, A., & Partridge, H. (2016). Making as learning: Makerspaces in universities. *Australian Academic & Research Libraries*, 1–17. https://doi.org/10.1080/00048623.2016.1228163

Yan, Z. Q., Zhou, C. Y., & Xu, T. (2013). Research on the structure of smart laboratory based on the internet of things technology. *Applied Mechanics and Materials, 427–429*, 2605–2608. https://doi.org/10.4028/www.scientific.net/AMM.427-429.2605

Yin, H., & Hu, J. (2019). SmartLab technologies. In Y. Zhang (Ed.), *Handbook of mobile teaching and learning* (2nd ed., pp. 371–386). Springer. https://doi.org/10.1007/978-3-642-54146-9_24

Nature and City: A Health Symbiosis

Marc Lostaló Vila-Trias and Joan Curós Vilá

Abstract

Cities should be the perfect symbiosis between urbanism and nature. The Anthropocene is evidence of the enormous impact that humans are having on the planet. In the context of cities, urban planning, architecture, and human development must interact with nature to offer a better quality of life to all citizens. Reflection and analysis are essential parameters for deciding which strategy is the most optimal in large and medium-sized cities so that they are correctly articulated and in harmony with their own interior and peripheral nature. This objective simultaneously entails the optimisation of our resources, from the most basic ones—air, water, and food—to the ecosystems and biodiversity that sustain us. This research, based on a literature review, studies some examples of cities where different strategies have been applied to achieve this regeneration in different contexts, such as the Asunción Plan in Paraguay, the 'Open Shore' project in West Palm Beach, Florida, the recovery of the natural geometry of the Aire River in Geneva, and the Cerdà superblocks in Barcelona. From this case study analysis, the conclusions point to the concealment of large infrastructures, the introduction of green corridors, the renaturation of river areas and waterfronts, the introduction of agriculture to the city, and the production of fresh water in coastal areas.

Keywords

Biophilic cities • Sustainable development • Urban environment • Urban sustainability • Healthy cities

M. L. Vila-Trias · J. C. Vilá (✉)
UPC (Catalonia Polytechnic University), Barcelona Tech, Barcelona, Spain
e-mail: joan.curos@upc.edu

1 Introduction: Nature as the Structure of the City

The growth of cities has been occurring in a way that is disassociated from our natural ecosystem, bringing with it a series of problems and consequences directly related to them, and how they feed off the territories beyond them, both in terms of people's health and environmental quality, and in terms of the anthropomorphisation of the natural environment, reducing resources and altering the water cycle, the balance of the ecosystem and biodiversity.

As a consequence, we have climate change is currently a threat to the planet, with rising ocean levels, reduced air quality, the frequent occurrence of extreme weather events, increased acidity of ocean waters, increased deforestation, reduction or loss of plant and animal species, inequalities between social classes, hunger, plutocracy, the increase and appearance of new diseases, the increase in world population and atmospheric pollution are all parameters that alert us and indicate the need to change our socio-economic model and our relationship with the environment and planetary resources (Jover & Wall, 2019).

Cities are centres of heat that contribute to rising temperatures, causing a greater demand for energy to maintain thermal comfort in living spaces. At the same time, they acquire higher levels of atmospheric, noise and light pollution that affect our health, impair rest and increase stress, affecting the circadian rhythms of people, plants and animals, as well as preventing the connection with the universe through the starry sky.

This number of problems is directly related to the growth of cities since the industrial revolution, in which their growth had a total dissociation from the natural environment, compared to a more organic development of the city where natural resources are at the basis of its development.

The current monocentric energy model contributes to the problem by requiring large energy infrastructures to feed the demand of cities. These large infrastructures require the

A. Battisti et al. (eds.), *Greening Our Cities: Sustainable Urbanism for a Greener Future*,
Advances in Science, Technology & Innovation, https://doi.org/10.1007/978-3-031-49495-6_28

occupation of a large surface area, an area that takes away from nature.

The economic subjugation of countries and people creates a plutocracy, which only seeks to increase its capital and people are only consumers. Health has become a business, so that having sick people will increase profits. Public services, which do not generate profits but provide welfare, are diminished in favour of private services that increase the profit of companies making people become only consumers of the product that is sold or offered.

In this socio-political framework, the need for change is evident and the main need is change in cities, in their development, in their relationship with the environment and with the citizen, in their energy and food dependence on the outside. It is in these aspects that architects and urban planners have much to say and contribute.

The total dissociation of cities from nature must come to an end, for if it does not, cities will suffer consequences in the face of the reality that the basis of everything is nature, the basis of scientific discoveries, the development of materials, medicines, the quality of air, water, food, etc. It is therefore necessary to remodel or restructure the urban planning of cities at different scales: planetary, continental, national, regional, city, and neighbourhood.

Urbature, as the urbanism of the future is called, encompasses the concept of urbanity and nature in a new word. In this sense, cities must be self-sufficient, polycentric, healthy spaces, and defenders of biodiversity. This is achieved by making them energy producers, in a decentralised model, where each contributing unit is self-sufficient, producing the energy it needs and sending the surplus to the grid. Because of this, buildings should act as small power plants by means of overhead turbines, photovoltaic panels, and geothermal generators.

The city can be endowed with the adaptation and transformation of spaces without a collective use, be it neighbourhood or urban, such as most flat roofs of buildings. The whole of these surface areas could be transformed into green spaces with small vegetable gardens or landscaped areas, which would help to reduce the temperature in the urban space. This transformation would mean the acquisition of an added value to currently residual spaces in terms of neighbourhood or community use, beyond the main function of protecting buildings.

With the emergence in 2020 of the covid-19 pandemic, it has become clear that citizens must have their own outdoor spaces to improve their well-being. It would be interesting for buildings to introduce nature into their interiors, such as interspersing high indoor spaces with vegetated outdoor spaces. This strategy leads to improved air quality and green spaces within the buildings themselves.

Polycentrism (cities having more than one centre) reduces the need for unnecessary travel, reducing CO_2 emissions and being more environmentally efficient. Decentralisation generates structure and urban fabric, linked to other urban elements (Ng, 2010).

Cities become healthy spaces when nature is introduced, which generates spaces of comfort for citizens. This means that space for vehicles, mainly private vehicles, is significantly reduced in favour of public or shared transport and zero or low-emission transport, especially underground transport.

The public space allocated to green areas should be increased to the point of creating a green infrastructure that links up with the rest of the urban space, creating a green network that crosses the entire city and connects with the forest, river and maritime ecosystems that delimit it. This leads to the continuity of the biophysical matrix, to the reduction of atmospheric pollution, to the improvement of urban quality, to the reduction of stress, and to the increase of curiosity and healthy stimulation of the citizens by maintaining a more closely linked relationship with nature.

This green infrastructure should be a main urban artery, linking all the elements necessary for the most basic needs, such as breathing, drinking, and eating, while simultaneously favouring the continuity of ecosystems, instead of segregating and sectoring them. Animals need large natural spaces, not isolated ones, for their survival and to maintain the balance of ecosystems and biodiversity. The partial or total dissociation of cities from nature, means that many children are unaware of the most basic species of animals or plants found.

2 Incorporation of Sustainability in Cities: Biophilic Cities

For the creation of a more resilient and sustainable future, it is worth considering the reflections of Leopold Aldo "*thinking like a mountain*" (Aldo, 1949, p. 128) where he discusses the geological time scale and its interrelationship with nature. In this sense, new urban strategies need to appreciate these two fundamental aspects, in order to create the balance between human development and the ecosystems that sustain it.

If we think like a mountain, we will realise that geological changes happen very slowly and last over time, having an impact on the distant future. Furthermore, the mountain maintains the balance between all its diversity of species, habitats, and ecosystems. From glaciers to its valleys and rivers, from predators, to small mammals, reptiles, amphibians, and small insects, all live in harmony and balance, interrelating with each other.

If we want to think and act like a mountain, energy production, agricultural production, and the relationship between urban development and forests need to be restructured. Agricultural areas need to be reforested with productive forests, with biodiversity supporting them. Monocultures deplete subsoil nutrients, especially nitrogen and phosphorus, and therefore such crops can be placed in other contexts, e.g. the introduction of agricultural uses on the roofs of buildings. All species are interrelated, maintaining a balance between them, so their combination is always favourable. The cultivation of native plants reduces the amount of nutrients and water needed, and consequently they are more efficient crops. Intensive monoculture agriculture, which allows for high food production, should be located in cities, contrary to current planning, giving up the space they currently occupy for reforestation, ecosystem restoration and biodiversity, in favour of a more resilient future (Davoudi et al., 2009).

It is not only agriculture that must conquer the city, but also the forests. In this sense, biophilic urbanism points in the right direction.

> Biophilic urbanism [...] is a city that puts nature first in its design, planning, and management; it recognizes the essential need for daily human contact with nature as well as the many environmental and economic values provided by nature and natural systems. (Beatley, 2011, p. 45)

Biophilic design, both in cities and buildings, aims to improve the environmental quality of spaces, connecting people with nature in a more direct way. This improves their health, creates a scenario of balance, peace and relaxation, improves air quality, reduces noise pollution, encourages creativity, etc. To do this, it uses vegetation inside and outside the spaces of buildings, in atriums, façades, corridors, courtyards, etc. Biophilic urbanism maintains the balance between nature and city. It is introduced by introducing an interconnected green infrastructure. The fusion of the city with nature allows to give continuity to the biophysical matrix, uniting instead of fragmenting. To this end, a series of strategies are applied that contribute to the continuity of the natural ecosystem.

It is advisable to study the possibility of superimposing a new fabric on the pre-existing urbanism, in other words, to give continuity to nature and to ensure that the urban fabric is not an interruption of the biophysical matrix, allowing it to connect with all the natural spaces outside the cities themselves. In this way, the different natural species would not be expelled from the cities but would have continuity throughout them. (Alberti, 2016).

It is important and necessary to revitalise the water of rivers, lakes, seas, and oceans, underlining that ¾ of our planet is water and this leads us to the basis of our ecosystem. It is also important and necessary to clean up the oceans with improved waste management, a better recycling chain and better treatment of human and animal waste. The elimination of plastic from food packaging and bags is another aspect to be managed in view of the danger it represents for marine fauna and flora. Other important considerations to be applied are the de-channelling of rivers, water purification, the prohibition of the dumping of chemical products, the renaturalisation of the space adjacent to rivers and lakes, and the creation of protected areas for marine fauna, such as the creation of new artificial reefs, with cavities that provide shelter for aquatic species.

Covid-19 has taught us that nature easily reconquers the city when human activity decreases and silence and tranquillity predominate, fauna has appeared in cities in search of a new habitat, which indicates that with green urban structures, fauna can coexist in cities and interact with citizens.

To establish the basis for landscaping and urban planning to create a more resilient and sustainable development, there are several elements and strategies suitable for achieving this goal. A distinction can be made between interventions in the existing city and areas of new development. In this regard, Singapore is a pioneer in biophilic urban planning. (Chan, 2012) states: "*[...] we want to emulate the forest. But the ideal would be not the city in a garden, a city in a forest [...]*". The idea is to emulate the city in the lush forest where plants, animals and humans coexist. This is the direction of Singapore's architectural and urban development, where nature is introduced into the roofs of buildings, corridors, atriums, facades, and streets, its network of parks and regenerating rivers are interconnected. It is an example to be considered and analysed in order to achieve this goal of nature conquering the city for the good of all citizens.

3 Methodology

The methodology applied for the research consisted of constructing a framework for discussion based on determining basic human needs, which are as follows:

- Oxygen.
- Water.
- Food.

Considering the application of these indicators, the cities themselves are unsustainable and need the support of non-populated areas. A citizen does not have trees in his immediate surroundings, except for those who live next to a forest. Some people do not have healthy drinking water naturally within reach, and in very few cases, rainwater is collected for drinking, and even less so in cities. In situations where people live near rivers or lakes, they usually cannot drink it. And of course, a citizen does not have an agriculture field at less than 500 m from his residence for self-consumption.

With these premises, we have observed the cities where different architectural projects and competitions have been executed, in which their actions have improved access to these needs to different degrees. In this way, a series of parameters, improved quality of the air, reduced acoustic pollution, and food self-sufficiency within the city, give continuity to the biophysical matrix, regenerate the river and maritime spaces, and give priority to the people, creating sustainable mobility, instead of the cars. It was established what cities should have in order to improve their sustainability based on human needs: A green infrastructure, a fabric that is superimposed on the pre-existing urban fabric, allowing the production of oxygen and local food, while at the same time fixing carbon dioxide in the soil. It also gives continuity to the biophysical matrix, allowing the flow of other living beings. The capacity to recover a denatured territory or to improve an urban space through architectural or urbanistic intervention was also considered a priority.

Finally, 4 projects have been selected that are committed to improving the quality of life of citizens in their cities. They are strategies to be analysed for the improvement of sustainability and urban efficiency.

The global nature of the existing problem of climate change has led to the selection of projects from different continents—America and Europe. This casuistry leads to the analysis and application to transform certain areas in different countries of the planet with a common objective, to transform large cities into more sustainable, bearable, and comfortable ones.

The similarity of the selected examples has a common denominator, which is to transform and increase the quality of life in large cities by optimising natural elements that coexist inside or on their periphery, such as the seas—West Palm Beach, rivers—River Aire Geneva, nature reserves—Plan de Asunción in Paraguay or the introduction of nature inside the city itself—superblocks in Barcelona. Another common aspect between them is that they are actions and results on a large scale, which means that the benefits obtained are felt by many inhabitants.

4 Cases Studies. Urban Strategies

4.1 Asunción's Master Plan. Paraguay

In Paraguay, we find an important biosphere reserve on the border of the city of Asunción and the river area. Near this reserve, there are a number of urban green spaces, with varying degrees of proximity and size, which are not connected to each other. In addition, the peripheral streets of the city give priority to the car instead of pedestrian and therefore to a less sustainable mobility.

In order to introduce nature—forest, vegetation, trees, etc.—into the city, it is necessary to create a new green fabric that is articulated with the existing urban fabric, complementing the isolated green spaces. A good example of how this fabric can be complemented in a pre-existing city is the Urban Ecosystem Master Plan project in Asunción (Paraguay) (Fig. 28.1). Through green corridors, which originate in a biosphere reserve and an active green front, adjacent to the existing city, nature penetrates the city, connecting with the green spaces as it passes through, establishing a primary relationship between nature and its landscaped spaces on the perimeter of the city. It is a project that seeks a balance through the introduction of natural spaces that are integrated into the urban space (Fig. 28.2).

This Master Plan uses a series of strategies to try to reconcile city and nature. To this end, it creates green, civic and dynamic corridors. To enhance these green and activity axes, it revitalises some urban areas, such as the Chacharita Alta, and also introduces a series of urban catalysts to give greater continuity to the axes and also favour the regeneration of the environment and facilitate development opportunities of different kinds. At the same time, they generate ecological and biodiversity reserves connected to these green axes.

These green axes are the most visible in the transformation of the city. The Cha Plan shows that there are two types of hierarchies of green infrastructure. On the one hand, and in the first place, there is a large green space which is the biodiversity reserve located on the banks of the Paraguay River. This space is of great importance, being the Banco de Sant Miguel together with the Bay of Asunción and having a surface area of several km^2. Secondly, there are the green corridors that arise from this nature reserve and go into the city, interconnecting different areas of urban parks and gardens, and disappearing in the middle of the city.

For the city of the future, a healthy symbiosis between nature and the city is necessary, and this necessarily involves establishing a green infrastructure superimposed on the current urban fabric, which will articulate between them. This network of corridors is hierarchical and has differences between its different elements. Just as a road network connects different cities, countries, and continents and is hierarchical with motorways, dual carriageways, national roads, regional roads, avenues, streets, and paths, this new green infrastructure must also be hierarchical and connect all the natural spaces with each other, not just going into the city and disappearing into it, but the biophysical matrix becomes a structuring axis of the city itself.

That is why green corridors are the main axes of this new urbature, the balance between urban and nature. The reason for this is that these arteries are the main connecting elements between the adjoining forests of existing cities, bringing natural spaces such as forests into the centre of

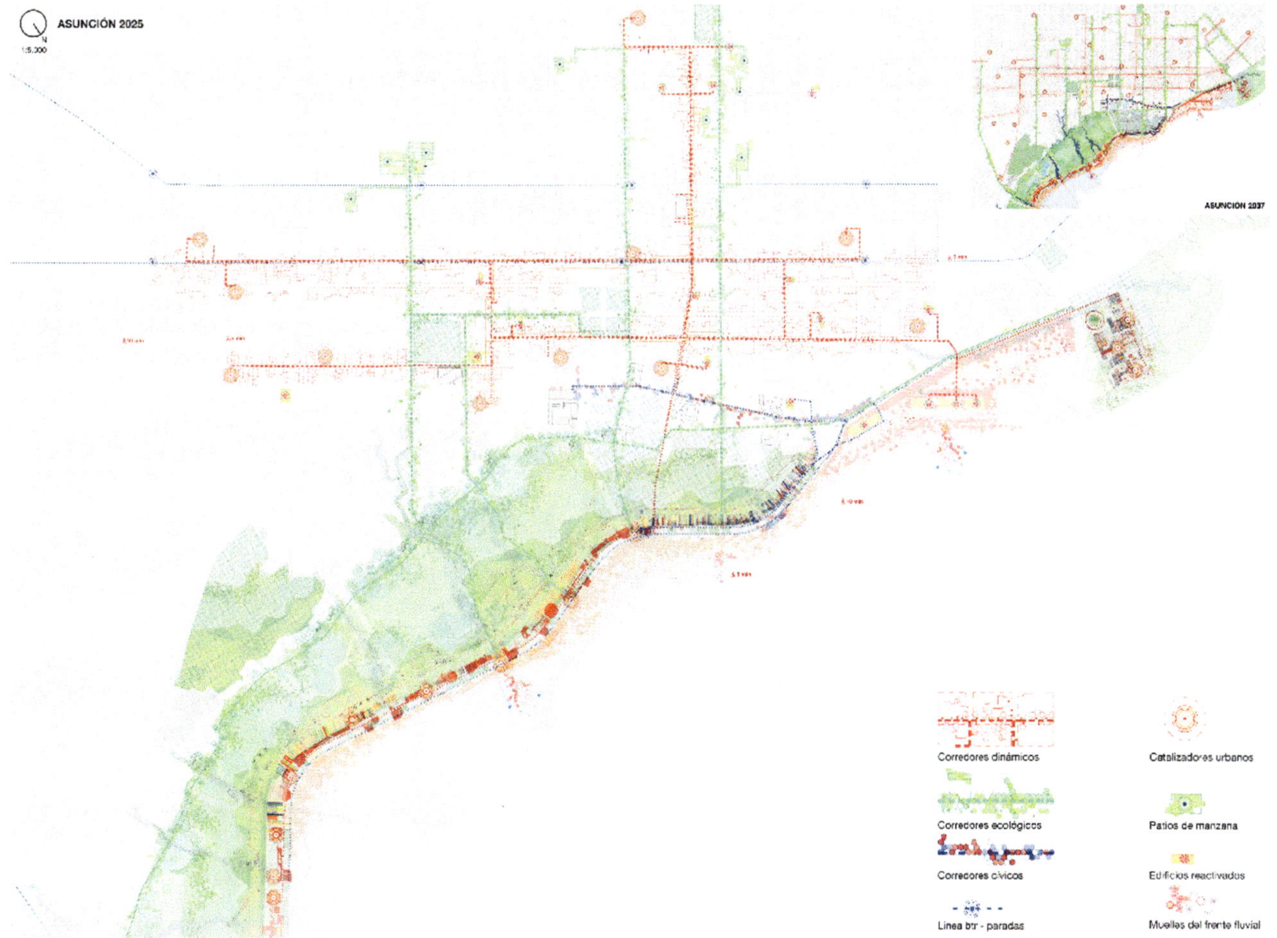

Fig. 28.1 Master Plan de Asunción, Paraguay. *Source* (Ecosistema urbano, 2019a, pp. 16–19)

cities and connecting ecosystems that have been isolated and fragmented by the rapid growth of cities, especially from the Second World War to the present day. These new green axes constitute a new forest and agricultural fabric that intersects with the existing fabric of the cities and restructures the internal traffic flows of the cities to facilitate the habitat or movement of animals. This system leads to a closer relationship between citizens and nature, whereby the new scenario restores natural origins. It allows children to learn by observation and direct experience by interacting with nature. These corridors should be wild nature in its purest state and different types and gradients can be categorised. Firstly, these corridors would be a combination of wilderness and riverine or maritime elements, which would allow the biophysical matrix to have full continuity, allowing terrestrial, aquatic and aerial species to move through these spaces. A second category would be a green corridor with wild nature, but without aquatic elements, allowing terrestrial and aerial species to move freely. And the third and last example is the green corridors, in which nature is more domesticated, where natural elements and human activities are combined. These would include parks, gardens, and urban agriculture, which would mainly facilitate bird flows, but could also be combined with small mammals, livestock, and insects (Ecosistema urbano, 2014).

To allow a better implementation of these corridors, it is advisable not to apply soil sealing but to allow water to filter into the subsoil, thus feeding the soil and aquifers and favouring the flowering of plants in these corridors. The layering of different natural elements will ensure that the filtered water is really clean.

The parks and gardens containing the third level corridors should be productive parks with the integration of food cultivation. There are two types: conventional gardens, but with the difference that the species planted are food producing, and agricultural parks that combine play and recreational activities with farming. In this way, cities move to being planned primarily for the ecosystem and for citizens, and finally for vehicles.

These new corridors also serve to reduce pollution and CO_2, improving air quality and emitting more oxygen, given that an average of 22 trees are needed for a person to

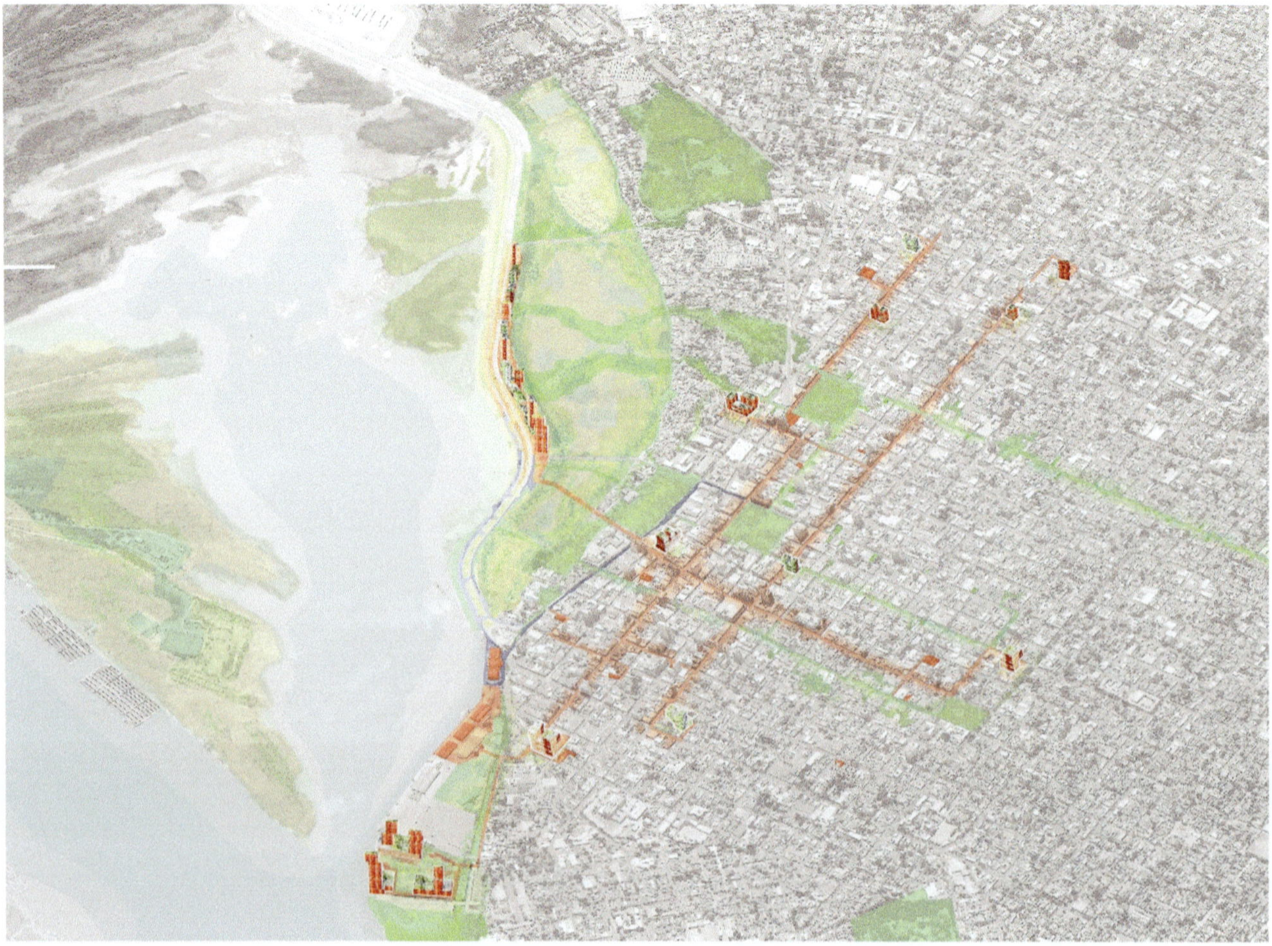

Fig. 28.2 Aerial view of Master Plan de Asunción, Paraguay. *Source* (Ecosistema urbano, 2014)

breathe. This value indicates the importance of reforestation, due to the increase in urbanised areas, and it is imperative that agriculture is integrated into the city, making the city productive and self-sufficient. In this respect, in addition to green corridors, any small area should contribute to food production. Buildings should also combine agricultural uses with other types of uses, providing green, productive, and recreational spaces between the buildings themselves to allow any citizen access to an outdoor community space.

Prime green corridors are those containing river or maritime spaces. In the case of already consolidated cities, where the seafront has been invaded by buildings, eliminating any trace of wild nature, fauna and flora should be reactivated and reintroduced in these contexts, giving them a sufficient length, minimum 200 m, to establish an ecosystem that serves as a link between aquatic and terrestrial ecosystems. To avoid beach erosion and favour the creation of protected underwater spaces, it is necessary to build new artificial reefs, between 500 m and 2 km away from the coasts, with these distances varying according to the type of sediments, seabed, or the type of section of the earth's crust at that point. This will allow the proliferation of marine plant and animal species, protected from fishing and ensuring the non-extinction of more species, adapting a space for them to regenerate and shelter.

In order to allow other animal species to use the beaches for their activities, it is necessary to provide them with a natural environment by establishing diversity protection areas and planting coastal plant species to provide them with shelter and protection.

In those urban areas where it is not possible to move back the building line, other strategies should be applied, ranging from the construction of sea walls to new areas located inland by means of artificial construction with rocks, stones, gravel, and sand, which would simultaneously generate other types of artificial reefs. In any case, these new marine areas must be sufficiently resilient to adapt to a possible rise in water levels. This new revegetation of maritime ecosystems will have to have continuity with terrestrial ecosystems, through the new fabric of green corridors that will connect the sea with the mountains.

Any maritime city must cease to be dependent on the under-supply of drinking water from rivers, lakes and glaciers. This requires the collection and storage of rainwater and the construction of new desalination plants, using passive solar energy to evaporate seawater, where it is combined with salt plants to produce salt and other necessary minerals. The plants should provide drinking water for the population, the new productive agricultural city and the parks and gardens, using the surplus to revegetate desert areas.

4.2 Open Shore. West Palm Beach. Florida. USA

In West Palm Beach, Florida, we find that the city abruptly ends at the waterfront, completely denaturing it and disassociating itself from the pre-existing natural ecosystem. There is no transition or filter space between the city and the ocean, where the asphalt ends, and the water begins. In this bordering space, we find minimal vegetation and the only small green space of note is the Meyer amphitheatre or the Great Lawn.

The 'Open Shore' project, West Palm Beach, Florida (United States), of Ecosistema Urbano (Fig. 28.3) contributes to the recovery of our coasts and their ecosystems, favouring the balance between human activities and marine ecosystems.

> The project envisions the future West Palm Beach as a dynamic and inclusive urban centre where activities and spaces are closely intertwined, providing access to a new urban waterscape, enhancing the city's connection to the lagoon and its endless possibilities for economic development, leisure and urban enjoyment. In this vision, the waterfront plays a key role as the most representative space... (Ecosistema urbano, 2016)

This is a good example of how a consolidated urban waterfront can be revegetated and simultaneously maintain urbanity by reclaiming land from the sea. On a large scale, the main strategy is to generate a series of themed squares closely linked to the tides and, consequently, to the existing natural ecosystem. These squares are arranged to allow the permeability of the dominant air currents and to favour visuals, without interfering with those of the existing urban front. At the same time, they facilitate the continuity of the existing urban fabric and brings it together in a new perpendicular promenade that limits the streets and the project space with the ocean.

The squares allow the development of different activities and atmospheres and introduce the changing dynamics of the tides and revegetate the area. The different microclimates generated in each square derive from bioclimatic canopies and mechanical elements that regulate them (Fig. 28.4).

These new green spaces with their corresponding vegetated buildings offer the possibility of practising different

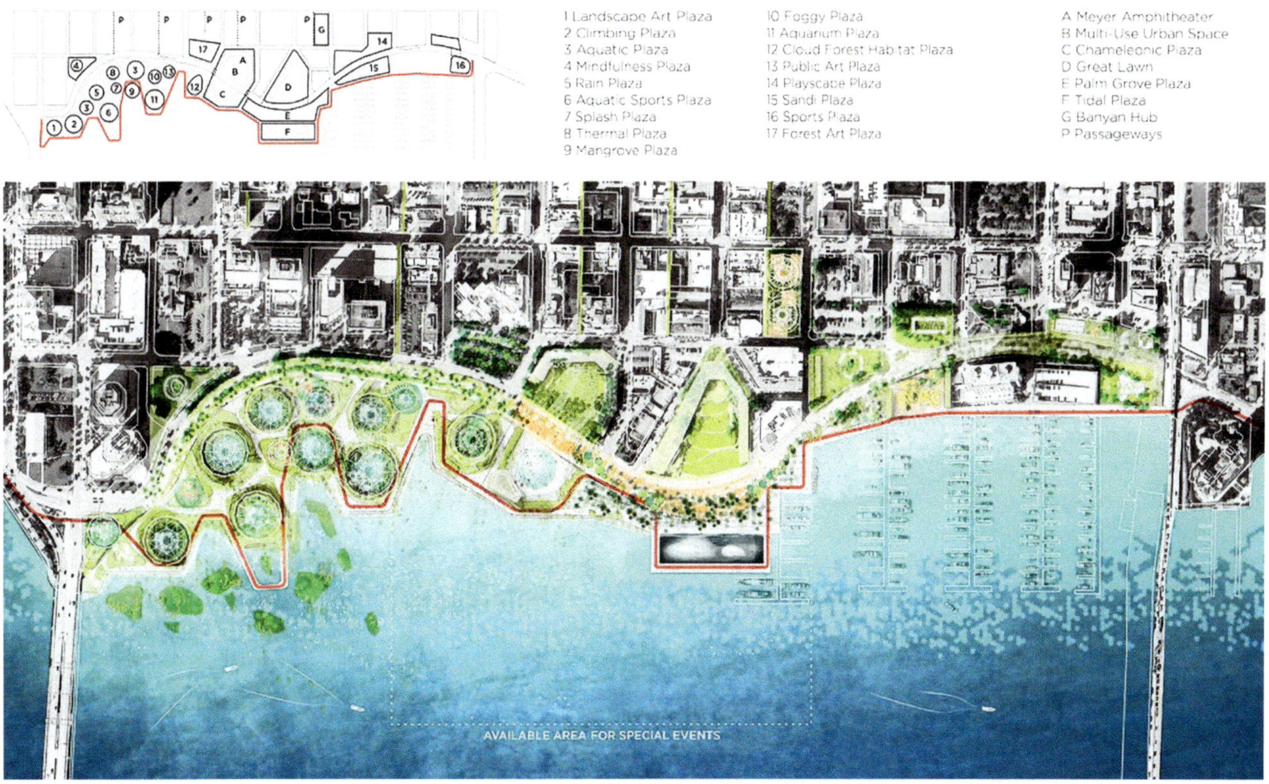

Fig. 28.3 Project 'Open Shore', West Palm Beach, Florida (United States). 2019b *Source* (Ecosistema urbano,, pp. 8–11)

Fig. 28.4 View from the water, 'Open Shore', West Palm Beach, Florida (United States). *Source* (Ecosistema urbano, 2016)

leisure activities, thus facilitating close contact between citizens in a more natural environment and at the same time providing public spaces at less than 100 m, which guarantees a much more frequent use than at greater distances.

The formal strategy of the squares consists of two permeable volumes that limit the space and configure it. They make it possible to differentiate the interior space, which lies between them, and relate it to the surrounding fabric. This space acquires different dimensions, depending on the activity it contains. These two volumes are topped by green roofs with their own topography, being an extension of the square, as they are born at ground level and rise to determine the volume.

Not only is the roof of the volume a vegetal space, but also the façades and the square is covered by a garden that is articulated with the canopy, giving different atmospheres and microclimates to the squares. The project also intervenes in the existing buildings and public spaces following the same philosophy of introducing nature into every corner of the city, improving its resilience in the long term, improving environmental quality, and in the short-term adapting to the different uses that occur during the day and night.

This intervention, on a smaller scale, has two main objectives. One is the reconversion of a parking area into a new urban core: the Banyan Hub (Fig. 28.5). The second is the restructuring of the existing Great Lawn urban space. The Great Lawn was already an important public space, but the new planning reinforces and enhances this idea. Two new green spaces, one more shaded and the other central with direct solar radiation, are generated by a new building that will form a new frontage to this space.

New urban elements are also added to facilitate different outdoor activities. Two new squares will be included that will maintain a relationship with the water, as is the case with the other new squares located in the new area above the lagoon.

Banyan Hub, for its part, create new green spaces, one in the interior and two on the rooftops. This new point of attraction and activity is enhanced by the integration of various public and private uses within the same volume, making it a new point of reference for its surroundings.

It combines residential uses, a sports area, a co-working space that will facilitate innovation by bringing people from different backgrounds into contact with each other, a YCMA centre, shops and a cultural centre (Ecosistema urbano, 2016).

4.3 General Plan of Renaturation of the River Aire. Geneva. Switzerland

In many parts of the world, rivers have been anthropomorphised, altering their geometry, either by channelling or redirecting them, or both simultaneously. This characteristic

Fig. 28.5 Banyan Hub section, 'Open Shore', West Palm Beach, Florida (United States). 2019b *Source* (Ecosistema urbano,, pp. 8–11)

alters the water cycle, affecting the ecosystem inside and outside. In the situation of the river Aire, in Geneva, part of the river itself had been channelled, in a rectilinear way, with the disadvantages that this entails.

The canalisation of rivers has several disadvantages because the velocity of water is not constant. In a natural river, current flows vary, not only along its course, but also in its width, providing different habitats for different species of living beings. Another disadvantage of increasing velocity is the reduction of water filtration and sediment input. Naturally flowing, fast-flowing rivers oxygenate the water and waterfalls and additionally generate ozone and negative ions. Viktor Schauberguer (1885–1958) studied and established a series of guidelines for revitalising water quality by creating whirlpools, spirals, and vortices.

The main characteristics for recovering river spaces and their natural forms are their simplicity and effectiveness, an example of which is the revitalisation of the river Aire in Geneva, Switzerland, designed by Georges Descombes, Atelier Descombes Rampini and his interdisciplinary team Superpositions (Rosenberg, 2019) (Fig. 28.6).

> Our project attempts to propose an alternate path, where the urgent ecological shifts are incorporated into a larger cultural change. The complex organization of the design associates the new river space and a linear series of gardens in the former canal. In reality the whole design becomes a linear garden" (ADR & Superpositions, 2016). "Facing the whole watershed, the original morphology of the mountains and the traces of human modifications, this long rivergarden organises the situations, views, confrontations, presences, aiming at introducing into this fragile and precious territory questions, worries, hopes. (ADR & Superpositions, 2016)

In its previous state, the architects find a river that has been channelled linearly by means of concrete elements. The strategy of interest is how the sinuosity of the river is recovered through a rhomboidal checkerboard, into which the flow of the river is introduced (Figs. 28.7 and 28.8). In this way, small meanders are generated, and the flow takes on different speeds at the same time as it favours the cleaning of the water, oxygenation, the contribution of sediments, and the generation of different microhabitats. It can be seen how, over time, the river has acquired the spiral geometry that Schauberger observed.

All natural spaces have been fragmented and separated, not only by cities, but also by infrastructures. The latter have not only constituted physical barriers to existing ecosystems, but have also fragmented urban spaces, often generating residual spaces, housing estates, or slums. These transects represent great spaces of opportunity not only to articulate and connect these urban spaces, but also to offer an opportunity to reconnect natural ecosystems, articulating and bringing nature closer to the urban space and its citizens, while simultaneously revitalising these spaces. There are

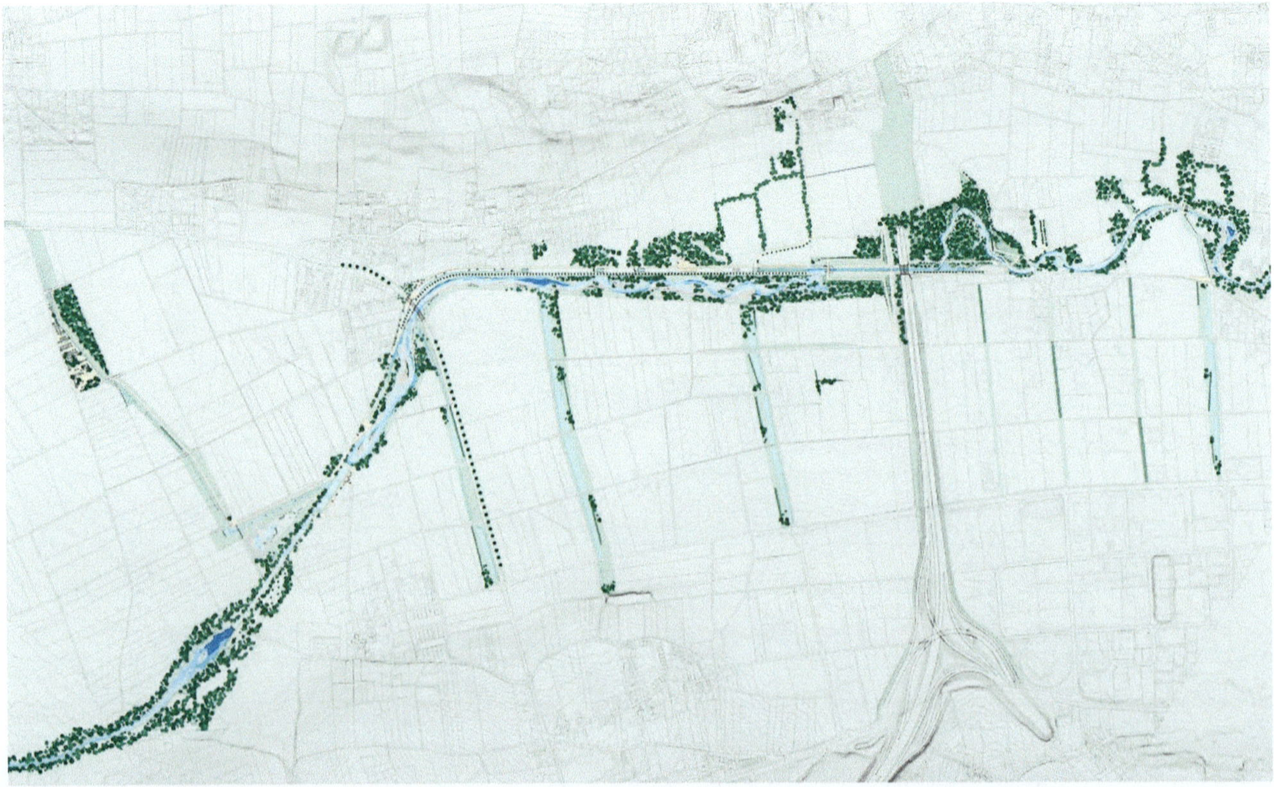

Fig. 28.6 General plan of renaturation of the river Aire, Geneva, Switzerland. 2016 *Source* (ADR & Superpositions,)

different techniques to connect two natural spaces through green corridors (Dramstad et al., 1996).

If the size of this corridor is larger, it allows for an increase in the number of species colonising it. Greater diversity will lead to greater sustainability, balance, and self-sufficiency, as each species brings its own benefits.

Vegetation must be sufficiently dense to conceal and provide shelter for larger species, as they will not use it when exposed to predators. The thickness and width of the corridor will act in the same way, providing shelter and security.

Another factor to consider is the number of alternative paths, or escape routes, available. If there is only one path, predators will be waiting there for their victims, so they will not cross it; conversely, if there are a greater number of possible routes, this will not happen.

All the elements of the network must have a certain continuity in order not to become isolated elements. If there is an isolated element, it will become an endangered habitat and may become an oasis of vegetation in the middle of nowhere. All animal species, even insects, need a certain continuity to be able to circulate within the ecosystem and therefore isolated elements must be articulated in this context in order to contribute to it.

Where isolated elements exist, adjacent spaces are colonised by means of green corridors and the occasional flowering of green areas. This establishes a continuity of the vegetation fabric and the ecosystem which allows species to circulate between the limits of the natural spaces that the fabric connects, defragmenting the natural spaces and at the same time constituting a larger forest area.

In order to improve this fabric, the boundaries of the connecting forests will also have to be treated. The basic strategy for this purpose will be to increase the contact surface, increasing in one way or another the roughness of this boundary (Fig. 28.9) (Dramstad et al., 1996).

When it comes to boundaries, green corridors or isolated plant elements that serve as connecting elements, species diversity should be encouraged (Fig. 28.10) either for self-sufficiency or for protection. There are species that favour each other and protect each other from diseases. Forests do not have monocultures, so one species protects the other, serves as food and provides nutrients (Dramstad et al., 1996).

Simultaneously, different types of plants provide different protection from direct solar radiation, which can be a cause of disease when it is excessive. If we look at the boundaries, tree trunks are protected by smaller trees and shrubs. It is worth noting that many species grow best under the shelter of a tree canopy which screens the light (Dramstad et al., 1996).

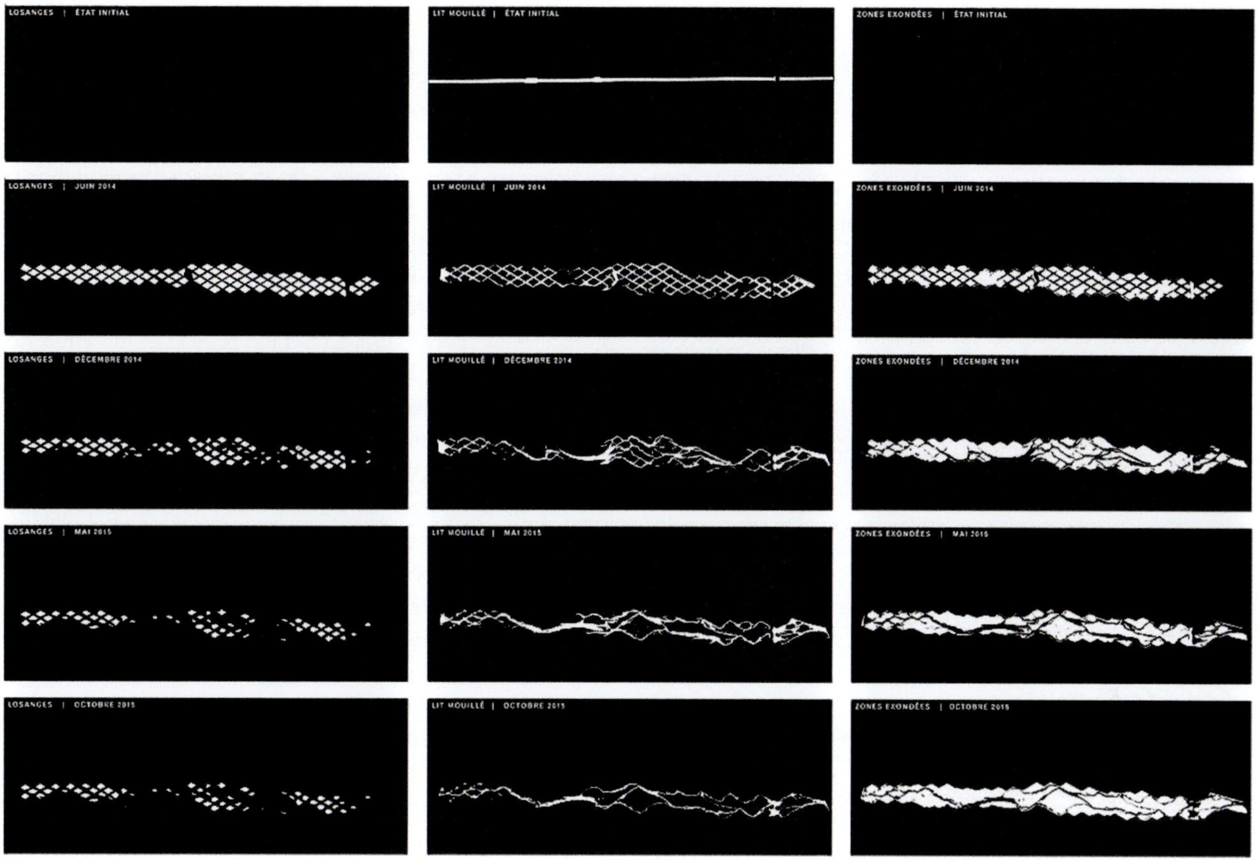

Fig. 28.7 Strategy evolution for the recovery of the natural geometry of the river Aire, Geneva, Switzerland. 2016 *Source* (ADR & Superpositions,)

Fig. 28.8 Evolution for the recovery of the natural geometry of the river Aire, Geneva, Switzerland. 2016 *Source* (ADR & Superpositions,)

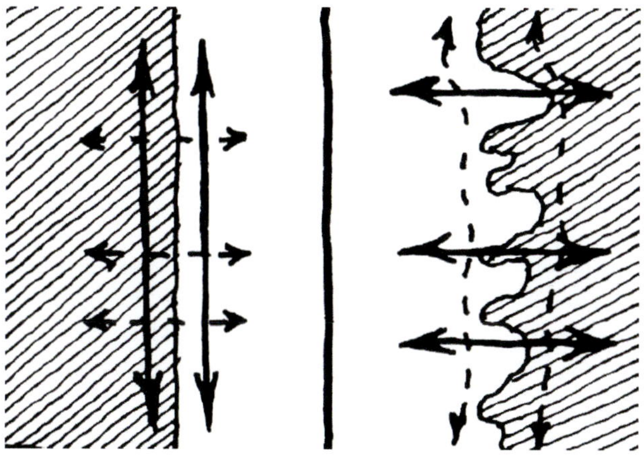

Fig. 28.9 Corridor boundary roughness. *Source* (Dramstad et al., 1996)

The need to cut distances between two points is met by means of corridors for the connection between two natural spaces, otherwise a set of elements constituting a network is necessary (Fig. 28.11).

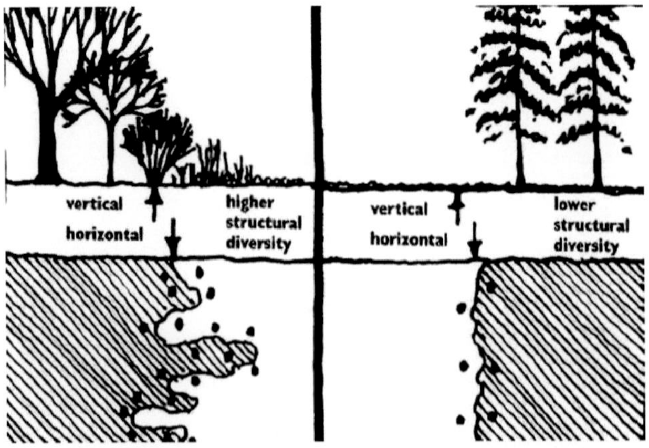

Fig. 28.10 Diversity and thickness of plant species along the boundaries. *Source* (Dramstad et al., 1996)

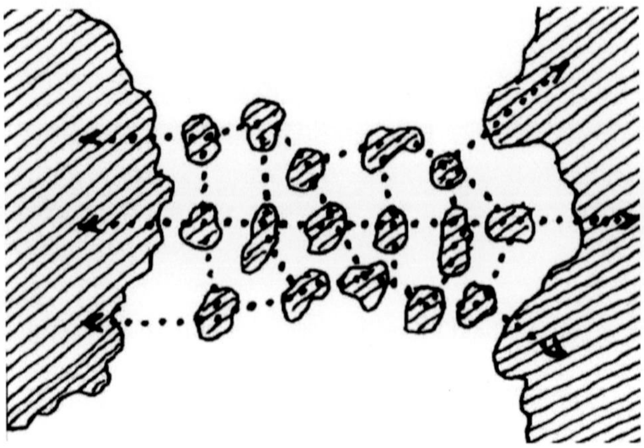

Fig. 28.11 Set of elements that establish a network connecting two vegetation areas. *Source* (Dramstad et al., 1996)

4.4 Superblock Model of Barcelona. Spain

In Barcelona, Spain, when the Eixample Cerdà was conceived, it was done, among other things, with the logic of the car in mind. With the idea of the superblocks, the aim is to improve the quality of life in the city, unifying nine blocks to create a superblock. Thus, new public spaces are created in the pre-existing road, creating spaces for interaction, relaxation, and social activity. It also establishes the tools for a more sustainable mobility, rationalising public transport, encouraging the use of bicycles and walking.

Not only should nature be introduced into cities, but cities should also be designed on a more human and citizen-friendly scale. Improving the environmental quality of urban spaces inevitably means reducing the number of vehicles, both to reduce environmental and noise pollution and to make streets more pedestrian-friendly. This situation leads to the promotion and increase of more sustainable mobility, which consists of increasing public transport (railways, subways, and buses), car sharing, bicycles and, more recently, electric scooters.

In the structuring of cities, until very recently, vehicles were given a lot of consideration and the citizen was in the background, who is the one who should really enjoy the city. In this sense, there is now a wave of reforms in urban spaces around the world where more sustainable mobility systems are being created, giving cities a more human scale. Some examples of this are Barcelona (Spain) with its Superblocks project, New York (USA), with the reform of the High Line and Times Square, Paris (France), with the reform of its most emblematic squares, such as Place de la République, Copenhagen (Denmark) reforming urban spaces, such as Nørreport station, or Israel Plads, which increases the number of kilometres of bicycle lanes and Singapore (Singapore) which favours more sustainable mobility systems and increases its pedestrian areas.

In the example of Barcelona, the Superblocks plan (Fig. 28.12) basically consists of grouping 9 blocks of L'Eixample into a superblock (Fig. 28.13). The streets inside this superblock adopt a more human-scale character, making them pedestrianised, restricting road traffic except in emergency situations, for services and for residents. In the transformation of these spaces, new plant species, spaces for recreational activities, and new urban furniture are introduced (Roberts, 2019). Public transport, buses, is further rationalised and organised in the remaining orthogonal grid. These new approaches contribute to improving the city by providing it with better air quality, both in terms of reducing noise pollution and creating healthier spaces. The more intense use of this new urban space contributes to the improvement of local economic activity and to the increase in consumption and improvement of outdoor living, which is so typical and common in Mediterranean cultures (Barcelona Architecture Walks, 2016).

For the time being, this strategy leads to traffic and parking problems, as the volume of cars remains the same and the public transport network has not been expanded to the extent that this urban transformation requires. Nor do citizens use and appropriate all of this new extension of public space, which means that this strategy is not yet at its best.

5 Results and Discussions

The growth and expansion of many cities has led to the isolation of large cities from nature, which means a reduced quality of life in them. To remedy this deficit, strategies are applied that revitalise or renaturalise existing physical elements such as rivers, canals, seas, corridors, large avenues, giving them a higher character than they currently have. This

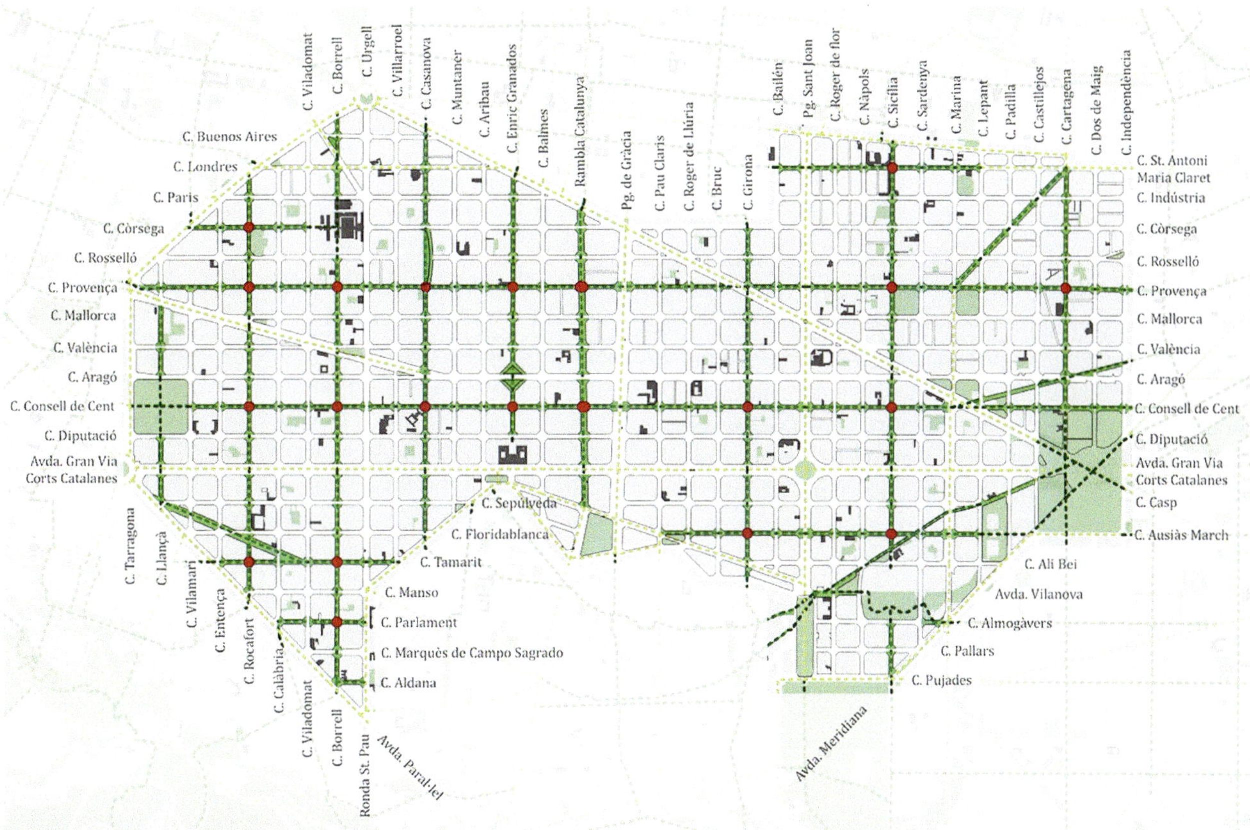

Fig. 28.12 Restructuring of the road model according to the Superblock model of Barcelona, Spain. Project not executed. *Source* Ajuntament de Barcelona

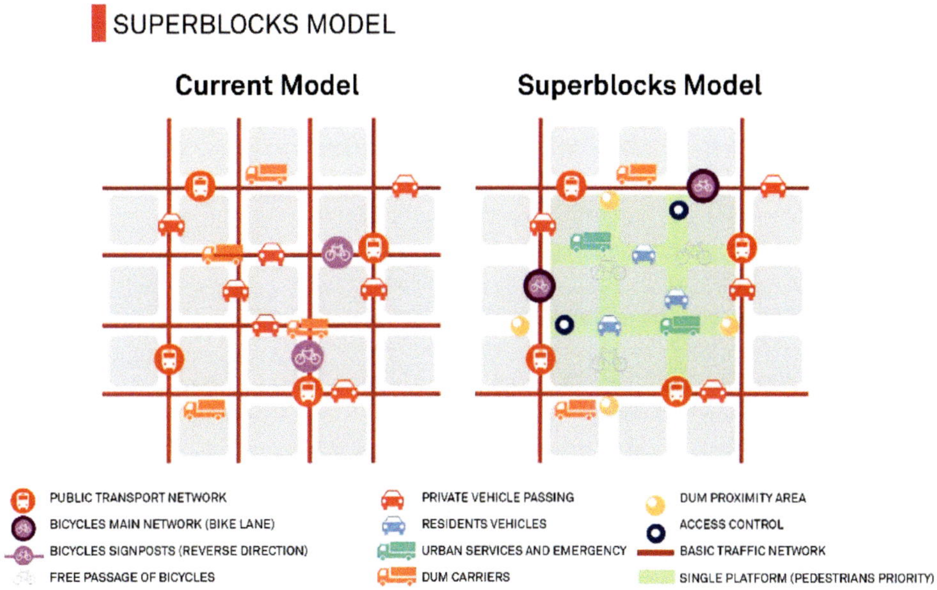

Fig. 28.13 Restructuring of the road model according to the Superblock model of Barcelona, Spain. Project not executed. *Source* Ajuntament de Barcelona (Barcelona Architecture Walks, 2016)

means increasing public space with nature as a protagonist to generate greater comfort and health for all its inhabitants.

The vision regarding cultural heritage issues has changed frequently, particularly since the 1960s. The notion of "a future for the past" in the 1970s changed to the concept of "*a past for the future*" in the 1980s and now the slogan "*making the past part of our future*" has been created (Habibi, 2002, pp. 3–5).

> The roots of sustainability with urban revitalisation are different but share commonalities. Revitalisation and sustainability have equivalent meanings and are often used to reduce the need to manage the world's properties: a) to ensure long-term harmony between man and nature / the built environment and b) to achieve a relentless improvement of the environment and the quality of human life. (Rodwell, 2007)

The Master Plan of Asunción in Paraguay offers us the ambitious strategy of uniting the different areas created due to the expansion of the city with the historic centre through ten actions, where natural elements acquire great importance both in the generation of new natural public space and in the construction of new buildings that act as catalysts. The proposal attempts to avoid the decentralisation of the historic centre due to the generation of new centralities or poles of attraction that are generated in the periphery with the Paraguay River as a limit and the Bay of Asunción with the José Asunción Flores road that separates the two.

> "At a time when many cities around the world are beginning to look for new ways to approach their development, Asunción's Master Plan aims to make the historic city centre a benchmark for other cities to follow as a pioneer in sustainable urban development. The idea is to create an "action plan" based on the best practices we have gleaned from hundreds of successful real-world initiatives around the world" (Ecosistema urbano, 2014). "In an effort to raise awareness and reconcile the city with its environment and natural heritage, our proposal aims to integrate sustainable urban development as a cross-cutting approach applied to all actions. Inspired by well-known initiatives, we propose the creation of a series of dynamic, green and ecological corridors, connected by urban catalysts: buildings with hybrid, mixed-use and flexible programming" (Ecosistema urbano, 2014). "The creation of an active green waterfront (floodable park) will respect the existing topography and become the lung of the city, using natural processes to facilitate air and water purification." (Ecosistema urbano, 2014)

Another example is in Florida West Palm Beach, where the beach is unified with the generation of a large natural public space that produces a blurring between the buildings and the sea. This large urban-natural area encourages citizens to walk and enjoy nature itself and its microclimates in an itinerary that connects the city with the beach. Previously, they had turned their backs on each other and therefore there was no interaction between the two, so this urban structure has achieved cohesion and unity.

> The new waterfront of the 'Open Shore' project, West Palm Beach, Florida hosts a wide range of outdoor activities and public spaces that enhance the urban experience and multiply the possibilities for interaction with the water. (Ecosistema urbano, 2016)

In the renaturation of the river Aire in Geneva, the main concern was to prevent future flooding and simultaneously generate natural spaces for social use along the course of the river that would structure the city and give the river itself a notoriety. In this way, the river becomes the backbone structure of the city with different scenarios along its course.

In the project, its authors Georges Descombes and Atelier Descombes and Rampini indicate the following parameters:

> "Our proposal had three objectives: flood defence, renaturation, and the creation of a new social space. We wanted to treat the flow of the river as natural architecture and the flow of the canal as designed architecture". It is a protection system, but we wanted to transform it into a kind of garden. (ADR & Superpositions, 2016)

The Barcelona superblocks project is being developed in the Cerdà urban expansion area and its objective is to eliminate space for vehicle traffic and adapt it as a public space for citizens, converting it into a green space that is as natural as possible. Therefore, certain streets in the Ensanche will be reconverted into green axes which will make access to the city by car more difficult. This will lead to a significant reinforcement of the public transport network in the different towns in the metropolitan area of Barcelona and in the cities which do not have railway lines.

> The Superblocks project in Barcelona is a territorial unit smaller than a neighbourhood but larger than a residential block, with quiet streets where urban and environmental activities take place. Compact, efficient spaces that have a positive impact on the lives of the people of Barcelona and are also self-sufficient in energy consumption. The aim is to improve the quality of life and the conditions of the residents, through the rational management of natural resources and public spaces, through the active participation of all those involved and by fighting against inequality and social exclusion. (Barcelona Architecture Walks, 2016)

With the four examples given above, there is a common denominator which is to increase the natural elements and enhance them in the interiors of the cities to provide a higher percentage of comfort and health to all their citizens. At the same time, and perhaps unconsciously, there is a buffer between the fabric of the city and its periphery, bringing into play and articulating the built and natural elements.

With the increase in public space shaped by natural scenery, new spaces for bird habitat are also being opened up. These newly applied strategies for cities will buffer both the expected future population growth and climate change. At the same time, these needs will optimise natural elements such as rivers, seas, forests or any geographical features within or on the periphery of cities.

6 Conclusions

In order to improve the well-being of cities today and in future, mainly due to climate change and global warming, it is desirable to implement strategies to increase their sustainability and resilience. These strategies are based on the application of a series of corrective parameters in architectural development strategies and the reconversion of the existing urban fabric. Transformation to this end is already being implemented in cities around the world. The strategies to alleviate and reduce heat, environmental and noise pollution in large cities, which are the most affected, are as follows:

a. Infrastructures: Covering, burying, or perforating large infrastructures. Eliminate the physical barrier they constitute, sewing and connecting the surrounding fabric and generating new public spaces. It is important to prevent these layouts from breaking up urban fabrics and segregating ecosystems and preventing the continuity of the biophysical matrix. If these barriers are hidden or transformed, or if pedestrian access spaces are generated that articulate both sides of the roads, the fauna habitat area will be expanded, not only in forested areas.
b. Green corridors: Generate a green fabric that connects with the adjacent forest areas based on different hierarchies of green corridors that intersect with the urban fabric and constitute a network that provides continuity to the biophysical matrix. In this way, natural ecosystems will have a natural continuity within cities. The fusion of nature with the city will lead to the achievement and enjoyment of healthier urban spaces, both in terms of air quality, the reduction of vehicle noise, the reduction of stress when listening to or observing nature, and the stimulation of our creativity when observing natural processes.
c. Rivers: Renaturalise river areas and their surroundings. Their routes will be a source of cohesion and centrality, offering a notable increase in the quality of life for citizens by connecting all the plant ecosystems (mountains, forests, and sea) and extending the radius of action of the biophysical matrix and its area of influence. It will provide the different animals with greater access to water, a basic element for life, and greater spatial freedom to feed and move around the territory. In addition, the fact of regenerating this type of space will allow some plant species to return to their place of origin, species which in some cases are medicinal, such as the Salix Alba, which will contribute to the recovery of biodiversity.
d. Sea fronts: Renaturalise the coasts and their nearby aquatic spaces by creating protection areas for aquatic and terrestrial ecosystems which will be articulated between them. It is necessary to build new natural reefs to prevent beach erosion and at the same time create aquatic areas protected from fishing activities and guarantee protection zones for marine ecosystems. These fronts should generate electricity through tidal power plants and should ensure the control of a possible rise in water levels. This will also ensure greater sustainability of marine ecosystems and the food they provide. By creating artificial reefs, protection zones will be created that will prevent fishing from depleting aquatic resources, thus promoting greater self-sufficiency and protection of marine biodiversity.
e. Agriculture: Conversion of agricultural land into forest land. Agriculture is introduced into the city and as a result, the emissions generated by transport are reduced, leading to the transformation of the city into a self-sufficient and productive type of city. All areas susceptible to generating food plantations will be reconverted to fulfil this function, such as streets, parks, gardens, and will be complemented with agriculture in height; façades, roofs, mezzanine spaces, terraces, balconies, patios will become productive urban spaces, which will be complemented with the rest of the activities and uses of the city. Simultaneously, the function of forests and their ecosystems can be restored with respect to the pressure that agriculture exerts on them and it will be easier for forests and their ecosystems to recover their natural space. With the application of these new strategies it will be easier for the biophysical matrix not to be reduced or confined to small territories, which will facilitate the recovery of different species, favouring the recovery and sustainability of biodiversity, bearing in mind that they are the source of our food, medicines and resources. It is necessary to safeguard and protect biodiversity to obtain a better future, and architecture and urban planning are very important actors in achieving this objective.
f. Water resources: Every coastal city must produce fresh water for human consumption, so that the city is self-sufficient in this aspect. Preferably passive systems will be used, e.g. by evaporation or in those systems that require the least energy input. Any energy input for water desalination should come from renewable energy sources. The main process for obtaining drinking water will be condensation through evaporation of water, produced by solar energy. The gaseous water will be collected and cooled, forcing it to condense and return to its liquid state, separating it from the solid particles dissolved in it. In this way, purified and potable water will be obtained without the need to introduce complex systems and large energy inputs, so that the energy requirements of the system are practically equal to the same amount as a booster pump.

The final conclusion is the integration of nature and the ecosystem that sustains us in the development of human activities, which will not depend on a single strategy or type of urban, landscape, building, or productive intervention, but rather on a series of elements and strategies that act jointly to guarantee the resilience and sustainability of urbanity and nature, guaranteeing the balance between both systems.

References

ADR-Atelier Descombes Rampini & Superpositions. (2016, June 30). *Renaturation of the River Aire*. LANDEZINE. http://landezine.com/index.php/2016/06/renaturation-of-the-river-aire-geneva/

Alberti, M. (2016). *Building cities that think like planets*. University of Washington Press.

Aldo, L. (1949). *A sand country ALMANAC*. Oxford University Press.

Barcelona Architecture Walks. (2016). *Superblocks: A sustainable strategy for regenerating the city*. https://barcelonarchitecturewalks.com/superblocks/

Beatley, T. (2011). *Biophilic cities*. Island Press.

Chan, J. (2012). *Singapore: biophilic city* [Video]. Youtube. https://www.youtube.com/watch?v=XMWOu9xIM_k

Davoudi, S., Crawford, J., & Mehmood, A. (2009). *Planning for climate change*. Earthscan.

Dramstad, W., Olson, J. D. & Forman, R. T. T. (1996). *Landscape ecology principles in landscape architecture and land-use planning*. Island Press.

Ecosistema urbano. (2014). *Plan Cha. Asunción (Paraguay)*. https://ecosistemaurbano.com/master-plan-for-the-revitalization-of-the-historic-downtown-of-asuncion/

Ecosistema urbano. (2016). *Open shore*. https://ecosistemaurbano.com/es/open-shore/

Ecosistema urbano. (2019a). Master Plan in Asunción (Paraguay). *Arquitectura Viva Proyectos, 92*, 16–19.

Ecosistema urbano. (2019b). 'Open Shore', West Palm Beach (United States). *Arquitectura Proyectos, 92*, 8–11

Habibi, M. (2002). *Urban regeneration*. Tehran University Publication.

Jover, M., & Wall, A. (2019). *Ecologies of prosperity*. Applied Research and Desing Publishing.

Ng, E. (Ed.). (2010). *Designing high-density cities*. Earthscan.

Roberts, D. (2019, April 9). *Barcelona wants to build 500 superblocks. Here's what it learned from the first one. Vox*. https://www.vox.com/energy-and-environment/2019/4/9/18273894/barcelona-urban-planning-superblocks-poblenou

Rodwell, D. (2007). *Conservation and sustainability in historic cities* (1st ed.). Wiley-Blackwell.

Rosenberg, E. (2019). *Before and after. Both. The revitalization of the Aire River, Switzerland*. Landscape Architecture Magazine. June 2019.

Gijrokastra Regeneration Through Smart Technologies

Luca Lezzerini

Abstract

The paper explores the use of digital tools in urban regeneration, the adaptive reuse of existing buildings, and the role of technology in co-living spaces. It is the evolution of previous research developed in 2021 and completes its original strategical picture with further considerations. The research question is how to reactivate Gijrokaster's economy and development by leveraging existing resources, applying modern smart technologies, and focusing on a holistic perspective driven by the urban design and planning of the city. The results of the research are the definition of realistic urban planning and design strategy that could be applied to Gjirokaster. Other results are the application of smart technologies into an urban development or recovery strategy, and the creative and adaptive use of existing buildings and space to introduce new factors in city development. Future results could be the abstraction of the used methodology to define a general approach to identify space for improvement or recovery in crumbling cities. The methodological approach of the paper starts from the analysis of the Gjirokaster urban context, its history and its potential, and provides a strategy to generate a virtuous loop that leads to Gijrokaster rebirth, leveraging both traditional resources (like tourism and agriculture) but also leveraging other local resources (university, hospital, industrial area, natural resources, …) to create a synergy capable to start, in a sustainable and manageable way, the rebirth of the city. In this approach, the urban layout and its adaptation to the strategy is a key element, but also how to gather starting equity and maintain an economic flux is another.

Keywords

Essence • Urban planning • Revitalization planning • Strategic planning • Smart technologies

1 Introduction

Albania is suffering a general shrinking of population. Recent statistics (INSTAT, 2022) evidence that deaths are increasing while the birth trend is slowing decreasing, the fertility rate is decreasing too, and life expectancy is shortening. Population structure is increasing over 65 and reducing in both ranges 0–14 and 15–64. Tiranë's prefecture is the most populated (about 33% of the overall population), followed by Durrës and Fier with 10% each. The least populated prefecture is Gjirokastër with 2.0% of the population.

The city of Gjirokastra is currently well known for its UNESCO accreditation as World Heritage Site. Today, most of its economy has been based on tourism, but results are not as expected: Gjirokastra is a shrinking city that needs a rebirth strategy or will die.

According to some studies (Bickert & Göler, 2012), tourism is not enough because tourists visit the city in one day and do not stay for the night and do not remain in the city for many days.

After analyzing the current state of Gjirokastra's economy, a strategy to let the city to rebirth has been defined, preserving existing economic positive factors and creating an economic flywheel that will attract money from both Albanian people and outside it.

The main idea behind this strategy is to focus on health, building in Gjirokastra a set of healthcare resources that can take advantage of existing accommodations infrastructure and touristic resources. To develop a such idea, the strategy focuses on residential kinds of healthcare like geriatrics and rehabilitation services. The healthcare services should be

L. Lezzerini (✉)
Polis University-Tiranë (AL), Tiranë, Albania
e-mail: luca_lezzerini@universitetipolis.edu.al

A. Battisti et al. (eds.), *Greening Our Cities: Sustainable Urbanism for a Greener Future*,
Advances in Science, Technology & Innovation, https://doi.org/10.1007/978-3-031-49495-6_29

designed to attract foreign people, generating a new kind of medical tourism. Albania is currently the target of medical tourism, mostly for dental (Nexhipi, 2018) and esthetic treatments, and this kind of tourism generates an economic positive flow because it requires the tourist to stay for many days in Albania.

Gjirokastra has an ideal configuration to support the development of a "distributed retirement house" for foreign people. The main idea is to attract retired people to spend their retirement living in Gjirokastra (Ewart, 2013; Gonzálezet al., 2009), like already happened, e.g., in the "sunset migration" for Portugal (King, 2013). The climate, the lower cost of living, and the natural and cultural resources can be very attractive for average-class retired people from middle and northern Europe. But attracting them is necessary to create good retirement services.

The most important of these services should be:

- Health services for older people (geriatrics and rehabilitation);
- Cultural opportunities;
- Free time management.

In this paper, a specific subset of them is analyzed considering the "distributed retirement house" (DRH) concept.

2 Research Question

Starting from previous research (Lezzerini, 2021) with the same context, the author has considered the possibility of developing a more general approach based on the concept of a smart city and deploying it onto Gjirokastra but also develop a creative process that can be applied to other similar cases.

The research questions that have driven the research were:

- How can a smart city approach be used to leverage existing resources in Gjirokastra and create a virtuous loop to let the city grow harmonically and in a sustainable way?
- The smart city-based approach cited above, can be used as a general approach for a general case of a city to be reshaped in a holistic perspective.

After the research, it has been clear that the proposed approach, as evidenced in the title, is mostly based on smart technologies that on a whole application of the smart city paradigm. As shown later, smart mobility and smart medicine are some of the keys to the proposed strategy. Nevertheless, the term smart city will be used because:

1. The used ontology has been designed to model a smart city or some portions of it (and this last one is the considered case);
2. Moving Gjirokastra toward a smart city will be easier and will also empower existing resources like tourism, internal market, university, and agriculture, so it should be considered a secondary objective.

For these reasons, in the text, the term smart city will be used, but describing in this paper only the first step which is based on some smart technologies, as evidenced in the title, than on the whole application of the smart city paradigm.

3 Expected Results

The expected results of the research were the definition of a strategic approach to drive the Gjirokastra renaissance and the development of a reusable approach, as more formalized as possible, to use the smart city as a tool for easing existing cities' development and growth, even if in this case the tool will be applied only partially, as explained above.

The developed approach can be generalized to become a general-purpose methodology applicable to any city that needs "improvement," to evaluate the feasibility of the smart city paradigm to "improve" it and, in the affirmative case, proceed with its planning and deployment. This general approach will only be outlined in this research because it is the subject of future research activity. Although in this paper, the smart city paradigm is used only for some smart technologies, and it can be considered the final objective (beyond this paper) and can be considered as a strategic approach to regenerating urban structures or whole cities. Moreover, the proposed approach based on Ramaprasad's framework can be applied to any other case of urban regeneration by "smartifying" the city.

In this paper, this "smartification" is done through the application of some smart technologies that are essential, in the specific case, to attract economical targets (elder people) giving them high-quality services, but this should be read as a first step in complete "smartification" of the city.

4 Methodological Approach

The methodology followed aims to the improvement of previous research (Lezzerini, 2021) redesigning its results through a more formal approach based on a smart city framework, also getting different, better results.

The methodology has been subdivided into the following steps:

- Literature review of smart cities frameworks;
- Literature research to evaluate forms of elder people organization in distributed accommodation;
- Analysis of the urban structure of Gjirokastra and of its neighboring;
- Structuring, through the selected smart city framework, the elements that should be considered at a high level;
- Identify available resources in Gjirokastra;
- Identify communication needs (information and transport needs);
- Identify suitable technologies (available and forthcoming);
- Layout the framework illustrative elements onto the Gjirokastra city structure;
- Define change proposals to the Urban Plan to fit the above items.

In the literature review of smart cities frameworks task, literature research has been done to find a suitable framework that can be used to formalize the approach to redesigning Gjirokastra city. No change has been done to such a framework: its improvement will be considered in further research. This framework can be applied to any smart city but can also be applied in this case, as explained later.

After finding the framework, second literature research has been conducted to identify and evaluate different forms of elder people organization in distributed accommodation, also considering the provision of required services.

An on-site survey has been done to develop an analysis of the urban structure of Gjirokastra and of its neighboring, identifying physical spaces and communication routes.

The following step was structuring, through the lens of the selected smart city framework, the elements that should be considered at a high level. These elements have been called, as stated in the selected framework, "illustrative elements" and will be discussed later.

Having focused on the illustrative elements of the smart city, a deep analysis has been done to identify available resources in Gjirokastra that could be used for them, also defining which resources are missing. Then communication needs have been identified, meaning with the word communication both information and transportation needs. Then, a list of suitable technologies (available and forthcoming) that could be used in illustrative elements implementation has been done.

Having defined all the above needs, the framework's illustrative elements have been mapped onto the Gjirokastra city structure to define different possible layouts to support the realization of illustrative elements.

The last step was defining change proposals to the urban plan to fit the above items.

5 Research Problem Elements

5.1 Gjirokastra's Rise and Fall

Gjirokastra has always been and is yet today, one of the most beloved cities in Albania (Fig. 29.1). The city is famous for its historical aspects, its peculiar traditions, and its enchanting landscapes that range from mountains to plains to hills.

Gjirokastra, since its foundation, has been one of the main settlements in southern Albania and has been representing one center of development for the entire region. Provided with important touristic places, a relevant industrial area, a university, and a hospital and located along an important communication path, it has been one of the most important cities of Albania. In the last thirty years, the city has fallen into a recession that has involved both the economy and the population. This dynamic has led the city to shrink, mainly following the decline of the industrial area but also enhanced by the flow of population toward bigger cities.

So, in the last three decades, Gjirokastra has entered into a vicious circle, typical of the so-called "shrinking cities."

The de-industrialization phenomenon that has hit the entire region has particularly injured Gjirokastra, driving it to starvation (Fig. 29.2). As a consequence of this situation, many people emigrated abroad (typically Greece, Europe, and the USA) while many others moved to the area of Tirana or other Albanian bigger cities. Existing villages of the area have been seriously emptied by this migration. Most of the emigrated people have been young. So, the result was a collapse in the economy and in demography.

The inclusion of Gjirokastra in the UNESCO list has been seen as an important step toward the rebirth of the city, and important efforts and resources have been invested in it, but the results are clear: they were not enough. Although investments in infrastructure, agriculture, and in tourism have been significantly increased since the UNESCO label, they are not enough and tourism itself is not capable to enable a real resurrection of the economy, currently the only perspective to re-attract young people. Today Gjirokastra is a well-known tourist target, but its economy struggles to restart.

The author has considered three different scenarios of rebirth: healthcare fueled rebirth, advanced agricultural pole, and modern tourism attraction. The three scenarios are subjects of three different research paths and will converge into a common vision. In this paper, only the first one will be approached, starting from already developed results (Lezzerini, 2021) and rewriting them according to a different approach based on the smart city paradigm, in a partial, initial, implementation, limited to allow distributed medical and geriatric services. Without the use of the selected smart

Fig. 29.1 Gjirokastra as a UNESCO World Heritage city (Image of the author)

technologies, it will not be possible to convince elder people to access the distributed retirement house, for example.

5.2 The Smart City Paradigm

Even if a precise definition of a smart city is yet to be agreed upon, many frameworks have been developed to describe the elements of a smart city. Investigating these frameworks and choosing one has led to the definition of a paradigm, in its sense of a framework of ideas, that can be used to drive the targeting of specific objectives.

Many smart city frameworks have been considered (e.g., Azizalrahman & Hasyimi, 2019), and other smart city concepts and business models have been evaluated (Townsend, 2015).

The chosen smart city framework was Ramaprasad's (Ramaprasad et al., 2017) because it is flexible and represents an ontology to describe elements of any smart city. Such a framework is a trial in defining a unified view of the smart city. In this paper, no changes have been done to this framework. In his vision, Ramaprasad states that the outcomes (i.e., the values provided by the smart city) are:

- Equity is the capability of the city to give each one the same opportunity, overcoming the differences between different people.
- Livability is a quality that if owned by a place implies that people will live and work there. It is usually a harmonic distribution of housing, services, healthcare, jobs, education, and a general sense of respect.
- Quality of life is when inhabitants of the city perceive that their needs, their goals, their expectations are provided by the cultural, values, and social context where they live.
- Resilience is the ability to return to a previous acceptable condition after an event that disrupted this state
- Sustainability is the ability of the smart city to indefinitely maintain itself operational at a predefined level

Fig. 29.2 Collapsing industrial area (Image of the author)

According to him, the smart city can be defined as a function of two elements, smart and city, that represent the ability to have "intelligent processes" (Smart) and the involved persons with their values and needs (city).

The city is then defined by Ramaprasad as composed of stakeholders and values (i.e., the outcomes above described).

For Ramaprasad, stakeholders have been classified as citizens, professionals, communities, institutions, businesses, and governments. The values, as already stated above, are the five outcomes. Even if Ramaprasad considers only these values, as will be seen later, it is possible to both use different values or define a deeper detail level.

The smartness of a city, according to Ramaprasad, can be defined as its "capability to be able to intelligent sense and respond through semiotics." Semiotics is defined by Ramaprasad as "an iterative process that generates and applies intelligence." Semiotics, in his vision, can be data, information, and knowledge. Data is the recording, in some encoding, of elements coming from the city and from the citizens. In Ramaprasad's vision, it is defined as the "symbolic representation of sensations and measurements." The term symbolic is referred to the chosen encoding. Information is the reduction or the complete removal of uncertainty which is based on data, but it depicts connections between data objects, connections that impact the uncertainty reducing or erasing it. For these reasons, Ramaprasad defines information as the "relationship among the data elements," relationships that contribute to the reduction/cancellation of uncertainty. But information is not enough. The information states a connection, but it does not discover the "meaning" of the connection. To make clear this meaning, the "sense" of the connection, knowledge is defined by Ramaparasad as the "meaning of the relationships among the data elements."

According to Ramaprasad, the "smart" is defined again as a function of four variables:

$$\text{Smart} = f(\text{Structure Functions Focus Semiotics})$$

where:

- The structure is defined by Ramaprasad as "the structure required to manage the semiotics." It is the set of elements that are needed to process, transfer, and store the semiotics. It can be further detailed as:
  - Architecture, that is the general architecture needed;
  - Infrastructure, that is the infrastructure required, considering both types of infrastructure, physical infrastructures and virtual infrastructures;
  - Systems, that can be based on computers, social behavior, and paper;
  - Services, that are provided by the above systems;
  - Policies;
  - Processes;
  - Personnel, i.e., people who act in the processes.
- Functions are defined by Ramaprasad as "the functions required to manage the semiotics". They are elements with an input and an output and can transform their input into something else (the output). Both inputs and outputs are semiotics. They are, according to Ramaprasad:
  - Sense is the ability to detect and gather semiotic elements;
  - The monitor is the capability to watch semiotic elements over time;
  - The process is the way of processing sensed elements;
  - Translate allows semiotic elements to be converted into actions (through actuators) of control (i.e., ensuring the persistence in a given state even if under perturbations;
  - Communicate is the ability to transmit, and exchange, semiotic elements
- Focus is the last independent variable and represents the aim, the target of the semiotic elements processing. Ramaprasad defines it as the "focus of intelligent sense and response," and this is the reason for the name. It represents different dynamics available in a city and can be further detailed as:
  - Cultural focus;
  - Economic focus;
  - Demographic focus;
  - Environmental focus;
  - Political focus;
  - Social focus;
  - Technological focus;
  - Infrastructural focus.

All the framework elements can be used as an ontology to describe the smart city in terms of "components" called "illustrative elements." These illustrative elements are generated by combining the various elements of the framework. In this way, it is possible to generate a high number of illustrative elements that represent single aspects of the smart city. A smart city will own (i.e., will have implemented) some or many, even all, of these illustrative elements.

A single illustrative element can be generated by combining the framework's elements according to the following schema:

<Structure > to <Functions > + <Focus > + <Semiotics > by/from/to <Stakeholders > for <Outcomes >

An example of an illustrative element obtained by applying the above schema could be "Services to Monitor Demographic Data from Citizens for Resilience."

The structure variable has 7 possible values, the functions variable has 5, the focus variable has 8, the Semiotics variable has 3, the Stakeholders variable has 6 and the Outcomes variable has 5. The possible combinations according to the illustrative elements generation schema is 7 * 5 * 8 * 3 * 6 * 5 which is 25,200 different illustrative elements. The real number is higher due to further possible combinations (e.g., the from/to alternative, different ways to implement the same variable, …).

6 Applying the Smart City Paradigm to Gjirokastra

6.1 The Previously Defined Strategy

As developed in previous research (Lezzerini, 2021), one flywheel for the Gjirokastra's economy could be the creation of a city focused on healthcare, especially on elder people's retirement, and on long time rehabilitation. This strategy has been derived by comparing the Gjirokastra reality with other similar cases (De Florio, 2007 and Rothuizen, 2011), and also sharing ideas with the well-established medical tourism (Kruja, 2016) that already happens in Albania.

According to this strategy, attracting elder people, especially retired people alone and with no children or with far children, will create a new economy that will input equity into the Gjirokastra economy and generate a virtuous loop also thanks to the university and the hospital.

Some urban planning changes have been proposed in this research and other considerations have been done.

This paper will start from such results and will demonstrate how the application of the smart city paradigm based on Ramaprasad's framework can be very effective in designing and further leveraging this strategy.

The idea that will be demonstrated in the next sections is that the previously defined strategy can be completely redesigned, keeping it intact in the contents but changing in the formal description, if the considered approach will be the one of a smart city. In this way, it is possible to define a two-step approach:

1. Implement the already-defined strategy using formal language (an ontology to be more clear) to describe it;
2. Upgrade it to a smart city (beyond the scope of this paper) using the same ontology, with the target to make Gijrokastra more efficient and more attractive.

What we will do in this paper, then, is to apply the smart city framework defined by Ramaprasad to check if it is suitable to describe even the first step.

6.2 Re-Discovering the Strategy

In this section, starting from the vision depicted in the above-cited research from the same author, the smart city paradigm will be deployed, using Ramaprasad's framework, and applied to define, first, the elements of the needed implementation and, then, the strategy to implement them.

The vision is a city where elder people from abroad can be attracted to live. Some factors are against this attraction: the language barrier, the foreign culture, the risk of solitude, and some others. To attract such people, the vision must enforce a high quality of life (QoL). Quality of life is one of the outcomes of Ramaprasad's framework and will be used, in this article, to show how to use it in designing both the elements and the strategy.

The first step is to detail the QoL into more specific factors. To do this, a wide literature review has been done to find a suitable model for quality of life that could be applied to Gjirokastra's case.

In our QoL modelling, we need to provide enough detail to define the following levels of QoL:

- QoL for elder individuals (needed to get maximum performance for each foreign resident elder);
- QoL for the community, to evaluate the overall improvement given by the smart city paradigm application;
- This second level will be considered, for the scope of this paper, with three different aggregations, but it can be also composed in other different ways:
- The foreign elder people community (to evaluate overall satisfaction);
- The original Gjirokastra's citizens (to evaluate the improvement in their QoL);
- The whole community (foreign elder people and Gjirokastra residents) to assess the global improvement in QoL.

Further different aggregations can be considered, and also additional metrics can be used to measure and evaluate the increase in QoL, but for the sake of simplicity, it will be not considered in this analysis.

After researching in literature, two models have been selected (Ratz et al., 2022; Ruzevicius, 2007). Although Ratz's model was more interconnected, Ruzevicius's has been chosen for its flexibility, which is needed to apply it to the described different community clusters.

Ruzevicius (Ruževičius, 2012) partitions the QoL into eight factors (Fig. 29.3):

1. Health;
2. Material aspects;
3. Spiritual and psychological;
4. Working life;
5. Life in the family/community;
6. Social relationships;
7. Leisure;
8. Environment (security, safety, legal, social, work, economic, political, and juridical environments).

These factors are depicted in the following image.

The eight factors have been considered how targets to be overcome or improved too, consequently, overcome or improve the overall QoL.

To design the implementation of these factors, the smart city framework has been applied. For space reasons, it is not possible to report all elements generated through the framework that can be used to design the QoL, so only a part of them will be provided, but this keeps valid the main reasoning.

For each factor, one or more illustrative elements have been generated through the framework.

Let us start, for example, with the health factor, which is "health quality of life" and can be understood as the physical state of the individual. It can be split into more detailed components like health, working load, physical endurance, and nourishment.

The first consideration is that some of these components do not apply to some cases. For example, an elder retired person will not be considered from the point of view of the workload, although some recreational activities can be compared to a workload.

The second consideration is that, again, some of these components can be more critical for some individuals than for others. For example, the health state is usually more critical in old people than in young people and will need more attention in the first group than in the second.

A third consideration is that the different levels of modelling for QoL (community, individual, …) described above will be analyzed starting from the group and then moving toward a more fine detail.

To manage, for example, the health factor of the individual, a set of illustrative elements can be designed through the framework. A result of such a design is given in Table 29.1.

In this example, a set of illustrative elements have been derived from the framework using its generations' formula.

Fig. 29.3 QoL model (*source* Ruzevicius, 2007)

Table 29.1 Example of QoL illustrative elements

Element	Description
Infrastructure to sense environmental (health) data from citizens for QoL	An infrastructure must be provided to measure the health parameters of the citizens (both resident and from abroad), to use them for QoL
Infrastructure to monitor environmental (health) data from citizens for QoL	An infrastructure must be provided to monitor the health data of the citizens (both resident and from abroad), to use them for QoL
Systems to monitor environmental (health) information from citizens for QoL	Systems are to be provided to monitor information about the health of citizens
Systems to process environmental (health) information from citizens for QoL	Systems are to be provided to process information about the health of citizens
Policies to sense/monitor/process/communicate data/information from citizens for QoL	Policies are needed regarding the privacy rights, reliability, and safety of the overall management system regarding health
Processes/personnel to monitor/process/communicate/translate environmental (health) data/information/knowledge from citizens for QoL	Processes and personnel are needed to actuate the health management system that handles data, information, and knowledge about health gathered from citizens
Architecture to process environmental (health) data/information/knowledge to citizens for QoL	Physical spaces and proper forms of the space must be arranged to process the data, information, and knowledge to provide healthcare to citizens
Architecture to communicate social data/information /knowledge to (between) citizens for QoL	Physical spaces where people can socialize and communicate. In this case, it is possible to evidence of the lack, in the framework, of a "between" preposition in addition to from/to/by

All these elements are needed to ensure an appropriate level of healthy quality of life for groups of citizens. Obviously, in these elements, the individual perspective is already present, even if they have been considered at the group level.

These elements are not all those needed but they are enough to evidence some issues.

The first issue is that the selected framework has a level too high: in fact, it has been necessary to adapt the environment variable, and specifying it was related to health aspects because there was no applicable focus in the original framework statements. This means that, even if in this case the framework has not been modified, an evolution of it is needed to better represent the possible focuses of a smart city.

A second issue is that semiotics is, in the framework, limited to data, information, and knowledge. But action can be associated with the semiotic element. For example, some semiotic elements are gathered, and others are acted, i.e., they are used to generate actions. These actions can be realized by actuators (mechanical, display, acoustic, …), and others are provided by personnel. The framework uses the translate function to state this concept because this function means the translation of the semiotics into control/action.

A third issue is about the "architectural" component of the framework. Really, this component should be able to map urban planning illustrative elements, and this means that, even if the term "architecture" should be considered in a wide sense, some other ontological elements are missing, e.g., "architecture to provide movement data/information/knowledge".

This last issue is the most important for the scope of this paper and will be further developed in the next section. But, before proceeding with it, it is important to explicitly state the roadmap to be followed.

After having defined all the illustrative elements, they are the detail of the vision, that is the final expected result, where all these illustrative elements will be implemented at the expected grade. To define the grade of the implementation (needed and then realized during the development of the smart city), an approach based on the customization of a formal descriptive theory for software engineering has been adopted. This standard is Essence 1.2, and it has been approved by the Object Management Group, one of the most important worldwide authorities on software and computer systems. The use of this standard is out of the scope of this paper and will be omitted, but it has been considered because, as an extreme simplification, a smart city can be considered a product of urban planning, architecture, and information and communication technology. Using Essence (Jacobson et al., 2019), in a customized application, customized through customization rules already present in the standard, has allowed the definition of a common representation that can bridge the three disciplines together in the description and progress measurement of a smart city development, in any instant of its life cycle.

Having a vision is then possible to define a strategy. The strategy is a sort of generic roadmap to arrive at the vision.

7 Creating a Detailed Design of the Strategy

7.1 The Urban Planning Elements Issue

For the scope of this paper, and for the sake of simplicity, the urban planning elements are mainly related to urban revitalization and economic development but also strategic UP is required.

To better structure the framework, further detail has been added to its outcomes. The chosen sub-elements of each outcome have been considered after a literature review. The resulting sub-elements (that will become the sub-outcomes) are listed in Table 29.2.

Considering the above table, to revitalize Gjirokastra, the most important (in an ABC approach) outcomes are quality

Table 29.2 Sub-outcomes

Outcome (value)	Sub-outcomes (sub-values)
Sustainability	Optimize current use of fossil fuels, eliminate waste, recycle, recover energy, save time, and reduce or eliminate pollution
Quality of life	Wealth, employment, the environment, physical and mental health, education, recreation and leisure time, social belonging, religious beliefs, safety, security, and freedom
Equity	Absence of unfair, avoidable, or remediable differences among groups of people. The groups are defined socially, economically, demographically, geographically, or among another dimension (sex, ethnicity, disability, …)
Livability	Safety, mobility options, employment and educational opportunities, public space, and political stability
Resilience	Local knowledge, community networks and relationships, communication, health, governance and leadership, resources, economic investment, preparedness, and mental outlook

of life and livability because they directly impact the increment of the resident population and related economic factors. In a second step, beyond the scope of this paper, equity and sustainability can be aimed. In a third step, also outside the purpose of this paper, the resilience outcome can be targeted.

Simply having defined this three-phase roadmap, the first kernel of strategy has been defined. Its first target is to improve the quality of life and livability to attract people and businesses. The second target is to improve sustainability (to optimize resource exploitation) and equity (to stabilize the overall smart city system of systems). Finally, the third target, manage changes by implementing better resilience.

To define strategic urban planning, moving from this strategic kernel, in this paper, we will focus on the QoL and livability only, to demonstrate the idea of the use of the smart city framework. The strategic urban planning process used has been explicitly designed to suit the framework and adapt to the smart city case. It has been named UPDRAFT and is a revision of the UN-HABIT strategic planning process (UN-HABIT, 2007). It has been conceived aiming to, at the same time, apply revitalization, strategic and development approaches in the context of Ramaprasad's framework. Any other UP process can, in any case, follow the same approach, *mutatis mutandis*.

The UPDRAFT process has four phases, as given in Table 29.3.

For the purpose of this paper, only the first two have been considered (and are the only two that have been adapted from the original process), but the approach can be used even in the other two.

For the first phase, the related tasks will be managed as in Table 29.4.

For the second phase, the related tasks will be managed as in Table 29.5.

Table 29.3 UPDRAFT phases and tasks

Phase	Tasks
Phase 1: urban situation and key resources analysis	Stakeholders analysis Urban situation assessment Local resources discovery Investment capacity assessment Consolidated urban diagnosis
Phase 2: urban development planning	Urban consultations Local resources exploitation strategy Investment attraction Drafting the strategic urban revitalization and development plan (SURDP) Approval and adoption of SURDP
Phase 3: sustainable action planning	Drafting action plans Local resource mobilization Public–private partnership
Phase 4: implementation and management of projects	Project design Management and coordination Monitoring and accounting Reporting

Table 29.4 UPDRAFT phase 1 tasks detail

Task	Execution
Stakeholders analysis	Stakeholders have already been identified by the framework and need only to be detailed
Urban situation assessment	This is a classical task in UP and will be conducted by interacting with all the stakeholders
Local resources discovery	All local resources, of any kind, are discovered and listed. The first set of categories of resources has been done during the research, and a subset will be used in the next section
Investment capacity assessment	This is another typical task of UP. It is split into two tasks (and this is the novelty of this process) the second of which is the next, economic flywheel and equity attraction
Economic flywheel and equity attraction	This is another peculiar step that emphasizes not only the possible investments but also defines the foundations for equity attraction in both forms of investment and paying users

According to the table above, the tasks that have been impacted by this research are: local resources discovery, economic flywheel and equity attraction, local resources exploitation strategy, and drafting the strategic urban revitalization and development plan (SURDP).

The proposed approach for strategy discovery can be used to support the strategic decision process. In a few words, the UPDRAFT process is activated by a strategy discovery process that supports the definition, enabling fast and reliable decision-making, of the strategy key points.

The local resources discovery has nothing special because it is a simple enumeration of all resources available in the area, with their stage of development. This can be done in various ways, from brainstorming to experience-based lists, but it is important to underline that this list can be improved during urban consultations. Each resource must be considered in relationship with the outcomes.

A second note is that it is important to remember that strategic urban planning is, usually, an iterative process that loops until a suitable SURDP has been developed. This is not evident from the above tables and, for such reason, has been underlined here.

7.2 Economic Flywheel and Equity Attraction

Having identified, from previous research, the possibility to attract foreign elder people as users of distributed retirement houses services, this will be considered a resource. This resource will also impact the development of industrial areas, accommodation services of hospital, and university development and more.

7.3 The Local Resources Exploitation Strategy

Let us show how it fits in the case of Gjirokastra. According to previously cited research (Lezzerini, 2021) and direct on-site surveys, Gjirokastra has the resources (called "original resources") given in Table 29.6 which is a traceability matrix about their relationship with elements of QoL:

The Use % has been roughly assigned (in this case, a precise value has not been demonstrated) and means the percentage of maturity of the resource to be exploited at its maximum capability. The trend of use means the trend to use more or improve the resource (+ means trend to improve or use more, − means trend to use less or degrade it, = means no meaningful changes).

For each element column, a + means a positive impact of the resource exploitation, − means a negative impact, and a blank cell means no significant impact. These values have been approximately defined to show the idea of the methodology.

There are also other resources that are those discovered in economic flywheel and equity attraction, and they are listed in Table 29.7.

These resources are of a different kind: they are resources that need to be built to implement the distributed retirement house and, so, they can be considered another type of sub-outcomes. They are, at the same time, both resources and outcomes because they are resources to be developed from scratch (or near from scratch). For this reason, they will be identified as "generated resources."

Some of these elements, both resources and generated resources, have relationships between them and with elements of other outcomes. For the sake of simplicity, this has been omitted.

Now, it is possible to use the smart city framework to define illustrative elements that can be generated by fixing the sub-outcomes. They are in the form (* means any of the possible values):

$$(*)\text{ to }(*)(*)(*)\text{ by/from/to }(*)\text{ for (chosen sub-outcome)}$$

An example can be infrastructure to sense cultural data from professionals for employment.

From this set of illustrative elements, the related original resources (related to the sub-outcome) can be considered, leading to the following formula:

$$(*)\text{ to }(*)(*)(*)\text{ by/from/to }(*)\text{ for (chosen sub-outcome)}\\ \text{using (all related resources)}$$

Table 29.5 UPDRAFT phase 2 tasks detail

Task	Execution
Urban consultations	This a typical step in strategic urban planning and is achieved involving all stakeholders
Local resources exploitation strategy	This is a peculiar step of this process and, starting from the list of resources defined in the previous phase, it associates to them a strategy to exploit them to achieve the desired outcomes (QoL and livability)
Drafting the strategic urban revitalization and development plan (SURDP)	This step is typical of a standard process, but it has some aspects that will be further described
Approval and adoption of SURDP	This is another typical step in a standard process

Table 29.6 Traceability matrix from original resources to sub-outcomes

Resource	Use (%)	Trend of use	Wealth	Employment	The environment	Physical and mental health	Education	Recreation and leisure time	Social belonging	Religious beliefs	Safety, security, and freedom
Cultural heritage	40	+	+	+		+	+	+	+		
Industrial area	10	−	+	+	−				+		−
River	10	=	−		+						
Airfield	40	=	+	+	−	+	+	+			−
Accommodation	30	+	+	+	−	+		+			
Hospital	75	=	+	+		+	+		+		+
University	75	=		+		+	+	+	+		
Stadium	75	=	+	+	−	+	+	+	+		
Hills	50	=	+	+	+	+	+	+			
Mountain	40	=	+	+	+	+	+	+			
Farmland	30	+	+	+	+	+	+		+		
Lake	30	=	+	+	+	+	+	+	+		
Roads	50	+	+	+	−	−					
Mosque, church	75	=				+	+		+	+	

Table 29.7 Traceability matrix from generated resources to sub-outcomes

Resource	Use (%)	Trend of use	Wealth	Employment	The environment	Physical and mental health	Education	Recreation and leisure time	Social belonging	Religious beliefs	Safety, security, and freedom
Private medical services for DRH	0		+	+	−	+					+
Sanitary services for DRH	0		+	+		+					
Accommodation services for DRH	0		+	+	−	+					+
Hospital for DRH	0		+	+	−	+	+		+		+
University for DRH	0			+		+	+	+	+		
Leisure services for DRH	0		+	+		+		+			
Cultural services for DRH	0		+	+		+		+	+		
Transportation services for DRH	0		+	+				+			+

An example can be infrastructure to sense cultural data from professionals for employment using [Cultural Heritage, Industrial Area, Airfield, Accommodation, Hospital, University, Stadium, Hills, Mountain, Farmland, Lake, Road].

At this time, all the resources that do not have a relationship with the selected focus and semiotics can be removed from the list. In the example, we will get infrastructure to sense cultural data from professionals for employment using [Cultural Heritage, University].

The same can be done in other cases as, for example, policies to communicate economic data from businesses for employment using [Industrial Area, Airfield, Accommodation, Hospital, Stadium, Farmland].

To discover the smart technologies to implement the distributed retirement house, the same method is followed but using, as outcomes, the generated resources, with a different formula:

$$(*) \text{ to } (*)(*)(*) \text{ by/from/to } (*) \text{ to provide (generated resource)}$$

For example, infrastructure to sense environmental (health) data from citizens for sanitary services for DRH. Another example could be architecture to translate social information to citizens for leisure services for DRH. Yet, another could be systems to communicate environmental (Health) information to professionals for private medical services for DRH.

After having done this, the impacted outcomes related to each generated resource (that are known thanks to Table 29.7—Traceability matrix from generated resources to sub-outcomes) can be added

$$(*) \text{ to } (*)(*)(*) \text{ by/from/to } (*) \text{ to provide}$$
$$\text{(generated resource) for (impacted outcomes)}$$

and so on. After having generated, manually or using a computer program, all the interesting combinations, and a set of illustrative elements will be available and can be used to define strategic planning using smart technologies.

For each original resource, its impact on outcomes will be quantified (e.g., counting their usage in the set, but in a more complex model, this can be done using a weight to quantify the impact, in this model the weight can be + 1 or – 1, respectively, the + or the – in the table).

For each generated resource, its smart elements (i.e., the illustrative elements generated with the generation formulas) will be known and further technology discovery can be made.

8 Discussion

According to the above results, the use of an ontology like Ramaprasad's framework for smart cities can be useful to generate all the possible elements needed to implement the smart city itself.

Using the urban planning process defined above, that is introducing the peculiar tasks of resource discovery and exploitation, it is possible to map original resources and generated resources on both the city outcomes and its illustrative elements. This will provide evidence of what can be provided and what needs to be implemented. Then, it is also possible to define the proper strategies for selecting, for each illustrative element (or group of them), the appropriate (smart but not only) technology.

With this approach, the urban plan will define a set of requirements in terms of illustrative elements that range from more abstract concepts, like sustainability, employment, and so on that belong to architectural or economic (but also to other disciplines) to technological requirements that are peculiar of the information and communication technology context.

In this way, it is possible to put together these requirements guaranteeing continuous traceability between all involved disciplines (architecture, economy, urban planning, ICT, robotics, …), ensuring all the advantages typically given by full traceability: coherence, change management, and reliability analysis.

Using the Essence 1.2, standard approach will introduce a unique language to describe all the illustrative elements. In addition, Essence will ensure the capability to set checkpoints (i.e., milestones) for smart city development and measure its progress even in terms of concepts that are external to the ICT world, like physical health or wealth.

Due to the high number of variables, the generated illustrative elements are tenth of thousands and this approach results difficult to be practically applicable in an easy way. This approach can be formulated as an optimization problem of nonlinear and mixed programming and can be approached using metaheuristics or unsupervised learning to dramatically improve its decision-support capability.

Even if, in this case, the use of smart technologies (i.e., some ICT technologies for medicine and smart mobility) had an essential role, this example demonstrates that Ramaprasad's framework for Smart Cities can be applied to implement a different strategic planning process for urban planning, in this case, used for revitalizing a city but that can be used for different purposes like, for example, planning from scratch or planning to empower.

9 Conclusions

As stated even in the title, this paper has considered an already-defined strategy, based on the use of smart technologies, to revitalize an Albanian city.

An important result of this paper is that it has derived (again) this strategy describing it in terms of elements of a smart city, putting the bases to develop a different urban planning process that used the generation of "illustrative components."

In practice, starting from an already existing result to apply smart technologies to the rebirth of Gjirokastra, an ontology to describe a smart city has been used in combination with a sample urban planning process to generate the strategic urban planning elements with a high detail level.

This new approach has been applied to this case because it can be considered an "in nuce" smart city, a seed that can grow, but starting from the use of ICT that is easily mapped onto Ramaprasad's framework.

An important challenge for future research, that is currently under study by the author, is the evolution of Ramaprasad's framework to be applied to any city or even at a lower scale. Developing this ontology in the general case will allow the use of a totally new approach in urban planning that, preserving planners' creativity and originality, will help them with specific software tools. In fact, although the proposed approach evidences the ability to discover all the elements that need to be implemented, the number of possible combinations is very high and needs specific further research development to express it in terms of a mixed programming optimization problem or unsupervised learning problem. Both approaches will be analyzed in successive research.

This kind of new process can benefit from specific decision-support ICT tools that could be designed starting from this case and that will be addressed in future research on which the author is yet working. The idea of these tools is to develop specific software to support urban planners to generate the illustrative elements of the smart city.

This approach allows continuous traceability from ideas to implementation. The author is further developing this capability that allows urban planners to trace their requirements up to implementation, and this can be a subject for other research aimed at tracing the evolution and the top-down decomposition of the original ideas through the city's life cycle. In this way, the stakeholders' values can be yet available even when looking at the implementation elements, allowing to remove of elements that do not match with them or to timely discover new stakeholders or values.

The use of a standard like Essence 1.2 will allow a common way of description valid for concepts coming from different disciplines like economics, architecture, urban planning, ICT, robotics, and more, providing at the same time a way to measure and plan their evolution during implementation.

The author is working on all these topics to provide a new, integrated, and holistic approach that allows urban planning communication with other disciplines (like engineering or sociology), also allowing the creation of new-generation software and methodological tools based on it.

References

Azizalrahman, H., & Hasyimi, V. (2019). *Towards a generic framework for smart cities*. INTECHOPEN

Bickert, M., & Göler, D. (2012). *Terra incognita oder Touristenmagnet? Einblicke in die albanische Welterbestadt Gjirokastra*. Hausdruckerei BBR.

De Florio, V. (2007) Participant: A new concept for optimally assisting the elder people. In *Proceedings of the twentieth IEEE international symposium on computer-based medical systems*, June 20, 2007.

Ewart, I. J. (2013). *Living 'from' home: Older people looking beyond the house*. Home Culture.

González, N., Kämäräinen, A., Luštrek, M., Gams, M., & Vélez, I. (2009). *From users' needs to system specifications: A care system supporting older people's independent living*. Centro de Estudios e Investigaciones Técnicas de Guipuzkoa, 20018 Donostia, San Sebastián, Spain.

INSTAT. (2022). *The population of Albania*. Retrieved from http://www.instat.gov.al/en/statistical-literacy/the-population-of-albania/. Last visited November 9, 2022.

Jacobson, I., Lawson, H., Ng, P., McMahon, P. E., & Goedicke, M. (2019). *The essentials of modern software engineering: Free the practices from the method prisons!* Morgan & Claypool.

King, R. (2013). Sunset migration. In *An introduction to international migration studies*. Amsterdam University Press.

Kruja, D. (2016). Innovative approach for tourism enterprise in Albania. *European Journal of Service Management, EJSM, 20*. ISSN: 2450-8535.

Lezzerini, L. (2021). *Rethinking Gjirokastra*. Observatory of the Mediterranean Basin UNECE/U_POLIS/Center of excellence, pp. 360–369.

Nexhipi, O. (2018). Medical tourism management challenges—The case of dental tourism in Albania. *European Journal of Interdisciplinary Studies, 4*(1).

Ramaprasad, A., Sánchez-Ortiz, A., & Syn, T. (2017). A unified definition of a smart city. In *16th international conference on electronic government (EGOV)*, September 2017, St. Petersburg.

Ratz, T., Michalkó, G., & Kovács, B. (2022). The influence of Lake Balaton's tourist milieu on visitors' quality of life. *Tourism: An International Interdisciplinary Journal, 56*(2).

Rothuizen, J. J. (2011). *Elder people learning to be mentors for young people*. VIA University College.

Ruzevicius, J. (2007). *Quality of life and its components' measurement*. Engineering Economics.

Ruževičius, J. (2012). *Management de la qualité. Notion globale et recherche en la matière* (432 p.). Maison d'éditions Akademinė leidyba.

Townsend, A. M. (2015). *Smart cities*. Norton.

UN-HABIT. (2007). *Enhancing urban safety and security, global report on human settlements 2007*. United Nations Human Settlements Programme.

Preservation and Conservation

Registering the Mediterranean Gradient: A Journey Along the Ebro River (Iberian Peninsula)

Caterina Anastasia and Alessia Allegri

Abstract

The research presented in this article originated within the 'MedWays' Research Project, an international research cluster of the Beniamino Segre Interdisciplinary Liceo Centre at the Accademia Nazionale dei Lincei (Italy) and the University of Trento (Italy). Following the invitation from MedWays, we responded to the Mediterranean topic with a research project (titled: 'Mediterranean Sections') conceived as a journey and, at the same time, as a collage. Our work gave rise to a representation of the forms of inhabiting the land–water interface of the Iberian Peninsula's Ebro River, narrated in graphic and chromatic registers. From the urban, architectural, and landscape-oriented points of view, we considered how to probe the specifically Mediterranean nature of a river that feeds the sea of this name. Using resources afforded by the Web, we assembled plans, sections, and sets of views into Mediterranean Sections—representations that travel the length of the river from its mouth to its source, from city to city, and field to field. Sectioning the river through units of length that serve as 'measures', we fix photograms by which to explore the variation of what we call the 'Mediterranean gradient' of the Ebro River. Beyond the simple geographical sense, we understand the 'gradient' as what this river gives and receives from its Mediterranean identity, with consideration of how this translates in terms of its forms. This article is one further step in this research. Using cross sections of the river, this work aims to construct photograms in sequences in which we can read the particular 'rhythm' of the fluvial landscape of the Ebro River. As was the case with the earlier Mediterranean Sections work, our notion of 'measure' is taken with reference to fields, urban blocks, and mobility infrastructures, documenting their variation as they develop along the river. Our visual register is sequentially elaborated in plan and in section, depicting the forms by which the Ebro's land–water interface is inhabited. These forms combine into the essence of its 'specific rhythm' (what we have termed the 'Mediterranean gradient') and provide the basis for its territorial project, which starts from the appreciation and preservation of its relationship with the (visible and invisible) water flowing across the territory.

Keywords

Land–water interfaces • Transect • Journey collage • Mediterranean gradient • MedWays

1 Introduction

This paper presents some insights into the final considerations of the *Mediterranean Sections* research project (Allegri & Anastasia, 2022). This study represents a new stage in the authors' reflection on the Mediterranean 'state' analysed through the land–water interface of the Iberic Ebro River.

Mediterranean Sections was born in response to an invitation to participate in an open reflection on the Mediterranean initiated by *MedWays*. *MedWays* (Italian name: *Le Vie del Mediterraneo*) is a research project funded by the Beniamino Segre Interdisciplinary Liceo Centre of the Accademia dei Lincei (the Italian National Academy of Sciences) and the University of Trento (Italy). The research aims to be a space for an open dialogue and comparison of the realities, studies, training, and exchange among those living close to (and reflecting on) the Mediterranean.

As defined by *MedWays* research project leader Mosè Ricci, the Mediterranean can be thought of as a reference

C. Anastasia (✉) · A. Allegri
CIAUD, Research Centre for Architecture, Urbanism and Design, Lisbon School of Architecture, Universidade de Lisboa, Lisbon, Portugal
e-mail: caterinaanastasia@fa.ulisboa.pt

A. Allegri
e-mail: alessia.allegri@fa.ulisboa.pt

A. Battisti et al. (eds.), *Greening Our Cities: Sustainable Urbanism for a Greener Future*, Advances in Science, Technology & Innovation, https://doi.org/10.1007/978-3-031-49495-6_30

manual for proactive regeneration and resilience. This 'manual' has its first stage in an open atlas of the Mediterranean reality (Ricci, 2022), a collection of the diverse research experiences responding to the invitation from *MedWays*. The *MedWays Open Atlas* is an 'open and potentially infinite interdisciplinary atlas of Mediterranean Routes: systems and material or immaterial relationships that somehow leave a trace in the Mediterranean landscape' (Ricci, 2022, p. 8). It aims to collect different point of view, narratives, and 'narrative devices' to explore, illustrate, and construct interpretative cartographies for the meanings, myths and, in a word, the 'nature' of the Mediterranean. Far from being exhaustive in its description of the Mediterranean reality, the *Open Atlas* aims rather to point towards the countless further ways it could be described.

Thanks to an agreement recently signed by the Sapienza University of Rome and the Municipality of Favignana (Sicily; Italy), *MedWays* has its physical core in the Favignana MedWays Centre (FMWC). The FMWC is a spin-off of the Accademia dei Lincei's funded research and is conceived as a place open to accommodate all interdisciplinary training and research initiatives concerning the environment and landscapes of the Mediterranean. Thanks to this agreement, the research activity carried out within the three-year (2019/2022) *MedWays* international research cluster can be continued. This article is intended as a starting point for a second research step of the *Mediterranean Sections* research project that could also be developed in the physical context of the FMWC.

Originally conceived as an in-person journey along the banks of the Ebro River in Spain, the pandemic emergency situation meant that *Mediterranean Sections* ultimately undertook this journey only through visual materials available on the Internet. This is how *Mediterranean Sections* became a voyage stimulated by vision, by the exploration of a territory through the eyes of others. *Mediterranean Sections* gives rise to a narration in graphic and chromatic registers of the forms by which the land–water interface of the Ebro River is inhabited.

Mediterranean Sections aspired to construct (or perhaps unveil) what we have termed the 'Mediterranean gradient' of the Ebro. This second step of the research aims to construct an image (observing a characteristic rhythm) of the Ebro's inhabited surroundings: what the river, as an ecological and cultural corridor, gives and receives from the Mediterranean Sea. This Mediterranean gradient may be read in the forms of inhabiting the soil of the river matrix.

Recording the specific forms of the Ebro involves registering its specific land–water interactions, especially including the way of inhabiting the ground along the water. Beyond this immediate area, a river also integrates fluxes of various natures across a much larger and more diverse environment that the channel itself; we know that even human activities far from the physical boundaries of the channel may strongly influence the river, and historical uses of water resources may continue to influence the contemporary river and its forms (Wolh, 2014). The same goes for settlements, uses, and crossings associated with the human presence. These originate from (Anastasia, 2013; George, 1961), depend on, and conform themselves in relation to the (visible or invisible) presence of river water.

Without wishing to be exhaustive, and referring to previous studies by the same authors of this article on recent research approaches on the theme of 'the city and the river', we highlight some perspectives towards an idea of the river as a planning and regeneration tool for urbanised territories. From the point of view of architecture and town planning, rather than landscape architecture, aware that the question of the city and its river always concerns a multidisciplinary field and that deepening one research approach does not mean forgetting the others, we highlight four study paths. On the one hand, there are research lines that emphasise a geographical (and geomorphological) or morphological perspective (in the broad sense of morphology). On the other hand, there are research lines that privilege a landscape approach, also understood as perception and collective imagination, and others that privilege a reading based on eco-efficiency and sustainability. Among the possible perspectives described above that place the river at the forefront of architectural and town planning considerations, urbanised river environments are often studied through drawing, also understood as a design tool (see: Anastasia, 2019).

Recording the specific forms of inhabiting the Ebro's land–water interface intends to follow lines of research that use the drawing, understood as an analysis and a project tool. Indeed, in the fields of architecture, town planning, and landscape, the study and design of urbanised river environments seems to require the use of graphic representation. In other words, if we think of the river (and its immediate surroundings), according to Besse's definition, as an 'inscription' that characterises human settlements (Besse, 2010), we know that, in order to be analysed and become a project ground, it needs to be represented.

In this line, authors suggest that such representation should be translated into a plan and cross-sectional drawing (Busquets & Correa, 2005; Mathur & Da Cunha, 2001; Viganò et al., 2016), others point towards new methods of representation (again Busquets & Correa, 2005) and new figures (Anastasia, 2013; De Wit, 2009; Mathur & Da Cunha, 2001; Motta & Pizzigoni, 2008; Motta & Ravagnati, 2008) capable of describing and making visible the specific dynamics of the river and, at the same time, the characteristic forms of the adjoining settlements (Anastasia, 2019).

The objective is to operate a 'description' through drawing. Description understood as an analysis that precedes diagnostics, forecasts, and intervention; also in the sense of a synthetic interpretation of places seen as matter and medium of ecological and social relations (Dematteis, 1995). Such a description is intrinsically evaluative and also implicitly project knowledge (Dematteis, 1995).

The formal reading of the territory, from an architectural and town planning perspective, relies on drawing as a research tool. As the main tool of architects and town planners, drawing reveals itself as a sort of device capable of inventing and proposing a reality while representing it, thus participating in the construction of the image of the territory it describes (Gregotti, 1966, 2014). To draw is to select, to select is to interpret, and to interpret is to propose; drawing is therefore to understand specific nuances and, at the same time, to stimulate and give life to their confrontation (LUB, 1981).

Drawing, as a tool for research and potential development of relationships between issues and interventions, is the means to grasp the intrinsic qualities of space. Even more so since we are dealing with water landscapes—so dynamic and elusive that they transcend different scales—the action of drawing is particularly stimulating (see Delta Urbanism Interdisciplinary Research Program—TUDelft Faculty of Architecture and the Built Environment, 2021).

Drawing represents, for the research behind this article, a process that aims to go beyond visualisation by making researchers, first and foremost, aware—awareness as perception + knowledge—of the river and the land–water interface as foundational elements of the natural and human construction of the territory and the essence of its image.

2 Plan and Cross-Sectional Readings: Narrating the Ebro Land–Water Interface

Starting from *Mediterranean Sections*' reading of the Ebro territory, the present work aims to explore and interpret the graphic materials elaborated therein. During the earlier search phase of *Mediterranean Sections*, three main types of 'form' were observed along the course of the river: fields, urban blocks, and mobility infrastructure crossing the waterline. The sequences constructed from these elements became the tools implemented to 'take the measure' of the Ebro River's land–water interface. In agreement with Matias and Castro's suggestion, our 'sense of measure' by which to observe and describe the Ebro started from the idea of a land–water interface conceived as a place that, while it fixes separate elemental types (and identities), also effectuates their integration—see Matias and Castro's definition of 'waterfront' (Matias Ferreira & Castro, 1999).

While *Mediterranean Sections* earlier ascertained the rhythm of the three chosen elements starting from a plan reading (see Sect. 3), this study constructs rhythmic sequences based on a cross-sectional reading. Our object of study therefore makes use of visual tools such as sections, cuts (portions of plans and views), and photosections (photographic sections) according to the definition of Mathur and da Cunha (2009): photographic sequences that cut through time and space with rhythm and purpose. The abstracted elements are composed into collages.

We use the term 'collage' in the sense of a 'collection' of things that may be similar or different and as a 'device' that brings 'order' and 'meaning' to an assortment of fragments that speak about a place. We offer a collage of elements (forms) that lie along a trajectory (a route), and then, we emphasise the concept of 'transect' (cut or transection) and its creative potential. According to Gandy's suggestions (Gandy, 2020), a 'transect' is a mobile form of data collection that involves tracing a line through a designed area, typically on foot, in order to systematically record what can be found. Although deprived of the physical act of walking—a familiar method in ecological surveys (Gandy, 2020)—our transects are still characterised by a mobile gaze that runs, crosses, cuts, and finally fixes frames, illustrating the entire length of the river through a few characteristic fragments.

After (re)tracing, (re)presenting, and (re)writing the Ebro through schemes of its land–water interface, in plan and in section, we then ask: what forms do fields, urban blocks, and mobility infrastructures crossing the water take as we move away from or approach the river? What forms do the same elements take as we move along the water from the source to the mouth and vice versa?

In this work, however, the specific characteristics of the riverbed guide the reading of the forms abstracted from the territory. As the shape of the riverbed (in plan and section) and the flow rate vary, we then ask: how do the forms of settlement, cultivation, and crossing vary? Going further, what is the rhythm (in our reading 'the gradient') that these forms imprint on the river matrix soil?

Fixing and analysing the characteristic land–water interface of the Ebro in parallel with the characteristic forms of its riverbed and valley, we then explore this image in its essence, the terrain on which to construct (or re-construct) its cultural and shared identity, its collective urban image, and that of the society that inhabits it, uses it, crosses it, and is reflected in its waters.

3 Forms of Inhabiting the Ebro River

The Ebro River flows through seven Spanish regions, arising in Cantabria and reaching the Mediterranean Sea in Catalonia; its catchment area drains territories of the Valencian Region and Castilla-La Mancha, including territories of Spain, Andorra, and France. The main riverbed crosses two Spanish regional capitals, Logroño and Zaragoza.

The Ebro is a mighty river with irregular flow. At the end of the summer, it has extreme low water levels throughout the basin, reaching up to a tenth of its average flow. The flow of the river is altered and regulated by numerous dams along its approximately 930 km, some of which create large reservoirs. The Ebro supplies urban and agricultural areas far away from its main channel, with two water transfers to Greater Bilbao and Camp de Tarragona. This water supplies hydroelectric power plants, irrigation systems, and areas dedicated to livestock farming. The reservoirs and confluence areas also provide recreational opportunities.

The characteristics of the river valley, the urbanisation of the riverbanks, and the use of the river water are the main factors that define the shapes of the main riverbed. Among these shapes are a deltaic area (the mouth of the river), river bars, river islands, areas of wide meanders, reservoirs, and tributary confluence areas. The non-inhabited lands contiguous to the river are protected natural areas and canyons.

Regarding the planimetric component, even with the conviction that 'the river is not a line' but a space defined by ever-changing boundaries, this research treats the axis of the river as a 'narrative axes'. The research performs a rectification of the fluvial course on the basis of the flow of the water, from which it reads the three chosen measures (fields, urban fabrics, and infrastructure along or crossing the river). The geographical orientation of the components is set aside, and the main reference becomes the river current (see Figs. 1 and 2).

From a plan analysis of the immediate surroundings of the river, we can apprehend the characteristics of the fluvial corridor.

From Fig. 1, we can note that cultivations occupy most of the banks. With special attention to the crops that generate geometries and are particularly evident in the aerial views, we read agricultural expanses and fields that mark extensive cultivation, orchards, portions of countryside in peri-urban areas, textures of fields dedicated to grazing, and cultivated flood areas. These are productive landscapes of rural character, green infrastructure supported by the river corridor and, at the same time, they are matrices of processes, marked by techniques, gestures, rituals, and practices in seasonal rhythms that punctuate the year.

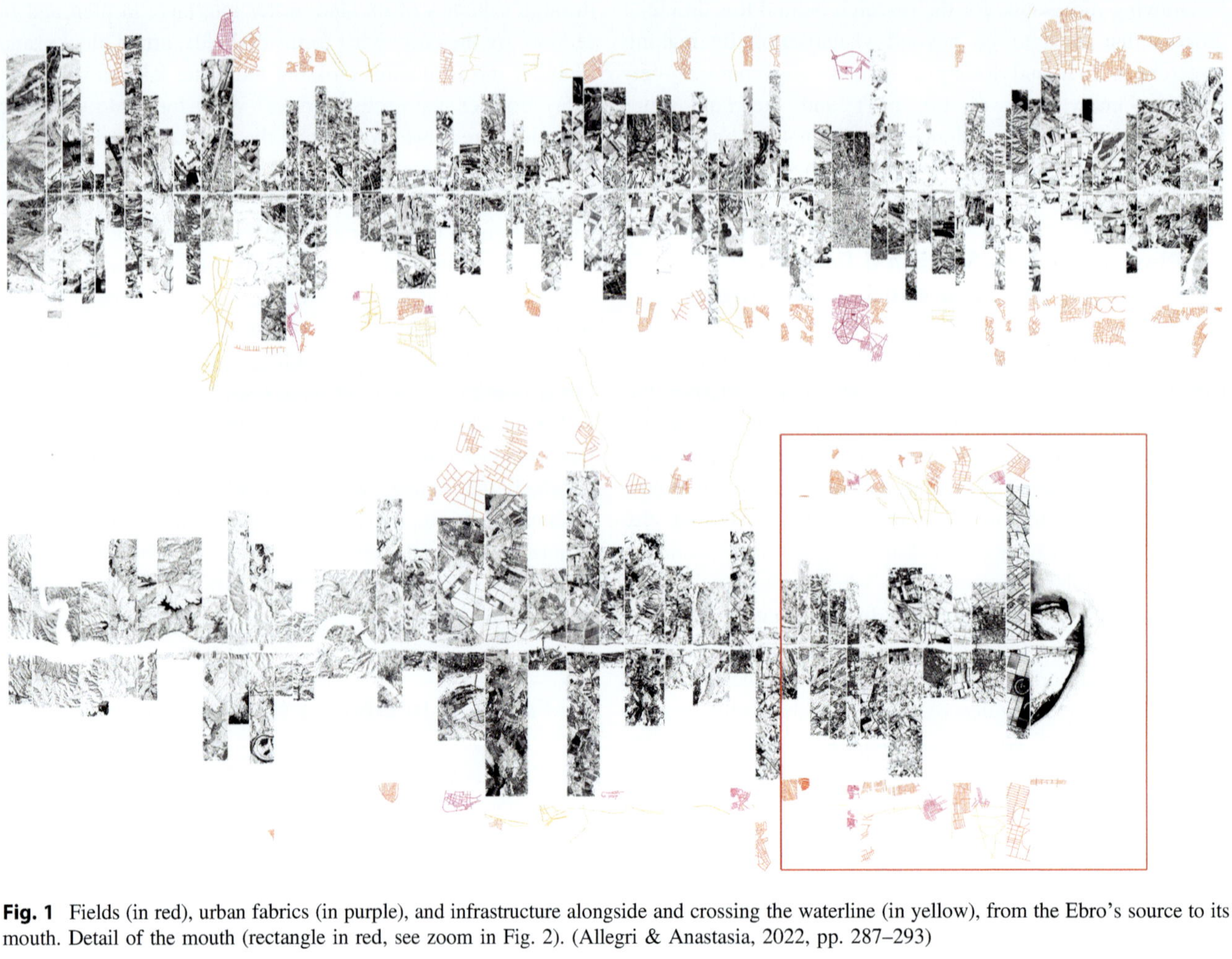

Fig. 1 Fields (in red), urban fabrics (in purple), and infrastructure alongside and crossing the waterline (in yellow), from the Ebro's source to its mouth. Detail of the mouth (rectangle in red, see zoom in Fig. 2). (Allegri & Anastasia, 2022, pp. 287–293)

Fig. 2 Fields (in red), urban fabrics (in purple), and infrastructure alongside and crossing the waterline (in yellow), detail of the Ebro's mouth (derived from: Allegri & Anastasia, 2022, pp. 287–293)

From Fig. 1, we can see that there is little urbanisation along almost the entire course of the river and there are no conurbations along the water, but rather inhabited cores that are not yet joined and are clearly compact. Urban cores, most of which are of reduced dimensions, tend to occupy only one river margin, with few bridges linking the urbanisation of both banks. Therefore, we find, opposite of the main urban core and across the river, a large number of urbanised areas more in the form of bridgehead towns. In other words, we can say that there are very few examples of balanced urbanisation on both sides of the river. As a result, the river corridor is rarely a space 'in the city'—it more often defines the urban limits or a 'second bank' that is the terrain of more recent urbanisation. According to a reading of the formal characteristics of the urban fabrics, we note that this 'second bank' is very often dedicated to monofunctional use.

From Fig. 1, we can finally draw information on the mobility infrastructure that follows the river corridor or crosses it with bridges. While national and regional roads tend to develop along the main riverbed, highways tend to build straight tracks that do not follow the meanders of the river. Minor routes, including local streets and maintenance paths, often follow the course of water derived from the river to feed the surroundings.

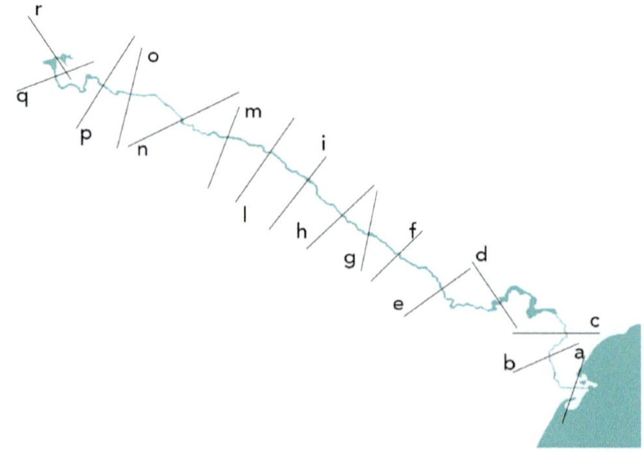

Fig. 3 Cross sections (from 'a' to 'r') alongside the Ebro River. Authors' elaboration

4 Forms of Inhabiting the Ebro Registered Through Cross Sections

In this chapter, the rhythms of the three chosen forms of inhabiting the land–water interface will be probed through cross sections of the river valley (transversal to the water current).

Reading a specific site through cross sections can be thoroughly informative. From Patrick Geddes in the early 1900s, to ecologists thereafter, and to architects and landscape architects using 'transecting as a line of investigation', the valley section has often been used as a tool to enquire about a terrain—especially if this is contiguous to water. From an ecological point of view, the distance between the two banks, the distance from the water on each bank, and the form of the valley section are important for the study of the characteristics of the corridor defined by the river.

In addition, we know that we cannot separate the hydraulic operation of the network systems from their adjoining surfaces; the riverbed is therefore inextricably linked to its adjacent banks (Anastasia, 2019; Rinaldo, 2009). For these reasons, the limit (and the link) that the riverbed establishes with the overflowing banks depends on the slightest topographical variation, which has to be evidenced, for example, through different abscissa and ordinate scales (e.g. elevation reported in metres and distance reported in kilometres, see Figs. 3, 4, 5, 6, 7, 8, 9, 10 and 11). In addition, the comparison of river valley cross sections at regular intervals makes it possible to organise and compare information that is more difficult to read in plan.

In Figs. 4, 5, 6, 7, 8, 9, 10 and 11, fields, urban blocks, and mobility infrastructure are schematised along the cutting axis of the section. From a first overall reading, the chromatic register of the forms reveals the presence of a lot of red, with less presence of yellow and violet, the latter often being sporadic. In the figure, the coloured lines have different thicknesses and, in the case of fields, also different fillings. Line thicknesses correspond to the width of fields, urban blocks, and infrastructure. The succession of colours reveals rhythms that are more or less tight (more or less 'slow').

Fields (in red) are the predominant texture in Figs. 4, 5, 6, 7, 8, 9, 10 and 11.

The plots are wider (their rhythm is slower)—the thickness of the red lines becomes larger—in the deltaic zone and in the flat areas of the open valleys. Here, the red becomes more intense. We therefore understand that fields are sizeable, regular rectangles of crops. On the contrary, in areas adjacent to the more or less tight river meanders, where the sediment supply makes the land fertile, we find long and narrow rectangles of characteristic tree plantations. As a result, the characteristic rhythm here is lighter, not continuous, and very often dotted. The dots may be thought to metaphorically represent tree plantations arranged in rows, in orderly groups, and woodland masses (in areas with steep slopes). The red in the figure has an intimate and obvious connection with the topography of the river valley.

Urban fabrics (yellow) and infrastructure (purple) are less present in Figs. 4, 5, 6, 7, 8, 9, 10 and 11 and are almost always associated with each other.

Thinner yellow and violet lines almost always chase each other, sometimes touching, sometimes tangent, and

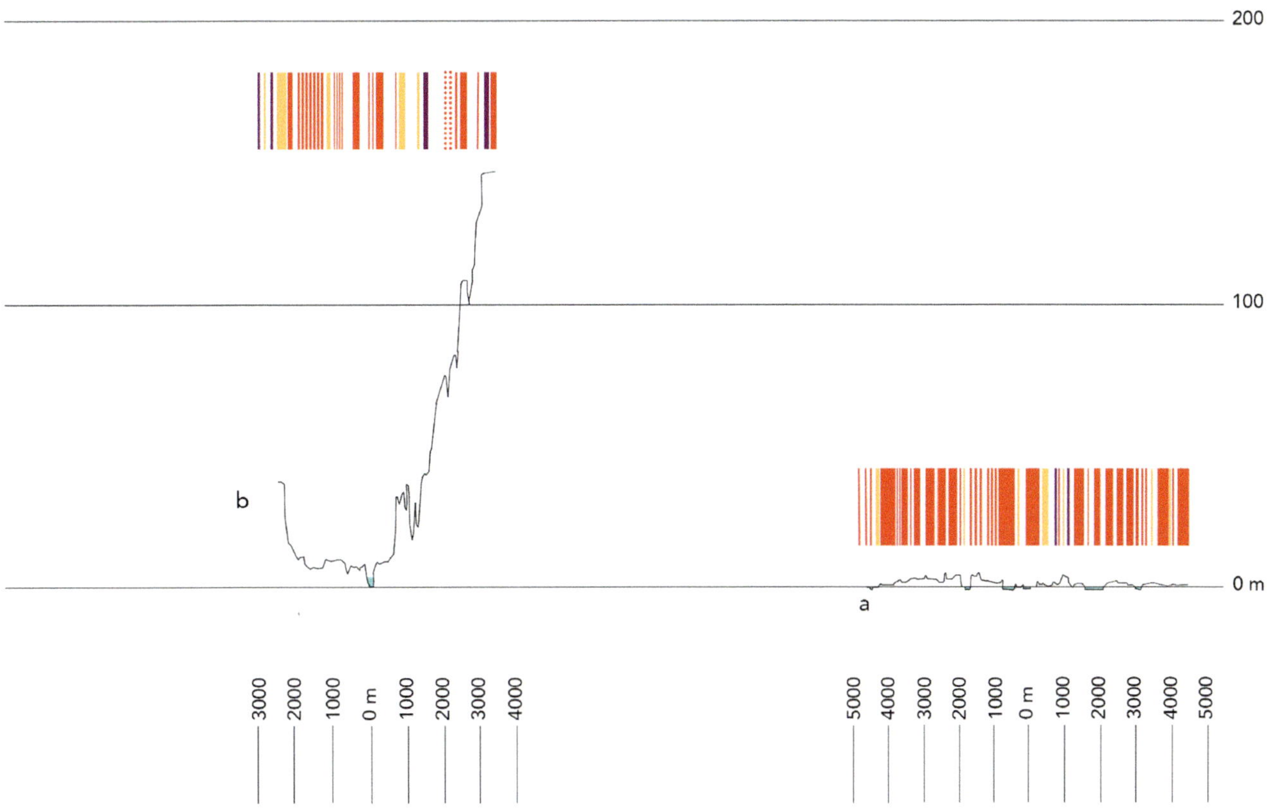

Fig. 4 Cross sections of the Ebro River (a–b) and the 'Mediterranean Gradient' from the Ebro's mouth to its source. Fields (in red), urban fabrics (in purple), and infrastructure alongside and crossing the waterline (in yellow). Elevation reported in metres and distance reported in kilometres. Authors' elaboration from Sistema de Información Geográfica Nacional (SIGN 2022)

sometimes overlapping. In lowland areas, violet and yellow are at tighter cadences and come to be perceived as orange, taking on strength and recognisability. In section, the rhythm confirms the reading in plan. Orange takes on strength and recognisability and, as we noted above, describes a frequency of few compact urban centres marked by the presence of one or more crossings (bridges). The rhythm with which yellow and violet enter the river corridor reveals that regional roads occupy both banks in wider valley stretches (such as near the delta or in the Zaragoza valley), and high-traffic infrastructure (e.g. motorways)—indicated by thick yellow lines—occasionally lead to urban centres.

5 Final Considerations

In the context of the extended research that gave rise to the *MedWays Open Atlas*, *Mediterranean Sections* aimed to record the forms of the land–water interface of the Ebro River through three element types that mark its territory. Based on these premises, the analysis through sections transversal to the Ebro's water current and the schemes of the associated 'gradient' is the step undertaken by this article.

In the framework of the Atlas, as a 'choral scientific enterprise with several coordinated and autonomous voices' and, 'as a random journey across the Mediterranean, a collection of meaning, landscapes, and social facts that explore the significance of the Middle Sea' (Ricci, 2022, pp. 8–9), this work takes its cue from the provocation inherent in the phrase 'Mediterranean Way', where 'way' may be understood in the double sense of routes and of habits. We therefore offer a narrative of the Ebro River rather than a description of it. Similar to the writers who also used the 'transect' as a way to structure a narrative around movement through space (Gandy, 2020, p. 164), we place it at the basis of our narrative in plan and section.

This is how we enter more deeply into the study of what we call 'the Mediterranean gradient', to get closer to understanding what a river gives and receives from its Mediterranean identity, from the river's source to its mouth and vice versa, and how this is translated in terms of 'form'.

In the belief that the inhabitation of the banks of the Ebro always comes from the combination of (at least) three elements (fields, urban fabric, and mobility infrastructure), we have constructed one possible characterisation. Inhabiting the Ebro produces a conceptual representation, and at the same time, it is the perception of a geometric and chromatic

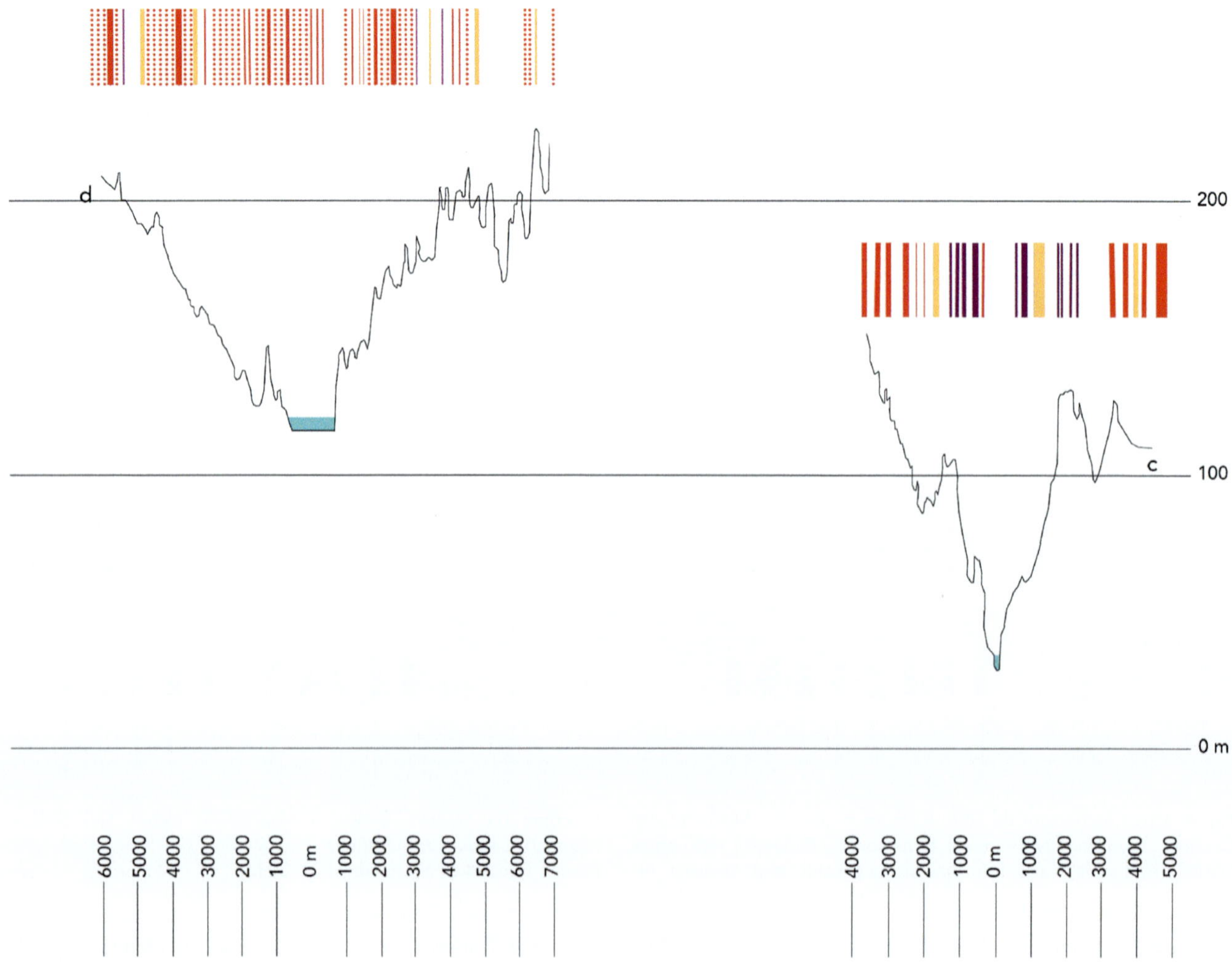

Fig. 5 Cross sections of the Ebro River (c–d) and the 'Mediterranean Gradient' from the Ebro's mouth to its source. Authors' elaboration from SIGN (2022)

succession that defines the observed landscape. What we call 'the gradient' could simultaneously be a transposition of the inhabitation of the shores into a scheme as well as the search for a means of representation.

The gradient is defined by a sequence of lines whose rhythm affords an intuition of its essence. The landscape becomes something written. The plans, and above all the 16 sections, tell a story with compositional autonomy. It is their sequence that gives them a narrative dimension, together telling the story of the inhabitation of the Ebro.

The Mediterranean Sections and gradient provide a morphological and phenomenal narrative of the geographical space of the Ebro Mediterranean Corridor. A collage of signs is given through rhythms of cultivating, living, and crossing that narrate a landscape, apprehending its identity. By linking forms of representation to perception, the gradient offers itself as a perceptual measure of reality.

First results of this research can be useful to the international scientific community, especially on three terms.

Firstly, this work constructs a reflection only from material obtained from the Internet. The pandemic has forced us to explore the possibilities of a non-physical but virtual journey. At the same time, it prompted us to discover and probe the potential of the Internet and the immense (graphic) material it can offer for research. Consequently, this study serves as a stimulus for research that cannot be carried out through the physical study of places (field analyses).

Secondly, from the reading (through drawing) and the subsequent graphic and chromatic register ('the gradient'), the research constructs a narrative of the river. As stated above, 'the gradient' could simultaneously be a transposition of the inhabitation of the shores into a scheme as well as the search for a means of representation. In this line, the work reinforces the idea of drawing and 'transect' as a tool of analysis towards the construction of a narrative.

Finally, what we have termed the 'Mediterranean gradient' becomes a 'narrative method' that can be used, and that we hope to use, in the analysis of other (Mediterranean and

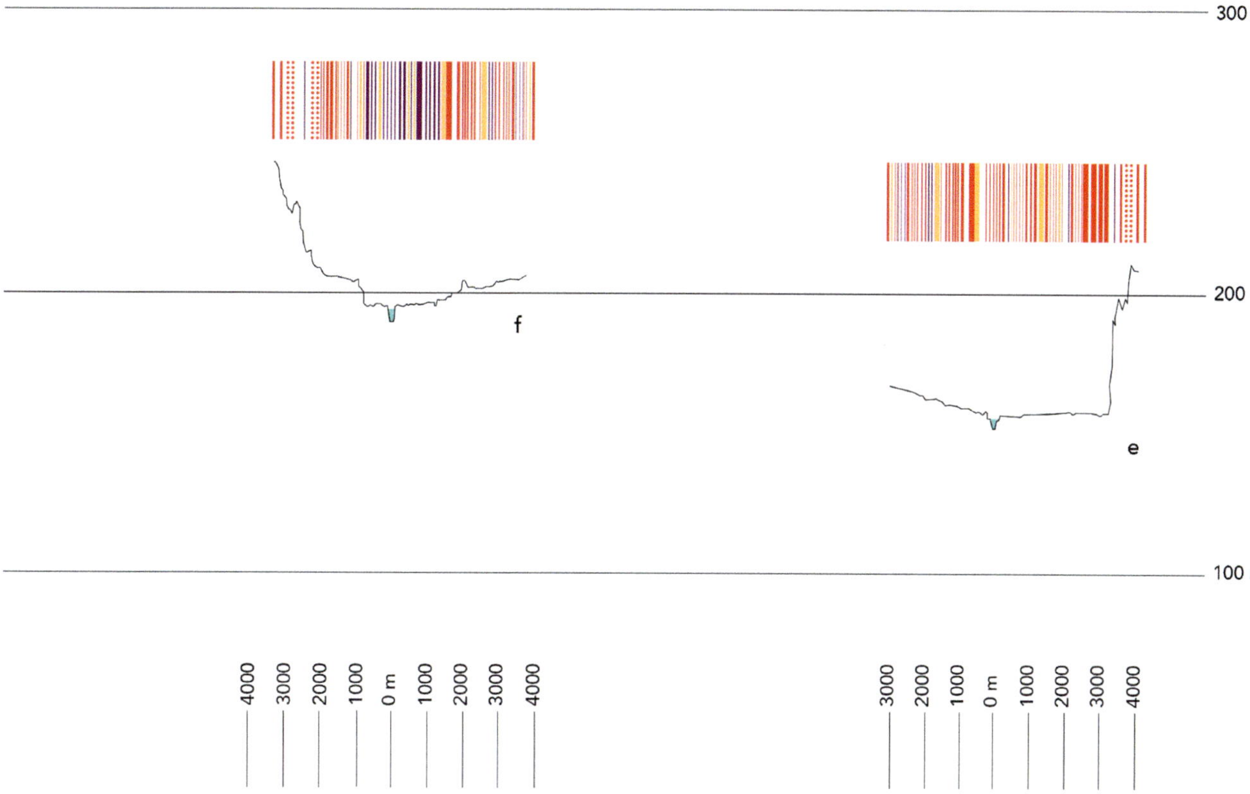

Fig. 6 Cross sections of the Ebro River (e–f) and the 'Mediterranean Gradient' from the Ebro's mouth to its source. Authors' elaboration from SIGN (2022)

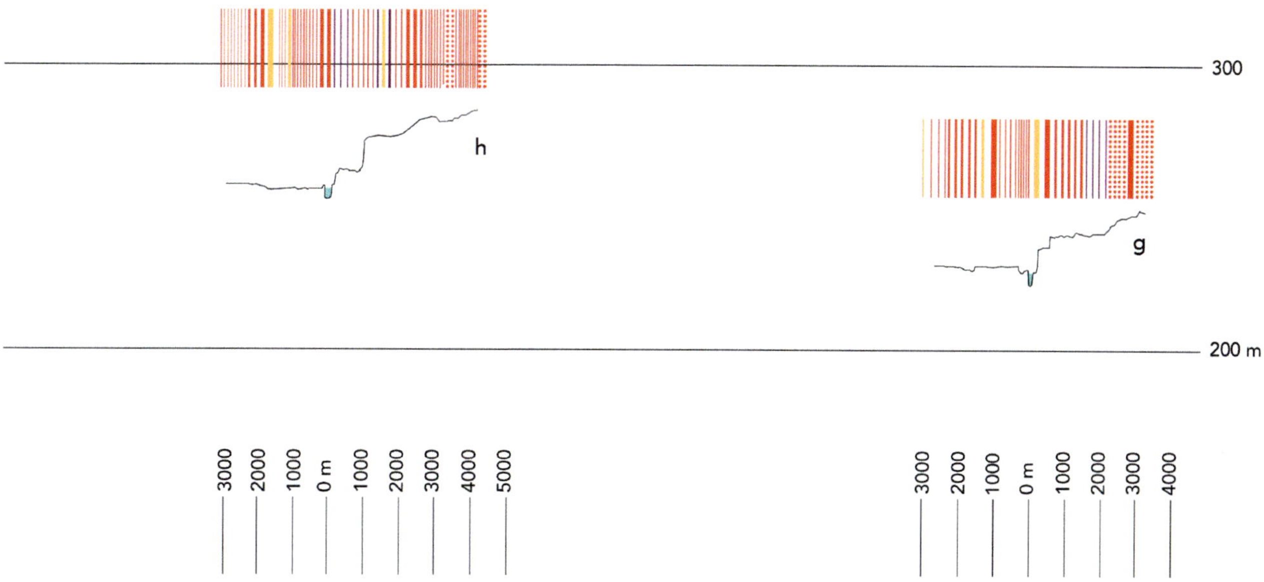

Fig. 7 Cross sections of the Ebro River (g–h) and the 'Mediterranean Gradient' from the Ebro's source to its mouth. Authors' elaboration from SIGN (2022)

non-Mediterranean) river contexts. A narrative that uses the 'transect' as a practice, particularly suited to the riverine context, which allows for the construction and comparison of representations of land–water interface forms.

Told through the traces, rhythm, density, and typological contrast, the Ebro's land–water interface, as a boundary and a shifting/thick edge encompassing contiguous spaces, can also tell us the basis of the interface's project. Paraphrasing

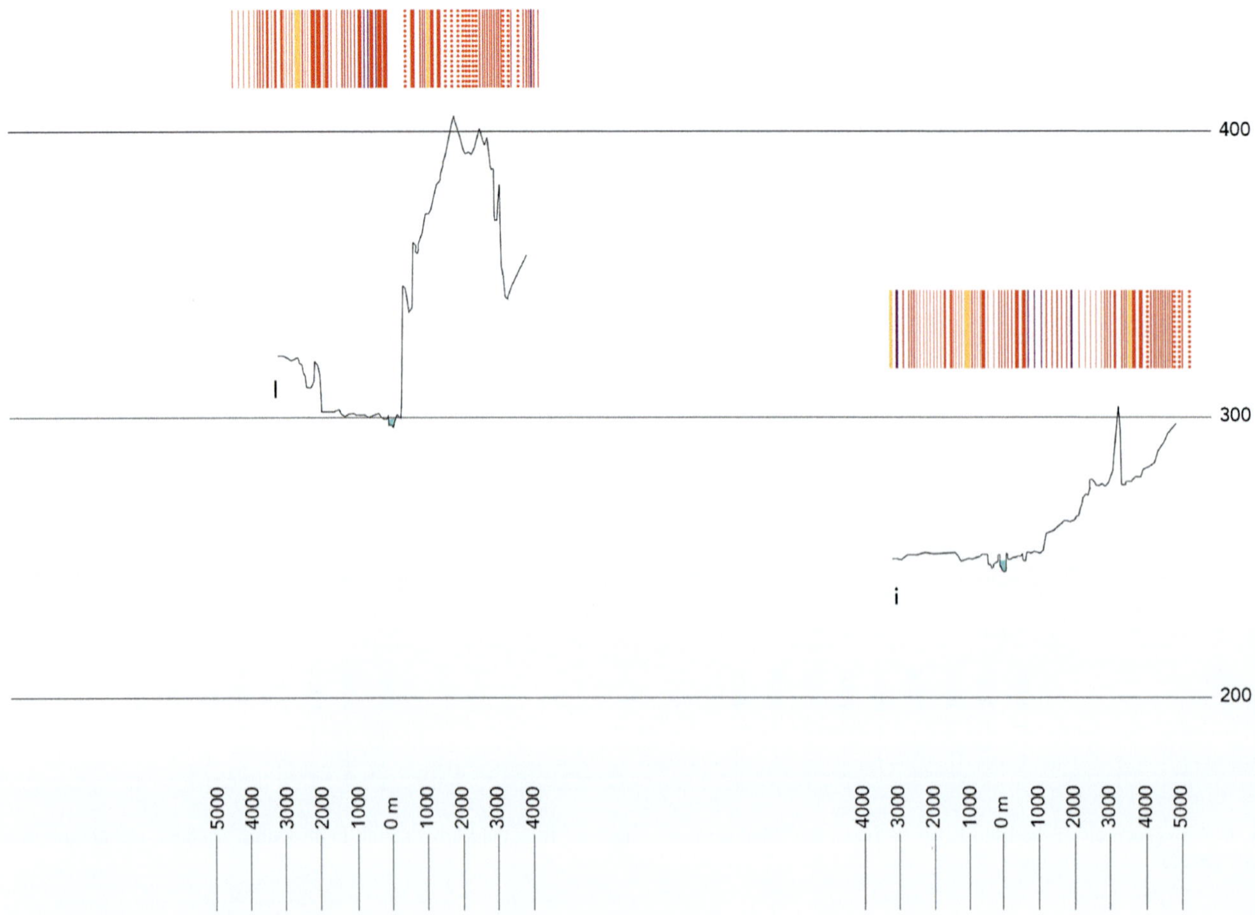

Fig. 8 Cross sections of the Ebro River (i–l) and the 'Mediterranean Gradient' from the Ebro's mouth to its source. Authors' elaboration from SIGN (2022)

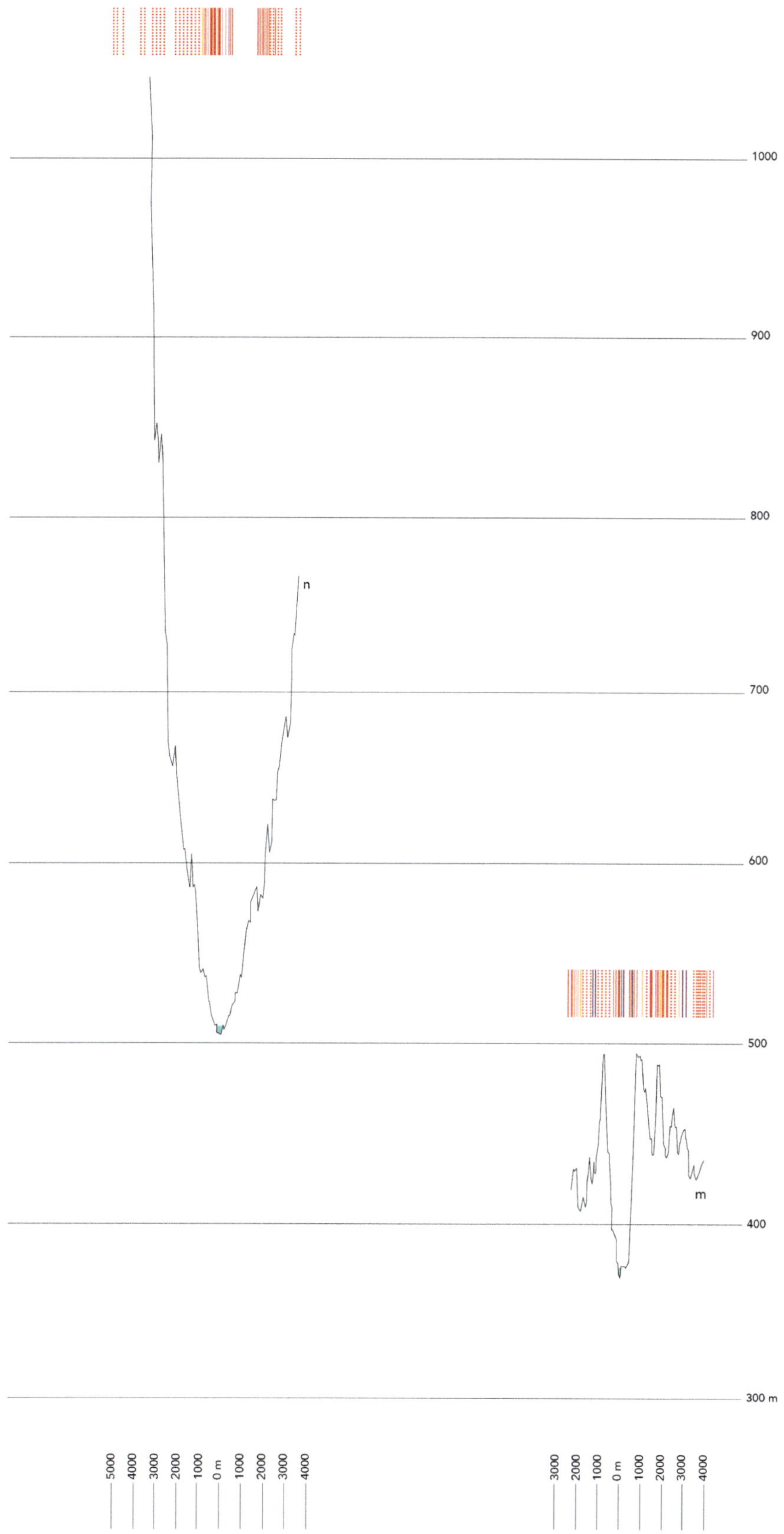

Fig. 9 Cross sections of the Ebro River (n–m) and the 'Mediterranean Gradient' from the Ebro's mouth to its source. Authors' elaboration from SIGN (2022)

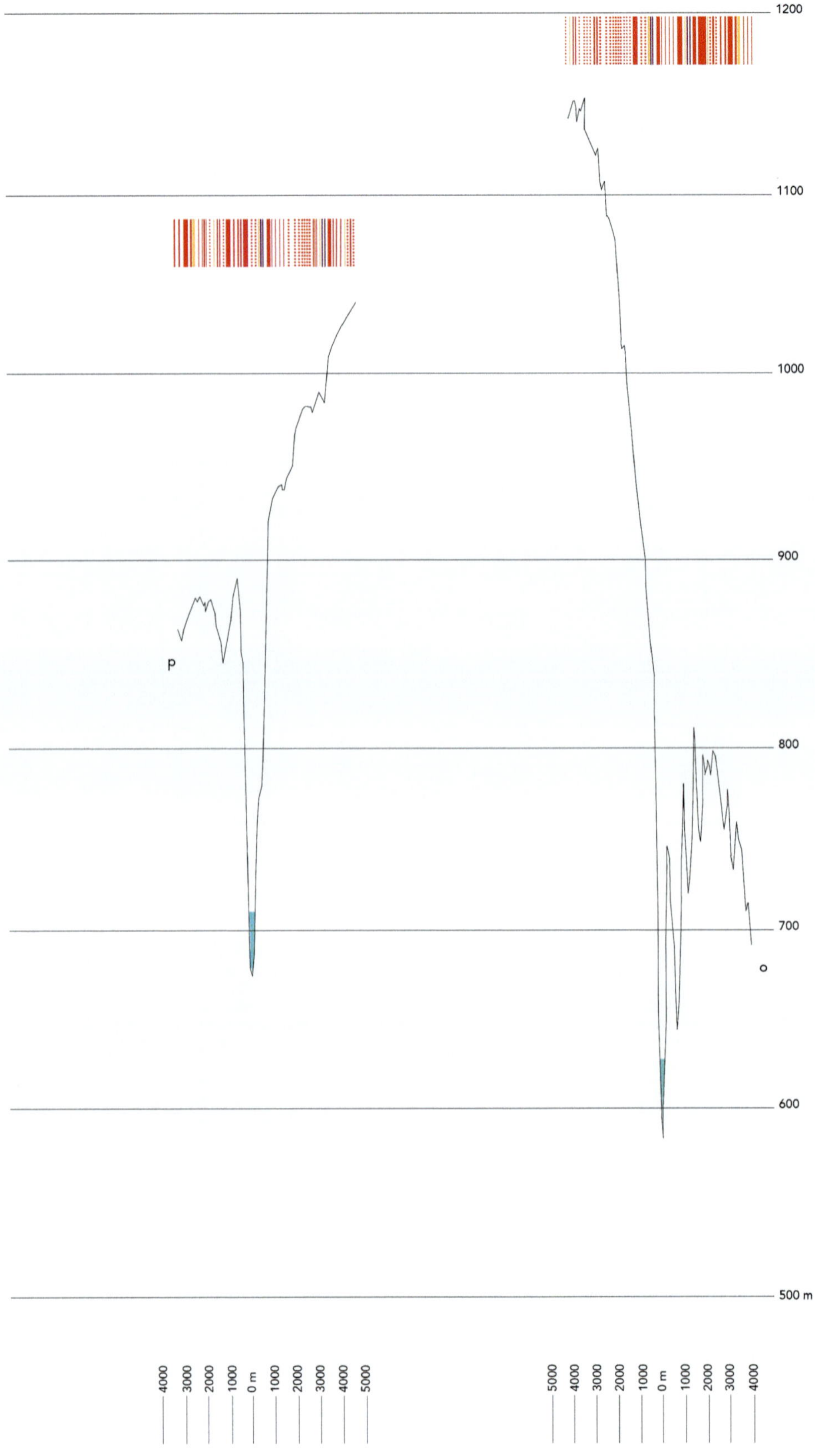

Fig. 10 Cross sections of the Ebro River (o–p) and the 'Mediterranean Gradient' from the Ebro's mouth to its source. Authors' elaboration from SIGN (2022)

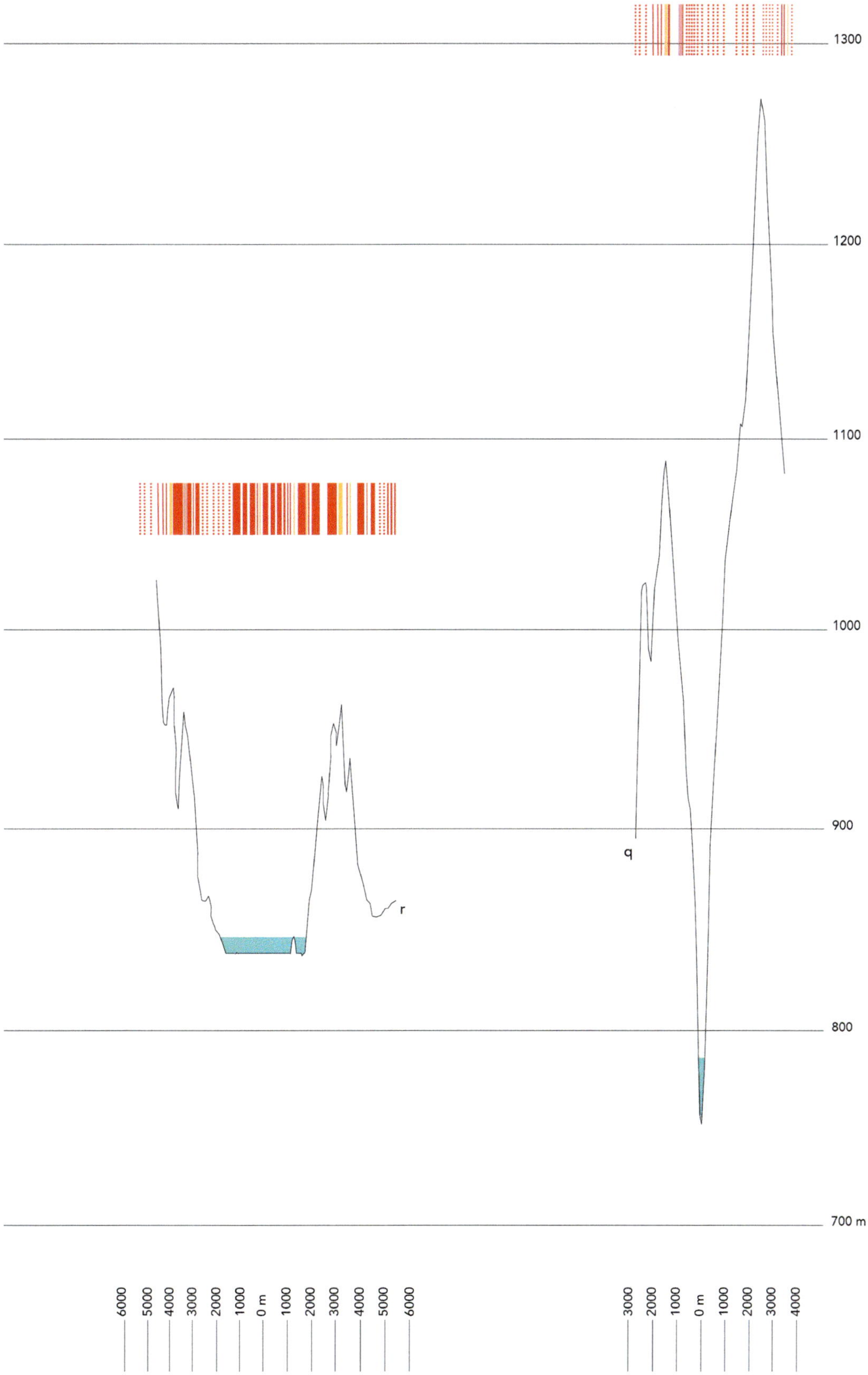

Fig. 11 Cross sections of the Ebro River (q–r) and the 'Mediterranean Gradient' from the Ebro's mouth to its source. Authors' elaboration from SIGN (2022)

what Orlando Ribeiro said about the city of Lisbon (Ribeiro, 2013), we emphasise the position of human settlements in relation to the river current as a way of explaining their fate.

Funding This work is financed by national funds through FCT—Fundação para a Ciência e a Tecnologia, I.P., under the Strategic Project with the references UIDB/04008/2020 and UIDP/04008/2020. Caterina Anastasia is currently working on a research project funded by FCT, 'Orçamento de Estado do Ministério da Ciência, Tecnologia e Ensino Superior'—and the European Commission/European Social Fund (ESF)—'ao abrigo do Quadro Estratégico Comum (2014–2020), através, nomeadamente, do Programa Operacional do Capital Humano', under the Postdoctoral Research Fellowship, individual postdoctoral grant (SFRH/BPD/116331/2016), carried out at CIAUD; Lisbon School of Architecture; Universidade de Lisboa, Lisbon (Portugal).

References

Allegri, A., & Anastasia, C. (2022). Mediterranean Sections: A journey along the Ebro River. In M. Ricci, S. Mannocci & M. Pasquali (Eds.), *MedWays Open Atlas* (pp. 286–297). Lettera Ventidue.

Anastasia, C. (2013). *Attraverso il fiume: Segni idraulici e territori abitati del basso corso del fiume Ter*. Universidad Politécnica de Cataluña.

Anastasia, C. (2019). Urban Regeneration Projects Bound to Water, along and towards the Tagus Estuary (Portugal). *Sustainability, 11* (23), 6578. https://doi.org/10.3390/su11236578

Besse, J.-M. (2010). *La Sombra de Las Cosas. Sobre Paisaje y Geografía*. Biblioteca Nueva.

Busquets, J., & Correa, F. (2005). *New Orleans: Strategies for a city in soft land*. Harvard University Graduate School of Design.

Delta Urbanism Interdisciplinary Research Program—TUDelft Faculty of Architecture and the Built Environment. (2021). https://deltaurbanism.org/about

Dematteis, G. (1995). *Progetto implicito. Il contributo della geografia umana alle scienze del territorio*. Franco Angeli.

De Wit, S. (2009). *Dutch lowlands: Morphogenesis of a cultural landscape*. Sun.

Gandy, M. (2020). Queering the transect. In M. Gandy & S. Jasper (Eds.), *The botanical city* (pp. 161–169). Jovis.

George, P. (1961). *Précis de géographie urbaine*. Presses Universitaires de France.

Gregotti, V. (1966). *Il territorio dell'architettura*. Feltrinelli.

Gregotti, V. (2014). *Il disegno come strumento di progetto*. Christian Marinotti Edizioni.

LUB—Laboratori d'Urbanisme de Barcelona. (1981). La identitat del territori català. Les comarques. *Quaderns d'arquitectura i urbanisme, Extra 1 and 2*

Mathur, A., & da Cunha, D. (2001). *Mississippi floods: Designing a shifting landscape*. Yale University Press.

Mathur, A., & da Cunha, D. (2009). *Soak: Mumbai in an Estuary*. Rupa & Co.

Matias Ferreira, V., & Castro, A. (1999). Cidade de água a lenta 'descoberta' da frente marítima de Lisboa. In V. Matias Ferreira, & F. Indovina (Eds.). *A cidade da EXPO '98. Uma reconversão da frente ribeirinha de Lisboa?* (pp. 17–40). Bizâncio.

Motta, G., & Pizzigoni, A. (2008). *La Máquina De Proyecto*. Universidad Nacional de Colombia.

Motta, G., & Ravagnati, C. (Eds.). (2008). *Alvei meandri isole e altre forme urbane*. Franco Angeli.

Ribeiro, O. (2013). O sítio e o crescimento de Lisboa. *Terra Brasilis, 2*. https://doi.org/10.4000/terrabrasilis.737

Ricci, M. (2022). The MedWays Open Atlas. In M. Ricci, S. Mannocci & M. Pasquali (Eds.), *MedWays Open Atlas* (pp. 8–27). Lettera Ventidue.

Rinaldo, A. (2009). *Il governo dell'acqua: ambiente naturale e ambiente costruito*. Marsilio.

Sistema di Información Geográfica Nacional (SIGN). (2022). https://signa.ign.es/

Viganò, P., Secchi, B., & Fabian, L. (Eds.). (2016). *Water and Asphalt: The project of isotropy*. Park Books.

Wolh, E. (2014). *Rivers in the landscape: Science and management*. Wiley.

Re-naturalizing to Regenerate the Piedicastello-Italcementi Urban System of Trento

Claudia Battaino and Maria Paola Gatti

Abstract

The regeneration of multi-layered urban structures involves complex theoretical, methodological, and technical issues that must be addressed to understand the reasons and means for a sustainable transformation that respects the distinctive characteristics of places and the demands of contemporary life. In its various forms, the need for regeneration concerns different types of urban fabric that have become dysfunctional, affected by abandonment and impoverishment, due to traditional functions succumbing to social and industrial transformation; in others, as in the case of historic centres, it corresponds to the physical abandonment of places where people struggle to survive. This paper seeks to outline possible strategies for the regeneration of multi-layered fabrics, starting from the present reality and from a deep knowledge that must be able to read and reinterpret the physical, social, and economic characteristics layered there. Controlled planning or any other way of reactivating the abandoned urban system, whether industrial or residential, can only be based on knowledge. The project, understood as a form of knowledge aimed at rethinking the natural and built landscape, proves to be an indispensable tool for creating a new, virtuous circle and creating better living conditions for the inhabitants (whether human, plant or animal). With regard to the Piedicastello district of Trento—the city's industrial and religious centre, while also being a mountainous area, leper colony, agricultural area, military centre, production centre, and residential district—it is essential to consider the district and the factors that have determined the evolutionary processes of this significant area of land, so that we may imagine the sustainability of an area that has undergone numerous and important transformations over time. Its various sections have also undergone radical changes in size and use, and over the course of abandonments and reuses they have remained viable. However, over the last 20 years, partly as a result of the loss of its twentieth century industrial purpose, and despite various regeneration proposals, the district, increasingly isolated and distant from the rest of the city's structure, has become less and less dynamic. This paper analyses the current fragile state of this area, which plays a crucial role in the physical, social, and economic development of the city of Trento, in its interrelated historical, cultural, and environmental aspects. Piedicastello-Italcementi is a significant case because it includes an old core, residential buildings, infrastructure, the large empty space in the industrial section, an unresolved waterfront on the Adige River; it will be considered in its entirety as an urban system. Starting from a consideration of development, critical issues and potential, underutilization, and dysfunction, this paper will outline a regeneration strategy based on re-naturalization. It will have the aim of redeveloping the urban district, emphasizing the identity of its parts and separate environmental settings, and rediscovering a relationship with the city and the river, the green agricultural areas, and the mountains.

Keywords

Town • Conservation • Urban regeneration • Smart cities • Re-naturalization

1 Introduction

Regenerating multi-layered urban structures is one of the most complex challenges, involving the intricate overlapping of objects and places of recognized value, elements of ordinary character, empty spaces and dense consolidated

C. Battaino (✉) · M. P. Gatti
Department of Civil, Environmental and Mechanical Engineering, University of Trento, Trento, Italy
e-mail: claudia.battaino@unitn.it

M. P. Gatti
e-mail: mariapaola.gatti@unitn.it

A. Battisti et al. (eds.), *Greening Our Cities: Sustainable Urbanism for a Greener Future*, Advances in Science, Technology & Innovation, https://doi.org/10.1007/978-3-031-49495-6_31

networks, modern road infrastructures, and pieces of nature that incorporate urban materials and/or are enclosed by buildings.

Regeneration processes in an existing city require inventive layering processes for these different fragments, reflecting on what is valuable and what is not, but could be, recognized as valuable.

In the contemporary dimension of the city, the full-empty relationship is undergoing an inversion. It is a common occurrence to find empty space as a resulting element, whether in built-up or vacant spaces.

Like Heidegger's "half spaces" (Heidegger, 1976), these vacant spaces have the potential to "re-measure" the existing city as a place where different things can meet and reunite, and they can therefore be very useful in the much-needed rebalancing between the natural, agricultural, and urban spheres. When transforming urban areas, some places have been abandoned but they have also been reused.

In the past, the abandonment was only temporary, with the search for a new use being a real necessity: land could not be taken away from agriculture, and buildings, the result of hard work, could not be abandoned. This process of reconversion/adaptation became an opportunity to improve and update buildings and spaces.

Over the course of the twentieth century, however, disused spaces gradually multiplied in parts of the city, sometimes built too hastily and without attention to the quality of the development. They were spaces to be reinterpreted, empty, and waiting.

In recent years, the issue has been occurring exponentially, with divestment in both residential (stable and temporary, such as hotels) and productive sections, as well as service and military facilities.

The first problem with regeneration is how to govern the transition of these spaces and how to reinterpret them so that they become resources that can make cities more sustainable.

The methodological process of the research began with knowledge, that is, the recognition of the layering of the physical traces of the place.

To explore the evolutionary process of the multi-layered urban fabric, historical archival material, iconography, and historical maps were collected and comparatively analysed.

Significant architectures were surveyed and studied, while the natural and artificial infrastructural system was analysed. The study of urban evolution also considered the ideas that have been imposed on this part of the city, including the urban planning and architectural projects that have been developed over time.

Literary works about the Piedicastello district and the city of Trento were also analysed. Maps and databases were compared to analyse the degradation of the Piedicastello area in different periods and the subsequent regeneration projects.

The analytical process made it possible to understand the transformation of the site from a completely natural landscape to an anthropic landscape.

The multi-scalar and multidisciplinary knowledge that guided this work provided the research with the broader vision needed to imagine future transformations.

To define the resilience of the area to be worked on, it is necessary to identify the urban design/construction tools, the economic and social planning tools, and more, to help deepen the analyses and thus the regeneration projects.

These regeneration projects should be approached with a multi-stakeholder approach to give cities more than just a new and competitive look.

Local identity needs to be revitalized, both in terms of landscape and architecture, to give it new cultural, economic, and social impetus, while taking account of environmental aspects.

2 Over the River, Piedicastello

On the orographic left of the Adige River, near a rocky hill, Doss Trento, at the confluence of several valleys, in an area sheltered by the river and steep mountains, the Rhaetians and then the Romans developed a small centre, at first temporary and then permanent (Fig. 1: The pictures illustrate the relationship between temporary and permanent settlement of Piedicastello and Trento).

The site was chosen because it ensured excellent defensive potential, maximum wind protection, good water availability, while also offering modest exposure to the sun, especially during winter.

Over time, the Castelliere developed into a small urban settlement with residential and productive buildings.

The need for a larger urban centre led to the transfer of the city centre to the other side of the Adige River, where the city of Tridentum, now Trento, was born.

However, the urban settlement of Piedicastello still continued to exist. The site's strategic position has ensured its intense vitality over the centuries.

According to some scholars, such as Nicolò Rasmo, Piedicastello reached the structure of a village under the rule of Theodoric (c. 450–526 AD), king of the Goths, who had a church built outside the settlement to serve his garrison.

The presence of a "refuge" castle is evidenced by Cassiodorus' letter (Granello, 2003) written during the rule of the Goths over the Adige Valley, hence the name of the settlement, although no trace of the castle remains today.

The development of Piedicastello was structured over time by the presence of the clergy, who reclaimed the land, gave it to the villagers, and built farmhouses and various religious buildings.

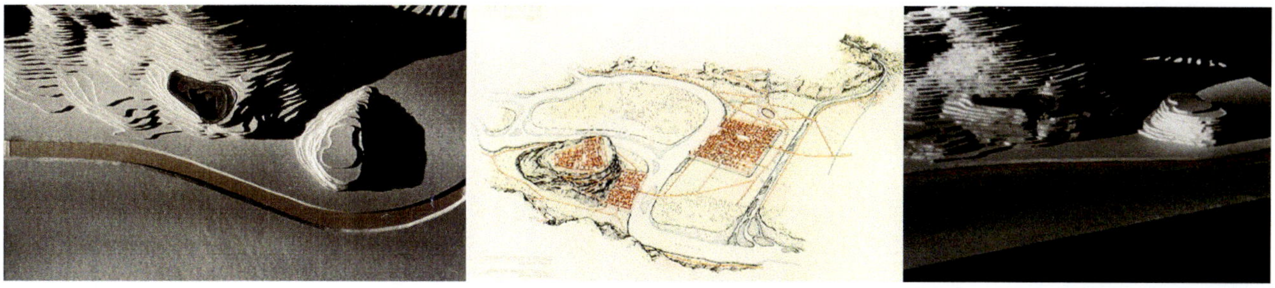

Fig. 1 Piedicastello between river and mountains (design table of the research FAMA)

After the year 1000, the village underwent a significant process of change, brought about by the acquisition of numerous lands by the Benedictine order of Vallalta.

Indeed, the confraternity received the entire marshy and uncultivated area of the valley floor that stretched from the slopes of Doss Trento, respectively, from Vela to the opposite side, at the foot of Calisio, where the village of San Martino now stands.

The vast plain was traversed by the Adige River, which divided it into islands with various branches, and by the Vela stream that flowed into the Adige near Sant'Apollinare.

The Benedictines (Iori, 1955) carried out timely land reclamation and land clearing, thereby increasing the amount of land used for agriculture. In addition, to avoid frequent flooding, they also built substantial embankment works on the river. A significant part of the land was leased to the villagers.

Improved agricultural techniques and a larger area of arable land made it possible to expand the settlement, with new buildings for residence and agriculture. Some of the village population began to make a living from fishing and craft activities.

The improved economic conditions of the settled population also allowed the construction of a new church during the twelfth century (between 1146 and 1180), entrusted to the Comacini masters, who were active in northern Italy.

The monks first settled in the buildings adjacent to the church of Sant'Apollinare (Brogiolo et al., 2013) (the parish priest's house and the administrative headquarters of the parish property).

Documents show that between 1237 and 1242, the village church, which had become the seat of the Benedictine abbey, was decorated with Gothic stylistic features from the French Cistercian area.

The original design, as noted by the scholar Rasmo, probably envisaged "a hall church, very high and bright because of the many Gothic windows open at the top of the nave" (Rasmo, 1966).

In 1236, after the demolition of several buildings around the church, the construction of the monastery began with a cloister, portico, dormitory, abbot's residence, and vegetable garden.

The small number of friars soon led to a change in the use of the buildings. The monastery turned into a manor house, fundamentally no different from that of the prince-bishops of Trento or the great feudal lords.

In the following centuries, the convent was demolished and rebuilt several times. The pillaging and consequent fire of 1444 was particularly devastating, requiring some artefacts to be reconstructed (the crenellated building, some of the service buildings, and the church).

Dürer's watercolour (Alberti, 1898) of 1494 is a detailed depiction of the urban layout of Piedicastello, displaying the consistency of the convent and the village surrounded by walls, with two gates which opened (Fig. 2: The pictures illustrate the Renaissance view of the village of Piedicastello. Piedicastello's relationship with the river and the city of Trento).

In the fifteenth century, the Benedictine fathers left Trento and Pope Martin V reduced the monastery estate to a prebend for the Provost of the Cathedral.

The urban development of Piedicastello depended on the road coming down from the mountain and crossing the wide Adige River. For many centuries, the connection between the two banks of the Adige was provided by barges.

At Piedicastello, a sort of harbour was built to serve the ferries connecting with the opposite bank, as well as for the vehicles that travelled along the entire length of the Adige. The presence of the harbour resulted in the construction of warehouses to store goods. The waterway was not only exploited for trade.

Only in the fourteenth century, in the narrowest bend of the Adige (north of the settlement) did the construction of a covered wooden bridge begin; on the right bank, a straight road parallel to the river led to the village.

Due to the presence of the water course and the road, the monastery and church maintained a high degree of individuality for centuries.

The wooden bridge (Mariani, 1673/1989) was destroyed several times by frequent floods and had to be rebuilt, while

Fig. 2 Dürer: views of Trento and Piedicastello, Landesmuseum, Hannover

in 1796, it burnt down, and in 1835, it was rebuilt for the last time.

In the sixteenth century, there was a great divide between city and countryside; while in Trento, the buildings were decorated, the areas and buildings in Piedicastello continued to retain a proto-urban structure, partly because the population was burdened with heavy imperial and bishop's taxes.

On the occasion of the Ecumenical Council in Piedicastello, the interventions proposed by Clesio in a letter dated 1533 "… to put the streets in order so that the roofs may be equally positioned, and to engage with those who may have the means to paint their façades." (Letter written by Clesio on 13 July 1533) were not implemented.

In 1545, with the opening of the Council, Piedicastello Abbey became the location for the clergymen's urban horseback rides (Cardinal Cervini, Pope Marcellus II) and the church and the adjacent abbey were refurbished.

Along the road that descended from the valley in the direction of the city of Trento, a terrace of buildings developed over time, consisting of several modules, bending around the shape of Mount Verrucca.

The continuous but broken structure allows the modules to be identified with a width between 4 and 5 m and varying depths (Fig. 3: The pictures illustrate the structuring and addition of building modules of the village of Piedicastello).

In the initial development, the depth was rather modest, as small vegetable gardens were built between the established terrace and the slope. Over time, the depth of the modules increased to 8–10 m and, through the tabernisation process, the transverse development led to the vegetable gardens being removed. Vertically, the modules developed on several levels.

The inclination of the hillside prevented the construction of completely underground floors, but the ground level on the uphill side was completely sunk into the ground.

The cellars, if they existed at all, occupied only a small part of the surface area of the constructed module. Originally, the constructed modules were built with one storey completely above ground level.

Over time, the buildings gained one or two storeys and today they have three or four storeys. The houses were simple, and the construction techniques were modest, yet

Fig. 3 Building units of the village of Piedicastello in 1920 and in a current aerial view (2022) (design table of the research FAMA)

this densely constructed ensemble presented an extremely well-preserved architectural appearance.

The group of buildings is made up of a series of characteristic elements that are organized in such a way as to form a balanced whole: the sloping tiled roof, the projecting wooden, the symmetrically arranged windows surrounded by stone, and the entrance door with a portal carved in stone.

Several sections were formed in the village: residential, religious, and productive, three loosely connected ensembles. Overall, however, the village was not redeveloped. Rather, it was decreed (Mariani, 1673/1989) to gather up the village beggars in the settlement.

In the seventeenth-century urban diagram drawn by Ludovico Sardagna (Chemelli, 1990), the village of Piedicastello is still encircled by the crenellated wall of the Abbey and continued to occupy the same area.

After the devastating plague of 1630, the Piedicastello suburb's isolated location from the rest of the city led to the invalid beggars and vagrants being located there.

The spread of the plague led the council of the Cà di Dio, a charitable organization of the city of Trento, to set up a lazaretto in the abbey meadow, which over time became a permanent structure.

Due to its position outside the walled city of Trento, the village also suffered from many fires and pillaging. The most devastating fire being that of 1703 caused by General Vendôme's French militia, but after the war the population proceeded to rebuild.

In the second half of the eighteenth century, as can be seen from the votive tablet dated 1753 kept in the Piedicastello rectory, the village still had walls, but the buildings appeared sparser, especially the buildings around the church.

It is therefore assumed that demolitions and alterations were performed on some buildings.

For example, looking at the Prepositura, the crenellations were removed and replaced with a simple four-pitch roof. Construction resumed in the second half of the eighteenth century, as evidenced by a 1760 plan showing some new buildings.

3 Transformations and Fragmentations

The enlargement and transformation of Piedicastello began at the beginning of the nineteenth century. In particular, a plan from 1822 shows the demolition of the entrance gate, and in later plans there is no trace of the wall.

In the second half of the nineteenth century, two interventions brought about a decisive transformation to the valley floor, namely the construction of the railway and the rectification of the Adige River.

In fact, the construction of the Verona–Bolzano railway line required a diversion of the river in order to avoid having to build two bridges (Fig. 4: The pictures illustrate the diversion of the Adige River in 1858 and the construction of the railway in 1859).

The rectification of the Adige, promoted by the Austro-Hungarian government (Austro-Hungary, 1883) and carried out in only four years, from 1854 to 1858, had a profound change on the urban layout, especially in the districts that overlooked the river.

The diversion of the river also made it possible to rehabilitate an area infested with miasma and malaria and was accompanied by the construction of embankments.

Fig. 4 Piedicastello. The diversion of the Adige River in 1858 and the construction of the railway in 1859 (design table of the research FAMA)

Once most of the original riverbed had been cleared, the land was reclaimed and used as farmland. The new infrastructural layout transformed the urban structure of the village and altered the connection with the city: the old road between the abbey of San Lorenzo and Piedicastello was occupied by the riverbed, and the bridge near the church of San Lorenzo was demolished.

Even though a new bridge was built, the canalization of the Adige and the railway further emphasized the peripheral character of the village, which was dedicated to agriculture and handcrafts.

In this period, Trento became increasingly bourgeois. The modernization project led to the construction of new buildings (courthouse, schools, leisure buildings, cinemas, etc.) and the construction of new water and electricity infrastructures.

During the nineteenth century, the development of the area was haphazard and sporadic.

The first industrial plants and workers' dwellings were built on the agricultural land, next to the old farms of the Giovannini, Manara, and Rossi families.

The reclaimed land owned by the municipal administration was used to build some public facilities or sold at low cost to private individuals, to build industrial plants and various warehouses and, later, the first working-class residential buildings. In Piedicastello, the largest cement plant in the province slowly took shape, occupying a large part of the Adige riverbed.

Trento became a fortified city, with the construction of barracks, military hospitals, military clubs, public baths, railway stations, restaurants, hotels, etc.

The construction of buildings caused the demand for building materials to explode. The timber merchant Antonio Frizzera (1827–1887), to "modernize" his business, combined the timber trade with the production of bricks, lime, and then cement (Mariani, 2011).

The small sawmill, located within Trento's city walls and having facilities near the Adige riverbanks, was moved in 1890 to a small plot of land in the San Nicolò area of Piedicastello, creating the timber depot, a building containing a Hoffmann kiln for baking bricks and a kiln for producing white lime.

The limestone was extracted from the mountain above, while the clay was taken from the quarries in the "Buco di Vela" area, in the municipality of Cadine, not far from the factory.

During the construction frenzy of the time, the potential of cement and the difficulties of its supply were immediately understood (for the main military works, cement had to be supplied by the Imperial factory in Kufstein), so the Frizzera family decided to produce only hydraulic cement (Gatti and Dalla Serra, 2019).

The Frizzera family bought new land in Piedicastello from the Municipality of Trento. The Municipality sold the land to the Frizzera at a low price but dictated the quantity of hydraulic cement to be produced, the construction of the factory, (within three years), of "workers' houses with dormitories and model refectories for unmarried workers, and with at least 12 quarters all corresponding to the provisions for buildings with workers' dwellings".

The city council required the Frizzera company to retain ownership and use for at least 25 years from the day of their completion.

From 1905 to 1912, the architect designed several buildings, including the volume to produce Portland cement, the large building containing dormitories and a refectory for unmarried workers, and the residential building with twelve units for employees (Fig. 5: The pictures illustrate Industrial Factory Frizzera—before— and Italcementi—after).

Paor designed modern buildings, both in terms of their plan and elevation; elementary artefacts that emphasized the value of industrial production in Trentino society at the turn of the century.

The solution was reminiscent of Maximilian Fabiani's (1865–1962) projects for Vienna, Ljubljana, Trieste, and nearby Bolzano. The productive buildings had rational floor plans based on the production cycle and refined elevations that structured the surroundings, giving the architecture a hierarchical role. The building containing the rotary kiln was innovative in terms of planimetry, demonstrating a synergetic design with the technicians.

The development presented innovative plant engineering solutions, such as the transport of wood and coal, and the drainage and water purification system.

In 1911, the sole proprietorship was transformed into a limited guaranteed company named the Domenico Frizzera First Trentino Portland Cement Factory ("Prima fabbrica trentina di Cemento Portland Domenico Frizzera").

The factory was extended and housing for workers and employees was designed and built on the adjacent land.

The project drawn up by Emilio Paor proposed that several buildings be extended and increased in height for the residential buildings, with four levels plus a basement and attic, as can be seen from the drawings preserved in the municipal archives of Trento.

The architect adopted a consolidated planimetric development, resolving the elevation with a refined and articulated formal structure, which was not typical of elementary and rational workers' architecture. This project was never realized.

A few years later, the Municipality built the workers' district on the eastern edge of the factory, south of the village and towards the river, with a simple urban design, with terraced modules, which still exists today.

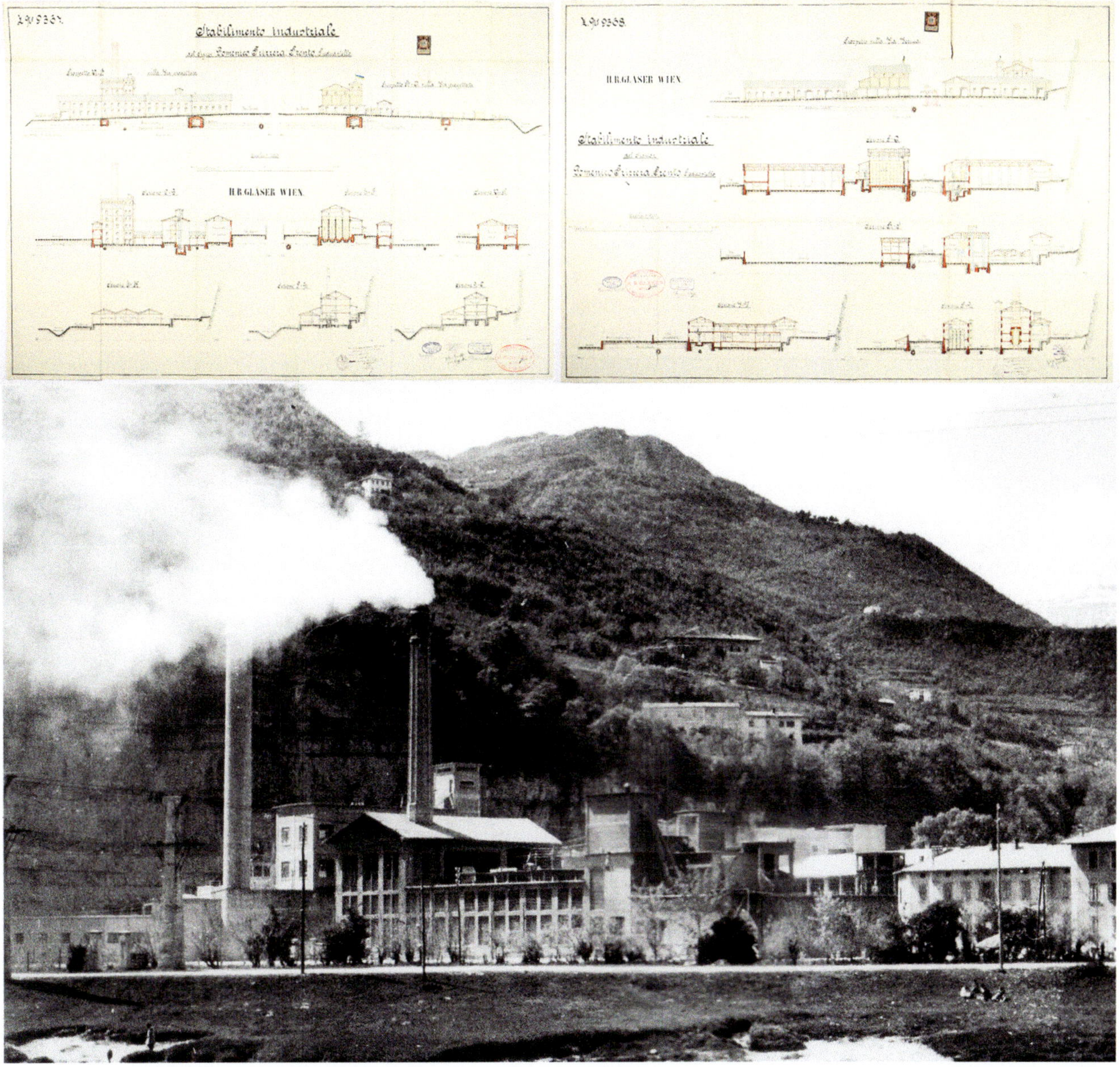

Fig. 5 Industrial Factory Frizzera (before) Italcementi (after). Original projects and photograph (1937) (design table of the research FAMA)

The development features a cross of tree-lined streets which provide different types of residences, surrounded by gardens and vegetable gardens.

In 1911, a chimney for the disposal of smoke, with a height of 50 m and a base diameter of 2.20 m, was built according to a design by the firm Alphons Custodis of Vienna, which adopted the Le Monnier system. The special nature of the construction required the technical supervision of Ottokar Culk, who directed the work.

The outbreak of the First World War brought production to a halt, and the factory was quickly stripped of all its technical and material equipment.

In 1919, the cement factory was acquired by the Bergamo Company (Società italiana e Società anonima Fabbriche riunite Cemento e Calce di Bergamo), owned by the Pesenti brothers, which was named "Italcementi" in 1926.

The factory enjoyed new growth, in part by taking advantage of the concession to use electricity supplies at favourable prices. The factory continued to expand, reaching its maximum workforce in 1939, with 240 workers.

Between the wars, the need to optimize production led to the demolition of buildings designed by Paor in order to construct new ones. At that time, civil and especially industrial buildings were not attributed any plant, architectural, or technological value.

The new buildings were designed by Italcementi's Technical Office, which planned and constructed functional buildings with a rationalist structure.

The formal and constructive simplification can be seen in the building of the porter's lodge, the refectory, and the workers' toilets (1934).

In the following years, the factory was continually updated, and in 1938, a cableway system was built to transport clay. At the end of World War II, the factory resumed production. At the same time, the restoration of some factory buildings and the renovation of the entire plant began. In 1947, some buildings were increased in height, and in 1954, new tanks were built.

The factory and its quarry, remained in operation, with alternating phases and changing owners, until the early 2000s: after a boom in the 1960s and 1970s, it suffered a slow decline in the 1980s, before closing in 1990, being abandoned, and demolished.

The plant, owned by Italcementi until 2005, was sold to the Atesino Development Institute which set itself the goal of implementing urban enhancement and redevelopment; the property was sold to Piedicastello SpA, and since 2015, the former production complex has been acquired by the Patrimonio del Trento real estate company.

In 2013, the asbestos was taken away, the factory buildings were demolished, and the large fuel tanks were removed. The entire area is now covered by one metre of soil (about 55,000 cubic metres) and a concrete slab. All that remains today are the soaring 64-m chimney stacks, built in 1950, which are under municipal protection.

In 2018, it was proposed to demolish them; structural damage was found, and the cost for their restoration was considered by many to be too high compared to their memorial value.

The risk of the rock face collapsing remains high, as to date nothing has been done. Italcementi is a symbol of the city of Trento, part of its collective imagination.

As the extraordinary images of the photographer Gabriele Basilico (Ferrari and Turella, 1990) document, the complex is a place-factory that encompasses not only the geographical but also the social context. (Fig. 6: The pictures illustrate the place-factory © Gabriele Basilico).

Fig. 6 Place-factory © Gabriele Basilico © Provincia Autonoma di Trento (Ferrari and Turella, 1990)

With the modern industrial plant, the large, simplified warehouses arranged at 90 degrees to the river, the chimney stacks contrasting against the rock face, the high fence, and the river, it is a place impervious to the rest of the city, introverted and closed.

It is grey location, yet one which tells of separation and negation; an empty space that establishes relationships with our memory and with the city.

New uses were allocated in an arbitrary manner, without any urban plan being drawn up. Only the road axes were defined, which partly traced the historical routes.

The large surface area made available due to the hydraulic regulation works carried out by the Austro-Hungarian administration in the previous century was divided into regular lots, with the surface area depending on the intended use and the economic availability of the purchasers.

On the Piedicastello lots, both the production and public buildings were constructed, including the nursery, primary school, embroidery school, and the first workers' residential building.

During the Fascism period, Piedicastello became a district serving workers. Through World War II, the village continued to be inhabited. The rocky hill protected it from bombing, although in 1943 a bomb exploded the bridge, breaking the connection with the city. For a few years, the Adige crossing was provided by a ferry. All these factors demonstrate Piedicastello's resilience.

Even when, in 1963, due to several landslides and detachments of boulders from the south face of Mount Verruca, the municipal administration forced the over 300 inhabitants to move away from the village, it was re-inhabited.

Piedicastello's industrial section has also displayed a slow and steady dynamism in its century of existence, even though this has altered the land in physical terms and caused a high level of environmental pollution.

4 Regeneration Strategies: Re-naturalizing Borders

Piedicastello therefore represents the initial nucleus of what over the course of history would become the city of Trento (Fig. 7: The pictures illustrate permanence and mutations: Piedicastello 1860–1970). It has undergone radical changes that have fragmented the area. The passage of the Brenner motorway and the bypass have divided the area into two parts: the neighbourhoods and the cement factory.

The medieval church of Sant'Apollinare has become a sort of island divided by the river, the village, and the factory.

Fig. 7 Permanence and mutations: Piedicastello 1860–1970 (design table of the research FAMA)

Fig. 8 Piedicastello-Italcementi: urban transformation 1971–2018. Year of the demolition of the Italcementi factory (design table of the research FAMA)

At the western edge of the urban structure, after the demolition of the cement factory, a large urban empty space was created, more than five hectares in size, enclosed on two sides by the mountain (Fig. 8: The pictures illustrate Piedicastello-Italcementi: Urban modifications 1971–2018).

This emptiness also raises questions about the values of the ancient village and the urban, social, economic, and landscape perspectives through which to imagine the regeneration of this layered part of the city.

The disappearance of the large production complex has left the area mutilated (Fig. 9: The pictures illustrate Piedicastello-Italcementi as urban void).

The cement factory, which occupied much of the site, was certainly a dominant and, in some ways, oppressive structure, and overly large compared to the old village.

However, it was also the village's vital element. The modern layering of the factory and its skyline are now replaced by a blank slate, locked in time both conceptually and physically, as has happened in other dysfunctional areas of Italian industrial history that are struggling to find a new identity or purpose.

The infrastructure has also been demolished; the bypass, which crosses the district and pierces through the Doss and passes under the hill, has cut it into two, adding to the fragmentation and separation from the rest of the city and the river.

Despite the former bypass tunnels being reused as a museum area and the spaces once occupied by the bypass junction being allowed to reconnect the village and the church, much remains to be done to improve the spatial relationship between its various parts and the rest of the city.

The Doss Trento hill, currently an urban park accessed by a walk that reaches the summit, is home to a mausoleum, an archaeological site, and a military museum, as well as a variety of flora and fauna that constitutes a valuable ecosystem.

However, due to the difficult connection between the park, the museum in the former tunnels, and the rest of the city, much potential remains untapped (Fig. 10: The pictures illustrate the urban system Piedicastello: Italcementi: Regeneration strategies. The images bring out Piedicastello's relationship with the city, highlighting how the waterway can become a unifying element. The images also clarify where action must be taken to regenerate Piedicastello. Extract from: Research project FAMA. Table elaborated during the 2018 winter school.).

Fig. 9 Piedicastello-Italcementi. Urban void (design table of the research FAMA)

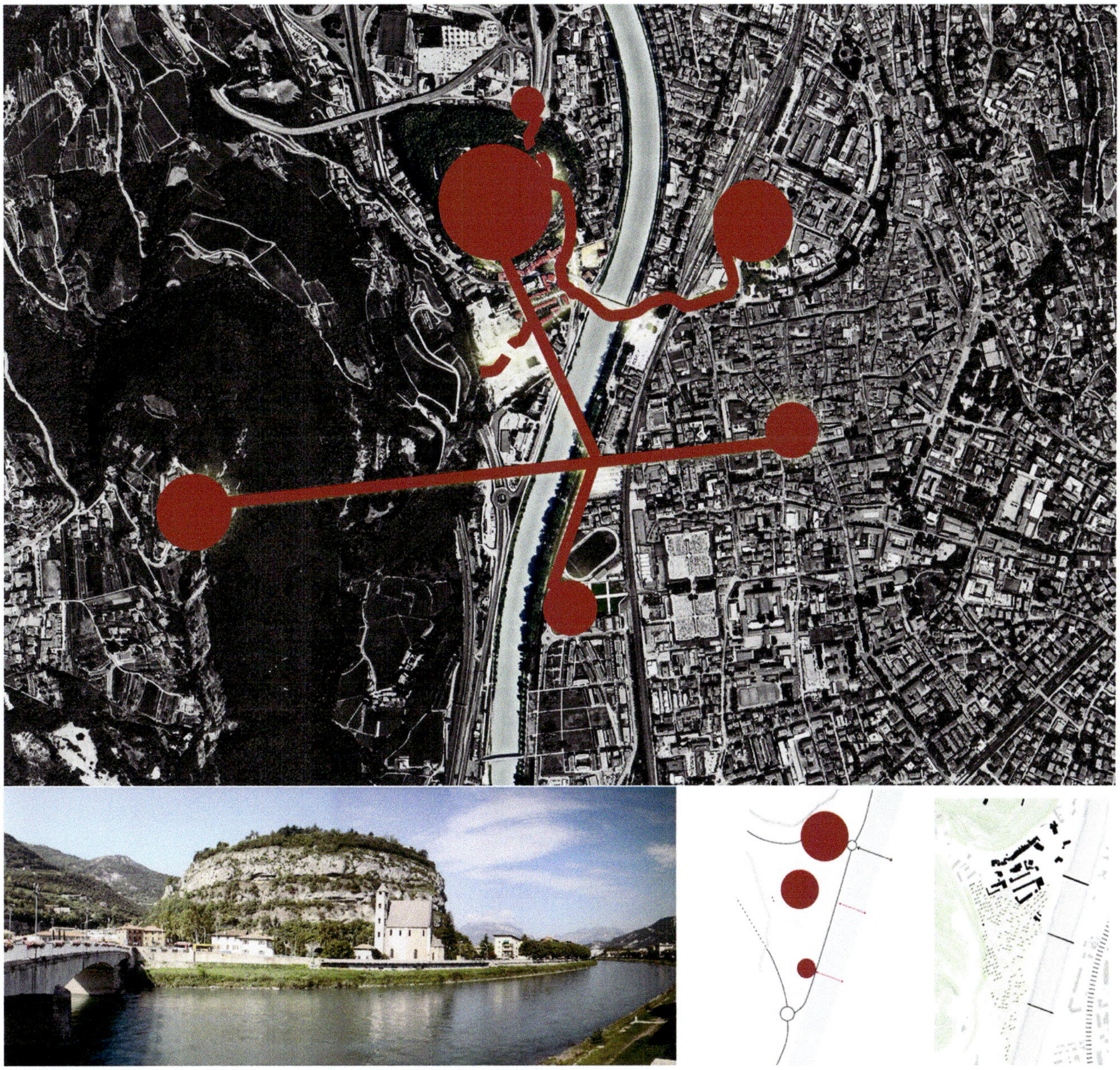

Fig. 10 Piedicastello-Italcementi. The dismission's landscape for regeneration strategies (design table of the research FAMA)

For Italcementi, even before production was ceased, many plans and ideas for reutilization were drawn up, but nothing was done.

For many years people have been looking for a strategy to reinterpret and reinvent the site, putting forward proposals to regenerate the city, and especially to regenerate with the city.

In 1992, the "New School Centre of Trento" competition was launched, and 154 different proposals were submitted. Some suggested demolishing the former factory structures in their entirety, while others proposed recovering and reusing the productive buildings (Cacciaguerra, 1993).

However, it was now clear that the area needed to be reclaimed, because it was invaded by pollutants and the mountain above it presented numerous problems.

After ten years of abandonment and lively discussions, ideas were periodically put forward for reuse, with ever different purposes: schools, research institutes, a shopping centre, a stadium, a trade fair-exhibition centre, a conference centre, residences with part of the volume destined for student accommodation, a car park, and even a large park. A masterplan was presented in 2020, which will fill the cement factory area substantially.

However, a reflection on the possible regeneration strategies for the urban system should have taken into account development undertaken in individual stages, and, in addition to the large empty space of the former Italcementi, should have considered the mountain and the geological instability, the water that periodically floods the area, the poor exposure to the sun, etc. In the collective memory, the Piedicastello-Italcementi system is still considered remote and isolated. This isolation is both physical and functional.

Natural elements such as the river, the lack of exposure to the sun, and the dominant mountains make it a sort of enclave in the city, which is further emphasized by the ways the land has been used—the lazzaretto (hospital), the Austro-Hungarian Armory, and lately the heavy industry—which have marginalized the site in terms of urban structure. All of these factors have made and continue to make it difficult to envisage solutions for regeneration.

Innovative planning is therefore necessary, taking into consideration a wider area, to include the areas of the motorway junction, the driver and vehicle licensing agency, the northern exit of the former tunnels, as well as the long strip on the right bank of the Adige River, where heavy infrastructure, pedestrian and bicycle paths, residential areas, and agriculture coexist.

The existing disorder, separation, and fragmentation could then be transformed into new order, into strengths of the urban system, and into an extraordinary mixture of different environments: mountain, plain, river, infrastructure, and city.

The great emptiness of the cement factory and the empty spaces of the former factory clash with one another, creating a sense of disorientation and making the district seem much more marginal and peripheral than it is, together with all the different elements that have been transformed or recycled in the meantime, without making the most of the possibilities offered by a complete rebirth.

The margins, whether natural or artificial, the boundary-spaces, the unresolved edges, have enormous potential for quality links and spaces.

Considering the separation of the disused and dysfunctional area from the historic village and the city, this should lead to various proposals for reuse. One of these is undoubtedly that of restoring the city boundaries with green elements.

Among the reflections undertaken, there is the opportunity to re-naturalize the Piedicastello-Italcementi urban system through a project that must be unitary in nature to overcome the current fragmentation and separation.

This process can be increased through re-naturalization of the location-factory, or at least of a substantial part of it, and through the creation of ecological corridors, connecting to Bondone and Doss Trento, which also act as a link between the two banks of the river.

The ecological network, by its very nature, has empty spaces, which in this case are the city and human activity, but these spaces cannot and must not cut connections.

The nature composition project in the city would help to breathe new life into the old village, which has also been flattened by the cement factory, restoring a more appropriate context to the terraced workers' houses from the early twentieth century, because both sites have real potential for receiving tourists.

This strategy must be linked to researching and imagining new types of green open spaces, alongside traditional gardens, and parks, going far beyond those proposed in green plans in the context of urban planning.

To qualify certain spaces, the links must be intensified between the parts, existing activities, connections, and remnant spaces.

At present, the district is a succession of isolated episodes in the urban plan; connections with the river are absent and circulation is no fluid, hindered by physical and visual barriers.

The district can be improved through urban "acupuncture" operations, with small interventions primarily targeting connections to the existing district and improving the green public spaces.

The district, together with the strip on the right of the Adige River, regenerated through connective green threads equipped with amenities, would bring hope for renewed quality and inclusiveness.

Finally, to solve some critical issues, it would seem appropriate to move back the road system that currently separates the district from the river, and to prepare new connections with the centre of Trento, which would re-emphasize the city's multi-centre structure with nuclei on the right strip of the Adige.

5 Conclusions

There has been a long succession of ideas and projects for the Piedicastello-Italcementi area, but these projects have not always respected the urban history and environmental peculiarities of the site.

Considering what has occurred in other parts of the city, such as the former Michelin area, the risk lies in not grasping the feeling and character of the pieces that make up the heterogeneous urban patchwork, and of emptying it only to fill it up with other elements, interrupting a consolidation and layering process that is necessary for a community to recognize itself.

Regenerating Piedicastello-Italcementi as an urban system requires a structure of layered arrangements to be recovered and looked at within a new system of values, one in which empty spaces can become vital economic, social, and environmental resources.

Nature is the main element for an adaptable and resilient urban requalification because it can enable objects and spaces between them to be reconverted and reconnected. The project is apparently in a fragile state, since it is conditioned by what is currently there.

However, by pursuing these objectives, the open space project can become a backbone to allow the large and empty space to take on a new, layered meaning. It would be a space that is cared for as part of ecological awareness, and a space that is understood to be more than merely green.

A space understood to be a common good, a resilient connective glue, which if well designed can also absorb future transformations.

Piedicastello-Italcementi in Trento needs this overall regeneration project/process, which is addressed as an urban system, rather than as occasional one-off projects. It is a project that must be founded on defining sequences and relationships of green and public space, for the creation of a friendlier urban landscape, with the possibility of implementing minimal interventions, reassembling existing empty spaces in search of a new meaning.

Taking these assumptions as a starting point, which obviously cannot disregard the expectations of the population living there, it comes down to designing an overall project that investigates innovative green typologies, for architectural/ecological configurations between full and empty spaces, to create unprecedented links between people, flora, and fauna.

Acknowledgements The present paper has been developed within the research activities of FAMA—Fragilità dell'Ambiente Montano e Alpino—University of 673Trento (https://projects.unitn.it/fama/it/).

References

Alberti, G. (1898). La più antica veduta di Trento. Acquarello di Alberto Dürer. Note e confronti con illustrazioni. Trento, Zippel.

Austro-Hungary. (1883). On the allocation of state funds to the Tyrol in the event of floods, Law n. 31, March 13, 1883. Available in: www.alex.onb.ac.at

Brogiolo, G. P., Cavada, E., Ibsen, M., Pisu, N., & Rapanà, M. (Eds.). (2013). Apsat 10. Chiese trentine delle origini al 1250. Mantova, SAP Società Archeologica.

Cacciaguerra, G. (Eds.). (1993). La scuola, il fiume, la città. Catalogo del concorso nazionale per il nuovo centro scolastico a Trento. Trento, Artigianelli.

Chemelli, A. (1990). Trento nelle Stampe d'Arte. Trento, Provincia autonoma di Trento, pp. 161–165.

Ferrari, E., & Turella, A. (Eds.). (1990). Trentino. Viaggio fotografico di Gabriele Basilico. Rovereto, Nicolodi.

Gatti, M. P., & Dalla Serra, D. (2019). Il cementificio di Piedicastello a Trento: la demolizione per la rigenerazione? In E. Garda et al. (Eds.), Ingegno e costruzione nell'epoca della complessità. Torino, Edizione del Politecnico Torino.

Granello, G. (2003). Note ad una lettera di Cassiodoro. *Studi Trentini di Scienze Storiche, 82*(1), 283–291.

Heidegger, M. (1976). Essere e tempo. Milano, Longanesi.

Iori, E. (1955). La Badia di San Lorenzo. Tempio Civico. Trento, Stabilimento Arti Grafiche Saturnia.

Mariani, M. (1989). Trento con il sacro Concilio et altri notabili. Libri tre. Commento e note A Chemelli. Trento, Lito Velox.

Mariani, R. (2011). Lo stabilimento per la produzione di cemento portland Domenico Frizzera. Le origini dell'Italcementi di Trento. *Studi Trentini. Storia, 90*(2), 361–400.

Rasmo, N. (1966). S. Apollinare e le origini romane di Trento. Trento, G B Monauni.

Urban Regeneration of Industrial Heritage, Tirana Case

Juljan Veleshnja

Abstract

The gradual delocalization of production activities, in the last decades of the last century, led to the progressive disposal of a considerable portion of the industrial heritage present in the periphery of the urban areas. Places originally characterized by a great productive fervor gradually have been transformed and now are presented in conditions of almost total abandonment and advanced degradation. Characterized by strong potential, many of the abandoned industrial sites were not perceived as resources to be used by their very existence. In the Albanian context, the industrial heritage is being completely or partially demolished, or alienated transformation, limiting the conservation interventions to a few isolated fragments, without any links with the context in which they are inserted. Tirana city nowadays is crowded with numerous used and unused buildings—such as former industrial buildings—which are the heritage of almost 50 years of the communist period. The process of de-industrialization and the late spontaneous development in Albania led to the depopulation and abandonment of entire industrial areas. With the phenomenon of Tirana city sprawl, former peripheral industrial sites are becoming degraded areas inside the city, now unable to communicate the relationships that originally existed between what remained and the disappeared activity, results meaningless today. This research aims to highlight the interest and the potential of the industrial heritage, with a focus on Tirana city, on the former industrial sites, with the challenge of how to develop and regenerate the former industrial sites. The assessment intentions at eventual acupuncture urban regeneration and to adapt and reuse these structures taking into consideration their former uses, for new future functions, and the influence these interventions have on a larger scale.

Keywords

Industrial heritage • Urban regeneration • Sustainable development • Communities • Social cohesion

1 Introduction

Albania has faced many transformations in the last decades caused by many political, economic, cultural, and social aspects that are still in continuous evolution (Barthelemy, 2016). Those transformations are characterized by spontaneous and informal development, creating different dynamic cities with good and bad urban realities. There is a lot of cultural and architectural heritage that is getting lost and demolished to free up space for new high-rise buildings.

Tirana, as the capital city, has faced an imponent urban sprawl, and the former periphery today is part of the urban fabric. The former industrial sites are under enormous pressure, dictating their decay and demolition as a unique answer to their transformation. The methodology of the research will be based on the desk research of the recent theoretical aspects, definitions, goals, models, or methodologies on how to regenerate the former industrial sites, and on the analyses of the best-practice cases of the international scene. This is to build strategies for urban regeneration of a specific former industrial site in Tirana, as an alternative to demolition and loss of this heritage.

2 Industrial Heritage

Cities are very dynamic organisms in continuous evolution, created by people for people, which encompasses cultural traces from the past and the present, shaping unceasingly for

J. Veleshnja (✉)
Department of Urbanism, Faculty of Architecture and Urbanism, Polytechnic University of Tirana, Tirana, Albania
e-mail: juljan.veleshnja@fau.edu.al

A. Battisti et al. (eds.), *Greening Our Cities: Sustainable Urbanism for a Greener Future*, Advances in Science, Technology & Innovation, https://doi.org/10.1007/978-3-031-49495-6_32

the future (Register, 2006). The city has to be understood, and for this, it is needed to determine its intrinsic and deeper values, the sense of its fundamentals, and the recognition of its important aspects (Roders & Bandarin, 2019). The development of the city should be outlined highlighting its divulged attributes to have a clear idea for its implied values, if those need to be preserved or transformed. The analyses and the inspection of the city are inseparable from the analyses of the society and the people who live in it, the social perception of the places as they are or as they should be, independently if those places are old or new, small or large, historic, industrial, underline that the city is the heritage of its inhabitants. The city is a visual and touchable aspect of our culture, generating cultural landscapes representing the sense of place evidencing the long relation of the people with the territory (Moore & Whelan, 2007).

The late development is threatening continuously the historic areas of the cities and the transformation of the landscape character, which are under pressure of the new constructions, but also in smaller settlements evidencing the diminishing of the local distinctiveness, that is why new strategies are needed to protect the culture heritage, which will benefit the communities in necessity to enhance the sense of place.

UNESCO defines that cultural heritage includes artifacts, monuments, a group of buildings and sites, and museums that have a diversity of values including symbolic, historic, artistic, aesthetic, ethnological or anthropological, scientific, and social significance (UNESCO, 2009). It includes intangible cultural heritage embedded into cultural and natural heritage artifacts, sites, or monuments, and it covers industrial heritage as well as cave paintings. The heritage regardless of being architectural, vernacular, industrial, etc., is a unique appearance of the wealth and multiplicity of common culture.

There is a large international debate about culture and heritage and their influence on the sustainable development of the territory, in parallel with the aspects of society, environment, and economy; culture and heritage are seen as the fourth pillar of sustainability (Meireis & Rippl, 2018). Industrial heritage as part of our culture needs attention and new ways to develop them.

Industrialization helped Europe to develop and change its face; this process generated a large industrial heritage, and a huge network of industrial sites was spread all over. Globalization, technological development, de-industrialization, industrial replacement, and economic re/adaptation have brought visible effects on the industrial sites (Loures, 2008). In cities that are developing, and facing the urban sprawl, the former peripheric industrial areas now are encompassed on the urban fabric, part of the city. In other areas that are facing the phenomenon of shrinking cities, the leftover industrial sites are transforming the natural landscape with their age and degradation. The former industrial sites with the late development gave birth to a massive collection of degraded and abandoned industrial facilities, with a negative impact on the environment and the quality of life in the surrounding (Goodman & Chant, 1999). The industrial landscapes are inseparably part of the image of the urban settlements and our cities, so part of our culture and heritage. Recently many researchers have focused on documenting, evaluating, and finding solutions for the further development of the industrial sites and society (Douet, 2016), underlying the importance of the former industrial sites and their role in strategic city planning. Former industrial heritage and the landscape created are a resource for the development of the city, as an essential part of our collective identity, to be examined in different perspectives, starting from the former function to culture, the layers of the history, and the human activity in time, in it and the surroundings (Orange, 2014). Before giving new attributes to all the aspects of the former industrial sites, it is necessary to define and categorize the material and immaterial traces of the industrial culture (Kahn & McComas, 2021). The late economic changes make the degraded former industrial sites be seen as no man's land, in between the new development in the surrounding and the former industrial buildings. The functional and physical decay of those sites is contributing more and more to creating a negative community perception about them (Ashworth & Graham, 2016). In 2003, the *Nizhny Tagil Charter* was created for the industrial heritage, which underlines the importance of the physical elements of the industrial buildings and structures, the processes and artifacts that take place in it and the surrounding urban areas, and all materials and immaterial aspects related to them (Parisi & Chimisso, 2021). Former industrial sites have to be seen as resources and assets, and they are the expression of the combination of the former natural territory and the industrial buildings defining a place. As historic sites, they give place to the creative practice in the conservation of the physical elements, to the design process of their transformation, and the urban regeneration of territory.

2.1 Theoretical Aspects of Urban Regeneration of Former Industrial Sites

The adaption and reuse of the former industrial areas should be done under the frame of urban regeneration, and for a better contextualization of the urban answers, the new development and the refurbishment of the former settlement should aim the sustainability. The urban regeneration process tends to bring sustainable and long-term improvements in time, in the aspects of economic, social, physical as well as environmental conditions of the entire area which is subject to transformation (Roberts et al., 2016).

Degraded and abandoned former industrial areas need responses to the opportunities that are present in them. The urban strategies and interventions can be done on different scales, from bigger, like on a regional or national scale, to smaller ones, like local or neighborhood scale. Many times, the smaller scale is easier to execute and achieve, to influence even the bigger scale, aiming to increase the quality of life of inhabitants affected by large-scale developments and demographic movement. The scales of urban interventions should be viewed with a bottom-up systematic perspective, and the macro-environment is viewed as a patchwork the sum of components in the smaller scale (Dopfer et al., 2004). The smaller scale interventions aim to repair the cracks of the degraded former industrial sites, rather than replacing the old with something new, but by filling the existing decay and problems with stimulating interventions (Stratton & Trinder, 2016). In many cases, interventions on the former industrial sites are not done appropriately, usually only in individual buildings alienating them from the idea of the harmony and unity of the whole industrial landscape, and other times the demolition of industrial constructions with important story and significance (Casella & Symonds, 2005).

Former industrial sites, as part of the city, have to be considered as assets and catalysts for urban regeneration, and they are a very pertinent part of the urban structure, the part that should be taken into consideration. Our heritage includes historical characteristics and cultural values that society must preserve, landscapes, buildings, collections related to identity, language, music, etc., all are part of our memory and belonging. This makes the place we live more competitive and underlines the distinctiveness, meaning, and quality, evidencing our identity and providing a sense of continuity, as a social and economic asset (Drury & McPherson, 2008). For the process of urban regeneration, former industrial sites are very important spots where to start from. There are different theoretical principles of urban regeneration, which aim to achieve a sustainable development of the degraded leftover urban areas.

2.2 Urban Regeneration Case Study, *Ex-INCET* Turin, Italy

In Europe, there are several initiatives on supporting, adapting, and reusing industrial heritage sites, like the European Route of Industrial Heritage (*ERIH*), which contributes to the tourism information network of industrial heritage in Europe, representing more than 1000 sites in 43 European countries (Grewenig, 2022). Industrialization has influenced visibly the development and the image of Europe, and there is a vast industrial heritage spread all over Europe. This heritage has to be repurposed and activated with new ideas, this is important, and that is what the European Route of Industrial Heritage is doing. The multiplicity of activities and initiatives are bringing back to life a vast network of former industrial sites, like "*The Emscher Landscape Park, Ruhr region re-naturalization*", or "*Officine Grandi Riparazioni (OGR) of Torino*", etc.

A particular case is the urban regeneration intervention of the former INCET area in Turin, which is located in the northern area of the city, in the district called *Barriera di Milano*, in an area affected by significant territorial transformation processes and expansion of the city (Camasso et al., 2013).

The former INCET factory (*National Electric Cable Industry*) is an industrial complex founded in 1888 by Vittorio and Giuseppe Tedeschi, and it was the seat of the Anonymous Society of Electric Cable Factory, thanks to the high-quality level of its products, mainly high-voltage underground cables tension, which made them obtain important orders in Italy and abroad; it recorded a significant increase in production and manpower. The company also had houses built for its workers on Corso Vigevano.

The production almost completely stopped after the bombing of 13 July 1943, and then resumed at the end of the conflict until the years after the Second World War when Pirelli, which had acquired the company keeping its name until 1968, abandoned the plant which became the property of the Municipality of Turin. A part of the site remained abandoned until 2010, while other parts underwent recovery interventions to house offices and commercial spaces of the FACIT clothing warehouses; other parts were demolished and replaced with new residential buildings.

Barriera di Milano, a working-class neighborhood subject to massive waves of inter-regional migration until the second half of the last century, has recently lost much of its original connotation; the territory is characterized, today, by conditions of physical and environmental degradation, phenomena of social exclusion, perception of insecurity and strong impact of the non-EU migratory phenomenon. From a social point of view, the area has almost 53,400 inhabitants, a varied social composition with increasing foreign residents and a strong incidence of the elderly. From a physical-morphological point of view, the intervention area was characterized by a densely inhabited central nucleus, a ring of abandoned industrial areas, and a strip of public housing.

The whole area was characterized by marked physical degradation, both in terms of private building heritage and in terms of public spaces. The area also had a strong lack of open urban spaces and a scarce number of square metres of greenery per inhabitant; moreover, there were no cycle paths or cycle-pedestrian paths as well as meeting spaces for cultural, social, and sporting activities.

From an economic point of view, the territory has a fragmented economic fabric, largely composed of small-scale entrepreneurial activities operating in the commerce sector. In recent years, we have also witnessed the weakening of proximity trade, a reduction in craft businesses, and the replacement of traditional activities in favor of low-skilled activities (e.g. money transfer, telephony, etc.) together with the establishment of large shopping centers.

Low levels of education of the resident population, compared to the city average, characterize a large number of employed people with limited professional qualifications. A high unemployment rate has meant that the area has significant critical elements, which have become even more acute with the economic crisis recently, and due to those aspects the area was placed in a weak position compared to the rest of the city.

In a context of this type, the need to redevelop numerous areas (primarily the former abandoned industrial areas) to be placed at the service of the district for social purposes has represented a strong request, together with the request for principals and functions aimed at dealing with the security problem (Bonini Baraldi et al., 2019).

The redevelopment and regeneration intervention project was started in the integrated development program of the former industrial site started in 2007. The interventions are divided into four aspects: physical requalification, support for the economy and employment, actions in the social and cultural sphere, and finally accompaniment and communication. Those aspects have specific objectives: regeneration of public space to promote social, economic, environmental, and cultural development; support the local economy and improve employability; strengthening social cohesion and the cultural offer of the territory; promoting the integration of interventions and guarantee the information and participation of local actors (IRES, 2017). At the former INCET, the regeneration project included the construction of the zonal headquarters of the Carabineer Command and a Multifunctional center of integrated services for the community. The implementation of the project was done in phases, in different periods, related to different public and private investments.

The former INCET area regeneration is about the different actions located within the physical-environmental aspect, aimed at the redevelopment, adaption, and reuse of the buildings. Of these actions, the first includes the reclamation, consolidation, structural recovery of existing buildings, and the creation of some spaces for local associations along the southern front of the complex of sheds located in the center of the intervention area. The second action is aimed at the construction of a multifunctional center of integrated services for the community and the superficial arrangement of the external areas, with the construction of level parking lots and a new road system inside the building. The third action has as its objective the construction of a new building, home to a creche (capacity 60 children) and a nursery school (capacity 75 children).

The other two actions are included in the socio-cultural aspects and are, respectively, the action called "Ex-INCET set-up and start-up", which provides for the preparation of the spaces intended for local associations, the covered gallery (Fig. 1) and the square in front of the Ex-INCET (Fig. 2), through the provision of structures, furnishings, tools, and technologies, to guarantee the full operation of the services offered and the best use of the places created, and the "Wireless Barrier" action aimed at making a wireless network available for the Internet connection of citizens in specific points of the neighborhood. The reconversion of the site from its original industrial vocation involved reclamation, disposal, and environmental safety operations which added the need to recover and consolidate the pre-existing buildings. The new Carabineer station has preserved the architectural outline and the historic facade of the pre-existing building on street Banfo. The station, in addition to the spaces of the carabineer, also includes city services. Three floors above ground for an area of approximately 2500 sqm, it is organized into three functional areas: the eastern part of the ground floor has the public functions, the western part hosts the common logistics services for operating and resident personnel, and the residences were arranged on the upper floors. The new Carabinieri station has been active since June 2014.

The multifunctional center is located in the complex of pre-existing industrial warehouses of which it has preserved the original volume, the architectural fronts, and the structural configuration. The building occupies a total area of about 4000 sqm and is divided into three parallel areas and completed on both ends by buildings that act as a unifying element. On the southern front, in October 2015, the Open INCET center was inaugurated, an Open Innovation space to promote business development and in particular youth entrepreneurship (Fig. 3). Open INCET hosts even the areas for the "Albertina Academy of Fine Arts—Turin" second level courses regarding workshop-courses of story and theories of scenography in theatres (Fig. 4). This aspect makes this place used by many artists and students, bringing life and ideas making it a more creative place.

The central area was converted to a public internal square, partially covered, and is connoted as a meeting space in the center. The west wing of the complex (two floors and a further elevation intended to house commercial, service, and residential activities for a total of 2950 sqm) was put up for sale by the Municipality and purchased, for almost 900,000 Euros, in February 2014 by a private company.

This implemented regeneration project case study can help in building new strategies and ways of valorizing the industrial heritage in different areas which presents physical decay, economic problems, and social pressure from

Fig. 1 Inner court ex-INCET (© photo Author JV)

low-income populations and emigrants, aspects that are similar in many other places, even in Tirana and Albania.

3 Albanian Context—The Late Development

Albania has faced different historical, cultural, social, and economic aspects that have outlined the development of the country, urban, rural, agricultural, and industrial areas. The former industrial sites' development in Albania seems to be not in comparison with the global discourse of sustainable development, exposing a particular development. For a better understanding of the Albanian industrial and former industrial contexts, it is necessary to analyze its development from the wide industrial era growth to centralized progress during socialist leadership to a long transitional laissez-faire situation of the post-socialist system. After a long period of Ottoman Empire invasion, Albania was presenting an agricultural development at the time. With the political changes during World War I and II, Albania faced the invasion of Fascism and Nazism later. The end of World War II signed the birth of a totalitarian regime in Albania "socialism". The development of the country was strictly designed, controlled, and dictated rigidly by the central government. The dogma of "*being all equal, of the socialist man*" made the government start a process of re-appropriation of all the goods to the state, private ownership and property was not allowed. The development of the territory and construction was centralized to the Central design research institute, as the only and main actor in the field of planning, design, and construction, which should stay in line with the ideology of the central government (Fischer & Schmitt, 2022). The controlled development was based on an anti-urban policy, distributing the population homogeneously in the territory and controlling the equilibrium between urban and rural areas. During the socialism 1944–90, the demographic movement of the people was controlled for maximizing

Fig. 2 Main façade ex-INCET (© photo Author JV)

the use of the whole territory, refurbishing the former urban areas and rural settlements, but also with the creation of new settlements near natural resources, forests, energy sources, industrial sites, etc. (Bërxholi, 2000). The construction of new industrial sites, during this period, was almost accompanied by new residential areas in proximity.

With the fall of the totalitarian regime in 1990, Albania faced a difficult transition period of deconstruction of the state and appearance of the democracy and the free market economy. The functioning, role, and power of the government were put into question, in strong contrast with a previously fully centralized one (King, 2005). The totalitarian regime evidenced visible progress in the isolated country where all the decisions were under the total control of the central government. The transition from isolation to an open, free and democratic society was a long, difficult, and transitional process. As an expected contra-reaction towards hyper-control, after the fall of the former system, the phenomena of spontaneous initiatives started to appear, guided by the market economic influence and individual interests (Pllumbi, 2013).

The difficulty of the government's instruments to control this transition phenomenon was evident in the development of the whole territory, where emerged the interest in the construction sector started from the high demand for housing (Velo, 1998). During this transition to the post-socialist system, there was an evident degradation of the industrial sites and surroundings, of the social, cultural, and natural sources. The country went through a period of multifaceted reforms aiming at the transformation of its economic system. During this period, there were different laws to regulate the private property of the former owner, taking it from the government. The reform in the privatization was accompanied by the diminution of the production and lack of the government to be competitive (Streule et al., 2019). On the other hand, the demographic shrinkage that accompanied the abandonment of the Albanian territory, in search of a better life, in urban areas or developed countries, makes the industry market more vulnerable. The transition and the spontaneous expansion brought two types of urban development, cities that faced sprawl and the ones facing urban shrinkage and degradation.

The industry passing from the central control of the government to the abandonment, brought in a large scale its physical decay and alienation (Fig. 5). This process influenced negatively even the surrounding residential areas. In some cases when the former industrial sites were

Fig. 3 Business start-up ex-INCET (© photo Author JV)

inside the urban centers, like Vlora city, they were fully demolished and transformed into residential areas erasing every trace of the former sites. In other cases, industrial sites were partly occupied by private activities alienating the former shape and function, and partly by communities with low incomes in search of shelter. During this transition, in many fields and directions, globalization and the individual interests brought about the loss of a good part of our cultural heritage and a large part of our industrial heritage.

The development of the industrial sites in Albania over the years is strongly related with the economic, political, cultural, and social aspects of the whole country (Parangoni, 2012). The first traces of industries came after the Independence of Albania in 1912, with the initial concessions which were issued to foreign companies like French, Austrian, and Italian. After World War II, the totalitarian regime was focused on relations with only a few states, and the first cooperation was with ex-Yugoslavia 1944–1949, as a neighbor country, in strong relation with Russia and socialist ideology, these cooperations gave birth to various factories and transport facilities. During the period 1949–61, international relations followed with the Soviet Union, which is also done by mutual trade exchange. This period was characterized by the industrialization of the former economy based on agriculture, and this brought the need for technical and professional skills in the labor force, which was possible by cooperation with the Soviet Union. In 1961, Albania and the regime of the time broke the relation with the Soviet Union. During the period 1961–1970, Albania formed an economic and political relationship with China. The period 1970–1975 signed the finalization of the "Steel of Party" metallurgical complex in Elbasan. Those were years when the totalitarian regime started closure and the doctrine to do everything ourselves to strengthen economic independence, improve the quantity and quality of industrial production, and also to increase agricultural production. The period 1976–1980 signed the last period of economic relations

Fig. 4 Scenography ex-INCET (© photo Author JV)

Fig. 5 Superphosphate factory, Laç (© photo Author JV)

with China and increment of self-production goal. The contributions of Russia and China was strongly related to the specific technological aspects of the industry and the architectural morphology of the buildings.

The period 1981–1990 evidence the use of intern assets (existing manufacturing capacities) and rupture of political and economic relations with all the other countries. The economic potential was low, and this was visible, very few industrial factories were constructed during this period. The last part of this decade coincides with the fall of the totalitarian regime, the standards of living very low, and as in many other sectors, most of the heavy industries showed to be not profitable. In this last period, the inner production was 40% less than previous periods.

With the changes of 1990, and the economic free market, the government tried to manage the situation, there were applied policies of privatization of public properties and state-owned enterprises, a phenomenon that involved the entire system of production and industries. Industrial production in this transitional period deteriorated, exports were negated, and imports consisted of small individual customer services and production.

These changes transformed the condition of industrial plants which were abandoned and in total decay status, and this affected even the surrounding areas. The spontaneous development enabled the different alienation of the industrial sites, from total degradation, to "tabula rasa", and partial and total transformation. The government year after year is growing its influence in the territory and this spontaneous development slowed. There are designed and approved Plans at different levels, starting from the National scale to general local plans, there are designed different development strategies, but none of these is focusing enough on the heritage, and especially on the industrial heritage.

4 Industrial Heritage in Tirana City

In almost 30 years, Tirana quadrupled the number of its residents and faced an imposing urban sprawl. Besides the center, which even if in a continuous process of change somehow has a consolidated urban form, the other parts of Tirana city have an impermanent urban structure (Bego, 2009). The palimpsest metaphor (Corboz, 1983) finds its best application in Tirana. The history of Tirana is the history of a city where the existing context never mattered.

Tirana become the core target for migration of the population from other small cities. For the public authorities, it was quite impossible to find a balance between the huge and fast needs of citizens and a fast transition process in a market economy. This kind of misconception was not only due to lack of information, but most of the time was a mindset issue that produced the perfect situation for people who took advantage of the confusion and made the first step in using public property without any legal provision. The government, during 1990–2010, was not able to control the situation to fulfil the accommodation needs and prevent informality (Xhindi, 2013), even in the following years. These new areas were not providing any services to the population, or any public space. Not any of the standards of urbanization were implemented (Kolevica, 2004).

The urban informality in Tirana finds different expressions in the built environment, and the urban informality can be divided into two main categories: *the urbanization of the surrounding agricultural land (sprawl)* and the *informal interventions in the urbanized area of the former city (additions)* (Veleshnja & Branko, 2018). This spontaneous and informal development has affected irrevocably even the industrial sites. Industrial constructions were built over the years expanding and adapting the production line, market needs as well as changing industry technologies.

These industrial sites, built mainly during the years 1945–1990, have had a special and very important role in the urban form of the areas they belonged (Parangoni, 2012). Industrial sites and their surroundings in the city were not just working places, but they were places where life was going on. "*The Combine (Kombinat - Textile Factory) more than a neighborhood is a story. Not all neighborhoods, including those of Tirana, have something to show or know how to show it. Kombinat has all the ingredients for a great story: beginnings, characters, heroes, conflicts, plots, activities, and conclusions. All this in a well-defined, easily recognizable, and describable space and location.*" (Za, 2011).

In Tirana, some of most the important industrial sites, which almost still have traces in the territory (Bushati et al., 2018) (Fig. 6):

Fig. 6 Former industrial sites in the updated schematic map of Tirana (Bushati et al., 2018)

Textile Factory "Stalin"—partly occupied/transformed/decay,
Plant "Enver"—demolished,
Wood processing factory "Misto Mame"—partly occupied/transformed/decay,
Mechanical plant "Partizani"—decay,
Food factory "Ali Kelmendi"—transformed function,
Porcelain factory—partly occupied/transformed/decay,
Mechanical plant "Dinamo"—partly occupied/transformed/decay,
Tractor plant "Enver Hoxha"—partly occupied/transformed/decay,
Chemical and food industries—partly occupied/transformed/decay.

The actual former industrial heritage in Tirana is not in good condition, but still with present traces. There are no proposals and reuses of them, but several demolitions of them, like the case of the *"Enver" plant*, which is demolished and new residential blocks have taken place, not any trace of the past, any residential and commercial areas on the ground floor. "*Dinamo*" *mechanical plant*, partly used by some artisans, and the rest storage and market, a degraded area with different interests to transform this area. *The tractor plant "Enver Hoxha"*, is partly used for some private leather production activities using the former building, in its shape, with new types of machinery, partly used as storage for other local public institutions, and partly appropriated and abandoned. The only site that has faced a project of transformation "*Reconstruction of Kombinat Neighborhood—KombinART*" is the *"Stalin" Textile Factory*, known by the name of *Kombinat*. As these few cases evidence, the vacuum of the legal frame and institutional agenda in valorizing the former industrial sites, unfortunately, none of them is part of the strategies of the General local plan Tirana2030.

5 Tractors Plant "Enver Hoxha"—Tirana, Regeneration Strategies

The tractors plant is located in the east area of Tirana city and represents one of the most important industrial sites in Albania. The area had its starting point as a small plant after World War II, producing simple mechanical parts, and then expanded during the financial plans 1966–1970.

During the 5-year financial plan in Tirana, 1971–1975, a wider territory was assigned to be built for the production of tractors, the east area of the former factory in Shkoza. There are says that this plant has a network of underground tunnels, which were built together with the buildings that were built after that period. These tunnels, eventually, would be put into operation in the case of an attack, whenever this could have happened, the necessary "war" tanks would be built and repaired (this was a common thought during the totalitarian regime, where the attack from the capitalism was an imminent supposed action).

The first Albanian tractor, type DT-54, according to the Chinese model of chain tractors of that time, was produced, released and officially inaugurated on 16 October 1978, exactly on the 70th anniversary of the birth of Enver Hoxha, dictator of totalitarian regime.

The first design was finalized by Albanian specialists because the Chinese aid was canceled due to the breakdown of relations between the two countries. The government used this situation to propagate that the first model was an Albanian achievement based on its forces. Apart from the political aspect, this work preserves invaluable memories of the years 1970 when the first auto tractors were produced in Albania. This tractor represented the spirit of Albanian agriculture, which dominated the country's rural landscape. The Tractors Plant occupies an area of 102,000 sqm; inside it, there are seven departments where about 3800 employees used to work. During that period, the surrounding area was not dense with residential buildings, the rural area of Shkozë where the industrial site took place was counting a few families, and this was a peripherical area of Tirana city at the time.

After the changes in 1990, many industrial sites stopped their activity as they were under public administration which was facing economic and political transition and difficulties. Since 1990, everything has stopped, from work to the dismissal of workers. In 1994, the departments that were still working were called the auto parts plant, with the main body being the mechanical and instrumental departments. A total of 94 workers were working in the plant, producing parts such as spare parts for cars, storage and maintenance. The Tractors Plant "Enver Hoxha" does not work nowadays, but some of the buildings have been adapted for residential purposes, the tailoring industry, premises rented to a private television in Tirana, and warehouses.

The Tractors Plant is divided into two parts, the east part which is more preserved, and in the west part which is located a few blocks below, where invasive and substitution changes in the dynamics and functionality of the building can be observed. The stoppage of the plant's function caused the area to lose the purpose it once had, which led to a large incursion of people moving into this area of abandoned buildings.

The eastern part of the plant is functionally and physically fragmented, giving parts of it to different private companies used with different functions over time, but still preserved former industrial building (Fig. 7). Some of the buildings on this part are occupied informally by some families with low income and without a place where to stay.

In the western part of the former industrial site, after the loss of functions in this part, the incursion and usurpation of

Fig. 7 View from tractors plant "Enver Hoxha", Tirana (© photo Author JV)

abandoned buildings by a small community of about 100 families took place, acting in total informality. From the architectural and morphological point of view of the buildings, a symbiotic coexistence is observed, where the residential buildings are attached to the walls of the factory building, becoming an integrated part of it. The informal development during the transition in this area shows a pronounced lack of infrastructure as well as a disorderly and deficient urban system, conditions which, in parallel with the alienation and degradation of the industrial area, present problematic environmental conditions. In the industrial area of the former tractor plant, various toxic wastes are observed, mainly oils and various chemical compounds that come as a result of the processing of various parts in the processing departments.

Despite the physical and environmental conditions, the population that came to this area shows a sense of community in certain segments. Since 2016 in Shkoza, there have been built 385 social housing units, placed at the disposal of the Roma community and those in need, as well as families that have expressed a need for social housing (Fig. 9). The presence of social housing, built in areas previously dedicated to industrial buildings, based "tabula rasa" phenomenon, significantly increases the number of residents in this area. The increase in the number of these residents also brings several social problems which significantly express of need for social cohesion.

This territory presents industrial and former industrial areas; educational institutions and recreation areas of them; health and services facilities; religious buildings; community buildings; public spaces and green areas; and natural areas in the surrounding. This former industrial area is close to a natural area of Farka (artificial) lake, still with vast natural areas and private residential villas. Actually, there are not many official documents with intervention to be done on former industrial sites. The only official document is the General Local Plan, Tirana2030, designed in 2016 and approved in 2017. This plan, for a polycentric development of Tirana, evidenced some poles, and one of these is the former industrial site of tractors plant "Enver Hoxha" and Farka Lake and surrounding. For this pole, the general local plan proposes functions of cultural facilities; residential; leisure/recreation; and natural oasis.

Close to the interested area, Lana River is passing by, which puts it in connection with Farka Lake, this can help emphasize the proposed "natural oasis". Another important urban element is the new greater ring road of Tirana city, which passes close to the area, becoming somehow a barrier between the former industrial area and the natural areas of Farka Lake (Fig. 8).

For the urban regeneration of the Tractors Plant "Enver Hoxha", in the General Local Plan, Tirana2030 is planned to be an urban pole for Tirana city. The former industrial site is proposed to increase the relation with Farka Lake and the natural area that surrounds it, and for this pole, there are proposed different main functions cultural facilities; residential; leisure/recreation; and natural oasis.

For the urban regeneration strategies in this research, after the desk research and analyses of the publications and

Fig. 8 Satellite image 2018, tractors plant TR, source ASIG State Authority for Geospatial Information

Fig. 9 "*Residents*", tractors plant TR (© photo Author JV)

official plan documents, there has been conducted several site visits and interviews with the different local actors, the workers of the industrial site, the inhabitants of the area (Fig. 9), and others. All those steps helped build some urban regeneration strategies in line with the General Local Plan strategies.

Food Market, Urban Agriculture, and Professional Building Capacities

On the site, it is important to propose an important economic pole, which dictated by the context can be strongly related to agriculture and natural areas around. One food market can be organized into the former industrial site, aiming for the involvement of the local people, who with their fields and production will be the first to sell their product and let people participate in the farming and agriculture processes. The market can be strongly related to different existing and new agritourism in the surrounding, which involves agriculturally based operations or activities that bring visitors to a farm; by combining agriculture and tourism, agritourism offers new sources of revenue.

On one of the other buildings of the former industrial site can be organized workshops which can help and involve local people to build and form professional capacities in different fields, but focusing more on agriculture processes, with urban agriculture, unemployed local people can practice cultivating, processing and distributing food around urban areas, and to the food market (Fig. 10). These activities will bring to a strong connection to the former tractors plant and the surrounding residential area with Farka Lake and natural areas around. (*Related to the General Local Plan, there involved: Economy + Education + Natural Oasis + Recreation.*)

Polyfunctional Cultural Center

The former industrial site has different buildings that can have new functions regarding cultural aspects, not only on the site scale but on the city scale. One of the buildings can be used as a polyfunctional cultural center, dedicated to theatre, exhibitions and other cultural activities. This will bring a creative community from Tirana city, which can help with the interaction and cultural exchange with local people.

The former industrial site has enough place for other economic services related to coffee-bars, restaurants, shops, libraries and others. (*Related to the General Local Plan, there involved: Cultural facilities + Leisure/Recreation.*)

Residential Buildings

The area of Shkoza has been proposed on the general local plan Tirana 2030, for the demolition of vast degraded areas, reconstruction and building of new residential blocks of collective and social housing. These new blocks can be combined with the natural parcels to give local people have possibility of having natural parcels of land with the possibility to cultivate and make agricultural processes. Those are some of the activities that help raise the economy, but although on help social cohesion. (*Related to the General Local Plan, there involved: Residential buildings.*)

Fig. 10 Courtyard—tractors plant TR (© photo Author JV)

Public Spaces, Green Paths

The existing residential areas need to be refurbished, trying to organize the infrastructure and give birth to new public spaces. The proposal of new residential blocks gives the possibility to propose public spaces for social and commercial activities. The network of public spaces has to be connected with the former industrial site hosting the food market, the polyfunctional cultural center, and other services, but although it is very important to build new natural paths and to connect those areas with the natural area of Farka Lake and the agritourism in the surroundings, and with Dajti mountain. This will make the area more walkable and cyclable, but although giving possibility to more activities for local people and visitors. (*Related to the General Local Plan, there involved: Leisure/Recreation + Natural Oasis.*)

6 Conclusions

This research aimed to underline the importance of former industrial sites as important starting points in the process of urban regeneration, and not only for the aesthetic aspect, but for all the complexities of those areas with a focus on human presence. Many former industrial sites in Tirana have a fundamental value in the regeneration process because this stock is part of our heritage and culture, and for this it needs to be protected and reused. Bringing the past to the future, to evidence the distinctiveness of the characteristics and "genius loci" of those sites can ensure the uniqueness of Tirana city.

With this research, there are given some main proposals for the former industrial area of the former Tractors Plant "Enver Hoxha" and the surrounding, aspects that surely needs to go deeper, to find the right partnerships, funding, and participation, and give specific answers to all the other problematics. These proposals try to follow the theoretical aspects of urban regeneration, giving answers to different aspects, economic, social, cultural and environmental ones, and going in the direction of the proposals of the general local plan. This is to give answers that the former industrial sites can be integrated into the future development without losing our heritage.

The recent development in Tirana city has increased the multicultural urban society, and this is evidenced of the need to increase and expand its cultural dimension. The former industrial stock can have a fundamental contribution to the cohesion of the communities and the creation of the shared local identity, that is why this stock can be seen as a source and their retrieval is a possibility to develop multi-functional areas (Magnaghi & Paloscia, 1992).

The development of Tirana city is evidencing the extinction of the former industrial sites, with the "tabula rasa" concept, that is why an urgent answer is needed not to lose the whole the former industrial stock. There is still a lot to be done by the government, developers, civil society, and local community, and the urban regeneration of the physical assets of the former industrial sites to achieve the goals, but it is an important starting point to raise questions and awareness.

References

Ashworth, G. J., & Graham, B. (2016). *Senses of place: Senses of time, heritage*. Taylor & Francis Group.

Barthelemy, M. (2016). *The structure and dynamics of cities: Urban data analysis and theoretical modeling*. Cambridge University Press.

Bego, M. (2009). *Skeda arkitekture 1965–2004 në kronikën e një jete të dallgëzuar: mbi degradimin urban sot*. DEA.

Bërxholi, A. (2000). *Regjistrimet e përgjithshme të popullsisë në Shqipëri*. Akademia e Shkencave.

Bonini Baraldi, S., Governa, F., & Salone, C. (2019). "They tried to make me go to rehab. I said, no, no, no". Representations of 'deprived' urban spaces and urban regeneration in Turin, Italy. *Urban Research & Practice*, 286–306.

Bushati, E., Thomai, G., & Mëzezi, I., 2018. *Industrial buidings in Tirana 1945–1990*. BOTIMET FLESH.

Camasso, M., Gron, S., & Vigliocco, E. (2013). *Gli spazi della costruzione nella ricomposizione urbana*. Celid.

Casella, E., & Symonds, J. (2005). *Industrial archaeology: Future directions*. Springer.

Corboz, A. (1983). The land as palimpsest. *Diogenes, 31*(121), 12–34.

Dopfer, K., Foster, J., & Potts, J. (2004). *Micro-meso-macro* (pp. 263–279). Springer.

Douet, J. (2016). *Industrial heritage re-tooled: The TICCIH guide to industrial heritage conservation*. Routledge.

Drury, P., & McPherson, A. (2008). *Conservation principles, policies and guidance—For the sustainable management of the historic environment*. English Heritage.

Fischer, B. J., & Schmitt, O. J. (2022). *A concise history of Albania*. Cambridge University Press.

Goodman, D., & Chant, C. (1999). *European cities and technology: Industrial to post-industrial cities*. Routledge.

Grewenig, M. M. (2022). *European route of industrial heritage—ERIH* [Online]. Available at: https://www.erih.net

IRES Piemonte. (2017). *L'intervento di rigenerazione urbana dell'area ex Incet a Torino*. IRES PIEMONTE - Regione Piemonte.

Kahn, M. E., & McComas, M. (2021). *Unlocking the potential of post-industrial cities*. Johns Hopkins University Press.

King, R. (2005). Albania as a laboratory for the study of migration and development. *Journal of Southern Europe and the Balkans, 7*(2), 133–155.

Kolevica, P. (2004). *Arkitektura dhe diktatura*. Logoreci.

Loures, L. (2008). *Post-industrial landscapes as renaissance locus: The case study research methods* (pp. 293–302). WIT Press.

Magnaghi, A., & Paloscia, R. (1992). *Per una trasformazione ecologica degli insediamenti*. FrancoAngeli.

Meireis, T., & Rippl, G. (2018). *Cultural sustainability: Perspectives from the humanities and social sciences*. Taylor & Francis.

Moore, N., & Whelan, Y. (2007). *Heritage, memory and the politics of identity: New perspectives on the cultural landscape*. Routledge.

Orange, H. (2014). *Reanimating industrial spaces: Conducting memory work in post-industrial societies*. Routledge.

Parangoni, I. (2012). *Industrial archaeology, an assessment of industrial heritage in Albania*. Albanian Heritage Foundation.

Parisi, R., & Chimisso, M. (2021). *La carta di Nizhny Tagil e la tutela del patrimonio industriale in Italia.* Rubbettino.

Pllumbi, D. (2013). Mirroring Tirana, reflections on Tirana's urban context and perspectives. *European Journal of Sustainable Development*, 73–84.

Register, R. (2006). *EcoCities: Rebuilding cities in balance with nature*. New Society Publishers.

Roberts, P., Sykes, H., & Granger, R. (2016). *Urban regeneration, second edition*. SAGE Publications Ltd.

Roders, A. P., & Bandarin, F. (2019). *Reshaping urban conservation: The historic urban landscape approach in action*. Springer.

Stratton, M., & Trinder, B. (2016). *Twentieth century industrial archaeology*. Taylor & Francis.

Streule, M., Sawyer, L., Karaman, O., & Schmid, C. (2019). Popular urbanization: Conceptualizing urbanization processes beyond informality. *International Journal of Urban and Regional Research, 44* (5), 652–672.

UNESCO. (2009). *UNESCO framework for cultural statistics.* ISBN: 978-92-9189-075-0.

Veleshnja, J., & Branko, I. (2018). *Urban development and the informality of Tirana city* (pp. 767–775). ISUFitaly.

Velo, M. (1998). *Paralel me arkitekturën.* Librat "Njeriu".

Xhindi, N. (2013). *Albania towards a sustainable regional development: The cases of Tirana, Shkodra and Kukes*. Universitätsverlag Potsdam.

Za, L. (2011). *Kombinat - Storia e vita quotidiana di un quartiere simbolo di Tirana*. Salento Books.

Sustainable Architecture Techniques in Albanian Traditional Houses

Etleva Dobjani, Ajda Çarçani, and Oltjon Luga

Abstract

The composition of an architectural space reflects the manner of life and culture of a society. In addition to building actual structures, architects also design spaces in which people might live. By using manual construction techniques and a lack of theoretical and in-depth technological knowledge, traditional homes usually reflect the society, culture, and technology of the era in which they were constructed. The boundaries between private and public spaces are typically merged in traditional homes, which frequently promote themselves as tiny communities. The development of the conventional home serves as evidence of logical semantic overlap since it is both a reflection of the collective symbolic universe and a manifestation of the dominant culture. The purpose of this article is to examine a variety of traditional homes in Albania and evaluate their plans, circulation patterns, accessibility, and relationship with light and the surrounding environment in light of their various geographic and cultural settings. The study's main objective, the examination of the dwellings, shows how traditional homes can implement sustainability principles that are solely based on craft knowledge, societal expectations, and meteorological data. These results may aid in creating a traditional architectural legacy sustainability index that can be applied to homes and enhance the process of creating sustainable structures based on traditional ones without abandoning cultural heritage practices, memories, or identities.

Keywords

Traditional architecture • Space composition • Cultural heritage • Sustainable buildings • Houses design

1 Introduction

The purpose of architecture is to improve people's quality of life. Today's topic focuses more on eco-friendly construction methods, biocompatible building materials, and sustainable design. The underlying principles of this philosophy have existed throughout human history since managing any kind of easily accessible and sustainable natural resource has always been a challenge. Sustainable architecture is a term used to describe design that is regarded to be environmentally beneficial. This process is dynamic and ongoing, thus it might occasionally be changed depending on the circumstances. These issues were not present in pre-industrial architecture due to the abundance of locally available resources and the architecture's significantly reduced environmental impact. Modern structures are the outcome of technological experiments made feasible by the industrial revolution with building materials, construction methods, related technologies, and technology itself. The majority of initiatives for pollution and environmental conservation began, particularly after the 1960s and 1970s. Sustainable architecture also results from the need to conserve electricity and the need to use renewable resources, which were both driven by the Yom Kippur War crisis of 1973, which has increasingly pushed national and international policies over the past few decades towards the introduction of alternative technologies for the use of renewable resources as well as regulatory.

Some works of the modern trend utilize new materials and designs that significantly distinguish them from earlier ones. New building trends are generally based on a positive view of how humans and the environment interact via the

E. Dobjani (✉) · A. Çarçani · O. Luga
Faculty of Applied Sciences and Creative Industries, Head of the Department of Architecture and Design, BARLETI University, Tirana, Albania
e-mail: arch.dep@umb.edu.al

A. Çarçani
e-mail: a.carcani@umb.edu.al

O. Luga
e-mail: o.luga@umb.edu.al

A. Battisti et al. (eds.), *Greening Our Cities: Sustainable Urbanism for a Greener Future*, Advances in Science, Technology & Innovation, https://doi.org/10.1007/978-3-031-49495-6_33

use of conservative-selective housing technology in light of new technologies for environmental management and regulation of energy and material fluxes between inside and outside. In addition to the three well-known principles of Vitruvius' architectural theory—venustas, utilitas, and firmitas—architecture aims to control comfort through the use of specific energy supply devices, resource exploitation, or maximizing the operation and production of renewable energy as well as the functionality of the building. As a result of its close ties to the nation's morphology, local resources, climatic, and environmental characteristics, as well as the socio-economic, cultural, and environmental needs for the protection of a specific community, Albanian vernacular architecture falls under the category of sustainable architecture and is the subject of this study. It also acts as a reminder of the nation's past and our cultural beginnings.

In Rudofsky's (1987) book Architecture without Architects, he writes that while it has been commonplace to reject primitive architecture as accidental, it actually appears to have developed as a result of distinctive and fairly sophisticated applications by people using basic equipment. The significance of regional natural and architectural resources is covered by Piesik (2017) in his study on vernacular architecture. Vernacular architecture is closely related to sustainable architecture. Vernacular architecture symbolizes the common practices and customs of a nation and defines various national principles beginning with local dispositions, climatic circumstances, local materials, etc. (Correia et al., 2014).

The research study of the paper examines the relationships between Albanian traditional architecture and passive tactics employed in sustainable architecture, as well as design principles, space composition, circulation, technological treatments, daylighting, and natural ventilation. This is achieved through critically analyzing Albanian traditional architecture. The idea is to assess the practicality and applicability of the typical project components using conventional systems in line with the goals of sustainability and minimizing the environmental effect of buildings. These findings may contribute to the development of a traditional architectural legacy sustainability index that can be used to evaluate the sustainability of new homes and improve the process of developing sustainable buildings based on traditional ones.

2 Methodology

The chapter uses descriptive, analytical, and comparative synthesis research methods to analyze the vernacular climate responsive strategies used in Albanian vernacular architecture. The program featured a variety of temperature- and location-varying scenarios and examples from numerous Albanian cities. Numerous passive climatic responsive solutions, with a focus on passive heating, lighting, and ventilation, as well as rainwater storage techniques, have been found in order to examine similarities and contrasts. The research indicates such passive strategies and offers guidance for potential future modern applications.

3 Traditional Urban Layout and Housing Models

Albanian cities saw various levels of growth during antiquity and the Middle Ages. First and foremost, the desire to protect it affected the decision of where to build the metropolis. According to the renowned scholar Karaiskaj (2011), there are two types of Albanian cities that were founded during the early medieval period and belonged to the slave-owning order. Walls surround the city in both types to provide security. The first category consists of cities having an acropolis, the inhabited center of which was typically located on the height's slopes. By merely enclosing the village with encircling walls, the second category of civic settlements—those without an acropolis—appears. These centers are protected by surrounding walls and are treated urbanistically. The primary road network is first connected to these features. Refering to Karaskaj, in (Meksi et al., 2016), the layout of the area inside the surrounding walls has a considerable impact on this network, which begins at the entry, or the main entrances, of the fortress (pp. 645).

The structures of the Christian worship, churches, which were devoted to various saints, held a significant role in the spaces of medieval Albanian cities. They are separated typologically into nave churches, cross-shaped basilicas with domes, and unusual situations.

One of the few cities where life practically remained unabatedly and developed for a very long period inside the castle's outer walls is Berat. The beginning of that significant process in the history of Berat and the end of the Albanian cities with the origin of the fortification, which is associated to the birth and development of the open city, is marked by the appearance of the inhabited center beyond the encircling walls. Far from the existing center, which from the perspective of urban planning became the core for the establishment of new residential areas, the construction of the Muslim cult's structures began. In the open city, craft trade hubs were expanding together with new residential districts. Gjirokastra stands out for its strong compositional unit, according to Meksi et al. (2011). The residential areas stand out for unique qualities that are primarily influenced by the STEEP TERRAIN on which they are situated.

The Rozafat fortification and the city of Shkoder are closely connected. There are still remnants of the Illyrian structures that once stood where Shkodra is now. Shkodra had an outlying city in addition to the residential core. The medieval city was relocated from the inundated plain to a higher site, clinging to the slope of the castle, during this time. The city's road system has the same lack of planning that is so typical of medieval spontaneity. According to the crafts, the bazaar was separated into many sections. Shkodra, a city in the plain region, was divided into a number of neighborhoods. Even in Shkoder, religious divisions have been evident. Shkodra's urban layout is among the most typical for urban areas, with many curves and bends.

Despite having a relatively long history of being a popular location for housing, Tirana's residential center serves as an important illustration of the gradual transformation from a rural to a civic center, even given the unique conditions of being Albania's capital at the turn of the twentieth century. Even Tirana is separated into two fundamental compositional units: commercial craft centers and residential districts. There have been many Muslim cult structures. The Lana bridge cut through Tirana, necessitating the construction of two stone bridges with three and five arches by the guilds of trays and blacksmiths in order for it to pass. The unique aspect of the urban development design is the extremely uncommon practice of constructing homes surrounded by relatively wide tracts of land, fruit trees, and vegetation. For Tirana, we can discuss a focus on the bazaar, the primary cult structures described by the linking arteries with neighboring cities, and the buildings that face the four directions of the horizon.

The city of Korce holds a distinctive place in the history of Albanian cities from the XV to XIX century because it is here that urban planning is first observed. The cadastral register from the years 1431 to 1432 describes Korca as an inhabited castle with 26 houses. The eighteenth century saw the greatest growth for the city of Korce. We now move on to the aspect of urban planning known as the hippodam system. This system entails establishing an approximately quadratic network of roads on equal or nearly equal-sized plots that served as building plots. The widespread use of corner houses is connected to the start of urban planning implementation in Korce.

With both general and unique characteristics, the urban and architectural values of the historical centers of Berat, Gjirokastra, Shkodra, Tirana, and Korce (Fig. 1) reveal us a common characteristic: a compact urban layout that actually reduces the number of surfaces exposed to the sun, narrow streets that shield pedestrians from harsh summer or winter weather.

Riza (2010, 2013) discusses the housing in his book, which covers Albanian housing from the 15th to the nineteenth centuries. He discusses how housing was defined at this time, which was different from how we currently define housing, as well as the portion of the courtyard and the structures included in it. If we consider the yard and the house as a single entity, there are some internal processes that take place that have an impact on how the yard and house interact with one another. Every change that occurs in the house has an impact on the yard and their interaction with one another. Visual connections between the inside and external spaces are strong, and social interaction with the intermediate space before moving on to the public space is highly valued. In addition, we will discuss the five basic types of housing and their compositional space, referring to the geographic map in Fig. 2.

These are the five typologies that this article examined:

- Fireplace house (Fireplace room)
- Hayat house (with a porch)
- Cardak House
- Fortified dwelling (Gjirokastra dwelling)
- Corner house (Qoshk house or Korca dwelling).

4 Characteristics of Traditional Houses' Development

The fireplace house has a two-story layout for the most part. The yard is directly impacted, and there are large gaps between the entryway and the house or hayat. The main gathering location for all other life-related activities in the fireplace house is the fire room. Figure 3 depicts the firehouse's various spatial distributions and size changes in relation to family size and socioeconomic class. The central fire room, which had an open height on both levels, served as the home's heating system. One room, two rooms, rooms

Fig. 1 Gjirokaster, Shkoder, Korçe, Tirana, Berat (BAÇE, et al., 1979)

Fig. 2 Geographical distribution of the traditional albanian dwellings. (*Source* Authors)

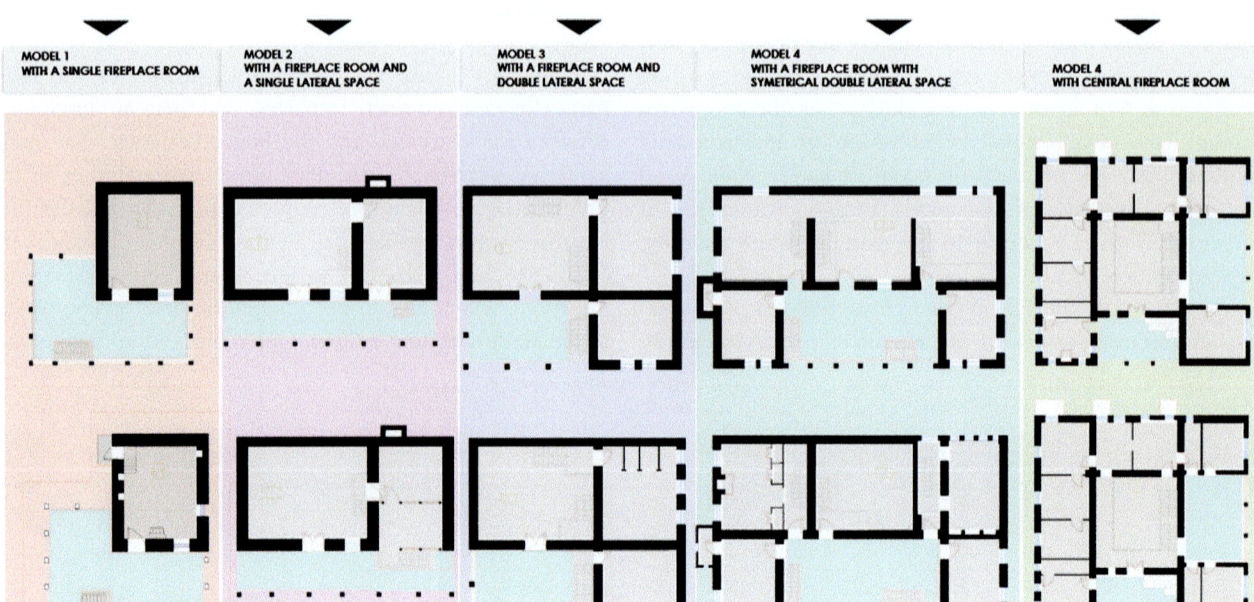

Fig. 3 Traditional two floor fireplace house. (*Source* Authors)

arranged in a U shape with the fire station in the middle, or a more complex layout with the fire station always in the center were all options for the house concept. This spatial design, which benefited from the fire and the heating of the beds on both floors, maintained the warmth within the home. Both the indoor stairs and the outside steps could be used for circulation (Fig. 4). The open roof above made it easier for smoke and natural CO_2 ventilation to circulate through the room. The living section is on the second level, while the service and storage areas are on the first.

The Hayat house (Fig. 5) was primarily constructed on a single floor. The spatial arrangement, in contrast to the fire house, mostly consisted of a main chamber and two ancillary rooms that were grouped on a single floor. The courtyard area, where the apertures are oriented, is where the main activities take place.

The cardak house (Fig. 6) has two storeys. The hayat home was the ancestor of this kind. In this case, there is a distinction between the first and second floors, with the first level functioning as storage and the second as a living area, and the courtyard is given clear emphasis. The cardak and hayati both have a direct line to the yard. Through the stairs and further away from the cardak, the generations converse.

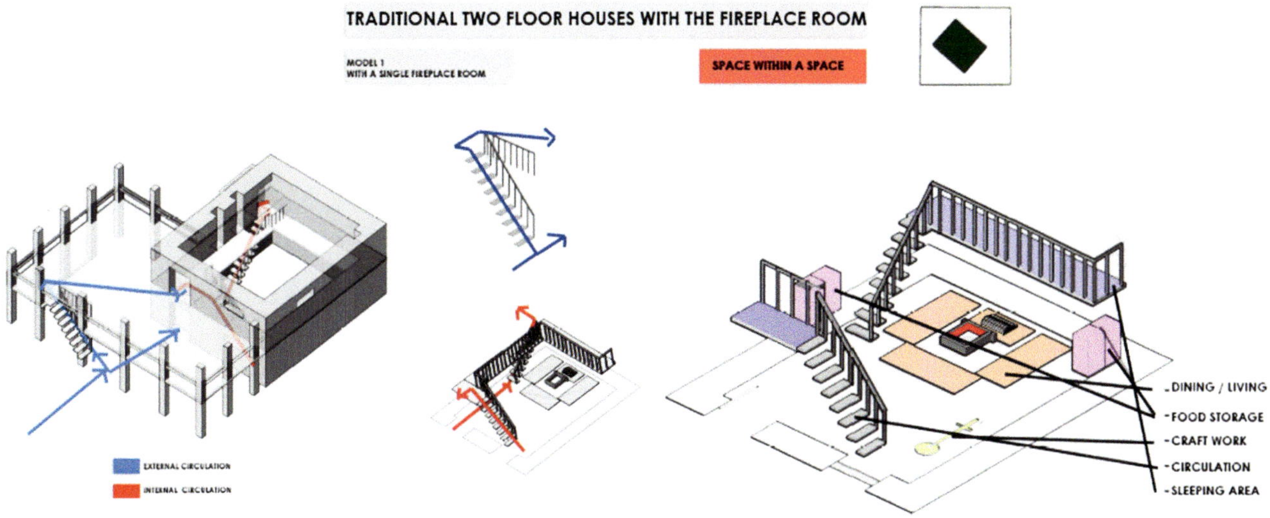

Fig. 4 Traditional two floor fireplace house and diagrams of external and internal circulation and space connection. (*Source* Authors)

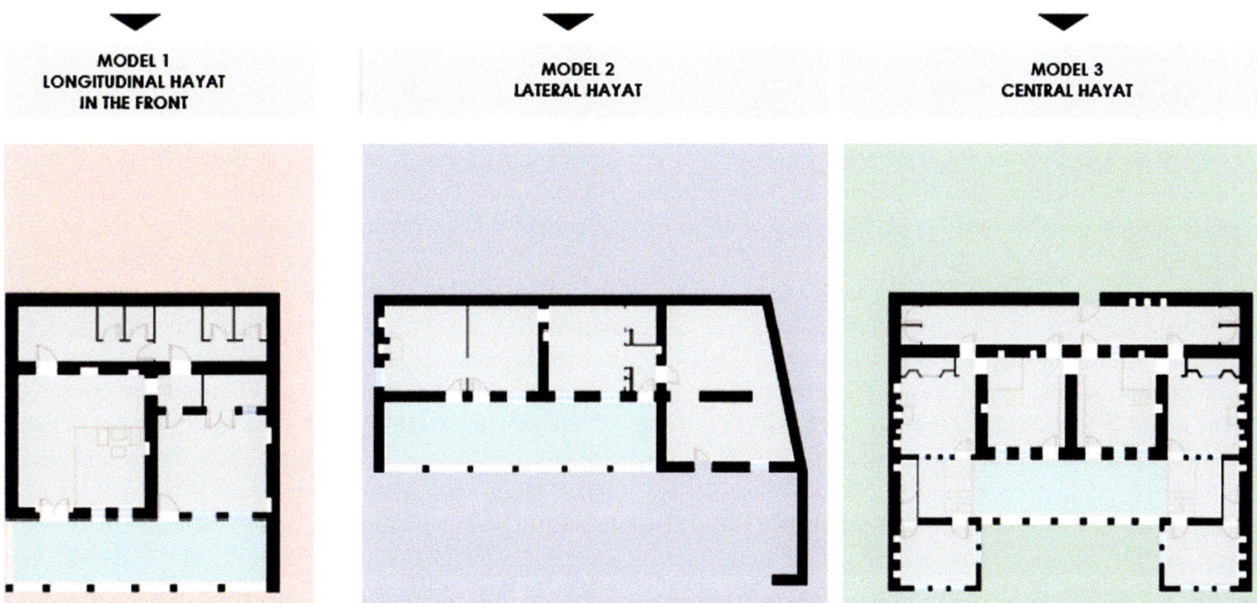

Fig. 5 Traditionaol hayat house. (*Source* Authors)

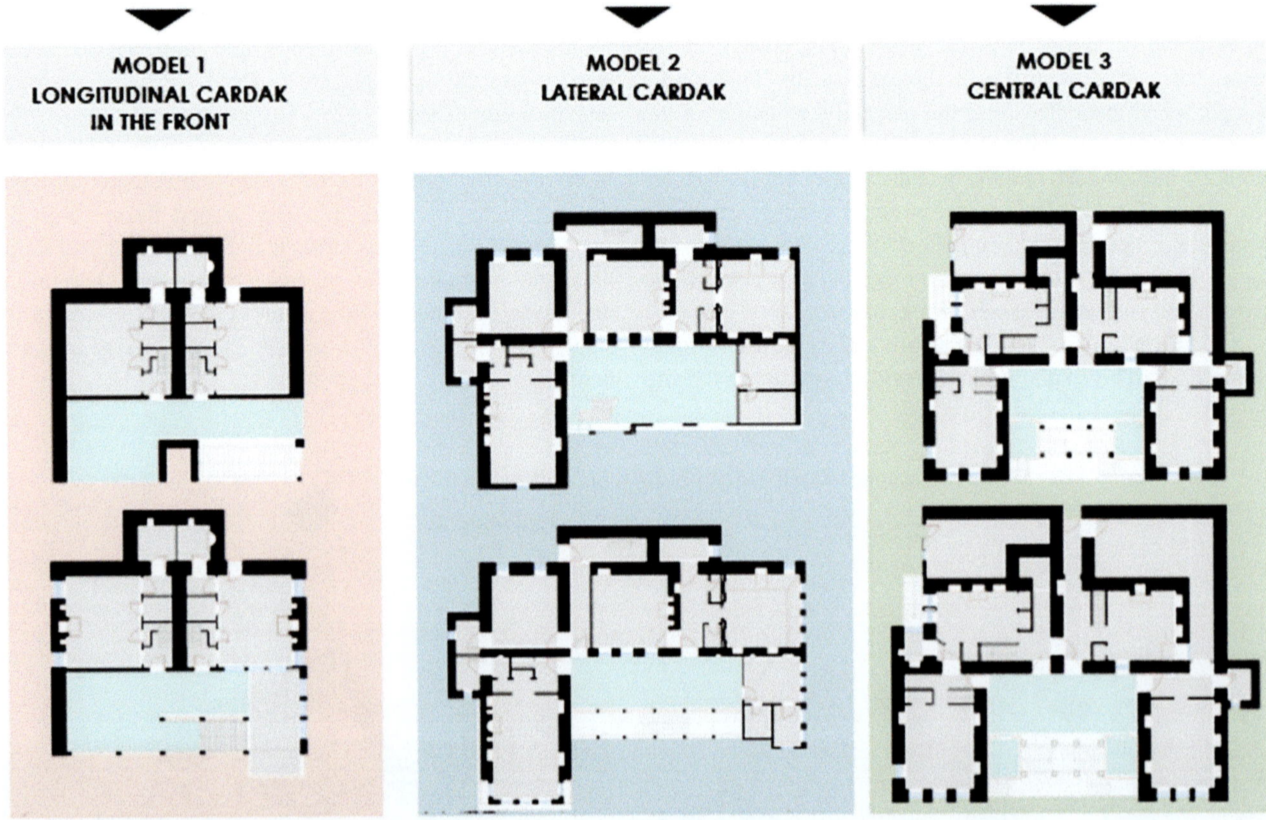

Fig. 6 Traditional çardak two floor houses. (*Source* Authors)

That cardak and hayati are in constant communication. The same compositional strategy is used on both levels, which merely happen to overlap in the floor plan and have the potential to work separately. This independence is guaranteed by the staircase's architectural feature, whose placement permits independent connection of the levels with one another. The yard and hayati and cardak share a close relationship.

The fortified home and the one with a porch can both be considered as ancestors of the corner house (Fig. 7). The phenomenon we encounter in this instance is the courtyard space being reclaimed by the circulating space with transverse orientation. With the placement of the internal corridor and the revolving internal relationship we are working with, the living rooms are lit in a particular direction. is constructed on a level surface. You walk straight into the house as you enter it.

The entrance to the Gjirokastra fortified home (Fig. 8) is made from below rather than from the outside (Fig. 9). In Gjirokaster, every house has an eastward orientation. We are in a place that develops on sloping terrain, which is the first feature of the terrain condition that shines out in the Girokastra fortified dwelling typology. Since the ground floor is treated as though it were a separate plot of land, it is paved with stones, while the first and second floors have wood pavements. Considering that the ground level is covered with stones and treated as though it were soil, it is crucial to consider whether we should think of the fortified structure as a three-story house or a two-story house with a basement. The most notable aspect of these flats, which were built when the city's water plant was in operation, is the layout of the large water tanks.

Water cisterns are an essential part of home construction, as Riza (1980) points out in the monograph "Qyteti Muze i Gjirokastrës." The ground-floor storage of rainwater in water chambers or cisterns allowed for the provision of water through a small sink that was positioned above the water line and close to the cistern. This alternative can be the best passive choice even in our day and age, especially during the hot summer months, for a select few activities like watering the garden, flushing the toilet, and mowing the lawn.

Another important element is the tightly packed design of the protected towers, which safeguards the internal energy gained from the heating. The winter rooms were built in a house that was divided into vertical zones and were located on the middle floor. The winter rooms' solid walls had small cracks in them. Their positioning, which was away from the floor and away from the ground, caused the interspace gaps

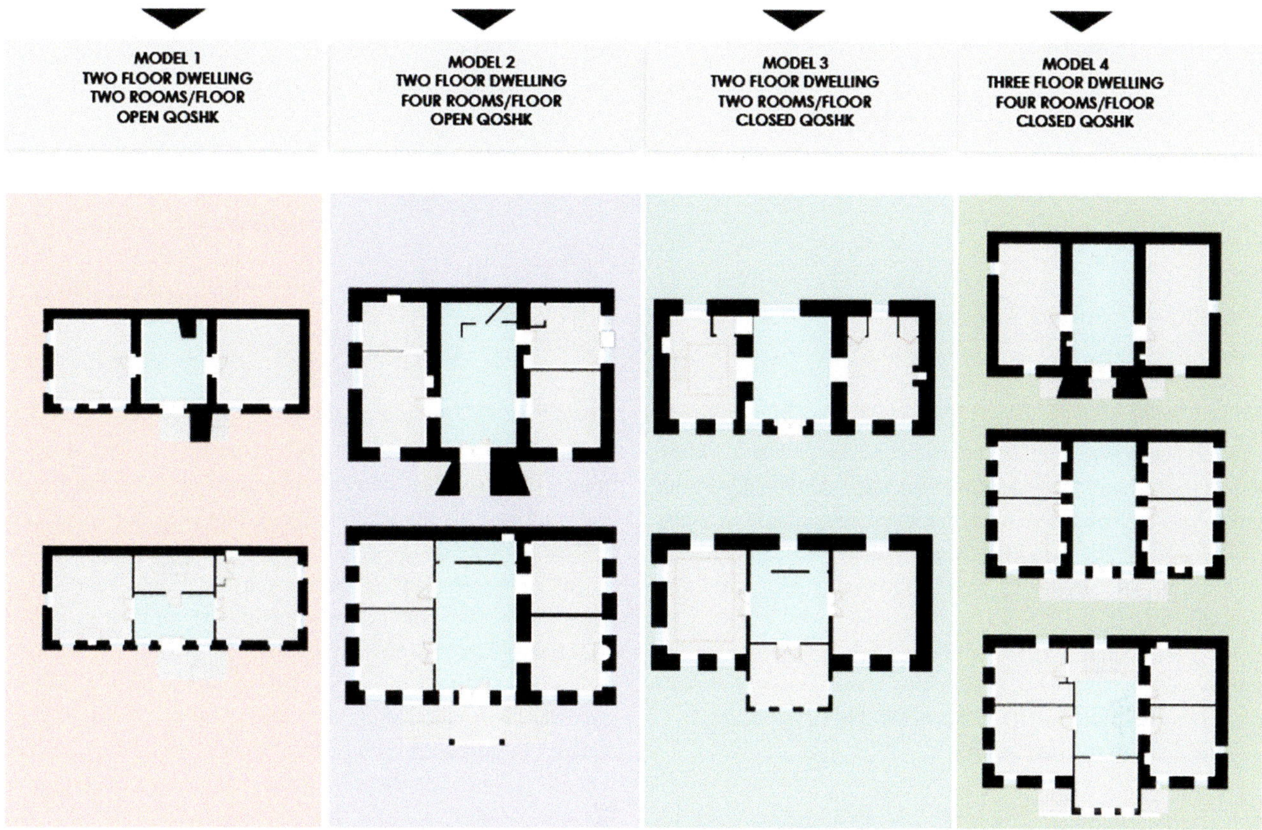

Fig. 7 Traditional two–three floor corner house (Qoshk House). (*Source* Authors)

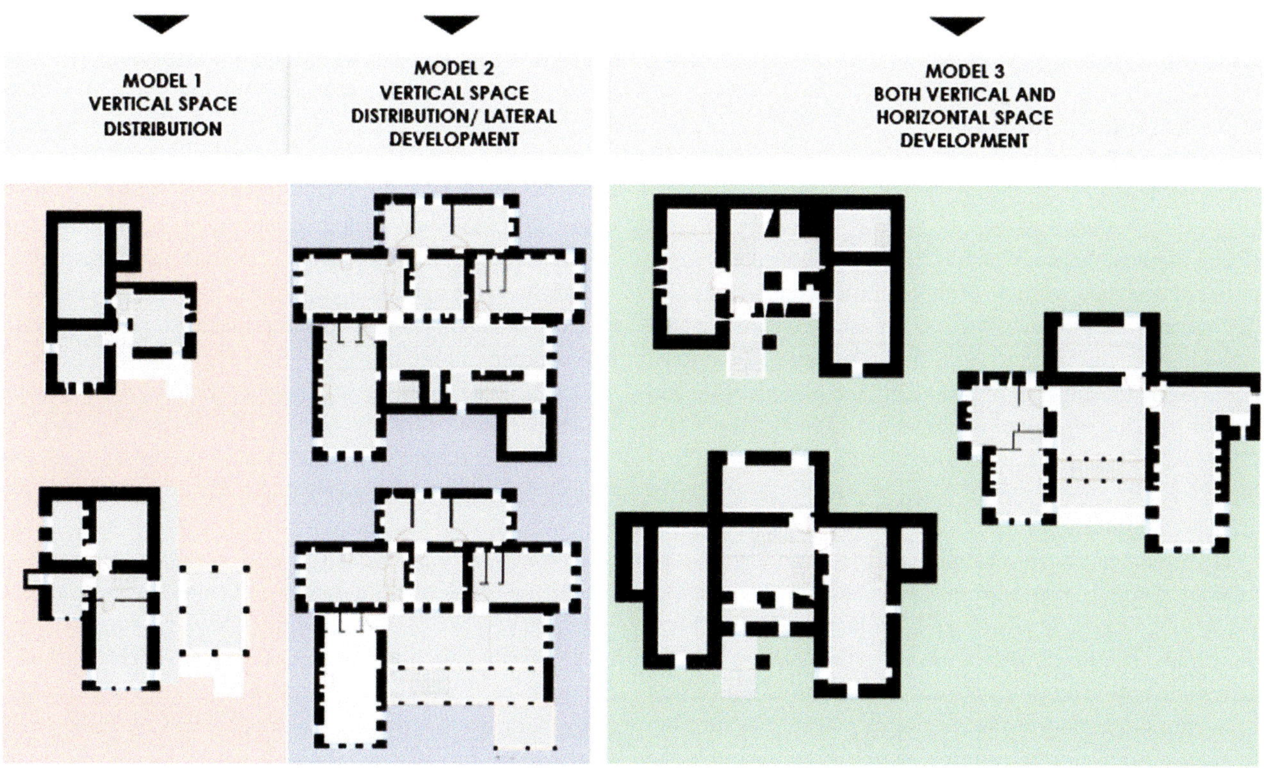

Fig. 8 Traditional fortified house (two and three floor). (*Source* Authors)

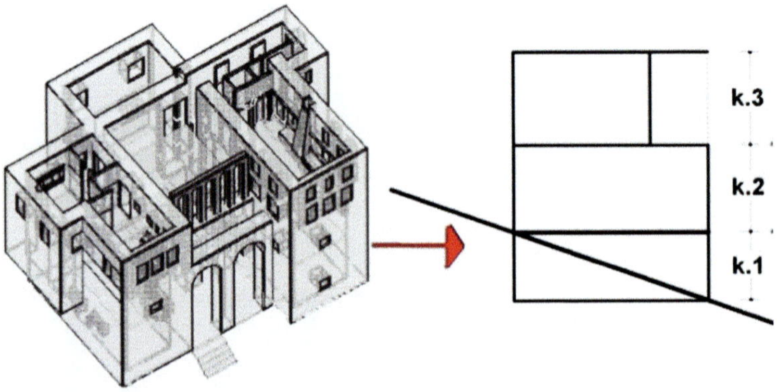

Fig. 9. 3d view and schematic section of the fortified house. (*Source* Authors)

between the lower and top cover. The room's few windows helped to maintain a consistent temperature. On the third floor, which was the top floor, were the summer quarters, though.

There were big windows, light wood walls, and natural ventilation. The ability to have cool rooms in the summer resulted from this.

5 The Relationship Between Natural Light and Space Hierarchy

Housing typologies can be split into two groups based on how their interior spaces are developed:

Horizontal and vertical development are both possible.

Those housing typologies whose functions evolve horizontally can be included in the first group; these would include the following typologies:

- Fire house with two floors
- Porch house (Hayat)
- Çardak house.

The fireplace house, which is normally two storeys high, is always designed with the fire room in the center (Fig. 10). The warmth of the house was maintained by this spatial arrangement. The interior stairs give the space a dynamic value. When the circulation is running, the space is split, and when it is halted, the space returns to its original state. When a dynamic element is in operation, the space becomes flexible since the element returns to its original place when the circulation is halted.

We also record an example of a subvariant of this building typology that takes place on two floors, which is precisely the example where we have half the hall and half the cardak. The house with a hayat, in its most basic form, has a main room and two secondary rooms, all of which take place on a single floor without rising in height, in contrast to the fire house, which takes place on two floors and uses the mafil for this.

Instead of a hayat that only connects the yard to the house, we have one like this one that joins the yard to the house as well as the interior space of the house (Fig. 11).

The interaction created by the light is directly impacted by the consequences of placing the functional bands in the house with the hayat. This inclination prevents both an orientation toward the hayat and the possibility of direct natural light illuminating the living quarters. (Fig. 12) This is important since it means that, in addition to serving as a means of circulation, the hayat in this form of dwelling also serves as the residence's main source of illumination. Because of this, we say that in a hayat house, rather than being created by means, the light used in the living areas comes straight from the outside.

The floor plan of the home with the cardak typology is divided, with the first level serving as storage and the second storey housing. It is also clear that the yard is important. For instance, the hayat and cardak are in close proximity to the yard. We observed a phenomenon in this type of housing where the second plan, which is the residential one, rises while the previous horizontal development, which was in the home with the hayat, is left on the lower floor. Communication with each of these bands is done through the hall and further from the cardak through the stairs. The hayat and the cardak, for example, are two levels that are directly related to one another.

The scheme in this case would be similar to a scheme in which two plans with the same floor plan overlap and may function independently of each other as a result of the outside degree. Each level, such as the hayat and the cardak (Figs. 12 and 13), has a direct link to the other (Figs. 14 and 15).

Through a transformation that changed the sections' height, the house with a cardak developed from the house with a hayat. We do not distinguish between the hayat- and cardak-containing homes. As it ascends in height, it takes on a new name, caradak, and serves the goal of connecting the entry with the yard after leaving the service zone on the lower level and passing through the residential zone one story above.

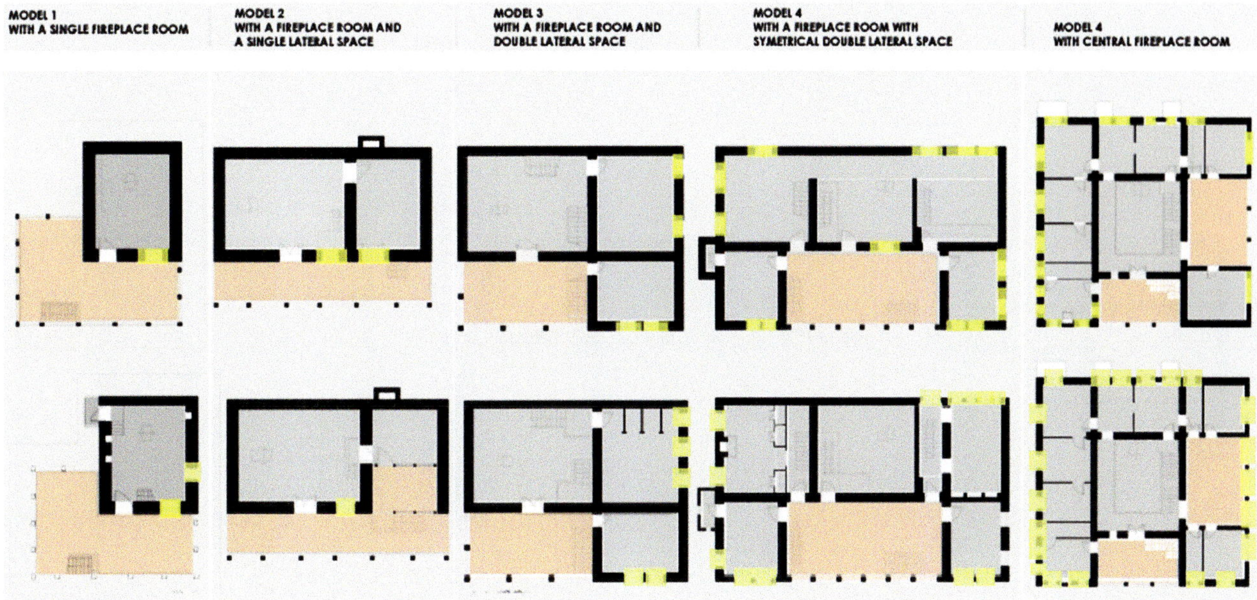

Fig. 10 Fire-house illumination toward the yard. (*Source* Authors)

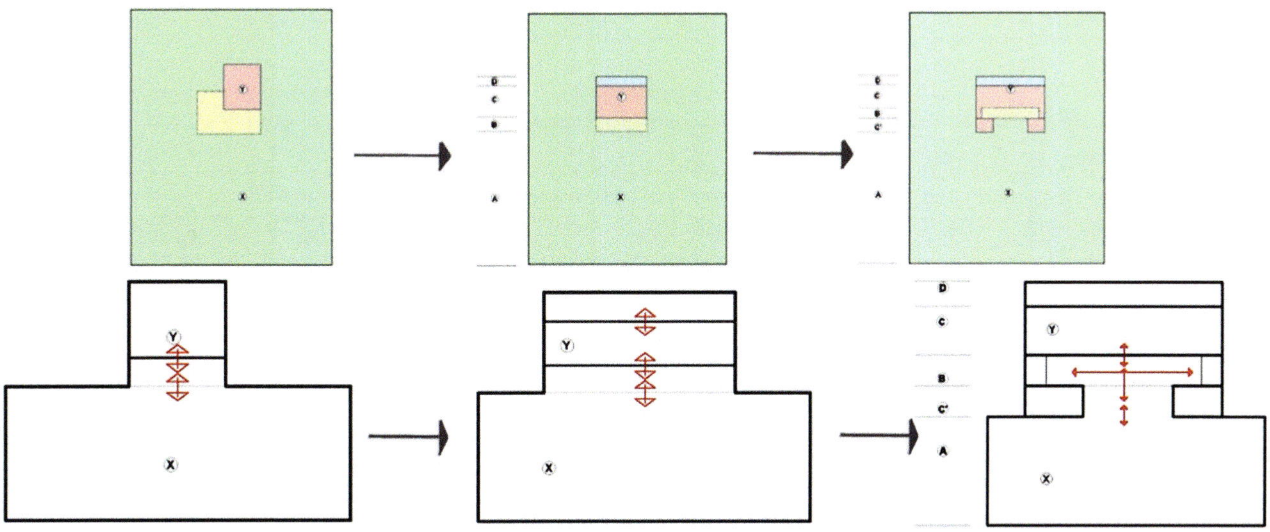

Fig. 11 Hayat house diagram in plan and the link between inside and outside. (*Source* Authors)

If the yard and the house were connected directly, without the use of the hayat and cardak as intermediary parts, the room would appear relatively small in compared to the yard. In this situation, the courtyard would strongly dominate the home without the hayat.

A 90-degree rotation of these belts to their orientation toward the yard marks the change from the hayat typology to the corner typology from the perspective of the belts (Fig. 15). And in respect to the light coming from the yard, the consequences of this shift are represented in the circulation-related elements.

The corner house interacts additively with the courtyard, creating a full-fledged entrance facade. The corner home arrangement eliminates the autonomous feature that was hayat. This idea states that when you enter the house, you are also entering the building. When it comes to building shape, this category is compact, and metropolitan areas frequently have courtyards. The buildings were located in the center of the courtyard.

The second group of houses are those that have developed vertically, but not just in terms of the fortress or the facades; rather, they have developed vertically in terms of

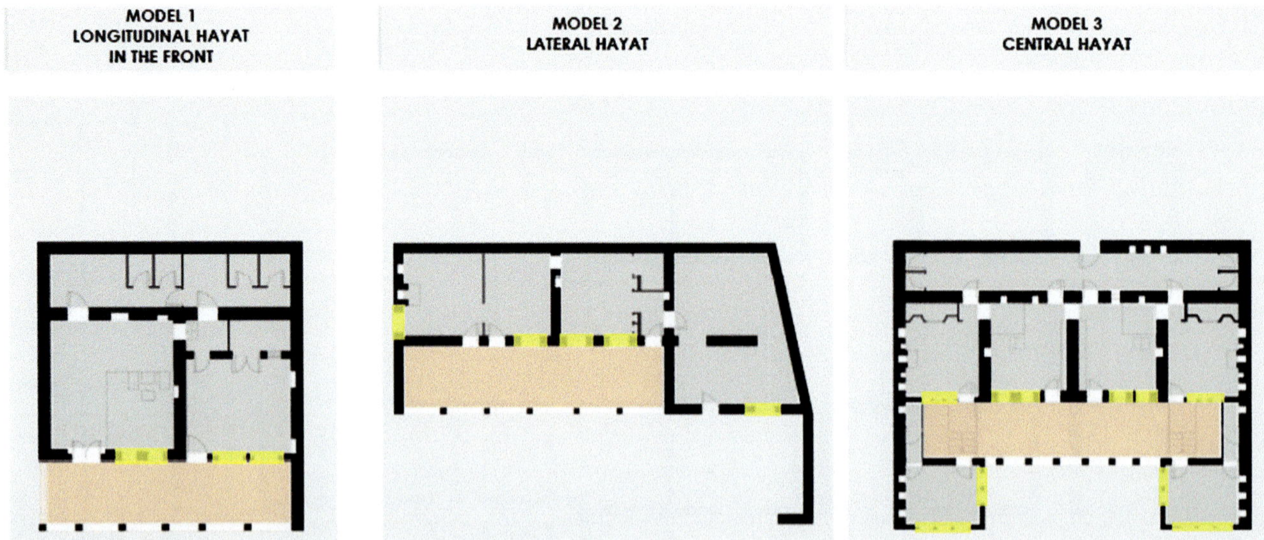

Fig. 12 Hayat house. (*Source* Authors)

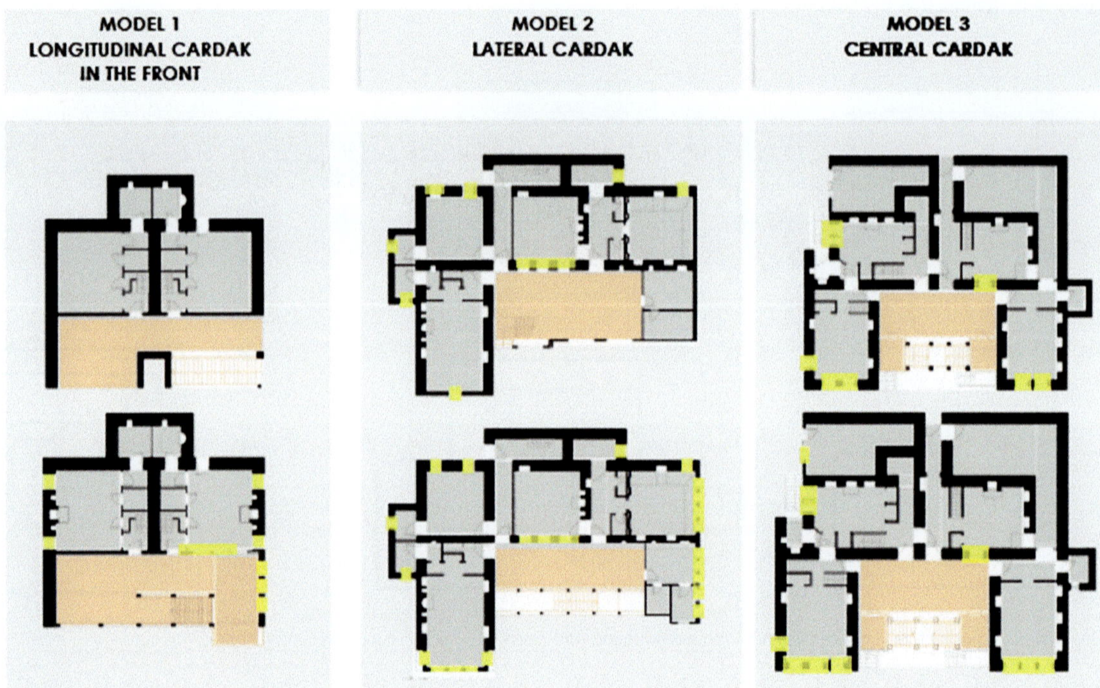

Fig. 13 Cardak house. (*Source* Authors)

how they organize the spaces and functions, which also has an impact on other compositional elements. The following homes will be included in these groups:

- Fortified home (Gjirokastra home)
- Corner home (Korca home).

With the inclusion of the element of height, the typology of the fortified house in Gjirokastra can be considered as an expansion of the cardak house typology. The fortified home of Gjirokastra lacks the spatial characteristic of the hayat, which is evident in dwellings with horizontal expansion. By removing this area and replacing it with just two or three feet of steps, the way people interact with the yard is altered (Fig. 17).

By removing this area and replacing it with just two or three feet of steps, the way people interact with the yard is altered. Within the fortified house, the connection to the yard

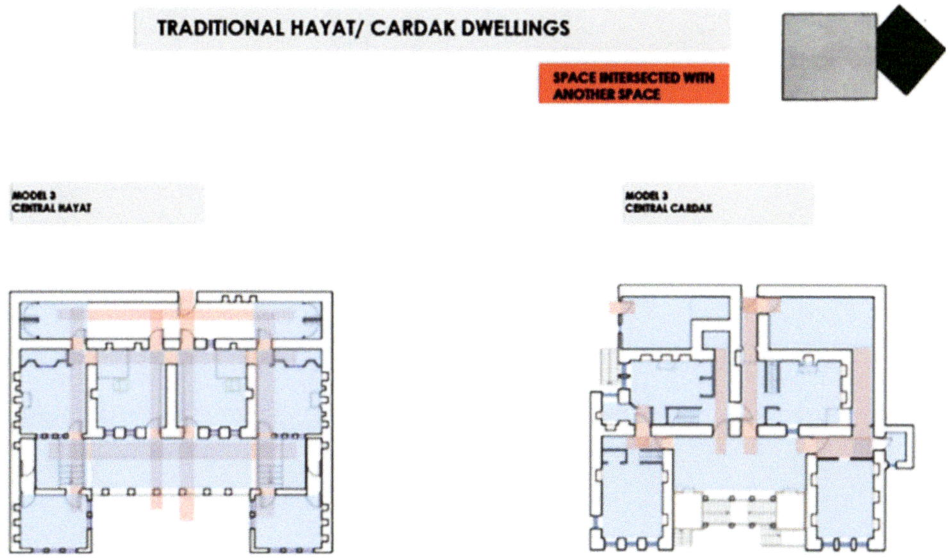

Fig. 14 Traditional hayat house. (*Source* Authors)

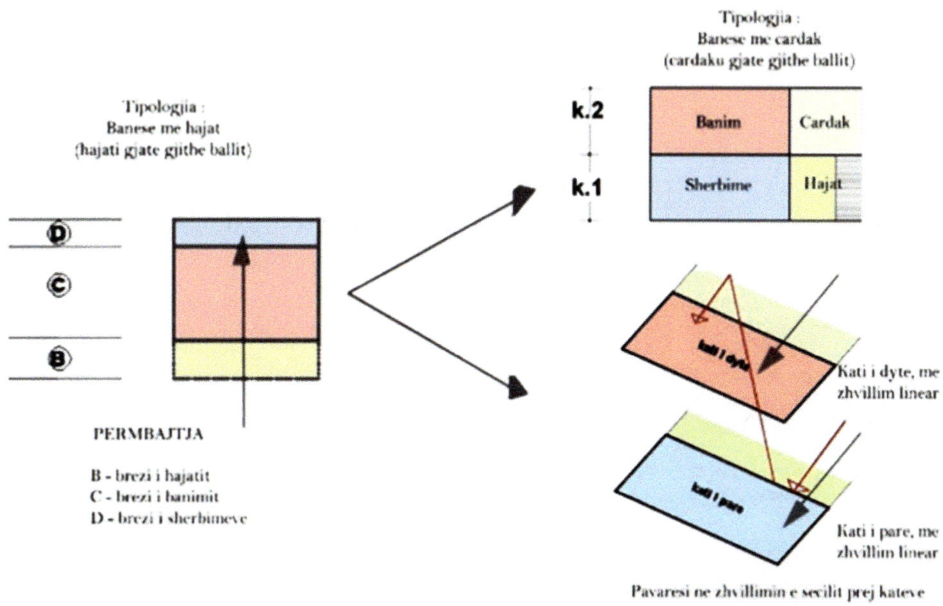

Fig. 15 Space connection and intersection. (*Source* Authors)

is broken. As a result, when the hall is excluded from the typology of the fortified dwelling, the connection between the courtyard and the building is broken. Instead of connecting to the yard as it did in the cardak home, the cardak in the Girokastra houses acts as a balcony, the last, nearly beautiful element that binds us to nature. The corner house is constructed on flat land and resembles a fortified dwelling (Fig. 18).

The area of the services on the second story, which pulls you upward in the corner house, is no longer an end space but rather a passing area. You must pass through this area to get to the top level, which is also the area with the emotional climax and was not designed in this way by accident.

We can see that the layout of rooms around the outside of the house provides for improved ventilation and natural light in both horizontal and vertical house growth.

6 Sustainable Aspects in the Various Case Studies

Local traditions are not lost but rather encouraged to flourish by elevating designers to the level of global aesthetic standards, as Benevolo also states in his book Introduction to Architecture (Benevolo, 2003). Designers are people with the baggage of their society and education, not abstract

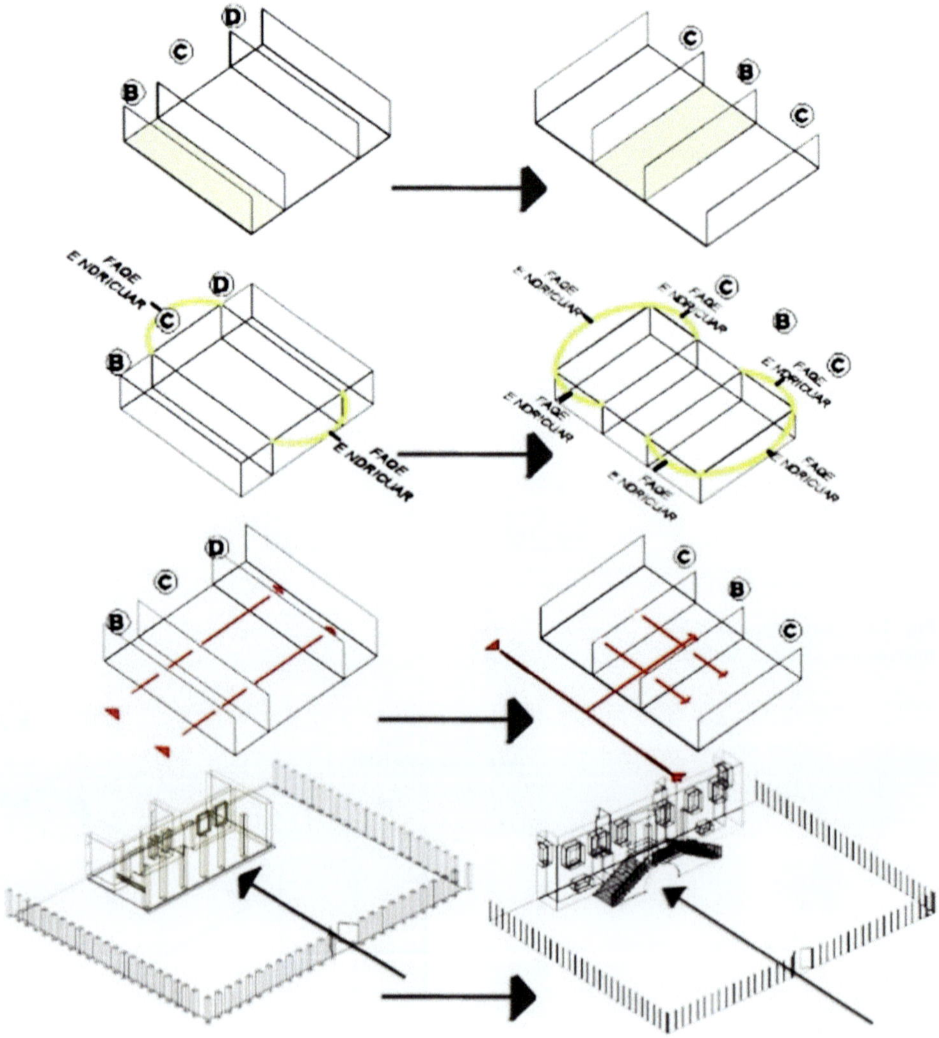

Fig. 16 Circulation-related elements and light study of the house. (*Source* Authors)

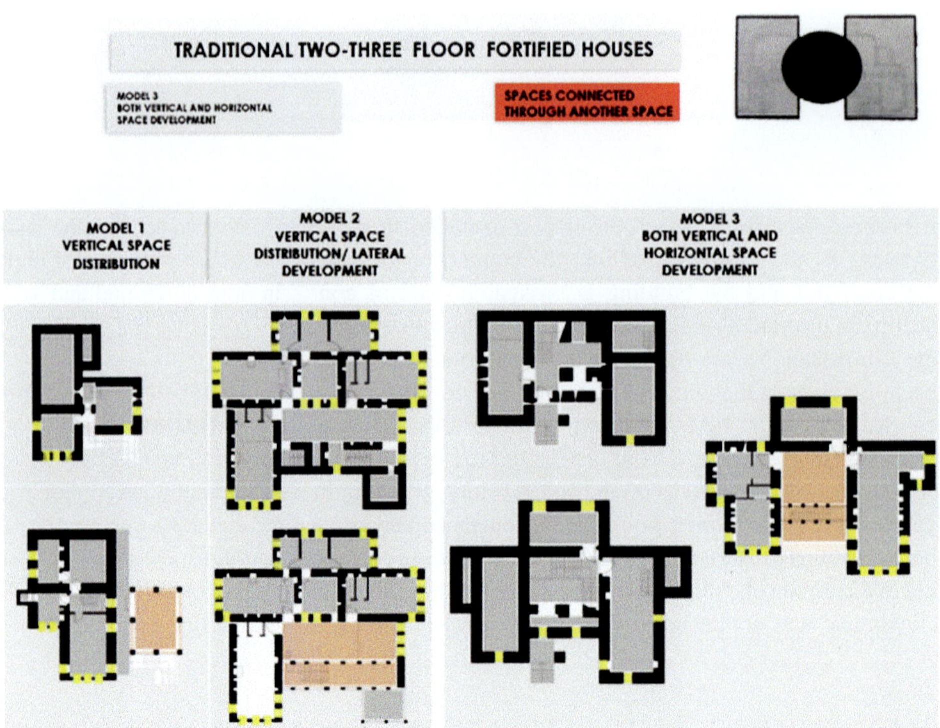

Fig. 17 Fortified house. (*Source* Authors)

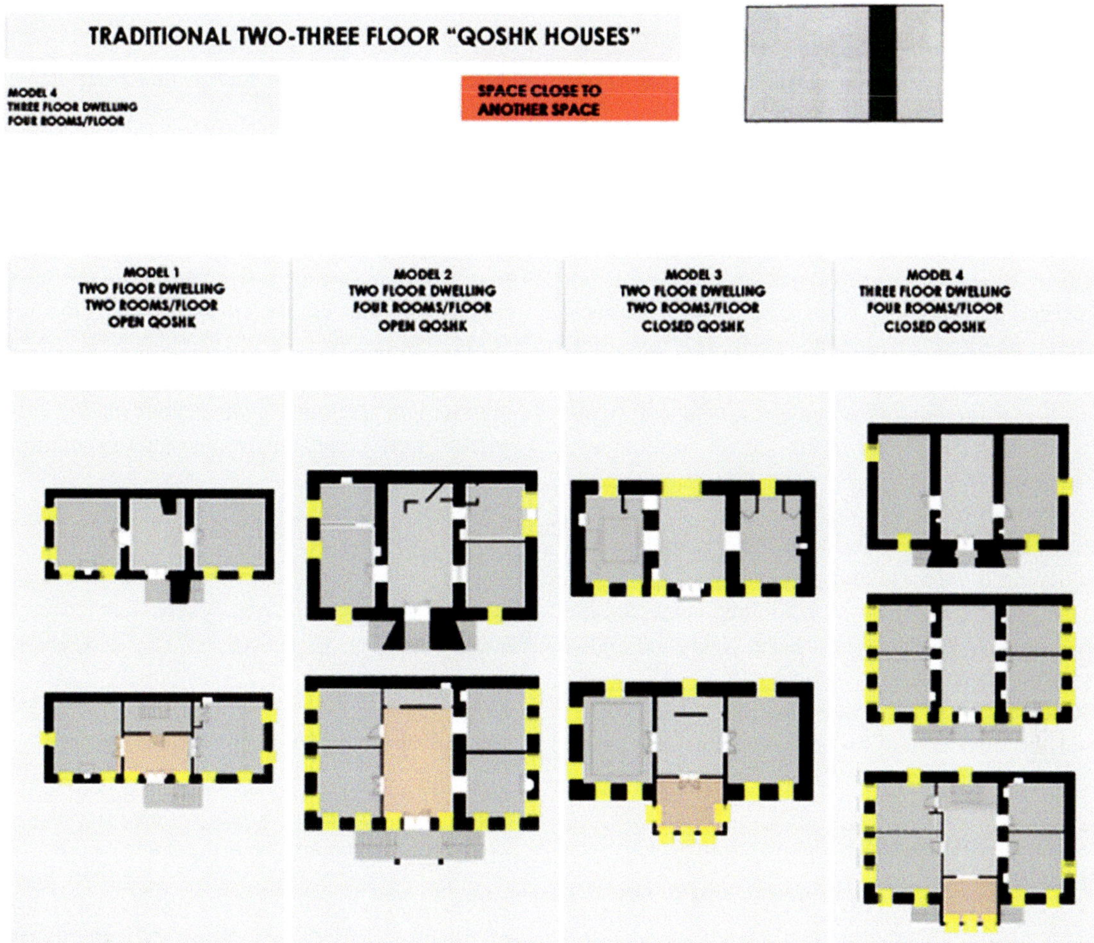

Fig. 18 Qoshk house. (*Source* Authors)

beings. Every nation has unique building methods, building materials, and natural and historical legacies that must all be taken into consideration. Heritage is being reevaluated through the lens of new historical and analytical research. The number of cities with comparable blocks that saw post-war growth during the building boom and left desolate landscapes in their wake is regrettable. The notion of an empty canvas also fails miserably. According to sociologist Lewis Mumford, the notion that the city is the end product of a difficult and drawn-out production process continuously shapes architects' thinking. (Saggio, 2010) As a result, the old are emphasized in this article as a heritage that not only has historical, memory, or identity values but is also revitalized by its solid and long-lasting architectural characteristics.

The typology of the traditional dwelling, which is precisely designed for the right place and period, symbolizes the rise of this type of commerce carried out in a more densely populated core (Riza, 2013). Since local design methods are the outcome of centuries of empirical logic, they produce hospitable environments that have withstood the test of time. (Simões et al., 2017).

Vernacular architecture provides knowledge that is essential to developing truly sustainable designs. Traditional Albanian homes have strong, thick exterior walls that block out summer heat while maintaining a cool, insulated interior. Strong wind and rain are better repelled by load-bearing walls and the compactness of medieval city plans. They are gated and walled for privacy. The walls are protected from rain and sunlight in the summer by the overhanging pitched roofs, which also create shadows. using largely local, organic resources like stone and soil.

Traditional homes relate to the building, the yard, the spaces between it and the structure, as well as its interior spaces, especially those with horizontal arrangements as described in earlier chapters. This indicated that social interaction with neighbors or passersby on the street was highly valued in Albanian culture and that there was a strong bond between the interior and the outside. In terms of building shape, it is compact, and courtyards are common in urban areas. The buildings were positioned in the center of the courtyard, especially in horizontal settlements, to allow for natural ventilation. The land's propensity to slant to one side in a mountainous area had an impact on where they were placed.

Better ventilation and natural light are made possible by the way the rooms are arranged around the outside of the building. Heat can spread over the entire area by placing the heating zone in the center. The fire room's atrium-like height, which is realized on two levels, heats both floors and directs smoke from the fire into the ventilated floor. The semi-open central circulation zones, which serve as a bridge between the apartment and the outdoor areas, get enough light and warmth to permeate the closed rooms. Small window surfaces in winter quarters and big window surfaces in summer rooms are examples of passive design.

The utilization of natural resources like rainwater and natural ventilation is encouraged by the collecting of rainfall for use as non-potable water in the reinforced home and the fact that no one lives on the bottom level. Using the higher levels warms the upper floors where people reside in chilly climates. Cooler spaces are made possible by the roof's shadow and the manner that its parts are angled to let in indirect light while blocking summertime bright sunshine.

7 Conclusion

This article looked at some traditional Albanian dwellings, and as a result, it can be said that traditional home designs have a lot of eco-friendly characteristics. When applied to new projects, such design concepts would be better preserved and might contribute to the creation of high-caliber, energy-efficient structures. For this project, traditional Albanian homes were chosen, and their major characteristics were identified as useful inspirations for the construction of modern housing buildings. The following are the key architectural characteristics of this study:

- The overhanging pitched roofs protect the walls from rain and lessen intense direct sunlight in the summer.
- The use of local, natural materials, mostly earth and stone; the positioning of the heating zone in the center to allow heat to circulate throughout the entire space;
- The design of the rooms around the perimeter of the home to allow for better ventilation and natural light.
- The fire room's two-level design creates an atrium-like height that heats both levels.
- Use of ventilated floor for the fire house.
- Small window surfaces in winter quarters and wide window surfaces in summer rooms. - A strong connection between inner and exterior space.
- The collection of rainwater for the fortified home's usage as non-potable water
- In frigid climates, the lower levels that are uninhabited by people use the higher levels to warm the upper floors where people live.
- A compact building shape reduces the amount of surfaces exposed to the sun. In the layout of old cities, narrow streets protect residents from sweltering summer days and chilly winter winds.
- Evaporative cooling (courtyard well).

References

Baçe, A., Meksi, A., Riza, E., Karaiskaj, G. J., & Thomo, P. (1979). Historia Arkitektures Shqiptare, Instituti I Monumenteve Te Kulturës, Tirane

Benevolo, L. (2003). Introduzione all'architettura, Laterza &Figli, 163

Correia, M., Dipasquales, L., & Mecca, S. (2014). *VerSus. Heritage for tomorrow. Vernacular knowledge for sustainable architecture*, Firenze University Press, Firenze available in https://www.researchgate.net/publication/310449583_VerSus_Heritage_for_Tomorrow_Vernacular_Knowledge_for_Sustainable_Architecture

Karaiskaj, GJ. (2011). *Kalaja e Gjirokastrës*. Revista Art-Trashëgimi, 23

Meksi, A., Baçe, A., & Riza, E. (2011). *Berati, historia dhe arkitektura*, Akademia e shkencave te Shqiperise, Tirane

Meksi, A., Baçe, A. Riza, E., Karaskaj, G. J, & Thomo, P. (2016). Historia e arkitektures ne Shqiperi, Ministria e Kultures, IMK, Tirane

Piesik, S. (2017). *Habitat: Vernacular Architecture for a Changing Planet*, Thames & Hudson Ltd, ISBN: 9780500343241

Riza, E. (1980). Qyteti Muze i Gjirokastrës, TIRANE

Riza, E. (2010). Qyteti dhe banesa qytetare Shqiptare e shek. XV-XIX, Dita 2000, 166

Riza, E. (2013). *Arkitektura popullore dhe vleresimi I saj*, Ministria e Kultures, 43, ISBN. 978-99956-92-83-4

Rudofsky, B. (1987). *Architecture without architects: A short introduction to non-pedigreed architecture*. University of New Mexico Press. ISBN-13: 0826310040-978

Saggio, A. (2010). *Architettura e Modernita. Dal Bauhaus alla rivoluzione informatica*, Carocci Editore, S.P.A

Simões, R., Cabral, I., Barros, F., Carlos, G., Correia, M., Marques, B., & Guedes, M. (2017). *Vernacular architecture in Portugal: Regional Variations*, 55

A Pragmatism Approaching of Demography: Documentation from North Mediterranean

Efthymios-Spyridon Georgiou

Abstract

Firstly, during the ten years, the demography in the Mediterranean countries changed. The purpose is the visualization of this change. The paper has goals for the various digital products of the maps, graphs, tables, and shapes that present simple, understanding, and intelligent ways. The urban planner uses Geographic Information Systems software and Microsoft Office applications for data management analysis. The methodology builds around the European statistics Web site. The official statistical Web site included statistical, vector, and raster datasets. In continuous, the urban planner uses the QGIS and ArcGIS Pro. The process builds on symbology, classification, and the construction map in ArcGIS Online. The steps of analysis are in detail in the manuscript. The number forming spatial data (polygon, lines, and points). The previous process is the methodology of the final products. The maps, graphs, and table present the population of the leading countries in the Mediterranean. Finally, the process moves into three levels: the G.I.S analysis, the maps, and the dataset design. The mosaic is the triptych of the project. It is essential to build a very high-quality mapping product. The broad needs to understand this demographic transformation.

Keywords

G.I.S • Maps • Statistics • Demography • Democracy

E.-S. Georgiou (✉)
Department GeoInformatics, University Salzburg, Salzburg, Austria
e-mail: efthymios.georgiou@stud.sbg.ac.at

1 Introduction

The main objects of the projects are the county's shape file, the population number, and the QGIS software. In addition, the lines of the work basing on the European Statistics Official Website (Database Eurostat 2023). This Web site is the primary source of the import data. The cartographer needs to be a mapping specialist. He could develop talents and hard and soft skills for designing and preserving the final product.

In conclusion, the demographic maps describe the ten-year situation, especially from 2012 to 2021. The maps export beneficial anthropological results. The populations increased/decreased in many countries. The products will combine with economic, environmental, and trade maps in future work. The benefits offer a holistic approach to spatial change. The mathematical form is an opportunity for model development and estimation in the following years.

2 The Vision

2.1 The Concept

The area is Mediterranean Basin. The strategic concept is the analysis of the progress of the population. The purpose of the work is to present the indexes of the inhabitants.

In details, the average population in Mediterranean Countries and the growth rate. The central questions are where focusing the significant percentage of the inhabitants? Which are the more important countries? Which are the decrease/increase in population? The goals are to present the data in an understanding, brilliant, and innovative way. Many tables create with flexible mathematical progress.

For example, the growth rate is a numeral equation of primary data. The paper summarizes the mathematical, statistical, and number dates in figures and tables. In addition, the data visualization in maps. Finally, the entire manuscript

A. Battisti et al. (eds.), *Greening Our Cities: Sustainable Urbanism for a Greener Future*,
Advances in Science, Technology & Innovation, https://doi.org/10.1007/978-3-031-49495-6_34

is a tremendous challenge for future work. It is possible to estimate the population condition in the next 30 years. The progress in the technological means applied to the field of protection and documentation of population. Among them are mentioned vector and raster datasets and excavation research, Geographical Information Systems and geometric documentation of mapping information, maintenance, restoration, and knowledge, etc.

At the same time, digital media, technical equipment, and applications that facilitate the above processes are multiplying, while programs and specialized software recognizing that are often used exclusively for the needs of mapping, such as:

(1) Geographic Information Systems (GIS),
(2) Advanced imprinting technology and registration of raster and vector datasets,
(3) Population pyramids,
(4) Computer-aided design software (CAD) at 2D/3D level, photorealism,
(5) Systems and special technical equipment for recording and monitoring population change,
(6) Electronic indexing databases.

2.2 Agenda Sustainable Development

The 2030 Agenda for Sustainable Development, adopted by all United Nations Member States in 2015, provides a standard plan for peace and prosperity for people and the planet today and tomorrow. The focus is on the 17 Sustainable Development Goals, an urgent call for action by all developed and developing countries in a global partnership. They recognize that eradicating poverty and other disadvantages must go hand in hand with policies that improve health and education, reduce inequality, and boost economic growth.

Sustainable development has a purpose in many ways. However, the most cited definition comes from Our Common Future, also known as the Brundtland Report: "Sustainable development is a development that meets its own needs without compromising the ability of future generations to meet their own."

The Sustainable Development Goals are a global agenda to end poverty, protect the planet, and ensure peace and prosperity for all. The definition comes from the joint edition of the International Union for Conservation of Nature (IUCN), the United Nations Environment Program (UNEP), and the World Wildlife Fund (WWF) and explains: "Sustainable development means improving the quality of life within the carrying capacity of the supporting ecosystems."

3 The European Union Policies

3.1 The European Policies in Spatial Planning

In the early European Union, the states in a period of economic abundance and strength designed spatial policies (NewTowns Policy) in Britain (Metropoles d'Equilibre) in France, which became the model for the "rival cities" policy for our country "at the end of the 70s, with the most characteristic, Five Fingers Plan in the greater Copenhagen area, Raumordnugns program in the then Federal West. Germany, Progretto in Italy, Facet Plan in the Netherlands" (Tsetsis, 1996). "The reasons that prompted the Community and the Member States to deal with issues of European spatial planning. These politics are related, in particular, to new interregional and intraregional inequalities.

The differences observed since the mid-80s go beyond national borders and the traditional distribution of territories in regions. "Phenomena such as the concentration of activities in central areas of the European Community (areas between London, Paris, Milan, Munich, and Hamburg or otherwise known as the "European Pentagon") and the subsequent marginalization of regions and isolated areas (e.g., islands and mountainous areas), the strengthening of metropolises and large urban centers and the abandonment of rural and old industrial region, unequal geographical access to infrastructure and knowledge, as well as pressures on cultural and natural heritage, were seen at the time as new threats to the balanced and sustainable development of the European space" (Giannakourou, 2008).

The approaches of an entire direction spatial policy that will be increasingly developed "from below", by local and regional bodies within the framework of the multi-level spatial governance long-term strategy, are expected to intensify in the future with the parallel enlargement and deepening of the Union. These approaches reflect not only the inability of the European Union to operate with the political criteria of a federal government but also the fact that, in reality, the dynamics of spatial development interconnect a set of local and regional actors with state and community actors in complex horizontal and vertical institutional negotiations to deal with the particular spatial problems of each country and region of the Union.

Thus, the future of spatial planning in Europe will depend primarily on the internal dynamics of member states and the ability of national spatial planning systems to adapt to a decentralized network model of horizontal and collaborative spatial planning within which the principles of consensus will prevail. Of synergy, partnership, and collective learning and action (Giannakourou, 2009). However, only some areas will have the same capabilities or know-how in spatial planning since there will be no central interventionist direction. Indeed, the regulation of space under the free market forces will create a variable geometry of spatial integration, in which some will be the winners and others—perhaps the political losers.

As Dawson (1993) prophetically observed, the dynamics of the European spatial economy shape new poles and axes of development and change the role of cities and regions in the European and international spatial division of labor. Also, it increases the comparative advantages of areas favored by the new axes of communication and new technologies while simultaneously exposing the disadvantages of slow-growing regions. It apologizes to them to the forces of relentless economic competition, a competition that becomes even more ruthless under the dynamics of globalization.

3.2 Similar Work

The work is a practical experience. Recently, I attended the ESRI User Conference in San Diego. The conference is one of the largest in the world in the Geographic Information Systems community. There are several valuable papers and presentations (LaLonde et al., 2022). The organization allowed me to participate remotely from my home in Athens. I was impressed by the content, the essentiality, the importance, the analysis, and the targeted presentations. Also, the excellent platform had online, recorded sessions.

Geographic Information Systems are a modern intersection of science. Cartography is a field that, in the era of the fourth industrial revolution, is constantly evolving. For example, the fourth transformation of map design from paper to digital format has significant benefits. In conclusion, the discussion can be about utilizing technological tools for better results. Storage, transfer, multi-user processing, dissemination, and attractiveness of visualization are significant benefits and changes. Increasing the profit of cartographers, G.I.S Analysts, and G.I.S designers is essential in professional planning. That is why a strategy, a system of action, and professional development are needed.

The growth and distribution of the human population on the planet's surface have profoundly shaped the characterization of "human civilizations" with varying levels of life. Alison Bashbord's book, Global Population concerns the size of the world's population. A scientific approach to the "population bomb" is not the beginning of 1968. It emerged after World War I and was considered a major ecological problem. The world population problem has affected both soil fertility and has always had analysis factors both "land" and "life". In Summarize, the book Global Population traces the demographic strategy of the world population problem as it developed from 1920 to 1960 (Bashford, 2014).

In short, natural scientists have found that population is as attractive a human dimension of environmental change because the data is readily available (in contrast to other human variables such as values, culture, and institutions). The projections are reasonably reliable analogously to all other quantitative variables. This has promoted a reductionist view of the interactions parallel to the population and the environment (Sherbinin et al., 2007).

4 Data Modeling

4.1 Statistical Data

An official Web site of the European Union is a significant source of statistical data (Data Base Eurostat 2023). The data is very reliable and valid. The Web site provides information from all European countries as a geographical and institutional entity—the European Union. Each European citizen, resident, and visitor has access to the data. Data from the Internet is accessible to every corner of the planet. The data present population, demographic, economic, and environmental elements.

They are the official data of each member-state and for states joining the EU. Finally, the importance of timelessness is essential. There is a great choice to download the data. There are choices such as a table, map, figure, and Excel file. The writer downloads the Excel file on a PC. The CSV file is accessible to working in each user. There are many excellent Web sites. The user can surf in web and look for an appropriate Web site. I see interest in the Web site of the Geographer Carlos Efrain Porto Tapiquen. The Web site is from University Autonoma de Barcelona. The Web site is a very respectable effort (Tapiquen, 2023).

The QGIS is valuable software (QGIS Training Manual, 2023). The benefits are the open-source function. The smart and details are progressing. The QGIS contains two tables. The table describes data such as the attribute table, in addition, the spatial table, such as area and shape file analysis. Finally, the QGIS has strong possibilities. The next step is to import the data. These two files are the only primary source. The same field joins the values from the attribute table to the shape file. This field will be the connecting link.

The spatial matrix is the processing and analysis space. To create a map, you first need a system. Each user must formulate a strategy for their construction. In addition, the design should be attractive and appealing. Finally, the cartographic representation with nomenclature, symbols, and colors is essential. Country shapefiles apply to all of Europe (Going Places with Spatial Analysis, 2022). The user selects the shape file and defines the Mediterranean countries. Finally, he is interested in the scale of the map of the Mediterranean basin. The symbology tool has the ability for attractive maps. Colors are a solution for proper geo-visualization. Also, the building maps in many ways: gradual, categorized, as a single categorization. Finally, suitable classes are essential. The result must be clear to a large audience (Getting Started with Data Management, 2022).

The datasets, which for a better understanding of the cartographic layer. Finally, the tag is in multiple fields, for example, country name, population, and capital. Sorting is a defining process. It is the factor that determines the final map. For example, a map with four classes has a different presentation than a map with eight categories. Finally, the classification is essential to be attractive and understandable. The information must be clear to the general public (Cartography, 2022).

The map printing function is the last step for the final product. Initially, there is a blank sheet. The sheet is adjusted based on the dimensions A4, A2, etc. Also, on the left, there are the map symbols. The main characters are as follows:

- The north symbol,
- The graphic,
- The written scale,
- The map coordinate system,
- The map title,
- The legend.

These commands determine the quality of the map. The map layout inserts into the white paper from the original project. Finally, it is essential to have margins evenly distributed.

4.2 The Demographical Indexes 2012 to 2021

Figure 1 refers to the size of the population from 2012. Figure 1 shows all countries near the Mediterranean. For example, Turkey has the most significant increase.

This phenomenon has a reason for economic problems for a new generation. The young person leaves from countries for a better business future. Finally, the new parents make a decision challenging to grow the relationship. Figure 2 shows the population in 2021. To summarize, the graphs refer to the biggest countries in 2012 and 2021.

The graph, Fig. 2 shows the population in countries. The average is approximately 20.000.000 inhabitants. A significant percentage focused on four countries. Turkey, Italy, Spain, Greece, and Portugal are the most prominent countries. Also, the lowest countries are Malta, Montenegro, Slovenia, Macedonia, and Albania. Finally, the growth rate decreased in more countries. The high rate per year presents Turkey.

The equation refers to the tempo of decrease/increase per year. Each country has a unique number of changes. Many counties have a low index of birth and a high death rate. The high increase has only a few countries.

Tables 2 shows the population's mean and sum in 2012 and 2021. Both indexes present an increase in ten years. This point means that the number of deaths is low than the number of births. Also, the factors of happiness increase the growth level. However, the balance between older adults over 65 years old and fewer than 23 old has inequality in old ages. In other words, it is the same. The population pyramid weights old age. Table 3 presents the difference from 2012 to 2021. The seven countries have increased, and seven counties have decreased. Turkey and France and Spain have the most significant increase. On the other hand, Bulgaria, Greece, and Croatia have the most significant decrease. The main reason for the raise in population is immigration from other countries.

In addition, prices for house rentals and Airbnb are increasing due to the increasing flow of tourists. The company's operations increased rents by 1.9%, transaction prices

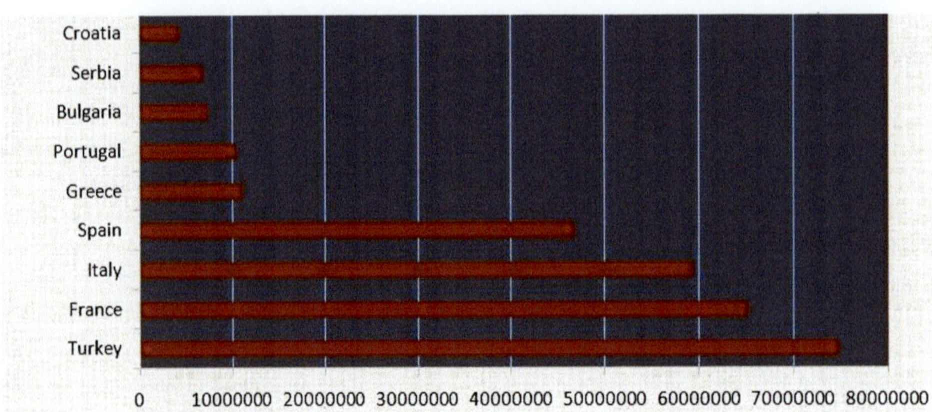

Fig. 1 Population 2012 in biggest countries. *Note* The data for the official Web site European Union {Eurostat: Documents] by Eurostat, 2023, @Copyright European Union Europe.eu (https://ec.europa.eu/eurostat/documents/2995521/5167222/3-20112013-AP-EN.PDF/73c338ef-13c4-42d2-99a5-179eabff244c)

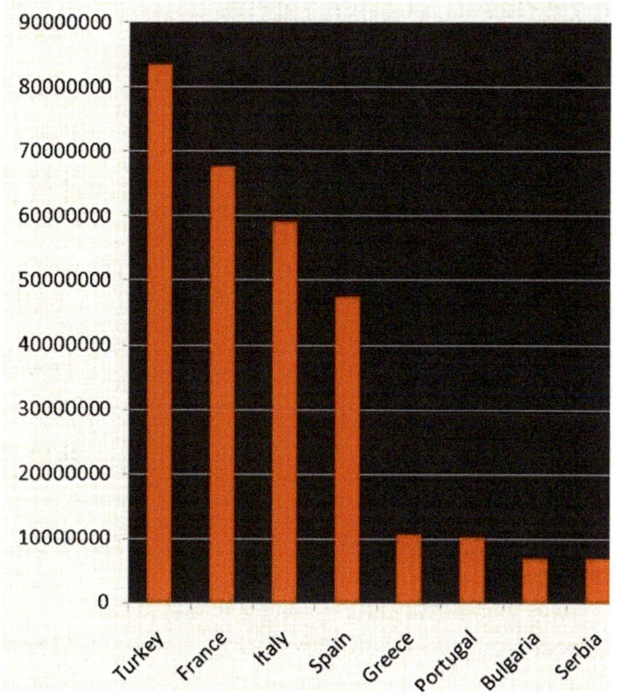

Fig. 2 Population 2021 in biggest countries. *Note* The data for the official Web site European Union {Statistical Database: Population and Social Condition] by Eurostat, 2023, @Copyright European Union Europe.eu (https://ec.europa.eu/eurostat/web/main/home)

Table 1 Countries in mediterranean (million)

Average 2021	21.778.4 inhabitants
Average 2012	19.884.5 inhabitants

Note The data for the official Web site European Union {Statistical Database: Population and Social Condition] by Eurostat, 2023, @Copyright European Union Europe.eu (https://ec.europa.eu/eurostat/web/main/home)
Note The data for the official Web site European Union {Eurostat: Documents] by Eurostat, 2023, @Copyright European Union Europe.eu (https://ec.europa.eu/eurostat/documents/2995521/5167222/3-20112013-AP-EN.PDF/73c338ef-13c4-42d2-99a5-179eabff244c)

Table 2 Countries in mediterranean (million)

Sum 2021	304.898.7 inhabitants
Sum 2012	294.428.5 inhabitants

Note The data for the official Web site European Union {Statistical Database: Population and Social Condition] by Eurostat, 2023, @Copyright European Union Europe.eu (https://ec.europa.eu/eurostat/web/main/home)
Note The data for the official Web site European Union {Eurostat: Documents] by Eurostat, 2023, @Copyright European Union Europe.eu (https://ec.europa.eu/eurostat/documents/2995521/5167222/3-20112013-AP-EN.PDF/73c338ef-13c4-42d2-99a5-179eabff244c)

by 4.6%, and record prices by 3.7% (Garcia-López et al., 2020). The population of metropolitan Barcelona and 26 other municipalities make up the city and is approximately 2.8 million. Finally, the metropolitan of Barcelona, which includes 162 municipalities, has around 4.2 million inhabitants (García, 2005).

Table 4 base on the mathematical equation: **Y = X (2012 + 2013 + ⋯ + 2021)/10.**

$X^{1+2+3+\cdots+n}$ = Years of population
Y = Result – Growth per Year
10 = Years from 2012 to 2021

5 The Process

5.1 Methodology of the Project

Spatial analysis is an innovative and understanding function. The cluster data technics basing on the software in ArcGIS Pro. The vector datasets create three maps of the population. The audience can observe the change and the size of the inhabitants. The crucial factors configure the map's benefits in the export of results and answer about the future of the North Mediterranean. The sea is often the physical border between the people but is an opportunity for transportation connection, market development, and financial strength.

The methods of cluster vector datasets develop under scientific rules. The user downloads the shapefiles from the recognition sites of Professor Efraim. In detail, it downloaded the shapefiles from countries in Europe. Also, the official Web site of the European Union is a very significant source. The Web site includes valuable information about the population from 2010 to 2021. The inhabitants of each county are download in an Excel file. The user joins the table of the inhabitants with the countries of Europe. The final step is the selection of the area of exciting counties. Finally, the process became in the software ArcGIS Pro. The three maps 2017 to 2021 exporting in the Layout menu as images. The final process is an attractive presentation of the population in the North Mediterranean basin.

5.2 Mapping Visualization

The spatial analysis shows the average population focuses in west Europe. Figure 3 is the map of the population 2017 presents the counties in the north Mediterranean. The two biggest counties and financial strength counties are France and Italy. Also, an important role is the Iberian Peninsula. The west entrance of Europe increases the population of South America. The reason is the common language background. Also, in Algeria, many people live in France and South Italy.

Table 3 Population 2012 to 2021 (million)

Countries	Population 2012	Population 2021	Increase/decrease
Turkey	74.724.7	83.614.3	8.890.0
France	65.327.8	67.656.7	2.328.9
Italy	59.374.2	59.236.2	−137.994
Spain	46.818.2	47.398.7	580.476
Greece	11.086.4	10.724.6	−361.807
Portugal	10.542.3	10.298.2	−244.146
Bulgaria	7.327.2	6.916.5	−410.676
Serbia	6.926.7	6.871.5	−55.158
Croatia	4.275.9	4.036.3	−239.629
Bosnia Herzegovina	3.839.2	No data	No data
Albania	2.903.0	2.829.7	−73.267
Macedonia	2.059.7	2.068.8	9.014
Slovenia	2.055.4	2.108.9	53.481
Montenegro	621.208	621.873	665
Malta	417.546	516.100	98.554

Note The data for the official Web site European Union {Statistical Database: Population and Social Condition] by Eurostat, 2023, @Copyright European Union Europe.eu (https://ec.europa.eu/eurostat/web/main/home)
Note The data for the official Web site European Union {Eurostat: Documents] by Eurostat, 2023, @Copyright European Union Europe.eu (https://ec.europa.eu/eurostat/documents/2995521/5167222/3-20112013-AP-EN.PDF/73c338ef-13c4-42d2-99a5-179eabff244c)

Table 4 Growth rate per year (million)

France	2.379.7
Turkey	8.890.0
Italy	−157.994
Spain	580.476
Greece	−361.807
Portugal	−244.146
Bulgaria	−410.676
Serbia	−55.158
Croatia	−239.629
Albania	−73.267
Macedonia	9.014
Slovenia	53.481
Montenegro	1.565
Malta	98.554

Note The data for the official Web site European Union {Statistical Database: Population and Social Condition] by Eurostat, 2023, @Copyright European Union Europe.eu (https://ec.europa.eu/eurostat/web/main/home)

In West Balkan, the phenomena of the previous two decades change the populations. The reason is a change of communist statement to capitalist economic status and the development of cooperation with the European Union (Fig. 4). This political decision has transformed the West Balkan. Places, for various reasons, have low growth; however, because it is near Italy and Austria, Croatia and Slovenia (Fig. 5) cutch the goals of sustainable development more quickly.

For instance, Dubrovnik, located in the Southern part of the Adriatic Sea, Dubrovnik is the most renowned tourist spot to situating in the Mediterranean Sea. The city owing to the city's picturesque architectural historical frame has grated a place on the UNESCO World Heritage Site list. In addition, the Peljesac Bridge serves as a connectivity path between the Dubrovnik exclave and the country.

Firstly, Table 5 shows the progress of the population from 2012 to 2021. The table presents the historic progress in central European countries. France is the only country which has an increase in inhabitants. Also, the four countries, Italy, Spain, Greece, and Portugal have a steady decrease in population.

5.3 The Multi-analysis of Demography

The Fig. 6a–c shows the comparative study in North Europe. The three maps examine the population from 2017 to 2021. The significant point is the destiny of the inhabitants in financial strength counties. Also, the countries such as the powerful administrative center, Austria, and the capital cities, for example, Athens, have increased populations.

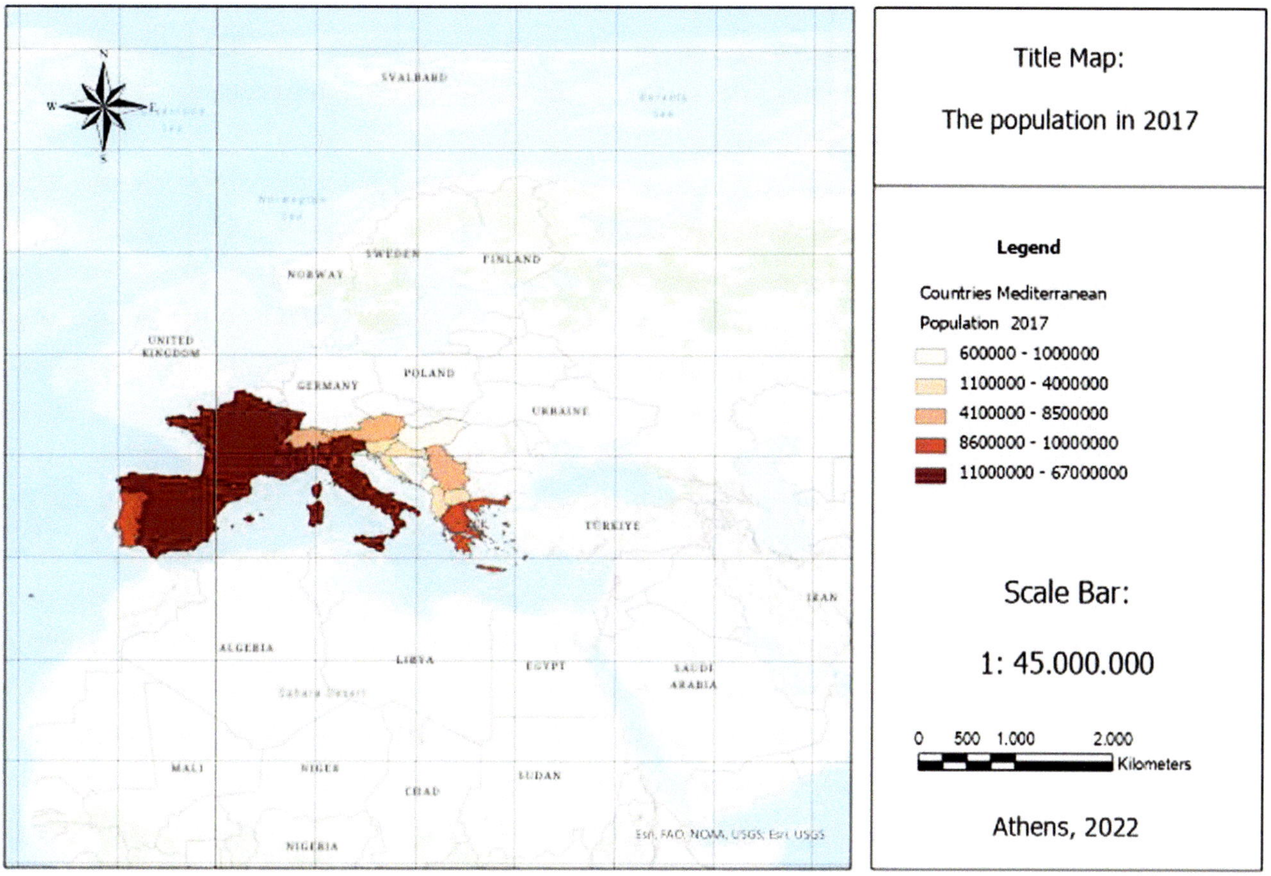

Fig. 3 Population in 2017. *Note* The data for the official Web site European Union {Statistical Database: Population and Social Condition] by Eurostat, 2023, @Copyright European Union Europe.eu (https://ec.europa.eu/eurostat/web/main/home)

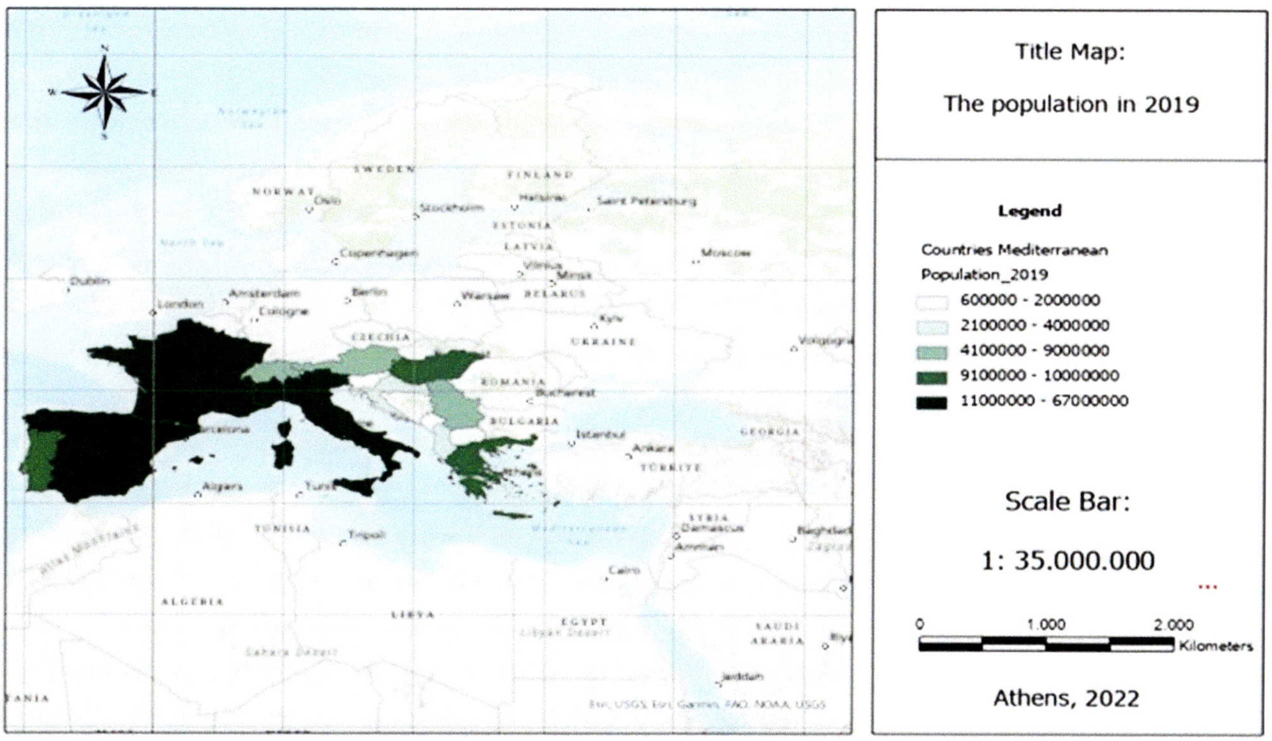

Fig. 4 Population in 2019. *Note* The data for the official Web site European Union {Statistical Database: Population and Social Condition] by Eurostat, 2023, @Copyright European Union Europe.eu (https://ec.europa.eu/eurostat/web/main/home)

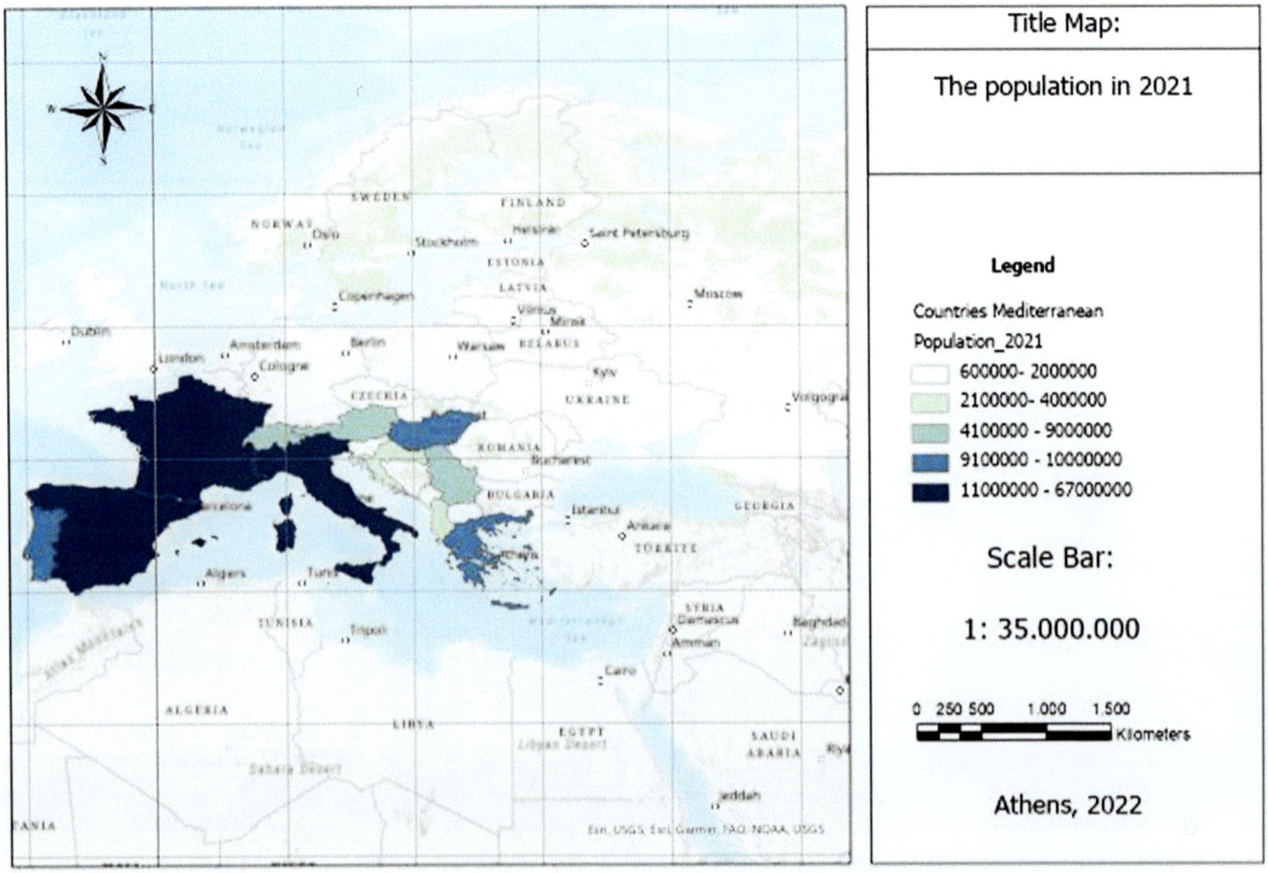

Fig. 5 Population in 2021. *Note* The data for the official Web site European Union {Statistical Database: Population and Social Condition] by Eurostat, 2023, @Copyright European Union Europe.eu (https://ec.europa.eu/eurostat/web/main/home)

Table 5 Population from 2012 to 2021 (million)

	Greece	Portugal	Spain	Italy	France
2012	11.086.4	10.542.4	46.818.2	59.394.2	65.277.0
2013	11.003.6	10.487.3	46.727.9	59.685.2	65.600.3
2014	10.926.8	10.427.3	46.512.2	60.782.7	66.166.0
2015	10.858.0	10.374.8	46.449.6	60.795.6	66.458.1
2016	10.783.7	10.341.3	46.440.1	60.665.5	66.638.4
2017	10.768.2	10.309.6	46.528.0	60.589.4	66.809.8
2018	10.741.1	10.291.0	46.658.4	60,484.0	67.026.2
2019	10.724.6	10.276.6	46.937.7	59.816.7	67.177.6
2020	10.718.5	10.296.0	47.332.6	59.641.5	67.320.2
2021	10.678.6	10.298.2	47.398.7	59.236.2	67.656.7

Note The data for the official Web site European Union {Statistical Database: Population and Social Condition] by Eurostat, 2023, @Copyright European Union Europe.eu (https://ec.europa.eu/eurostat/web/main/home)

Synoptic, the multi-factors that have significant meaning in an increase of inhabitants are as follows:

(1) Financial strength,
(2) Tourist development,
(3) The common language,
(4) The common historical background,
(5) The level of capitalist growth.

These factors increase happiness, long-term life, and the number of children. The paper's analysis focuses on the index, the reason, the political decisions, and the results. The

a: 2017 b: 2019 c: 2021

Fig. 6 Population in three periods. *Note* The data for the official Web site European Union {Statistical Database: Population and Social Condition] by Eurostat, 2023, @Copyright European Union Europe.eu (https://ec.europa.eu/eurostat/web/main/home)

future needs to configure for all. The 17 goals of sustainable development need to gain from each country and all of Europe. The people from the south Mediterranean search for a better future in the United States of America and Europe. Adaptation to culture and language is not essential, but people must keep excellent and good living standards. Figure 6a–c presents the population in the Meditteranean basin. The RGB colorful maps show each chronological period.

In short, scientists have found that population is attractive as a human dimension of environmental change because data is readily available (as opposed to other human variables such as values, culture, and institutions). The forecast models are reasonably reliable, and residents can process all other quantitative variables in analog models. This process has fostered a reductionist view of population–environment interactions.

6 Conclusion

Geographic Information Systems are a modern intersection of science. The science of cartography is a field that, in the era of the Fourth Industrial Revolution, is constantly evolving. For example, the fourth transformation of map design from paper to digital format has significant benefits. The storage, transfer, multi-user processing, dissemination, and attractiveness of visualization are benefits and changes. The development profit of cartographers, G.I.S Analysts, and G.I.S designers is essential in professional planning. That is why a strategy, a system of action, and professional development are needed (LaLonde et al., 2022).

According to Alison Bashbord's book Global Population (Bashford, 2014) concerns about the size of the world's population did not begin with the 1968 "population boom" but after World War I, which sees the situation as a problem. The main pillars are the significant ecological, agricultural, economic, and geopolitical consequences. The world population problem analyzes soil fertility and affects both land and life. Also, Global Population traces the synthesis of a global population problem as it developed from the 1920s to the 1960s.

References

Bashford, A. (2014). Global population. In *Global population*. Columbia University Press.

Cartography. (2022). ESRI Academy MOOC. Remotely.

Dawson, A. H. (1993). *A geography of European integration*. Belhaven Press.

García, M. (2005). The case of Barcelona. In *Metropolitan governance and spatial planning* (pp. 351–372). Routledge.

Garcia-López, M. À., Jofre-Monseny, J., Martínez-Mazza, R., & Segú, M. (2020). Do short-term rental platforms affect housing markets? Evidence from Airbnb in Barcelona. *Journal of Urban Economics*.

Getting Started with Data Management. (2022). ESRI Academy MOOC. Remotely.

Giannakourou, G. (2008). The institutional dimension of territorial cohesion. In search of a new form of territorial governance in Europe (Dedication: Reform Treaty and Territorial Cohesion, International and European Politics. Quarterly Political and Economic Review).

Giannakourou, G. (2009). The Territorial Cohesion of the European Area: A New Policy Field for the European Union (25 Texts on Spatial Planning and Development, Collected Volume). Thessalias.

Going Places with Spatial Analysis. (2022). ESRI Academy MOOC. Remotely.

LaLonde, A., Jacods, J., & Mesimer, C. (2022, July). Presentation getting started with data management. In *ESRI User Conference*. San Diego, CA.

Navigation. Database—Eurostat. (n.d.). https://ec.europa.eu/eurostat/data/database

QGIS Training Manual—QGIS Documentation documentation (2023). https://docs.qgis.org/3.28/en/docs/training_manual/index.html

Sherbinin, A. D., Carr, D., Cassels, S., & Jiang, L. (2007). Population and environment.

Tapiquen, C. E. P. (2023). Europe shapefiles (*.shp). Geografía, SIG y Cartografía Digital. https://www.efrainmaps.es/english-version/free-downloads/europe/

Tsetsis, S. (1996). Towards a European Spatial Policy, Urban Transformations in an Evolving Europe. Papazisi.